From
Atoms
to
Galaxies

From
Atoms
to
Galaxies

A Conceptual Physics Approach to Scientific Awareness

Sadri Hassani

CRC Press
Taylor & Francis Group
Boca Raton London New York

CRC Press is an imprint of the
Taylor & Francis Group, an **informa** business

CRC Press
Taylor & Francis Group
6000 Broken Sound Parkway NW, Suite 300
Boca Raton, FL 33487-2742

First issued in paperback 2019

© 2010 by Taylor & Francis Group, LLC
CRC Press is an imprint of Taylor & Francis Group, an Informa business

No claim to original U.S. Government works

ISBN-13: 978-1-4398-0849-8 (hbk)
ISBN-13: 978-0-367-38411-1 (pbk)

Library of Congress Cataloging-in-Publication Data

Hassani, Sadri.
   From atoms to galaxies : a conceptual physics approach to scientific awareness / Sadri Dean Hassani.
     p. cm.
   Includes bibliographical references and index.
   ISBN 978-1-4398-0849-8 (hardcover : alk. paper)
   1. Physics--Philosophy--Textbooks. I. Title.

QC6.H3187 2010
530--dc22
                                     2010006622

Visit the Taylor & Francis Web site at
http://www.taylorandfrancis.com

and the CRC Press Web site at
http://www.crcpress.com

To Sarah

Contents

Preface

This book is as much about atoms and galaxies as it is about scientific awareness. The appearance of the latter in the subtitle does not imply any reduction of its significance. I could have easily titled the book, "SCIENTIFIC AWARENESS via a tour of conceptual physics." In fact, my primary reasons for writing this book and developing the courses out of which it grew, were the disheartening status of the scientific literacy of students, the teaching of pseudoscientific ideas in some departments on many university campuses, and lack of evidence for the effectiveness of what has been done to raise the scientific awareness of students and the general public.

The exponential growth of physics, chemistry, and biology in the nineteenth century brought science to such a level of prominence that the educators and intellectuals of that era could no longer ignore it, and felt an obligation to incorporate it in school curricula. The emergence of different philosophies of education and science shaped the nature and variety of this incorporation, and provided a platform on which the intellectuals of the period could debate. One of these intellectuals was Henry Adams, who wrote:

> The future of Thought, and therefore of History, lies in the hands of the physicists, and ... the future historian must seek his education in the world of mathematical physics. A new generation must be brought up to think by new methods, and if our historical departments in the Universities cannot enter this next phase, the physical departments will have to assume this task alone.[1]

Henry Adams was a nineteenth century American historian, journalist, and philosopher, who grasped the value and power of a solid scientific education for all citizens, *including* the teachers of humanities in the universities. Unfortunately his inspiring voice was drowned by the raucous row of relativism, neo-positivism, progressivism, constructivism, and a host of other antiscientific philosophical "-isms" pursued on campuses and inculcated in our students.

Although most of the faculty in the humanities and social sciences do genuine, useful research, some scholars cross the boundaries and step into controversial, often pseudo- and anti-scientific, areas. Scholars like the educational psychologist who thinks that students need to learn how to decide which life insurance to buy and how interest rates determine home mortgages, rather than acquire scientific and mathematical skills; the sociologist who proclaims that the *contents* of scientific theories are dependent on the social conditions in which they were discovered; the Freudian who touts the pseudoscience of psychoanalysis as science, and to prove his point, claims that Möbius strip "explains many things about the structure of mental disease;" the medical doctor who, in his attempt at unifying quantum physics and Ayurveda, finds a sugar molecule intelligent; the mystagogue who "discovers" a parallel between the occult Eastern philosophy and modern physics; and the historian of

---

[1] *The Degradation of the Democratic Dogma*, 1919 (New York, Macmillan).

science who gives equal weight to quantum theory—whose validity has been tested in a multitude of harsh experiments—and Aristotelian "dynamics"—which is a collection of (mostly false) philosophical statements—thereby devaluing *all* scientific theories and claiming that "knowledge is relative" and "any knowledge is true."

College students have become alarmingly illiterate in science. In a recent survey of 439 college students[2] in the United States, 56% believed in psychic or spiritual healing; 40% believed in haunted houses; 40% believed that people can be possessed by the devil; and 39% believed in ghosts and the return of the spirits of the dead. And the number of years a student spends in college does not improve his/her literacy. In fact, as if proving the effectiveness of the anti-science scholars' teachings, there is an *increase* in the paranormal belief from freshmen to upper-level students: 23% of freshmen believe in the paranormal compared to 31% of seniors and 34% of graduate students!

Physics has a unique place in the community of sciences: post-Renaissance physicists (re)discovered the scientific methodology; physics is the science that investigates the fundamental questions concerning space, time, and matter, and as such, deals more directly with questions having a philosophical overtone; the instrumentation developed by physicists is indispensable in the breakthroughs of other sciences; and many results obtained by physicists become cornerstones of other sciences. Because of this unique position, physics can play a significant role in improving scientific awareness. For this to happen, we first have to equip the public with a good appreciation of the natural laws and how they govern the orderly phenomena of the universe. Through a detailed examination of the physical ideas (conceptual, theoretical, experimental, historical, philosophical, etc.) using clear explanations, numerous examples, and repeated usage, I have tried to accomplish this task.

But this detailed treatment by itself is not sufficient, as many educators have realized. The general public mostly (although not exclusively) wants physics to relate to its everyday life. Unfortunately, this relation has been equated to *relevance*, which nowadays gets easily translated into technological wizardry and entertainment. This tendency has become so predominant among the advocates of relevance in physics education that any idea that is not "fun," or "hands-on," or is not relevant to the operation of a TV, a CD player, a computer, or the special effects in a movie, is discarded.

The overemphasis on entertainment and technology has two detrimental consequences. On the one hand, *students* are willy-nilly, led to expect high quality entertainment from the physics educators. They are bound to compare any technological presentation of physics with the myriad technological presentations to which they are constantly exposed. They will set the simple animated simulation of the application of the Bernoulli principle in the flight of an airplane against the meticulous and highly sophisticated animation of the flight of a fighter jet in *Pixar* movies. And it is clear who the loser is in this comparison!

On the other hand, the minimally qualified physics *teachers*—of whom there is no shortage, especially in the rural districts—see only the last word in the highly overused phrase "science and technology," and make a very narrow-minded interpretation of it. Combined with "relevance" and entertainment, this interpretation turns a high school physics class in the American Midwest into a showing of the movie *Forrest Gump* because of its highly sophisticated use of computer graphics *technology*!

There is another aspect of "relevance" not explored by the educators of physics: the exposure of students to pseudoscientific ideas. Such an exposure is very "relevant" to the formation of the intellect of the exposed, and to the society in which they live. A society, some of whose institutions of higher learning teach its citizens that chiropractic is a viable alternative to scientific medicine, that psychoanalysis is science, that scientific truth is not different from religious and philosophical truth, and that objectivity does not exist in science, is in danger of entering a Dark Age. I have diligently pursued this aspect of "relevance" throughout my book: in the last chapter of each part, in Chapters 2 and 35, in

---

[2]Source: *Skeptical Inquirer*, Vol. 30, No. 1, Jan/Feb 2006, p 37.

the entire Part IX, and in discussions designated by the symbol 📖 **Food for Thought** 🌱 .

The dominant educational philosophy in America, pragmatism, coupled with the inherent difficulty of physics, has *forced* the well-trained and qualified teachers, and *encouraged* the ill-trained and unqualified teachers of physics to overemphasize its *utility* in their instruction. Even the pupils are trained to ask, "What is physics good for?" although the same question could be asked of many other disciplines, including music, art, literature, and poetry.

There were many occasions in writing this book on which I was tempted to expound on the everyday application of physics. But I resisted that temptation, not because there are already numerous books devoted to that subject, but, more importantly, because I think that the reader should be given the opportunity to appreciate physics, not for its pragmatic utility, but for its inner beauty. Instead of asking for a practical application of a newly discovered particle, the reader—it is my hope—will be enlightened by the power of the theory which predicts the particle (along with a detailed specification of all its properties), and the ingenuity of the experimentalists who will ultimately discover it.

The question of what the best solution is to the urgent problem of raising the scientific awareness of the public has no easy answer. Some educators believe that feeding the public a lot of practically oriented *facts* about such topics as health, climate, environment, and technology will make the citizenry more scientifically literate. Others, at the other extreme, believe that the teaching of the scientific *method*, of "how science works," is more effective. Whether any of these attempts has produced any results is debatable. The polls do not show any improvement in the scientific literacy of the public. In fact, if anything, they indicate an opposite trend.

If the belief in, and practice of, pseudoscience and anti-science is a symptom of scientific illiteracy, then controverting this belief and practice is the first step in the direction of raising scientific awareness. Many excellent books, aimed at general readers, have been written to address this issue. However, a classroom textbook, which teachers can use not only to cover scientific topics, but also to use those topics to disprove unscientific claims, has been lacking. This book is a modest attempt at filling the void. Because it is a first attempt, the book has much room for improvement, and I'll be deeply indebted to those readers and adopters who send me their comments and suggestions.

What would be the appropriate content of a physics textbook which aims at raising the scientific awareness of the untrained reader? It would lay the foundation of the laws that govern our universe in a nontechnical way, at the same time that it provides a healthy dose of technical detail for those who have sufficient background and desire to see it; it would emphasize the topics that excite the mind; it would, whenever possible, use physics to expose the sand upon which most unscientific beliefs stand; it would outline the methods developed by physicists for scientific investigation of nature, and contrast them with those developed by the outsiders who claim to be the owners of scientific methodology; finally, it would convince the reader that *science* is the only element capable of advancing *technology*, at the same time that it distinguishes between the two. Here is how I have tried to accomplish these goals:

- **Modern physics**: Over half of the book is devoted to quantum physics, relativity, nuclear and particle physics, gauge theory, quantum field theory, quarks and leptons, and cosmology. The coverage of these topics not only excites the mind, but is also necessary for uncovering the abuse made of them by the gurus of the New Age Movement and some academic scholars.

- **New Agers' mysticism**: The New Age Movement has become a viable alternative to the traditional spirituality, and because it has no rigid authority, it appeals to the more educated public. To make it even more appealing, the New Age gurus borrow words from modern physics to cloak their unfounded beliefs in scientific garments.

I have carefully analyzed some of these beliefs, and shown how a good understanding of physics reveals their baselessness, regardless of the amount of modern-physics terminology used in them.

- **Emphasis on concepts**: Technical and mathematical details have been avoided and conceptual understanding emphasized, so that readers with little math background, the intended audience of the book, can appreciate the basic principles of physics.

- **The enclosed CD**: Nature speaks in the language of mathematics. The more mathematics you know, the better you can understand nature. Conceptual physics is a "translation" of this language, and, as such, is limited in its capacity to convey the true meaning and beauty of the poetry of this language, physics. While this limitation is unavoidable for the maximally non-technical readers, it is possible to communicate some of this poetry using high school algebra. The accompanying CD, referred to in the book as *Appendix.pdf*, provides a great deal of algebraic details for the readers who have both the background and the desire to see them.

- **Science and technology**: In advancing the scientific awareness of the public, it is absolutely vital to make a clear distinction between science and technology. And this is not simply because the popular media identify the two and limit their coverage almost exclusively to technology, but also because many academic scholars who study the impact of society, politics, economics, and culture on *science* and vice versa make the same identification and come up with faulty conclusions. The last part of the book is devoted to the characteristics of science in general, and the distinction between science and technology, in particular.

- **Philosophy of science**: To the scholars of humanities, philosophers of science are the arbiters of scientific methodology. The medley of the philosophies of science, which often contradict each other, makes the message delivered by the philosophers utterly confusing, and allows the scholars of humanities to draw erroneous conclusions about science. Scattered throughout the book (mostly in *Food for Thought* and *Epilogue* segments) are in-depth discussions of these messages, clearly pointing out their errors.

- **Pseudoscience**: Whether it is psychoanalysis, chiropractic, or alternative medicine, pseudoscience has crept into the college campuses at an unprecedented rate. An exposure of these disciplines is essential in making students scientifically aware, and the book contains many detailed discussions of them in *Food for Thought* segments and the last chapter.

In promoting scientific awareness in this book, I have differentiated between the unscientific or even antiscientific beliefs of an individual and the publicized and published ideas degrading science. The individual beliefs may (and sometimes do) have constructive results for the believer. I do not look down on anybody's submission to acupuncture, chiropractic, psychoanalysis, faith healing, or any religious, spiritual, and New Age beliefs.[3] But when these beliefs are publicized as alternatives to science, I have had no hesitation in the book to expose them. It is one thing for an alternative medical doctor to treat patients with herbal drugs, voodoo, or shamanism; it is quite another to write a book about the practice, universalize treatments that work on some people (as much as placebo works on them), and attack scientific medicine. It is one thing to go to a church, synagogue, mosque, or New Age retreat to attain inner peace; it is quite another to advertise "intelligent design"

---

[3]The use of the label designating a group in my critical analyses should not be interpreted as a critique of the individuals in that group. The labels, whenever mentioned in the book, refer to the leaders, publicizers, and "theorists" of those groups. For example, "New Agers" refers to the authors and leaders of the New Age Movement, not to individuals who believe in and practice its teaching.

as an alternative to the theory of evolution, or to muddle up modern physics and Eastern mysticism.

This book is the product of many years of intensive work and research, which would not have been possible without the support of my family: my wife Sarah, and my children Dane and Daisy, and I thank them for it. I would like to especially thank Sarah for her selfless encouragement, for the countless sacrifices she has made over the years we have been together, and for reading the entire manuscript meticulously and giving numerous helpful comments and suggestions. I would also like to thank my editor at T&F/CRC, John Navas, for noticing the potential of the book, providing much valuable advice, and seeing through its seamless production.

Sadri Hassani
hassani@phy.ilstu.edu

Note to the Reader

This book goes against the philosophy that the only physics worth learning is that which can be applied to the making of a cell phone, high definition TV, computer, SUV, or movie filled with graphics trickery. It does not glorify the pragmatic aspect of physics. Rather, it talks about some of the great ideas that have defined our era; about the inner workings of that magnificent enterprise of our race, science; about the power of theoretical prediction and the delight of experimental confirmation; about the lives of the people who discovered the path to truth; and *most importantly* about how science, in general, and physics, in particular, can identify and combat the irrationality that is sometimes cloaked in scientific garment.

The collection of detailed explanations using occasional formulas, and a multitude of explanatory line-drawn figures and examples in the text, plus glossaries, review questions, conceptual exercises, numerical examples, and mathematical notes at the end of each chapter or in appendices, qualifies this tome as a *text*book. However, by moving the (more difficult) numerical examples, numerical exercises, and mathematical notes to appendices—and using the results obtained there only when necessary—I have made the book more easily accessible to a general reader, who has a background in mathematics at the level of high school algebra, with the caveat that sections whose headings are highlighted are a little more technical than the rest of the book, and can be skipped on the first reading.

Although the highest level of math used is high school algebra, the book can be read at several different levels. If you like math and want to see the technical reasoning behind ideas introduced in the text, refer to the Math Notes, Numerical Examples, and Numerical Exercises in the appendix provided on the enclosed CD. If mathematics comes fairly naturally to you but you are not a math buff, go over the mathematical discussions while reading the text and look at some of the numerical examples, spending as much time on them as you please, but avoid getting frustrated, as the text material is independent of the level of your understanding of the technical stuff. If you have absolutely no inclination toward technicalities, skim through formulas, numerical examples, mathematical discussions, etc. Regardless of your level of mathematical sophistication, however, I urge you to pay attention to *Food for Thoughts*, the epilogues at the end of each part, the nontechnical chapters such as Chapters 2 and 35, and the entire Part IX. These are the discussions that you normally don't see in any other physics book and may shed a new light on how you view the world.

I hope that after you read the book, you will not only understand the rational laws governing our universe, past and present, but also be able to identify the unreason that surrounds us in various shapes and forms, and consumes a substantial portion of our global resources.

Note to the Instructor

The level—and therefore the content—of "science literacy" courses differs greatly from instructor to instructor and class to class. Even if we agree on the *theme* of the course, some of us may want to stress the technical and mathematical aspect of concepts at one extreme, while others may find it more useful to emphasize the social, philosophical, and methodological implication of concepts at the other extreme. I have written the book in such a way that these two extremes, as well as the approaches in the middle, can be accommodated with no difficulty.

The theme of this book, as outlined in the preface, is the beliefs that contradict science and are often offered as replacements for science. Within the framework of this theme, I can envision at least three types of courses for which this book may be suitable:

- A discussion/reading-laden course, which puts little emphasis on physics at the same time that it spotlights the role of physics in disproving many unfounded pseudo- and anti-scientific beliefs propagated by the laymen as well as academic professionals. For such a course, I recommend a fairly detailed coverage of the entire Part I, since it examines the emergence of science and the scientific methodology. Once this part is covered, a sample of *Epilogue* and *Food for Thought* writings, accompanied by as much physics as the instructor feels necessary, could be explored. This sample may be chosen from any parts of the book, with the caveat that some discussion of the *physics* of those parts may be necessary. I also highly recommend Chapter 35, as it is a fairly elaborate refutation of some trendsetting abuses of modern physics by the gurus of the New Age Movement. Once these topics are covered, the three chapters in the last part of the book would be ideal to end the course. Although this course would be best suited for a small class, using the on-line teaching tools and software, it could also be adopted for much larger classes.

- A course in which physical concepts, especially those of modern physics, are studied with a little mathematical rigor. Students in this course learn the concepts and apply them to simple new situations by doing some conceptual as well as numerical exercises. Although the emphasis is slightly shifted from the course described in the first bullet, this shift of emphasis should not result in a huge reduction of discussion/reading. For this course, I recommend a fairly detailed coverage of Parts I and II, selections from Parts III and IV, and some in-depth coverage of one of Parts V to VIII. The instructor can assign readings from *Epilogue* and *Food for Thought* segments, as well as the nontechnical Chapters 2, 35, 41, 42, and 43. These readings can be material for short quizzes at the beginning of each session. Ideally, these quizzes are taken with "clickers," which are becoming increasingly popular.

- The course described in the second bullet could be made more rigorous by the implementation of some mathematics (not surpassing high school algebra) and using the CD provided. In this course, a good amount of emphasis would be placed on the technicalities behind the concepts, more numerical examples and exercises would be assigned, and the instructor would spend more time on the mathematical details. However, I do not recommend completely abandoning the reading/discussions described in the first and second bullets. To remain true to the subtitle of the book, the instructor is encouraged to cover some *Epilogue* and *Food for Thought* segments, and nontechnical chapters.

There is sufficient flexibility in the topics and style of coverage to allow instructors to design courses that match the need and technical level of their students. A physics course, concentrating entirely on physical concepts with no coverage of the "awareness" component of the book is also possible. However, the alarming statistics given in the preface should be sufficient motivation for the instructor to season that course with at least some readings from the nontechnical material of the book.

Pedagogy is an important part of our job as instructors of physics. The assortment of pedagogical approaches to the teaching of physics, some of which have entered our discipline from outside, can deter the honest effort of some instructors and reduce the effectiveness of their teaching. That is why, on occasions, I have engaged in discussions of the merits of some pedagogical approaches and faults of others. I hope that such discussions will stimulate fruitful dialogues at least among the instructors using this book, if not among the larger teaching community.

As an instructor of a course taught out of this book, you have access to some useful resources. A **test bank** with well over 1500 class-tested multiple choice questions facilitates designing quizzes and exams. A set of animated interactive movie **presentations**—covering a wide range of topics, including the emergence of science in the ancient civilizations, astronomy, motion, electromagnetism, and quantum physics—arouses the interest of students in learning. Among these presentations are *The Energetics of Santa Claus*, where the implausibility of Santa Claus's motion is demonstrated using energy conservation; *Gravity*, where the universal gravitational constant of a hypothetical universe can be measured; and *Kepler's Laws*, where Kepler's three laws of planetary motion are effectively demonstrated and his third law can be fairly accurately verified with the whole class participating. My students have invariably found these presentations the most favorite part of the course. **Web quizzes** consisting of a huge number of questions, which students can take multiple times and at their own pace—each time with a different (random) content—enhance students' learning ability through repetition. **Clicker quizzes**, drawn primarily from the "awareness" component of the book, are fast and efficient tools of ensuring that students become aware of the nature, relevance, and importance of science. Students are given a reading assignment (typically a page or two from the *Food for Thought* segments, *Epilogues*, or the last three chapters of the book) a couple of days in advance, and later tested using clickers, or any other means deemed appropriate by the instructor.

Part I

In the Beginning . . .

Chapter 1

Science Kindles

1.1 The Beginning

Science has its origin in the foggy horizon of time. Ever since the dawn of civilization humankind's curiosity has led them to observe, ponder, speculate, induce, and deduce. This curiosity tried to make sense out of the objects of their experience: trees, animals, rivers, forests, mountains, ... Sun, Moon, the stars, ... rain, wind, tornado, blizzard, storm....

In their primitive classification scheme, humans may have subconsciously divided all such things into three large categories: terrestrial, celestial, and "in-between." The first category consisted of objects and creatures that they could touch, feel, see, and understand. The second category was even more understandable because there was a conspicuous regularity in it which made the humankind comfortably knowledgeable of the sky, and, thus, at ease with it. Everything in this category seemed to be moving; and any motion requires a mover, a *live* mover. Therefore, the sky must be brimming with creatures!

It was the third category that caused discomfort and uncertainty. And since unpredictability interfered with their essential day-to-day activities, humankind sought ways of controlling it. After some unsuccessful attempts at solving the problem head on (for example, by climbing mountains and physically and verbally challenging the source of the uncertainty), they probably decided to open a dialog—a channel of communication—with the forces of the whimsical third category. The occurrences of this category seemed to originate from the sky, and the sky appeared to be the seat of regularity. Therefore, humankind saw some hope in taming the unpredictable by communicating with the predictable. But how?

A direct communication seemed ineffective as the sky was too far away and no detectable source of the unpredictable seemed in sight. The next best idea was to create *images* of the heavenly "creatures" and communicate with those images. The physical appearance of the images was left to the imagination of humankind: a horse with the head of a man, a woman with many hands, a four-legged animal with the head of an eagle, or a man with the head of an animal. The ineffectiveness of communication with the images did not cause the abandonment of the entire idea; it only led to changing the images and the content of communications, hence the variety of idols and prayers.

> Science starts in the temples, and scientists start as priests.

Communication with the creatures in the sky through their images on Earth required an extraordinary imagination and an unusual mental capability. Thus the brightest and most devoted people of society were given the task of communicating with the idols, and they would receive the support and encouragement from the rest of the community. This division of labor created the first priests whose job was not only to develop the best way to communicate with the images, but, as it was soon realized, to observe the seat of their

Astrology is but a form of superstition.

source, the sky. So, the priests were also the first astronomers. However, the ultimate purpose of sky-watching was to control the events on Earth, and in this capacity, astronomy and **astrology** were the two sides of the same coin.[1]

What do you know? 1.1. If you wanted to go to "college" in ancient Egypt and Babylon, to what institution would you go?

In their attempt to understand the gods in heaven the priests must have discovered some connections between the events in the sky and certain occurrences on Earth. For example, the ancient Egyptian priesthood noticed that the annual flooding of the Nile coincided with the rising of the Star Sirius. Such astronomical observations were also prompted by terrestrial needs such as the prediction of the flood, harvesting seasons, seasonal changes in the climate, et cetera. Thus, early astronomy was intimately mixed with the spiritual curiosity and the practical motivations of the ancient priesthood.

Looking at the sky, the priests could not help but notice the two most conspicuous objects: the Sun and the Moon. The daily regularity of these objects could not have escaped the attention of even the earliest priests. That is why calendar is as old as civilization itself. This regularity, together with other attributes, such as the warmth and the luminosity of the Sun and its influence on life and seasonal changes, bestowed upon it a unique place in some myths and religions of the world. In Egypt, for instance, where, for centuries, the dry land has been baking beneath a hot cloudless blue sky during the day and shivering in a bitter cold dark night in which stars glitter like shards of ice, the day has the warmth and the bustling of life, the night, the coldness, and the silence of death. It is thus only natural that in the Egyptian pantheon, the Sun is elevated to the rank of God itself in the person of Ra, the Sun God.

What do you know? 1.2. There is one thing that connects human activity to the occurrence of celestial events. What is that?

In Babylon, on the other hand, where the silt of Tigris and Euphrates deposited at the mouth of the Persian Gulf pushed the shores of the civilization further and further into the capricious sea, and unpredictable rain flooding of the rivers destroyed entire civilizations in a seemingly chaotic fashion, the supreme being is capable of infinite wisdom and calming spells. Thus, Ea, the supreme god of Mesopotamia, "... skillfully made his overpowering sacred spell. Reciting it, he cast it on the water (on Apsu), poured slumber over him, so that he soundly slept."[2]

Somewhere far removed from today's world of specialization, in the valley of Tigris and Euphrates in the ancient Babylon, at a time when priests were scientists and scientists were priests, the monks of the temple of Ea resolved that to understand their god, they needed to study His habitat, the sky. And the science of Babylonian astronomy was born. Babylonian clergy developed a sophisticated catalog of the motion of the Sun, the Moon and the planets, so that they could not only predict the location of these bodies among the stars, but also the recurrence of lunar eclipses, without, as far as we know, any kind of geometrical theory of the celestial motions.

---

[1]Astrology, the claim that earthly happenings are related to, and can be controlled by, heavenly motions, was the start of astronomy, the science of the observation and theoretical understanding of the celestial objects. Astrology had supporters among respected scientists such as Kepler well into the seventeenth century. However, today it is only a superstition practiced by fortune tellers and palm readers, and has absolutely no sound foundations in spite of the fact that many people, including some noted leaders and politicians, base their action on it.

[2]From *Enuma elish*, the grand Mesopotamian cosmology, which describes the Babylonian perception of the Universe and how the present world order was established.

Farther north, the Egyptian monks had developed their own version of astronomy. They imagined the whole universe to be like a large box. The Earth formed the bottom of this box with Egypt at its center. The sky stretched over it like a ceiling, its earthward face sprinkled with lamps lit at night and extinguished during the day. This ceiling was supported by four lofty mountain peaks connected by a continuous chain of mountains. On the other side of this mountain chain, hidden from us, a great river flowed. The Nile was a branch of this river, turning off from it at its southern bend. The river carried a boat in which was a disc of fire, the Sun, the living god *Ra*. This boat was occasionally attacked by a huge serpent during the day, whereby the Sun was eclipsed for a short time.

The two cradles of civilization, Babylon and Egypt, began the observational astronomy, and the cradle of the Western civilization, Greece, embarked on the geometrical explanation of the subject. Not unlike the two older civilizations, early Greek cosmology is rooted in mythology, especially as described in such epic poems as *Iliad* and *Odyssey* by Homer who is said to have lived in the eighth century BC Although alterations were made to this cosmic mythology by the later generations, no "scientific" theories or explanations were forthcoming until the middle of the sixth century BC when the giant of early Greek philosophy, Pythagoras, stepped onto the stage.

Food for Thought

FFT 1.1.1. The ancient unity of science and religion has become a *modern* source of confusion and misuse. Certain philosophers and followers of the Eastern mysticism argue that, because science and religion were one and the same in early civilizations, they should be reunited now as well. They lament the separation of modern science from religion and advise the scientists to change course and methodology to make science more amiable to religion. They beguile their readers into thinking that there is a connection between modern science and the old mysticism (see Chapter 35 for a detailed examination of this beguilement).

The truth is that the connection between science and religion existed up to about 600 BC *only because science was at its infancy.* Starting with Socrates and Plato in Greece (and apparently *only* in Greece), religion came under rational scrutiny, and from a philosophical and logical perspective, found to be full of contradictions. And from that point on, the gap between religion and science *inevitably* grew wider and wider.

While the Egyptian and Babylonian priests attached deities to the motion of celestial bodies, Greek astronomers sought a purely logical explanation for the motion of planets and stars, engaging and inventing more and more mathematics as they went along, and disengaging the Greek gods farther and farther from the objects of their study. Only such a strong separation between the gods and the stars made the progress of astronomy and science possible. This separation was the strongest in Greece, making the scientific progress uniquely Greek. No other ancient civilization made this separation, and therefore, no other civilization became capable of producing abstract thoughts such as geometry and theoretical astronomy.

The insistence that modern science and religion unite makes absolutely no sense and is tantamount to the abandonment of science altogether. The recent acceptance by the Catholic church of the theory of evolution is not to be considered a unification of science and religion; it made neither a science out of Catholicism nor a religion out of evolution. It was a concession forced on a pragmatically responsive church by the weight of countless bits of evidence accumulated over 150 years. The more idealistic sects cannot (and will not) find any compromise between the science of evolution and the religious dogma of "intelligent design."

While the Catholic church accepts the theory of evolution, some scientists, especially physicists, are lured to a prize worth over a million dollars and given annually to scholars who can find a connection between science and religion. The **Templeton Prize** was initially the *Templeton Prize for Progress in Religion*, and the first winner in 1973 was Mother Teresa. Since then winners have included Billy Graham, Charles Colson (of Watergate fame), and believers of other faiths. But in 1999 Ian Barbour, a student of Fermi and a professor of physics and theology at Carleton College, was the recipient. He was given the award for initiating a "dialog between science and religion." Templeton, a billionaire who believes that God has chosen him to show the world that theology and science are "two windows on the same landscape," admired Barbour, and coveted his dialog. So Templeton changed the name to the **Templeton Prize for Research or Discoveries about Spiritual Realities**. It is the largest prize for intellectual accomplishment in existence, chosen to be bigger than the Nobel. Since that time, seven of the last ten winners of the Templeton Prize have been physicists!

Many scientists are practitioners of some form of religion, but they keep their profession apart from their faith, because otherwise the obvious contradiction between the two can cause a lot of mental pain for them. Abdus Salam was a devout moslem and a great physicist, a Nobel Laureate. His profession never interfered with his faith and his religion never determined the course of his research. The only major influence of his faith on his profession was the establishment of the International Center for Theoretical Physics in Trieste, Italy, where physicists from developing countries are given a chance to do first-hand research. This is a far cry from "faith-based" physicists, who, to win the million-dollar prize, advertise a fake relation between religion and science. They do more damage to science than anti-science community because they are usually well-known scientists in whom people have invested a trust.

1.2   Early Greek Astronomy

Of the several schools of philosophy in the earlier era of the Greek civilization, that of Pythagoras is unique in many respects. He is said to have traveled a good deal to Egypt and Babylon, and to have been indebted for much of his knowledge of science to what he had learned during his travels.

Pythagorean philosophy is based on the primacy of numbers and the interpretation of nature. All the followers of this philosophy appear to have been devoted to the cause of science. To them, numbers were everything; they did not merely represent the relations of the phenomena to each other, but were the substance of things and the cause of all phenomena of nature. Pythagoras and his followers were led to this assumption by observing how everything in nature was governed by numerical relations: the regularity of the celestial motions, and the harmony of musical sounds.[3] Pythagoreans believed that the world was ruled by harmony, all the different heavenly revolutions producing different tones.

As for the heavenly bodies, Pythagoras coined the word "cosmos," was the first to recognize that the morning and the evening stars were the same body, that the Moon was a mirror-like body, and that the planets moved in separate orbits inclined to the celestial equator. This is all that is attributed to Pythagoras himself. The Pythagorean school, of course, flourished far beyond the time of Pythagoras, and the theory of the cosmos taught by his followers developed into one of the most ingenious theories of antiquity.

The first scientific theory of heavens.   The leading idea of the Pythagorean system of the world is that *the apparent daily rotations of the starry heavens and the daily motion of the Sun are caused by the spherical Earth being carried in twenty-four hours round the circumference of a circle.* The Pythagoreans thought that the nature of the Earth was too gross to make it fit the exalted position occupying the center of the universe. In this commanding position they placed the *central fire* or the *hearth of the universe* round which the Earth and all the other heavenly bodies moved *uniformly* in *circular* orbits.

Remarkably, this first "scientific" theory of the heavens had the two main characteristics of a good theory even by modern standards: **simplicity** and **symmetry**. One can say, in fact, that Pythagoreans started the trend. Being well trained in the mathematics of their time, they saw in spheres (or circles) a most symmetric and flawless geometric figure, the only figure matching the perfection of heaven. Similarly, uniformity constituted the simplest assumption concerning the motion of the cosmic objects. Fortunately for Pythagoreans, the

---

[3]They were the first to discover the numerical relation between various octaves.

nonuniformity of these objects (and thus, the discrepancy with their theory) was discovered much later.[4]

The central fire was not the only source of light and heat in the universe. The Sun received its heat and light partly from the central fire and partly from the fire that surrounded the sphere of the fixed stars. Sun then scattered its light in all directions including to the Moon, which in turn, reflected that sunlight in our direction. The faint glow on the entire surface of the new Moon was a reflection of the light of the central fire.

BIOGRAPHY

Pythagoras (ca. 580–ca. 500 BC) was born in Samos, settled in southern Italy about 540 or 530, and died there about 500. A few years after the tyrant Polycrates seized control of the city of Samos (about 540 BC), Pythagoras went to Egypt, where he visited many of the temples and took part in many discussions with the priests. In 525 BC Cambyses II, the king of Persia, invaded and conquered Egypt. Pythagoras was taken prisoner and sent to Babylon. While he was in Babylon, he reached the acme of perfection in arithmetic and music and the other mathematical sciences taught by the Babylonians. Pythagoras left no writings of his own, but his school attracted a great deal of attention not only because its members formed a religious-like brotherhood with unorthodox opinions, but also because they became involved in politics, thereby subjecting themselves to persecution in Southern Italy.

The Pythagorean system was a very ingenious model of the workings of the cosmos. It was also revolutionary in its assumption of the sphericity of the Earth and in the fact that the Earth was moving just like the rest of the heavenly bodies. Unfortunately, this model did not win any adherents outside the Pythagorean schools due to its philosophical roots, and, in particular, the almost religious significance the Pythagoreans attached to numbers. Plato and Aristotle popularized a different kind of astronomy. They placed the Earth motionless at the center of the universe, and let everything else move around it. They thought that each heavenly body was attached to a *crystal* sphere that rotated uniformly about the Earth: one sphere for the Moon, one for the Sun, one for each of the five planets, and one for the stars, making a total of eight heavenly crystal spheres.

Influence of Plato and Aristotle popularizes the Earth-centered astronomy.

Plato even gave a mystical flavor to his astronomy in that he believed the cosmos to be alive and to have a soul! Aristotle, Plato's favorite student, contributed very little to astronomy, and was of the same opinion as his mentor without the latter's mysticism. This motionless-Earth model was to influence—with very few exceptions—all astronomers for almost 2000 years.

The subsequent generations of astronomers accumulated new data, and reexamined the old Egyptian and Babylonian records. The result was the revelation of new features in the sky, and the first astronomer associated with these new features is **Eudoxus** of Knidus (ca. 408 BC–ca. 355 BC). Eudoxus was only 23 when he went to Athens to attend Plato's lectures. However, he soon discovered Plato's overemphasis on the mind, while Eudoxus was convinced that experience could have powerful influence on mind's invention. So, he set off to Egypt, where he received instruction from a priest of Hellopolls concerning the planetary motions, and where he came across indications of irregularities in the motion of Mars whereby it was seen that sometimes it moved backwards. This phenomenon, which came to be known as the **retrograde motion** seemed to topple the simple Pythagorean idea of uniform motion.

Signs of nonuniformity and irregularity in the motion of celestial bodies appear.

What do you know? 1.3. Did Pythagoras know about the retrograde motion of Mars and other planets?

---

[4]Egyptian priests knew of the nonuniform motion of Mars, but this knowledge was apparently not passed on to the Pythagoreans.

To account for this nonuniformity while keeping simplicity and symmetry of the theory, Eudoxus assumed *multiple spheres* for each celestial body. According to this model, each planet had several spheres, which were situated one inside the other, each rotating with uniform speed about its own axis, which was oriented in a direction different than the axes of the other spheres. Because all these spheres were assumed to be concentric with the Earth, the theory long afterwards became known as the **homocentric model**. The planet was supposed to be situated on the equator of the innermost sphere which revolved with *uniform speed* round its two poles. A variation of this model became the starting point of the geocentric model to which we shall come back later.

1.3   The Age of Measurement

With the advance of geometry at the end of the fourth century BC, the time was ripe in the third century BC for astronomers to pose new questions whose answers utilized this new science. Astronomers were no longer content with mapping the location of the Sun and the Moon in the sky. They wanted to see how far away they were. They were no longer satisfied with the simple fact that the Earth was a sphere, but how *big* a sphere it was. Two prominent figures of this period are Aristarchus and Eratosthenes.

1.3.1   Aristarchus' Measurements: Heliocentrism

Motion is a very intricate concept that humankind could not unravel until the end of the seventeenth century. One of the most difficult aspects of the concept of motion is the fact that it is relative. Due to the ubiquity of motion in the modern life we have all experienced this relativity when, stopping at a corner waiting for the green light, we startle to see our car move backwards. To prevent a collision with the car behind us, we slam on the brakes only to realize that it was not our car that was moving but the car next to us that started to move forward slowly.

Such a routine experience was nonexistent 2300 years ago, and it required a magnificently creative imagination to conjecture the relativity of motion. As early as the fifth century BC, there were philosophers, who, on purely hypothetical grounds, proposed that it was the Earth which turned on its axis from west to east in twenty-four hours, and this motion appeared as the daily rotation of the heavens from east to west. They also proposed the revolution of some planets around the Sun.

While this old heliocentric model was based entirely on conjectures, it was **Aristarchus** of Samos, flourishing about 280 BC, who based his reasoning on the *quantitative* observation of the heavenly bodies. From the shadow of the Earth on the Moon in a lunar eclipse, Aristarchus could estimate the diameter of the Moon to be a third of the Earth's diameter (see Figure 1.1). This estimate was later refined by Hipparchus to the more accurate value of 0.27.

> **What do you know?** 1.4. The following figure shows three hypothetical eclipses of the Moon. Which one of the three pictures represents the smallest Earth? The largest Earth?
>
>

Having found the size of the Moon, Aristarchus could now calculate the Earth-Moon distance by measuring the angular size of the Moon (see Appendix B for details). He estimated the Earth-Moon distance to be 25 Earth diameters; an estimate that was later

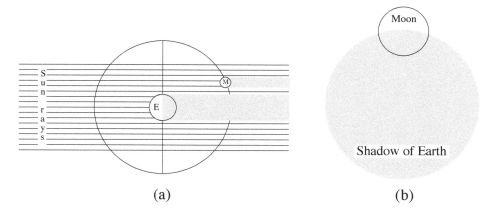

Figure 1.1: (a) The Moon is about to enter the shadow of the Earth. (b) As the Moon is entering the shadow of the Earth, the image of the Earth on the Moon can be compared with the Moon itself, and the size of the Moon relative to the size of the Earth can be estimated.

improved by Hipparchus, who obtained 30 Earth diameters, very close to today's accepted value.

Aristarchus' next task proved to be very difficult. The calculation of the Earth–Sun distance is complicated by its sheer enormity. Nevertheless Aristarchus' procedure was so ingenious, so simple, and so original that even a brief description of it can be testimony to the magnificent power of the human mind and its intimacy with the universe.

There are two half-moons in a single revolution of the Moon around the Earth. These are called *quarters* because it is really a quarter of the Moon surface that is visible to us.[5] One is called the *first quarter* and the other the *third quarter* (see Figure 1.2). By measuring the time the Moon takes to go from the first quarter to the third quarter and comparing it with the time from the third quarter to the first quarter, one can determine the angle α in the figure, which in turn can be used to find the Earth-Sun distance in terms of the Earth-Moon distance. This idea is very clever theoretically, but in practice, extremely difficult to implement, because the difference between the two times is immeasurably small, which, in turn is due to the fact that the Sun is very far (compared to the Earth-Moon distance). Nevertheless, Aristarchus' measurements led him to conclude that the Sun is about 20 times farther away from Earth than the Moon (see the **Math Note E.1.1 on page 75** of *Appendix.pdf* for details).

What do you know? 1.5. The following figure shows three hypothetical locations of the first and third quarters of the Moon on its orbit around Earth. Which one of the three pictures represents the farthest Sun? The nearest Sun? Is the Sun on the left or on the right?

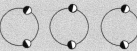

Aristarchus then argued that, since the angular sizes of the Moon and the Sun are almost the same, the Sun must be twenty times bigger that the Moon. He had already measured the diameter of the Moon to be about one-third that of the Earth. So, he estimated the

Aristarchus calculates the diameter of Sun to be seven times that of Earth's.

---
[5]Remember that a full Moon shows only *half* of its surface!

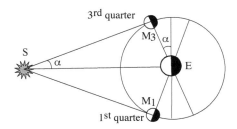

Figure 1.2: From the difference between the arc lengths $M_1 M_3$ and $M_3 M_1$, one determines α, which gives $\overline{SE}$ in terms of $\overline{ME}$. The Moon is assumed to be moving uniformly counterclockwise.

diameter of the Sun to be about seven times that of the Earth. He then argued that *a small object is more likely to go around a bigger object rather than vice versa*. And this is how he came up with the heliocentric model of the solar system!

The very few and scanty references to the system of Aristarchus by classical authors indicate that the system was never favorably received. Without a doubt, the principal reason for the fall of the heliocentric idea was lack of evidence for Earth's motion around the Sun. Astronomers expected to be able to detect the effect of this motion by observing the angular displacement of stars in much the same way that we detect the angular displacement of distant objects when we drive on a road. As we drive past a distant mountain on a highway, we notice a change in our line of sight: the mountain is first ahead of us, say to our right, and after some time—the farther the mountain, the longer this time—it "moves" directly to our right. When no such displacement of stars was observed, heliocentrism died. Aristarchus correctly reasoned that the stars must be much farther away than the astronomers of the time believed. But his reasoning fell on deaf ears, as the stars were thought to be only slightly farther away than the Sun.

BIOGRAPHY

Aristarchus of Samos (ca. 310 – ca. 230 BC) was born on the island of Samos, in Greece. He was influenced by the Pythagorean Philolaus of Kroton, but in contrast to Philolaus he identified the central fire with the Sun and put other planets in correct order from the Sun. The only extant work attributed to Aristarchus, *On the Sizes and Distances of the Sun and Moon*, is peculiar in that it assumes the Sun's diameter to be 2 degrees, rather than the correct value, 1/2 degree. The latter diameter is known from Archimedes to have been Aristarchus's actual value. Though the original text has been lost, a reference in Archimedes' book *The Sand Reckoner* describes another work by Aristarchus in which he advanced an alternative hypothesis of the heliocentric model. The geocentric model, consistent with planetary parallax, was assumed to be an explanation for the unobservability of the stellar parallax. The rejection of the heliocentric view was common, as the following passage from Plutarch suggests *On the Apparent Face in the Orb of the Moon*: "Cleanthes [a contemporary of Aristarchus and head of the Stoics] thought it was the duty of the Greeks to indict Aristarchus of Samos on the charge of impiety for putting in motion the Hearth of the universe [i.e. the Earth], ... supposing the heaven to remain at rest and the Earth to revolve in an oblique circle, while it rotates, at the same time, about its own axis."

1.3.2   Eratosthenes and Size of Earth

When Alexander the Great died in 323 BC at the age of thirty three, he left a vast empire to his feuding successors, who finally divided it into three parts, one of which was Egypt which fell under the rule of Ptolemy I, who founded a dynasty in the new port of Alexandria. Once Ptolemy established himself in the new country, he erected a magnificent library and a museum on the advice of one of Aristotle's pupils. Although no trace of either building exists now, we know that about a hundred scholars from all over the Mediterranean area lived

A magnificent ancient library with no extant trace!

in the museum, and their work was generously supported by Ptolemy and his successors. Teaching was relegated to second place and priority was given to research. Dinners or drinking parties, where solutions to problems were given and many literary prizes were awarded, were another regular feature. In addition to excellently furnished quarters, the Museum contained a communal dining hall, a covered colonnade with seats where the lectures and discussions could take place.

When Ptolemy I died and his son Philadelphus succeeded him, a more vibrant director was appointed, who further strengthened the library and museum. Under the new directorship, the museum attracted many learned men, among them Euclid whose famous treatise *The Thirteen Books of the Elements* contained so sound and logical an exposition of geometry that it formed the basis of the teaching of the subject up to the beginning of the twentieth century. It is Euclid who, during one of the museum's discussions, is credited with having replied to Ptolemy I who enquired about an easier way of learning geometry, with the quip "there is no royal road to geometry."

"There is no royal road to geometry."

Euclid

Astronomy and mathematics prospered at the museum, most notably by the efforts of its director, **Eratosthenes**, whose interests included not only those subjects, but also geography and calendar studies. But his real claim to fame rests on his excellent determination of the circumference of the Earth. At first glance this measurement may sound an incredible achievement for anyone in 240 BC, but in fact measurements of a similar kind had been made before and were being carried out at the same time. The point about Eratosthenes' measurement, and the reason why it is still remembered, is that his result is almost exactly the same as our present-day value! Eratosthenes' method is so simple and elegant that it is well worth our effort to get acquainted with it.

BIOGRAPHY

Eratosthenes (276–197 BC) was born in Cyrene on the southern shores of the Mediterranean. It was here that he received his early education before being attracted to Athens, where he studied for several years at the Academy and the Lyceum, the school founded by Aristotle. Besides measuring the circumference of the Earth, Eratosthenes made many other major contributions to the progress of science. He worked out a calendar that included leap years, and he laid the foundations of a systematic chronography of the world when he tried to give the dates of literary and political events from the time of the siege of Troy. He is also said to have compiled an astronomy catalogue containing 675 stars. By the time he was twenty-nine he had already gained some reputation and was invited by Ptolemy III to go to Alexandria as tutor to his son. There he stayed for the rest of his long life, becoming director of the museum ten years later and dying at the advanced age of eighty.

In Figure 1.3 you see two objects—they could be anything: trees, buildings, people, wooden bars, and so forth—casting shadows at noon, which make the easily measurable angles α_1 and α_2 as shown. Furthermore, the figure is drawn in such a way as to help convince you that the angle θ is the difference between α_1 and α_2. With θ so determined, you can calculate the circumference of the Earth from a measurement of the distance $\overline{P_1P_2}$. In fact, the ratio of the Earth's circumference to $\overline{P_1P_2}$ is the same as the ratio of 360 degrees to θ. Eratosthenes compared the angles from vertical at Syene and Alexandria, and noted that they differ by 7.2 degrees. The north-south distance between the two cities was 500 miles. So, he found the circumference of the Earth from

Eratosthenes' method of measuring the radius of the Earth.

$$\frac{\text{circumference}}{\overline{P_1P_2}} = \frac{360}{7.2} \quad \text{or} \quad \text{circumference} = 500 \times \frac{360}{7.2} = 25,000 \text{ miles},$$

which is remarkably close to today's accepted value of 24,850 miles.

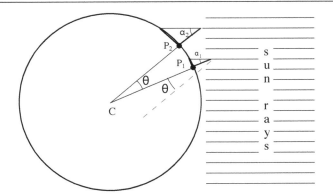

Figure 1.3: In the northern hemisphere, objects in the north cast a longer shadow than those in the south. The Sun rays, coming from an extraordinarily long distance, are parallel. The heights and their shadows are enormously exaggerated for clarity.

What do you know? 1.6. The following figure shows Sun rays hitting the surface of Earth at two different locations in the northern hemisphere. Which location is farther north? What if the locations were in the southern hemisphere?

What do you know? 1.7. Omaha is 1600 miles directly north of Houston. You measure the shadow angles in both cities. Which one is bigger? By how many degrees?

1.4   The Geocentric Model

The library and museum flourished under Eratosthenes' directorship, which continued through three reigns, and attracted for a time **Apollonius**, a great mathematician and astronomer who was some ten years Eratosthenes' junior. Apollonius was a brilliant geometer, and most of his education, as his work itself, was carried out in Alexandria. It was here that he discovered a novel way of combining circular motions to provide a real alternative to the spheres of Eudoxus. The motivation for this endeavor? The desire to invent a theory that agreed with the new observations contradicting the homocentric theory.

A closer scrutiny of various planets, especially Mars, revealed that in their retrogression, *they shine more brightly*. Astronomers attributed this to the approach of the planet towards Earth. Homocentric theory could not account for this approach, as it affixed the planets to spheres *concentric* with Earth.

In homocentric theory, the planets are attached to a sphere concentric with Earth. Therefore, the distance of the planet to Earth cannot change.

Apollonius invents the notion of an epicycle.

What Apollonius suggested was the brilliantly simple idea of making up a planet's motion from two or more circular motions built, as it were, on top of each other. Each planet was considered as fixed to the outside of a small circle, which spins about its center, and at the same time moves around the edge of a larger circle. The small circle carrying the planet was known as the **epicycle** (i.e., the upon-circle) and the larger one with the Earth as its center as the **deferent** (the carrier). Figure 1.4 depicts a case in which the planet M moves four times on its epicycle as the center of the epicycle completes a single revolution on the deferent. Some snapshots of the motion of M around Earth are shown. As the planet moves on the epicycle and the center of the epicycle on the deferent, the planet describes a path which includes retrograde motion as well as the change in the brightness of the planet

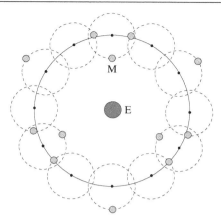

Figure 1.4: The planet M moves four times on its epicycle as the center of the epicycle completes a single revolution on the deferent. Twelve snapshots of the planet, epicycle (smaller circle), and deferent (larger circle) are shown. The combination of the two motions of the epicycle on the deferent and the planet on the epicycle results in the retrograde motion of M and the change in its brightness during the retrograde motion.

during this motion. Further details of the motion of planets on epicycles and deferents can be found in **Math Note E.1.2** on **page 75** of *Appendix.pdf*.

> **What do you know? 1.8.** When the center of the epicycle is at the 11 o'clock position of the deferent, the planet is at the 2 o'clock position of its epicycle. At what o'clock of its epicycle is the planet when the center of the epicycle is exactly half way between 10 and 11 o'clock of the deferent? Exactly half way between 2 and 3 o'clock of the deferent?

The Alexandrian research center flourished until about 220 BC, when the Ptolemies became embroiled in political strife in Egypt, continuing for more than seventy years, during which time many of the scholars in the museum left the country for Pergamum or other centers of learning such as Athens, Antioch, or Rhodes. Among those who fled Alexandria in the great upheavals of 150 BC was **Hipparchus of Nicaea**, who is primarily remembered for his observational work. Following Aristarchus, he too tried to measure the relative distances of the Sun and the Moon, although he was well aware that his precursor had been dissatisfied with his results a century before.

However the fame of Hipparchus rests not on his measurement of the distances of the Sun and the Moon, but on a much subtler kind of measurement, which he undertook successfully. This arose out of his plan to make a catalogue of the positions of the stars, a task which he accomplished after he left Alexandria and moved to the island of Rhodes. This catalogue was the first full list of viable stars ever to be compiled. Hipparchus was not content merely to list the stars and their positions, he also specified their brightness, dividing them into six ranges of intensity thus introducing a method which is fundamentally the same as that still used today.

As far as we know, Hipparchus died on the island of Rhodes, perhaps in 120 BC, and with him there ended, for over two centuries, the Greek tradition of astronomical observation and speculation. Politically Rome was gaining in ascendancy, and it was not until well into the early days of the Christian era that **Claudius Ptolemy** began his work and the next (short) scene in the play of humankind's quest to fathom the mystery of the Universe.

Ptolemy's astronomical efforts are enshrined in an immense tome called *Almagest*, which

Hipparchus makes a detailed catalogue of stars to be used by posterity.

Lacking any observable evidence for Earth's motion, Ptolemy accepts Aristotle's Earth-centered universe.

is regarded as the authoritative exposition of the **geocentric model** of the solar system. In it, we find a description of the full stature to which Greek astronomy attained and due credit is given not only to Hipparchus but also to many other astronomers of earlier times.

Beginning with broad outlines, Ptolemy sides with Plato and Aristotle and accepts a spherical Earth immovable in the center of a spherical Universe. This does not mean that he casts away a Sun-centered Universe without due reflection; on the contrary he considers the matter carefully but concludes that no observations of the stars reveal any of the effects to be expected if the Earth does, in fact, move in space.

BIOGRAPHY

Claudius Ptolemaius (85–165 AD) or Ptolemy, was born in the Greek city of Ptolemais. We know from his writings that he flourished in Alexandria between AD 127 and 151. This was a time when the Roman empire stretched over the cultured Mediterranean world. Hadrian was emperor, and with his love of culture inspired by Greek traditions, the intellectual climate was not unfavorable to scholarship. The library at Alexandria was now flourishing but scholarship consisted primarily of collating and assessing the achievements of previous generations. This intellectual climate is reflected in Ptolemy's works. Ptolemy wrote not only on mathematics, but also on optics and astrology. But it is for his achievements in geography and especially in astronomy that he is remembered, and it is in these two fields that his influence on later generations was to be so profound.

Having given his readers the basic facts and then equipped them with a basic knowledge of solid geometry, Ptolemy sets out to describe the motion of the Sun and the theory of the Moon's orbit. His work on the Moon is one of his highest achievements, for the problem is essentially a difficult one as the Moon's path in the sky is subject to so many varying forces. Using epicycle and deferent, Ptolemy was able to account fairly accurately for its observed motion provided he did not place the Earth exactly at the center of the Moon's path, a legitimate mathematical loophole, which he was unable to improve upon, as hard as he tried.

To account for the observed motion of Moon, Ptolemy displaces Earth from center of Moon's orbit.

The two final sections of the *Almagest* deal with the stars and the planets. The stars are catalogued à la Hipparchus, but their total is increased from 850 to over a thousand, showing that Ptolemy was also a very competent practical observer. As far as the planets are concerned, he used a deferent and epicycle to account for their observed motion, but in order to deal with the complications of their movements, Ptolemy had not only to offset the Earth a little from the center of things, as he had been obliged to do for the Moon, but also to imagine that the epicycles performed small oscillations as they circled on their way.

To account for observation, Ptolemy offsets Earth from center and ascribes small oscillations to motion of planets.

This highly intricate and complicated theory of the motion of heavenly bodies survived a millennium of scientific apathy of the Dark Ages well into the Renaissance, until a Polish canon introduced a simpler and more elegant theory. How could the complicated geocentric theory last for so many years? Why did nobody question the theory? What happened to the Greek-like curiosity? The answer is perhaps in the Roman civilization, the subject of the next chapter.

1.5   Wonder of Ancient Greece

In concluding this chapter on the kindling of science it is worth paying tribute to the people who, in the course of about three centuries advanced human knowledge to a peak never achieved by our race either before or for many generations after. It is remarkable that a tiny region on the globe centered around Athens and stretching only a few hundred miles could produce so many great men of science. It is true that the Greeks imported the raw material of science from Egypt and Babylon, but the refined abstractness with which they molded that knowledge, and the theoretical exactness to which they excelled it was only a Grecian trademark.

The love for abstraction and intellectual adventure is best described by Edith Hamilton in the following passage:

> Once upon a time—-the exact date cannot be given but it was not far from 450 BC—an Athenian fleet cast anchor near an island in the Ægean as the Sun was setting. Athens was making herself mistress of the sea and the attack on the island was to be begun the next morning. That evening the commander-in-chief, no less a one, the story goes, than Pericles himself, sent an invitation to his second in command to sup with him on the flag-ship. So there you may see them sitting on the ship's high poop, a canopy over their heads to keep off the dew. ...Pericles bethinks [the young boy filling the cups] of the poets and quotes a line about the "purple light" upon a fair young cheek. The younger general is critical: it had never seemed to him that the color-adjective was well chosen. He preferred another poet's use of rosy to describe the bloom of youth. Pericles on his side objects: that very poet had elsewhere used purple in the same way when speaking of the radiance of young loveliness. So the conversation went on, each man capping the other's quotation with one as apt. The entire talk at the supper table turned on delicate and fanciful points of literary criticism. But, nonetheless, when the battle began the next morning, the same men, fighting fiercely and directing wisely, carried the attack on the island. [Ham 93, p. 65]

Whether this exchange of discourse between two warriors actually took place is not known. But the very fact that it has come down to us points to the remarkably advanced intellectual capacity of the Greek citizenry. No such anecdote exists between Caesar and his second in command; nor between General Grant and General Sherman. Such a discussion of color-adjectives between generals could take place only in ancient Greece.

We could only ponder as to why and how the process of the intellectualization of the populace occurred in Greece. Was it because of the detachment of Greek science from religion? Was it because of the rise of secular philosophy in Greece? Was it because of the way the Greeks educated their youth? Each of these factors may have had a hand in the development of knowledge in Greece, but they only displace our lack of knowledge about the real cause of this development, because they leave unanswered the question of the roots of the factors: Why was *Greek* science detached from religion? Why did secular philosophy rise in *Greece*, and nowhere else? (And if it rose in other regions of the globe, why did abstract science not develop in those regions?) How did *Greeks* come to educate their children in such a way as to develop them into abstract thinkers?

No other ancient civilization beside the Greeks developed abstract science.

We may never know *why* the Greek civilization turned out the way it did. One thing, however, is certain: In no other ancient civilization has history encountered such an advanced achievement in philosophy, astronomy, geometry, and physics. Other civilizations such as the ones in ancient China or in central America may have had means of predicting the motion of some celestial bodies, or an eclipse of the Sun or the Moon; nevertheless, their astronomy was purely observational much like the Egyptian and Babylonian astronomy of the first and second millennia BC. There is no indication that any of these civilizations discovered trigonometry, or proved any theorem in geometry, or had any theoretical framework for their astronomy. As far as we know, abstract science in antiquity was purely a Greek phenomenon. And the rebirth of scientific thought during the Renaissance started at precisely the point where the ancient Greeks left off.[6]

Each great civilization of the past has left a wonder to remind the posterity of the glory that once was. The Egyptian Pyramids tell of the commanding power of Pharaohs over thousands of slaves who carried one boulder after another to the top of each pyramid over a period of many years. The Hanging Gardens of Babylon convey the magnificent achievement of a nation at the peak of the agricultural revolution. The Colosseum in Rome tells the story of an engineering marvel built for a nation subsumed in a gladiatorial entertainment of the most brutal kind. The wonder of Greece, however, is not found in the temple of

---

[6]In acknowledging the contribution of Islamic scientists of the middle ages, we have to point out that *their* starting point was also the works of the Greek scholars such as Euclid and Archimedes.

Athena *Parthenon* nor in the monumental temple of Zeus ruined in the battles of the ages. Greek wonder is in *The Elements* of Euclid, the *Method* of Archimedes, and in many other pyramids and colosseums of thought destroyed by the raging winds of time. And these wonders truly surpass all others.

1.6   End-of-Chapter Material

1.6.1   Answers to "What do you know?"

1.1. You would go to the temples. That's where the priests were and that's where higher knowledge was learned and taught.

1.2. Calendar!

1.3. No, he could not have. If he had known about it, he would have had to incorporate it into his theory. But his theory assigned a *single* sphere to each planet, and the sphere was in *uniform* motion.

1.4. The left picture represents the smallest Earth; the middle, the largest.

1.5. The closer the quarter moons are to the top and bottom of the Moon orbit, the farther the Sun is. Therefore, the middle picture represents the farthest Sun; the left picture, the nearest. The Sun is on the right, because the right half of Moon is illuminated.

1.6. Figure 1.3 shows that as you get closer to the north pole *in the northern hemisphere*, the larger the angle the vertical makes with the Sun rays. So, the location of the picture on the right is farther north than the picture on the left. It is also clear from Figure 1.3 that as you get closer to the *south* pole in the southern hemisphere, the larger the angle the vertical makes with the Sun rays. So, the location of the picture on the left is farther north than the picture on the right.

1.7. Omaha and Houston are both in the northern hemisphere, and Omaha is north of Houston. Therefore, the shadow angle there is larger. The ratio of the distance between Omaha and Houston to Earth circumference must equal the ratio of the angle difference to 360 degrees. So, if θ denotes the angle difference, we have $1600/25000 = \theta/360$ or $\theta = 23$ degrees.

1.8. The planet moves a third of its epicycle from one snapshot to another. So as the center of the epicycle moves from 11 to 10 o'clock, the planet moves from 2 to 10 o'clock of its epicycle (see Figure 1.4). So, for half way between 10 and 11 of the deferent, the planet reaches half way between 2 and 10, i.e., 12 o'clock. Similarly, as the center of the epicycle moves from 3 to 2 o'clock, the planet moves from 6 to 2 o'clock of its epicycle. So, for half way between 2 and 3 of the deferent, the planet reaches half way between 6 and 2, i.e., 4 o'clock.

1.6.2   Chapter Glossary

Astrology The study of the sky with the purpose of understanding and controlling what happens on Earth. Astrology is not a science, because there is absolutely no connection between what humans do on Earth and what happens in the sky.

Astronomy The *science* of the study of the objects in the sky, without any intention of relating them to what is happening on Earth.

Deferent A large circle centered at the center of Earth on which the center of another circle, called epicycle, moves.

Ea The Babylonian supreme god of Mesopotamia.

Epicycle The circle (or sphere) on which a planet moves. The center of this circle moves on a larger circle. In the simplest model, this larger circle, called *deferent*, has the Earth at its center. The combination of the two motions results in a retrograde motion as well as a change in the brightness of the planet (due to its approach to Earth).

Geocentric Model A model of the solar system according to which the Earth is at the center and the Moon, the Sun, all the planets, and the stars move around it.

Heliocentric Model A model of the solar system according to which the Sun is at the center and all the planets move around it.

Ra The Egyptian Sun God.

Retrograde Motion The slow-down, reversal of direction, another slow-down, and another reversal of the direction of motion of planets. This is most conspicuous for Mars.

1.6.3   Review Questions

1.1. When did science start? Name some of the objects of humans' experience out of which they tried to make sense.

1.2. What are the three large categories into which humans divided the world around them? Which category was subject to close scrutiny, and therefore, understandable? Which was regular, and what characteristic of it prompted humans to associate it with creatures?

1.3. Why did the third category cause discomfort? Why was is related to the second category? What did humans do to control the uncertainty associated with the third category? How did they communicate with the sources of the third category?

1.4. Who were responsible for communicating with the creatures in the sky? Why do we call them the first scientists?

1.5. What is the difference between astronomy and astrology? Which one is a science? What is the main characteristic of astrology?

1.6. Why did Sun have a special place among Egyptian gods? What is the name of the Egyptian Sun God? How is the Babylonian supreme god different from the Egyptian Sun God?

1.7. Where did the observational astronomy begin? Where did the geometric explanation of astronomy start?

1.8. Who proposed the first astronomical theory? Where did he get the knowledge on which he based his theory? What kind of philosophy did he initiate? What was the role of numbers in this philosophy?

1.9. State the Pythagorean system of the world. What is at the center and what goes around it? Where does Sun get its light from? Where does Moon get its light from?

1.10. What are the two characteristics of a good theory? Who introduced these characteristics in their theory? How do these characteristics manifest themselves in that theory?

1.11. Why was the Pythagorean astronomy abandoned? What replaced that astronomy?

1.12. What is retrograde motion, and which planet exhibits it most conspicuously? Who came up with a theory that explained the retrograde motion without sacrificing simplicity and symmetry? How did he do it? What is his theory called?

1.13. What is the difference between Aristarchus' heliocentrism and the older one? How did Aristarchus determine the size of the Moon compared to Earth? What did he get for the size of the Moon in terms of Earth size? How did he measure the Earth-Moon distance? What is this distance in multiples of Earth diameter?

1.14. Describe the method Aristarchus used to estimate the Earth-Sun distance. How many times the Earth-Moon distance was his estimate? What did that say about the size of Sun compared to size of Moon? Compared to Earth size? How many times was Sun bigger than Earth? For which one does it make more sense to be going around the other?

1.15. In what city of Egypt was the center for higher learning and research built? What did this center consist of? Name one famous person who worked at the center, and a famous director who contributed to the prosperity of astronomy and mathematics at the center.

1.16. Describe the method that Eratosthenes used to measure the circumference of Earth. How does his measurement compare with today's measurement?

1.17. What was the new feature of the retrograde motion of planets that the homocentric model could not explain? How was the homocentric model modified to account for this feature? What name is associated with the new theory?

1.18. Who made the first catalogue of the stars? Whose name is associated with the geocentric model due to the authoritative book he wrote on the subject? What is the name of that book?

1.6.4   Conceptual Exercises

1.1. The central fire in the Pythagorean astronomy is not visible. How could you explain this fact by considering the spin of the Earth in relation to its orbital rotation around the central fire?

1.2. Consider the shadow angles used in measuring the circumference of Earth in the *southern* hemisphere. Which angle is larger, the one in the south or the one in the north? See Figure 1.3 for help.

1.3. Santa Rosa and Cordoba are two cities in Argentina. Cordoba is 350 miles north of Santa Rosa. The shadow angle in which city is bigger?

1.4. Figure 1.5 shows snapshots of the epicycle of planet M as it goes around the Earth E counterclockwise. Suppose that M moves around the small circle three times as the center of the small circle moves on the big circle once. M starts at the position shown in the figure; call it the first snapshot.
(a) On which small circle will you find M after its first revolution? Draw M on that circle.
(b) On which small circle will you find M after its second revolution? Draw M on that circle.
(c) What fraction of its epicycle does M cover from one snapshot to the next?
(d) Starting with the second snapshot, draw the location of M on all the remaining small circles.

1.5. A planet M moves 6 times on its epicycle while the center of the epicycle goes around Earth once on the deferent. M starts at the 6 o'clock position of its epicycle when the center of the epicycle is at 12 o'clock position of the deferent. All motions are counterclockwise.
(a) When the center of the epicycle reaches the 11 o'clock position of the deferent, at what position of its epicycle is M?
(b) When the center of the epicycle reaches the 7 o'clock position of the deferent, at what position of its epicycle is M?

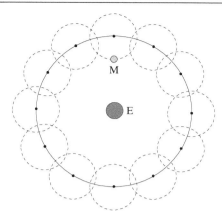

Figure 1.5: Planet M moves on the little circle four times while the little circle moves once on the big circle.

(c) When the center of the epicycle reaches the 2 o'clock position of the deferent, at what position of its epicycle is M?

(d) When M is at 3 o'clock position on its epicycle during its first revolution, at what position on the deferent is the center of the epicycle?

(e) When M is at 12 o'clock position on its epicycle during its second revolution, at what position on the deferent is the center of the epicycle?

(f) When M is at 9 o'clock position on its epicycle during its third revolution, at what position on the deferent is the center of the epicycle?

(g) When M is at 3 o'clock position on its epicycle during its fourth revolution, at what position on the deferent is the center of the epicycle?

(h) When M is at 9 o'clock position on its epicycle during its last revolution, at what position on the deferent is the center of the epicycle?

1.6.5   Numerical Exercises

1.1. On planet Middle Earth, the city of Shire is 50 miles directly north of the city of Mordor, both cities lying in the northern hemisphere of Middle Earth. The angle of the shadows in Shire is 9 degrees larger than those in Mordor.

(a) What is the circumference of Middle Earth?

(b) What is its radius?

1.2. Johannesburg and East London are two cities in South Africa. Johannesburg is 475 miles north of East London. The shadow angle in which city is bigger? By how many degrees?

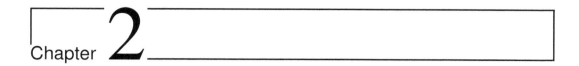

Chapter 2

A Dark Interlude

2.1 Roman Civilization: ca 250 BC–ca 400 AD

Ptolemy was the last of only a handful of scientists who flourished in the three centuries following Archimedes's murder. The Greek science was fatally wounded when in 212 BC, the sword of a Roman soldier ran through the delicate body of Archimedes, the prince of all Hellenic geniuses. After centuries of diligent observation, monumental discoveries, and ingenuous theoretical findings, the Greek political and economic system could no longer support science. However, unlike the earlier period, when the setting of the scientific Sun in the river civilizations of Egypt and Babylon heralded its rise in Hellas, Greek science could not find a rising horizon. The torch that Pythagoras and his forbears brought from Egypt and Babylon and passed on to such champions as Aristarchus, Euclid, and Archimedes, remained smoldering in the hands of Ptolemy until it burned itself out.

The center of political and economic power was clearly shifting from Athens to Rome, but, alas, science did not survive this shift. Romans were a pragmatic conquering people with very strong military and political establishments, but lacking in culture. And when they came in contact with the advanced Hellenistic civilization to their east, which had reached to their footsteps in Sicily and Syracuse, they absorbed its culture like a sponge, albeit selectively. Aside from the artistic treasures that decorated the mansions of the rising aristocracy in Rome, much of the "culture" that Romans took from the Hellenistic milieu was only of practical utility.

The pragmatic Roman citizenry showed little interest in the abstract and theoretical achievements of the Hellenistic civilization. From among such magnificent scientific treasures as Euclidean geometry, Epicurean atomic theory, and the sophisticated astronomy of giants like Aristarchus and Hipparchus, the Romans selected some unknown practical manuals on agricultural management; and in the education of Roman children, science and mathematics had a vanishingly small role.

The rapid economic growth of the Roman Empire led inevitably to the stratification of the population. The aristocracy was reaping the fruits of prosperity and living in luxurious mansions built on enormous estates. The artisans in the cities and the farmers in the countryside were being taxed more and more to meet the expenses of the army and the bureaucracy. In the absence of a collective societal interest in the rationality of philosophy, mathematics, and science, the stress of economic and cultural poverty of the general public found two outlets: entertainment and religion.

2.2   Entertainment in the Roman Empire

The word "gladiator" is synonymous with the Roman Empire. Gladiators were not simply a coincidence which could have popped up in any other civilization. They were at the root of the Roman culture and social structure. The Flavian Amphitheater (Colosseum) at Rome had a seating capacity of 50,000. Comparable amphitheaters were built in major cities in Italy, France, Yugoslavia, and North Africa. Smaller amphitheaters holding 20,000–25,000 were numerous. Even small towns in the western provinces had their own amphitheaters, albeit quite small ones. "They came in all sizes, down to the little arenas hollowed out of the hillside beside military camps, which served the garrison for weapon drill, as well as for the occasional gladiatorial show" [Wel 84, p. 272].

The construction of the great amphitheaters required a tremendous amount of labor, materials, and engineering skill that were not easy to come by. The design of the Colosseum and similar amphitheaters were very sophisticated in terms of crowd control, ease of access and evacuation for so many people, and arrangements for the delivery and storage of props and performers, animal and human. Underneath the spectators was an elaborate labyrinth of corridors and dens in which the wild beasts were kept in cage and later released into the arena untouched by human hand.

There is no doubt that gladiatorial shows were the most important events in the Roman culture. In most cities and towns, the amphitheater was the biggest building, its only rival being the circus—another institution of entertainment—if there was one. It is clear that Romans enjoyed public slaughter of men and beasts to the point that it became a fundamental institution and a social ritual, which had to be properly accommodated, and for which the society was willing to set aside a good fraction of its resources.

The justification for these brutal shows was that—as Pliny the Younger describes in his famous *Panegyric 33*—they "inspired a glory in wounds and a contempt of death, since the love of praise and desire for victory could be seen, even in the bodies of slaves and criminals" [Rad 75]. The games had their origins in the Roman belief that human blood propitiated the souls of the dead. Therefore, people sacrificed prisoners of wars and slaves of "poor quality" bought specifically for this purpose.

The person in charge of the show boasted of his cruelty. One such "sponsor," a certain magistrate, has on the base of his statue, "Over four days he showed eleven pairs [of gladiators]: from these eleven of the best gladiators of Campania were killed, with ten bears killed cruelly" [Wel 84, p. 275]. Other reports describe equally hideous shows of "the half-eaten beast-fighters, a mass of wounds, dust and blood, who even so beg to be preserved until the next day, although they will be flung again to the same teeth and claws" [Cha 61, p. 109].

The amphitheater was the stage on which the terror of pain and death, the fragility of life, and the arbitrariness of the power of the ruling class unfolded. Those who died in the arena died for the established social order. The show was a "terrifying demonstration of what could happen to those who failed to please their masters, who failed to conform to the established order.... A spectator who was witty at Domitian's expense was dragged out and thrown to the dogs in the arena. Commodus, a handsome and ruthless emperor of Rome with enormous athletic ability who had a hostile relation with the senators, walked towards the senators' seats holding in one hand the head of an ostrich which he had just sacrificed and in the other the sacrificial knife. The threat was blatant. Dio records how he himself chewed on a laurel leaf from the wreath on his head to stop himself giggling, presumably from sheer terror" [Wel 84, p. 275].

2.3   Religion in the Roman Empire

Contrary to the Hellenistic society, where rational philosophers fought superstition and controlled its unbridled spread, the Romans were thrilled by mysticism and magic. In fact, many of the "philosophers" in the Roman society took the philosophy of Plato, and introduced so many supernatural and mystical elements into it that it finally turned into various forms of religious cults.

Philosophy turns into various cults in the hands of Roman intellect.

2.3.1   Rome, a Thriving Marketplace for the Supernatural

The Roman society, like many other before it, had gods of its own, with Jupiter as the dominant deity. Other deities such as Mithras of Iran and Isis of Egypt were also worshiped in Rome, but they were done privately. The need for deity was enhanced by the uncertainty in the daily life. All the sacrifices and prayers of the citizens were their part of an invisible contract to which the gods were supposedly bound. And if the gods failed to perform their part of the contract, their alters would be overturned, their images dishonored, and their temples vandalized. This mob action could be very dangerous politically. So the state imposed severe penalties against those who either enticed or participated in such riots. It was therefore natural for the populace to gravitate to magic and cults.

While the same tendency may have existed in the earlier Greek civilization, the smaller size of the city-state and its accompanying sense of community, *together with the constant opposition of anti-superstition philosophers* such as Socrates and Plato, retarded the growth of cultism and mysticism in ancient Greece. In contrast, in the vast reaches of the Roman Empire, and in the absence of the rational guidance of philosophers, the individual felt detached from the society and took refuge in the teachings of charismatic "holy men."

The word "holy man" often refers not only to the familiar religious activists such as Jesus Christ, but also to such cult leaders as Apollonius of Tyana, who claimed to be the reincarnation of Proteus, a minor sea god (servant of Poseidon) who could change his appearance at will; to the puzzling Peregrinus Proteus, who immolated himself after an Olympic event; or to Simon Magus, reputed to have had a bad fall when he tried to fly [And 94, p. 4]. The Roman Empire was a huge marketplace for holy men, claiming to be associated with supernatural powers. Each warned of the coming of a "doomsday," offered salvation in a virtuous life, in which all ties to the "evil" material objects were severed, and this salvation came only if men adhered to *his* principles.

2.3.2   Emergence of Christianity

The free marketplace of holy ideas lasted for a long time. After the impact of Christianity and the crucifixion of Christ, and before the consolidation of the authority of the church, the holy men comprised of converts who offered their own interpretations of Jesus Christ's teachings. The focus of these interpretations was the notion of the Messiah—and at first, the only notion—in which the followers of Christ disagreed with their fellow Jews. However, the cleavage between the new group and their Jewish brethren was to widen to the point that many Christians were persecuted for blasphemy in the hands of the Jews.

A leading figure in this period, with a genius for organization, was a Jew named Saul (later to be known as Paul), who was at first a vehement persecutor of the followers of Christ. Then, on his way to Damascus, he allegedly had a vision of Christ, and from that moment on, he devoted his energies to the service of Christianity. After a period of meditation, Paul engaged in passionate and illuminating missionary activities in Syria and eastern Asia Minor. For his intense devotion to his mission Paul is considered the second founder of Christianity. He was instrumental in establishing Christianity as a gentile faith with its own non-Jewish calendar, for instance. Paul, however, was not the only missionary. Another noted disciple, who went to Rome before Paul and founded a church there was Peter.

After Peter and Paul, the churches were still small and suffered greatly from the hatred of pagans and Jews alike. During the next century, many of the qualities that characterized

Christianity during the Roman empire and even later, took form. The church in each city[1] was independent, and, much like a synagogue, it had a fairly democratic organization under a board of elders and deacons. However, as time passed, and the church felt more and more under attack both from outside (the pagans and the Jews) and inside (the heretics), the need for the concentration of power appeared inevitable. By the third century, the bishop of each city, though still elected, was the spokesman for, and director of, his congregation.

The relation between the church and the state was strained, to say the least. No Roman ruler tolerated any cult which instigated popular discontent. At the time of Nero and the Flavias, simple membership in a church was sufficient reason for execution. However, the firm belief in the teachings of Christ and the promise of salvation in the afterlife were strong incentives for the oppressed public to adhere to the doctrine. So, although the Roman state took every opportunity to persecute the leaders of Christianity, the news of their martyrdom and the episodes of their embrace of death to "join their Father in Heaven" traveled very fast and very effectively to the ears of the masses who were looking for a way to emancipate themselves from the tyranny of the emperors.

Ineffective persecution of Christians by the Roman Empire.

The end of the third century brought about fundamental changes to the political and social structure of the Roman Empire. The consolidation of power in the hands of the emperor accompanied more oppression and economic decay, bringing with it an unprecedented internal dissidence. And, the external threats by Sassanian Persia and the Germans added to the weakening of the state.

These political changes in the state were followed by equally important reorganization of the church. In 284, when Emperor Diocletian came to power, the number of Christians had grown to about 10 percent of the Empire, including members of the imperial government and the army. Diocletian, although tolerant of political foes, showed no enthusiasm for intellectual deviations. And on February 23, 303, he focused his intolerance on the Church and carried out the worst persecution ever on Christianity in which the church at Nicomedia was invaded and burned, and the Holy Scriptures within it were deliberately given to the flames. Diocletian's successor carried out similar persecution until 311, when he finally issued a decree of tolerance.

The accession of Constantine brought an incredible change in the relation of the Church and state. During one of his campaigns in 312, Constantine is said to have had a dream in which he was allegedly told to mark a Christian emblem on the shields of his soldiers. His victory in that campaign sealed his belief in the power of the Christian God, and he never wavered in his support of Christianity. The first great Christian basilicas were now built in Rome; Constantine's mother Helena built the church of Nativity in Bethlehem; and the new capital of Constantinople was deliberately designed as a Christian city. The power of jurisdiction was granted to episcopal courts, confiscated properties were returned to the Church, and the clergy were exempted from pagan public rituals. Constantine himself was not baptized until he lay on his deathbed, partly to "wipe away as many of the black sins in his ruthless life as possible, partly because the great majority of his subjects were still pagan" [Sta 98, p. 680].

Roman Empire gives in; Christianity becomes the state religion.

Slowly, paganism gave way to Christianity. In 382, the pagan symbol of the Roman Senate House, which was previously removed by one emperor and restored by another, was again taken away. This action brought protests from some pagan senators, which were countered by the impassioned arguments of Bishop Ambrose of Milan. In 392, Theodosius banned all pagan sacrifices, and by his death in 395, Christianity was the state religion.

2.3.3   Philosophization of Christianity

Now that the struggle for survival was over, Christianity could concentrate on intellectual activities and theo*logy*. The Christian converts of the earlier period, who had a background

---

[1]The hubs of Christianity were the cities.

in classics and philosophy, recognized a need for clarification and extension of Christian theology. The most influential theologian after Paul is Augustine (354–430), who, in his youth at Carthage, was so immersed in the worldly pleasures as to have an illegitimate son. But his search for a spiritual base led him through Manicheism, skepticism, and Neoplatonism, until he came under the influence of Bishop Ambrose at Milan and was converted to Christianity—the religion of his mother—in 386.

St. Augustine attacks rationalism.

In his widely known book, the *Confessions*, Augustine very effectively and persuasively tells of his own search for meaning in life and how he found safe mooring in the Church. He attacks pagan pride in reason and the resulting elevation of man as an end in himself. He warns that man could not penetrate the unknown beyond a certain limit "lest freedom of thought beget impiety of opinion." In his letters, he writes more bluntly, "God and the soul, that is what I desire to know. Nothing more? Nothing whatever!" [Sta 98, p. 690].

St. Augustine incorporates Plato's philosophy and astronomy into Christian doctrine.

Augustine was familiar with some of Plato's writings and commentaries on them, which had been written in Roman times and later. From these he concluded that the Platonic conception of eternal truths was consistent with the Christian theology and formed a foundation on which knowledge of the Universe could be built. In particular, Plato's cosmology with the Earth at the center and the Universe filled with the eternal "soul" blended very nicely with the biblical story of creation.

As the Roman Empire declined, the central power of the Emperor was divided into the hands of a few land owners whose *villas* or country houses became the local center of power. The rising popularity of the Church and its prestige in the imperial family placed it in a unique position to share this power during the following Dark Ages. The Church, whose base started in the cities, now moved to the countryside, where the mode of life was reverting toward the Neolithic level of simple food-production in a network of stable, independent cells, which could survive the decline of the imperial machinery and the disappearance of a money economy.

2.4   Education in the Roman Empire

The unique position held by entertainment, politics, and religion in the Roman society could be better understood by examining the experience of its youth as the bearer of the Roman tradition. A significant part of this experience is contained in the way the youth was prepared for its future responsibility. This preparation was the task of the Roman educational system, which, like other aspects of the Roman culture, was taken from the Greeks. How did the Romans adapt the Hellenistic education to their own needs? To answer this question, we need to examine the Greek education.

2.4.1   The Greek Roots

The scientific and philosophical discoveries of the sixth century BC by such great thinkers as Pythagoras stimulated an unprecedented intellectual activity in ancient Greece. However, the general impact of such activities was not felt until a century later when democracy began to take hold of the society, and a greater number of citizens participated in political life. This in turn created a demand for education, which would train the citizens for effective and efficient government.

By the time of Plato (around 400 BC), there were several methods of educating the youth, of which two stood out. The **sophists** were professional teachers charging fees for instruction, and therefore, could be afforded only by the wealthy. As the sophist movement started with the democratic tendencies in Greece, their main concern for education was to inculcate in students the ability to debate in a group with differing viewpoints. This was of course necessary for democracy especially for those students who would become leaders in the emerging government. Thus the entire effort of the sophist was concentrated on

Sophists emphasized rhetoric much like today's educators emphasize "communication skills."

rhetoric, although other subjects such as animal husbandry, veterinary science, agriculture, and household medicine were, to a much lesser degree, discussed.

The first and most respected sophist, *Protagoras*, emphasized the teaching of poetry, music, and gymnastics in his elementary education. Mathematics and astronomy were taught in so far as they had *practical* use in the affairs of the state. For his more advanced instructions, Protagoras did not believe in a "more intensive study of the content of the elementary curriculum, for he specifically rejects 'calculation, astronomy, geometry, and music'. But for the study of poetry at the higher level he expresses great regard" [Bec 64, p. 170]. He had a special interest in logic and dialectic, but this interest was a basis for the teaching of rhetoric.

The **Platonic** education was in complete contrast to the sophists'. In his *Republic*, Plato describes an ideal society as one that is governed by an elite, philosophical, governing class,

Plato's emphasis on mathematics and its consequence.

and his educational theory is based on the fulfillment of this ideal. The curricular content of Plato's elementary education consists of literature, art and music, gymnastics and sports, practical mathematics, and mathematics as a discipline. It is this emphasis on mathematics that sets Platonic education apart from the sophists'. This emphasis helped plant the seed of rational, logical, and mathematical thinking in the minds of so many bright scientists, and nourished such great politicians as Ptolemy I who built the Alexandrian Museum, in which such great scientists as Euclid, Aristarchus, Eratosthenes, and many others were trained.

In the *Republic*, Plato assumes some teaching of elementary mathematics, not only as a basis for the higher studies to be undertaken later, but also for its practical use in making citizens and soldiers more efficient, as well as for its general educational merit. He "refers to arithmetical knowledge as 'universally useful in all crafts and in every form of knowledge and intellectual operation—the first thing everyone has to learn .... Arithmetic, then geometry and all branches of the preliminary education which is to pave the way for Dialectic should be introduced in childhood'" [Bec 64, p. 209].

Plato does not confine the mathematics of elementary educational curriculum solely to

Plato's belief in mathematics as a test of intelligence.

practical matters.

> [Mathematics] plays its part in character development in providing a stimulus to effort. It forces the mind to reason and stirs the dull youth to become quick, retentive, and shrewd 'beyond his natural powers'. It awakens the power of thought, disciplines and trains the mind, preparing for any kind of further study. In fact, the power of calculation is identified with the operation of the rational element in the soul.
>
> Not only does mathematics train the mind towards the natural limits of its capacity, but it also serves to determine the relative capacities of different minds. ...That is, mathematics provides a means of selecting those intellects which will benefit from advanced study. It acts as a sort of intelligence test. [Bec 64, p. 210]

In fact, Plato suggests building a strong character not only by confronting it with hardship, danger, and temptations of pleasure, but also [Bec 64, p. 210]

Box 2.4.1. *by trying it "in many forms of study to see whether it has the courage and endurance to pursue the highest kind of knowledge [i.e., mathematics], without flinching as others flinch under physical trials."*

Plato's higher education consists entirely of math and science.

Plato's secondary education consists entirely of mathematics and science, and serves to select those who are to be admitted to the higher study of philosophy. Plato includes all branches of mathematics in his secondary curriculum. These are arithmetic, geometry, astronomy and harmonics, and even the "new" science of solid geometry.

> In the *Laws* there is a similar emphasis on mathematical studies in the higher education of the members of the Nocturnal Council. But there Plato believes that insight into

the principle of unity may be obtained through a study of the orderly movement of the stars and the planets. Thus astronomy becomes the major study to which the other branches of mathematics essential for an understanding of the mathematical relations involved in the movements of the celestial bodies are subservient. Astronomy performs in the *Laws* the same function that dialectic performs in the *Republic*. [Bec 64, p. 214]

Plato's secondary education, while encompassing the current practice of mathematics, severely excludes that other study, which was the most important part of influential sophists like *Isocrates*, and which was to dominate much of ancient education for centuries—the study of rhetoric. His objection to it was simply that

Box 2.4.2. *Rhetoric is incapable of unraveling the truth, because the orator must always seek to please and persuade by conforming to the mood of the public.*

Plato compares a sophist to the keeper of a beast whose mood is to be studied, calling "good" what it enjoys, and "bad" what it dislikes. In another place, Plato compares the rhetorician to a chef whose job it is to please the palate of his client, as opposed to a doctor who prescribes what he thinks is good for the patient. Plato naturally condemns the skill that seeks to distort facts simply to persuade an audience, and he indignantly rejects rhetoric as an instrument in the search for truth.

Greek cultural thirst for knowledge helped establish places in which the philosophical, mathematical, and scientific talent of the Greek youth could blossom. These places were not just the primary and secondary schools, but also institutions of higher learning such as Plato's *Academy*, Aristotle's *Lyceum*, and Epicurus' *Garden*, in which philosophers and scientists could exchange ideas, transfer knowledge, and create new knowledge. Greek aristocracy, themselves being a product of these institutions and an intelligent society in which they were raised, donated generously to the building of these centers.

The Library and Museum built in Alexandria by Ptolemy I, one of Alexander's successors, is a testament to the love of the Greek aristocracy for science and mathematics. So many talented scientists gathered together in these centers, so much genius diffused in every corner of those buildings, and so much knowledge was created there that at no other time in history did humanity witness such an outpouring of intelligence, except in post-Renaissance Europe. Alas, when Rome became a superpower and conquered Egypt, the Library and Museum received several destructive blows from zealots raised in the Roman-Christian culture. What was left of the Library and Museum received a final blow from another zealot, who thought that either the books in the buildings taught the same lessons as his holy book, in which case they were redundant, or something opposed to it, in which case they were heretical; in either case, the books had to be burned.

2.4.2   The Roman Adaptation

The creative genius that embellished the Hellenic society in all forms of human expression—art, literature, philosophy, mathematics, and science—was transformed into a prosaic pragmatism in the hands of the Romans. And in their adaptation of the Greek culture, the Romans took that system of education which suited their style of government: the sophist's system. The path that the Hellenistic studies took in Rome was grammatical and philological, and in higher schools, it was rhetorical. If there was any interest in philosophical issues, it was from a practical angle: to equip the orator with effective intellectual ammunition.

In the Roman secondary school, the school of grammaticus, a pupil would learn Greek, literature, grammar, spelling, writing, music, and geography with varying degrees of emphasis. Arithmetic was taught; however, neither in the secondary nor in higher education was it the theoretical arithmetic of Plato, but mere calculation.

After the secondary school, the youth's further education depended on his future occupation. Those intended for a farmer's life ended up at some farm station; those intended for the army went into the service; those intended for public life or for jurists, went to the rhetorical schools. In short

> **Box 2.4.3.** *The only institutions of higher education in the Roman Empire were the rhetorical schools.*

Geometry, astronomy, and mathematics were only lightly touched in Roman higher education.

The following passage nicely sums up the Roman higher education [Lau 70, p. 341]:

> In the rhetorical schools the young men studied rhetoric and all the arts which could make an effective orator. Cicero tells us that in the last century of the Republic "no studies were ever pursued with more earnestness than those tending to the acquisition of eloquence." ...In the schools the youths wrote declamations on prescribed themes ...and delivered them with proper accent and articulation. ...The analysis of language with a view to mastering all its forms was studied. Mathematics, philosophy ...and law, as well as literature, entered more or less into this higher curriculum; *but the three former seem to have been studied under specialist teachers, and did not form an essential part of the higher instruction with the majority of students* [our emphasis]. ...In short, we may say that in the higher education of youths who aimed at some form of public life—as all the ambitious among the well-to-do did—the two words "law" and "oratory" practically summed up their studies. Philosophy and geometry, which, along with astronomy, included in those days the whole of physical science, were merely touched ....

Lack of math and science education in Rome gave way to a burgeoning of sorcery, magic, and astrology.

A cursory comparison of the Roman with the Platonic education reveals the antipodal relation between the two. As much as Plato insists on mathematics, astronomy, and geometry, the Romans stay away from them. The Roman youths inherited a state for which they were equipped only militarily and "oratorily." Instead of doctors for social diseases, the Roman educational system produced chefs that, with their oratory delicacies, pleased the palates of society.

2.4.3   The Vanished Library

The only sign of scientific activity and the haven for the very few rational thinkers who were pursuing mathematics and astronomy was in the great library of Alexandria. Here, built by Greeks, filled with Greek books, and administered for a long time by the descendants of the Hellenistic glory, the dying heart of reason and science was feebly pulsating.

Of this library not a stone now remains. The beautiful but ruthless Syrian queen Zenobia gave the library its first destructive blow in AD 270. In the next century the library was deserted by many of its remaining scholars who could no longer endure the theological wrangling which by then had begun to split the city. The library continued to function, until Cyril was elected patriarch of Alexandria in 411.

Cyril burns the Alexandrian library and brutally murders its director Hypatia, a female mathematician.

Cyril was typical of one faction of the Christian Church at that time, which thought its duty was literally to fight and to destroy all that was in disagreement with it, and believed that knowledge for its own sake had little to offer Christianity. As soon as he was appointed to his see, Cyril began to attack anything which he considered unorthodox or irregular; he evicted one Christian sect from their homes and deprived its bishop of his personal belongings, drove the Jews from their synagogue and allowed his followers to plunder them, and eventually turned his attention to the library. In the belief that it was the repository of heathen teaching and pagan ideals, Cyril incited mob violence to such a degree that the place was set on fire and the director **Hypatia**, a scholarly woman mathematician, was murdered with unnecessary brutality in 415.

Some two hundred odd years later, a different zealot, arriving from far away was standing at the doorsteps of the library. This new visitor was Amrou Ibn el-Ass, caliph Omar's military commander who had just captured Alexandria. He had written to Omar about the thousands of volumes of books that the library was holding, and was awaiting the caliph's order as to the fate of the books and the library. When Amrou Ibn el-Ass opened Omar's message, he read thusly in it [Can 90, p. 98]:

> ...As for the books you mention, here is my reply. If their content is in accordance with the book of Allah, we may do without them for in that case the book of Allah more than suffices. If, on the other hand, they contain matter not in accordance with the book of Allah, there can be no need to preserve them. Proceed, then, and destroy them.

These words of Omar, echoing those of Augustine uttered some three hundred years earlier: "God and the soul, that is what I desire to know. Nothing more? Nothing whatever!", destroyed the last stronghold of the once glorious age of reason and science.

2.5   The Aftermath

Archimedes' murder by the sword of a Roman soldier symbolizes the end of the Greek ideology and the beginning of the long period in which the Roman ideology ran supreme. The global influence of Rome and its doctrine of law, politics, rhetorics, and oratory ushered the human race into a long era of mental void. The decay of the cities led to the concentration of local political power in the countryside, where the serfs produced huge quantities of food for their feudal landlords, who had absolutely no interest in philosophy, astronomy, mathematics, or science. To them the maintenance of their territory, the subordination of their peasants, the collection of taxes, the abundance of crop yields, the passing of all the privileges to their children, and, of course, a good relation with the church were the ultimate goals of existence.

Long were gone the competent generals, who on the eve of a major battle were engaged in a heated philosophical discussion of a poet's use of color-adjectives (see Section 1.5). Eventually no one asked questions; no one was interested in the philosophical investigation of nature as was so intensely pursued at the time of Plato and the generations after him, as if the collective mind of the civilized world had gone into a coma. The landlords and the Church found a symbiotic relation whereby the Church assured the aristocracy of a good life in heaven, and the aristocracy assured the Church of a good financial life on Earth.

The entire intellectual activity fell into the hands of the clergy, who, in the footsteps of St. Augustine, adopted those fragments of Plato's and Aristotle's philosophy, which perfectly filled the Christianity's void of the worldly rationality. And for well over a millennium the entire intellectual activity of the intelligentsia (the bishops, the monks, and all the clergy) was to interpret and reinterpret Plato and Aristotle in light of the Bible.

In the meantime life was moving on; the feudal lords had plenty of food, luxurious mansions, comfortable households, and other landlords to assist them in ruling the peasantry. In their magnificent cathedrals, the clergy gained enormous political power and control over the spiritual life of the masses. The masses were content with the food and shelter provided them by the landlords. "Who needs mathematics, astronomy, and geometry? How can science be relevant to an agricultural society whose only tools are plows and oxen?" Indeed, there were tremendous advances in the art of plowing since the time of the Romans, and the cathedrals built during the Dark Ages testify to the creativity of the Christian architects of the time. *They* did not use any Platonic science to build those cathedrals! Medieval society, like any other stable society at any other period of human history, was the most advanced society of the day, and the general public was content with what it had. Only in retrospect, and with over a millennium of hindsight, are we affixing the adjective "Dark" to this period. To the typical ordinary man of the time, his days were as "bright" as they could ever be.

And so the landlords came and went; the kings and queens came and went; the bishops, and the popes, and the peasants and the serfs came and went. And they fed themselves, and housed themselves, and the popes and the bishops blessed the landlords and their household, and warned the peasants of the fires of hell. And everything seemed very ordinary and as it should be. And no question was asked; and no curious mind-eye looked at the heaven; and everybody thought that all answers were in the Bible and in the works of Aristotle and Plato. And this went on for over a thousand years.

A minor exception was the people of what is now called the Middle East. With the birth and rapid expansion of Islam in the seventh century, the center of research moved to the East. The voluminous works of Aristotle were translated into Arabic accompanied by new interpretation and outlook. Even in the East, Aristotle became God's intellectual gift to man and the representative of the Divine revelation about the natural Universe. The Islamic scholars were mostly concerned with the rehashing of the Greek mathematics, although some very novel ideas—such as the invention of algebra—originated from their work. And they definitely had very little interest in observation.

Islamic observations of the heavens were confined primarily to a re-determination of quantities like the size of the Earth, the precession of the equinox discovered by Hipparchus, and the position of the Sun's path among the stars, while the movement of the Moon was also studied with great care because of the demand for its use in the Islamic lunar calendar. Ptolemy's system of epicycles and deferents was used as a basis for preparing tables setting out the future positions of the planets, the most notable of which were the Toledan tables which were calculated by Jewish scholars working in the eleventh century under their Arab masters in Toledo, and in the first hay of the fifteenth century a catalogue of stars was prepared at Samarkand, the first to be compiled since Ptolemy's thirteen hundred years before.

2.6   A Lesson from the Past

After two thousand years, we still wonder what happened to the great Hellenic civilization. Why a great culture that produced Pythagoras, Epicurus, Euclid, Aristarchus, and Archimedes turned intellectually so dry. Why, out of all people, the Romans succeeded in leading the humanity for almost half a millennium. There is no simple answer to these questions. However, one thing is clear: rational thinking and reason took a back seat in the Roman empire.

Reason takes a back seat in the Roman empire.

The most influential Roman thinker, Cicero, dismisses the study of nature (science) as either too hard or irrelevant. As the intellectual voice of the Roman society, which was based on a pragmatic ideal, he preached the importance of the practicalities of Roman life such as politics, law, and statesmanship, and the cornerstone of his educational philosophy was rhetorics and oratory. Cicero's teachings produced great politicians, lawyers, statesmen, engineers, and military generals, but no scientists or mathematicians.

Two thousand years later, at the dawn of the twenty-first century, Howard Gardner, a leading American educational psychologist tells the readers of *New York Times* of March 3, 1998 (our emphasis):

> ...Half a dozen years ago, when our economy was languishing along with our test scores, it was easy to blame our poor schools and to push for better results from our students. Now the United States stands at the top of the world economically, but our students are still scoring at the bottom on international math and science tests. *Since high scores on these tests obviously aren't crucial to our economic success*, we need to decide what kinds of tests matter in helping form the kind of citizen we want to have.

...But most standardized tests ask fact-based questions that sample a wide range of topics in a somewhat superficial way. Students who score well on these tests are like *well-trained athletes or musicians*: through practice, they have become proficient at a certain skill—in this case, they have done extensive problem sets in many different "content areas" and can move quickly from one question to another.

These tests are helpful in the real world, especially in high school and college. But they simply do not show whether a student can think seriously about a scientific issue. We could drill our students with problem sets and raise their test scores, but still be left with a population that remains scientifically and mathematically challenged.

After all, students should be able to apply scientific and mathematical concepts to the world around them. *As adults, they need to know how to decide which life insurance to buy, how pesticides affect their food and how interest rates determine home mortgages.* Citizens also need to be able to decide whether cloning research should be banned, whether more money should be poured into studying global warming and whether there should be a national health care plan.

This quote is brimming with pragmatism! As the author of the "theory" of *multiple intelligences*, which equates the intelligence of a rap dancer (bodily-kinesthetic intelligence) with that of a mathematician (logical-mathematical intelligence), Howard Gardner cannot be expected to differ significantly from Cicero when addressing issues of mathematics and science. Gardner's satisfaction with the US economy in 1998 even though "our students are still scoring at the bottom on international math and science tests," is typical of many American educators. The very scientific and mathematical skills that Gardner dismisses because they make students "like well-trained athletes or musicians," are driving the US economy which "stands at the top of the world." This economy, whose basis is electronic and information technology, is driven mostly by Indian, Chinese, Korean, and other foreign engineers, "who score well on these tests," and were educated in schools which "through practice ... [taught them to] become proficient at a certain skill."

The idea that science should not be taught as a skill, but as a "way of thinking" goes back to John Dewey, the father of American education. As one of the creators of American philosophy of pragmatism, Dewey abhors a theory which has no practical application:

John Dewey's opinion on theories.

> The value of any fact or theory as bearing on human activity is, in the long run, determined by practical application—that is, by using it for accomplishing some definite purpose. If it works well—if it removes friction, frees activity, economizes effort, makes for richer results—it is valuable as contributing to a perfect adjustment of means to end. If it makes no such contribution, it is practically useless, no matter what claims may be theoretically urged in its behalf. [Arch 74, p. 195]

Dewey wrote this in 1895, when the precise theory of electromagnetism was 30 years old, electromagnetic waves had been discovered, industrial electric generators were being introduced, and the possibility of instantaneous transmission of information across the globe was looming in the horizon. The cell theory of living things, the germ theory of diseases, the advance of organic chemistry and the synthesis of the organic substances dealt a death blow to vitalism, and established science as the only viable vehicle of knowledge and its utility for the betterment of the human life.

While Cicero could proclaim the science of his time as either too complicated or irrelevant and therefore practically excluded it from his curriculum, many educators of the late nineteenth and early twentieth centuries were facing a science that was becoming more and more "relevant." Thus, they perceived a need to incorporate it in their curricula. But how could a pragmatic educator incorporate science in his curriculum? Note how Dewey echoes Cicero in the following passage written in 1910:

> The infinitely extensive character of natural facts and the universal character of the laws formulated about them is sometimes claimed to give science an advantage over literature. But ... this presumed superiority turns out a defect; that is to say, so long

as we confine ourselves to the point of view of subject-matter. Just because the facts of nature are multitudinous, inexhaustible, they begin nowhere and end nowhere in particular, and hence are not, just as facts, the best material for the education of those whose lives are centered in quite local situations and whose careers are irretrievably partial and specific. If we turn from multiplicity of detail to general laws, we find indeed that the laws of science are universal, but we also find that for educational purposes their universality means abstractness and remoteness. [Arch 74, p. 184]

Dewey is saying that science has too many facts (Cicero said it is too complicated), and its universal laws are too abstract and remote (Cicero said they were not relevant). However, as we mentioned above, he cannot exclude it from education, as Cicero did. So, he finds a compromise: he excludes it as a subject-matter, but retains a phantom, a ghost, a shadow, a haze which he calls "scientific method" [Arch 74, p. 183]:

...science has been taught too much as an accumulation of ready-made material with which students are to be made familiar, not enough as a method of thinking, an attitude of mind, after the pattern of which mental habits are to be transformed.

Dewey must have heard of the two pillars of modern physics—relativity and quantum theory—in 1910, and must have known of the abstractness and the mathematical character of those theories. And as a staunch pragmatist, who wants to find practical and useful results quickly and with the minimal mental effort, he found no need to initiate learning the subjects. Instead, he might have asked himself "What are these good for in students' everyday lives?" And readily found the answer: "For nothing." Thus, in Dewey's curriculum, science has no place as a subject-matter; only its "method of thinking, an attitude of mind" is to be taught.

Dewey's pragmatic approach to teaching was the basis of what he called the progressive movement in education. That movement was consistent with democracy, the antithesis of autocracy, which was associated with the traditional teaching techniques in which teachers were the authority figures. Just as in democracy individuals have equal rights and responsibilities, progressive teachers were considered members of the group (the class). But if teachers are only members of the group, who will be teaching the kids?

If it is asked how the presentation of such bodies of knowledge would differ from the standardized texts of traditional schools, the answer is easy. In the first place, the material would be associated with and derived from occupational activities or prolonged courses of action undertaken by the pupils themselves. In the second place, the material presented would not be something to be literally followed by other teachers and students, ... [Arch 74, p. 179]

This approach is called **student-centered** teaching today. The curriculum is revolved around the activities "undertaken by the pupils themselves," and not around what they ought to learn. As a result, there is no set curriculum, and teachers present their material based on the need of students in their "group."

Student-centered teaching is 100 years old!

If there is no preset curriculum, teachers cannot teach out of textbooks, because, by definition, textbooks contain preset material. So, how do pupils acquire knowledge? To acquire knowledge, students have to think. Thus, to Dewey, the *process* of thinking becomes more important than the acquisition of knowledge:

Since the situation in which thinking occurs is a doubtful one, thinking is a process of inquiry, of looking into things, of investigating. *Ac*quiring is always secondary, and instrumental to the act of *in*quiring. It is seeking, a quest, for something that is not at hand. We sometimes talk as if "original research" were a peculiar prerogative of scientists or at least of advanced students. But all thinking is research, and all research is native, original, with him who carries it on, even if everybody else in the world is sure of what he is still looking for. [Dew 87, p. 148]

This is the **inquiry-based** method of teaching, and it has become a mania among many science educators. It is the complement of student-centered teaching: If the curriculum revolves around the activities undertaken by the pupils themselves, and the knowledge gained by the previous generations can only be transferred by teachers or textbooks, which is not allowed in progressive teaching, then the only way pupils can gain knowledge is by discovering it themselves through inquiry.

Inquiry-based teaching is 100 years old!

How does this work in physics and math? Any new knowledge in these fields is crucially dependent on the previous knowledge. Take one of the most fundamental concepts in physics, acceleration. Let the pupil watch a ball fall and note that its speed increases. Tell the pupil that this change in speed is called acceleration, and ask him to relate speed with acceleration. He might say that acceleration is how speed changes with distance, and he would be following the mistaken footsteps of many physicists and mathematicians that came before Galileo, who discovered the correct definition of acceleration as the way speed changes with *time*, not distance.

Newton said, "If I have seen further than others, it is by standing upon the shoulders of giants." One of the giants on whose shoulders Newton stood was Galileo. If Newton had not learned that acceleration was the rate of change of velocity with time, he would not have discovered the laws of motion, which are the cornerstones of all physics. Inquiry-based teaching takes away the giants from under the pupil's feet. She can no longer see further, and is left alone to discover the knowledge whose discovery required the genius of many generations of scientists.

I have spent more time than I intended on the contentious topic of the pedagogy of science, a topic on which hundreds of books have been written and thousands of educators are in heated discussion. In all likelihood, no consensus will ever come out of all these discussions. However, history is always a good teacher, and our examination of the education of the two great civilizations of antiquity can be a good guidance. So, I end this chapter with a question. In preparing our youth to lead our society, do we want to follow Cicero who said "The investigation of nature seeks to find out either things which nobody can know or things which nobody needs to know," and whose teaching produced not a single Roman scientist or mathematician, or to follow Plato who said

> **Box 2.6.1.** *Mathematics plays its part in character development in providing a stimulus to effort. It forces the mind to reason and stirs the dull youth to become quick, retentive, and shrewd "beyond his natural powers." It awakens the power of thought and disciplines and trains the mind.*

and whose teaching produced the likes of Euclid, Aristarchus, Eratosthenes, and Archimedes?

2.7   End-of-Chapter Material

2.7.1   Review Questions

2.1. When was Archimedes murdered? Who killed him and where was he from? What does this murder symbolize?

2.2. Which superpower came after the Greeks? How would you describe the citizens of this superpower? What kind of heritage did they take from the Greeks?

2.3. Name a few Greek scientists, mathematicians, astronomers, or philosophers. Name a few Greek emperors. Name a few Roman scientists, mathematicians, astronomers, or philosophers. Name a few Roman emperors. This should tell you the difference between what Greek and Roman cultures emphasized.

2.4. From reading Section 2.2 which one of the following areas would you say the Romans were good at? Mathematics, science, politics, sports, engineering, showmanship. What is the popular name for the Flavian Amphitheater?

2.5. What kind of a building existed in every Roman city, and what happened in that building? What was the origin of gladiatorial shows?

2.6. What happened to Plato's philosophy in the hands of the Roman philosophers?

2.7. Besides entertainment, what other human activity thrived in Rome?

2.8. Who were the "holy men?" What did they warn their listeners of? Name a few of them, and what they claimed and accomplished.

2.9. What was the reason that Saul, the Jew, converted to Christianity? What did he change his name to?

2.10. Where were the hubs of Christianity? How were the churches run at the very beginning? What happened to them later? By the third century, who was the spokesman for the congregation of a city?

2.11. What percentage of the Roman population was Christian when Emperor Diocletian came to power? What did he do in 303 which was significant in the history of Christianity?

2.12. Who was Constantine? What city was named after him? When did he become a Christian? When did Christianity become the state religion?

2.13. Who combined philosophy and Christianity? How would you describe his attitude towards science and reason? Whose philosophy did he recognize as consistent with Christianity?

2.14. Who were the sophists? What subject did they emphasize the most? Name a famous sophist, and the subjects he chose to teach in his elementary education. What was his attitude towards mathematics and astronomy both in his elementary and higher education?

2.15. Compare Plato's educational philosophy with the sophists'. What did Plato's elementary education consist of? What subject(s) did he emphasize? What is the first thing everyone has to learn according to Plato?

2.16. Describe some of the effects of mathematics on students according to Plato. What kind of a test would Plato give to select students for advanced study?

2.17. What subjects are taught in Plato's secondary education? What subjects does Plato emphasize in the training of the members of the Nocturnal Council?

2.18. What does Plato think of sophists? What is his attitude towards rhetoric in the search for truth?

2.19. Which system of Greek education did Romans adopt? What was the content of the Roman higher education? What did students study in rhetorical schools?

2.20. In what city was science and astronomy pursued? Who is Zenobia, and what did she do in AD 270?

2.21. Who is Cyril, and what was his attitude towards knowledge? What did he do to those who disagreed with him? What did he think of the Library in Alexandria? What was his role in setting the Library on fire? Who was the director of the Library at the time, and what happened to her?

2.22. Who delivered the last blow to the Library? Where did he come from?

2.23. What was Cicero's attitude toward science? What kind of professionals did Roman society produce under Cicero's teaching?

2.24. What factor is important to Howard Gardner in judging the effectiveness of high scores in math and science tests? To what does he liken students who score well on math and science tests? What does he think our adult population ought to know, and what does this have to do with teaching science and math to our youth?

2.25. How does Dewey view a theory? Did he follow Cicero or Plato in teaching science? Could he completely exclude science in his curriculum as Cicero did? How was science included in Dewey's curriculum?

2.26. What is student-centered teaching? Who introduced it in the American education?

2.27. What is inquiry-based teaching? Does it encourage the transfer of knowledge from one generation to the next? Is the acquisition of knowledge more important or the process of thinking? Does the process of thinking necessarily end with the acquisition of knowledge?

2.28. What did Newton say about his discoveries? Could he have discovered the laws of motion had he been taught physics by the inquiry-based method?

Chapter 3

Science Rekindles

3.1 Renaissance and the Copernican Revolution

The passage of time weakened the social bond that held the feudal landlord, the clergy, and the serfdom together. The need of the growing society which the peasantry could not meet by itself any longer, gave rise to a new class of citizenry. The task of manufacturing such items as clothing, building material, furniture, and certain agricultural products fell into the hands of *artisans*, who, because of their dependence on one another, clustered together in communities which later developed into cities.

The rise of the cities led to new commerce, new social relations, and new ideologies. More and more peasants migrated to the urban communities, and through contact with their rural relatives, transferred the new ideologies to the countryside. The old feudal bondage started to shatter; peasants started to revolt; the foundations of the Church, which acted as an agent of peace and quietude of the peasantry, started to crack from within. By the middle of the sixteenth century, monks such as Thomas Müntzer and Martin Luther of Germany, and Huldrych Zwingli and John Calvin of Switzerland had stood up against the corrupt and aristocratic church and demanded a separation of church and state. The *Renaissance*, the new intellectual life was born.

Although the Renaissance is said to have started in the fourteenth century Italian literature, the true Renaissance, the one that lifted the fallen icon of mankind and put it on the pedestal in the Greek tradition, was in science and mathematics. Awakened by the sound of history, our race began to ponder, scrutinize, question, and speculate again, not merely for pragmatic ends (as the Romans did), but for the sake of discovery (as the Greeks did). Pragmatism leads only to the fulfillment of the most rudimentary desires of our race such as the production of food and shelter, as in the Dark Ages. However, the *homosapien* curiosity and inquisitiveness drive the wheels of evolution, as in Babylon and Egypt (with the invention of hieroglyphics and cuneiforms), in Greece (with the formalization of mathematics and astronomy), and in the Islamic world (with the invention of algebra). Now, this same *homosapien* trait was rejuvenated in Western Europe in the form of the revival of interest in science and mathematics.

Human race starts to think again.

Nothing can illustrate the homosapien desire to know better than Copernicus' attempt at explaining the motion of planets around the Sun. Unlike the old theories which kept changing due to the force of observation, his was completely a product of imagination, a response to the need for simplicity, and a reaction to the growing complexity of an intricate mammoth-like theory, on which, over the years, observation and accurate measurements had placed the unbearable weight of dozens of epicycles and numerous planetary oscillations.

Copernicus's theory is purely and completely a product of his fertile mind.

Copernicus assumed that the Sun was at the center of the solar system and all planets, including Earth, revolved around it with different speeds on perfect circles. His theory is called the **heliocentric model** (from the Greek word *Helios* meaning the Sun). The different size of the orbits and different speeds of the planets cause them to move away and towards each other during the course of their orbital motion. In particular, looking at a planet such as Mars from Earth, Mars possesses an irregular motion, sometimes approaching Earth and sometimes moving away from it (causing a change in the brightness of Mars); sometimes moving in one direction, other times in the opposite direction (the retrograde motion of Mars). **Math Note E.3.1** on **page 76** of *Appendix.pdf* explains the details of how a planet appears to be moving relative to Earth, and how the heliocentric theory leads to all the observed features of the planetary motion without the complications of the geocentric model.

BIOGRAPHY

The first notable new thinker, who questioned the old wisdom in astronomy was **Nicolaus Copernicus** (1473–1543). Nicolaus was brought up by his maternal uncle, who was consecrated Bishop of Ermland, a post which carried with it a palace, a cathedral at Frauenberg, and the power of civil rule over the area. At the age of 18 Nicolaus was dispatched to Kracow University to broaden his mind. University education at Krakow was, Copernicus later wrote, a vital factor in everything that he went on to achieve. There he studied Latin, mathematics, geography, philosophy, and, of course, astronomy, in which he developed a keen interest. Around 1514 he distributed a little hand written book to a few of his friends who knew that he was the author even though no author was named on the title page. This book, usually called the *Little Commentary*, set out Copernicus's theory of a universe with the Sun at its center. This theory, however, was not published until the last year of his life upon the persistent urging of one of his protestant friends, when he wrote *De Revolutionibus*.

What do you know? 3.1. Suppose a hypothetical planet was farther from the Sun than Earth and made its revolution around the Sun in exactly the same time that Earth did. How do you describe the motion of this planet relative to Earth? Ignore the daily spin of Earth and consider Earth and the planet as points.

The Copernican revolution spread very quickly in Europe. Italy, the bedrock of Renaissance embraced Copernicanism with exceptional fervor. The heliocentric model did not simply change the science of the time; it undermined the very foundation of the philosophy and the intellectual life of the period. Once the Earth was removed from the center of the universe, endless philosophical conclusions could be drawn from it which clashed with the view of the Church. And the Church was determined to crush such heretical philosophies. One man paid dearly to this antiheresy hysteria; his name was **Giordano Bruno** (1548–1600).

Bruno left his home town of Nola to travel to nearby Naples when he was 14 years old to study there. He attended lectures on humanities, logic, and dialectics in Naples and it was at this time that he was influenced by one of his teachers towards Averroism. This was Christian philosophy based on an interpretation of Aristotle's works through the Muslim philosopher Averroes. Its basic belief was that reason and philosophy are superior to faith and knowledge founded on faith. This radical view attracted Bruno to Copernican heliocentric theory and, after reading Lucretius' poem *On the Nature of Things*, to the materialistic philosophy of the Italian Renaissance. His fertile mind was his laboratory in which the Epicurean notion of an infinite universe, combined with the Copernican heliocentrism, led him to believe in the existence of infinitely many solar systems, each having its own Earth, its own Garden of Eden, its own Adam and Eve, and its own fall from grace.

What do you know? 3.2. Why is a heliocentric model of the solar system conducive to the belief that there are more systems like it in the infinite universe while a geocentric model is not?

A passionate debater who spoke his mind with the honesty and purity of a child, he refused to recant his views even after serving a prison term from 1593 to 1600. Instead, during his Inquisition trial, he gave his final and most impassioned speech, in which he harangued his judges before hearing his death sentence: "Perhaps you tremble more in pronouncing the sentence than I do in hearing it." He was burned at the stake on February 17, 1600. His statue has now been erected at the exact spot where he was burned.

3.2   New Observations: 15 Centuries after Ptolemy

Copernicus drew the attention of humanity to the simplicity and elegance of his *theoretical* model of the solar system. But simplicity and elegance, as essential as they seem to be, are only of aesthetic quality. The real test of a theory is observation, an activity forgotten for nearly 1500 years. It was a Dane by the name of Tycho Brahe, who revived this activity, and put the Copernican theory to the harshest test.

In the process of observing the *nova* of November 11, 1572, Tycho introduced *error analysis*—a method in use even today—based on the assumption that any act of observation or measurement introduces an error irrespective of the capability and the precision of the instrument. He therefore set about determining the errors inherent in his equipment before he used it. In this way he was able to attain a degree of accuracy never before reached and, although he still had no telescope to help him, his measurements were five times more accurate than those of Hipparchus, the greatest observer of antiquity.

Tycho invents method of error analysis; makes measurements 5 times more accurate than Hipparchus'.

BIOGRAPHY

Tycho Brahe (1546–1601) was born three and a half years after the death of Copernicus and the publication of his *De Revolutionibus*. His uncle (and his guardian) wanted him to be a statesman, and decided that he should gain experience abroad. In February 1562 Tycho set off with a traveling companion to go to the University of Leipzig. Astronomy was not officially part of his studies, which included classical languages and culture. However, he had brought with him the astronomy books that he had purchased earlier. While still at the University of Leipzig, Brahe began making observations and keeping a record of them. The second observation he recorded was a conjunction of Jupiter and Saturn which proved significant for Tycho's subsequent career. Neither tables based on Copernicus nor on Ptolemy gave the correct date for the conjunction, Ptolemy's being out by nearly a month and Copernicus' by days. Tycho, with the confidence of someone not yet seventeen, thought he could do better, and he later proved himself to be right! He designed and built a large number of precision astronomical instruments that helped him measure the location of planets with unprecedented accuracy. Tycho showed that comets are celestial objects whose appearance must follow the "shattering" of the heavenly crystalline spheres. And since no such shattering was observed upon the appearance of a comet, crystalline spheres must not exist.

Tycho spent a long time considering the nature of the planetary system since he was not satisfied by the proposals of Copernicus for two reasons. In the first place he could find no observational evidence of the Earth's motion, and as he was an extremely careful observer he felt sure in his own mind that the lack of any evidence for the shift of nearer stars against the background of more distant stars (much like the movement of the trees on the side of a road—on which a car moves—relative to more distant objects in the background) was a strong argument against a Sun-centered universe. Secondly, he was a Protestant and to him the Bible was divinely inspired in every word and phrase and he well knew that the Old Testament was very definite in its description of the Earth as fixed in space.

Tycho's greatest contribution to astronomy is his most accurate observations of planets and the Moon.

Without any doubt Tycho's greatest achievement is his observation of the Moon and planets. In addition to observing with amazing precision, Tycho instituted the procedure of observing these bodies wherever they might be in the sky. Hitherto the positions of the planets and even of the Moon, had only been determined at astrologically important moments. Tycho, believer in astrology though he was, cast his observational net as wide as possible and noted positions as often as weather and visibility permitted. The results of this new approach were soon to be felt and the procedure has been followed ever since.

3.3   The Fall of the Spherical Dynasty

Tycho's accurate observations disagreed with both Copernican and Ptolemaic models. Two theories, one extremely complicated and awkward, one simple and elegant, confronted the harsh reality of observation, and they both failed. This failure prompted the downfall of a 2000-year-old cherished idea introduced by none other than the giant intellectual of antiquity, Pythagoras.

The collapse of the spherical dynasty came very slowly and very reluctantly. It was one of those moments in the history of science, when the harsh reality of observation pushed aside the long-held prejudices stored in the mind of humanity. Unlike the dogmatic belief in the words of an "authority," a practice common to creationists, intelligent designers, psychoanalysts, chiropractors, and other alternative medical doctors, Johannes Kepler, the German physicist, astronomer, and mathematician, put observational evidence ahead of any prejudices, and cast aside—albeit hesitantly—the notion that celestial bodies move on spheres.

Kepler's basic interest lay in uncovering the mysteries of the planetary system. After Tycho's death, a collection of observations of unparalleled accuracy came into Kepler's possession and he set himself two tasks. On the one hand, to complete a volume of calculations begun by Tycho showing the future positions of the planets and, on the other, to examine Tycho's very complete observations of the planet Mars with a view to determining exactly how it moved.

BIOGRAPHY

Johannes Kepler (1571–1630) was a premature baby and a very delicate child who was brought up by his grandparents. After elementary and secondary schooling, Kepler entered Tübingen University to become a Protestant minister. At Tübingen, Kepler was taught astronomy by one of the leading astronomers of the day, Michael Maestlin (1550–1631). The astronomy of the curriculum was, of course, geocentric astronomy. At the end of his first year Kepler got "A"s for everything except mathematics. Probably Maestlin was trying to tell him he could do better, because Kepler was in fact one of the select pupils to whom he chose to teach more advanced astronomy by introducing them to the new, heliocentric cosmological system of Copernicus. It was from Maestlin that Kepler learned that the preface to Copernicus's book, explaining that this was "only mathematics," was not by Copernicus. Kepler seems to have accepted almost instantly that the Copernican system was physically true, and from then on, astronomy and mathematics became his passion. Kepler also worked and wrote a book in optics, in which he used the idea of a "ray of light" for the first time.

Kepler concludes that the orbit of Mars must be oval.

Kepler chose the observations of Mars because, of all the planets, it seemed the most difficult to fit in with the theoretical motions accepted at that time, and he hoped that Tycho's masterly observations would help him solve the problem. By the end of 1604 he was partly successful for he had come firmly—but reluctantly—to the conclusion that the movement of Mars could not properly be accounted for by assuming the principle of uniform *circular* motion, either about the Sun or the Earth. He had, in fact, arrived at the monumental discovery that the planet moved round the Sun in an *oval* orbit. This was a complete break with authority, an utter repudiation of a tradition which had been held without question since the days of Pythagoras more than two thousand years before.

What do you know? 3.3. Some philosophers (even some physicists) believe that the human mind can create knowledge by examining the existing knowledge. How does Kepler's work fit in this belief system?

Kepler's discovery was not published until 1609, when it appeared under the title of *The New Astronomy*. Important though the new discovery was, Kepler was unable in 1609 to prove that it applied to *all* the planets, and he was also trying to find an explanation of why an oval orbit was the kind of path which Mars described. He believed that some force from the Sun was the cause, and he even went so far as to suggest that the Sun emitted rays which moved round as it rotated and so drove the planets along in their orbits.

Kepler extended and developed his ideas of oval orbits, applied them to the newly discovered moons of Jupiter and, between the years 1618 and 1621, published his *Epitome of Copernican Astronomy* as well as what he himself considered to be his greatest work, *The Harmony of the Universe*. Throughout his researches he had been seeking a harmony in the Universe; his discovery of oval orbits had forced him to reject the Greek theory of uniform circular movements and he continually sought for some uniformity of motion in an oval path in the firm belief that God must have created a Universe with the beauty of regular motion at some point within its structure. He partly reached his goal when he discovered that a degree of mathematical regularity could be discerned in a planet's motion within its orbit, however oval it might be. The fulfillment of this goal came in the *Harmony*, in which he gave a precise relationship between the size of the orbit of a planet and the time it took to complete its circuit.

Kepler discovers the gravitational pull of the Sun ... Well, almost!

Kepler discovers a harmony in the universe, not in the form of a uniform motion, but in a mathematical statement.

What do you know? 3.4. How crucial was the role of new observation in the theoretical model proposed by Copernicus? How crucial was it in Kepler's theory?

Kepler's discovery of the nature of the planetary motion was a cornerstone of the seventeenth century in more ways than one. Not only did it break with the 2000-year-old tradition of Pythagorean obsession with sphericity, but it also laid the foundation of the universal law of gravitation to be discovered by Newton.

He summarized his findings in three laws. The first two laws are statements about the shape of the orbit of a planet and its speed. However, the pride of Kepler's life and work was his third law in which he claimed to have found a harmony of the universe. Here are **Kepler's three laws**:

Box 3.3.1. (Kepler's three laws of planetary motion)
First Law: Each planet moves around the Sun in an elliptical orbit with the Sun located at one of its foci.
Second Law: The line joining the Sun to the planet sweeps out equal areas of the ellipse in equal times. Hence, the planet moves faster when it is closer to Sun.
Third Law: The square of the period T of revolution of each planet is proportional to the cube of its semi-major axis a. In symbols this is written as $T^2 = ka^3$, where k is the constant of proportionality.

To understand the three laws of planetary motion, we need to know some properties of an ellipse. Figure 3.1 shows an ellipse with its two foci F_1 and F_2. The long axis drawn from A to B is the major axis denoted by $2a$, so that the semi-major axis is a. Pick two arbitrary points P and Q on the ellipse, draw one line segment from each to the foci. The

Properties of an ellipse.

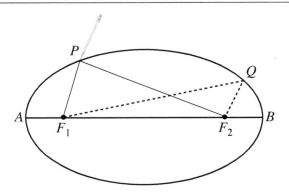

Figure 3.1: The two points F_1 and F_2 are the foci of the ellipse. The line segment $\overline{AB}$ is the major axis. The sum of the length of the two line segments drawn from P to the foci is the same as those drawn from Q or any other point.

sum of the length of the line segments in each pair is the same: $\overline{PF_1} + \overline{PF_2}$ is the same as $\overline{QF_1} + \overline{QF_2}$.

In fact, this is how you draw ellipses: Tie the two ends of a piece of string to two fixed points (the foci); place a pencil next to some point of the string and push it so the string is taut; holding the string taut, move the pencil. The resulting curve is an ellipse. It is clear that the smaller the distance between the two foci is, the more circular the ellipse will be. When the two foci coincide, we'll have a circle, and the major axis becomes the diameter of the circle. For most planets, the ellipse is very nearly a circle.

A consequence of the second law is that the planet moves faster when it gets close to the Sun. This is because, as Figure 3.2 shows, to cover equal areas, the planet must traverse a longer arc of its orbit at P than at Q. Kepler's third law is demonstrated numerically in **Example D.3.1** on **page 13** of *Appendix.pdf*.

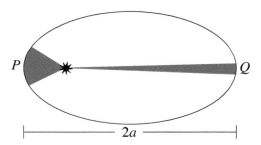

Figure 3.2: The shaded areas are assumed equal. Kepler's second law says that the planet moves faster at P than at Q. The major axis of the ellipse is $2a$.

3.4   End-of-Chapter Material

3.4.1   Answers to "What do you know?"

3.1. The planet would not have any motion relative to Earth. It would appear fixed in the sky.

3.2. Once Earth becomes "just another planet" and not the "center of the universe," it is much easier to think of other planets like it in the infinitude of the cosmos. An Earth located at the center of the entire universe is, by definition, unique.

3.3. Kepler could not possibly have discovered the elliptical orbits without a detailed analysis of the observational data. In fact, the circular path was so ingrained in the mind of astronomers (Kepler included), that any departure from it was inconceivable. Only observation could force the ellipticity of the orbits into the psyche of the seventeenth century astronomers.

3.4. Copernicus was not responding to any new observation when he proposed the heliocentric model. In fact, there *were no* new observations at his time. On the other hand, Kepler studied Tycho Brahe's new improved observations with utmost care, and was practically *forced* by Tycho's data on Mars to abandon the idea of a circular orbit and replace it with an elliptical one.

3.4.2   Chapter Glossary

Focus One of the two points of an ellipse having the property that if you connect any point on the perimeter of the ellipse to these points, the sum of the resulting line segments is a constant, independent of the point chosen on the perimeter (see Figure 3.1).

Heliocentric Model A model of the solar system in which the Sun is assumed at the center while the planets, including Earth, revolve around it on circular (later modified to elliptical) orbits. Although Copernicus is associated with heliocentrism, Aristarchus, a third century BC mathematician was the first to propose the model based on his measurement of the sizes and distances of Moon and Sun.

Period The time it takes a planet to make a complete revolution around the Sun.

Semi-Major Axis Half of the major axis of an ellipse. The major axis is the length of the longer axis of the ellipse (see Figure 3.1).

3.4.3   Review Questions

3.1. Name some monks who rebelled against the established church.

3.2. Describe the attitude of the Greeks and Romans toward knowledge. Which attitude advances science and mathematics? Which attitude advances only engineering and technology?

3.3. What was Copernicus' motivation for his heliocentric model? Did any particular observational evidence have a role in Copernicus' proposal for heliocentrism?

3.4. Describe Copernicus' heliocentric model. Did it agree with observation when Copernicus proposed it? Did it explain the retrograde motion of planets and the change in their brightness during this motion?

3.5. Who was Giordano Bruno? Who influenced his thinking? What was the connection Bruno made between the Biblical stories and the universe? How did he die?

3.6. Who is Tycho Brahe? What kind of contribution did he make in astronomy? What did his observation of the planets say about the geocentric model? About the heliocentric model?

3.7. What planet did Kepler study carefully? What was his conclusion about the shape of its orbit? When did he publish his findings about this planet? How long did it take him to apply his conclusions to all planets?

3.4.4   Conceptual Exercises

3.1. You saw that in Figure 3.1, the sum of the length of the line segments drawn from any point on the ellipse to the foci is a constant independent of the point. Now apply this to point A.
(a) How is $\overline{AF_1}$ related to $\overline{BF_2}$?
(b) How is the sum of the length of the line segments from A to the foci related to the major axis?
(c) How is the sum of the length of the line segments from *any* point to the foci related to the major axis?

3.2. A planet is at some point P of its orbit. A few months later it is at Q, closer to the Sun than P.
(a) Is the planet moving faster at P or at Q?
(b) Consider the line joining the Sun to the planet. This line sweeps out an area in one second at P. The line also sweeps out an area in one second at Q. Which area is bigger?

3.3. Planet M, located farther from Sun than Earth, moves around Sun once as Earth moves around Sun 4 times. Both move counterclockwise on circles starting at the 12 o'clock location.
(a) When Earth completes its first revolution, at what position of its orbit is M?
(b) When Earth completes its third revolution, at what position of its orbit is M?
(c) When Earth is at 4 o'clock during its second revolution, at what position of its orbit is M?
(d) When Earth is at 8 o'clock during its third revolution, at what position of its orbit is M?
(e) When Earth is at 6 o'clock during its last revolution, at what position of its orbit is M?

3.4. Planet M, located farther from Sun than Earth, moves around Sun once as Earth moves around Sun 6 times. Both move counterclockwise on circles starting at the 12 o'clock location.
(a) When Earth completes its first revolution, at what position of its orbit is M?
(b) When Earth completes its fifth revolution, at what position of its orbit is M?
(c) When Earth is at 3 o'clock during its second revolution, at what position of its orbit is M?
(d) When Earth is at 6 o'clock during its second revolution, at what position of its orbit is M?
(e) When Earth is at 9 o'clock during its fifth revolution, at what position of its orbit is M?

3.4.5   Numerical Exercises

3.1. It takes Comet Halley 76 years to go around the Sun.
(a) What is the semi-major axis of the orbit of the comet?
(b) Suppose that it almost grazes Earth's orbit when the comet approaches Earth. What is the farthest Comet Halley gets from Earth?

3.2. The star Herates has only two planets Neemaz and Lahoz, both of whose orbits are circular. Neemaz is nine million km away from Herates and has a period of 50 days.
(a) Find the k in Kepler's third law for this star system in scientific units.
(b) If Lahoz is 25 million km away from Herates, what is its period in days? Use (a) to answer this question.
(c) Use proportion to find Lahoz's period.

3.3. A star has two planets, both of whose orbits are circular. The first planet is 5 million km away from Herates and has a period of 2 days.

(a) Find the k in Kepler's third law for this star system in scientific units.
(b) The period of the second planet is 8 days. How far is it from the star? Use (a) to answer this question.
(c) Use proportion to find the second planet's distance from the star.

Chapter 4

From Heaven to Earth

Today astronomy and astrophysics are subbranches of physics. But earlier in our history, they were completely separate. In fact, when astronomy was well on its way to becoming an exact science in antiquity, physics was nothing but a (false) description of the most obvious phenomena. The reason was twofold: the primacy of the mind and the loftiness of heaven.

Plato's influence on the intellect of Hellenistic scholars and his canonization of the mind disparaged any attempt at studying objects by experimentation. To him and his followers human mind—and only human mind—was capable of understanding the universe. Plato showed a great deal of contempt to those who tried to understand geometry by drawing figures on sand. Geometry and mathematics were reserved for the pleasure of the mind and for the study of heaven, where, according to Plato, the Soul (Idea, Form) resided.

4.1 The Ancient Physics

Astronomy was no doubt the heart of the Greek science. In fact, for over three centuries, it was the only science studied by Greek scholars. After all, the Greeks received their knowledge from Babylon and Egypt, and in both civilizations astronomy was the only scientific enterprise for its relation to gods and climate change. This connection between astronomy and gods held a sacred position for that science among human knowledge even for the secular Greeks.

Physics was not worthy of attention because it was inherently experimental, and as such, required the "dirtying of one's hand." In the dominant Platonic school of thought, nothing was more intellectually demeaning than to lower oneself to the point of experimenting with worldly objects with the purpose of gaining knowledge. That is why Archimedes, the greatest mathematical physicist of antiquity, was hesitant to publish his worldly observations, and only reported the final mathematical products. In contrast, Aristotle, a pupil of Plato, reduced physics to a few personal experiences and a lot of (wrong) philosophical speculations.

4.1.1 The Aristotelian Dynamics

Aristotle was a *philosopher*; and whatever he wrote on physics was from a philosopher's perspective: a great deal of emphasis on argumentation and logical inference, with insignificant reliance on observation. This made his statements either so general as to render them useless, or when he was specific, his statements were usually wrong. Take his dynamics. He states that vertical motion is a *natural* motion in which some objects such as smoke and fire rise, and other objects such as stone and water fall. This is so general a statement that it

"Aristotle maintained that women have fewer teeth than men; although he was twice married, it never occurred to him to verify this statement by examining his wives' mouths."

Bertrand Russell

teaches us nothing. It is similar to the statement: "The Earth population consists of males and females." While the statement is true, it adds nothing to our knowledge.

As soon as Aristotle tries to be slightly more specific, he fumbles. Consider his "law of motion." Book H of his Physics starts with the statement: "Everything in motion is necessarily being moved by some thing" [Apo 69, p. 127]. This appears so "evident" that Aristotle sees no need to explain it any further. After all whatever one sees in motion on the streets and roads inside and outside Athens, is either being pulled, or pushed, or carried, or thrown.[1]

Aristotle then tries to quantify his law, and he comes up with the following statement:

> ... if A is the mover, B the thing in motion, S the length over which motion has occurred, and T the time taken, (1) in time T a force equal to that of A will cause a thing which is half of B to move over length $2S$, and (2) it will cause it to move over length S in half the time of T; for thus there will be a proportion. And (3) if the force of A causes B to move over the length S in time T, it also causes B to move over half of S in half the time of T, and (4) a force equal to half of A causes a thing equal to half of B to move over a length S in time T. [Apo 69, p. 146]

The entire quotation above can be summarized in the concise mathematical statement $A/B = k(S/T)$ where k is a constant of proportionality. When we recognize that S/T is simply the speed v of the object in motion, we may rewrite the mathematical statement as $A = kBv$. In particular, if $A = 0$, i.e., if there is no force acting on the object in motion, then $v = 0$, i.e., there will be no motion!

This Aristotelian law of motion hits a brick wall when it is applied to the motion of an arrow. As long as the arrow is in the hand of the thrower, the motion appears to obey the law, with the thrower providing the necessary push. However, as soon as it leaves his hand, the law breaks down, because now there is no provider of any push or pull. Aristotle himself argued that the air must provide this missing force. That this is completely wrong could have been "discovered" by Aristotle if he had consulted a horseback rider, who could have told Aristotle about the enormous *retarding* drag he felt every time he set his horse galloping.

Aristotle "reinterprets" his law in light of observation.

That $A = kBv$ is completely wrong can be gleaned from the observation that some (very heavy) objects will not move even if they experience a substantial amount of force. Aristotle was, of course, aware of this situation, and remedied his law by arbitrarily adding the proviso that the proportionality holds in the increasing direction: If the law holds already, then doubling A doubles the speed v for a given B. However, halving A may not result in halving v, because one half of A may not be enough to move B in the first place. This constant change in the interpretation of the formula is one of the main tenets of the pseudoscientific methodology: If observation disagrees with your theory, do not change the theory; reinterpret it!

The problem with Aristotelian dynamics is that it is based entirely on argumentation, and no reference is made at all to observation. Had Aristotle performed some simple experiments with motion, he would have discovered the fallacy of his law. But in his time, pure thought and the force of logic ruled the world of philosophy, and any knowledge that used "manual" activity was considered base and unworthy of academic pursuit.

4.1.2   The Archimedean Physics

In the golden age of Hellenistic rationalism, and before the political and military pragmatism of Rome bulldozed the ancient civilizations into the Dark Ages, in the city of the Pythagoreans, there lived an intellectual giant, a rare breed of scholars who could blend

---

[1]Aristotle actually divides motion into several categories such as pull, push, carry, and throw. He then goes into details of what the difference is between all these categories. Categorization was an ecstasy for ancient philosophers.

theoretical reasoning of a skilled mathematician with the dexterity of an artisan, an intellect matched in genius only two millennia later in the person of Isaac Newton. This intellectual giant was **Archimedes of Syracuse.**

BIOGRAPHY

Archimedes of Syracuse (287–212 BC) spent some time in Egypt early in his career, but he resided mostly in Syracuse, the principal city-state in Sicily. He is considered one of the greatest mathematical physicists of all time. He not only excelled in the mathematics of his time, but invented the method of infinitesimals which was later used by the seventeenth century mathematicians in their development of calculus. Archimedes was also a great mechanical inventor. He constructed such effective war machines for the defense of Syracuse against the siege of the Romans in 213 BC that the capture of the city was delayed for a long time. But Syracuse eventually fell in the autumn of 212 or spring of 211 BC. Archimedes, absorbed in his intellectual activity, did not notice a Roman soldier entering his house. Upon the command of the soldier to accompany him, Archimedes asked for some time to finish his problem. The enraged soldier drew his sword and ran him through; Roman militarism conquered the crown of the Greek rationalism.

Roman militarism (in the person of a soldier) conquers Greek rationalism (in the person of Archimedes).

According to Plutarch, Archimedes had so low an opinion of the kind of practical invention at which he excelled and to which he owed his contemporary fame that he left no written work on such subjects. While it is true that all of his known works were of a theoretical character, his interest in mechanics deeply influenced his mathematical thinking. The story that he determined the proportion of gold and silver in a wreath made for King Hieron by weighing it in water is probably true, but it may not be true that he actually leaped from the bath in which he supposedly got the idea, and ran naked through the streets shouting "Heureka!" ("I have found it!").

> **What do you know? 4.1.** What was wrong with Aristotelian dynamics?

Archimedes' most important work is *Method Concerning Mechanical Theorems*, in which he describes the process of discovery in mathematics. It is the sole surviving work from antiquity and one of the few from any period that deals with this topic. In it Archimedes recounts how he used a "mechanical" method to arrive at some of his key discoveries, including the area of a parabolic segment and the surface area and volume of a sphere. Archimedes emphasizes that, though useful as a heuristic method, this procedure does not constitute a rigorous proof. Nevertheless, it comes as close as one can get to the modern scientific process (originated by Galileo) of making theories based on observational measurements.

Archimedes uses modern scientific methodology 1800 years before Galileo rediscovers it!

> **What do you know? 4.2.** What was right with Archimedean physics?

Archimedes' mathematical proofs and presentations exhibit great boldness and originality of thought as well as extreme rigor, meeting the highest standards of contemporary geometry. While *Method* shows that he arrived at the formulas for the surface area and volume of a sphere by "mechanical" reasoning involving infinitesimals, in his actual proofs of the results in *Sphere and Cylinder* he uses only the rigorous methods of passage to the limit that had been invented by *Eudoxus* in the 4th century BC. These methods, of which Archimedes was a master, are the standard procedure in all his works on higher geometry that deal with problems of integration.

The greatest effect of his work on mathematicians and physicists—of post-Renaissance Europe—came with the printing of texts derived from the Greek, and eventually of the Greek text itself, the *editio princeps*, in Basel in 1544. The Latin translation of many

Mathematicians and physicists pick up where Archimedes left off.

of Archimedes' works in 1558 and his complete works in 1615 contributed greatly to the spread of his ideas, which were reflected in the work of the foremost mathematicians and physicists of the time, including Johannes Kepler, Galileo, René Descartes, and Pierre de Fermat. Without the background of the rediscovered ancient mathematicians, amongst whom Archimedes was paramount, the development of mathematics in Europe in the century between 1550 and 1650 is inconceivable.

4.2   Galileo's Study of Motion

Copernicus' heliocentric astronomy was indeed a revolution. It was a locomotive that set in motion the train of human intellect, having gathered dust and rust for over 1400 years. The degree of its impact on the mind of mankind can be measured by the number of great scientific scholars it produced in the generation immediately after Copernicus: Johannes Kepler (1571–1630), René Descartes (1596–1650), Giordano Bruno (1548–1600), and Pierre de Fermat (1601–1665) just to name a few. These great thinkers, as radical as their points of view were, were still confined to the study of heaven and the rules of logic, i.e., astronomy and mathematics. The progress of science required a new way of studying nature, a new methodology. This came about by studying motion, and it was done by Galileo Galilei.

BIOGRAPHY

Galileo Galilei (1564–1642) may have been influenced by his father's line of experimentation in music theory. As a student at the University of Pisa in 1583, he discovered the constancy of the period of a pendulum while watching the oscillations of a lamp at the cathedral of Pisa. Although he attended the University of Pisa, Galileo left it after four years without a degree. And to support himself, he gave private lessons in mathematics. He started to investigate some problems of physics a la Archimedes rather than Aristotle. Following Archimedes, he introduced experimental procedures in the study of motion. He is credited, among other things, with the first observation of the sky through a telescope, the invention of the first thermometer, and the first law of motion. Galileo's views on Copernican theory were condemned by the church, which put him under house arrest until his death.

Galileo's new methodology of the natural philosophy truly revolutionized the discipline, and transformed it into a science in the modern sense of the word. He brought the study of motion from the heavens down to Earth, and made it the first *experimental*—as opposed to merely observational—science. In this respect he was a follower of Archimedes at a time when, except for a handful of new visionaries, the entire academic community followed Aristotle's philosophy. Archimedes himself did not apply his *Method* to the study of (earthly) motion, although he applied it to many other phenomena including hydrostatics. This is partly because motion was not as common a phenomenon as it was in Galileo's time that witnessed global journeys on an almost daily basis.

Galileo uses inclined planes to study motion.

In his study of motion, Galileo used an **inclined plane**. Place a block of wood B on a plank IP laid horizontally on the floor. When you raise one end of the plank, you create an inclined plane with an inclination angle denoted by θ (Figure 4.1). Keep raising the end of the plank until the block starts to move. Call this angle $\theta_{\min}$, and note that if you lower the plank below $\theta_{\min}$, the block stops.

Now change the block, or the inclined plane, or both, and try to determine $\theta_{\min}$ for the new set-up. You will see that $\theta_{\min}$ will be different for different Bs and IPs. In particular, there are some combination of Bs and IPs for which the angle $\theta_{\min}$ is smaller than others. Let's call these *smoother* Bs and IPs. Repeat the experiment for different levels of smoothness. These experiments will show us that *the smoother the surface of contact between the block and the inclined plane the smaller the angle θ_{min}.*

The foregoing statement is very much in tune with our common sense and intuition. And if we stop here, we will have added very little to our knowledge of motion. However, Galileo

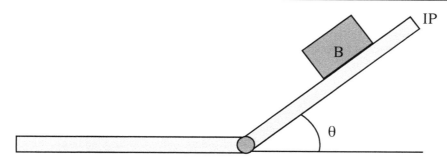

Figure 4.1: A block B is resting on an inclined plane IP with an inclination angle θ.

did not stop with this statement. Upon a stroke of genius, he asked a deeper question: What θ_{min} corresponds to an *infinitely smooth* surface of contact? Now, the concept of infinity is purely a mathematical concept, one that, analogous to the mathematical concept of a point, is *not* realizable in physical terms. However, just as the *concept* of a point can be *approximated* by actual physical objects,[2] so can the concept of an infinitely smooth surface be approximated by planks that are planed or filed by finer and finer instruments.

What do you know? 4.3. In an inclined-plane experiment, for one surface the angle at which the block starts to move is 30°, and for another surface it is 25°. The smoothness of a third surface is between the first two. What are the upper and the lower limits of the size of the angle corresponding to this surface?

Human mind is not fettered by the limitations of the instruments with which human hands occupy themselves. While the "hands" are extremely important in showing the direction in which human mind is to engage itself, it is the (trained) mind that takes us to the boundaries of knowledge and points us to the direction of nature's hidden possibilities.

The role of human mind in science.

Although infinitely smooth surfaces do not exist in our immediate regions of contact with nature, their *conceptual* framework could (and did) lead us to its hidden secrets. So, what *if* the surface of contact is infinitely smooth? What kind of θ_{min} would we be getting? Since the minimum angle decreases with smoothness, we expect an *infinitely small* angle for an infinitely smooth surface. But what measure corresponds to an infinitely small angle? The answer is surprisingly obvious: zero! So, you need not raise an infinitely smooth surface to move the infinitely smooth block resting on it! "You need not raise" is too strong a statement. A better statement is to say "You need to raise the plank by a *nonzero* angle." This angle could be a billionth of a trillionth of a quadrillionth of ... a degree! A more meaningful way of stating the conclusion is to say that once in motion on a perfectly smooth *horizontal* surface, you cannot stop the block by manipulating the plank, because you can't lower the angle below zero! In other words

Galileo discovers the first law of motion.

Box 4.2.1. *On an infinitely smooth surface, an infinitely smooth block, once set in motion, will move on its own without any assistance from outside.*

This is a restricted version of the **first law of motion**, or the **law of inertia**. It is in complete contrast with the Aristotelian law of motion whereby "any moving object must have a mover." It is also in complete contrast with common sense. But what is common

---

[2]Such as the mark made by the tip of a sharpened pencil on a sheet of paper.

sense but a collection of "prejudices we accumulate by the age of eighteen."[3] And if the fundamental laws of physics were to have no dissonance with our common sense, they would be incapable of revealing to us the hidden secrets of the universe, which by their very nature escape our common senses.

🧺 Food for Thought 🍸

FFT 4.2.2. There are two ways of looking at Galileo's statement of the first law of motion. One way is to appreciate it as a product of the human mind and, ideal as it may be, seek instances of the actual motion that approximate the ideal. This is what I call the Greek way. The other way is to argue that since there is no infinitely smooth surface, the statement cannot be real, and therefore has no use. I call this attitude the Roman way.

The Greeks were fond of knowledge for its own sake, and it was not uncommon to see two generals discussing philosophical issues over a drink on the night before a battle. The Greeks were

Greek and Roman ways of looking at idealizations.

not solely after a knowledge that had utility. Their philosophy was anything but pragmatic. Their geometry was based on the concept of infinite and zero. In fact, the notion of a point as something that has infinitely small length, height, and width is very similar to the notion of the infinitely smooth surfaces in Box 4.2.1. And if the Greeks had abandoned the notion of an ideal point because it could not be realized in actual situations, they would have had to abandon geometry as a whole, and we would have been deprived of a priceless treasure.

The Romans were the complete opposite. If a notion had no practical utility, they would not "waste" their time on it. Since studying science was, in the eyes of the Roman educators, either hard or useless, the Romans abandoned it altogether. If a geometric point cannot be actually realized, so the Romans argued, abandon the point and with it the whole geometry. And they did! As the primary impetus in the western civilization, the Roman ideology led humanity into almost two millennia of intellectual darkness.

The lesson to learn is not to look at the first law of motion (or any other physical law) as a statement that can be exactly realized in nature, but as an *idealized* statement to which any practical situation is an approximation. Furthermore, under controlled laboratory environments situations could be created that approximate the ideal better and better. However, this simple and significant lesson is dismissed by some modern philosophers of science just as it was dismissed by the Romans.

In essence, these philosophers are saying that since the idealized conditions necessary for the fulfillment of the laws of physics are not realized in nature, *the laws themselves must be wrong*! A good example, to which we shall return later (see Food for Though 20.2.1), is the rejection of the law of gravity by some philosophers of science because of its inability to predict the fall of a piece of paper in air! Now, the mathematical law of gravity applies to two completely isolated objects which exert *only this force* on one another: if a piece of paper were *only* under the influence of the gravity of the Earth, the mathematical law would describe its behavior exactly. The presence of the atmosphere, of course, introduces other forces on the piece of paper, which alter its motion drastically. Yet some philosophers of science argue that, because we cannot get rid of the atmosphere, the law of gravity must be wrong!

Along the same line, consider the statement: "Grass is green." This statement assumes that you look at the grass with your naked eye. Now a gravity-denying philosopher of science comes along and says that the statement is wrong. Her reason is: "I always wear brown Sun glasses and I don't see green; I see almost black; and I can show you many others who wear brown glasses, and they always tell me that grass is not green when they have their glasses on. Therefore, your statement is incorrect." This kind of reasoning may appear bizarre, yet it is identical to the argument against the mathematical law of gravity.

This kind of thinking is behind the educational philosophy which relies solely on "inquiry-based" methods, and disposes of any idealization that cannot be achieved by such methods. It is also behind the pedagogical philosophy that teaches only those parts of mathematics and physics which answer the question "Is it good for some practical application?" in a resounding affirmative.

---

[3]This is a rephrasing of a famous quote by Einstein.

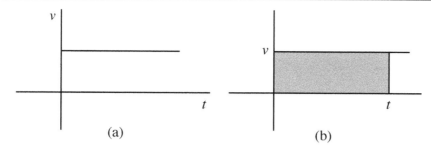

Figure 4.2: (a) The graph of v versus t for a uniform motion is simply a horizontal line. (b) The area under the graph is the distance.

4.3   Rectilinear Motion

A general study of motion (that which takes place in space) will be taken up in Chapter 6. In the present section, we want to study a simpler kind of motion; the one that takes place on a straight line, the **rectilinear motion**.

One of the major contributions to the study of motion, by which Galileo clarified some fundamental concepts—and along the way discovered a major formula in mechanics—is **uniformly accelerated motion** (UAM). It is remarkable that his study of this specific motion led him to some basic notions of calculus just as Newton's study of the general properties of motion led him to some advanced ideas of calculus.

4.3.1   Uniform Rectilinear Motion

To understand the uniformly accelerated motion, we need a little background in the simpler *uniform motion* on a straight line. This is a motion in which the speed of the moving object does not change. Denote the speed by v (which stands for velocity, and to which we shall return later), distance by x, and time by t, and get $v = x/t$, or

$$x(t) = vt, \tag{4.1}$$

which is the mathematical expression for the familiar idea that distance equals speed times travel time. Here—and in what follows—the notation $x(t)$ emphasizes the fact that distance is a function of time.

For our purposes, it is convenient to draw a graph of speed versus time—as opposed to the more common graph of distance versus time. For uniform motion, this graph is as shown in Figure 4.2. Since the value of speed does not change over time, the graph is simply a horizontal line [Figure 4.2(a)]. Figure 4.2(b) shows the area of a rectangle of length t and width v. This area is equal to vt, which is the distance traveled! Thus, for uniform motion, the area under the graph of velocity versus time is the distance traveled.

Area under a v-t plot is the distance traveled.

Some remarks concerning physical quantities and their measurement are in order here. Whenever we measure a physical quantity, we should end up with two attributes: a number *and* a **unit**. The number by itself has absolutely no meaning. If I say that the distance between this point and that point is 12, you (should) immediately say 12 what? Is it 12 inches (the two ends of a ruler), 12 meters (the two ends of a lecture hall), 12 miles (the two gates of a city), or 12 light years (the distance from the Sun to one of the closer stars to it)?

Role of units in physics.

Therefore, the study of physics should start with a convention regarding units. Most physical quantities can be expressed in terms of three fundamental units: the unit of **length**, the unit of **time**, and the unit of **mass**. Most physicists of the world have agreed to measure length in **meters**, abbreviated as m, time in **seconds**, abbreviated as s, and mass in **kilogram**, abbreviated as kg; and these are the units we shall use in this book. As you will discover in the sequel, most other physical quantities are measured in units that are derivable from these fundamental units. For example, velocity or speed, which is distance divided by time, has the unit or meters per second, abbreviated as m/s. For a full discussion of units see Appendix A.

Example 4.3.1. As a very simple example of the uniform motion, let's calculate the conversion factor for going from miles per hour (mph) to meters per second (m/s). A mile is 1610 meters, and an hour is 3600 seconds. If an object travels at the rate of one mph, it covers a distance of 1610 meters in 3600 seconds. Therefore, 1 mph = $\frac{1610}{3600}$ or 0.4472 m/s.

1 mph is 0.4472 m/s.

Whenever you want to convert a speed measured in mph to m/s, simply multiply the number by 0.4472. For example, the speed limit on highways—which is 65 mph—is $65 \times 0.4472 = 29.07$ m/s. Similarly, a speed given in m/s can be converted to mph by *dividing* the number by 0.4472. For example, sound travels with a speed of 340 m/s. This speed is $340/0.4472 = 760$ mph, which is also called **Mach** one. ∎

Mach as a unit of sound speed.

4.3.2 General Rectilinear Motion: Distance

Now suppose that the moving object has a nonuniform speed such as the one depicted in Figure 4.3(a). How do you find the distance traveled by the object? Although the speed changes considerably during the entire motion, for short enough time intervals, the speed of the object is approximately constant. For example, as you travel on a straight highway, you will have to change your speed according to the flow of the traffic and possibly the speed limits of construction areas. For long travel times, you will get a curve similar to the one in Figure 4.3(a), for which the speed changes dramatically. However, if you concentrate on short intervals, say five seconds, your speed will not be changing appreciably. Suppose, for the sake of calculation, we approximate the speed *during* those five seconds to be the speed *at the beginning* of the interval. Then, you can say that the distance traveled during those five seconds is still the (initial) speed times the travel time. During the second five-second interval, your speed, which is equal to the speed at the beginning of the second interval, will have changed slightly, but the distance during the second interval is still (the new) speed times the time interval. To find the total distance you add all these small distances.

On the graph, this procedure is tantamount to drawing rectangles whose heights are initial speeds during certain (small) time intervals, and whose widths are those time intervals. The distances are the areas of these rectangles such as those shown in Figure 4.3(b). Note that in this figure, the height of a rectangle is the speed at the beginning of the given time interval. If the time intervals are not sufficiently small, you will obtain only a very rough estimate of the total distance traveled, i.e., the sum of the areas of the rectangles will be only a rough approximation of the total distance. However, by taking the time intervals small enough, it is possible to get a more and more accurate measure of the distance [Figure 4.3(c)]. And the sum of the areas of the rectangles gets closer and closer to the actual total distance traveled. In the limit of infinitesimally small time intervals, the sum of the areas of the rectangles will coincide with the area under the velocity curve. This leads to

> **Box 4.3.2.** *The area under the speed-versus-time curve, bounded by the vertical v-axis (corresponding to $t = 0$) and the vertical line at t, is equal to the total distance traveled during the time t. More generally, the area bounded by the curve and the vertical lines $t = t_1$ and $t = t_2$ (with $t_2 > t_1$) is equal to the total distance traveled during the time interval $t_2 - t_1$.*

Example 4.3.3. Figure 4.4 shows the graph of the speed (measured in m/s) of a bicycle versus time (measured in seconds). Some general description of the motion can be deduced from the

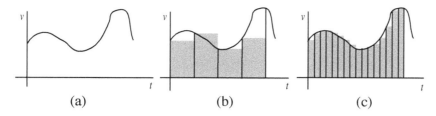

Figure 4.3: (a) The graph of v versus t for a general motion. (b) The sum of the areas of the rectangles very roughly approximates the area under the graph. (c) The sum of the areas of the rectangles more accurately approximates the area under the graph.

plot. Figure 4.4(a) shows, for example, that the bicycle is traveling with a nonzero speed at $t = 0$. This does not mean that the bicycle magically acquires a speed in no time! It means, rather, that we *started* to observe the motion of the bicycle while it was moving. The bicycle is slowing down for a while (for approximately 10 seconds); and then speeds up. About 10 seconds after the start of observing the motion (between approximately 9 and 11 seconds after timing the motion), the bicycle is moving almost uniformly (the speed is almost constant). A little later, about 15 seconds after timing the motion, the speed starts to increase sharply for about one second, and then increases less sharply after that. All these statements can be stated somewhat more precisely by scrutinizing Figure 4.4(b).

To find the distance traveled, refer to Figure 4.4(b). Since v is in m/s and t is in seconds, the area (product of v and t) will be in meters. Each vertical line represents one second and each horizontal line represents one m/s. So, the bicycle travels about 7 meters in the first second (the area between the first vertical line and the v-axis). In the second second, the bicycle covers a distance of about 6 meters, totaling approximately 13 meters 2 seconds after the start of observation. Three seconds into the motion, the bicycle has traveled about 18 m, and after 6 seconds the bicycle has covered a distance of 29.5 m. The reader is urged to make a table, the first column being t with entries 0 sec through 15 sec. The second and third columns are to represent, respectively, speed (in m/s) and distance traveled (in m). Entries of the second column can be read directly from Figure 4.4(b). For example, the initial speed (at $t = 0$) is about 7.5 m/s, while a second later it is a little below 6.5 m/s. The last column is the hardest one, because one has to add up all the squares under the curve up to the time given in the first column. ■

> **What do you know? 4.4.** Figure 4.4 shows the velocity of a car as a function of time. What is the velocity of the car (a) 4 seconds after the start of motion? (b) 8 seconds after the start of motion? (c) How far does the car travel during this time?

4.3.3   General Rectilinear Motion: Acceleration

The graph of speed of an object A versus time contains a great deal of information about the motion of that object. You learned how to calculate the distance A travels in time t by measuring the area under the v-t curve. Now you will learn how to find the **acceleration** of A from the same graph.

The acceleration a of A at time t is defined as the change in velocity divided by time, when the time is very very small. To measure the (instantaneous) acceleration at time t, find the velocity at t (label it $v(t)$); find the velocity a little later at t' (label it $v(t')$); subtract $v(t)$ from $v(t')$ and divide the difference by $t' - t$; make sure that t' is as close to t as possible. The smaller the difference $t' - t$, the more accurate the acceleration. In symbols,

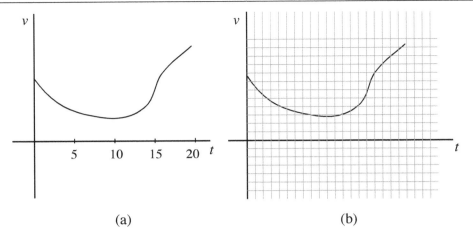

Figure 4.4: (a) The graph of v (in m/s) versus t (in seconds) for a general motion. (b) The grid can help us find the distance traveled.

$$a = \frac{v(t') - v(t)}{t' - t}, \qquad t' \text{ should be very close to } t, \tag{4.2}$$

where $v(t)$ is the speed of A at t and $v(t')$ its speed at a later time t'.

Note that acceleration, being (the change in) velocity divided by time has the unit of meters per second divided by second or (m/s)/s, or m/s^2. Note also that a could be positive [if $v(t') > v(t)$, corresponding to speeding up, or *acceleration*] or negative [if $v(t') < v(t)$, corresponding to slowing down, or *deceleration*].

> Acceleration has the unit of m/s^2 and can be positive or negative.

Suppose you are given a v-t graph such as the one shown in Figure 4.5(a), and you are asked to find the (instantaneous) acceleration at some time, say 12 seconds after the motion starts. How can you determine acceleration from the v-t graph? Here's how: Read off the speed at $t = 12$ s (approximately 6 m/s), read off the speed a little—say 2 seconds—later at $t' = 14$ s (approximately 8.25 m/s), and use the definition of the acceleration given above, to find $a_{\text{avg}} = \frac{8.25-6}{14-12} = 1.125$ m/s^2. The subscript points to the fact that you have calculated the *average* acceleration during the 2 seconds of its motion. As a rule, we designate the acceleration as *average* when the time interval is large (two seconds *is* large for this discussion).[4] For future reference, note that this ratio is simply the slope of the line passing through points P and Q of the curve in Figure 4.5(a).

There are two ways to improve the accuracy of this calculation. First, you can improve the calculation of the slope of the line through P and Q by reading larger numbers. For instance, the length of the dashed line segment (vertical increment of 10.5 units) over the corresponding horizontal line segment (increment of 9.5 units) will give a value of $a_{\text{avg}} = 10.5/9.5 = 1.10526$ m/s^2 for the slope. This value is more accurate than the previous one, because we could read the large increments much more clearly and accurately than small values.

The second improvement is more substantial, because it has to do with the very notion of "instantaneity." Recall that in Equation (4.2), t' is to be as close to t as possible. Points P and Q are separated by 2 seconds, which we decided was too large. So, draw lines from P to points closer and closer to it, and number these lines. In Figure 4.5(b), line number 1 is the line passing through P and Q, lines number 2 through 4 pass through P and points that are closer and closer to P. The last line (number 4) passes through a point so close

---

[4]If the difference between the speeds at the beginning and the end of the time interval is not (very) small compared to either of the two speeds, then the time interval is large.

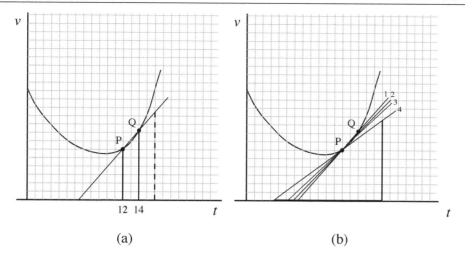

Figure 4.5: The graph of v (in m/s) versus t (in seconds). (a) The first estimate of the acceleration. (b) A more accurate way of calculating the acceleration.

to P that you cannot distinguish it from P. Such a line is called the **tangent** line, and it is *the* line whose slope gives the most accurate value for the acceleration. You can verify from the graph that the slope (average acceleration) of lines 2 is $a_{\mathrm{avg}} = 11/11.2 = 0.982$, for line 3 is $a_{\mathrm{avg}} = 11.5/12.75 = 0.902$, and for line 4 it is $a = 9.5/13.5 = 0.704$. The last acceleration has no subscript because it is the slope of the *tangent* line at P. We therefore conclude that the (instantaneous) acceleration of A, 12 seconds after the start of motion, is about 0.7 m/s^2.

In general,

What is a **tangent** line?

> **Box 4.3.4.** *On a velocity-versus-time graph, the instantaneous acceleration at a given time is the slope of the tangent to the curve at that time.*

To draw the tangent line at a given time, locate the point—call it P—on the curve corresponding to that time. Align a straight-edge so that it passes through P and another point that is as close to P as you can make it. Now draw the line. Tangent lines making an acute angle with the horizontal axis have positive slopes (positive accelerations); those with obtuse angles have negative slopes (negative accelerations).

How to draw the line tangent to a curve at a point.

4.3.4   Uniformly Accelerated Motion (UAM)

The case in which the v-t curve is a straight line is of considerable importance. As you know, the slope of a straight line is the same for all points of the line. What this means is that for such a v-t curve, the acceleration does not change during the motion of the object. Such a motion is called a **uniformly accelerated motion**.

> **What do you know? 4.5.** Figure 4.5 shows the velocity of a car as a function of time. What is the acceleration of the car (a) 7 seconds after the start of motion? (b) 9.5 seconds after the start of motion?

Galileo was the first person to clarify the concept of acceleration and to discover the mathematical formula for the UAM. You can easily find the velocity as a function of time if you remember how to write the equation of a straight line. For an xy-plot, it is $y = mx + b$, where m is the slope and b is the y-intercept. For a vt-plot, the slope is the acceleration a, and the v-intercept is the initial velocity, which we denote by v_0. Therefore, the equation of velocity as a function of time becomes $v = at + v_0$, which we rewrite as

$$v(t) = v_0 + at \qquad (4.3)$$

The graph of the speed versus time for a uniformly accelerated motion is shown in Figure E.4(a) of **Math Note E.4.1** on **page 78** of *Appendix.pdf*. The distance turns out to be

$$x(t) = v_0 t + \tfrac{1}{2}at^2, \qquad (4.4)$$

as shown in that Math Note.

Positive and negative acceleration and their relation to the direction of motion.

The acceleration in either (4.3) or (4.4) could be positive or negative. To determine whether a quantity is positive or negative, you have to designate a *direction* as positive. It is convenient to *take the direction of motion to be positive*. So, "acceleration points oppositely to the direction of motion," means that acceleration is negative. Similarly, positive acceleration means that "acceleration is in the direction of motion."

Example 4.3.5. A train starts from rest and accelerates uniformly to 80 mph in 2 minutes. We want to find the acceleration, and the distance traveled in those 2 minutes. First convert the (final) speed v_f to m/s using the conversion factor found in Example 4.3.1: $v_f = 80 \times 0.4472 = 35.8$ m/s. We also need to convert time to seconds: $t = 2 \times 60 = 120$ s.

For uniform acceleration, the time interval in which the change in velocity is to be calculated need not be small, because the slope of the $v - t$ plot is constant. Thus, we can take t' to be the final time and t to be the initial time, which is taken to be zero. The acceleration is now calculated from its definition $a = (35.8 - 0)/120 = 0.298$ m/s^2. The distance can be easily found from Equation (4.4) with $v_0 = 0$ (because the train starts from rest), $a = 0.298$, and $t = 120$ s:

$$x = v_0 t + \tfrac{1}{2}at^2 = 0 \times 120 + 0.149 \times (120)^2 = 2145.6 \text{ m},$$

which is about 1.33 miles. ∎

A naturally occurring UAM is **free fall**, by which is meant the fall of objects under the influence of gravity when the retarding effects of air are neglected. Under such a circumstance, all objects fall with identical acceleration of approximately 9.8 m/s^2. Galileo was the first to recognize this, and is said to have performed an experiment in which he dropped two dissimilar objects from the leaning tower of Pisa and noted that they landed at the same time. Sometimes this acceleration is denoted by g rather than a to emphasize that it has a *g*ravitational origin.

Free fall acceleration is 9.8 m/s^2.

Example 4.3.6. As just noted, an object, dropped from a height, accelerates downward with a uniform acceleration of 9.8 m/s^2. Taking *down* to be the positive direction, you end up with a positive acceleration. Since in a drop—rather than a downward *throw*—the initial velocity is zero, Equation (4.4) becomes $x(t) = \tfrac{1}{2}(9.8)t^2 = 4.9t^2$. Thus, after 1 second the object falls 4.9 m; after 2 seconds, it falls $x(2) = 4.9(2^2) = 19.6$ m; and after 3 seconds it falls $x(3) = 4.9(3^2) = 44.1$ m, etc. You see that in each second, the object covers a larger distance than in the previous second. This is, of course, because the object gains speed as time passes.

Q: How long would it take the object to fall to the ground from the top of the Sears Tower in Chicago?

A: The Sears Tower is about 500 m high. So, $500 = 4.9t^2$, or $t = \sqrt{102.04} = 10.1$ s. This is an underestimate, because the air drag—which increases with speed—can have a substantial retarding effect on the motion of the object, causing it to slow down, and increasing the time of the fall. ∎

Example 4.3.7. A ball is thrown vertically up with a speed of 100 mph. The acceleration of gravity is 9.8 m/s^2 *downward*.

Q: How high does the ball rise?

A: There are only two formulas related to uniformly accelerated motion. These are Equations (4.3) and (4.4). If you know the time of flight, the second equation can immediately give you the distance (height) by substitution. But this time of flight can be obtained from the first equation, because, you know (or should know!) that at the maximum height, the speed will have to be zero (otherwise, the ball will pass the maximum height!). Since you know v, v_0, and a, you can immediately find t.

First, convert the initial speed to m/s. This gives $v_0 = 100 \times 0.4472 = 44.72$ m/s. Next, note that since the acceleration is opposite to the direction of motion, *it carries a negative sign*. Therefore, (4.3) gives $0 = 44.72 - 9.8t$, or $t = 4.56$ s. The distance can now be easily calculated from Equation (4.4) with $v_0 = 44.72$, $t = 4.56$, and $a = -9.8$:

$$x = v_0 t + \tfrac{1}{2}at^2 = 44.72 \times 4.56 - 4.9 \times (4.56)^2 = 102 \text{ m}.$$

This result can be generalized to a formula connecting distance and speed directly, as shown in **Math Note E.4.2** on **page 78** of *Appendix.pdf*.   ∎

Here is an interesting application of both uniform and uniformly accelerated motions to a life-and-death situation! You are driving on a quiet street. Suddenly a cat jumps in front of you several meters away. You "immediately" slam on the brakes. Do you run the cat over or don't you?

The word "immediately" is in quotation marks for a reason. Nothing in nature is immediate. Every process takes a finite amount of time. The time may be very small—a very small fraction of a millionth of a second. But it is never zero. Consider the process of slamming on the brakes. First you have to spot the cat, i.e., the light from the cat must reach you eyes. This takes a while ... a very short while, but nevertheless a while. For example, if the cat is 300 m away, it takes light a millionth of a second to reach you.[5] And if you are moving with a speed of 20 m/s, the car will move 0.00002 m, or 0.02 mm by the time the light from the cat reaches your eyes. Since this distance is so very short we can ignore it.

What we cannot ignore is what happens *after* you receive the cat's light signal. Your eyes communicate the signal reception to your brain (that takes time); your brain analyzes the situation and decides what to do (hopefully to convey to your leg the instruction of slamming on the brake pedal), which also takes time; then your leg muscles, which up to this time have been relaxed, initiate their motion and indeed slam on the brakes, which also takes time. All this may add up to 0.1 to 0.2 second, in which time the car moves 2 to 4 meters ... not an insignificant distance. **Example D.4.1** on **page 14** of *Appendix.pdf* peppers this discussion with some numbers.

4.4   End-of-Chapter Material

4.4.1   Answers to "What do you know?"

4.1. Aristotelian dynamics was not based on precise observation. It was a collection of philosophical arguments with a little rudimentary (and wrong) observational input.

4.2. Archimedean physics was based on observation and experimentation. This is the right way of doing science.

4.3. Since the smoothness is between the first two surfaces, its corresponding angle must lie between the two angles given. Therefore, the third angle is less than 30° and greater than 25°.

4.4. (a) 4 m/s. (b) 3 m/s. (c) Approximately 13.5 m.

4.5. (a) I used a ruler, and made it tangent to the line at $t = 7$. The ruler crossed the v-axis at 8.5 and the t-axis at 23. The acceleration is, therefore, $-8.5/23 = -0.37$ m/s^2.

---

[5]The speed of light is 300 million m/s.

Acceleration is negative because the slope is negative. (b) The tangent line is (almost) horizontal. So, there is no acceleration, i.e., acceleration is (almost) zero at 9.5 seconds.

4.4.2   Chapter Glossary

Acceleration The change in velocity divided by time. Velocity is speed to which is assigned a direction. So a change in velocity can be made by changing its magnitude as well as its direction (see Chapter 6).

Aristotelian Dynamics A set of philosophical statements laid out by Aristotle without any experimental verification of those statements. Do not confuse the word "dynamics" used here with the same word used in today's physics where observation and experimentation is crucial.

First Law of Motion One of the three laws of motion, a limited version of which was discovered by Galileo. He discovered that on an ideal infinitely smooth surface, an object keeps moving without having to be pushed.

Free Fall Any motion that is caused by gravity *and gravity alone*. A projectile or an object dropped from a height is a good approximation to free fall. The reason that it is not *exactly* a free fall is the presence of atmosphere which affects the motion of most (but not all) objects only slightly. A small stone dropped from a height is almost in free fall, but a parachute is not.

Uniform Motion A motion in which velocity of the moving object does not change. This motion necessarily takes place on a straight line.

Uniformly Accelerated Motion A motion in which acceleration of the moving object does not change.

4.4.3   Review Questions

4.1. What prevented the development of physics in antiquity? Whose influence caused this retardation?

4.2. What did Aristotle say about the number of teeth in a woman's mouth? Was his statement scientific? What did Aristotle say about the connection between the speed of an object and the force acting on it? Was he correct? Did he make the statement based on actual experimentation?

4.3. How do you compare Archimedes' physics with Aristotle's? Why did Archimedes have such a low opinion of his practical inventions?

4.4. What is the role of Galileo in physics and science? Was he a follower of Aristotle or Archimedes?

4.5. What is the statement that summarizes the *actual* observations of the inclined plane experiment? What statement goes beyond the actual observations? What is the role of the human mind in the later?

4.6. What is a uniform motion? How are velocity and distance related in such a motion? How do you find distance on a plot of velocity versus time for a uniform motion? How do you find distance on a plot of velocity versus time for *any* kind of motion?

4.7. How do you find acceleration on a plot of velocity versus time for a general motion? What is the name of the line used to determine the instantaneous acceleration? What property of this line gives the instantaneous acceleration?

4.8. Why can't you say that distance is speed times time in a uniformly accelerated motion?

4.4.4   Conceptual Exercises

4.1. Suppose you want to find the acceleration of a car as it is moving. In one case, you wait one second and measure the change in the car's speed and get an acceleration of 3.5 m/s^2. In the second case, you wait half a second and measure the change in the car's speed and get an acceleration of 4 m/s^2. Which value do you accept?

4.2. Throw a ball vertically up with a very large speed. It moves up while slowing down until it reaches its maximum height at which point it stops and starts moving down. Where does it have the largest acceleration, at the beginning or at its maximum height?

4.3. Drop a ball from a tall building. Do you expect the distance covered by the ball during the fifth second of its fall to be (a) the same, (b) more, or (c) less than the distance covered during the sixth second?

4.4. Skyscraper A is twice as tall as B. You drop a watermelon from each. Ignoring air resistance, is it true that the A's watermelon takes twice as long to reach the ground?

4.4.5   Numerical Exercises

4.1. Figure 4.6 shows the plot of the velocity of an object versus time. The units on the time axis are seconds and those on the velocity axis are m/s.
(a) What is the initial velocity of the object?
(b) What is the velocity after 4 seconds?
(c) What is the velocity after 8, 10, 12, and 14 seconds?
(d) Describe qualitatively the motion from the beginning to the end.
(e) Is the acceleration constant during the first 8 seconds of the motion? If so, what is the acceleration (include sign)?
(f) What is the acceleration between 8 and 14 seconds (include sign)?
(g) Is the acceleration constant between 14 seconds and the end of the motion? If so, what is the acceleration (include sign)?
(h) What is the distance traveled in the first 4 seconds?
(i) What is the distance traveled in the first 8 seconds?
(j) What is the distance traveled between 4 and 8 seconds?
(k) What is the distance traveled between 8 and 14 seconds?
(l) What is the distance traveled between 14 seconds and the end of motion?
(m) What is the distance traveled during the entire motion?

4.2. A car accelerates uniformly from 45 mph to 65 mph in 2 seconds.
(a) What is the acceleration of the car in m/s^2?
(b) How far does the car travel in the process?
(c) If the car traveled with the same acceleration for another 5 minutes, what would its speed be in m/e? In mph?
(d) How far would it travel in those 5 minutes?

4.3. A ball is thrown vertically upward with a speed of 150 mph.
(a) How long does it take the ball to reach its maximum height?
(b) What is this maximum heigh?

4.4. A car is moving at the rate of 20 m/s on a street. A cat suddenly jumps in front of the car 26 m away. The driver slams on the brakes "immediately" with a reflex time delay of 0.1 second. The brakes cause a deceleration of 8 m/s^2. Is the cat dead or alive? To answer this question
(a) find how far the car travels before the brakes are *actually* applied. (b) Plug in all the known quantities in the two kinematic formulas for uniformly accelerated motion. Pay

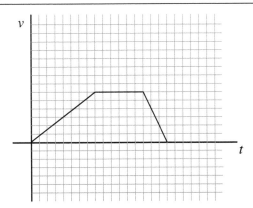

Figure 4.6: The plot of velocity versus time. Each tick on the horizontal axis represents one second. Each tick on the vertical axis represents one m/s.

attention to the *sign* of a!

(c) Determine how long it will take for the car to come to a complete stop after the brakes are applied.

(d) Knowing the time of decelerated travel, calculate the distance the car travels while decelerating to a complete stop.

(e) Add the two distances.

4.5. Estimate the time it takes a watermelon to hit the ground when dropped from the third floor (about 10 m) of a building.

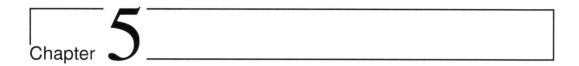

Chapter 5

Epilogue: Underdetermination?

5.1 Lessons from Astronomy

From the early days of its inception in Babylon and Egypt to Kepler's discovery of the sophisticated mathematical relations governing the planetary motions, astronomy underwent a substantial metamorphosis. Let us analyze the causes of this change through an examination of the procedures used by the astronomers who contributed to the field.

Looking back at the development of astronomy, we see that **observation** played a crucial role. This observation was quite different from casual "observations" every human being makes of the sky. The observations of the Egyptian and Babylonian priests were **quantitative** and **recorded**. They charted the sky; gave names to certain conspicuous heavenly bodies; kept accurate tracks of their locations; and recorded their findings. Somehow, they knew that unraveling the secrets of the universe was not the task of a single generation; that the observations of one generation had to be passed down to the next, so they did not have to start anew.

Quantitative and recorded observation is the beginning of science.

As the number of quantitative observations grew, and as patterns were discovered in the motion of heavenly bodies, **theories** were proposed to summarize these observations and patterns. At first, theories were no more than myths adapted to the cultural settings of the theory makers: the stories of Ea in Babylon, of Ra in Egypt, and of Zeus in Greece. Later, myths were replaced by secular stories describing the motion of *material* objects in the sky. Pythagorean astronomy was of this form. It was, nevertheless, a story that attempted to encompass the entire cosmos in one general swoop.

Materiality of objects of investigation in science.

This generality hampered the advancement of knowledge, and after a few generations, astronomers came to the conclusion that it would be best if they concentrated on the motion of some *specific* objects. It soon appeared that a few objects had a pronounced motion against the background stars; and so the attention was shifted from general cosmos to the five planets, the Sun, and the Moon. This reduction in the number of objects had a profound effect on the development of astronomy. With only a few objects to keep track of, astronomers were able to chart their location in the sky, watch closely their luminosity, their color, and any other feature that was easily missed in the necessarily diluted observation of the entire cosmos.

Specificity is crucial for the advancement of science.

This reduction, for instance, led to the discovery of the retrograde motion, and generated some of the most intense mental activities of ancient Greece, activities that eventually produced Eudoxus's brilliant homocentric theory. And when astronomers discovered the small increases in the brightness of a specific planet such as Mars, the idea of epicycles came into being.

As the drive for the quantitative measurement of physical properties of celestial objects intensified, so did the concentration on very specific aspects of those properties. Instead of holistically lumping the entire cosmos into crystal spheres which moved uniformly around the Earth, Eratosthenes wanted to know *specifically* how big the Earth was, while Aristarchus was interested in the very "narrow" question of the Earth-Sun and Earth-Moon distances. No more lofty ideas of the motion of the entire universe for which there was no measurable answer, but a very specific question about a very specific measurement for which there was a possibility of an answer.

The achievements of these scientists are among the most notable of ancient Greece. Eratosthenes gave us a sense of the size of our home. While humans were aware of the enormity of Earth, they had absolutely no idea of *how* big it was. It could have been ten times the distance between Athens and Sparta, or a hundred times, or even a thousand times. With Eratosthenes' discovery, mankind now had an accurate picture of how big the Earth was, and how small our entire species appeared next to it.

Specificity of Earth-Sun measurement leads to a **universal** shift in astronomical viewpoint.

Aristarchus's measurement of the Earth-Moon distance dwarfed our entire civilization to a tiny speck in the cosmos. His next measurement, although very inaccurate, foretold of an entirely different universe. Aristarchus's *specific* measurement of the Earth-Sun distance showed that the Sun was seven times bigger than the Earth. Although grossly underestimated, the enormity of the Sun prompted Aristarchus to question the prevailing geocentric model of the universe. How can the Sun, being so much bigger, go around the Earth? It *must* be the other way around. And he stumbled on the heliocentric model of the universe. But if Earth is moving, why can't we detect its motion against an ocean of fixed stars? His answer? The stars must be much farther away from us than previously thought! A remarkable *universal* statement coming from two very *specific* measurements! It took over seventeen centuries for mankind to grasp the value of Aristarchus' contribution.

The most valuable lesson of this part.

Perhaps the most valuable lesson we learn from this part of the book is the value of abstract and scientific thought. Abandoning them plunges the human race into darkness. When the Romans took over the rein of humanity from the Greeks, they substituted pragmatism for the abstract thought, rhetoric for mathematics, law for science, and politics for philosophy. The result was a burgeoning of magic, cults, and entertainment, with science and mathematics fading into the horizon of the Roman empire, being completely lost in the ensuing Dark Ages.

Cicero's evaluation of science.

Some may argue that such a claim is pure speculation; it may well be. But the evidence of history is overwhelming. History has not recorded a single Roman scientist, Roman mathematician, or Roman astronomer;[1] at the same time, it is full of Roman warriors, Roman engineers, Roman politicians, Roman orators, and Roman gladiators. The battle of Plato against rhetoric and his emphasis on mathematics (see Section 2.4.1) may not be the only cause of the scientific achievements in Greece, but nobody can deny the great influence of Platonic thought on the development of the Golden Age. Compare Plato's remarks on mathematics and science with his Roman counterpart, Cicero, who thought that the investigation of the universe seeks either what nobody can know or nobody needs to know. If Cicero, the great Roman intellectual, degrades pure thought to such a level, is it surprising that the seed of science and mathematics found no Roman young mind in which to grow?

During the Middle Ages, when the eclipse of knowledge darkened the scientific landscape of Europe, a few Islamic astronomers, prompted mostly by the need of their religious festivities—which were based on the motion of the Moon—were polishing the Ptolemy's *Almagest*. Their contribution was mostly theoretical, and although they advanced spherical geometry and trigonometry—by-products of astronomy—to a considerable extent, they did not give astronomy the impetus needed to bring it to its new level of development.

Such an impetus required new *observation*. As long as one relied on the ancient observations of Hipparchus and Ptolemy, the complicated geocentric model or the newly proposed

---

[1]The few who happened to *live* in the Roman empire had a distinctly Greek background.

heliocentric model of Copernicus were adequate. But when Tycho Brahe charted the motion of the planets and stars to unprecedented accuracy, a new explanation became urgent. And this explanation was discovered not by a holistic study of the entire cosmos, nor of the solar system, nor of the entire collection of the planets, but by a very careful analysis of a *specific* planet over a period of eight years.

Importance of observation.

Kepler was obsessed with the planet Mars, because it defied all explanations, geocentric and heliocentric. He noted that without Tycho's measurements, Copernicus's heliocentric model with circular orbits for all planets, was capable of explaining the solar system. However, when planets were placed on their paths as measured by Tycho, discrepancy arose. This discrepancy was particularly conspicuous for Mars. Kepler decided to tackle the Mars problem, and after eight years of careful geometrical analysis, he concluded that the orbit of Mars had to be an ellipse with the Sun being at one of its foci. It was now natural to *generalize* this specific finding to other planets in the form of Kepler's first two planetary laws.

Understanding the motion of a single planet yields a universal truth!

What we learn from this discussion of astronomy is that

> **Box 5.1.1.** (1) *Observation is crucial in the development of astronomy*
> (2) *Only concentration on a **specific** question can lead to general universal statements.*

This was as true for the astronomy of 300 BC as it was for that of AD 1600. The general statements such as those made by Pythagoras can never uncover the scientific truth. But some very specific questions such as "How far is the Sun?" or "How far is the Moon?" can lead to a general (and truer) picture of the cosmos, in which planets are moving around the Sun, and the stars are much farther from us than previously thought. Similarly, the very specific question "What is the shape of the orbit of Mars?" can unravel the mystery of the entire solar system.

5.2   Birth of Scientific Methodology

Under the reign of Plato and Aristotle, nothing but pure thought (Idea, Form) was worthy of philosophical attention. And since physics was part of philosophy, purely speculative and argumentative methods were the only way of investigating nature. This methodology was well-suited to the ideology of the Church during the Dark Ages; however, it was so powerfully ingrained in the psyche of mankind that even some of the free thinkers of scientific Renaissance could not abandon it. Take Descartes, for instance. While he believed in the importance of matter and the fact that the universe was like a machine obeying knowable rules and laws, he thought that the power of the mind was sufficient to unlock all the mysteries of the universe. He was a skeptic of the first rate, and like Plato, distrusted the senses. However, while he was aware of the importance of the sense impressions, he argued that the "truth" lies beyond these impressions, and through the power of the mind and logical and mathematical analysis, one can obtain the truth. He was contemptuous of Galileo because of the latter's interest in the experimental side of science.

5.2.1   Experimentation

Galileo, unlike Descartes, was a follower of Archimedes, who over 18 centuries earlier was using the experimental methods in hydrostatics to calculate the volumes of solids of different shapes. The great contribution of Galileo was to apply the same method to the study of *motion*. In Galileo's own words [Bro 62, p. 143]:

> My purpose is to set forth a very new science dealing with a very ancient subject.
> There is, in nature, perhaps nothing older than motion, concerning which the books

written by philosophers are neither few nor small; nevertheless, I have discovered by experiment some properties of it which are worth knowing and which have not hitherto been either observed or demonstrated. ...

It has been observed that missiles and projectiles describe a curved path of some sort; however, no one has pointed out the fact that this path is a parabola. But this and other facts, not few in number or less worth knowing, I have succeeded in proving; and what I consider more important, there have been opened up to this vast and most excellent science, of which my work is merely the beginning, ways and means by which other minds more acute than mine will explore its remote corners.

How prophetic was the last sentence! Only 22 years after Galileo's death, a 22-year-old student of Cambridge explored and discovered one of the remotest and most hidden corners of this "vast and most excellent science."

Prior to Galileo, the scientific investigation of motion had always been confined to the objects in heaven. Observation, that crucial first step, had always been directed at the sky. The "physicists" of antiquity—with the exception of Archimedes—considered earthly objects too menial to be "observed." It was only after the liberation of the human mind from the medieval bondage that the study of motion descended from the sky to terrestrial laboratories. This was done by Galileo, who showed us that the "dirtying of one's hands" is not only not menial, but quite *necessary* if we are to unravel the secrets of the universe. He introduced a methodology whereby the act of observation became just as integral a part of terrestrial science as it was of celestial science.

This methodology—the placement of primacy on observation—did not come about easily. When Roger Bacon (c.1220–1292), the most celebrated European scientist of the Middle Ages, sought "to work out the natures and properties of things"—which included studying light and the rainbow and describing a process for making gunpowder—he was accused of black magic [Boo 83, p. 401]. He failed to persuade Pope Clement IV to admit experimental sciences to the university curriculum; he had to write his scientific papers in secrecy, and was imprisoned for "suspected novelties."

Galileo was not the only one to depart from the Aristotelian doctrine of the primacy of mind over experimentation. The wave of renaissance propagated into many areas of human curiosity, and rekindled an interest that had been suppressed for centuries. It was in this spirit that, as we saw, Tycho Brahe turned the eye of humanity once again toward heaven. It was also in this spirit that Gilbert started his crude, but fundamental, experiments with magnets and electricity. But Galileo's experiments in mechanics was by far the most significant of all, because it opened not only the path of the most fundamental branch of physics, but also reopened the road to the science of mathematics, a road that was, for the most part, closed for over 16 centuries. Galileo's legacy was to apply the methodology that was so successful in the advancement of astronomy to the study of motion. What are the characteristics of this methodology?

5.2.2 Specificity

Unless you concentrate on specifics, you cannot discover general laws.

Instead of making—a la Aristotle—a holistic statement that some objects fall and some rise, Galileo asked the *specific* question of how a *specific* object, such as a block of wood, slides on another object, such as an inclined plane, having *specific* properties—such as smoothness. Just as some ancient Greeks achieved immense success by asking the *specific* questions: "How big is Earth?" and "How far is Moon?", and even discovered a *universal* world view—that the Sun must be the center of our cosmic neighborhood—by asking the *specific* question "How far is Sun?"; and just as Kepler unraveled the *universal* law of the motion of planets around the Sun by asking the *specific* question "What is the precise path of Mars?", so did Galileo discover the *universal* first law of motion by concentrating on the *specific* set-up of a block and an inclined plane.

5.2.3   Quantification

Measurement is at the heart of any scientific knowledge. This could be as simple as Eratosthenes measuring the angular shadows of buildings in two different cities (thus determining the circumference of the Earth), or Galileo measuring the angle and the smoothness of an inclined plane (thus arriving at the first law of motion). Measurement by itself is not sufficient to make an investigation scientific. It has to be done in conjunction with an **experiment** designed to answer a specific scientific question. Experimentation was not new to Galileo; Archimedes was a master at this method of investigation almost two millennia before Galileo. What Galileo did for the first time in history was to apply the method to the study of *motion*.

The act of measurement is necessarily accompanied by **control**. Once you concentrate on a specific question and design an *experiment* to conduct your investigation, you have the freedom to *control* various parameters of the experiment, thereby discovering the relation between those parameters. For example, by controlling the smoothness of the surfaces in contact and the angle of the inclined plane, Galileo discovered that on smooth surfaces, the block started to slide down the inclined plane at smaller angles.

5.2.4   Extrapolation and Idealization

The specificity of the question, the quantification of the investigation, and the control of the parameters are very crucial in gaining scientific knowledge. By its very nature, any experiment is limited by the amount of time and effort the experimenter puts in it. For example, in the inclined plane experiment, one is limited by the number of smooth surfaces and their corresponding angles. If one is lucky, one can obtain 10 or 20 surfaces, and therefore, 10 or 20 angles. To conclude from this the statement "The smoother the surface, the smaller the angle at which the block starts to slide" requires a leap of faith. But it is a leap that is essential in the development of science. It says that if the surface S_1 requires a $30°$ angle for the block to start sliding, and surface S_2 requires a $40°$ angle, then a surface whose smoothness is somewhere between S_1 and S_2, will require an angle between $30°$ and $40°$. One need not do the actual experiment to verify this, although one could. This is called *extrapolation*.

What is even more essential is what we may call **idealization**. This is the extremely important step that takes the science from the boundaries of the experimental laboratory to the realm of the universal laws. Although in extrapolation one could in principle verify the correctness of the conclusion, such a direct verification is impossible in idealization. In the case of the inclined plane, the idealization is the statement "For *infinitely* smooth surfaces, the angle at which the block starts to move is *zero*." There is no way that we can test this assertion *directly*, because no *infinitely smooth* surfaces exist in nature.[2] We shall see that all future scientific knowledge is based on this idealization. It is the process that connects experiments with laws and theories.

How science moves from the boundaries of laboratories to the realm of universal knowledge.

5.3   Underdetermination?

Extrapolation and idealization are indispensable for scientific progress. Any experimenter deals with only a *finite number* of trials, and any scientific conclusion is *necessarily* based on a finite number of experimental "points." This is how science has been ever since its inception in Egypt and Babylon, where the paths of stars and planets were charted from only a *finite number* of sightings. However, philosophers of science argue that because of the finiteness of the number of experimental trials, no general conclusion ought to be drawn from those trials. They call this argumentation the *underdetermination* of theories.

---

[2]We can test it *indirectly* by noting that as we increase the smoothness of the inclined plane and the block, the angle required to start the slide of the block decreases.

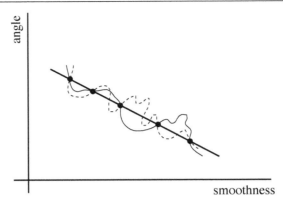

Figure 5.1: The five points represent the outcomes of five possible experiments with an inclined plane. The straight line represents what Galileo (a physicist) would draw as the theoretical statement summarizing the results of the experiments. The other two curves represent two of an *infinite number* of curves that Thomas Kuhn would draw to make his underdetermination argument.

Based on "underdetermination" argument you cannot say "the smoother the surface, the smaller the angle at which the block starts to slide," because you have not tested the statement for *every possible* angle and smoothness; that even if physicists successfully test one thousand different surfaces with varying degrees of smoothness against the statement, still they are not allowed to jump to the conclusion that the statement holds for all surfaces.

Underdetermination is the basis of the philosophers' lack of interest in observation and experimentation. To them, observations can at best give you a finite number of examples of a physical process. It is the theory that claims to be universal, and thus, worthy of philosophical consideration. In fact, Thomas Kuhn, the noted American philosopher of science, points out that given any set of observations, there are infinitely many theories that can fit those observations; therefore, he goes on to say, theories are chosen not by experimental confirmation, but by social persuasion! This is akin to the mathematical fact that given a finite number of points (representing experimental results), there are infinitely many curves (each curve representing a theory) that can be constructed to fit those points (see Figure 5.1); therefore, philosophers of science argue, why even bother to pick one curve as *the* curve. This argument ignores the fact that, in physics, future adds many (albeit a finite number of) points to the set, and usually it is the unique curve rightly chosen, on which these points will fall. But underdeterminists are not satisfied by these future predictions, because the predictions confirm the theory only a finite number of times!

Let us see what sort of a worldview we would have if underdetermination argument were taken to its extreme logical boundaries. Consider the statement "The Sun rises from the east." This is something that everybody accepts and its validity is taken for granted. But underdetermination tells us that since the rising of the Sun from the east has been tested only a finite (albeit a very large) number of times, it should not be generalized. "How can you be sure that the next time the Sun rises it won't be from the west (or the north, or the south, or ...)?" Of course, no philosophers of science would submit to this underdeterminational argument for the rising of the Sun, although the logic of their argument applied to scientific theories is identical to the one applied to the sunrise.

Consider another statement "All human beings will die before their 1000th birthday." Again, the underdeterminationists w(sh)ould argue that although the statement may have been tested on all human beings that ever walked on this planet, you cannot generalize it to all humans; that some of us now living on Earth may live to be 1000 years old. Imagine how much harm this kind of argument inflicts on rational thinking. "If the educated philosophers of science tell me to accept underdetermination," a person on the street may think, "then I

Underdetermination and failure of Sun to rise from the East!

Underdetermination and the 650-year-old woman who gave birth to a healthy child!

should probably believe what I read in the tabloid papers the other day about a 650-year-old woman giving birth to a healthy child!"

5.4   The Role of Mathematics

When the Hellenistic rationalism gave way to the Roman pragmatism; when the interest of the intellect shifted from questions about nature and universe to questions about law and politics; when "What is the diameter of the Earth?" was displaced by "What is the best way to build a colosseum?"; and when mathematics was conquered by rhetoric, humankind entered the most barren desert of the landscape of its history.

Upon its second awakening in the sixteenth century, mankind rediscovered the power of mathematics and rationalism. And this time it was no less authority than nature itself that sounded the wake-up call. As soon as humankind started to study nature again— either experimentally as Galileo did, or observationally as Kepler did, or philosophically as Descartes did—mathematics was proven to be the only means of communication. Nature was telling us once more that if we want to understand her, we had better equip ourselves with this powerful tool of communication as our forefathers did centuries ago in the golden age of Hellas. It was in this spirit that Galileo in Italy said:

Once again Nature demands mathematics from mankind.

> Philosophy is written in that great book which ever lies before our gaze—I mean the universe—but we cannot understand it if we do not first learn the language and grasp the symbols in which it is written. The book is written in the mathematical language, and the symbols are triangles, circles and other geometrical figures, without the help of which it is impossible to conceive a single word of it, and without which one wonders in vain through a dark labyrinth.

And Descartes in France echoed back:

> If we possessed a thorough knowledge of all the parts of any animal (e.g., man), we could from that alone, by reasons entirely mathematical and certain, deduce the whole conformation and figure of each of its members, and, conversely if we knew several peculiarities of this conformation, we would from those deduce the nature of its seed.

And Kepler in Germany resounded:

> The chief aim of all investigations of the external world should be to discover the rational order and harmony which has been imposed on it by God and which He revealed to us in the language of mathematics.

This time around, however, the demand of nature was much more forceful. While at the time of Greeks, the study of nature was confined to geometry and the *static* properties of objects, the task of analyzing *motion* brought forth the notion of time. And for this, mankind had to learn (discover) new vocabulary of the language of nature. Thus Galileo and his students were forced to discover some rudimentary properties of calculus, and by the time of Newton—who unraveled the complete set of the laws of motion—it developed into a new branch of mathematics and theoretical physics.

"The latest authors, like the most ancient, strove to subordinate the phenomena of nature to the laws of mathematics."

Isaac Newton

Part II

Newtonian Era

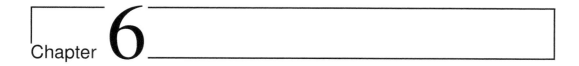

Chapter **6**

Kinematics: Describing Motion

Chapter 4 taught us some basic knowledge of motion along a straight line. This chapter generalizes the concept of motion and introduces some fundamental quantities necessary for its description.

6.1 Position, Displacement, and Distance

What is motion? Vaguely speaking, it is the change in the state of an object. More precisely, it is the change in the *position* of an object with *time*. Still more precisely, we must speak of the motion of an *object* relative to an *observer*, although the "object" could be a person, and the "observer" a thing.

To analyze the motion, draw an arrow from the "observer" O to the "object" A, and call the arrow the **position vector**.[1] The very definition of the position vector assumes that both O and A are points. The position vector, denoted commonly by $\mathbf{r}$, determines the *instantaneous* position of the object A [Figure 6.1(a)] relative to an observer O. The word "instantaneous" is important because the position of the point object A is, in general, constantly changing. If we were to take snapshots of A at various times, t_1, t_2, t_3, etc., and label the corresponding points at which A is located by A_1, A_2, A_3, etc, we would have a situation depicted in Figure 6.1(b), with position vectors $\mathbf{r}_1, \mathbf{r}_2, \mathbf{r}_3$, etc. In this figure, only three out of an infinitude of possible snapshots are shown. Bear in mind that every directed line segment from O to a point on the curve in Figure 6.1 is a possible position vector.

Position vector.

> **What do you know? 6.1.** Can you say that the observer—as defined in the description of motion—does not move?

If the point A does not change, i.e., if the position vector $\mathbf{r}$ does not vary with time, we say that the object is **stationary** relative to O. You have to make a clear distinction between the *distance* between O and A, which is the length of $\mathbf{r}$, and the position vector, which is the directed line segment $\mathbf{r}$. This distinction becomes essential in situations where the distance does not change while A is clearly not stationary (for instance, when O is at the center of a circle while A moves on the circle).

All objects/observers that are stationary relative to one another constitute a **reference frame** (RF). For example, all students in a lecture hall form a RF, because an arrow drawn from one student to another will not change (as long as all students are sitting still in their

Reference frame.

---

[1]Appendix C summarizes some rudimentary properties of vectors at a level needed for this chapter.

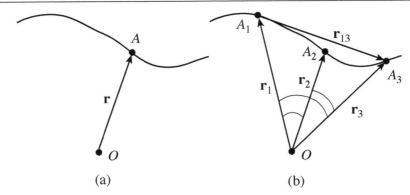

Figure 6.1: (a) The instantaneous position vector of A, and (b) some position vectors of A at different times with respect to the observer O.

chairs). This RF also includes all the chairs, the desks, the equipments, etc. The lecturer will be part of this RF only if he/she is not moving in the room. Similarly, all the sitting passengers as well as all the chairs, tables, equipments, etc. in a moving train form a RF. This RF does not include objects outside, nor people that are walking in the train.

> **What do you know? 6.2.** One car is following another on a highway. Both are moving at 65 mph north. Which of the following is true? (a) All passengers of the car in front are in the same reference frame. (b) All passengers of the car in the back are in the same reference frame. (c) All passengers of both cars are in the same reference frame. How does your answer change if the cars move in opposite directions?

Because motion is relative, different observers perceive the motion of the same object differently. For one observer the motion may be very simple, for another it may seem very complicated. You have already seen that the motion of planets is very simple—very nearly circular—as seen from the Sun. However, these same planets appear to have complicated motions, including retrogression and change in brightness, etc., when viewed from Earth (see Chapter 3).

> **What do you know? 6.3.** A car is moving on a highway at 60 mph. Can we just say that the car is moving?

Example 6.1.1. Your teacher paces the lecture hall in front of you. In the back, there is a clock mounted on the wall. As she talks about the relativity of motion, she decides to designate herself as the observer and the clock as the object in motion. She starts at one end of the lecture room, stretches her arm to constantly point to the clock in the back, and tells the class to carefully watch her arm, which is in the direction of the *position vector* of the clock, as she walks to the other end of the room. By the time she reaches the other end, her arm has shifted in direction. The position vector of the clock (the object) has changed relative to the teacher (the observer). Therefore, *the clock was moving relative to the teacher*! ∎

Displacement.   As objects move (relative to some observer), they are *displaced*.

> **Box 6.1.2.** *The **displacement** of A moving relative to O from an initial point A_1 to a final point A_2 is the directed line segment from A_1 to A_2.*

Displacement is denoted by $\Delta\mathbf{r}$ or $\mathbf{r}_{12}$, or (rarely) $\mathbf{d}$, and is usually measured in meters. In terms of the position vectors $\mathbf{r}_1$ and $\mathbf{r}_2$, we have $\Delta\mathbf{r} = \mathbf{r}_2 - \mathbf{r}_1$, and this *defines* the difference between two position vectors.

What do you know? 6.4. Karl and Emmy don't move relative to one another. They look at a car move from an initial point to a final point. Does Emmy see the same initial position vector for the car as Karl? Does she see the same final position vector as him? Does she see the same displacement for the car as him? What if Emmy moves relative to Karl?

You have to differentiate between distance and displacement. **Distance** is the *length of the path* that an object takes in going from an initial point to the final point. For example, in Figure 6.1, the displacement between A_1 and A_3, denoted by $\mathbf{r}_{13}$, is simply the directed straight line segment from A_1 to A_3, while the distance is the length of the curved path joining the two points.

Position is only one of the ingredients of motion. Another ingredient is time. The concept of time is much harder to explain than position. Even an operational definition of time is not straightforward. To define time operationally, you need to be able to *measure* time. **What is time?** This involves a measuring device (a clock), just as the measurement of position involves a pair of measuring devices (a meter stick to measure length and a compass to measure direction).[2] However, although it is conceptually easy to construct a meter stick and a compass, it is difficult to come up with a good clock. Ironically, it is only *after* physicists mastered the concept of motion that they were able to determine what a good clock is. We shall not dwell on this subject any further and simply assume that there exist good clocks. (However, you may find Food for Thought 18.1.1 relevant to the present discussion.)

Example 6.1.3. Suppose that observer O_1 watches a driver move north 50 miles, then east 100 miles and finally south 50 miles where he stops (see Figure 6.2). The distance the driver traveled is simply the length of the path taken, i.e., the sum of three sides of the rectangle: distance $= 50 + 100 + 50 = 200$ miles.

What do you know? 6.5. Karl is driving on a highway. He says "The trees are moving." (a) He is crazy! (b) He is dead wrong! (c) He is right.

On the other hand, the displacement is simply the (directed) line segment from i to f whose length is obviously 100 miles and whose direction is eastward. This last specification is important, as displacement always requires a designation of direction. So, we write

$$\text{displacement} = \Delta\mathbf{r} = 100 \text{ miles, eastward}$$

It is clear from the figure that, although different observers such as O_1 and O_2, *who do not move relative to one another*, measure different initial and final position vectors, the displacement will be the same for both. That is why, when finding displacements, you usually do not need to specify the observer. ∎

6.2   Parallax

When an object moves relative to an observer, the position vector of the object at any particular time makes an angle with the position vector at an earlier time. As the object moves, the angle changes [see Figure 6.1(b)]. This angle change is called the **parallax** of **Parallax defined.**

---

[2]Direction could also be measured astronomically, as the ancient civilizations did.

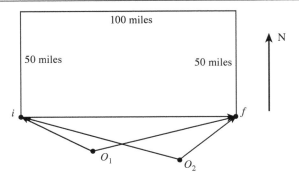

Figure 6.2: The driver moves from the initial point i to the final point f. Note that although the initial and final position vectors are different for observers O_1 and O_2, the displacement vector is the same for both.

Transverse motion.

the motion. If the object moves along the line joining it to the observer (line of sight), the motion will not have a parallax. Thus, parallax is an indication of a motion that has a perpendicular component to the line of sight. Such a motion is called a **transverse** motion.

If you draw a position vector from yourself to the window of your room as you are sitting at your desk, and see that that position vector does not change, then the parallax is zero and clearly the window is not moving relative to you. Therefore, parallax is an indication of (transverse) motion of an object relative to the observer. A zero parallax indicates no transverse motion.

> **What do you know? 6.6.** You draw a line from yourself to an object. Then you start to move *perpendicular* to that line, and you don't see a parallax. What can you conclude?

As you move on a highway, looking at the trees on the side of the road, it is the effect of parallax that turns your head to keep track of a single tree. Figure 6.3(a) shows a car moving on a highway looking at a (stationary) tree. To the people in the car, the tree moves in the opposite direction [Figure 6.3(b)], and describes a parallax. The figure also shows that the farther an object is from the car, the smaller the parallax (angle change) it describes. That is why the foreground objects alongside a highway seem to be moving faster than the background distant objects.

One of the strong oppositions to the ancient heliocentric model was the lack of any observable parallax. If the Earth is moving around the Sun, and therefore in the vastness of the universe, then we should be able to see *some* parallax for the stars. In fact, the Greeks thought that the stars were not too much farther from the Earth than the Sun was. So, if the Earth moved around the Sun, it had to cover such enormous distances as to make any parallax easily detectable. Failure of such detection was convincing evidence that the Earth was stationary. Of course, now we know that even the closest star is thousands of times more distant than the Sun, and its parallax is immeasurably small. **Example D.6.1** on **page 15** of *Appendix.pdf* calculates the parallax of Epsilon Eridani, one of the closest stars to the solar system, and shows that it is a mere 0.000169 degree.

The smallness of parallax explains why the Moon "follows" us as we drive at night. Let's go back to Figure 6.3 and replace the tree with the Moon. As the car moves from its initial position to its final position, a very "small" parallax will be formed due to the enormous distance of the Moon. How small? Suppose we look at the Moon, then drive a mile and look at it again. What is the angle between the two position vectors? The angle in radian

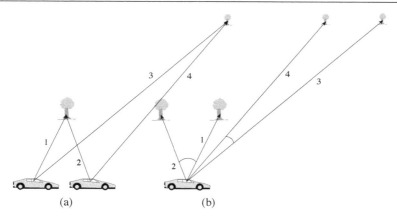

Figure 6.3: A car in motion detects a parallax of a tree. (a) To an outside observer the car goes from position 1 to position 2. (b) To the people in the car, the tree goes from position 1 to position 2. These two positions define a parallax. Positions 3 and 4 correspond to a distant tree, and describe a smaller parallax.

is one mile divided by the Earth-Moon distance. The latter being about 400,000 km and one mile being 1.61 km, we get 1.61/400,000 or 4×10^{-6} radian, or about 0.0002 degree! This angle is so small that the two position vectors appear to be parallel (see Figure B.4 of Appendix B). Figure 6.4(a) shows these two position vectors. The result is that the Moon appears to be "attached" to the car with a very long rigid and immobile (and invisible) rod, dragged by the car wherever it goes [Figure 6.4(b)].

Parallax explains why Moon follows you.

6.3   Velocity and Speed

Motion is related to change. The continuous change in the instantaneous position vector is what we call motion.[3] We can represent this change by a directed line segment from the initial position to the final position, i.e., by the displacement. It is clear that objects covering the same displacements in *different times* describe different motions. What distinguishes these motions is velocity.

Average velocity and average speed.

Box 6.3.1. *The **average velocity** $\mathbf{v}_{\mathrm{avg}}$ of a moving object is its displacement divided by Δt, the time elapsed:* $\mathbf{v}_{\mathrm{avg}} = \dfrac{\Delta \mathbf{r}}{\Delta t}$. *The **average speed** v_{avg} of a moving object is the distance it travels divided by the time elapsed Δt:* $v_{\mathrm{avg}} = \mathrm{distance}/\Delta t$.

Note that although the letter "v" symbolizes both the average velocity and average speed, the *magnitude* of these quantities are completely different as the following example shows.

Example 6.3.2. Let us return to Example 6.1.3 and assume that the south-north and the north-south legs of the trip take one hour each, and the west-east leg three hours. Then the elapsed time is five hours. The average velocity is

$$\mathbf{v}_{\mathrm{avg}} = \frac{\Delta \mathbf{r}}{\Delta t} = \frac{100 \text{ miles, eastward}}{5 \text{ hours}} = 20 \text{ mph, eastward,}$$

---

[3] The position vector is drawn relative to an observer, of course. In the following discussion, all the properties of motion are described relative to an observer, although the observer is not identified specifically.

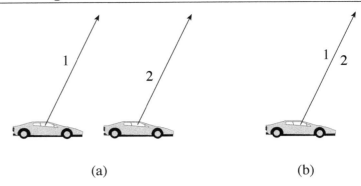

(a)   (b)

Figure 6.4: (a) As the car goes from position 1 to position 2, the lines to the Moon are parallel. (b) To the people in the car, the Moon appear to be "attached" to their car, and dragged by it.

while the average speed is

$$v_{\text{avg}} = \frac{\text{distance}}{\Delta t} = \frac{200 \text{ miles}}{5 \text{ hours}} = 40 \text{ mph}.$$

Speed requires no specification of direction. You see that contrary to the ordinary interchangeable use of the two words, the (magnitude of the) average velocity is not equal to the average speed. For another illustration see **Example D.6.2** on **page 16** of *Appendix.pdf*. ■

Assume that object A moves on a complicated path as shown in Figure 6.5(a), and you take snapshots of the motion. If the time intervals are too long,[4] the object may have gone through a multitude of intervening points before the second snapshot is taken. Thus, if A_1 represents the initial point on the photographic plate, and a long time later the particle is recorded at A_2, the change in the position vector (the displacement) appears to be $\Delta \mathbf{r}$. You have lost a considerable amount of information about the motion of the object by waiting too long.

To avoid this loss of information, make the time interval between successive observations smaller and smaller. Then the point A_2 will be very close to A_1 and the displacement $\Delta \mathbf{r}$ becomes arbitrarily short. How are we going to deal with such small directed line segments? By defining the instantaneous velocity! It is true that $\Delta \mathbf{r}$ becomes shorter and shorter as the time interval between t_1 and t_2 becomes smaller and smaller, but if you divide $\Delta \mathbf{r}$ by the small time interval, you will get a sizable number.

Instantaneous velocity defined.

Box 6.3.3. *Locate the moving particle at time t and call its position vector $\mathbf{r}$; locate it at a later time very close to t, say t' with position vector $\mathbf{r}'$; find the displacement $\Delta \boldsymbol{r} = \mathbf{r}' - \mathbf{r}$, as shown in Figure 6.5(b); divide this displacement by $t' - t$. In the limit that t' gets arbitrarily close to t, the ratio is called the **instantaneous velocity** $\mathbf{v}$ at time t. Thus, $\mathbf{v}_{\text{avg}}$ becomes $\mathbf{v}$ when Δt is very small.*

We measure velocity and speed in units of *meter per second* abbreviated m/s. However, mile per hour (mph) is also used sometimes. A few remarks are in order.

Instantaneous speed.

- The instantaneous velocity is also a *directed* line segment, whose length v is called the **instantaneous speed** of the object. Instantaneous velocity tells us how fast the object is moving at time t, and its direction gives the instantaneous direction of motion of the object at that time.

---

[4]The phrase "too long" has different meanings for different motions. While minutes are fairly large in the analysis of a racing car's motion, hours, even days, are small when analyzing the motion of some planets.

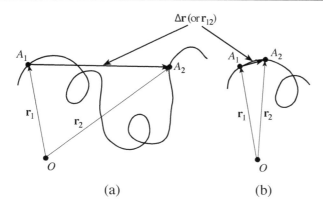

Figure 6.5: (a) The average velocity coming from this displacement does not contain the detailed information of the motion of the object. (b) To find the *instantaneous* velocity, the change in position vector and the time interval must be small.

- That instantaneous speed is indeed the speed of the object as defined earlier (distance over time) can be seen by noting that contrary to the *average* velocity and *average* speed, which involve the unequal *displacement* and *distance*, in the present situation, the two are (almost) equal. This is evident from Figure 6.5(b) where the arc between A_1 and A_2 (the distance) is very nearly equal to the straight line segment (the magnitude of the displacement) connecting them. The closer A_2 is to A_1, the "more equal" the length of the arc and the length of the straight line, and the closer you get to the ideal definition of instantaneous speed and velocity.

  Although the magnitude of $\mathbf{v}_{avg}$ is not equal to v_{avg}, the magnitude of $\mathbf{v}$ is equal to v.

- At the (instantaneous) location of the particle, the instantaneous velocity is tangent to the curve describing the path of the particle.

- In a general motion, the instantaneous velocity continuously changes both in length and in direction.

What do you know? 6.7. Which member of each of the following pairs has a bigger value?
(a) Average velocity or average speed.
(b) Instantaneous velocity or instantaneous speed.

6.3.1   Some Common Kinds of Motion

Let us now look at some of the more common motions that exist in nature. The simplest kind of motion is called **uniform motion** [Figure 6.6(a)]. In a uniform motion, the object moves in the same direction on a straight line with constant speed. Equivalently, in a uniform motion the velocity is constant.

Uniform motion.

   Next in complexity is the rectilinear—straight line—motion of an object thrown straight up with a velocity $\mathbf{v}_0$, or dropped from a height. The length of the velocity vector, i.e., the instantaneous speed, decreases on the way up, becomes zero at the very top, and increases on the way down [Figure 6.6(b)].

   The third type of motion is that of a **projectile** in which an object is thrown up at an angle. This is a generalization of the motion discussed immediately above, but now the velocity $\mathbf{v}_0$ is tilted. Notice the decrease in speed of the projectile on the way up,

Projectile.

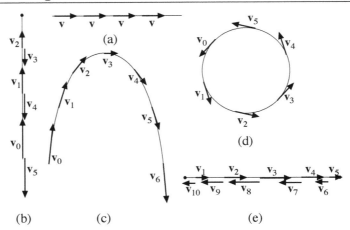

Figure 6.6: The most common types of motion. (a) The velocity does not change in uniform motion. (b) An object thrown up slows down (uniformly) until it stops momentarily at a maximum height, after which it falls down. (c) A projectile describes a parabola. (d) In a uniform circular motion the speed does not change, but the direction of velocity keeps changing uniformly. (e) An object in simple harmonic motion moves back and forth between two points.

the attainment of a (horizontal) minimum at the top, and the increase on the way down [Figure 6.6(c)].

Uniform circular motion.

The fourth example is the **uniform circular motion** in which the object moves with uniform *speed* on a circle. Note that although the speed, the length of the velocity, is constant, the velocity *changes* continuously [Figure 6.6(d)]. The motion of artificial satellites is well approximated by a circular motion. Similarly, the motion of the Moon around the Earth, and that of the Earth around the Sun, are almost circular.

Simple harmonic motion.

Finally, a **simple harmonic** motion is a rectilinear motion, in which the moving object repeats its movement (also called *oscillation*). The maximum speed occurs at the midpoint of the range of oscillation, while at the end points, where the particle comes to a momentary rest, the speed is zero [Figure 6.6(e)]. A concrete example of simple harmonic motion is that of a mass attached to one end of a spring whose other end is fixed. When the mass is displaced slightly, it will undergo a simple harmonic motion.

6.4   Acceleration

The final concept to be considered in kinematics is **acceleration**. The common meaning of acceleration describes only the *increase* in the *speed* of an object. The scientific definition of acceleration is a generalization of the common meaning. It is simply the *change in the velocity* of the moving object per unit time interval. To be precise:

Acceleration defined.

> **Box 6.4.1.** *Measure the velocity—both speed and direction—of the object at time t and call it* **v***; a short time later, say at t', measure the velocity again, and call it* **v**$'$*; subtract the initial velocity from the final velocity; divide the result by $t' - t$. In the limit that $t' - t$ becomes infinitesimally small, the ratio will be the* instantaneous *acceleration at time t. In symbols,* **a** $= \Delta \mathbf{v}/\Delta t$*, where* $\Delta \mathbf{v} = \mathbf{v}' - \mathbf{v}$*,* $\Delta t = t' - t$*, and Δt is taken to be very small.*

If the acceleration is constant, i.e., if neither its direction nor its magnitude changes, then the time interval $t' - t$ need not be small. The unit of acceleration is meter per second per second, abbreviated m/s^2.

What do you know? 6.8. Is it true that acceleration means speeding up?

The following remarks should elucidate the concept of acceleration.

- In the procedure described above, it is necessary to find the difference between the final and the initial *velocities* as discussed in Appendix C.

- Acceleration, calculated from the change in the velocity *vector*, is itself a vector quantity. Thus it has a length (the magnitude of acceleration) and a direction.

- A source of common mistake is to confuse acceleration with velocity. Although the direction of the velocity vector is the same as the direction of the instantaneous motion of the particle, the direction of the acceleration vector *has nothing to do with this motion*. (See the following examples.)

- Any change in velocity represents an acceleration. This change could come from a change in the speed with no change in the direction, or a change in direction with no change in speed, or changes in both direction and speed.

- A motion whose acceleration remains unchanged, *and whose initial velocity is in the same direction as the acceleration*, is called a **uniformly accelerated motion** (UAM). Since acceleration always points in the same direction, all changes in velocity will take place along the line defined by the initial velocity (or acceleration). So, such a uniformly accelerated motion is necessarily rectilinear. This is the motion studied by Galileo, who discovered the mathematical formulas that give the velocity and distance as a function of time (see Subsection 4.3.4).

- When the direction of the acceleration is opposite to that of the velocity, it is sometimes called *deceleration*.

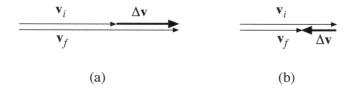

       (a)   (b)

Figure 6.7: (a) The change in the velocity of the car as it *increases* its speed is in the direction of motion. (b) The change in the velocity of the car as it *decreases* its speed is opposite to the direction of motion.

What do you know? 6.9. A car is moving north at 65 mph. A little later it moves at 64 mph. What is the direction of the change in velocity? What is the direction of the acceleration of the car?

Example 6.4.2. This example will shed some light on the sign convention used for acceleration in Section 4.3.

Suppose a car, moving exactly east, increases its speed. Then as Figure 6.7(a) shows, the change $\Delta\mathbf{v}$ in the *velocity* of the car is also due east. This change is obtained by drawing the initial *velocity* $\mathbf{v}_i$ and the final velocity $\mathbf{v}_f$ from the same point (in the figure the two vectors have been displaced

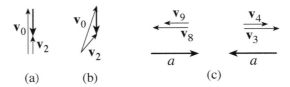

Figure 6.8: The change in velocity for both (a) rectilinear and (b) projectile free fall points down. (c) The change in velocity—and thus acceleration—varies in direction as well as magnitude for a simple harmonic motion.

so they can be clearly identified) and connect the tip of $\mathbf{v}_i$ to the tip of $\mathbf{v}_f$. Since acceleration and $\Delta\mathbf{v}$ have the same directions and the latter is in the direction of motion, we conclude that the acceleration has the same direction as motion itself. If we take the direction of motion to be positive, this acceleration is also positive.

Now suppose that the car decreases its speed (decelerates). Then as Figure 6.7(b) shows, the change $\Delta\mathbf{v}$ in the *velocity* of the car is due west. We conclude that the acceleration is due west, and therefore opposite to the direction of motion. Since the direction of motion is positive, acceleration is negative. So deceleration is the same as negative acceleration.   ■

> **What do you know? 6.10.** A race car is moving on the straight leg of a race track at 200 mph. A bicycler is moving at 10 mph as she turns left on a corner. Which has more acceleration?

Let us qualitatively analyze the acceleration of the kinds of motion discussed so far. Bear in mind that to find the true instantaneous acceleration, one has to compare the velocity vectors infinitesimally close to one another; however, to make drawings clear, we use the initial and final velocity vectors which are reasonably far apart. In Figure 6.8, we have shown the accelerations for some of the motions of Figure 6.6. In a uniform motion, the velocity vector is constant. Hence, the change in velocity is zero, and when you divide this change by Δt, you get zero for the acceleration: *the acceleration of an object in uniform motion is zero.*

In the motion of an object that is falling or is thrown straight up [Figure 6.8(a)], as well as in the projectile motion [Figure 6.8(b)], the change in the velocities is *always downward*. Hence, the acceleration always points down. Furthermore, a careful analysis of these motions shows that the magnitude of this acceleration is constant. This constant value, which happens to be approximately 9.8 m/s^2, is called the **gravitational acceleration** and is denoted by $\mathbf{g}$. Close to the surface of the Earth, all objects moving freely, i.e., not influenced by non-gravitational forces (such as air drag), have this downward acceleration. Such objects are all said to be in **free fall** (see also Section 4.3.4).

Finally, the acceleration of an object in simple harmonic motion is linear (along the line of motion), but periodically changes direction. Figure 6.8(c) compares two velocity vectors on each side of the motion ($\mathbf{v}_8$ and $\mathbf{v}_9$ on the left and $\mathbf{v}_3$ and $\mathbf{v}_4$ on the right). It is clear that $\mathbf{v}_9 - \mathbf{v}_8$ points to the right, and $\mathbf{v}_4 - \mathbf{v}_3$ points to the left. Therefore, the acceleration vectors are as shown in Figure 6.8(c).

Example D.6.3 on **page 17** of *Appendix.pdf* illustrates a numerical calculation of the acceleration of a car moving with constant speed on a circle. **Math Note E.6.1** on **page 79** of *Appendix.pdf* generalizes this and shows that the acceleration for a uniform circular motion *points towards the center of the circle*, and its magnitude is given by

Formula for centripetal acceleration.

$$a = \frac{v^2}{r} \tag{6.1}$$

where v is the (constant) speed and r is the radius of the circle. Such an acceleration is called a **centripetal acceleration**.

> **What do you know? 6.11.** A car is moving north *at constant speed* on a circle as shown below. Is it accelerating? If so, what is the direction of its acceleration?

Example 6.4.3. A car is moving on a curved road in the shape of an arc of a circle of radius 100 m. Suppose that the car is moving northward momentarily with a constant speed of 50 mph, and that the road curves westward.

Q: What is the magnitude of the centripetal acceleration of the car at that moment?

A: Equation (6.1) gives the magnitude of the acceleration. To use that equation, you have to convert the speed from mph to m/s. Using the conversion factor 0.4472, you obtain $v = 50 \times 0.4472 = 22.36$ m/s. Thus, $a = 22.36^2/100 = 5$ m/s^2, using Equation (6.1).

Q: What about the direction of the acceleration?

A: Recall that the centripetal acceleration of an object always points toward the center of the circle on which the object is moving. In this case, the center happened to be to the left of the car at the moment that it moves northward. Therefore, the direction of the acceleration would be to the left of the car, or westward.   ■

6.5   End-of-Chapter Material

6.5.1   Answers to "What do you know?"

6.1. Strictly speaking, yes! Because motion is determined only relative to an observer, once the latter is specified, it (he, she) does not move (relative to itself, himself, herself).

6.2. When the two cars move in the same direction, all three statements are correct. The first two are obvious. The third becomes transparent after you draw a position vector from any object in one car to any object in the second car and ask "Is this position vector changing?" The answer is no, because the distance between the two cars does not change. If the cars are moving in opposite directions, then (a) and (b) are correct but (c) is not.

6.3. No! Because you have not specified an observer. If the observer is assumed to be the ground, then the car is moving *relative* to the ground. But if the observer is the driver of the car, then the car *is not moving* relative to the driver. In fact, the ground is moving relative to the driver!

6.4. The initial and final position vectors are generally different for two different observers (unless they are next to each other). However, the displacement is the same for both *as long as they don't move relative to each other*. If Emmy moves relative to Karl, then all three quantities will be different. For example, if Emmy is the driver of the car, then all three quantities are zero for her!

6.5. None! Motion is relative! So, he should specify relative to whom are the trees moving.

6.6. Since you are moving perpendicular to the line of sight, your motion is transverse; and if the object is near, you should see a parallax. The only conclusion, therefore, is that the object is very far from you.

6.7. (a) Average speed, because distance is larger than displacement. (b) They are equal, because infinitesimal distance is equal to infinitesimal displacement.

6.8. No. Speeding up is only one form of accelerating. Any change in *velocity* accompanies an acceleration.

6.9. Change means final quantity minus initial quantity. It is what you add to the initial to get the final. Since in this case final is smaller than initial, we need to add a vector pointing south to the initial velocity to get the final velocity. So, the change in velocity points south. Acceleration, therefore, also points south.

6.10. The race car has no acceleration because neither its speed nor its direction is changing. The bicycler has acceleration because she is changing the direction of her velocity. So, the bicyle has more acceleration.

6.11. Since the *direction* of velocity is changing (even though its magnitude is not), the car is accelerating. In fact, it has a centripetal acceleration, with direction pointing toward the center of the circle. Therefore, the acceleration must be pointing west.

6.5.2 Chapter Glossary

Average Speed Distance traveled in some time interval divided by that time interval. Do not confuse average speed with average velocity.

Average Velocity A vector quantity defined as displacement of an object in some time interval divided by that time interval. Do not confuse average velocity with average speed.

Centripetal Acceleration The acceleration of an object moving *with constant speed* on a circle. Its direction is toward the center (thus the name *centr*ipetal).

Displacement A directed line segment (arrow) drawn from the initial position of an object in motion to its final position. The initial and final positions are determined by the beginning and end of a time interval.

Distance The length of the path taken by an object in motion in some time interval.

Instantaneous Speed Distance traveled in some time interval divided by that time interval when the time interval is taken to be as short as possible.

Instantaneous Velocity A vector quantity defined as displacement of an object in some time interval divided by that time interval when the time interval is taken to be as short as possible.

Observer A point with respect to which the motion of an object is considered.

Parallax The change in the angle of the line of sight of an object in motion relative to an observer.

Position Vector A directed line segment (arrow) drawn from the observer to the object in motion.

Reference Frame The collection of all objects (including people) which do not move relative to one another; i.e., the position vector of each object relative to any other object does not change.

6.5.3 Review Questions

6.1. What is motion? When we say that a car is moving, what (who) is the observer?

6.2. What is a reference frame? Do all objects (people) in a lecture hall constitute a reference frame? What about the lecturer who is pacing the width of the hall?

6.3. Is it possible for one observer to detect a simple motion for an object while a second observer detects a complicated motion for the same object? Give an example.

6.4. Is it possible to say that the clock fixed on the wall is moving? Why or why not?

6.5. What is displacement? How is it different from distance? Under what circumstances is displacement equal to distance?

6.6. Define parallax and state whether you observe it for all kinds of motion. What is transverse motion? What does parallax have to do with the demise of the old heliocentric models?

6.7. What is the definition of average velocity? Of average speed? Are the two quantities equal in magnitude?

6.8. What is the definition of instantaneous velocity? Of instantaneous speed? Are the two quantities equal in magnitude? Explain why the direction of motion is the same as the direction of instantaneous velocity.

6.9. What is the definition of acceleration? Is deceleration a kind of acceleration? Can you have acceleration without a change in speed?

6.10. What is centripetal acceleration? Does the speed of an object change when it has a centripetal acceleration?

6.5.4   Conceptual Exercises

6.1. Is it possible for the distance between the observer and the object to remain unchanged while the object is moving relative to the observer?

6.2. You are put to sleep on a spaceship with a window and sent to outer space away from any visible objects except distant stars. You wake up. Can you tell if you are moving? Explain!

6.3. Using parallax, explain why when you look at objects on the side of a highway, you see the near objects move relative to the distant ones.

6.4. Using parallax, explain why you can detect the motion of a *near* object that moves directly away or directly towards you (along the line of sight), but you cannot detect such a motion for a distant object like a star? If a distant object has exactly zero parallax, can you say that it is not moving relative to us?

6.5. You are moving on a country road when suddenly a deer jumps in front of you. To avoid an accident, you slam on the brakes and turn the steering wheel. Are you accelerating at that moment? What if you just slammed on the brakes? What if you just turned the steering wheel? Explain!

6.6. Give an example of a motion in which the acceleration is constant but it is not in the direction of motion (i.e., instantaneous velocity).

6.5.5   Numerical Exercises

6.1. Vladimir, who lives in Chicago, has a job interview in New York City 900 miles away. He takes an airplane at 6:00 am from O'Hare and lands at JFK Airport after two hours. The location of his interview is 30 miles from JFK. So he takes a cab which takes him to the job interview in 45 minutes. The interview lasts for one hour, after which Vladimir takes a cab to the airport, which because of a traffic jam, does not get there until one hour later. He waits another hour in the airport for his return flight, which brings him back to exactly the same spot at O'Hare at 2:00 pm.
(a) How many miles did Vladimir travel for his entire trip?
(b) How long did the entire trip take?
(c) What is the average speed in mph? In m/s?
(d) What was the average speed of the first cab in mph? In m/s?
(e) What was the average speed of the second cab in mph? In m/s?
(f) What was Vladimir's displacement for the entire trip?
(g) What was Vladimir's average velocity for the entire trip?

6.2. A car is moving with constant speed on the semicircle of Figure 6.9 from A to B in one hour.
(a) What is the displacement of the car?
(b) What is the distance traveled by the car?
(c) What is the average velocity of the car?
(d) What is the average speed of the car?
(e) What is the direction of the instantaneous velocity of the car at A, B, C, D, and E?
(f) What is the magnitude of the acceleration of the car?
(g) What is the direction of the acceleration of the car at A, B, C, D, and E?

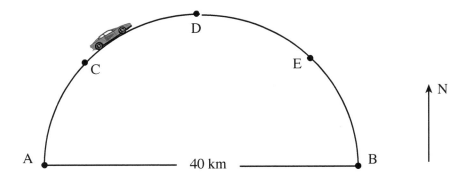

Figure 6.9: The car moving on a semicircle.

Chapter 7

Dynamics: Cause of Motion

In his study of motion, Galileo clarified many useful kinematical concepts such as velocity and acceleration. He also discovered (see Box 4.2.1) that, contrary to Aristotle's theory, an object moving on an infinitely smooth horizontal surface requires no "movers." Once in motion, the object moves forever on the horizontal surface with constant velocity. Galileo, moreover, emphasized the role of mathematics in studying motion.

The seeds of the scientific methodology of Galileo and his emphasis on mathematics found a perfect soil to grow in the person of Isaac Newton. Never before since the time of Archimedes had there been such a harmonious blend of mathematics and nature as in the work of Newton. In him converged and climaxed the forceful tide of rational science that started with Copernicus. New societies of science, for the first time, were exposing observations and presenting theories and discoveries for discussion, criticism, endorsement, correction, and diffusion. For a quarter-century as president of the Royal Society in London, Newton made it an unprecedented center of publicity and of the power of science.

7.1 The First Law of Motion

A simple form of the first law—also known as the law of inertia—was given by Galileo (see Box 4.2.1). Quite surprisingly, Descartes, contemptuous of experimentation, discovered a more general version of this law:

First law stated.

> **Box 7.1.1.** *An isolated body at rest will remain at rest, and an isolated body in motion continues its motion with constant velocity along a straight line.*

This is in complete contrast to everyday experience. After all, we seldom see an object in eternal straight-line motion without some agent "driving" it! Without a driving force, all objects will eventually stop. Box 7.1.1 resolves this apparent paradox by the adjective "isolated." Objects of our everyday experience are *not* isolated, but in contact with another object. And this contact introduces **friction**.

Both Newton and Galileo were aware of the existence of friction. They also knew that reducing the friction brings the motion of the object closer to the ideal straight-line uniform motion. The first law is simply the extrapolation of this experimental fact to an idealized situation in which friction is completely absent. Such an idealization occurs frequently in science. In fact, one is safe to say that in general, theories, laws, and principles are almost impossible to formulate without idealization.

BIOGRAPHY

Isaac Newton (1642–1727) had a very rough childhood. His father died before his birth; his mother remarried and left him with his grandmother on a farm; when his stepfather died, his mother moved in with three children. Newton attended the Free Grammar School in Grantham, five miles from his home. His school reports described him as "idle" and "inattentive." His mother, by now a lady of reasonable wealth and property, thought that her eldest son was the right person to manage her affairs and her estate. Newton was taken away from school but soon showed that he had no talent, or interest, in managing an estate. Newton's uncle persuaded his mother that Isaac should finish his schooling and go to the university. At 19, older than the other undergraduates, Newton entered Trinity College, Cambridge. Upon his graduation in the summer of 1665, he had to leave Cambridge due to the widespread plague in the London area to spend the next 18 months in the quiet of his family farm at Woolsthorpe. These 18 months were the most productive of Newton's (as well as any other scientist's) life. He developed calculus, discovered the laws of motion, and found the mathematical formulation of the law of gravity.

It is a pleasing irony of scientific investigation that to comprehend a *real*, seemingly inexplicable case, one has to abandon the case by idealizing and simplifying it, unravel the idealized version, and only then come back and explain the original case. The first law is a simple example. A real motion, with friction, could not produce the first law. By idealization to a frictionless environment, not only do we fathom the concept of motion, but we obtain a fuller understanding of what it was that made the analysis of the motion difficult in the first place. Friction becomes that external agent which prevents the object from obeying the (idealized) first law of motion.

7.1.1   Momentum

In the study of the "cause" of motion, two important concepts ought to be understood clearly. These are *mass*, the quantity of matter, and *momentum*, the quantity of motion. To introduce these concepts, start with the first law and replace it as follows:

Concept of momentum introduced.

> **Box 7.1.2.** *There is a vector quantity denoted by* **p** *associated with the motion of every physical system called* **total momentum,** *and if the system is completely isolated, this total momentum will not change.*

The following remarks should help clarify the concept of momentum:

- An isolated system is one on which no *external* agent acts.

- A system may consist of more than one part.

- The *total* momentum of a system is the vector sum (see Appendix C) of the momenta of its parts.

- The parts may exert influences on each other. Such influences will not affect the *total* momentum of the system as long as the external influence is absent.

- The word "system" is completely arbitrary. In principle many separate systems can be combined to form a new larger system. The choice is dictated by the convenience achieved in specific situations.

- The momentum at rest is zero.

- There may be more than one external agent acting on a system. As long as their net effect is zero, the first law holds.

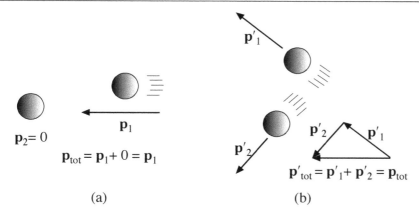

$\mathbf{p}_2 = 0$

$\mathbf{p}_1$

$\mathbf{p}_{\text{tot}} = \mathbf{p}_1 + 0 = \mathbf{p}_1$

(a)

$\mathbf{p'}_1$

$\mathbf{p'}_2$

$\mathbf{p'}_2$

$\mathbf{p'}_1$

$\mathbf{p'}_{\text{tot}} = \mathbf{p'}_1 + \mathbf{p'}_2 = \mathbf{p}_{\text{tot}}$

(b)

Figure 7.1: The total momentum before collision (a) is equal to the total momentum after collision (b).

What do you know? 7.1. A system is seen to move on a circle. Is the system isolated?

Let us look at some real situations involving systems and their momenta.

Example 7.1.3. Consider two billiard balls, one moving on a straight line towards a second one, which is at rest. Let us define our system to be these two balls taken together. Ignoring the small friction of the table and the smaller air resistance, you can assume that the ball in motion moves on a straight line with constant velocity. You can, therefore, conclude that there is no external agent disturbing this ball. The second ball is also undisturbed by any external agent, because it is not moving. Thus, as a system, there is no external agent acting on the two balls. The first law, then, says that the total momentum of this system will not change. This is obvious as long as the two balls have not collided [Figure 7.1(a)]. What happens *after* collision? Clearly the two balls may wander off in two arbitrary directions [Figure 7.1(b)], but if you add the momenta of the two balls, so says the first law, you will get the same vector as before collision. Note that momentum for each part is *defined* so that equality and constancy in the absence of external agents hold. Note also that each ball will exert an influence on the other during collision, but this influence is internal. ■

What do you know? 7.2. Two clay balls approach each other head-on with the same momentum. What is the total momentum of the system consisting of the two balls? What happens if they stick to each other after collision?

Example 7.1.4. A block of wood is resting on a smooth frictionless table while a bullet is approaching it with momentum **p** [Figure 7.2(a)]. The bullet subsequently collides and penetrates the block and comes to rest inside it while the system of block and bullet moves [Figure 7.2(b)]. The momentum of the combination will be the same as before collision.

Imagine that you are standing on a boat which is at rest. As you start to move in one direction, the boat moves in the opposite direction. Initially the total momentum of the system consisting of you and the boat is zero (neither you nor the boat is moving). After you start to move with a given momentum, the boat moves with exactly the same momentum in the opposite direction to render the total momentum the same as before. This happens because there is no external agent.

A more dramatic illustration of this is the motion of a rocket. In this case, the system consists of the rocket plus its fuel. After ignition, the fuel gushes down while the rocket goes up. Again the momentum of the rocket exactly balances the (opposite) momentum of the exhaust fuel.

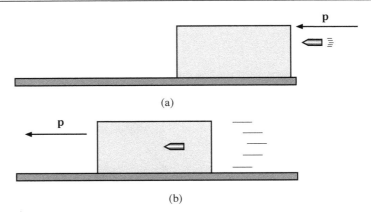

Figure 7.2: The bullet and the block constitute a system. (a) The block is resting while the bullet carries all the momentum of the system. (b) The bullet is inside the block, and they move with the same momentum as before collision.

A firecracker explodes into many fragments, each going in a different direction. Since the force of the explosion is an internal influence, the total momentum after the explosion—obtained by vectorially adding the momenta of all fragments—is exactly the same as the momentum of the firecracker before it explodes. ■

> **What do you know? 7.3.** A billiard ball, moving to the left, hits 8 other billiard balls. When you add up the momenta of the 8 balls, what do you get? (a) You get zero. (b) You get a momentum pointing to the left. (c) The answer depends on where the original ball is going.
> Answer the question for all 9 balls.

The foregoing examples illustrate one of the most important laws of physics, namely the law of **momentum conservation**, which is equivalent to the first law of motion as stated earlier. However, it is instructive to rephrase the latter as follows:

Principle of momentum conservation stated.

> **Box 7.1.5.** *The total momentum of an isolated system is constant.*

7.1.2   Mass

We have talked about momentum, we have given examples of its conservation, but, we have not as yet defined what momentum *is*. Surprisingly, there is no definition of momentum! We must accept momentum as given (or defined) by the first law and, when appropriate, define other quantities in terms of momentum (see Food for Thought 7.2.2 for some insight).

One of the most important quantities defined in terms of momentum is mass.[1] Consider two blocks A and B initially at rest on a perfectly smooth surface [Figure 7.3(a)]. Between A and B there is a small compressed spring, which can be thought of as part of—and attached to—one of the blocks, say A. The two blocks are held together by a string which can be very gently severed (for example, very slowly burned) to minimize the effect of

---

[1]It is possible to keep mass undefined and define momentum in terms of it. This, in fact, is done in most textbooks. We prefer momentum because not only mass but also force can be defined in terms of it, as you shall see shortly.

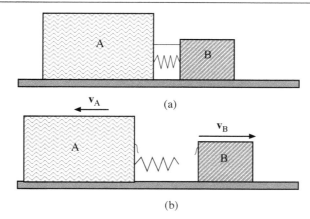

Figure 7.3: As the two parts of an isolated system separate, the "bigger" one moves more slowly. Mass is defined in terms of momentum and velocity.

external influences. After the string is severed, the blocks will move in opposite directions [Figure 7.3(b)]. The system $A + B$ is an *isolated* system. Although the burning is done by an external agent, the motion of the system is not affected by such an agent because both A and B remain motionless during the burning process until the very last fiber of the string is burned. But right after the last fiber is burned and before the blocks start to move, the only influence acting on the blocks is caused by the compressed spring which is *internal* to the system. (One can make the whole process more ideally isolated by replacing the spring-string combination by a miniature time bomb.)

Assuming an ideal case in which the external influence can be completely ignored, the first law can be put into action, and the conclusion can be reached that A has as much momentum (but in opposite direction) as B. It is clear that the velocities will not be the same. The "heavier" object will move slower than the "lighter" object. We make the vague notions of "heaviness" and "lightness" more precise by the concept of mass: Physicists *define* the ratio of the mass of A to that of B to be in inverse proportion to their speeds: $\frac{m_A}{m_B} \equiv \frac{v_B}{v_A}$. Thus, the product of mass and speed for the two blocks will be the same: $m_A v_A = m_B v_B$

Mass and momentum.

This procedure defines only the ratio of masses. To find the mass of an object, we assign the value of one unit to an arbitrary but standard object and find the ratio of any other mass to this standard mass by the experiment described above. The standard unit of mass that we use is **kilogram** (kg) which is approximately the mass of one liter of water. Since 1889, the kilogram is defined to be equal to the mass of the *international prototype of the kilogram*, which is made from an alloy of platinum and iridium of 39 mm height and diameter, and is kept at the *Bureau International des Poids et Mesures* (International Bureau of Weights and Measures) in Paris. Official copies of the prototype kilogram are made available as national prototypes, which are compared to the Paris prototype (*"Le Grand Kilo"*) roughly every 10 years.

Unit of mass is kilogram (kg).

Now that the concept of mass is defined, we can obtain a better insight into the nature of momentum by working backwards. You saw above that the product of mass and speed is the same for the two objects A and B. Invoking the first law now, we identify the product mv as the magnitude (or length) of the vector quantity, momentum. Proceed one step further and write

$$\mathbf{p} = m\mathbf{v}. \tag{7.1}$$

It is then clear that $\mathbf{p}_A + \mathbf{p}_B = m_A\mathbf{v}_A + m_B\mathbf{v}_B = \mathbf{0}$ because $m_A\mathbf{v}_A$ has exactly the same length as $m_B\mathbf{v}_B$, but its direction is opposite, and the sum of any two vectors which have equal length but opposite directions is always zero. Thus, defining $\mathbf{p} = m\mathbf{v}$ ensures the conservation of momentum: zero momentum before burning; zero momentum, $\mathbf{p}_A + \mathbf{p}_B = \mathbf{0}$, after burning.

What do you know? 7.4. In Figure 7.3, A moves twice as fast as B. If A is 10 kg, what is the mass of B?

It turns out that measuring the mass is much easier; all you need is a scale! In general, however, the direct measurement of momentum is difficult. Hence, by measuring the mass and the velocity of an object and using the relation (7.1), you can calculate its momentum.

Example 7.1.6. An astronaut, holding a wrench in his hand, is detached from his spaceship and moves away from it. The astronaut and the wrench constitute an isolated system having some momentum which points away from the spaceship. The only way he can get back to the ship is by using the fact that the momentum of the system does not change. He decides to throw the wrench as hard as he can. But which direction does he throw it? Suppose that he throws it *towards* the spaceship. Since the momentum of the system was away from the spaceship initially, the momentum of the astronaut has to be *away* from the spaceship with a magnitude equal to the sum of the initial momentum and the momentum of the wrench. Therefore, he should throw the wrench *away* from the spaceship. This, nevertheless, will not guarantee that the astronaut will acquire a momentum *towards* the spaceship. Only if the momentum of the wrench is *larger* than the initial momentum of the system will the astronaut acquire a momentum towards the spaceship. **Example D.7.1** on **page 18** of *Appendix.pdf* adds a numerical flavor to this situation.   ■

What do you know? 7.5. Why does a gun recoil when fired? Would it recoil more or less or the same if the bullet were heavier?

Example 7.1.7. In the olden days, before the age of laser and computer technology, in the ballistics laboratories, the speed of a bullet was measured using the constancy of the total momentum of an isolated system. The bullet would be fired into a block of wood resting on a surface. The (small) velocity of the subsequent motion of the block with the bullet inside would be measured. Knowing the masses of the block and bullet, one would calculate the momentum of the system. This momentum is of course the momentum of the initial fired bullet. With the mass of the bullet at one's disposal, the initial speed of the bullet could be determined. See **Example D.7.2** on **page 19** of *Appendix.pdf* for a quantitative analysis.   ■

What do you know? 7.6. In the motion of a rocket, the amount of fuel that is ejected per second is the same throughout the initial stage of the motion of the rocket. The speed of the fuel coming out is also the same. Using these two facts, compare the momentum of the rocket at the beginning of the initial stage and at the end of the initial stage. What about its speed? Hint: At the beginning, about 90% of the mass of a rocket is its fuel.

7.2   The Second Law of Motion

The first law applies only to situations in which there are no external agents acting on the object in motion. What happens if there is an external influence? The second law answers this question.

The second law is the key to the secrets of motion. Applied to the motion of any system, it gives a full description of the future of that system. It includes the first law as a special case, and—as we shall see—explains many phenomena that were unexplained before. The second law is so universal that most of the physics and mathematics of eighteenth and nineteenth centuries were discovered by applying it to diverse physical systems.

We are now ready to state the **second law of motion**. Again, like all great laws, it carries a definition with it. Let us state the law first, and then discuss it:

Box 7.2.1. (**Second Law of Motion**) *The net instantaneous force acting on an object is precisely the instantaneous change of its momentum per unit time. In symbols, the second law can be written as*

$$\mathbf{F}_{\text{net}} = \frac{\Delta \mathbf{p}}{\Delta t} \qquad \textit{with } \Delta t \textit{ very small.}$$

If the mass of the object remains constant during its motion, then $\Delta \mathbf{p} = m\Delta \mathbf{v}$ and $\mathbf{F}_{\text{net}} = \Delta \mathbf{p}/\Delta t = m(\Delta \mathbf{v}/\Delta t)$. Thus, recalling that acceleration is the change in velocity divided by time, in this special (but very useful) case one writes

$$\mathbf{F}_{\text{net}} = m\mathbf{a}, \tag{7.2}$$

which is the version of the second law most often used in introductory discussions of dynamics.[2] Equation (7.2) implies that force and acceleration point in the same direction. This relation also suggests the unit of force. It is simply the product of the unit of mass (kg) and the unit of acceleration (m/s^2). Thus, the unit of force is $\text{kg} \cdot \text{m/s}^2$ which is called **Newton** and abbreviated as N.

*Unit of force is **Newton** (N).*

🧺 Food for Thought 🌱

FFT 7.2.2. Henri Poincaré, the great French mathematician of the late nineteenth and early twentieth century has pointed out that every fundamental law of physics has a definition built into it. The law and the definition come together. This seems a bit unconventional (logically), and a logician would frown on such a procedure: "How can you make a statement about something that is not known to you?" Nevertheless that is how physics develops ideas and new quantities, and in many cases physicists cannot assume that the fundamental law should be a statement about predefined quantities.

The first law of motion is a statement about momentum, which is undefined in that statement. The second law of motion is a statement about force, which is *defined* in that law as the rate of change of momentum. You may ask "This procedure of making a statement about concepts that are defined by that statement makes scientific statements quite arbitrary. I can write an arbitrary equation, and *define* the quantities involved at my will. Would that make me a great physicist?" This is a legitimate question that requires clarification.

The power of physical laws lies in their agreement with observation, and the form which these laws take is the outcome of many trials and errors sometimes spanned over many generations of physicists. You may define "force" as, say, velocity times momentum, or something else, but your definition will be useless because, when confronted with observation, it will give the wrong answer. In fact, this is how some concepts were developed: by trying different (wrong) definitions until the right definition was discovered. For example, because most objects either slow down or speed up as they cover more and more distances, many physicists defined acceleration as the rate of change of velocity with *distance*. Galileo was the first to come upon the right definition of acceleration as the rate of change of velocity with *time*.

Physics and other sciences are *empirical*. Logic and mathematics are not. That is why in the latter fields one *has to* define new quantities and concepts in terms of the old; otherwise the fields become a collection of random statements. The empirical nature of physics allows statements involving new and old concepts *simultaneously*. Through trials and errors and comparison with

---

[2]The subscript "net" is usually suppressed. However, the reader should always keep in mind that it is the *net* force that enters the second law (see also the remarks made below).

observation, the correct statements will be picked out, and in time become part of the scientific vocabulary.

The second law is the most important of the three laws, as the following discussion will reveal. Hence, a good understanding of its content is essential. The following remarks and examples are intended to elucidate this law.

- The second law *defines* force in terms of the instantaneous change in momentum. Since momentum is a vector quantity, force is also a vector quantity. As defined by the second law, it replaces the inexact notions such as "influence," or "agent." Thus, *force is anything that can cause a change in momentum.*

- The second law implies the first law: If there is no external agent, there is no force; hence, there is no change in momentum.

- In situations in which two or more agents apply forces on a moving object, the total force causing the momentum change will be simply the vector sum of individual forces. This vector sum is precisely $\mathbf{F}_{net}$.

- $\mathbf{F}_{net} = m\mathbf{a}$ becomes invalid in situations where mass varies. For such cases the concept of force as the change in momentum per unit time must be used.

- Being vectors, both $\mathbf{F}_{net}$ and $\mathbf{a}$ can be decomposed into components. Once this is done, there is one equation for each component. The most common decomposition is horizontal and vertical. This decomposition is useful in the study of some practical motions. For instance, **Example D.7.3** on **page 19** of *Appendix.pdf* shows that a projectile has the shape of a parabola, as Galileo pointed out (see Section 5.2.1), and that the maximum range of a projectile is obtained when the angle of firing is 45°.

A freely falling body has a constant acceleration $\mathbf{g}$ of 9.81 m/s². Thus, according to the second law, there must be a force, $\mathbf{F} = m\mathbf{g}$ acting on the body. This force, usually denoted by $\mathbf{w}$ is called the **weight** of the object. Strictly speaking, there is a difference between

Concept of weight clarified.

mass and weight, although, in everyday language one speaks of the "weight" of an object being "one kilogram." As long as we are on the surface of the Earth, and ignore minor variations in $\mathbf{g}$ at different locations, weight and mass are equivalent. One has to bear in mind, however, that on the surface of the Moon, for example, the weight of an object is less than its weight on the surface of the Earth while the mass is exactly the same.

Now let's consider some examples of the second law. In all cases, assume that the mass of the object does not change so Equation (7.2) is valid.

Tension in a rope.

Example 7.2.3. A mass is attached to the end of a rope and everything is at rest. Since the mass is not moving, it has no acceleration, and the net force acting on the mass is zero. Gravity pulls the mass down. So, there must be another force balancing the force of gravity. This is called the **tension** in the rope. Tension is always directed along the rope. ∎

What do you know? 7.7. Tarzan is clutching the top of a rope which can support a maximum tension equal to Tarzan's weight. The loose end of the rope is above a river bed known to be infested with crocodiles, some of whom are circling right under the rope. He starts to climb down the rope slightly. (a) Right after he starts to climb down, what is the direction of the net force on Tarzan? (b) At that moment, which force is bigger, Tarzan's weight, or the tension in the rope? (c) Was it smart of him to move downward?

Example 7.2.4. A person starts to walk; a car takes off; a train starts to move; a plane takes off. In all cases the velocity changes. Thus the person, the car, the train, and the plane all have accelerations. Newton's second law, then, implies that there are forces causing these accelerations. Where do these forces come from? Stay tuned! ∎

Example 7.2.5. A box of mass m rests on a rough floor and is pulled by too small a horizontal force to move the box. Let us analyze the "motion" of this box.

If the box does not move, it has no acceleration. Vertically, there is the force of gravity pulling the box down. But since the vertical acceleration is zero, the second law tells us that there must be another force balancing the gravitational force. This force **N**, exerted by the floor on the box, is called the **normal** force.

Normal force.

Horizontally, the acceleration is also zero. Therefore, the applied force must be balanced by another force. This, of course, is the force of friction. It is interesting to note that *as long as the box does not move*, the frictional force "adjusts itself" to the applied force! If the applied force increases, but remains smaller than the *maximum* force of friction, the latter also increases. The force of friction can, of course, never overpower the applied force. Otherwise, you would encounter the *spooky* situation in which the box moves to the left while being pulled to the right!

A spooky motion!

As the applied force increases, a point will be reached at which this force will overcome the maximum frictional force and the box will start to accelerate. By measuring the acceleration and the mass of the block, one can determine the frictional force between the box and the floor. **Example D.7.4** on **page 20** of *Appendix.pdf* adds some numbers to this exercise. ∎

> **What do you know? 7.8.** Tarzan of **What do you know? 7.7.** is clutching the end of his rope which can support a maximum tension equal to his weight. Should Tarzan climb up the rope or stay put, while the crocodiles are circling below?

Example 7.2.6. When you jump down from a height, you have a natural tendency to bend your knees. Why? The second law of motion explains. In jumping from a height, you pick up momentum on your way down, and just before you hit the ground, you carry a considerable amount of it. When you hit the ground your momentum reduces to zero. So there is a big change in your momentum. Box 7.2.1 tells us that the larger the time interval Δt is during which this change takes place, the smaller the force. You instinctively continue the motion by bending your knees thereby increasing the time interval. A stiff-kneed jump causes the momentum to come to a sudden halt, with Δt considerably reduced. This can yield a large enough force to cause an injury.

Jumping and the second law of motion.

Air bags are also based on this principle. In a collision, our head is thrown forward with a large momentum. If it hits a solid object, it will stop very quickly (Δt is very small), and by the equation of Box 7.2.1, the force **F** will be large, causing fractures, or even death. An air bag increases the stopping time considerably, thereby decreasing the force.

Air bags and the second law of motion.

Diving into water, jumping on a trampoline, or falling in a haystack increase the stoppage time, and therefore, decrease the potentially damaging force. ∎

An important example of motion, discussed earlier, is the uniform circular motion, in which the acceleration points to the center—centripetal acceleration. Newton's second law now implies that there must be a net force which also points to the center. This force is called a **centripetal force**. Every object moving on a circle with uniform speed experiences a centripetal force. For a stone attached to the end of a string and whirled on a circle, the centripetal force is the tension in the string. For the Moon or a satellite going around the Earth, the centripetal force is gravitation. For an object in a centrifuge, the centripetal force is the reaction force of the spinning container in which the object is placed. What is it for a car rounding a curve?

Centripetal force.

> **What do you know? 7.9.** What is the origin of the centripetal force for a car rounding a curve?

Example 7.2.7. In the fall of a parachutist, two forces are present, the weight and the force of air drag. The weight is, of course, constant but the air drag increase with speed. If the parachutist opens her parachute early, the weight is larger than the air drag, and the parachutist accelerates downward, her speed increasing. As the speed increases, so does the force of air drag. This process

Terminal velocity of a parachutist.

continues until the force of air drag equals the weight. From this point on, the speed remains constant. This constant speed is called the **terminal velocity**.

If the parachutist opens her parachute later, much later, the weight is *smaller* than the air drag, and the parachutist decelerates, her speed decreasing. As the speed decreases, so does the force of air drag. This process continues until the force of air drag reduces to the weight. Again, from then on, the speed remains constant. **Example D.7.5 on page 21** of *Appendix.pdf* adds some numbers to the motion of a parachutist. ∎

7.3   The Third Law of Motion

The first and second laws concentrate on the motion of a single object (or a system of objects). But they don't give a complete picture of reality. In fact, we would have a very chaotic world if only the first two laws of motion were ruling it. The ground in Example 7.2.4 would "decide" when, where, and in which direction to push on people, cars, and trains! Fortunately, there is a third law to prevent this haphazard reality.

The third law pays attention to both the agent and the receiver of the force. Stated briefly it says: To every action there is an opposing reaction. Put more elaborately—and more precisely—it says:

Box 7.3.1. (Third Law of Motion) *If object A exerts a force* $\mathbf{F}_{AB}$ *on an object B, then object B exerts a force* $\mathbf{F}_{BA}$ *on object A such that* $\mathbf{F}_{AB} = -\mathbf{F}_{BA}$.

Thus the two forces are in opposite directions.

It is extremely important to *separate* the agent of the force from its receiver. Failure to do so is the cause of utter confusion, even for more experienced students. The classic example of such a confusion is that of the horse pulling a buggy. By Newton's third law, the buggy also exerts a force on the horse which is equal in magnitude but opposite in direction. Hence, the total force, which is the vector sum of these two forces, is zero. The second law now says that the acceleration must be zero. How, then, can the horse pull the buggy from rest? The answer is, as mentioned earlier, in separating the horse from the buggy. To analyze the motion of the buggy, we must look at all the forces that act *on the buggy*, only one of which is the force exerted by the horse. The force exerted by the buggy on the horse is not among these forces, so it cannot cancel the force by the horse. Similarly, to analyze the motion of the horse, we look at the forces that act *on the horse*, one of which is the force exerted by the buggy. You'll see the analysis of this motion further in Example 7.3.4.

What do you know? 7.10. A person is standing on a weight scale in an elevator that has just started moving up. (a) What direction is the acceleration pointing? What is the direction of the net force? (b) Beside gravity pointing down, what other force is acting on the person? In which direction? (c) What is the reaction to this force? Is the reading on the scale larger or smaller than the actual weight of the person?

Example 7.3.2. In Example 7.2.4 we concluded that on a person who just starts to walk (on a flat surface) there acts a force due to the second law.

Q: Where does this force come from?

A: As you walk, you push on the floor (next time you walk, try to detect this push). A lot of this (pushing) force is vertical, some of it is horizontal. For the force to have this crucial horizontal component, the floor must have friction or traction. Newton's third law implies that there is a reaction from the floor onto the person to the horizontal component of this push. Newton's second law now implies an acceleration. Everybody has experienced the difficulty of walking on a slippery ground. Walking on a flat floor would be impossible if the traction were absent!

What about the other forces? Clearly, traction is not the only force acting on a walking person. Do we not have to take these other forces into account? Indeed we do. However, it turns out that the vector sum of all the other forces is zero. For instance, the weight of the person tends to accelerate him downward, and if the floor were not there to stop him, he would move vertically down. Since he has no downward acceleration, there must be a force by the floor on the person which exactly balances his weight. Thus the ground exerts not only traction, but also the force that balances gravity (the normal force). ∎

What do you know? 7.11. A person is standing on a weight scale in an elevator moving down. Suppose that the acceleration of the elevator is g, the gravitational acceleration. What is the reading on the scale?

The motion of a person standing on a scale while riding an elevator is more fully discussed in **Example D.7.6** on **page 22** of *Appendix.pdf*.

Example 7.3.3. We also considered the acceleration of a car in Example 7.2.4 and concluded that there must be a force acting on the car. A misconception is that it is the engine that pushes the car. To see that the engine alone is incapable of moving the car, lift it off the ground. No amount of power delivered by the engine can move the car! As in the case of walking, through the property of traction between the tires and the road, the car pushes on the ground. By the third law, the ground pushes back on the car, causing its acceleration.

It is not the engine that pushes the car forward!

It is worth mentioning that the engine does exert a force on the rest of the car in a complicated way. But in order to move the car *plus the engine*, an external agent must provide a force. This external agent is the road. It is also important to note that the force of traction becomes meaningful only when the surfaces in contact *tend to move* relative to one another. Although there is traction between a standing car and the road, the *force* is activated only when the car starts to move. ∎

What do you know? 7.12. In some roller coaster rides the roller coaster goes on a vertical loop. Why don't passengers fall down? Hint: The speed at the top of the loop should be so large that the centripetal force on a given passenger is larger than the weight of the passenger.

Example 7.3.4. Let's return to the motion of the horse and buggy [Figure 7.4(a)]. First concentrate on the buggy [Figure 7.4(b)]. The buggy will start to move as soon as the force of friction between its wheels and the ground is overcome. Thus, the horse must be able to exert on the buggy a force at least as large as this frictional force. What about the horse? The buggy pulls the horse back by a force which, in magnitude, equals the force exerted by the horse on the buggy [Figure 7.4(c)]. If the horse is to move, there must be another force acting on the horse. The horse's stamping, kicking, and pushing on the icy ground in winter, is its honest struggle to create a *traction* between its hooves and the ground! It is this traction *on the horse* that causes the whole motion. ∎

Horse and buggy paradox explained.

The motion of a roller coaster that loops vertically is discussed in **Example D.7.7** on **page 23** of *Appendix.pdf*.

Example 7.3.5. In a tug-of-war contest, one kindergartener is competing with a team of (American) football players. The team first pulls the rope very gently so the kindergartener is not dragged. In this case, the force exerted by the team on the kindergartener is equal to the force exerted by

A tug of war between a football team and a kindergartener.

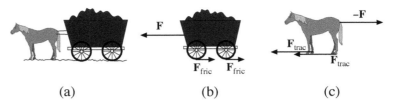

(a)   (b)   (c)

Figure 7.4: (a) The horse and the buggy. (b) Analysis of the motion of the buggy. (c) Analysis of the motion of the horse.

the kindergartener on the team. Now the team exerts a larger force, causing the kindergartener to move towards the team. Is the kindergartener exerting a force on the team equal to the force of the team on the kindergartener? The tendency is to say "No," because of the apparent lack of symmetry. However, even in this case the forces are equal!

How the kindergartener can win the tug-of-war.

The reason that the kindergartener moves is because the net force on *him* is nonzero: the force by the football team is larger than the traction (friction) between the kindergartener and the floor. The reason that the team does not move is because the kindergartener's force on the team, *which is equal to the team's force on the kindergartener*, exactly balances the traction between the team and the floor. If the kindergartener were standing on a rough floor and the team on an (almost) infinitely smooth surface, then the kindergartener would win the competition, because now even the slightest pull by the kindergartener can overcome the very tiny traction between the team and the surface! ■

The three laws of motion, discovered by Newton, and described in this chapter, are the three pillars upon which most of what is known as "classical physics" is built. Special emphasis must be placed on the second law because it relates the source of motion, force, with a kinematics variable, acceleration.

The task of physics, for about two hundred years after Newton, was to apply, test, and mathematically re-express these laws in as many novel situations as possible. In all cases these laws proved to be true. The only problem facing physicists was the formulas expressing forces. Once the force became known, the second law would take over and predict the motion of particles under the influence of the given force. No wonder classical physics and "Newtonian physics" have become synonymous.

It was only at the beginning of the twentieth century, when all attempts at applying Newtonian physics to the domain of atoms, and also to the domain of extremely fast objects failed, that new theories, quantum theory for atoms and relativity for the fast objects, were forced on the physics community.

7.4   End-of-Chapter Material

7.4.1   Answers to "What do you know?"

7.1. Remember that momentum is a *vector* quantity. This means that you can change it by changing its *direction*. Since the direction of the momentum of a system (which is the direction of the system's motion) changes if it moves on a circle, the momentum itself changes. Therefore, the system cannot be isolated.

7.2. Since the two momenta are equal and opposite in direction, they cancel to zero. After they stick, the momentum is also zero. So, the two stuck balls do not move.

7.3. (c) Unless you know where the original ball is going, you can't say anything about the motion of the other 8 balls, because it is the *total* momentum that does not change. On the other hand, when you include the original ball, then you can say that the momentum of all 9 balls add up to a momentum pointing to the left. Not only that, but that this total momentum is equal to the initial momentum of the original ball.

7.4. In the experiment of Figure 7.3, the ratio of the masses is equal to the inverse ratio of the speeds. So, if A is moving twice as fast as B, it must be half as massive as B. So, B is twice as massive as A. Since A is 10 kg, B must be 20 kg.

7.5. The bullet in the gun and the gun form an isolated system, which initially has zero total momentum. When the bullet comes out, it has a momentum in the forward direction. Therefore, the gun must have a momentum in the backward direction. That's the kick! If the bullet is more massive and comes out with the same speed, then its momentum is larger, and the kick should be stronger.

7.6. At the beginning, a little fuel comes out carrying a little bit of momentum. The rocket has the same momentum but with all the fuel inside, it can move very slowly. As the fuel is ejected, the rocket becomes lighter and, even though the momentum of the ejected fuel and the rocket are the same, the rocket moves faster than before.

7.7. (a) Right after Tarzan starts moving down, the acceleration is downward, so the net force is also downward. (b) The weight must be bigger than the tension to give a net downward force. (c) No! Because if he stops, i.e., if he decelerates, the acceleration will be *upward*; so the tension will be larger than Tarzan's weight. But the maximum tension the rope can support is his weight. So, the rope will sever!

7.8. If Tarzan starts moving up, he will be accelerating upward, so the net force will also be upward. This means that the tension will be bigger than the weight, and the rope will break! So, he has a better chance of survival if he stays put.

7.9. It is the traction (friction) between the tires and the road. To see that it is indeed the traction that keeps the car on the circle, imagine what would happen if the road were icy (with little traction).

7.10. (a) Since the elevator has just started moving up, acceleration is upward. The net force is also upward. (b) The scale is pushing the person upward (the normal force). And this force is larger than the weight of the person because the net force is upward. (c) By the third law the person is pushing back on the scale with a force equal to the force of the scale on person. Thus, the reading of the scale is larger than the actual weight of the person.

7.11. The net force on the person is his weight (his mass times g) pointing down and the force of the scale on him, pointing up. The net force is his mass times the acceleration of the elevator. Since this acceleration is g, the net force is the same as his weight. Therefore, the force of the scale on him is zero. The reaction to this force, the force he exerts on the scale, is also zero. So, the reading of the scale is zero! He is weightless!

7.12. If the speed is large enough, the centripetal force (which is the net force, because it is mass times the centripetal acceleration) on any given passenger at the very top of the loop can be larger than his/her weight. This means that the track must be pushing the passenger downward. Therefore, the person (really the car in which he/she is sitting) must be pushing *up* on the track. If it is pushing up on the track, it cannot fall down!

7.4.2   Chapter Glossary

Centripetal Force The force on any object that moves on a circle. It is the force that keeps the object on the circle.

Force Anything that changes the momentum of a system. Everything else being equal, the force that changes the momentum faster is stronger.

Friction A force that acts against the motion of an object. It is present whenever the surface of the object is in contact with another surface.

Inertia The property of an object that maintains its state of motion. Objects with large inertia tend to resist any change in their motion more than objects with smaller inertia.

Kilogram (kg) The scientific unit of mass.

Mass Measures the inertia of an object. The ratio of the speed of the standard of mass (kept in Paris) to the speed of the object whose mass is to be determined, in an experiment in which the standard mass and the object are at first stationary and then fly apart due to an internal mechanism.

Momentum Also known as the *quantity of motion*, is a vector quantity in terms of which the first law of motion is stated. An isolated system retains its momentum forever.

Momentum Conservation The essence of the first law of motion. The momentum of an isolated system is conserved (i.e., it does not change).

Newton (N) The scientific unit in which force is measured.

Normal Force The force felt by any object that is placed on a surface. It is usually perpendicular to the surface of contact.

Tension (As in a rope) is the force that a rope sustains when it is stretched. The harder the stretching, the larger the tension.

Weight The force of gravity exerted on an object is its weight.

7.4.3   Review Questions

7.1. State the first law of motion in terms of an object being at rest or having a velocity.

7.2. State the first law of motion in terms of momentum. What is an isolated system? What is its momentum in terms of the momenta of its parts?

7.3. Before two billiard balls collide, how many balls carry momentum? What about after collision? How does the total momentum of the two balls compare with the momentum of the initial moving ball?

7.4. Before a bullet hits a block of wood on a smooth floor, what carries the momentum of the bullet-block system? After the bullet penetrates into the block, what carries the momentum of the system? Which momentum is bigger: the momentum before the bullet hits the block or after?

7.5. State the principle of the conservation of momentum. How is it related to the first law of motion?

7.6. What is mass, and how is it defined in terms of the first law of motion?

7.7. How is momentum related to mass and velocity?

7.8. What is the connection between fundamental laws of physics and definition of physical quantities?

7.9. What is the difference between physics and mathematics in the context of definition of terms they use?

7.10. What is force and how is it related to momentum? How is it related to acceleration?

7.11. One force changes the momentum of a system in one second. A second force changes the momentum of the same system by the same amount in half a second. Which force is bigger?

7.12. Which is more general: the first law of motion or the second law?

7.13. What is weight? If you go to the Moon, does your weight change? Your mass?

7.14. What is the normal force? Would there be a normal force if there were no gravity, say in outer space?

7.15. When you start to walk, do you accelerate? If yes, what force causes your acceleration? What is the source of this force?

7.16. You push on a very heavy box resting on a rough floor, and it does not move. How does friction compare with your push?

7.17. Increase your push on the heavy box (of the previous question), which still does not move. Does friction increase or remain the same as before?

7.18. Why do we bend our knees instinctively when we jump from a height? What effect does this have on the force of impact?

7.19. What quantities do air bags affect, the change in the momentum of our head or the time in which this change takes place? What effect does this process have on the force applied to our head?

7.20. What is a centripetal force? Give two examples in which centripetal forces are operative.

7.21. What is a the terminal velocity? What property of the force of air drag results in a terminal velocity?

7.22. What is the precise statement of the third law of motion?

7.23. A horse pulls a buggy. The buggy pulls the horse by the same amount of force, but in the *opposite* direction. So the two forces cancel. How can the horse move the buggy then?

7.24. When you start to walk you accelerate because the floor pushes on you. How does the floor know when to push on you?

7.25. Is it the force of the engine that drives the car or something else? If something else, what?

7.4.4 Conceptual Exercises

7.1. Can you answer the question "Which has more momentum, a moving truck or a moving car"? Why or why not?

7.2. A time bomb motionless in midair explodes into many pieces. What do you get if you add the momenta of all the flying pieces?

7.3. A moving billiard ball hits an identical stationary one head on and stops.
(a) Does the second ball remain stationary?
(b) If it moves, does it move faster, slower, or with the same speed as the first one before hitting?

7.4. A ping pong ball hits a block of wood resting on a frictionless floor and bounces back with the same speed. Will the block acquire (a) the speed of the ping pong ball, (b) half the speed of the ping pong ball, (c) twice the speed of the ping pong ball, (d) the momentum of the ping pong ball, (e) half the momentum of the ping pong ball, (f) twice the momentum of the ping pong ball?

7.5. When you jump from a height, you stop. How do you reconcile this with momentum conservation?

7.6. You are driving on a straight highway at constant speed. What forces are acting on your car? Do they cancel each other? Now you enter a curve. While on the curve, what forces are acting on your car? Do they cancel each other?

7.7. A race car is moving at 100 mph on a straight part of a track. It then accelerates to 150 mph and moves at this new speed on another straight part of the track. On which track is the net force on the car is bigger?

7.8. An object has no acceleration. Can you say that no forces are acting on the object? Explain.

7.9. Use the laws of motion to explain why it is safer to land on a haystack than on a concrete floor when jumping from a tall building. Why is a thicker haystack safer than a thin one?

7.10. Why are elastic ropes which stretch a lot favored by mountain climbers? Explain using the laws of motion.

7.11. Use the laws of motion to explain why firefighters use trampolines for people jumping from a tall building.

7.12. A 100-kg mountaineer slides down a rope at constant speed. What is the net force on the mountaineer? What is the force of friction on the mountaineer?

7.13. Consider a bucket of liquid as shown in Figure 7.5. Concentrate on the portion **A** of liquid shown in Figure 7.5(a).
(a) What is the net force on **A**? Draw on Figure 7.5(a) an arrow representing the weight of that portion of liquid.
(b) What other force is acting on **A**? How big is this force compared to the weight of **A**? Which direction is it pointing? Draw it on Figure 7.5(a). What is the source of this force?

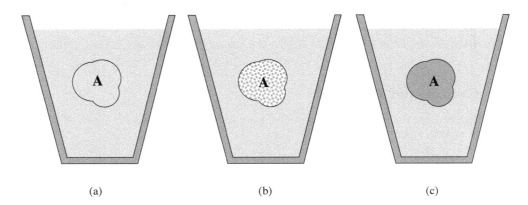

(a)   (b)   (c)

Figure 7.5: The bucket of liquid and the Archimedes' principle.

7.14. Now suppose we suck the liquid out of **A** and immediately fill it up with a metal made precisely in the shape of region **A** as shown Figure 7.5(b). Draw on Figure 7.5(b) an arrow representing the weight of the metal.
(a) What other force is acting on the metal? How does it compare with the weight of the metal? Which direction is it pointing? Draw it on Figure 7.5(b).
(b) How does this force compare with the force on the liquid of region **A** in the previous question? What is the source of this force?
(c) Which direction does the net force on the metal point? In what direction does the metal start to move?

7.15. Instead of metal, we fill region **A** with wood. Draw on Figure 7.5(c) an arrow representing the weight of the wood.
(a) What other force is acting on the wood? How does it compare with the weight of the wood? Which direction is it pointing? Draw it on Figure 7.5(c).
(b) How does this force compare with the force on the metal of the previous question, and on the liquid sucked out of region **A**? What is the source of this force?
(c) Which direction does the net force on the wood point? In what direction does the wood start to move?

7.16. Based on your experience with the last three questions, try to describe the force exerted on an object immersed in a liquid in terms of the volume of the liquid displaced by the object.

7.17. A monkey is hanging motionless on a rope.
(a) Is there a net force on the monkey?
(b) What forces are acting on the monkey, in which direction?
(c) How does the tension in the rope compare with the weight of the monkey?

7.18. Now suppose that the monkey starts to climb up the rope.
(a) Is there a net force on the monkey?
(b) What forces are acting on the monkey, in which direction?
(c) How does the tension in the rope compare with the weight of the monkey?

7.19. After reaching the speed of 1 m/s, the monkey moves at that speed for a while.
(a) Is there a net force on the monkey?
(b) What forces are acting on the monkey, in which direction?
(c) How does the tension in the rope compare with the weight of the monkey?

7.20. After reaching a certain height, the monkey slows down to stop.
(a) Is there a net force on the monkey?
(b) What forces are acting on the monkey, in which direction?
(c) How does the tension in the rope compare with the weight of the monkey?

7.21. When you jump vertically upward,
(a) what is the net horizontal force on you right before taking off?
(b) Which force is bigger, your weight or the force exerted on you by the floor?
(c) How does the floor know when to push on you?
(d) Which law of motion gives you the answer to the last question?

7.22. A crate sitting on the floor experiences two forces, gravity and the normal force which act in opposite directions. Are these an action-reaction pair? If not, what force pairs up with each to form an action-reaction pair?

7.23. You push on a heavy crate standing on a rough floor and it does not move.
(a) Does the crate push back on you? How do you compare the crate's force on you with your force on the crate?
(b) You push harder so the crate starts to move. Does the crate push back on you now? Is the crate's force on you smaller than your force on the crate?

7.24. You and a 5-ton crate are standing on a frictionless floor. A rope is attached to the crate and you hold its loose end. You pull the crate. Does the crate move? Do you move? If both you and the crate move, which moves faster?

7.25. Put four identical 1-kg cubic blocks next to each other in a row on a frictionless surface. Number them 1 through 4 starting at left. Push the left end of 1 with a force of 4 N.
(a) What is the acceleration of each block?

(b) How hard is 2 pushing on 1? How hard is 1 pushing on 2?
(c) How hard is 3 pushing on 2?
(d) How hard is 4 pushing on 3?

Losing weight!

7.26. Consider the downward motion of a person standing on a scale in an elevator. Concentrate on the motion at the very beginning, i.e., when the elevator starts accelerating downward with an acceleration that is less than the gravitational acceleration.
(a) Is the upward force of the scale on the person larger or smaller than the weight of the person? What is the reaction to this force?
(b) Does the scale read the weight of the person or something different? Is it larger or smaller?

Gaining weight!

7.27. Consider the downward motion of a person standing on a scale in an elevator. Concentrate on the motion at the end, i.e., when the elevator starts slowing down.
(a) Is the upward force of the scale on the person larger or smaller than the weight of the person?
(b) What is the reaction to this force? Does the scale read the weight of the person or something different? Is it larger or smaller?

Falling up!

7.28. Consider the downward motion of a person standing on a scale in an elevator. Concentrate on the motion at the very beginning, i.e., when the elevator starts accelerating downward with an acceleration that is *larger* than the gravitational acceleration.
(a) Which direction is the force other than gravity?
(b) Could the source of this force be the scale? If not, what is the source of this force?

7.4.5 Numerical Exercises

7.1. An 80-gram bullet traveling with the speed of 100 m/s hits a 5-kg block of wood which is already in motion in the same direction on a smooth floor with a speed of 1 m/s (see Figure 7.6). The bullet sticks to the block.
(a) What is the initial momentum of the bullet? Draw an arrow!
(b) What is the initial momentum of the block? Draw an arrow!
(c) What is the momentum of the system before the bullet hits the wood? Draw an arrow!
(d) What is the momentum of the system after the bullet hits the wood? Draw an arrow!
(e) What is the speed of the block with the bullet inside?

Figure 7.6: The bullet is about to hit the block that is moving in the same direction.

7.2. An astronaut with a total mass (astronaut plus wrench) of 90 kg is detached from her spaceship at a distance of 30 m and is moving with a speed of 0.1 m/s away from the spaceship. The commander tells the astronaut to throw the 0.5-kg wrench she is holding as hard as she can. The astronaut follows the order, throwing the wrench at a speed of 30 m/s. Drawing arrows for momenta will help!
(a) What is the appropriate system and what does it consist of?
(b) Which direction does the astronaut throw the wrench?
(c) What is the momentum of the system before she throws the wrench?
(d) What is the momentum of the system after she throws the wrench?

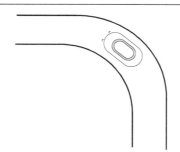

Figure 7.7: The car in uniform motion on a curve.

(e) What is the momentum of the wrench? Draw an arrow!

(f) What is the momentum of the astronaut? Draw an arrow!

(g) What is the speed of the astronaut?

(h) Will the astronaut get back to the spaceship? If so, how long does it take her to reach the spaceship?

7.3. Consider the motion of a parachutist. Suppose that the force of air drag can be written as $F_{\text{drag}} = 500v$ where v is the speed of the parachutist, that the mass of the parachutist plus the parachute and everything else in motion is 250 kg, and that the parachutist does not open the parachute for a while. He falls down freely until his speed reaches 20 m/s. Now he opens the parachute.

(a) What is the net force on the parachutist right after he opens the parachute? Which direction is it pointing?

(b) What is the acceleration of the parachutist right after he opens the parachute? Which direction is it pointing? Is he speeding up or slowing down?

(c) A little later the speed of the parachutist is 10 m/s. What is the net force on him now? Which direction is it pointing?

(d) What is the acceleration of the parachutist now? Which direction is it pointing? Is he speeding up or slowing down?

(e) At what speed do the weight and the drag force become equal? Once the parachutist reaches this speed, can he speed up or slow down? Explain!

7.4. Jack pushes on a 200-kg block sitting on a floor with a force of 150 N to the right. He notices that the block acquires an acceleration of 0.5 m/s^2.

(a) What is the net force on the block?

(b) What is the force of friction?

7.5. A 1200-kg car is moving around a curve at a speed of 50 mph as shown in Figure 7.7 (northwestward at the moment depicted). The radius of the curve is 100 m.

(a) What is the acceleration of the car?

(b) What is the net force acting on the car? Draw an arrow on the car indicating the direction of the net force.

(c) What applies this force on the car?

Chapter **8**

Further Topics on Motion

The universality of Newton's laws of motion invites their application in a wide variety of situations. Newton himself applied these laws to the solar system and obtained the entire collection of results previously discovered by other means, including Kepler's three laws of planetary motion. The post-Newtonian physicists and mathematicians applied the same set of laws to the motion of objects on Earth, and a burgeoning of ideas and techniques—ideas and techniques that created entirely new branches of mathematics and physics, not to mention other sciences—ensued.

The development of mathematical tools, in conjunction with the variety of application of the laws of motion led to the introduction of new concepts in mechanics. Concepts which, although taken from ordinary imprecise experiences, were given precise mathematical meaning. Some of these concepts proved to be extremely helpful in solving some difficult dynamical problems. Two such concepts are work and energy with which this chapter starts.

8.1 Work and Energy

By now the reader has noticed that the words used in everyday life acquire special, precise—and sometimes strange—meanings in the hands of physicists. This applies not only to physics, but to all sciences and especially in mathematics. For instance, take the word "function," whose ordinary meanings include such notions as "normal or characteristic actions," "a special duty," "occupation or employment," and "a formal ceremony." In mathematics, function becomes a specific association between two sets of objects. Such a precision and specificity is at the root—and a prerequisite—of the development of science. This is particularly pronounced in physics, nicknamed "the queen of sciences" and even more so in mathematics, whose first encounter with precision goes back to 300 BC when Euclid axiomatized geometry.

The concepts of work and energy are no exceptions. For a physicist a weight lifter sweating under 250 lb of weight is not doing any work as long as he does not move the weight! This may appear cruelly ungratifying, but physics demands precision not emotional justification. It has been found—by numerous historical trials and errors—that no motion implies no work. The situation is worse than this. Even if there *is* motion, there is no guarantee that work is being done!

Work defined.

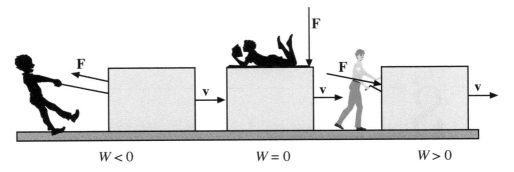

Figure 8.1: The work done on an object can be negative, zero, or positive depending on the angle between the force and displacement.

Box 8.1.1. *Work (W) done by a force F is displacement (d) times the component of force along that displacement, and is thus nonzero only if the force that causes the work is not always perpendicular to the instantaneous velocity. We write $W = F_{||}d$, and interpret $F_{||}$ as the component of the force along the displacement.*

What do you know? 8.1. Tie a stone to one end of a string. Hold the other end and whirl the string above your head.[a] Three forces act on the stone: its weight, the tension in the string, and the air drag. The work of which force is positive, negative, or zero?

[a]Make sure nobody's head is on the path of the stone!

Work can be positive or negative.

If the force makes an acute angle with velocity (it reinforces the motion), the work is *positive*, otherwise the work is *negative* because an obtuse angle hinders the motion (Figure 8.1). The unit of work is N·m, which is also called **Joule**, named after James Joule whose careful analysis of heat led to the equivalence of heat and energy (to be discussed later in the book).

8.1.1   Conservation of Mechanical Energy

The second law of motion tells us that a (net) force acting on an object changes its momentum, and therefore its velocity.[1] This change in momentum accompanies a displacement of the object. Hence, a net work is done on the object (assuming that the net force is not perpendicular to the displacement). Therefore, when a net work is done on an object, its speed changes. A quantitative analysis of the dynamics of the object shows that a quantity can be *defined* so that the change in this quantity during the action of the force is precisely the work done on the object. This quantity is called the **kinetic energy** (KE) of the object in motion:

$$\text{Work} = \text{Change in KE} \quad \text{or} \quad W = \Delta(KE). \tag{8.1}$$

Kinetic energy defined.

If this is to hold in Newtonian mechanics, then KE has to be defined as (see **Math Note E.8.1** on **page 80** of *Appendix.pdf* for details)

$$KE = \tfrac{1}{2}mv^2, \tag{8.2}$$

---

[1]As usual, we consider mass to be constant.

where m is the mass and v the speed of the object. Since work is related to (the change in the) KE, the latter must have the same unit, Joule.

Unit of work and energy is **Joule** (J).

Example 8.1.2. A 5-kg block, moving with a speed of 10 m/s, comes to a stop after covering a distance of 5 meters.

Q: What is the work done by the force of friction, and how big is that force?

A: The change in KE, i.e., the final KE minus the initial KE, is equal to the work done by friction. But $\Delta KE = \frac{1}{2}mv_f^2 - \frac{1}{2}mv_i^2 = 0 - \frac{1}{2} \times 5 \times 10^2 = -250$ J. So, the work done by friction is -250 J. Furthermore, $W = F_{||}d$, or since force of friction *is* parallel to the direction of motion, $-250 = F_{\text{fric}} \times 5$ yielding -50 N for the force of friction. The minus sign tells us that the friction opposes the motion. ∎

> **What do you know?   8.2.** Consider the motion of a parachutist who opens the parachute from the very beginning. Is the work done by her weight positive, negative, or zero? Is the work done by air drag positive, negative, or zero? Is the work done by the net force positive, negative, or zero? Is the change in kinetic energy positive, negative, or zero?

An omnipresent force is gravity. So, it is worthwhile to apply Equation (8.1) to the force of gravity. Figure 8.2 shows an object fall from an initial height h_1 to a final height h_2. The force of gravity (or the weight of the falling object) is mg and the parallel displacement is $h_1 - h_2$. Thus, the work done by gravity is $mg(h_1 - h_2)$. The change in KE is $KE_2 - KE_1$. Equating these two quantities and rearranging terms, we obtain $mgh_1 + KE_1 = mgh_2 + KE_2$.

This is a very interesting and important result. It states that the quantity $mgh + KE$, where h denotes the height, is the same for both initial point (point 1) and final point (point 2). It is customary to define the **gravitational potential energy** as

Gravitational PE.

$$PE = mgh. \tag{8.3}$$

Then the equation above can be written as $PE_1 + KE_1 = PE_2 + KE_2$, which is the statement of the conservation of *mechanical* energy. It says that if one adds the potential energy at a point P_1 to the kinetic energy at P_1, one obtains the same numerical value as the corresponding sum at any other point P_2. Since P_1 and P_2 are arbitrarily chosen, one can state

> **Box 8.1.3. (Conservation of Mechanical Energy)** *The total **mechanical energy** $ME = KE + PE$ of a system is a constant during the entire motion of the system.*

> **What do you know?   8.3.** Consider the motion of a parachutist who opens the parachute much later. Is the work done by her weight positive, negative, or zero? Is the work done by air drag positive, negative, or zero? Is the work done by the net force positive, negative, or zero? Is the change in kinetic energy positive, negative, or zero?

The name "potential energy" was not given capriciously. To lift an object to a new height, we must do work on the object. The object in return acquires the capacity or the "potential" to do useful work for us: If we let it go, it will pick up (kinetic) energy, which could be used to drive a wheel, for example. The turbines of a hydroelectric power plant use the potential energy of the water dropped from a height to produce electricity.

Using Box 8.1.3 in hydroelectric power plants.

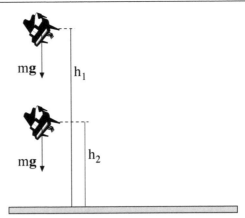

Figure 8.2: As an object falls, gravitational force performs some work.

Example 8.1.4. How much work is required to lift a car of mass 1 metric ton (1000 kg) to a height of 5 meters? Assume that the applied force always balances the weight of the car.

Applied force is mg and displacement is 5 m. So

$$W = \underbrace{1000 \times 9.81}_{F_{||}} \times \underbrace{5}_{d} = 49,050 \text{ Joules},$$

which is stored in the car as the potential energy. That there is indeed energy stored in this lifted car is evident from your hesitation to stand right under it! ∎

Gravity is not the only force leading to potential energy. The following example discusses another common PE.

Example 8.1.5. Consider a block attached to one end of a spring whose other end is held firmly to a wall (Figure 8.3). The block is allowed to move on a perfectly smooth horizontal surface. Initially the block is at rest at point O. Now stretch the spring to a position A and let go of it. The subsequent motion is a succession of compression and elongation of the spring. Whenever the spring is either compressed or stretched it has the potential of doing work, thus it has a potential energy. This potential energy is maximum when the spring is maximally stretched or compressed, i.e., at points A and B. Since the spring is neither stretched nor compressed at O, we say that the potential energy is zero there. At any other point between A and B, the potential energy acquires a value between zero and its maximum value. On the other hand, the kinetic energy is zero at A or B because the block comes to a temporary rest at these two points. The maximum of the KE takes place at O where the block moves at the greatest speed. ∎

Height is measured relative to a convenient (and arbitrary) reference level.

Equation (8.3) contains h, height. But height relative to what? Somebody on the first floor of the Science Building wants to measure height relative to the ground, which is a few meters below the ground of another person, who happens to be on the second floor. So, which floor does one have to use for measuring height? The answer: it doesn't matter! It turns out that it is the *difference* in PE that is of physical significance, and this difference is independent of the "floor."

Place a ball at the edge of a table. Now jerk it horizontally so that when it leaves the table, it has some horizontal speed. As it hits the floor, its motion is no longer horizontal, but makes an angle with the floor. It is possible to break the motion of the ball into vertical and horizontal components and, using the second law of motion, find the horizontal and vertical velocities of the ball, its speed, and the angle it makes with the horizontal as it hits the floor. This is rather complicated, and requires some trigonometry. However, if we are interested only in the speed of the ball, we can easily obtain it using the principle of the

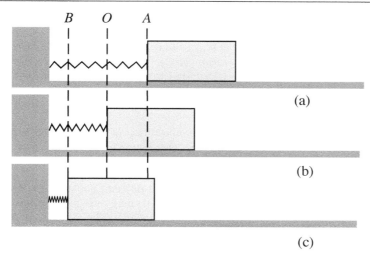

Figure 8.3: The mass-spring system oscillates between points A and B, at which the potential energy is maximum. The potential at O, where the spring is neither compressed nor stretched, is zero.

conservation of mechanical energy. **Example D.8.1** on **page 24** of *Appendix.pdf* shows how to do this.

A roller coaster starts from a very high altitude (with no kinetic energy but a lot of potential energy). As it reaches lower altitudes, its potential energy is partially converted to kinetic energy. When it climbs back up a hill, its KE is partially converted to PE. The continuous conversion of PE into KE and vice versa is what makes the motion of a roller coaster exciting. **Example D.8.2** on **page 25** of *Appendix.pdf* calculates the speed of a particular passenger of a roller coaster at various points of its track.

Example D.8.2 analyzes the motion of a roller coaster for a combined mass (car plus passenger) of 100 kg. What happens if the passenger is heavier (or lighter)? The values of the forces and the energies will change. However, the values of the speeds had better not change! If it did, some cars will be going faster than others, and will be pulling on the latches connecting them to the slower cars, causing them to snap! The interested reader is urged to go through the example using a different mass to see that this is indeed the case. **Math Note E.8.2** on **page 81** of *Appendix.pdf* proves this claim rigorously.

Speed of cars of a roller coaster is independent of their masses.

We have to emphasize that both KE and PE are *defined* in such a way that the foregoing conservation law becomes valid. It is also important to note that the conservation of mechanical energy does not always hold. For instance, if the system of the block and spring is laid on a rough surface, the block will come to rest after a while, losing all of its KE and PE. Later in the book you will see that this decrease in the mechanical energy is not a loss, but a transformation into another form of energy, heat. A system for which the conservation of mechanical energy holds is called a **conservative system**.

Conservative systems.

🛒 Food for Thought 🌱

FFT 8.1.6. The word "energy" has been unfairly subjected to a lot of abuse in the hands of alternative medical practitioners, where it has been given a stature equivalent to "aura," "soul," and the Chinese mystical "Qi." This is mainly because of the misconception that energy is "non-material." In the community of alternative medicine, where pseudoscientific ideas are the basis of the enterprise, where spiritual entities are believed to be the cause of diseases, energy—and its cousin, *field*—become an attractive companion to those mystical entities.

Contrary to the misguided misconception of the alternative medical community, energy is a

Energy is a property of only *material* objects.

property of only *material* objects. Furthermore, it is a quantity that has demonstrable effects on—and interaction with—matter. When we talk about potential energy, we not only think of the energy of *a material object* such as a ball or a body of water, but this energy will eventually manifest itself through *interaction with material objects*. This potential energy will eventually turn into kinetic energy, and the latter will be transferred to another *material* object (e.g., a turbine) upon impact, and this transfer of energy can be measured and recorded on instruments. An energy that can only be felt by the "hands of a healer," or the "soul of a preacher," or the "mind of a Qi master" does not exist.

As you will see later, potential energy is a manifestation of a *field*, which is as material as a tennis ball, a car, or a train. Just as a ball, a car, or a train can make itself felt when it collides with a racket, another car, or a truck, so do the electromagnetic fields make themselves felt when they "collide" with the antenna of a radio, a television, or a cellular phone.

Not only is energy a manifestation of solely material objects, but modern physics has shown that energy itself is *material*. The fields causing potential energy are nothing but a huge army of *material* subatomic particles. In the case of the electromagnetic fields, these particles are called photons.

8.1.2 Power

The rate at which energy is produced or consumed is called **power**. If E is the energy produced or consumed in time t, then the power P is

$$\text{power} = \frac{\text{energy}}{\text{time}} \quad \text{or} \quad P = \frac{E}{t}. \tag{8.4}$$

Unit of power is **Watt**.

This definition gives the unit of power as Joules per second which is given the name **Watt** in honor of the British inventor James Watt who improved the steam engine.

We can rewrite Equation (8.4) as energy = power × time or $E = Pt$, suggesting the definition of Joules as Watt·second. In fact, there is a unit of energy in commercial use which is based on this idea. Let Δt be one hour (3600 s) and with power in units of Watts, define a unit of energy called Watt·hour, where

$$1 \text{ Watt·hour} = 1 \text{ Watt} \times 1 \text{ hour} = 1 \text{ Watt} \times 3600 \text{ s}$$
$$= 3600 \text{ Watt·s} = 3600 \text{ Joules}.$$

Kilo Watt hour in terms of Joules.

The more common unit of energy used commercially is 1000 times this and is called kilo-Watt-hour, abbreviated **kWh**. Thus, 1 kWh = 3.6×10^6 J.

As an example, consider a 100 Watt light bulb that is on 8 hours a day. The daily energy consumption of this light bulb is 100 Watts × 8 hours = 800 Watt hours, or 0.8 kWh. At 15 cents per kWh, the daily cost of this bulb would be 12 cents.

What do you know? 8.4. In your favorite fitness club, you lift 50 kg half a meter 30 times, and it takes you 2 minutes to do this. What is your power output?

Horse power, another unit of power.

Watt, the scientific unit of power, is not the common one used in industry. The energy output of large machines is measured in **horse power**, the British unit of power. At the dawn of industrial revolution, when heavy machines such as steam engines were being invented, the tendency was to compare them with the "natural machines" in use, namely the beasts of burden. Although the strength of horses varied widely, it was not hard to average over this variety and come up with an ideal average horse and quantify the output of such a horse.

A horse power was soon to be defined as the amount of work an ideal horse performs in one minute. By studying various horses, it was decided that the ideal average horse can lift

3300 lb to a height of 10 ft in one minute. To see how many Watts there are in a horse power, note that there are 2.204634 lb per kilogram. So, 3300 lb is $3,300/2.204634 = 1,496.85$ kg. Multiply this by 9.8 to get the weight, i.e., the *force* associated with 3300 lb. This yields 14,669.13 N, and if a horse lifts this weight to a height of 3.048 meters (each foot is 0.3048 m), then it will be doing $14,669.13 \times 3.048 = 44,711.5$ J of work. Since this work is done in 60 seconds, the power of an ideal horse will be $44,711.5/60 = 745.2$ Watts. Officially, a horse power is equivalent to 745.7 Watts.

8.1.3 The Physics of Santa Claus

First let me assure the reader that the title of this subsection is not a joke! Physics has a lot to say about Santa Claus, especially when it comes to his consumption of energy. This subsection is devoted to an analysis of the amount of energy Santa needs to make all his deliveries as well as the consequences of his stops and starts. Because of the very difficult task Santa is facing, in the following analysis I shall make life as easy as possible for him by estimating quantities enormously in his favor.

Start by estimating the number of houses Santa has to visit. There are approximately 2.1 billion Christians in the world, and—estimating drastically in Santa's favor—let us assume that eligible children constitute only 7 percent (or 0.07) of this population. Thus, the number of children he has to visit is 2.1 billion times 0.07 or approximately 150 million.

The number of houses Santa has to visit depends on how many children live in each house: The more children live in a typical house, the fewer houses he will have to visit. To make life really easy for Santa, let's assume a (very high) concentration of 6 children per house. This brings the number of houses down to $1.5 \times 10^8/6 = 2.5 \times 10^7$, or 25 million chimneys to plunge down and climb up!

> Santa has to visit approximately 25 million houses!

Santa cannot afford to spend too much time in each house, because he has only 24 hours to deliver all the toys he is carrying.[2] On the other hand, he cannot spend too little time moving through a chimney, because then his speed may be so huge as to require too much energy to attain. As **Math Note E.8.3** on **page 81** of *Appendix.pdf* shows, the optimal time spent per chimney—the time that minimizes Santa's energy consumption—is about 26 microseconds;[3] so Santa covers over 38,000 chimneys every second! Since this does not violate any laws of physics, let's accept it. After all, we are all on Santa's side!

For 25 million chimneys, he needs $25,000,000 \times 26 \times 10^{-6} = 650$ seconds. Since there are 84,600 seconds in 24 hours, only 85,750 seconds are left for Santa to hop from chimney to chimney to chimney ... all 25 million of them. Ignore the time he spends in each house finding the appropriate place for the gifts, taking them out of his bag, and placing them down, although these are legitimate concerns. Nevertheless, can he make all the deliveries? It depends on how fast he will be moving from house to house.

To further help Santa in his seemingly impossible task, we make the houses very small, bring them next to each other, and place them side by side on a straight line. Suppose that all this sets the distance between consecutive chimneys to be merely 50 ft or 15 meters, with a total distance of $15 \times 2.5 \times 10^7 = 3.75 \times 10^8$ m. Since he has 85,750 seconds to travel this distance, Santa's speed will be $3.75 \times 10^8/85750 = 4373$ m/s. This is larger than half the speed of the fastest humanly possible vehicle, and covers the NY-LA distance in less than 20 minutes! Although *implausible*, let's assume that it is *possible*.

> Santa has to travel with a speed of 10,000 mph to deliver all the toys!

It appears that we have allowed Santa to accomplish his task. Or have we? Any travel requires fuel (or energy). Now let's estimate the energy consumption of Santa's journey, which is assumed to be used simply to speed him (and his toys) up.[4] Every

---

[2] Actually, because of the Earth's rotation, Santa has a little more than 24 hours, but the difference is not essential.

[3] A microsecond is a millionth (or 10^{-6}) second.

[4] Again, to give Santa the benefit of the doubt, we ignore such hurdles as the drag and friction forces, which are responsible for almost all the fuel cost for ordinary people.

time Santa increases his speed, there is a corresponding increase in his kinetic energy, which requires fuel. To find Santa's kinetic energy, we need his total mass. Assuming an average mass of 2 kg per toy—and only one toy per child—Santa's *initial* cargo mass is $2 \times 1.5 \times 10^8 = 3 \times 10^8$ kg. Since this mass decreases as Santa delivers the toys, take the effective mass to be the average of this initial mass and the final mass (zero), or 1.5×10^8 kg. [It turns out that this average mass gives the same result as exactly calculating the kinetic energy of individual hops and adding them.] Therefore, the attainment of a speed of 4373 m/s from rest requires an energy of $KE_{\mathrm{hop}} = \frac{1}{2}(1.5 \times 10^8)(4373)^2$ or 1.43×10^{15} Joules. Thus, every time Santa hops from one chimney to the adjacent one, he speeds up from zero to 4373 m/s, and in the process, consumes 1.43×10^{15} Joules. For all the 25 million chimneys he needs 3.6×10^{22} Joules.

Santa needs over 1430 trillion Joules just to hop from one chimney to the next!

This is one part of the energy Santa needs for his mission. Remember his plunging and climbing through the chimneys? That requires energy too. First, what is his speed in the chimneys? Assuming a (short) chimney length of 4 meters, the round trip distance is 8 meters, which Santa has to cover in 26 microseconds. This gives a speed of $8/26 \times 10^{-6} = 307692$ m/s.

Santa's speed in the chimney is 688,000 mph!

Suppose that Santa leaves his cargo on the roof of each house he is visiting, and carries only the toys through the chimney. The energy required for his trip down—for a *thin* Santa of only 100 kg—is $\times \frac{1}{2}(112)(307692)^2 = 5.3 \times 10^{12}$ Joules, where the extra 12 kg in the mass accounts for the 6 toys he is carrying. For the trip up, the energy is $\frac{1}{2}(100)(307692)^2 = 4.7 \times 10^{12}$ Joules. Thus, the total chimney energy per house is the sum of these two, or 10^{13} Joules. The chimney energy for the 25 million houses is $(2.5 \times 10^7) \times (10^{13}) = 2.5 \times 10^{20}$ Joules. This is less than 1% of the energy needed for hopping from chimney to chimney. So, let's take the latter to be the total energy.

The hopping energy, 3.6×10^{22} Joules, is huge. Nevertheless, it is the minimum amount he can spend. But what does it really mean? To get a feel for its magnitude, you have to compare it with another huge quantity of energy. For example, how does it compare with the yearly energy consumption of the world? The entire annual energy consumption of the world was about 4×10^{20} Joules in 2001 [DOE 01]. This included not only the typical residential usage such as heating, lighting, cooking, commuting, and entertainment, but also the large scale industrial, agricultural, and transportation consumption; and 4×10^{20} Joules is typical of the recent annual consumptions. Santa uses $3.6 \times 10^{22}/4 \times 10^{20} = 90$ times the annual world energy supply *in one day*! Thus, the entire world must stop using any form of energy for 90 years so that Santa can deliver his toys in one day! Clearly Santa's visit cannot be an annual event. At best it can be a centennial event, for the preparation of which the whole population of the world (Christian and non-Christian) must stop consuming any form of energy for the entire century!

Santa uses 90 times the annual world energy supply *in one day*!

Although extremely difficult, the people of the world might be willing to tolerate all the harshness caused by Santa's trip were it not for the revelation that behind his jolliness and smile there is destruction. Huge destruction in the form of massive explosions! Of course, Santa's explosions are not intentional. He just can't help exploding houses as he visits them! How can that be?

An explosion is simply the release of a large amount of energy in a small time period. Take Santa's plunge down a typical chimney. We found that he has a kinetic energy of 5.3×10^{12} Joules when he reaches the bottom of the chimney. This energy turns into heat when he brakes to a complete stop. For comparison, the heat produced in the explosion of a ton of TNT is about 4×10^9 Joules. Thus, Santa releases the equivalent of $5.3 \times 10^{12}/4 \times 10^9 = 1325$ tons of TNT when landing in the house, and almost the same amount when he climbs up the chimney and stops at the roof! And a time interval of 13 microseconds for each of these releases is short enough to qualify them as explosions.

Santa's plunge heat is tons of TNT!

But the real killer is the explosion caused by his landing at the chimney as he comes from the previous house. His KE as he lands at the chimney was calculated to be 1.43×10^{15} Joules, and this energy is turned into heat in less than the time it takes to go from one

Santa's hopping energy compared to the Hiroshima bomb!

chimney to the next—which, by the way, is $85750/2.5 \times 10^7 = 0.0034$ second, qualifying this release of energy as an explosion as well. How many tons of TNT is this equivalent to? $1.43 \times 10^{15}/4 \times 10^9 = 357{,}000$ tons of TNT! Suffice it to say that the destructive power of "Little Boy," the bomb that was dropped on Hiroshima, was a "mere" 15,000 tons of TNT. Every time Santa lands at a chimney, he detonates about 24 Hiroshima-type bombs, and he is at ground zero of every blast!

8.2 Rigid Body Motion

One of the difficulties with applying Newton's laws of motion is that ordinary objects are not points; yet the laws can only be applied to points. The very essential act of constructing the position vector of a train relative to a person becomes a confusing issue: it is not clear which part of the person is to be connected to which part of the train to construct the position vector. This leads to the technique of breaking up objects into infinitesimal pieces and applying the laws to those pieces, and finally somehow summing up the motion of all individual "point" pieces. Two simple cases of the application of this technique—a rigid body and an incompressible fluid—are important. The simpler case of a rigid body is the subject of this section.

 Food for Thought

FFT 8.2.1. The conclusion of the preceding analysis would be a devastating blow to the spirit of a three year old. However you, the reader of these lines, have no difficulty accepting the force of logic and the laws of physics, and convincing yourself that the story of Santa Claus is simply a superstition. In fact, arguments at a much lower level than the analysis above have no doubt convinced you of the nonexistence of Santa Claus. But have you rid yourself of other, more sophisticated, forms of superstition?

There are various levels of superstition. Santa Claus and his cousins, the Easter bunny and leprechauns, are "introductory" superstitions, out of which you grow by the time you reach junior high school. At the "intermediate" level, we have astrology, "scientific" creationism, psychic and palm reading—to name a few—to which a large fraction of our adult population subscribes. It does not take much to "convince" even a college student that UFOs exist, that certain people have the ability to predict the future, that it is possible to heal the body by the power of the mind.

"Advanced superstition" consists of such delicate and scientifically cloaked disciplines as the alternative medicine, chiropractic, and psychoanalysis, which are the subject of academic "research" at our institutions of higher learning. When college students see their professors specializing in psychoanalysis conduct "research," just as chemists do, write textbooks, which are adopted by many universities, just as chemists do, and attend conferences to present papers and exchange ideas, just as chemists do, it becomes very hard to convince students that there is a substantial difference between chemistry and psychoanalysis. Alas, no simple formula can be written down to refute the discipline of psychoanalysis as there is to refute the existence of Santa Claus.

The heart of the problem is differentiating between proof and plausibility. We have not proved that Santa Claus does not exist. We can never prove that, just as we cannot prove that Elvis does not show up at midnight in Graceland (see Section 43.1.3). It is impossible to prove a negative statement. However, the implausibility of Santa's trip is so strong that it can almost replace the proof of his nonexistence.

Unfortunately, we cannot say the same thing about other "higher superstitions." They require a much more sophisticated critical examination. Their implausibility will be revealed only to those who are exposed to the inner workings of natural sciences. Only then can they see the universality (and limitation) of the laws governing the natural phenomena, and judge for themselves that certain phenomena are simply so implausible that they can be called impossible.

Implausibility of higher (and more subtle) superstitions.

A powerful factor that prevents children from getting rid of superstition is the influence of the adults. We cannot ignore the crucial role of parents and society on the upbringing of a citizen. The reason that we no longer believe in Santa Claus by the time we finish elementary school is not that we have been exposed to the type of argument presented above. It is that our parents, and

Role of parents in getting rid of superstition.

all the adults with whom we come in contact during our adolescence, give us the message that it is time we left our sweet childhood belief in Santa behind us. The same parents and the same adult population do not give us the message that psychics do not have any special "power" or "gift," that palm readers cannot give us any useful advice, that horoscopes will not tell us about our future, and that psychoanalysis, alternative medicine, herbology, etc. are as effective as drinking a glass of water.

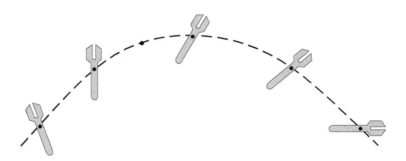

Figure 8.4: The motion of a rigid body is the superposition of a translation of the center of mass and a rotation about the center of mass. The picture describes the motion of a wrench after it is thrown. Notice how the center of mass of the wrench (the dot on it) follows the path that a point particle would describe.

As a car moves on a highway—ignoring certain internal relative motion of its parts—it always remains in one piece, i.e. the distance between its various parts remains the same regardless of the motion it is subjected to. An object with this property is called a **rigid body**. While the car is rigid, the smoke that comes out of its exhaust pipe is not, because the distance between various smoke particles changes as the smoke diffuses into the atmosphere.

Defining a rigid body.

What do you know? 8.5. A bucket is half filled with water. Does the distance between various parts of the water change? Why is water not rigid?

8.2.1   Center of Mass

Although a rigid body may appear complicated at first glance, its motion turns out to be a superposition of two *simple motions*. The overall motion of a rigid body is described by the *translation* of a single point, called the **center of mass** (CM) of the rigid body, plus a *rotation* about the center of mass (Figure 8.4).

What do you know? 8.6. How does the motion of the wrench of Figure 8.4 appear to you if you were riding along with its center of mass?

We shall not dwell on the concept of the center of mass. Suffice it to say that for *uniform* symmetric geometrical figures it coincides with the center of symmetry. Thus, the geometric center of a uniform cube or sphere is also its CM. Operationally, CM, sometimes called **center of gravity**, is determined to be the point on which the object can be balanced. The basic property of the center of mass is that the overall translation of the rigid body due to *external* forces can be determined by assuming that all its mass is concentrated at CM and applying the second law of motion to it.

External force determines motion of **center of mass**.

8.2.2   Angular Momentum

Once the translation of the rigid body is determined as above, one can concentrate on rotation by situating oneself at the CM. Then the rigid body will appear to be only rotating. The simplest kind of rotation is that about a fixed axis, called the *axis of rotation*. Such a rotation is characterized by the property that the perpendicular distance from any point of the rigid body to the axis of rotation remains unchanged during the course of the motion. Thus, the top in Figure 8.5(a) rotates about the axis Oz making an angle α with the vertical.

Rotation about an axis.

Another noteworthy example of rotating objects is that of the Earth as it moves around the Sun. The center of the Earth translates on the Earth's orbit making a full circle in approximately 365 days. Superimposed on this motion is the rotation of the Earth about an axis connecting the north and the south poles. The latter motion is called *spin* while the former is the *orbital* rotation. The axis of spin is slightly tilted with respect to the plane of the orbit, giving rise to seasons.

Rotations are also described in terms of **angular momentum**. For the simplest kind of rotation, namely rotation about a fixed axis, angular momentum is a vector **J** along the axis of rotation whose direction is determined by the sense of rotation. Specifically, if an ordinary (so-called right-handed) screw is allowed to turn in the direction of rotation, it will advance in the direction of angular momentum. An equivalent way of finding the direction of angular momentum is using the so-called *right-hand rule* (RHR): Curl the fingers of your *right hand* in the direction of rotation, the direction of the angular momentum vector will be defined by your stretched thumb [Figure 8.5(b)].

Angular momentum.

The magnitude of **J** in this simple case can be written as $J = I\omega$ where ω is the speed of rotation, called *angular velocity*, and I is called the **moment of inertia**, or rotational inertia. In a qualitative sense, the moment of inertia is a measure of the resistance of the rotating object to a change in its state of motion. Thus, moment of inertia is to rotation what mass is to translation. It is possible to write a mathematical formula relating I to the mass and the "extension" of the object. We shall forego this relation, but simply point out that

Moment of inertia.

> **Box 8.2.2.** *The farther the mass concentration is from the axis of rotation, the larger the moment of inertia.*

If a single (point) object is rotating on a circle about an axis perpendicular to the area of the circle, then the magnitude of the angular momentum of the object is defined as the product of its (linear) momentum p and the radius of the circle.

$$J = rp = rmv \tag{8.5}$$

It can also be written as $I\omega$, where $\omega = v/r$ and $I = mr^2$ for a single point object.

The word "momentum" used in christening J is not without purpose. Recall that ordinary momentum was conserved when there were no external agents. The same is true for angular momentum. The law of angular momentum conservation states that

> **Box 8.2.3. (Conservation of Angular Momentum)** *The angular momentum of an isolated system remains constant.*

This means that both its direction and its magnitude will remain unchanged. The following examples illustrate this law.

Example 8.2.4. A pirouetting ice skater opens up her arms to slow down. This is because by opening her arms, she sends some of the mass of her body away from the axis of rotation (running

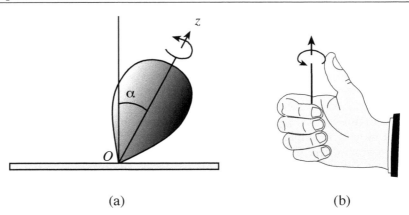

Figure 8.5: (a) Rotation about an axis. (b) Angular momentum and its relation to the direction of rotation via the right-hand rule (RHR).

vertically through her body), causing an increase in I. Since **J** is constant, the angular speed ω must decrease to keep the product $I\omega$ unchanged. Similarly, by bringing her arms in, she can speed up because the folding of the arms decreases I.

Divers crouch in midair to reduce their rotational inertia. This increases their angular speed, allowing them to perform more somersaults in the little time available to them from the diving board to the surface of the water.

A bicycle at rest cannot stand. As soon as it starts to move, the wheels acquire an angular momentum in a horizontal direction (remember the right-hand rule), which resists any change. As long as there is no severe push or pull to the sides, the conservation of angular momentum ensures the constancy of the rotation (and thus the motion) of the bicycle wheels.

Bullets are given a little angular momentum by having them traverse a helical path in the barrel of a gun. This gives a little more stability to the bullet's motion, because any external agent trying to alter the motion of the bullet encounters an additional resistance due to the angular momentum of the bullet similar to the resistance of a bicycle to falling when it is in motion.

Artificial satellites are given a little rotation to provide their motion with an extra stability. With angular momentum added to their motion, it will be harder to alter their course.　■

> **What do you know? 8.7.** Why do tightrope walkers carry such a long pole as they walk on the rope?

Just as a general rigid body motion is a superposition of a translation—of the center of mass—and a rotation (about the CM), *total angular momentum* is a superposition of orbital angular momentum and spin. For instance, in the case of the Earth, the orbital angular momentum is associated with the revolution of its CM around the Sun, while the spin measures that part of the (total) angular momentum arising from the rotation of the Earth around its own axis, causing alternate succession of days and nights.

The concepts of spin and orbital angular momentum become particularly important in the quantum theory of subatomic particles. It is shown in quantum theory that spin, orbital, and total angular momenta are all quantized. Furthermore, particles with different spins have fundamentally different properties. More on these later.

> **What do you know? 8.8.** A person is sitting on a swivel chair holding a spinning bicycle wheel. If she changes the orientation of the wheel's shaft, the chair starts to rotate. Why?

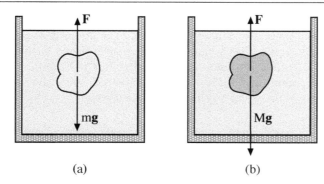

Figure 8.6: The origin of the buoyant force. (a) A "piece" of the fluid. (b) A piece of an immersed object.

8.3 Mechanics of Fluids

Newton's laws of motion can not only describe the behavior of point particles and rigid bodies, but also the more complicated objects called **fluids**. These are substances which change their shapes easily. Common examples of fluids are liquids and gases.

8.3.1 Hydrostatics

First let's study fluids at rest. In this subsection assume that all parts of the body of fluid under consideration are motionless as in a quiet pond, or a glass of water sitting on a table.

Buoyancy

Now consider any "piece" of the fluid. Since the piece is at rest, its velocity must be zero. In addition, there is no change in this velocity. Therefore, its acceleration is also zero.[5] The second law of motion now says that the net force on the piece must be zero. In Figure 8.6 a small piece of a fluid, say water, is isolated to analyze its motion. This piece is part of a larger body of water which could have any size or shape. We know that gravity is acting downward in the form of the weight of the piece and that the total force is zero. So, there must be a second force exactly equal to the weight, but acting upward. The source of this force can be nothing but the *area of the remaining* liquid in contact with the given piece.

   Now suppose you replace that piece with something else, say a metal, cut in exactly the same shape, so that it fits snugly in place of the piece. Clearly the gravitational force on the metal will be different from that on the replaced liquid, but since the shape of the contact area between the metal and the rest of the water is exactly the same as before, the upward force will remain the same. This upward force is called the **buoyant force**, and was first discovered by Archimedes, who upon such discovery is said to have run out of his bath naked and shouted: $E\nu\rho\epsilon\kappa\alpha$! This discovery is called **Archimedes' principle**:

Newton's laws of motion imply Archimedes' principle.

Box 8.3.1. (Archimedes' Principle) *An object in a fluid is pushed up by a force equal to the weight of the fluid it displaces.*

[5]In general, a zero velocity does not imply a zero acceleration. For example, a ball thrown vertically up comes to a momentary stop at its highest point, but the acceleration is not zero there. For the acceleration to be zero, the velocity should *remain* zero.

Flotation.

Submersed objects displace as much liquid as their own volume. However, a floating object is immersed only partially. Thus the volume of the liquid it displaces is not equal to its total volume. Nevertheless, the weight of the displaced liquid is equal to the weight of the object, making the total force zero, and the object in equilibrium. As one adds weight to the floating object, more and more liquid is displaced, with the weight of the newly displaced liquid always being equal to the newly added weight.

What do you know? 8.9. How does the weight of a floating submersed submarine compare with the sea water it displaces? How do you explain this, knowing that a submarine is made of heavy metals?

Liquid Pressure

To study fluids in more detail, we need to learn a new concept, **pressure**, denoted by P, which simply stated, is the amount of force exerted perpendicularly to a unit area. If F is the force exerted (in perpendicular direction) to a surface of area A, then the pressure is

Pascal is the scientific unit of pressure.

$$P = \frac{F}{A}. \tag{8.6}$$

The unit used in measuring pressure comes directly from its definition: N/m^2, which is also called **Pascal**.

What is density?

We most often speak of the pressure of a gas or a liquid. For such substances, pressure is connected to another property of the substance, **density**. In general, density is the amount of mass contained in a unit volume. So, if the mass of a quantity of matter is m and its volume is V, then its density ρ is given by $\rho = m/V$, and measured in kg/m^3. The densities of some common substances are given in Table 8.1.

Substance	Density (kg/m^3)	Substance	Density (kg/m^3)
Air	1.2	Iron	7,860
Aluminum	2,700	Lead	11,300
Copper	8,920	Uranium	18,700
Gold	19,300	Water	1,000

Table 8.1: The densities of some common substances.

Consider the column of water (or any other liquid) shown in Figure 8.7(a). This column is stationary. Thus, its acceleration is zero, implying that the total force on it is also zero. There is one obvious force acting on the column, the omnipresent gravitational force, mg, pointing downward. Since the total force on the column is zero, there must be a second *upward* force. This is provided by *the remaining body of water* in the form of pressure applied at the bottom surface of the column.[6]

Now suppose we make the column longer. Then its mass and, therefore, mg, will increase. To keep the column in equilibrium, the pressure at the bottom must increase as well. This increase is independent of the width of the column. The quantitative analysis of **Math Note E.8.4** on **page 82** of *Appendix.pdf* shows this independence clearly. The only thing that determines the pressure in a liquid is depth and the liquid density. Thus, we speak of pressure *at a point* inside liquid. Since the pressure at this point is determined only by the depth, we conclude that

---

[6]Although there is pressure on the lateral surface, the force producing it is perpendicular to that surface; hence, it does not act to balance gravity.

Box 8.3.2. *The pressure of a fluid increases with depth. All points at the same depth in a given liquid have the same pressure.*

We are situated at the bottom of a deep sea of air, and by the very argument given above, there is a pressure at such a depth. This is precisely what is called **atmospheric pressure**. Thus, in the analysis above, a third force is also present, the force of atmosphere exerted on the top surface of the column. This pressure results in a force that adds to the downward strength of gravity.

Atmospheric pressure.

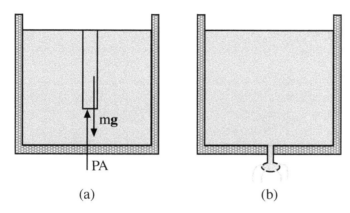

Figure 8.7: (a) Analysis of the origin of fluid pressure. The upward *force* balancing the weight of the column is PA. (b) Direction of fluid pressure.

Let us call the pressure at the surface of the liquid P_0. Usually this is just the atmospheric pressure as discussed in the previous paragraph. However, it is possible to apply extra pressure on the surface of a liquid, for instance, using a piston. Then the downward force on the column in Figure 8.7 increases and—since the liquid is assumed motionless—the pressure at the bottom must increase by exactly the same amount. This is true for columns of all sizes and shapes. Therefore,

Box 8.3.3. *Increasing P_0, the pressure at the surface, increases liquid pressure by the same amount everywhere.*

In which direction does the pressure in a liquid apply? In all directions! To see this refer to Figure 8.7(b) and note how the water gushes out in all directions due to this pressure. Besides, we just mentioned that liquid pressure is defined at a point, and a single point cannot determine any preferred direction for the pressure to point to.

Figure 8.8 shows a container with different vases all connected at the bottom. At the same depth in any part of the liquid, the pressure is the same. In particular, the surface of the liquid must be at the same level in different vases, because otherwise, at a given common depth, the pressure of the vase with higher liquid surface would be more than that of other vases. Therefore, all our conclusions and results are independent of the shape of the liquid, and consequently, of its container.

Consider the following experiment. Take two identical thick glass tubes each one meter long closed at the bottom, and open at the top. Fill them both with water and plug the tops with stoppers. Now flip the tubes and immerse the open ends in a vessel full of water

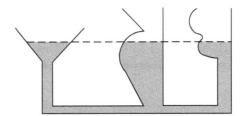

Figure 8.8: The shape of the vessel is irrelevant in the analysis of hydrostatics.

and then unplug the stoppers. As you move the tubes up and down in the vessel, they will remain full. In particular, it is possible to obtain a configuration shown in Figure 8.9(a).

> **What do you know? 8.10.** How do you resolve the paradox that the water levels of the tubes in Figure 8.9 are at different heights while we just showed, using Figure 8.8, that they should be at the *same* height?

Example 8.3.4. Instead of two identical tubes, take glass tubes of varying length as shown in Figure 8.9(b). Start with the length of 1 m, fill it with water and immerse the open end of it in the water vessel. Next take a 2-m tube and do the same thing. Then 3 m, 4 m, etc. Doing the experiment may become more and more awkward, but it stimulates the question: How long should the tube be before it stops remaining full once immersed in the water from the open end? This is a legitimate question, because points A, B, C, D, and E are all at the same level, and therefore, must have the same pressure, the atmospheric pressure. If I keep increasing the length of the tube, the pressure due to the weight of the water column will eventually reach the atmospheric pressure. So, for a sufficiently long tube, water will only partially fill the tube [Figure 8.9(c)], and for all tubes longer than that, the height of the water will remain the same, with the top of the tube containing *nothing*, a vacuum. This means that at the surface of the water in the long tube at this height, the pressure is (almost) zero. One can say that the atmospheric pressure is equal to the pressure of such a height of water. What is this height?

Barometry started by the Florentine well diggers and a student of Galileo.

Surprisingly, this height was discovered by Florentine well diggers who reported to Galileo that their suction pumps could not raise water to more than about 10 m. Suction pumps work by creating vacuum which reduces the pressure at the pump, causing the air pressure to push the water up the pipe. But if one places the pump at an elevation of 10 m or more, the atmospheric pressure cannot push the water up any further. To make things more manageable, Evangelista Torricelli (1608-1647), a student of Galileo, used mercury instead of water. Since mercury is about 13.6 times heavier than water, the column should be proportionately smaller. So, he expected the height in his tube to be about 73.5 cm. He took a tube 1 m long; filled it to the rim with mercury; held the open end of it while immersing it in a vessel full of mercury, and noticed that mercury dropped to a height of 76 cm (or about 30 inches), the remaining 24 cm (or about 9 inches) was empty. This was the beginning of barometry. ∎

Mercury barometer.

Figure 8.10 shows a crude picture of a **hydraulic jack**. Since pressure is the same at the same height, a little bit of force at A can support a huge weight at B. For instance, suppose the area at A is 400 cm^2 (a 20 cm by 20 cm square) and that at B is 4 m$^2 = 40,000$ cm^2. Then a girl weighing 50 pounds at A can support 5,000 lb, or about 2.5 metric tons at B! However, displacement of objects becomes inconvenient at B, as the ratio of displacements is the reciprocal of that of forces.[7] Thus, to raise the weight at B by a mere 1 cm, the girl at A should be pushed down a distance of 100 cm, or 1 m!

Hydraulic jack.

---

[7]This is due to the energy conservation: The work done by the force at A must equal the (negative of the) work done by the force at B.

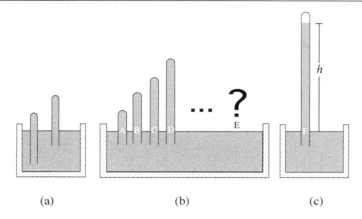

Figure 8.9: (a) The water levels of the two closed tubes are different. (b) A tube of water remains filled as long as it is not too long. (c) For a sufficiently long tube, the height h of the water produces a pressure at F equal to the atmospheric pressure. In the empty space on top of the tube, the pressure is (almost) zero.

8.3.2   Hydrodynamics

The application of the laws of motion to moving fluids can predict their behavior and properties. Instead of applying them directly, however, it is more economical to use the work-energy version of those laws. For simplicity, let's ignore hydrodynamic friction—also called viscosity—and study the flow of an incompressible liquid moving horizontally. The motion of such fluids is governed by what is known as the **Bernoulli principle**, whose details can be found in **Math Note E.8.5** on **page 82** of *Appendix.pdf*.

Bernoulli principle states that when the velocity of a fluid increases, its pressure decreases and vice versa. Thus by creating a velocity gradient, you can create forces that can do work for you. This is the principle behind airplane flight, where the wings are designed to have an air flow of higher speed on top and lower speed at the bottom [see Figure 8.11(a)]. The pressure at the bottom of the wing will be greater than at the top and the force larger, causing a lift.

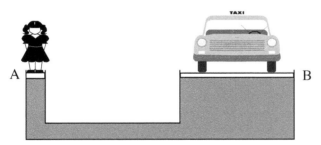

Figure 8.10: Schematic of a simple-minded hydraulic jack.

Example 8.3.5. This example looks at some of the most notable applications of Bernoulli principle.

A *horizontal* wind can turn an umbrella inside out [see Figure 8.11(b)]. As the wind blows, the air on top of the umbrella moves faster than the air under it, which, because of the curvature of the umbrella, is almost stationary. Because of the lower air speed underneath, the pressure there is higher than the pressure above. The high pressure is sometimes enough to turn the umbrella inside out.

Why umbrellas turn inside out in high winds.

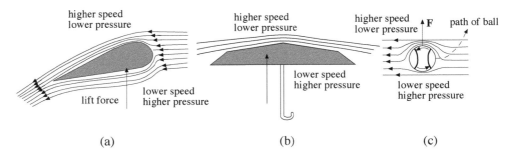

Figure 8.11: Examples of Bernoulli principle in action. (a) Airplane take-off. (b) An umbrella about to be turned inside out. (c) A spinning ball moving on a curved path.

Why spinning balls move on a curved path.
As a spinning ball moves through the air, it curves. To see why, refer to Figure 8.11(c), in which the motion is analyzed in the ball's reference frame. In this reference frame, air flows past the ball opposite to the direction of motion of the ball (remember the relativity of motion from Chapter 6). Let us assume that the ball rotates counterclockwise when looking at it from the top. Because of the friction between the ball and the air in contact with it, a layer of air is dragged along the ball. The motion of this layer is opposite to the air flow at the bottom, and in the direction of the air flow at the top. This causes the net velocity of the air flow to decrease at the bottom and to increase at the top. Bernoulli principle now says that the pressure should be larger at the bottom than at the top, causing the ball to curve upward. ∎

8.4   End-of-Chapter Material

8.4.1   Answers to "What do you know?"

8.1. The weight and the tension in the string are perpendicular to the circular path of the stone; therefore, the work of each of these two forces is zero. Air drag is opposite to the direction of motion; hence, it does negative work on the stone.

8.2. At the beginning the air drag is small; so the work done by it is also small and *negative*, because air drag points opposite to the direction of motion. The work done by the weight is larger than the drag work and it is positive. The net force (weight minus drag) points downward; so the net work is positive. Therefore, the change in KE is positive (parachutist speeds up).

8.3. At the beginning the air drag is large now; so the work done by it is also large and *negative*. The work done by the weight is smaller than the drag work and it is positive. The net force (drag minus weight) points upward; so the net work is negative. Therefore, the change in KE is negative (parachutist slows down).

8.4. The work done in each lift is *weight* of the load times displacement. Weight is $50 \times 9.8 = 490$ N. So, the work done in each lift is $490 \times 0.5 = 245$ J. Total work done is $245 \times 30 = 7350$ J. This is done in 120 seconds. Therefore, your power output is $7350/120 = 61.25$ Watts. Later, we will learn that each dietary calorie is 4180 Joules. So, each second you are losing $61.25/4180 = 0.0147$ calories! To lose 147 calories, you'll have to exercise for 10,000 seconds or about 3 hours![8] It is not easy to lose weight!

8.5. As long as the water is stationary in a bucket, the distance between its various parts does not change. But that does not imply that water is rigid. Rigidity requires the constancy of the distance between various parts when the system is subjected to *any* kind of motion. To see that water is not rigid, turn the bucket upside down and follow the motion

---

[8]This is an exaggeration, because it does not consider the other forms of calorie loss, such as internal metabolism.

of various parts of the water.

8.6. The wrench would be simply rotating about its center of mass.

8.7. With a long pole (especially if a weight is added at each end), the tightrope walker acquires a large moment of inertia, and it will be harder to rotate him (or tip him over).

8.8. The chair, the person, and the bicycle wheel constitute a system with an initial angular momentum concentrated in the wheel. Reorienting the shaft changes the direction of the angular momentum of the wheel. Since the *total* angular momentum is conserved, the system starts to rotate in such a way as to compensate for the change in the angular momentum caused by the reorientation of the shaft.

8.9. The weight of the submarine equals the weight of the water it displaces. Although the body of a submarine is made of heavy metals, inside it is filled with air, which is much lighter than sea water. The overall weight can be adjusted to be equal to the sea water displaced. This is done by injecting or ejecting water from its reservoir.

8.10. The water, which tries to reach the height of about 10 m, encounters the resistance of the closed end of the tube; it exerts an upward force on the tube, and the tube, obeying the third law of motion, reacts by exerting a downward force on the water. This force translates into a pressure, which, when added to the pressure of the column of water, produces the atmospheric pressure at the water level [points A, B, etc. in Figure 8.9(b)]. The shorter the tube is the larger the pressure at the closed end, so that the total pressure at the water level is always the atmospheric pressure.

8.4.2 Chapter Glossary

Angular Momentum The analog of ordinary momentum in rotational motion. One can call it the "quantity of rotational motion." If the rotation takes place about a fixed axis, then angular momentum of a system is the sum of the product of the momentum of each constituent of the system and its distance from the axis.

Archimedes' Principle A principle—derivable from Newton's laws of motion—stating how a liquid exerts an upward force on an object immersed in it.

Atmospheric Pressure The pressure at some point on the surface of the Earth due to the weight of the column of air (extending over the thickness of the atmosphere) located at that point.

Bernoulli Principle A principle in fluid dynamics relating the pressure of a fluid to its speed.

Center of Gravity Same as center of mass.

Center of Mass A point of a rigid body whose motion under the influence of the external forces describes the motion of the rigid body if the entire body were concentrated at that point. Sometimes called **center of gravity** because it is the point at which you can balance the rigid body on a pivot.

Conservation of Angular Momentum A principle similar to the first law of motion. It states that if a system is isolated, then its total angular momentum does not change.

Conservation of Energy A deep fundamental principle of physics, which states that energy cannot be created or destroyed. When restricted to kinetic and potential energy, it becomes **conservation of mechanical energy**, which states that, in the absence of other forms of energy, the sum of the potential and kinetic energies is constant.

Conservative System A system for which conservation of mechanical energy holds.

Density Mass divided by the volume it occupies. Density measures the concentration of matter in a region.

Horse Power A unit of power defined as the amount of work an ideal horse performs in one minute. Officially, a horse power is equivalent to 745.7 Watts.

Hydrodynamics The study of fluids (liquids and gases) in motion.

Hydrostatics The study of fluids (usually liquids) at rest.

Joule Scientific unit of work and energy, named after the English physicist who showed that heat is a form of energy.

Kilo Watt Hour A unit of energy used commercially, especially by electric power companies. It is the energy consumed or produced in one hour by a consumer or producer whose power is 1000 Watts.

Kinetic Energy The energy associated with the motion of an object. It is *defined* in such a way that, when the work of the force on the left-hand side of the second law of motion is calculated, the right-hand side gives a change in the kinetic energy.

Mechanical Energy Energy associated with the mechanical motion and forces. It is the sum of the potential and kinetic energies.

Pascal Unit of pressure. Same as N/m^2.

Potential Energy Energy which has the "potential" of turning into kinetic energy. Potential energy is defined for the so-called conservative forces, of which gravity is one.

Power The rate of production or consumption of energy. The number of Joules produced or consumed in a second.

Pressure Normal force exerted on a surface divided by the area of that surface.

Right-hand Rule (RHR) A rule that associates a straight-line direction with a rotational motion: Curl the fingers of your right hand along the direction of rotation, your thumb points in the direction of the straight line. For example, when rotation is about a fixed axis, the rule gives a direction along that axis.

Rigid Body An object for which the distance between its constituents does not change when (mild) forces are applied to it. Stated differently, a rigid body does not change shape under the application of (not too strong) forces.

Spin The angular momentum or the rotational motion of a rigid body about an axis (usually an axis of symmetry) passing through the body.

Watt Unit of power, named in honor of the British inventor James Watt who improved the steam engine.

Work A physical quantity defined for a force. It is the product of that force and the displacement of the object on which the force acts.

8.4.3   Review Questions

8.1. What is the definition of work in physics? What is the unit of work? Whom was it named after?

8.2. Can work be negative? How? Is it possible for a force to be displaced without doing any work? Explain!

8.3. What is kinetic energy? How is it related to mass and velocity? How is it related to work?

8.4. What kind of energy is associated with the work done by gravity? How does it depend on height? Does it matter where one measures the height from?

8.5. Is potential energy conserved? Is kinetic energy conserved? Is total mechanical energy conserved?

8.6. Give a practical application of the conservation of mechanical energy. What happens to the kinetic energy when potential energy increases?

8.7. What is power? Is kilo Watt hour a unit of power or energy? What is horse power? How many Watts are there in a horse power?

8.8. Would you be able to see Santa Claus as he moves down and up a chimney? As he moves from one house to another? Why or why not?

8.9. Approximately how many times the annual world consumption of energy does Santa use for his one-night trip?

8.10. What is an explosion? How strong are Santa's explosions as he goes down and up the chimneys? How do these explosions compare with the explosion caused by his landing at a chimney? How does the latter compare with the atomic bomb dropped on Hiroshima in WWII?

8.11. What is a rigid body? What point of the body describes the translational motion of the body?

8.12. What is the simplest kind of rotation? What is angular momentum for a rotation about a fixed axis? How is its direction determined? What is the right-hand rule?

8.13. What is moment of inertia, and how is it related to angular momentum? What happens to angular velocity if you increase the moment of inertia of a rotating system?

8.14. What is a fluid? What branch of fluid mechanics studies fluids at rest?

8.15. State Archimedes' principle. What is the name of the force involved in that principle?

8.16. What is pressure? What is its unit? How is the pressure of a fluid related to the depth in that fluid? What is the atmospheric pressure?

8.17. What branch of fluid mechanics studies fluids in motion? State Bernoulli principle.

8.18. How can a *horizontal* wind turn an umbrella inside out? What does Bernoulli principle have to do with flight of airplanes?

8.4.4 Conceptual Exercises

8.1. You will learn later that in a gas mixture, all molecules have the same KE. Which molecules move faster, the lighter ones or the heavier ones?

8.2. If you double the velocity of an object, by what factor does its kinetic energy increase? Its momentum? If you double the mass of an object, by what factor does its kinetic energy increase? Its momentum? If you double both the velocity and the mass of an object, by what factor does its kinetic energy increase? Its momentum?

8.3. The magnetic force on a moving electrically charged particle is always perpendicular to the velocity of the particle. Explain why magnetic forces cannot change the kinetic energy of such particles.

8.4. Recall that the centripetal acceleration of an object in uniform circular motion points toward the center of the circle and that the velocity vector is tangent to the circle. Now explain why the force of gravity of Earth does no work on a satellite moving on a circular orbit.

8.5. Which is easier to stop, a bicycle or a car moving with the same speed?

8.6. Which requires more work to stop, a bicycle or a car moving with the same momentum? Hint: The KE of an object of mass m and momentum p is given by $KE = p^2/2m$.

8.7. Why do long-range cannons have longer barrels? Hint: Think of the exploding gas pushing with a constant force on the cannon ball and having no way of escaping except at the opening of the barrel.

8.8. Consider a pendulum consisting of a string tied to a ceiling at one end with a bob attached to the other end. As the bob moves, its KE changes. Which force is responsible for this change in KE, the weight of the bob or the force of the string?

8.9. A real roller coaster has 150,000 J of potential energy at the highest point where it starts from rest, but only 145,000 J of KE at its lowest point. What happened to the other 5000 J?

8.10. Is it possible for a roller coaster to have a summit higher than the very first summit on which it starts?

8.11. Consider an ideal world in which there is no friction, no air drag, no retarding forces whatsoever.
(a) Do you need any fuel to speed up your car from rest to some given speed?
(b) Do you need fuel to keep the car moving at that given speed?
(c) Why do you need fuel to keep the car moving at constant speed in the real world?
(d) Does this contradict the fact that the net force on the car is zero, and therefore, that the net work is also zero?

8.12. In the absence of air resistance, a ball thrown upwards reaches a certain ideal maximum height. When air resistance is present, does the ball, thrown with the same speed, reach a height that is larger, smaller, or equal to the ideal height?

8.13. Take two identical balls and go to the roof of the Sears Tower. Throw the first one downward. Throw the second one upward with the same speed. How do you compare the KE of the two balls when they both land on the street? Hint: What is the speed of the second ball when it reaches the rooftop on its way down?

8.14. A car changes its speed from 30 mph to 45 mph. A second car, identical to the first, changes its speed from 40 mph to 55 mph. Do they change their momentum by the same amount? If not, which car's momentum changes more? Do they change their KE by the same amount? If not, which car's KE changes more?

8.15. A stationary firecracker explodes into pieces. The total momentum of the flying pieces is zero (right?). Is the total KE of the pieces also zero? Why or why not?

8.16. Two identical lumps of clay move with the same speed in opposite directions. They stick together after they collide.
(a) Is momentum conserved? If not, where did the difference go?
(b) Is KE conserved? If not, where did the difference go?

8.17. What happens if out of the allowed time per house, Santa spends too much in the chimney? What if he spends too little time in the chimney?

8.18. We have assumed that Santa reaches his plunge, climb, and hopping speeds instantaneously. What kind of acceleration does this correspond to? Would it change the amount of energy consumed if acceleration were smaller?

8.19. A rigid body is rotating about an axis passing through it.
(a) Do all parts of the body move with the same speed?
(b) Do they move with the same *angular* speed?
(c) What can you say about the speed of the parts of the rigid body as you move away from the axis of rotation?

8.20. Which object has a larger moment of inertia, a disk or a spoked wheel of equal mass (ignoring the mass of the spokes)?

8.21. Take a very thick and viscous liquid and rotate it about an axis passing through the liquid.
(a) Do all parts of the liquid move with the same speed?
(b) Do they move with the same angular speed? Does this clarify the difference between a liquid and a rigid body?

8.22. The water of some lakes has a large concentration of salt, to the extent that people don't sink in the lake. Explain this using Archimedes' principle.

8.23. Using Archimedes' principle explain why a helium-filled balloon rises, but a balloon filled with air falls.

8.24. Why is the bottom of a dam thicker than its top?

8.25. Why does a 300-lb male not make a dent in the linoleum of a kitchen, but a 100-lb female on spike heels does?

8.26. Fill a one-meter thin glass tube with mercury and invert it into a vessel full of mercury. The height of the mercury column in the tube will be about 76 cm. Now fill a huge glass tube, the diameter of whose cross section is one meter, and do the same thing. Would the height of the mercury column be 76 cm or less?

8.27. Take a sheet of paper, hold it vertically and blow on one side of it parallel to the sheet. The paper bends toward the side you are blowing. Why?

8.28. When taking a shower, have you noticed that the curtain moves toward the water coming out of the shower head? What is the reason?

8.4.5 Numerical Exercises

8.1. A car has a total mass of 1500 kg and is moving with a speed of 40 mph.
(a) What is the kinetic energy of the car?
(b) How much work is done by air drag and friction (net force on the car) to stop the car?

8.2. A roller coaster starts from rest at point A. The mass of the car plus the passenger is 250 kg (see Figure 8.12). Assume that $h_1 = 80$ m, $h_2 = 65$ m, and $h_3 = 30$ m.
(a) What is the total mechanical energy at A?
(b) What is the total mechanical energy at B? What is PE at B? What is KE at B? Speed at B? What happens to the KE and speed at B if you assume that the reference level is 10 m below B?
(c) What is the total mechanical energy at C? What is PE at C? What is KE at C? Speed at C?
(d) What is the total mechanical energy at D? What is PE at D? What is KE at D? Speed at D?
(e) Redo the problem for a mass of 300 kg. Are the new answers consistent with your expectation? (Think of what would happen if different cars moved with different speeds!)

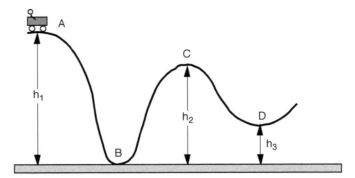

Figure 8.12: The roller coaster rolls and coasts!

8.3. A 100-gram bullet moving at 80 m/s hits a stationary 2-kg block of wood and penetrates into it. Use conservation of momentum to find the speed of the block (with the bullet inside it). From this, find the KE of the block (with the bullet inside it), and compare this KE with the initial KE of the bullet. Why are they different?

8.4. Suppose that there are only 5 houses that Santa has to visit, and to each house he has to deliver 10 kg of toys. Assuming that Santa's speed is to be 100 m/s, calculate the total hopping kinetic energy by adding the KE for each house. What kind of effective mass would yield the same value for the total KE? How close is this effective mass to half the initial mass? If you increased the number of houses to 10, how close would the effective mass be to half the initial mass? Can you see why when the number of houses increases the effective mass becomes close to half the initial mass?

8.5. Two trucks 80,000 lb (36 metric tons) each, moving at 75 mph collide head-on, with the wreckage being at rest after collision. Each kilogram of TNT has 4 million Joules of destructive energy.
(a) What is the KE of each truck?
(b) What is the total KE?
(c) How many kg of TNT is the destructive energy of collision?

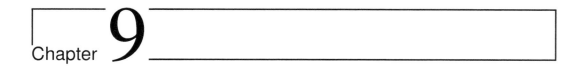

Chapter **9**

Gravitation

There are four fundamental forces in nature: gravity, electromagnetism, strong nuclear, and weak nuclear. Of these four forces, gravity is the first to be discovered and the least understood. Although we have a good understanding of the behavior of the gravitational force on a large scale, i.e., as applied to celestial bodies, when we try to unravel the mysteries of this force in the domain of the atoms and subatomic particles, we fail. This failure manifests itself most clearly in the attempts to reconcile gravity with the theory that most successfully describes the behavior of the subatomic world, the quantum theory. The process of reconciling a classically observed force with the quantum theory is called quantization. It has been possible to reconcile electromagnetism with the quantum theory with great success (as Chapter 32 testifies), so we say that electromagnetism has been quantized.

The nuclear forces have no classical analog. They live only in the quantum domain, inside the nucleus of atoms. Therefore, if there is any successful theory for the nuclear forces, it has to be a quantum theory. In fact, beginning in 1973, a very successful theory started to emerge which could explain the behavior of nuclear forces. This so-called **standard model** of electromagnetic, weak, and strong forces (which is the topic of Chapter 34) has passed all the rigorous tests of experimental scrutiny with flying colors. The standard model also has the extra aesthetic appeal of unifying the electromagnetic force with the weak nuclear force.

Despite the invention of many ingenious techniques, and the attempts by some of the greatest minds of contemporary physics, a satisfactory quantized theory of gravity is still lacking. Such a theory is actively pursued for many reasons. It is believed, for instance, that a knowledge of gravity on a subatomic scale will lead not only to a better understanding of gravity itself, but also to a more complete description of the other three forces. In fact, there are many experimental and theoretical indications that the four forces are but different manifestations of a single "super force." To discover this super force, we must first unravel gravity at the subatomic level. Moreover, the theory of the origin of the universe, the **Big Bang theory** (the topic of Part VIII of this book), will not be complete unless we know how the fundamental particles, the ingredients for the formation of the universe, interacted with one another right after the bang when the universe was infinitely hot and infinitely small. Such a knowledge requires the quantization of gravity.

This chapter deals with gravity on a large scale. There are two theories in use: Newton's theory of gravity which is adequate for all "ordinary" applications such as the motion of planets, satellites, meteors, etc., and Einstein's **general theory of relativity** (GTR) which is appropriate for all ordinary as well as extraordinary applications such as black holes, big bang theory, and GPS (global positioning system), in which colossal masses, very large speeds, and extremely high clock precision are to be dealt with. Only Newtonian theory is discussed in this chapter. The general theory of relativity is the topic of Chapter 29.

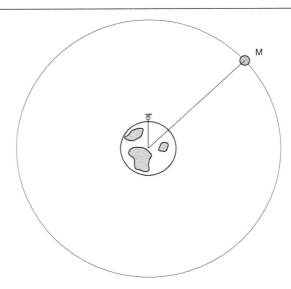

Figure 9.1: The distance of the apple from the center of the Earth is (very nearly) equal to the radius of the Earth, i.e., 6400 km. The Earth-Moon distance is 384,400 km. Distances (and sizes) are not drawn to scale.

9.1   The Universal Law of Gravitation

Gravity and electricity are the only two of the four forces that had displayed themselves on this planet continuously long before the oldest human civilization, long before the oldest known human species, even long before the first living cell started to scull in the rough waters of the primitive Earth. The fall of free objects above the ground, and the phenomenon of lightning are the two manifestations of gravity and electricity which are as old as the Earth itself. The irregularity of lightning hampered our understanding of electricity for over a century longer than the unraveling of gravity by Sir Isaac Newton in 1665.

Newton *did not* discover gravity!

It is said that Newton discovered gravity when he saw an apple fall from a tree. This is a wrong statement. Newton did not discover gravity! Gravity was discovered long before Newton. Aristotle, for instance, talks of two forces, gravity and levity, with the former being the tendency to fall and the latter, the tendency to rise. What Newton did, was to discover the *mathematical law* governing the gravitational force. Whether he used an apple to discover this law is not important. What is important—and this requires a genius and an incredible insight—is that Newton saw a Moon in an apple and an apple in the Moon. The following example explains.

Example 9.1.1. To an ordinary person, the motions of an apple falling to the ground and the Moon circling Earth 60 times farther than the apple, have nothing in common (Figure 9.1). But Newton saw the connection between the two in that they are both moving under the influence of gravity alone. Having brilliantly discovered this connection, Newton calculated the acceleration of the Moon using the formula $a = v^2/r$ for the centripetal acceleration. The speed v is simply the circumference of the Moon's orbit divided by its period, both of which had been known ever since the Greeks' golden age. Newton discovered that the Moon, at a distance 60 times larger than the apple's distance from the Earth's center, had an acceleration that was about $3600 = 60^2$ times *smaller* than the apple's. He concluded that the acceleration due to gravity must decrease as the square of the distance from the center of the gravitating object. **Example D.9.1** on **page 26** of *Appendix.pdf* supplies the numerical details. ∎

Newton might have come up with his universal law of gravitation through an argument similar to the one given in the foregoing example, and in the process, he must have had the keen insight to detect the sameness of the acceleration of the Moon and that of an apple.[1] Any similarity between the fall of an apple directly downward, and the motion of the Moon thousands of miles away circling the Earth, goes undetected by ordinary people. Yet this very similarity is the first giant step toward unraveling the mysteries of gravity, a step that only a devoted genius such as Newton could take.

If the gravitational acceleration caused by Earth is proportional to the inverse of the square of the distance, then the gravitational force on an object of mass m, which by the second law is $F = ma$, must be proportional to $m/(distance)^2$. Now let M denote the Earth's mass. Then the third law of motion implies that m applies the same force on M. Thus, switching the role of M and m, we conclude that the gravitational force must also be proportional to M, i.e., proportional to $mM/(distance)^2$. Supplying a constant of proportionality, denoted by G, Newton arrived at the law of gravitation between the Earth and a terrestrial object, $F = GMm/r^2$, where M is the mass of the Earth, m that of the object pulled by the Earth, and r the "distance" between them. Newton realized that this law could be generalized to act between *any* two objects. Calling the masses of the two objects m_1 and m_2, the **universal law of gravitation** becomes:

An important application of the third law of motion.

Universal law of gravitation.

$$F = \frac{Gm_1m_2}{r^2} \quad \text{and} \quad F = \frac{GMm}{r^2}, \tag{9.1}$$

where the second equation is usually used when a celestial body of mass M influences the motion of an ordinary object of mass m.

The constant G, called the **universal gravitational constant**, was not determined until 1798 when the British physicist Henry Cavendish measured it. The reason for such a lapse of time is that to measure G, one had to measure the two masses m_1 and m_2, the force F, and the distance r between them. The celestial determination of these quantities was impossible (one had to measure the masses of, say the Sun and the Earth) while the terrestrial determination of the quantities proved to be very elusive due to the extremely small force of gravity between ordinary objects (see Example 9.1.3), which in turn was due to the fact that G itself is too small.

How big (or small) is G?

Example D.9.2 on **page 26** of *Appendix.pdf* shows that, in the scientific system of units in which length is measured in meters, time in seconds, and mass in kilograms, the value of G must lie somewhere between 1.82×10^{-11} and 3.65×10^{-10} based on some very general observation. But, with extreme care and extraordinary patience, Cavendish was able to measure the actual value of G. His value, in the scientific units we use in this book, was slightly different from today's accepted value of 6.6742×10^{-11}.

🧺 Food for Thought 🎋

FFT 9.1.2. Some philosophers believe that there is something fundamentally wrong with science because its theories and its observations are not independent. They dismiss any kind of observation as *theory-laden*. They think that if an observation is not completely independent of the theory it is testing, then both the theory and the observation are invalid. And if they are right, then the determination of the universal gravitational constant G is one of the biggest blunders in the history of science, because Cavendish *assumed* that Equation (9.1) was correct, and based on this assumption, he measured m_1 and m_2 (using a scale), r (using a tape measure), and F (using a newly invented instrument called a *torsion balance*). Then he *used* Equation (9.1) to *calculate* G.

And Cavendish was not alone in intertwining observation with theory. In fact, *no observation or experimentation is done that does not incorporate one or more theories*. Take the simplest experimentation, the measurement of the distance between two points. When you use a straightedge to do so, you are *assuming* that only one straight line segment can be drawn between the two points,

A simple distance measurement is "theory-laden."

---

[1] And to speculate that the Earth's mass could be thought of as concentrated at its center so that the "distance" from the Earth to the apple could be compared to that of the Moon. This speculation led Newton to the invention of integral calculus.

and that you are measuring the length of that unique line segment. This assumption is at the heart of one of the oldest theories in existence, the Euclidean geometry.

Modern experimental physics is even more intertwined with theory. When an X-ray machine is used to examine an atom, the boundary between observation and theory becomes completely blurred. On the one hand, the behavior of an atom obeys the laws of quantum theory, and the examination of an atom is ultimately a test of this theory. On the other hand, the construction of an X-ray machine is entirely based on the quantum theory.

The inseparability of observation and theory is the key to the advancement of science. It is always breakthroughs in theoretical understanding of nature that suggests new observations for whose rendering the accumulated theoretical knowledge of humanity becomes indispensable. And the breakthroughs in observation bring about conditions under which new theories could be born.

All observations are theory-laden. This is how science works!

So, if Cavendish *used* Newton's universal law of gravitation to *measure* the value of G, he was following the most fundamental principle of scientific investigation. The important point is that all such observations and theories fit nicely into a framework in which a myriad opportunities exist to check and recheck the experimental and theoretical results. Example D.9.2 shows how methods completely independent of those used by Cavendish—i.e., the size and the density of the Earth—put stringent upper and lower limits on the value of G.

People don't attract people ... gravitationally!

Example 9.1.3. The gravitational force between two ordinary objects is extremely small. For example two cars each having a mass of 1 ton (1000 kg), whose centers are separated by a distance of 1.5 m exert a gravitational force of only 2.96×10^{-5} N, as can be easily calculated using Equation (9.1). Compare this with the Earth's gravitational pull on a penny which is about 0.03 Newton! That is why people are not attracted to each other ... not gravitationally, any way! ■

The gravitational force due to the Earth on an object of mass m is $F = GmM/r^2$. This force, like any other force, is related to the acceleration by $F = ma$. These two equations imply that $ma = GmM/r^2$ or $a = GM/r^2$. It is more customary to denote the *gravitational acceleration* by g. Therefore, we write

$$g = \frac{GM}{r^2}. \tag{9.2}$$

Although this equation was derived for the Earth, it is easily seen that the derivation is completely general and applies to the acceleration caused by an arbitrary (spherical) mass M. This result is a profound property of nature which, as we shall see, is at the heart of one of the most beautiful theories ever discovered, the general theory of relativity. The profound property is

> **Box 9.1.4.** *The acceleration due to gravity is independent of the mass of the object being accelerated.*

A grain of sand, a penny, a cannon ball, and a satellite all accelerate at the same rate—if they are at the same distance, of course—towards the gravitating body. Such an implication has been tested again and again. The first test was performed by Galileo in his famous experiment at the leaning tower of Pisa. The most famous—and still one of the most accurate—experiment was performed in 1889 by Eötvös, a Hungarian physicist, in which he showed that the variation in the gravitational acceleration of various objects is less than a few parts in a billion! More recent measurements set an upper bound for this variation in the acceleration at values between 10^{-11} and 10^{-13}.

No other fundamental force of nature has the property stated in Box 9.1.4. For instance, accelerations due to the electrical forces are sensitive to the mass of the accelerated object. In fact, one can say that gravity is *that* fundamental force whose acceleration is insensitive to variations in mass.

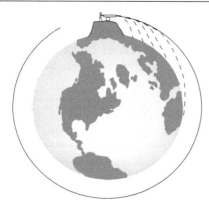

Figure 9.2: By constantly increasing the horizontal speed of the projectile, one can eventually reach an orbiting speed.

Another important property of gravitation is the fact that it is *always attractive*, unlike the electric force, which can be attractive or repulsive depending on the sign of the electric charges. This is because the *source* of gravity, mass, is always positive, while the source of electricity, electric charge, can be both positive and negative.

What do you know? 9.1. Go to Mt. Everest and drop an apple. Fire another one horizontally fast enough to orbit the Earth. Which apple has a larger acceleration?

Example 9.1.5. We can now "weigh" the Earth as follows. At the surface of the Earth $g = 9.81$ m/s^2 and $r = $ radius of the Earth $\approx 6,400$ km. Thus, using Equation (9.2), we obtain $9.81 = 6.67 \times 10^{-11} \times M/6,400,000^2$, from which we can calculate M. The value we obtain is $M = 6 \times 10^{24}$ kg. Notice how kilometer had to be converted to meter in Equation (9.2). This is something you should keep in mind whenever you do a numerical exercise: Any formula used in this book requires that length be measured in meter, time in second, and mass in kilogram. ∎

Weighing the Earth!

What do you know? 9.2. Inflate Earth so its radius is doubled. What is the acceleration at the surface of the new Earth?

9.1.1 Satellite Speed

Newton saw the connection between a circulating Moon and a falling apple, and discovered the law of gravity. He then proposed the possibility of "apples" circulating the Earth at low altitudes. He imagined a very tall mountain peak, on which was mounted a cannon, firing cannon balls horizontally with varying speeds (Figure 9.2). As you increase the speed of the projectile, its range increases. If the speed is large enough, the range will be so huge that the ball will miss the Earth completely. We then have an orbiting cannon ball.

We can find a general expression relating the speed and the orbital radius of a circulating object going around a celestial body of mass M. On the one hand, we have Equation (9.2) which gives the acceleration of the object in terms of the mass of the celestial body. On the other hand, we know that this acceleration, being a centripetal acceleration, is given by $g = v^2/r$. Equating these two forms of the same acceleration yields $GM/r^2 = v^2/r$ implying $GM/r = v^2$, or finally

$$v = \sqrt{\frac{GM}{r}}. \tag{9.3}$$

Equation (9.3) shows that the farther the object is from the gravitating mass, the slower it has to move to stay in orbit. Low-altitude satellites (low compared to the Earth radius) have a speed of about 8 km/s or approximately 18,000 mph. **Example D.9.3** on **page 27** of *Appendix.pdf* finds the speed of a satellite at an altitude of 100 km.

> **What do you know? 9.3.** If you want a satellite to orbit Earth at an altitude of $3R$ above the Earth's surface, how fast should it be moving?

> **What do you know? 9.4.** Is it possible to have a satellite circling Earth at 10 km/s? Why or why not? Remember that the radius of Earth is 6400 km and its mass is 6×10^{24} kg.

9.1.2 Kepler's Third Law

The Newtonian laws of motion and the theory of gravity, when applied to the solar system, yields Kepler's three laws of the planetary motion. To derive the first and second Kepler laws from Newton's law of gravity requires higher mathematics (differential equations). However, the third law—for the special case of a *circular* orbit—can be derived with little (mathematical) effort, as done in **Math Note E.9.1** on **page 84** of *Appendix.pdf*. The result is:

$$T^2 = \frac{4\pi^2}{GM}r^3. \tag{9.4}$$

So the k in Box 3.3.1 is $k = 4\pi^2/GM$.

🧺 Food for Thought 🍂

FFT 9.1.6. Look at Equation (9.4)! Look at it carefully! It is a collection of letters, numbers, and symbols. It contains the strange symbol "=" and a horizontal line with some "stuff" on top and bottom of it. It is a product of the human mind, which also invented the letters and the numbers that go into the equation. It is the collective invention of generations of thinkers working in their solitary rooms day after day and night after night, guided by the information collected by generations of observers, looking painstakingly at heaven and Earth, and gathering glimmers of knowledge from them. In short, Equation (9.4) is one of the collective masterpieces of our race.

It also contains "G." A letter. But what's in the letter? A message. A universal message encrypted in the cosmic dance of the Moon, planets, solar system, and beyond. To decipher this message you need training, devotion, and perseverance . . . lots of all three! You also need a torsion balance. But that comes in the package if you are trained and devoted beyond imagination and persevere equally. All this converged in the person of Henry Cavendish, who, in his laboratory, on a tiny speck of the planet Earth called London, deciphered the secret universal message hidden in G. Once this secret was revealed, humans acquired a cosmic arm with which they could reach into the farthest corners of the cosmos and *touch* the planets, stars, and galaxies out there.

Contact with the cosmos is not a new phenomenon. Ancient astronomers *looked* at the heavens, collected information from it, and when the information fell into the hands of the Greek scientists, they perfected the pidgin in which Egyptians and Babylonians communicated with nature, into an abstract language. Mathematics became the means of communication with heavens, and the cosmos taught the Greek scientists the basics of this new language.

How remarkable that Aristarchus put this new language to use in his measurements of the sizes and distances of the most conspicuous objects in heaven! Theoretical astronomy extends our sense of vision: the local earthly sizes and distances, which we can estimate using our eyes, are extended to heaven by merging the use of our vision with mathematics. But theoretical astronomy, by itself, cannot go beyond that. To extend our other senses, the tactile sense, for example, a more earthly science and a more sophisticated language is needed. This earthly science started with Galileo and

Newton, and its more sophisticated language was developed by Newton and his contemporaries. This earthly science is physics.

Equation (9.4) is more than just Kepler's third law. It contains the mass M of the Sun. Therefore, knowledge of the period of a planet and its distance from the Sun can lead to a measurement of the Sun's mass! **Example D.9.4** on **page 28** of *Appendix.pdf* uses the Earth's period and its distance from the Sun to weigh the Sun. The mass of the Sun turns out to be 2×10^{30} kg or about 330,000 times the mass of the Earth.

Weighing the Sun!

Another useful application of Equation (9.4) is in telecommunication. Communication satellites are most useful if they remain stationary relative to the Earth, in which case they are called **synchronous satellites**. The orbit of the satellite is a circle centered at the center of the Earth. The satellite must move with the surface of the Earth to stay stationary relative to it. But all points on the Earth move on circles whose planes are perpendicular to the axis of rotation of the Earth (the axis connecting the north to the south pole). Thus, the plane of the orbit of the satellite must also be perpendicular to this axis. The only plane that is perpendicular to the axis of the Earth and goes through its center is the plane of the equatorial circle. Therefore, the satellite must go around a circle concentric with the equator. If we want to receive signals from such satellites, we must point our satellite dishes to the sky above the equator. In the northern hemisphere, this means that our dishes must point to the southern sky. In the southern hemisphere, the dishes point to the northern sky. **Example D.9.5** on **page 28** of *Appendix.pdf* shows how to find the altitude of a synchronous satellite.

Synchronous satellites used in communications.

9.1.3   Dark Matter

Equation (9.4) or (9.3) ties three quantities together: mass M of the gravitating body, speed v (or period T) of the circulating object, and the distance r of the circulating object from the gravitating body. Knowing any two of these quantities determines the third one. For example, a knowledge of the distance and the mass determines the speed; this is how we calculated the speed of a satellite circling the Earth (see Example D.9.3). Similarly, the period and the distance of Earth from the Sun determined the mass of the Sun (Example D.9.4). The application of the (equivalent) formulas (9.4) or (9.3) to the cosmos has led to one of the most exotic and enigmatic conclusions in the history of cosmology.

Example 9.1.7. Suppose you look at a star, and based on certain of its visible properties, you determine its mass. Now, you look at one of its planets, determine its distance from the star and the period of its revolution around the star. You plug this information in Kepler's third law and note that it doesn't work; that the period is much smaller than it should be. Since Kepler's third law is on a firm theoretical and observational footing, we have no choice but to accept it. The value of G has been determined to a very high accuracy; so we have no room to fudge it. We are therefore left with r, M, and T. The first and the last quantities can be measured directly and accurately. The period can be measured by charting the planet's position in the field of a super-powerful telescope (such as the Hubble Space Telescope). The distance can be measured accurately by finding the angular separation of the star and the planet once the distance of the star from Earth is known.

The only alternative left for explaining the discrepancy is to assume that the mass of the star is not necessarily its *visible* mass. If we assume the validity of Kepler's third law, and the accuracy of the distance and period, we must conclude that the actual mass of the star is more than the visible mass and that much of the mass of the star is *invisible*! **Example D.9.6** on **page 28** of *Appendix.pdf* adds some numerical flavor to this discussion. ∎

The invisible mass of the foregoing example has come to be known as **dark matter**. There is no dark matter surrounding a star; but the essence of the argument for the existence of dark matter is as outlined in that example; the details are slightly different. Instead of a planet moving around a star, one has to focus on stars in the galactic arms moving around the core of a galaxy.

Dark matter explained.

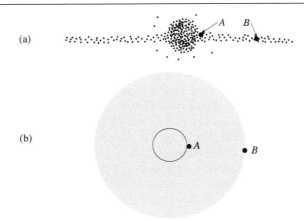

Figure 9.3: (a) A typical galaxy and two stars moving around its huge core. (b) Star A experiences only the force of gravity of the mass inside the small sphere, while star B experiences the force of gravity of the entire mass of the large sphere.

According to Equation (9.3), the farther a star is from the mass M of the visible part of a galaxy, the slower it should move. Observation, however, indicates that

Box 9.1.8. *Beyond a certain distance from the core of the Milky Way, all stars move at the same speed* regardless of their distance *from the center of the galaxy.*

Squaring (9.3), rewriting the result as $M = (v^2/G)r$, and noting that v is a constant, we have to conclude that M increases as r increases! The visible stars contained in the sphere of radius r cannot account for this increase.

To augment your understanding of dark matter refer to Figure 9.3. Two stars A and B revolve around the massive core of a galaxy. Neglecting the thin arms of the galaxy, star B, being four times farther away than A from the core, must move at half the speed of A according to Equation (9.3). To move as fast as A, B must experience the force of gravity of four times more mass than A is experiencing.[2] Therefore, ***the distance between the center of the galaxy and B must be filled with an invisible mass, i.e., dark matter.***

You may ask "Why does A not experience the gravity of this mass?" **Math Note E.9.2** on **page 84** of *Appendix.pdf* shows that *the gravitational force inside a spherical cavity dug within a spherical distribution of mass is zero.* Now, think of the mass in the large sphere of Figure 9.3(b) as consisting of a solid sphere in the middle (extending right up to A) surrounded by a thick spherical shell. Then A, being in the cavity of the spherical shell will not experience any gravitational force due to the shell; only the small sphere in the middle attracts A.

The gravitational force inside a spherical cavity is zero.

9.2   Gravitational Field and Potential Energy

Newton's universal law of gravitation states that a mass exerts a force on another mass located at a distance from the first. This is called **action-at-a-distance** and embodies a frame of mind in which it is possible, without an intermediary agent, for an object to influence another distant object. Since there is no agent to "transmit" the force, it is

---

[2]So that the 4 of M in the numerator of (9.3) cancels the 4 of r in the denominator.

understood that a change in the state of the source of gravitation is instantaneously reflected in the force on the second mass.

9.2.1   Gravitational Field

The "telepathic" transmission of information alluded to above was indigestible to the nineteenth century physicists, especially when they tried to apply it to the electrical phenomena. Thus, in the hands of the English physicist Michael Faraday, and the Scottish physicist James Clerk Maxwell, the untouchable action-at-a-distance was replaced by the more tangible notion of **field**.

Very generally speaking, any force has a field associated with it. We speak of gravitational field, electric field, magnetic field, and nuclear fields. If, through experimentation, we find that there is a gravitational *force* $\mathbf{F}_g$ acting on a (small) mass at a point P, we say that there is a gravitational field at P. In fact, the gravitational field $\mathbf{g}$ is defined to be the ratio of the gravitational force on a (small) object divided by the mass m of the object:

Gravitational field.

$$\mathbf{g} = \frac{\mathbf{F}_g}{m} \quad \text{or} \quad \mathbf{F}_g = m\mathbf{g}. \tag{9.5}$$

It follows that the gravitational force and field have the same direction because m is always positive. Since the gravitational force is always attractive (the force pulls m towards the center of the gravitating body),

> **Box 9.2.1.** *The gravitational field of a gravitating body always points toward that body. If the body is spherical, the field points toward the center of the sphere.*

As Equation (9.5) shows, the gravitational field is nothing but the gravitational acceleration. The second equation in (9.5) is the same as the weight of the object; it confirms the fact that weight is the force of gravity. This equation also shows the equivalence between the force concept and the field concept: Knowing the field at a point, we can find the force at that point and vice versa. Other fields are defined similarly.

Gravitational field is gravitational acceleration.

The concept of the field *localizes* the concept of a force. Whereas in action-at-a-distance theory a force *here* is caused by an object (source) *there*, in field theory a force *here* is caused by a field *here*. In field theories, the primary objects are fields rather than sources. Of course, any field has a source, and to determine the value of a field at a point, theoretically, one has to know the exact location and status of the source. This connection between the source and the field is what ties together the action-at-a-distance theory and the field theory.

The field concept was invented merely as a "tangible" device to render action-at-a-distance more "concrete." Whether a field had any physical and measurable property that would give it an independent identity was irrelevant. However, further developments, especially in the hands of Maxwell and Einstein, showed that not only do fields have independent existence of their own, but they are the fundamental quantities on which (modern) physics is to be based.

One may very well ask: Given that there is a force at a point, say P, how do we know what *kind* of a field is present there? Is it gravitational, electrical, or nuclear? The answer is: by experimentation. By varying certain properties of the "test" object, we can determine the kind of field present at a point. For instance, if by varying the amount of charge on the test object, the force does not change, we can eliminate the electric field which is sensitive to the charge. Similarly, by varying the mass—while keeping other properties intact—of the test object and observing the change in the force, we can determine whether a gravitational field is present at P.

One of the most conspicuous effects of gravitational field, next to the fall of objects, is the phenomenon of **tide**. Consider the Earth-Moon system as depicted in Figure 9.4. The

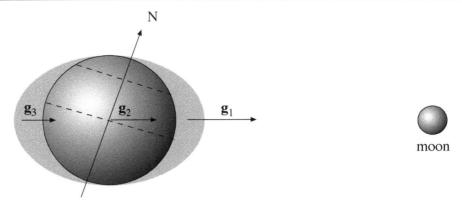

Figure 9.4: Tides are created because the gravitational field of the Moon is different at the three regions of the Earth shown.

Tides explained. ocean waters closer to the Moon feel a gravitational field (or acceleration) $\mathbf{g}_1$. The Earth as a whole feels a smaller gravitational field $\mathbf{g}_2$ at its center, and the ocean waters at the far end of the Earth feel the weakest gravitational field $\mathbf{g}_3$. Therefore, the waters on the right want to "escape" *from Earth* with a relative acceleration of $\mathbf{g}_1 - \mathbf{g}_2$. That is why a hump is formed on the side closer to the Moon. Why is there a second hump at the far end? Since the Earth as a whole feels an acceleration $\mathbf{g}_2$ which is larger than $\mathbf{g}_3$, the whole Earth is running away from the waters on the left with a relative acceleration of $\mathbf{g}_2 - \mathbf{g}_3$. Thus, these waters are "left behind" in the form of a second hump. Of course, all the waters are prevented from escaping by the strong gravitational field of the Earth. The formation of the two humps explains why tides occur twice a day.

9.2.2   Gravitational Potential Energy

The gravitational PE defined earlier as mgh is valid as long as the height h is much much smaller than the radius of the gravitating body, such as the Earth. This is because, as long as we stay close to the surface of the gravitating body, the weight $m\mathbf{g}$ is very nearly constant and the work done is simply this weight times the displacement (change in h). When the displacement from the center of the gravitating body is large, the force (weight) does not remain constant during the displacement. In such a case, one must use the techniques of integral calculus to find the exact formula for the gravitational potential energy. It turns out that, if one chooses the reference height not at the surface of the body but *at infinity*, *Gravitational PE for the case when reference "height" is at infinity.* then at a distance r from the center of the gravitating body, the potential energy would be

$$PE = -\frac{GMm}{r}. \tag{9.6}$$

Note that as r goes to infinity (becomes larger and larger), the potential energy also gets larger and larger, going to zero at infinity, as it should.

What do you know? 9.5. How would you change Equation (9.6) to make the surface of the gravitating body of radius R the reference height, i.e., where PE is zero? Recall that PE is defined to within a constant.

We now have two ways of finding the potential energy of an object at a height which is small compared to the radius of the Earth: The exact way given by Equation (9.6), and the

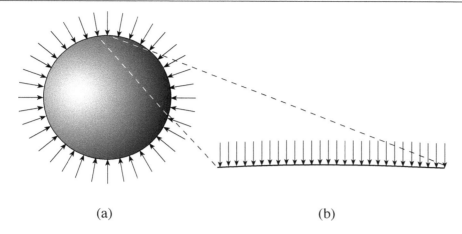

(a)   (b)

Figure 9.5: The gravitational field points (a) toward the center, and (b) "down" perpendicular to the surface of the celestial body. Note that (b) is a magnified version of a small part of (a).

approximate way given by mgh. Do these formulas agree? **Example D.9.7** on **page 29** of *Appendix.pdf* shows that if you use accurate values for the mass and radius of the Earth, they indeed do!

It should be clear from our discussion so far that (a) the gravitational field always points toward the mass (source) producing it (see Box 9.2.1), and (b) that, as we move *opposite* to the direction of the field (i.e., away from the gravitating body), the potential energy increases. Let us summarize this as follows:

> **Box 9.2.2.** *Points that are farther away from the gravitating body are at a higher potential than closer points. An object released in a region of gravitational field has a natural tendency to move from regions of high potential to regions of low potential, i.e., in the direction of the field.*

The law of conservation of energy is particularly important when applied to the motion of an object moving at cosmic distances under the influence of the gravitational force of a celestial body:

Conservation of mechanical energy at cosmic distances.

$$E = KE + PE = \tfrac{1}{2}mv^2 - \frac{GMm}{r}, \tag{9.7}$$

where v and r denote, respectively, the speed and the distance of the object from the center of gravitating body. In Chapter 8 we used ME for the mechanical energy to distinguish it from other forms of energy—such as heat—that may be present. Here E stands for the (total) mechanical energy, because in the discussion of fundamental forces such as gravity, it is a more common notation than ME. The quantity E can be either positive or negative, because the particle may have more KE, in which case E will be positive, or more (negative) PE in which case E will be negative.

Just as in the motion of a roller coaster, the total energy E will not change during the entire motion of the object [see Box 8.1.3]. In particular, its sign will not change: If it is negative at one point of the motion of the object, it will be negative all the time; and if it is positive or zero, it will remain so during the entire motion of the object.

> **What do you know? 9.6.** Explain the qualitative form of Kepler's second law (i.e., that planets speed up when getting close to the Sun) using Equation (9.7).

What goes up need not come down!

Now suppose that at some point in its motion (for instance at the beginning, when the object is set in motion), the object has an energy that is either positive or zero, i.e., $E \geq 0$. Then the object will never stop. Why? What would happen if it did stop? Its speed would be zero; thus its KE would be zero; and its entire energy E would be its PE, which is *negative*. That is impossible. Thus the old adage "What goes up must come down" is no longer valid! If you fire a projectile with a large enough speed so that its KE is larger than its potential energy at the surface of the Earth, then it will never return to Earth.

> **What do you know? 9.7.** A deep space probe has a positive total energy E. What happens to its PE as it gets farther and farther? What happens to its KE as it gets farther and farther? What will its KE equal eventually?

Similarly, if $E < 0$, the object cannot go on moving away from the gravitating center forever, because the PE term keeps decreasing in magnitude, and will eventually be dominated by the KE term, leading to a positive (or zero) E.

> **Box 9.2.3.** *If the total energy E of an object moving away from the mass M is positive or zero, the object will move away from M forever. If $E < 0$, the object will come to a stop at some distance from M, and start its descent toward M.*

Example D.9.8 on **page 30** of *Appendix.pdf* illustrates the statement of this box.

Gravitational potential.

A quantity related to the gravitational potential energy and sometimes used in the discussion of gravity is **gravitational potential** denoted by Φ. **Math Note E.9.3** on **page 85** of *Appendix.pdf* shows that for points close to a gravitating body (a planet, a star, etc.), $\Phi = gh$, with g the gravitational acceleration (field) of the gravitating body, and for points far away (at a distance r from the center of the gravitating body), $\Phi = -GM/r$, with M the mass of the gravitating body.

> **What do you know? 9.8.** Dig a hole that goes through the center of Earth and comes out on the other side of the globe. Drop an apple in the hole. Describe the subsequent motion of the apple. Where does it have the largest speed? The smallest speed? Where does it stop?

9.2.3   Escape Velocity

Escape velocity.

We have seen that when $E \geq 0$, the projectile will never stop. The size of E is determined by the "initial" KE—and therefore, the initial speed—of the projectile. What is the smallest speed with which the projectile can be launched at a distance r from the center of the gravitating body of mass M so that it never returns? This clearly corresponds to the smallest KE, and therefore, smallest E. But the smallest E is, of course, zero. Setting $E = 0$ in Equation (9.7) and multiplying both sides by 2, we obtain $0 = mv^2 - 2GMm/r$ or $mv^2 = GMm/r$ or $v^2 = 2GM/r$. Taking the square root, and calling this smallest velocity the **escape velocity**, we have $v_{\text{esc}} = \sqrt{2GM/r}$. The escape velocity usually refers to the minimum speed with which a projectile is to be fired *at the surface* of a spherical celestial body—where r is equal to the radius R of the body—for it never to return. For such a projectile, we obtain

$$v_{\text{esc}} = \sqrt{\frac{2GM}{R}}. \tag{9.8}$$

Example 9.2.4. By substituting the mass and the radius of the Earth in Equation (9.8) we can find the escape velocity of Earth. It turns out to be about 11 km/s. Similarly, the escape velocity of the Moon can be calculated. It is about 2 km/s. **Example D.9.9** on **page 31** of *Appendix.pdf* explains how we got these numbers.

One of the reasons that the Moon has no atmosphere is because of its low escape velocity. The average thermal molecular speed of most gases is comparable to the escape velocity of the Moon. Therefore, in their constant random motion, they will eventually move away from the Moon, and never return. ■

What do you know? 9.9. Inflate the Earth by a factor of a million, so that the inflated Earth has radius one million times larger than the present Earth. What would be the escape velocity at the surface of the inflated Earth? If you throw a ball as hard as you can, will it escape the inflated Earth?

Example 9.2.5. A satellite is circling at a certain distance from a planet.[3] From Equation (9.3), the velocity and KE of the satellite can be calculated. You can also find the potential energy of the satellite. Now add them and find that the total energy of the satellite is *negative*.

This is a characteristic of a **bound system**: the satellite is *bound* to the planet, and cannot escape from it (without the injection of some energy from outside). Box 9.2.3 has already told us that when $E < 0$, an object, initially moving away from a mass M, will eventually return to it, i.e., it is *bound* to M. The circling satellite is trying to move away from M, but *instantaneously* is drawn back to M (see Figure 9.6).

The minimum energy that the satellite needs to "unbind" itself from M is that which makes the total energy zero (in which case the satellite reaches the escape velocity). This is called the *binding energy* E_b, and is simply the negative of the total energy (or the total energy without regards to its sign). See **Example D.9.10** on **page 31** of *Appendix.pdf* for the numerical details ■

A bound system and its binding energy.

9.2.4 Black Holes

A remarkable property of the escape velocity in Equation (9.8) is that it is independent of the mass of the projectile. This means that a bullet, a cannon ball, a missile, and a gigantic space probe all require the same speed to escape a celestial body. Taking the opposite limit, if the mass of the escaping object becomes smaller and smaller, its escape velocity does not change. In particular, if a particle has *no mass at all*, its escape velocity will still be the same. There *is* a particle with no mass. It is called **photon** and it is what light is made of.

We now pose the intriguing question: Is it possible for a celestial body to prevent light from escaping? A glance at Equation (9.8) answers this question in the affirmative. What is needed is a high enough mass, or a small enough radius to make the fraction under the square root sign large enough. Specifically, this ratio should be so large as to make the left-hand side of the equation greater than the speed of light. Denoting this speed by c (a universally accepted notation for the speed of light), we get $c < \sqrt{2GM/R}$ or $c^2 < 2GM/R$, which after dividing both sides by $2G$ yields

$$\frac{M}{R} > \frac{c^2}{2G}. \tag{9.9}$$

Thus, if a celestial body has a mass-to-radius ratio larger than $c^2/2G$, then light will not be able to escape that body. Such bodies exist, and are called **black holes**.[4] Although black holes were suggested in the 1960s on theoretical grounds, no *direct* evidence for them was

---

[3]The satellite could be a moon, a meteor, a man-made object, or anything else. The planet could be any celestial object.

[4]The existence and properties of black holes require the full machinery of the general theory of relativity. It is remarkable that the result obtained so straightforwardly in Equation (9.9) agrees completely with the exact result.

found for a while. The launch—and the correction in the optical system—of the Hubble Space Telescope saw a black hole approximately 3 billion times more massive than the Sun in 1994.

Example D.9.11 on **page 31** of *Appendix.pdf* shows that if the Sun—which can house over a million Earth s in its volume—were to become a black hole, it would have to shrink to a sphere with a radius of only 3 km. This is less than the size of even a small city! This should give you an idea of how dense a black hole is.

Food for Thought

FFT 9.2.6. You say "No one can see a black hole or dark matter; so how can you even say that they exist." I say "Have you flown on a jet airplane? The pilot cannot see through the clouds; cannot see distant objects—which are rapidly approaching the plane due to its enormous speed; cannot see all the planes that are landing at the destination airport at about the same time. But the 'eye' of the airplane, the radar device mounted on it, can."

The accuracy and the speed with which a radar equipment can detect an approaching object is far superior to any human's vision. We don't see radar; we don't see the objects from which radar gets reflected; we are completely blind to radars! But science (physics) is not! Physics predicted the existence of microwaves when no human could even dream of such a thing. Then physics *produced* the first (artificial, of course!) microwave in the laboratory. Later physics also created devices that were sensitive to microwaves.

After many years of successful and failed experimentations with microwaves, they were used for detecting the motion of approaching objects. These experimentations were not unlike those of a baby, who tries to discover the world around her: "these two blurry 'objects' that the two crude light detectors on my face keep detecting, the same two objects that feed me when I'm hungry, clean me when I'm dirty, comfort me when I'm uncomfortable, spoil me for no reason at all, and keep forcing me to say 'papa' and 'mama,' must be special in my life." So, when the device was perfected to the point that it could use microwave to detect motion very accurately, it became known as radar. And when scientists used radar over and over again (mostly in the military) and saw that it worked, they started using it in airplanes. The baby grew up and her vision sharpened!

Every piece of physical knowledge, when used and reused with successful outcome, becomes an extension of our senses. And for the same reason that we do not doubt the existence of a tree when our eyes detect a green object with a brown stem, we should not doubt the existence of an object if the physical knowledge in our possession "detects" that object. If Kepler's third law, which has been tested over and over again with success, tells us that the speed of stars in a galaxy indicate a larger mass than the visible mass of the galaxy, we should infer dark matter. And if the universal law of gravity, which has been relentlessly tested by astronomers and physicists, tells us that there exist objects in our galaxy, which have such an enormous density as to capture light and trap it permanently, we should accept black holes.

Just as the eyes of a baby are trained to recognize its parents and the world around it once it becomes an adult, so are the immature inventions of science trained to recognize the universe (mostly invisible to us), once they become mature. So, yes, black holes and dark matter are as real as trees and tables!

9.3   Weightlessness

We have already mentioned that all objects fall at the same rate in a gravitational field. Consider now dropping a box with a cat inside it. Since the box and the cat fall at the same rate, the cat will appear to be floating in the box. By the same token a man in an elevator whose cables are severed, i.e., is in free fall, will be floating in the elevator. Thus, he feels weightless. (See also the discussion at the end of Example D.7.6.)

If, instead of dropping the box and the elevator with their occupants straight down, we give them an additional horizontal speed, the situation will not change. In this latter case, both the container and its content will acquire the *same* horizontal speed, and again they will move together. Hence, the cat and the man will feel weightless in the new motion as

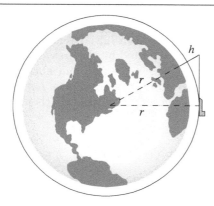

Figure 9.6: A space shuttle is constantly pulled back by the gravitational force of the Earth. If gravity were turned off for a small amount of time, the shuttle would move on a straight line, increasing its height by a small amount. The reason that the shuttle does not increase its height is because gravity pulls it down by exactly the same amount.

well. To elucidate the argument first imagine throwing two marbles at the same time with one hand. The two marbles will move together until they land. Now replace one marble with a small box. Once again, the box and the marble will move together, because they have the same horizontal speed and the same vertical acceleration. In the third experiment put the marble inside the box. Since the relative position of the marble and the box is irrelevant, the two will move together, and if you look carefully inside the box, you will see the marble floating. All the above experiments can be repeated for various sizes of marbles and boxes with the same result. In particular, a 200 lb marble inside a room-sized box will give the same result. Replacing marbles with cats and people will not change the outcome.

Now imagine standing on top of an incredibly high mountain peak and dropping the box (or the elevator). Figure 9.2 shows that if the horizontal speed is great enough, the container will eventually orbit the Earth. Thus any *orbiting* container is in *free fall*, and all objects inside it appear weightless.

An orbiting space shuttle is in free fall!

To further appreciate the seemingly puzzling fact that orbiting objects are in free fall, imagine what would happen if the gravitational field of the Earth were turned off. The object would move away from the Earth on a straight line with constant speed due to the first law of motion. The gravity, on the other hand, pulls the object down, i.e., makes the object fall from the potentially long distance it would travel if gravity were absent. This process of pulling down takes place constantly and instantaneously. The following example puts a little quantitative touch on the discussion.

Example 9.3.1. A space shuttle or a satellite has a circular orbit with radius r slightly larger than the radius of the Earth (see Figure 9.6). Previously, we calculated the speed of the satellite to be about 7900 m/s. To see that this motion is a free fall motion, imagine that a cosmic switchboard operator turned off the Earth's gravity for only one second. Then, according to the laws of motion, the satellite would continue on a straight line as shown in the figure, traveling 7900 m in one second. This distance is exaggerated in the figure for clarity. We can readily calculate the increase h in height using Pythagorean theorem and the fact that $h << r$ [see **Math Note E.9.4** on **page 86** of *Appendix.pdf* for details]. The result is $h = 4.9$ m. Thus, with gravity absent, the satellite would move away from the center of the Earth only 4.9 m in one second. But, as shown in Example 4.3.6, this is precisely the distance an object falls freely toward the center in one second. We therefore conclude that all objects circling the Earth are pulled back to the Earth precisely at a rate to keep them in their orbits, no more and no less! This pull-back is what we call a free *fall* motion.   ■

An astronaut in a space shuttle performs a physics experiment to test the validity of the first law of motion. He releases a ball in midair, being careful not to impart any initial speed to it, and notes that it stays afloat in midair. He concludes that the first law holds. In a second experiment he gives the ball an initial velocity and notes that it travels on a straight line with constant speed as expected by the first law. He comes down on Earth and performs the same experiment. He releases a ball and notices that it falls to the ground. He gives it an initial velocity and observes that it does not follow a straight path, but a curved one terminating on the surface of the Earth. He then concludes that either

- the first law is wrong, or

- there is a force acting on the object, or

- the Earth is a peculiar reference frame in which the first law does not hold.

As you shall see later, the choices above differentiate between Newton's theory of gravity and Einstein's general theory of relativity (GTR). Newtonian theory accepts the first law, and the fact that there is a force, called gravity, and rejects the peculiarity of the Earth's reference frame. Einstein, on the other hand, rejects the existence of a gravitational force, and blames the violation of the first law on the peculiar nature of the Earth reference frame. This basic difference in outlook leads to two different descriptions of gravitation to which we shall return later in the book.

9.4   End-of-Chapter Material

9.4.1   Answers to "What do you know?"

9.1. Both apples are the same distance from the Earth center. Therefore, they must have the same acceleration because of Equation (9.2).

9.2. Equation (9.2) indicates that the acceleration is a quarter of the previous acceleration. Numerically, the new acceleration is one fourth of 9.8 or 2.45 m/s^2.

9.3. Equation (9.3) shows that the speed of a satellite circling any celestial body decreases as the square root of the distance from the center of the body. For Earth, at one radius (low altitude), the speed of the satellite is 8 km/s. The *altitude* of $3R$ corresponds to a distance of four radii from the Earth's center. With 4 times the distance of the low altitude satellites, this satellite must have $1/\sqrt{4}$, or half the speed of a low altitude satellite. Thus, the speed is 4 km/s or 9000 mph.

9.4. If you plug in the values of speed (10,000 m/s), G, and M in Equation (9.3), you get 4,002,000 m or 4002 km for r, the orbit's radius. This is *less than* the Earth's radius! So, no such satellite can exist for Earth.

9.5. If the radius of the gravitating body is R, then add the constant GMm/R to Equation (9.6).

9.6. Decreasing r, the distance from the Sun, decreases $-GMm/r$. So the KE must increase, i.e., the speed of the planet must increase.

9.7. The potential energy increases with distance (the magnitude decreases, but since PE is negative, it increases). Since total energy is constant, KE must decrease. Eventually, when the probe is almost infinitely far away, PE is almost zero. Hence, KE becomes equal to the total energy E.

9.8. The apple only feels the gravity of the sphere right beneath it. At the beginning it is the entire globe. As it moves toward the center, the gravitating sphere becomes smaller and smaller (see Figure 9.3 and page 138) and the force of gravity weaker and weaker, until it reaches the center. All this time the apple has been accelerating. So, it has the maximum speed when it reaches the center. As it passes the center, the force of gravity has a retarding effect, dragging the apple toward the center. The apple slows down, until it reaches the other side of the globe. The maximum speed is reached at the center, because all the initial potential energy (which is the total energy, as the apple is assumed to have no KE at the beginning) gets converted into KE at the center. When it reaches the other end, it will stop, because the PE there is identical to PE at the beginning, where KE was zero.

9.9. Since the escape velocity is proportional to the inverse of the square root of the radius, for the inflated Earth it will be $\sqrt{1,000,000}$ or 1000 times smaller than that of the current Earth. Example 9.2.4 gives the Earth's escape velocity as 11000 m/s. So, the inflated Earth would have an escape velocity of 11 m/s or about 25 mph. A hard throw has more speed than this.

9.4.2 Chapter Glossary

Action at a Distance The notion that a body exerts a force on a test object instantaneously, no matter how far the object is from the source of the force.

Big Bang The initial moment at which the entire universe, including matter, space, and time was created under extreme conditions of temperature, pressure, and energy. See Part VIII for a full discussion.

Binding Energy The absolute value of the total energy (which is negative) of a bound system.

Black Hole A celestial body which captures light, when the light passes the body at a close enough distance. A black hole is characterized by the fact that its escape velocity is larger that the speed of light.

Bound System A system consisting of two or more bodies held together by the force of gravity. A characteristic of a bound system is that its total energy is negative.

Dark Matter An abundant *invisible* constituent of the universe whose existence is vindicated through its gravitational effect on the *visible* part of the universe.

Escape Velocity The minimum velocity—associated with a celestial body—which, when given to a projectile, sends the projectile farther and farther away from the body in such a way that the projectile never returns.

Gravitational Constant The universal physical constant G setting the scale of the strength of the gravitational force.

Gravitational Field The physical entity surrounding any massive body. A test object introduced in this field experiences the gravitational force exerted by the massive body.

Gravitational Potential Energy The energy related to the work done by gravity. More precisely, the work done by gravity in taking a test object from an initial position to a final position is the difference between the potential energies at those two points.

Quantization The process of reconciling a classically observed force with the quantum theory.

Standard Model The theory that explains three of the four fundamental forces of nature: electromagnetism, weak nuclear force, and strong nuclear force.

Super Force The Holy Grail of fundamental physics. It is the single force which encompasses all the fundamental forces of nature, including gravity.

Universal Law of Gravitation The mathematical formula giving the gravitational force between two objects in terms of their masses and the distance between them.

Weightlessness The condition prevailing in a freely-falling enclosure under which objects in the enclosure float.

9.4.3   Review Questions

9.1. Name all the fundamental forces of nature. How well is gravity understood? At what level?

9.2. Did Newton discover gravity? If not, why is his name associated with gravity?

9.3. What is common between an apple falling down and the Moon going around the Earth? Who saw this commonality?

9.4. Who measured the universal gravitational constant? When did he do it? How many years after Newton discovered the law of gravity? Why was the measurement touted as "weighing the Earth?"

9.5. What is dark matter? Can we see it? If not, how do we know that it even exists? Which law is used to confirm the existence of dark matter? Where do we find it?

9.6. What is action at a distance? What is a field? How is it related to force? How does field overcome the shortcomings of action at a distance? What is the difference between gravitational field and gravitational acceleration?

9.7. Explain how tides occur. Why are there two tides per day?

9.8. What is escape velocity? What is the total energy of an object moving at its escape velocity? Does escape velocity of an object depend on its mass?

9.9. Is it possible for a celestial body to have an escape velocity larger than the speed of light? If so, what is this celestial body called?

9.10. What is weightlessness and how can it be achieved? Under what conditions do you feel weightless in an elevator? Explain why objects in an enclosure falling freely (not necessarily straight down) under a gravitational force experience weightlessness.

9.4.4   Conceptual Exercises

9.1. What happens to the gravitational force between two objects if you double the distance between them? If you triple the distance between them?

9.2. A satellite circling the Earth is gravitationally attracted to Earth. Why doesn't it crash into Earth?

9.3. A satellite is circling the Earth with a certain speed. If you were to keep it on an orbit at a higher altitude, do you have to speed it up or slow it down?

9.4. You are in an airplane holding a box. You drop the box, and note that it crashes to Earth. Now you are holding a box in a space shuttle orbiting the Earth. You drop the box, does it crash to Earth?

9.5. A low-orbiting satellite is circling the Earth at 8 km/s relative to the Earth. It projects a capsule backwards at 8 km/s. What will be the path of the capsule?

9.6. By observing Io, one of Jupiter's moons, you can determine how long it takes Io to make a complete revolution around Jupiter. If you have a good telescope, you can also estimate the distance from Io to the center of Jupiter. From these two pieces of information, can you estimate the mass of Jupiter? If so, which law do you use?

9.7. When you use $PE = mgh$, the potential energy increase with height. Is this conclusion also true when you use Equation (9.6)?

9.8. When you fire a projectile from Earth's surface at the escape velocity of 11 km/s, it will never return, but it keeps slowing down. What is the speed at very very (almost infinitely) far away?

9.9. At very very (almost infinitely) far away from Earth, you let go of a projectile. It moves towards Earth, assuming only Earth's gravity is at work. When it crashes to Earth, what speed will it have?

9.10. Can the total energy of an object under the influence of the gravity of a celestial body be positive? Negative? Zero? Can the kinetic energy of an object under the influence of the gravity of a celestial body be positive? Negative? Zero? Can the potential energy of an object under the influence of the gravity of a celestial body be positive? Negative? Zero?

9.11. An elevator in free fall eliminates gravity inside. So, if the elevator falls at the rate of 9.8 m/s^2, no gravity exists inside the elevator. Suppose the elevator keeps falling at the rate of 9.8 m/s^2, but somehow magically the Earth's gravity disappears. What happens to objects in the elevator? Does this suggest a way of producing artificial gravity in space?

9.4.5   Numerical Exercises

9.1. A planet in a remote star system is seen to move around its star once every 50 days (the period of the planet) on a circular orbit. The distance of the planet from the star is 30 million km.
(a) What is the distance covered by the planet in one period?
(b) What is the planet's period in seconds?
(c) What is the speed (in m/s) of the planet as it moves around the star?
(d) What is the (centripetal) acceleration of the planet?
(e) From the knowledge of the centripetal acceleration, find the mass of the star.
(f) Using Kepler's third law, find the mass of the star.

9.2. Moon has a radius of 1740 km and a mass of 7.35×10^{22} kg. What is Moon's escape velocity?

9.3. To how small a radius should the Earth shrink for it to become a black hole?

9.4. The Large Hadron Collider (LHC) is a colossal machine that is designed to look inside matter at the smallest scale. For this, it needs to accelerate particles to enormous speeds and energies. Because of the size of its energy, some people are afraid that it might create a black hole which will gobble up the Earth. Some groups are even suing CERN, the institution building and operating the LHC. From formula 9.9, estimate the mass needed to create a black hole the size of the nucleus of a hydrogen atom, 10^{-15} m. Compare this with the mass of a few grams used in a typical run of the LHC.

Chapter 10

Epilogue: Determinism

Galileo predicted that in the hands of the future generations of physicists, his methodology would open completely new vistas of the universe. Newton was the first and the greatest personification of this prediction. He carried out Galileo's methodology to its most prolific limits, thereby climbing to the highest peak of the scientific terrain in history.

10.1 Newton's Methodology

Once the Aristotelian holistic approach to the investigation of nature was toppled by Galileo, the gate to the garden of knowledge opened up to mankind. Newton inherited, among other things, the indispensable scientific methodology introduced by Galileo, and applied it very effectively to the physics of motion, discovering its three fundamental laws. Later generations of mathematicians and physicists applied these laws successfully to such a wide variety of situations that for over two centuries the words "physics" and "Newton" were indistinguishable.

For over two centuries, "physics" and "Newtonian physics" become synonymous.

10.1.1 Specificity

In his analysis of motion, Newton adhered to the Galilean philosophy of concentrating on a *specific* aspect of a *specific* moving object. If we are interested in the properties of motion *in general*, Newton argued, we had better concentrate on the motion of a ball, for example. The Aristotelian holistic statement that "some objects fall, and some rise," and that "apple belongs to the first category" is a completely useless statement, Newton might have said. We gain knowledge about motion only if we concentrate on a moving apple.

His concentration on the motion of specific objects led Newton to a precise definition of velocity and acceleration and the (co-)invention of calculus. To find the velocity of an apple, it is not sufficient to simply divide the distance it travels by the travel time, because the apple may change its velocity from moment to moment. Thus, Newton argued, we have to speak of the *instantaneous* velocity of the apple, whereby apple's travel time is shrunk to an infinitesimal. Similarly, one has to speak of the *instantaneous* acceleration, because the rate of change of the velocity of an apple may be different at different parts of its motion. These *instantaneous* quantities were the paradigms of the general concept of the *derivative*, a fundamental notion in calculus.

Analysis of a *specific* motion leads to the *universal* concept of the derivative.

Newton's further analysis of the motion of an apple led to the question of why it is accelerating. His answer was that the apple must be influenced by a force and that the acceleration of the apple must be directly proportional to the force exerted on the apple. And only after analyzing the *specific* motion of an apple did he arrive at the general statement that the acceleration of *any object* must be directly proportional to the force exerted on that object. Thus the (*universal*) second law of motion came about as a result of analyzing the motion of a specific object, in exact analogy to Kepler's discovery of his three planetary laws by investigating the motion of Mars.

Newton's discovery of the *universal* law of gravitation was also a result of the analysis of the *specific* motion of the Moon and an apple: By comparing the accelerations of a *specific* object in free fall on Earth with the acceleration of the Moon as it circles the Earth, he was able to arrive at his mathematical statement of the law of gravitation.

10.1.2   Idealization/Isolation

Earth, Moon, and apple are very complicated structures. The water covering about three fourths of its surface, the atmosphere hovering above its surface, the mountain ranges, the forests, the life, and a myriad other complications make Earth anything but simple. Similar complications exist for Moon. The non-uniformity of its internal parts, the existence of a stem with a dimple on the skin, etc., make an apple also a complicated structure. Yet Newton, following Galileo's advice, ignored all such complications and *idealized* the three structures to geometrical points. This took a tremendous amount of insight and abstraction, especially since Newton was the first to implement the notion of idealization in this particular fashion.

A holistic philosopher would have nauseated at the thought of such a breach of the apparent reality. On the other hand, a holistic philosopher would have never discovered the mathematical statement of the law of gravity. A holistic philosopher sees the complexity of a "whole" physical system and gets lost in the resulting imbroglio. In contrast, a scientist, following in the footsteps of Galileo and Newton, sees the simplicity in the *relevant* properties of an apple, or a Moon, or an Earth, ignores the complications, and discovers laws governing the behavior of not only the system at hand, but other systems that may be out of his/her reach. In fact, it was Newton himself, who after shrinking the Earth, Moon, and the apple to a point and discovering the universal law of gravitation, came back and applied the law to *different parts* of the Earth, and successfully explained the occurrence of the tides.

Newton elevated the process of idealization to a new level, which we can call **isolation**. Although he was aware of the influence of the Sun and the rest of the planets on the motion of the Moon and the apple, *he ignored them*! As revolting as this procedure may have appeared to a contemporaneous holistic philosopher, it was the right thing to do. It was the scientific thing to do. It still is the scientific thing to do! These philosophers have a tendency to emphasize the *existence* of the "influence," but conveniently ignore its *magnitude*! If A and B both influence C, but A's influence is a thousand times stronger than B's, is it not reasonable to ignore the latter? After all, our measuring devices are not infinitely accurate, and we may not be able to even detect B's influence! **Example D.10.1** on **page 31** of *Appendix.pdf* compares the forces of gravity on an apple and the Moon exerted by the Sun and the Earth. In it, you will see that the Sun pulls an apple with a force that is over ten million times smaller than the Earth's force; and the force of the Sun on the Moon, although relatively larger, is still only about one percent of the Earth's.

Did Newton do the wrong thing in ignoring the Sun's gravity? Did he have to bog himself down with the Sun's gravitational effect on the apple even though it is less than one ten millionth of the Earth's effect? Or its effect on the Moon even though it is only about one percent of the Earth's effect? All scientists say "no" to all these questions, while all holistic philosophers of science say "yes." The latter abhor specificity, and as such, are against the scientific methodology of Galileo, Newton, and all the other scientists that came after them.

A holistic philosopher would have never discovered the law of gravity.

10.1.3 Continuity

"If I have seen further it is by standing on the shoulders of giants," said Newton of how he made his enormous contributions to physics and mathematics. With this, not only was he acknowledging his gratitude to others, but also codifying the very nature of *scientific* progress. The noted scientists before Newton were the initiators of Renaissance science, and as such, were the *creators* of new knowledge. Copernicus started his revolution without ammunition from previous generations of astronomers;[1] Galileo had very little to go by in his study of motion; Kepler had access only to a collection of *raw* (albeit, very accurate) *data* left to him by Tycho Brahe, but to no systematic *knowledge* except the Copernican astronomy.

The most decisive litmus test for scientific nature of a discipline is to see how much the discipline incorporates its existing knowledge in its new discoveries.

Newton set a trend in that he recognized the essentiality of using the existing knowledge to gain further understanding of nature. He realized that, contrary to Aristotle and other holistic philosophers, the universe was too big for a single man to understand. That is why he so beautifully summarized this aspect of science:

> **Box 10.1.1.** *"To explain all nature is too difficult a task for any one man or even for any one age. 'Tis much better to do a little with certainty, and leave the rest for others that come after you, than to explain all things."*

The scientific nature of a discipline is determined by the extent to which its new discoveries use the accumulation of the existing knowledge. Any discipline whose practitioners propose "theories" only to be ignored by the future practitioners, who in turn come up with their own "theories" to be ignored by the next generation, is not a scientific discipline. In fact, one can go so far as to say that *the* most decisive litmus test for scientific nature of a discipline is to see how much the discipline incorporates its existing knowledge in its new discoveries.

10.1.4 Use/Creation of Mathematics

The title of Newton's opus magnum is *Philosophiae Naturalis Principia Mathematica*, translated as *Mathematical Principles of Natural Philosophy*. His use of the word *Mathematica* in the title of the book signifies another milestone he laid for the investigation of the physical universe.

Galileo was the first scientist to discover the "language of Nature," but Newton became the most fluent speaker of this language. As he investigated motion, Nature whispered to Newton that he needed to learn and perfect the "dialect" of differential calculus. And when he was working on his gravitational theory, he discovered that Nature was chatting with him in the new parlance of integral calculus.

The use and creation of mathematics resulting from the communication with Nature was not new to Galileo and Newton. In fact, Nature started to whisper into Egyptian and Babylonian ears when they were compelled to employ mathematics in their day-to-day activities. They then passed on their acquired pidgin to the Greeks, who in turn transformed the primitive language into the most sophisticated tongue of the time, Euclidean geometry.

However, Newton's conversation with Nature included a new part of speech, time. Although the ancients, including the Greeks, were aware of the notion of time, and although the idea of a derivative and an infinitesimal were known to Newton's contemporaries (even to the ancient mathematicians), it was Newton who combined the two concepts and introduced the idea of infinitesimal rate of change or derivative with respect to time, an idea that proved to be crucial for the development of not only mechanics, but the entire field

---

[1]It is not clear whether Copernicus knew of the heliocentric model of Aristarchus.

of physics. After discovering the mathematical expression for the law of gravity, Newton employed the notion of derivative with respect to time in conjunction with that law to come up with the first **differential equation** in the history of mankind.

10.2   Determinism of Newtonian Physics

It is a fact of nature that physical quantities do not change abruptly and suddenly with time or motion.[2] Any force applied to a moving object, changes smoothly and slowly. For example, gravity decreases with distance and it does so slowly and smoothly. The same smoothness, or **continuity**, applies not only to all forces but also to all physical quantities. Even artificial abrupt application of forces such as the force of hammer on a nail only *appears* as abrupt. On a microscopic level, the atoms of the nail "feel" the approach of the atoms of the hammer at a distance because of electrical forces that act between the two sets of atoms.

As **Math Note E.10.1** on **page 86** of *Appendix.pdf* illustrates, this principle of continuity, combined with Newton's laws of motion, leads to the fundamental philosophical concept, **determinism**. As applied to the motion of a particle, determinism states that

> **Box 10.2.1.** *If the initial velocity and position of a particle are given, the subsequent motion of the particle can be predicted exactly, once the force acting on the particle is known during its entire motion.*

10.2.1   Calculating Earth's Orbit

Let us apply Newton's laws to the motion of Earth around the Sun to illustrate determinism. The line connecting the Earth to the Sun and the line along which Earth moves *initially* form a plane which we take to be the xy-plane. Since the gravitational force has no component perpendicular to this plane, Earth will always be confined to the xy-plane. Choose the x-axis to be the line joining the Sun to the *initial* position of Earth, and assume that Earth moves with a speed of 30,000 m/s (the approximate value of the average speed of Earth around the Sun) in the positive y-direction. The subsequent motion can be broken up into an x-motion and a y-motion.

From the initial coordinates and velocity, as well as the laws of motion, we can calculate the coordinates and the components of the velocity a little later. For the interested readers, this is discussed in great detail in **Math Note E.10.2** on **page 87** of *Appendix.pdf*, where "a little later" means 60 seconds later. From these newly obtained data and the laws of motion *as applied to the new location and velocity* of Earth, we can find the coordinates and the components of velocity another minute later. Continuing this process, we can chart the location of Earth around the Sun for a complete revolution as shown in Math Note E.10.2. Thus the initial location and velocity of Earth *determines* its motion for eternity. This is a simple example of determinism.

10.2.2   The Solar System

What if we have more than one particle? Can the laws of motion, combined with the theory of differential equations, predict the behavior of an assembly of particles? Yes. Of course,

---

[2]Here "abrupt" and "sudden" mean "in no time," i.e., in *zero* time in the strictest mathematical sense. The notions of "suddenness" and "abruptness" are relative. In the same time interval that it takes you to "suddenly" apply the brake pedal of your car as soon as you see a danger, the radio waves of your favorite station travel about 20,000 miles, and millions upon millions of cosmic rays pass through your body! There is nothing abrupt for them!

we must know the initial positions and velocities of all particles, as well as the interacting force between each pair of particles in the assembly.

The solar system was a great testing ground for Newtonian physics. The law of gravity, in the form of a mathematical formula for the force between any pair of bodies in the solar system, was known. Furthermore, the theory of differential equations was greatly developed by the end of the eighteenth and the beginning of the nineteenth century. Thus, starting with an observed configuration of the solar system, the mathematical physicists of that era could predict the configuration, i.e., positions and velocities of the planets in the solar system at any later time. The incredible agreement between theoretical predictions and observations compelled almost all physicists of that time into a semi-religious faith in Newtonian physics. There were of course instances in which the observation disagreed with the theory, and those of weaker faith started to doubt Newtonian physics. However, the true believers found other ways of bringing the theory into agreement with observation without abandoning the cherished laws of physics. One such instance deserves closer examination.

Sir William Herschel (1738-1822), the German-English astronomer, discovered the planet Uranus in 1781. This discovery brought a tremendous wave of excitement to the tranquil field of astronomy of the solar system which had studied only six planets ever since the Greeks. Now the astronomers, both theoretical and observational, had to take into consideration the motion of the new intruder when studying other planets and their satellites.

Two astronomers, John C. Adams (1819-1892) from England and Urbain J. J. LeVerrier (1811-1877) from France were particularly interested in the motion of Uranus itself. As they scrutinized its motion, they noted that Uranus did not behave according to Newton's theory of gravitation. Being firm believers of this theory, they both assumed (Adams in 1845 and Leverrier in 1846) the existence of yet another planet whose presence beyond Uranus could explain the anomalous behavior of the latter. Adams sent his findings to Airy, the astronomer royal and a nonbeliever, who discarded it, being sure that the anomaly of Uranus' motion was due to the imperfections in the theory of gravitation. Leverrier, on the other hand, was lucky. He sent his findings, along with the approximate location of the new planet in the sky, to Johann G. Galle (1812-1910), the director of the Berlin Observatory. Upon the receipt of LeVerrier's letter, Galle started to look for the new intruder, and within hours could spot the planet that we now call Neptune.

Newtonian physics predicts the existence of Neptune.

It was such flawless predictions of ghostly celestial bodies, plus countless other terrestrial and celestial predictions by Newtonian physics that gave it an awesome veracity. Put yourself in the position of the nineteenth century theoretical physicists, who, merely on the basis of a small disturbance in the orbit of one planet, and working with a mathematical equation written down more than 150 years earlier, could forecast the existence of a whole new world in the solar system. Imagine the awesome joy they felt when observation confirmed their predictions, and the impulse toward a religion-like faith in the theory that predicted the observations!

10.2.3   Determinism

If Newtonian theory works for the solar system with such precision, why not apply it to everything? Find out the equation for interactions among any number of particles and the machinery of differential equations can predict the behavior of these particles once their initial configuration is given! Of course, the determination of the initial configuration and the subsequent propulsion of the machinery of differential equations may not be an easy *practical* task. In practice, even the problem of three particles interacting gravitationally has not been "solved." In principle, however, there is no difficulty imagining a supercomputer capable of crunching an enormous quantity of numbers corresponding to the initial configuration and the differential equations manipulating them, and giving out the configuration of the system at a later time. The existence and uniqueness of the configuration at any specified time are proven mathematical facts.

Determinism is a result of such forceful mathematical statements concerning the existence and uniqueness of subsequent configuration given an initial configuration. Apply this statement to the collection of all atoms and molecules in the universe, and you have
Determinism stated. excluded any kind of free will and chance. Determinism states that

> **Box 10.2.2.** *If the configuration of all the atoms in the universe is given at an initial time, the atoms will evolve in a predetermined way for the rest of eternity.*

According to the deterministic philosophy, the formation of galaxies, the emergence of stars in galaxies, the appearance of planets, the evolution of life, the development of human beings, and the subsequent human history are all predetermined by the initial configuration of the atoms in the universe. An omnipotent supreme being that chooses the initial configuration, simultaneously fixes the future of the universe for the rest of eternity. In a very real sense, even the supreme being loses control of the subsequent development of the universe. Little wonder Einstein, a determinist, said "I wonder if God had any choice in creating the universe."

The world is not deterministic! But the world is *not* deterministic! The advent of the quantum theory in 1925–1926 paved the way for an indeterministic universe, and by 1930 every evidence pointed to a highly unpredictable atomic world. According to this theory, of which we shall have a lot more to say in Part V, it is impossible to predict the behavior of atoms and subatomic particles deterministically as Newtonian physics suggests. In fact, it was due to failing attempts to reconcile the deterministic Newtonian physics with the behavior of subatomic particles that the indeterminism of the microworld was forced on the physicists of the 1920s and 1930s.

Part III

Waves and Electromagnetism

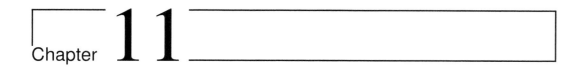

Chapter 11

Waves

When he was 23, Newton bought a glass prism "to try therewith the phenomena of colour." And with this the science of optics started. He probably made many fundamental discoveries in that field during the next three decades of his life. However, sometime in February, 1692, a light, left burning in his room while he was away, started a fire that eventually destroyed his papers including a large work on optics describing the results of his research of twenty years. Nevertheless, he managed to reconstruct most of his experiments, and published his results in *Optick* in 1704.

In one of his experiments, Newton showed that lights of different colors, when sent through a prism, bend (or *refract*) at different angles. And when he sent sunlight through the prism, he noticed that it dispersed into different colors. He thus concluded that white light is simply a mixture of lights of different colors. In another experiment he passed light through some combination of thin lenses and discovered what is now called *Newton's rings*, a succession of bright and dark circles formed when the lenses were illuminated with light.

Both of the phenomena mentioned above—especially the formation of Newton's rings—can best be explained by assuming that light is a wave. However, Newton vehemently opposed the idea that light can be a wave, and he got into bitter argument with those who advocated this idea, in particular the Dutch physicist, Christian Huygens. Newton thought of light as a beam of tiny *particles* streaming through space. Huygens, on the other hand, believed that light is a wave; and although the wave theory explained the dispersion of sunlight and the formation of Newton's rings more effectively than the particle theory, it did not account for lack of evidence for the "bending" of light around edges as sound (which all agree to be a wave) does. This debate continued until 1801 when Thomas Young, a British physicist, showed unequivocally that light is indeed a wave.

With the unprecedented advance of mathematics—resulting mostly from the application of Newtonian physics to different media—in the eighteenth and the beginning of the nineteenth centuries, it was realized that *mechanical* waves appear in many different situations. In fact, with the exception of light itself, all other waves could be treated as mechanical waves, arising from a mechanical motion of some kind of elastic medium. Furthermore, all these waves obey the same mathematical equation, known as the **wave equation**. The universality and the simplicity of this equation makes it truly a piece of (mathematical) art, which—to emphasize its affinity to other forms of art—we display in Figure 11.1.

In the wave equation, the symbol Ψ represents the displacement of the medium from its equilibrium position, for example, the up-down displacement of water in a quiet pond when a stone is thrown in it. The symbol ∇ represents how Ψ changes in space, and $\frac{\partial}{\partial t}$ how it changes with time. Thus, Ψ is a function of space and time: At any fixed point of space (e.g., a point in the pond) the displacement changes with time (the water moves up and down), and at any fixed time (e.g., a snapshot of the pond), the displacement Ψ changes

Figure 11.1: The wave equation—a masterpiece of theoretical physics.

over space (at some points it has a crest, at others a trough). The wave equation connects the space and time variations in a particular way. The symbol v in the wave equation is the speed with which the wave travels.

We shall not deal with the mathematical aspect of the wave equation at all. However, because of its importance in the subsequent discussions, we shall talk about waves themselves, and as a paradigm, concentrate on mechanical waves produced in elastic media by mechanical oscillations, keeping in mind that almost all properties discussed here carry over intact to considerations involving other kinds of waves. First let's understand the notion of oscillation.

11.1   Oscillation

If a mass is attached to one end of a spring, whose other end is fixed, it performs a repetitive motion when the spring is stretched out and released. In general, the repetition may be complicated. However, *if the amount of stretch is small,*[1] the motion of the mass-spring system will be simplified considerably. This motion is called a **simple harmonic motion** (SHM) and whatever performs such a motion, a **simple harmonic oscillator** (SHO). In Figure 11.2 we have shown the (small) displacement of the mass-spring system as a function of time.

Period defined.   An important characteristic of an SHM is its **period** T defined as the time required for the SHO to return to its "original" position, which could be any position during the course of its motion. Figure 11.2 shows the SHM of a mass attached to a spring. Imagine the spring oscillating vertically while a scroll of paper moves uniformly under a marker fastened to the mass. Then the horizontal distance on the paper represents time (e.g., if the paper scrolls at the rate of one centimeter per second, then each centimeter of the paper represents one second), and the curve produced by the marker describes the motion of the mass as a function of time. On such a graph, you can easily read off the period of the SHM: It is the horizontal distance between any position of the mass and the next identical position. In Figure 11.2 three arbitrary positions, P_1, P_2, and P_3, are shown. At a later time, the mass will be executing identical motions at the corresponding positions, P_1', P_2', and P_3'. It follows that any of the horizontal distances, $\overline{P_1P_1'}$, $\overline{P_2P_2'}$, or $\overline{P_3P_3'}$ is one period.

The motion undergone by an SHO during one period is called a **cycle**. An important

---

[1]Compared to the unstretched (and uncompressed) length of the spring.

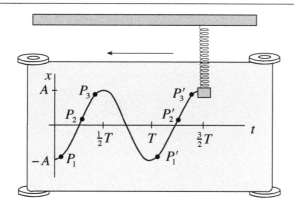

Figure 11.2: Simple harmonic motion is the simplest kind of repetitive motion. As the paper scrolls to the left, the marker on the mass traces a simple harmonic curve.

quantity given in terms of cycles is the **frequency** f which is defined as the number of cycles performed by the SHM in one second. If the period is T seconds (say $T = 0.1$ second), then the number of cycles repeated in one second would be $1/T$ ($1/0.1 = 10$). Thus, one can write

$$f = \frac{1}{T}. \tag{11.1}$$

Cycle and frequency defined.

Since T is measured in seconds, frequency is measured in units of s^{-1}, or cycles per second (cps), which is also called **Hertz** and abbreviated as Hz.

> **What do you know? 11.1.** The frequency of an oscillator is 500 Hz. What is its period?

Example 11.1.1. A pendulum—a heavy bob hanging from one end of a very light string whose other end is fixed—describes SHM when its initial displacement is small. For such a pendulum the period in seconds is given approximately by $T = 2\pi\sqrt{l/g}$, where l is the length of the string in meters and g is the gravitational acceleration (equal to 9.8 m/s^2 at the surface of the Earth). If $l = 5$ cm, the period will be approximately 0.449 second and the frequency $f = 1/T = 1/0.449 = 2.23$ Hz. The same pendulum on the Moon, with a gravitational acceleration of 1.62 m/s^2, will have a period of 1.1 seconds and a frequency of 0.9 Hz. ∎

The significance of SHM lies in the fact that any other repetitive motion can be represented as a superposition of (infinitely) many SHMs with different frequencies. In some sense, by studying SHM, you become familiar with the basic building block of all repetitive motions. The construction of a general repetitive motion out of SHMs, however, is a complex procedure beyond the scope of this book.

11.2   Mechanical Waves

Any disturbance in an elastic medium has a tendency to propagate in that medium. For instance, a stone dropped in a quiet pond creates a disturbance in the form of a small circle with center at the point where the stone strikes the water; and this disturbance propagates (spreads) through the pond in the form of circles increasing in size. Similarly, a disturbance at one end of a rope propagates to the other end.

Disturbances which take place only once (such as a single jerk at one end of a rope) propagate in the form of a **pulse**. On the other hand, if the source keeps disturbing the

Pulse and wave differentiated.

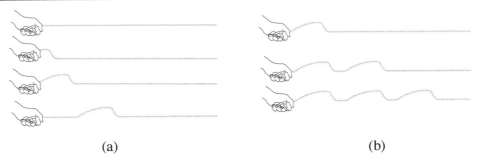

Figure 11.3: (a) A pulse on a rope, and (b) a repetition of it producing wave.

medium continuously, the resulting effect propagates in the form of a **wave**. Figure 11.3 shows a pulse and the corresponding wave on a rope. In Figure 11.3(a) the shape of the rope carrying a pulse is shown at various times. In Figure 11.3(b) the source keeps producing disturbances on the left which move to the right in the form of a train of pulses. This train of moving pulses is called a wave.

> **What do you know? 11.2.** Create a pulse on a rope by jerking its end. Half a second later, create another pulse. How far ahead of the second pulse is the first pulse if the speed of the disturbance is 0.6 m/s.

The shape of the wave depends on the motion of its source. Only waves produced by a source in SHM will be discussed here. As mentioned above, any other wave is a superposition of these simple waves. Suppose that the source that produces the complicated wave pattern on the rope of Figure 11.3(b) is replaced with a SHO. Then the shape of the wave will be simple and smooth as shown in Figure 11.4. Such a wave is called a **simple harmonic wave**. The shape of the rope is shown at times equal to multiples of $T/2$ where T is the period of the oscillation of the source. Before going any further let us make sure that we know how to differentiate Figure 11.4 from Figure 11.2. In the latter, the displacement of the *source* (e.g., the mass in the mass-spring system) is shown as a function of time. Figure 11.4, on the other hand, is a snapshot of the actual *shape* of the rope at various times.

Simple harmonic wave.

Wavelength defined.

An important characteristic of a simple harmonic wave is its **wavelength**. As shown in Figure 11.4, it is the distance the wave front moves in one period. Thus, the distance between the open circle at $t = 0$ and the one at $t = T$ is the wavelength λ. Similarly the front corresponding to the open triangle is produced at $t = T/2$ and will have moved a distance λ when the time is $t = 3T/2$. The figure does not show when the open square was formed. (It was produced at $t = T/4$, for which no snapshot is shown.) However, we can identify the point as the peak immediately following the open circle. In fact, in a general case, where the extent of the wave may be too large—because its production started too long ago—to allow the identification of its front, the wavelength is determined by measuring the distance between two "similar" points, such as two crests or two troughs, of the wave train.

Motion of wave is not the same as motion of rope as a whole!

If you examine the rope carefully, you'll see that, as a whole, it is not moving. Concentrating on an arbitrary point on the rope, you'll note that it simply moves up and down, i.e., it performs a SHM. If the rope as a whole is not moving to the right, and its parts are not moving to the right, then what *is* moving to the right? It is the *disturbance* caused by the source which, from piece to piece, from molecule to molecule, and from atom to atom is transferred to the other end. As a small piece of the rope moves up, it drags the

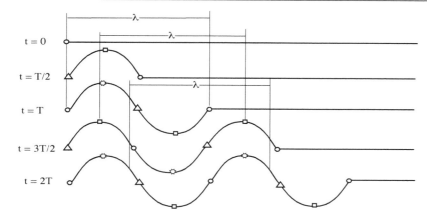

Figure 11.4: A simple harmonic oscillator produces a simple wave.

neighboring piece behind it. The dragged piece drags its neighboring piece which in turn drags its neighboring piece and so on. The overall effect is the production of wave trains moving along the rope. Since the disturbance carries energy, we say that it is the *energy* that moves along the rope. This is particularly noticeable in the up-and-down motion of a boat caused by water waves produced by other boats. That the wave carries energy, is clear from its capacity to bob the boat. That it is not the water that moves as a whole, is clear from the fact that the boat is not carried away.

We can calculate how fast the wave—the disturbance, the energy—moves: It travels a distance of λ in a time interval T. Hence, its speed v will be $v = \lambda/T$. But since $1/T = f$, the frequency, we can write

$$v = \lambda f. \tag{11.2}$$

How wavelength, frequency, and speed are related.

This is a very important relation and applies to all waves, not just waves on the rope. If you know any two of the three quantities v, λ, and f, you can calculate the third one. To determine the speed of the (mechanical, such as water) wave, you can measure the frequency f, for instance, by counting the number of wave crests arriving each second, and the wavelength λ (by measuring the distance between successive crests) and multiplying them together. The following examples illustrate and clarify these concepts.

Example 11.2.1. A fisherman notices that the water wave crests are 3 meters apart and that 10 crests arrive at his boat each second. What is the speed of the water waves? Here $\lambda = 3$ meters and $f = 10$ Hz. Hence, Equation (11.2) immediately yields $v = 3 \times 10 = 30$ m/s.

Sound travels with a speed of 340 m/s in air. The concert A note on a piano has a frequency of 440 Hz. Thus, the wavelength of concert A in air can be determined from Equation (11.2): $340 = \lambda \times 440$. The result is $\lambda = 0.77$ meter or 77 cm.

An FM radio station transmits a radio wave of frequency 100 MHz (M stands for **mega** meaning a million). It is known that a radio wave travels at the speed of light, i.e., 300,000 km per second, or 3×10^8 m/s. We can find the wavelength of the FM wave: $3 \times 10^8 = 100 \times 10^6 \lambda$ or $\lambda = 3 \times 10^8/10^8 = 3$ m. AM stations have frequencies in the range of 535 to 1605 kHz (kilohertz or 1000 Hz). Thus, the wavelength for a station broadcasting at 600 kHz is determined from $3 \times 10^8 = 600,000\lambda$ or $\lambda = 3 \times 10^8/600,000 = 500$ m. ∎

Wavelength and frequency are the precise quantitative analogues of every-day notions used to describe our perceptions of sound and light. The pitch of a sound is described by its frequency, the higher the pitch of a sound wave, the larger its frequency, or the smaller its wavelength. Normal human ear can detect frequencies as low as 20 Hz and as high as 20,000 Hz.

Frequency describes both the pitch of sound and the color of light.

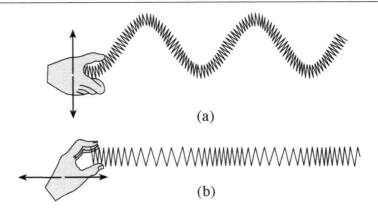

Figure 11.5: (a) Transverse and (b) longitudinal waves made by a slinky.

Color to light is what pitch is to sound. Thus, various colors of light correspond to various wavelengths. The wavelength of the *visible* light extends from 4×10^{-7} m, the wavelength of violet light, to 7×10^{-7} m, the wavelength of red. There are other (invisible) lights, generally called *electromagnetic waves*, to which the human eye is not sensitive, and whose wavelengths are either shorter than 4×10^{-7} m (ultraviolet light, X-rays, and gamma rays), or longer than 7×10^{-7} m (infrared light, microwaves, and radio waves). Chapter 14 talks about electromagnetic waves.

Energy of a wave is proportional to the square of its amplitude.

Another correspondence between quantitative properties of waves and their every-day counterparts is the wave **amplitude** which is related to loudness for sound and brightness, or intensity for light. Amplitude is simply the maximum displacement of the medium (the rope, say) from its undisturbed position. Thus, huge water waves have larger amplitudes than small ripples. Big water waves, with big amplitudes, are more capable of moving a boat than small ones. Therefore, a wave with a larger amplitude carries more energy than the one with a smaller amplitude. In fact, it is shown in mechanics that[2]

> **Box 11.2.2.** *The energy (or intensity) of a wave is proportional to the square of its amplitude.*

Based on the relation between the oscillation and propagation directions, waves are divided into two categories: If the oscillation of the parts of the medium is at right angle to the direction of the propagation of the wave (as in a rope or in water), the wave is said to be **transverse**. A wave whose propagation is in the same direction as the oscillation of the medium is called a **longitudinal** wave. A prime example of a longitudinal wave is the sound wave. Figure 11.5 shows these two kinds of waves.

Transverse and longitudinal waves.

Consider the two waves of Figure 11.5 produced on the slinky. Suppose we pass the slinky through a long rectangular slit (whose width is only slightly bigger than the diameter of the slinky) cut out in a wood plank. If the rectangular slit is positioned vertically and the slinky is given vertical transverse waves, then the wave will pass through the slit, oscillating up and down along the slit. Now suppose we turn the plank 90 degrees. Then, the slit will be horizontal while the transverse slinky waves are vertical. The slit will stop the waves. On the other hand, a longitudinal slinky wave is not sensitive to the orientation of the slit. This is how transverse waves are distinguished. Due to this property, transverse waves are said to be **polarized** or to have **polarization**. Polarized Sun glasses take advantage of the transverse nature of light waves.

Polarization is a property of transverse waves only.

---

[2]We shall learn later that a light wave consists of *quanta*, carrying energy which is proportional to the

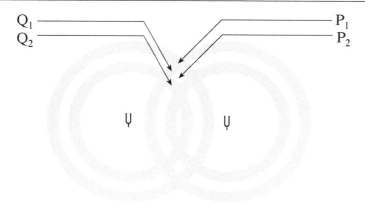

Figure 11.6: By superposition principle each wave acts as if the other were not present. The overall effect is the algebraic sum of the two waves. In this figure, the gray circles represent crests, and the white circular regions between them, troughs. At P_1 and P_2 the amplitudes add, while at Q_1 and Q_2 they subtract.

11.3   Interference and Diffraction

When two waves in a quiet pond reach a common region, they will "meet" for a while and then go their own separate ways. This is a consequence of an important property of waves called **superposition**, which simply states that when two waves reach a common point of the medium in which they are propagating, the total disturbance of that point is the algebraic sum of the two disturbances and the presence of one wave does not affect the properties of the other (see Figure 11.6).

Superposition principle.

Let us take a closer look at the superposition principle. Recall that when a wave reaches a point P of the medium, it tends to start a local oscillation of that point. If two waves reach P at the same time, the total oscillation of P is simply the algebraic sum of the oscillations corresponding to the two waves. If both waves have a tendency to move P up—as the two waves in Figure 11.6 tend to do to P_1—then both oscillations are positive and the total oscillation will be the sum of the two oscillations. Similarly, if both waves have a tendency to move P down—as the two waves in Figure 11.6 tend to do to P_2—then both oscillations are negative and the total oscillation will be the (negative) sum of the two oscillations. Points such as P_1 or P_2 where the waves reinforce each other are called points of **constructive interference**. In contrast, the two waves of Figure 11.6 reaching Q_1 or Q_2 have opposite signs because one wave tends to move Q_1 up (and Q_2 down) while the other wave wants to move Q_1 down (and Q_2 up). The result is the difference between the two waves. Points such as Q_1 or Q_2 where the waves cancel each other are points of **destructive interference**.

Constructive and destructive interference.

> **What do you know? 11.3.** How many points of constructive interference do you see in Figure 11.6? How many destructive?

Now suppose that we have two sources which produce waves in such a way that at certain points in the medium the positive oscillation of the wave coming from one source is *always* canceled by the negative oscillation of the other wave. Then at these points we have minimum or no oscillation. This means that a detector of wave placed at these points will detect no wave at *any* time during which the two sources above are operating.

frequency of the light wave.

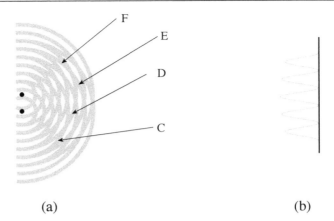

(a)   (b)

Figure 11.7: (a) A spot such as half way between C and D oscillates with maximum amplitude, while C, D, E, and F have no oscillation at all. This figure is very similar to the original drawing by Thomas Young. (b) Plot of intensity as a function of (vertical) position on the screen.

In contrast, there are other points in the medium for which the two waves always arrive with oscillations of the same sign. Such points oscillate strongly and a wave detector can detect maximum oscillation at all times. Such a consistent cancellation or reinforcement is possible only because the motion is periodic. If the two waves cancel (or reinforce each other) the first few times, then they will cancel (or reinforce each other) repeatedly because the same motion gets repeated over and over again. For example, half way between C and D in Figure 11.7(b) has maximum oscillation at all times because it is always the meeting ground of two wave crests (or two troughs). Point C or D, on the other hand, has no oscillation, being where a crest and a trough meet.

Interference pattern.

Not just two slits, but also multiple slits produce interference.

A pair of sources capable of producing the foregoing effect are called **coherent** sources. The effect itself is called the **interference**, and the pattern of maxima and minima produced is called **fringes** or the **interference pattern** (Figure 11.7). Since the points of maxima and minima remain so for the coherent sources, the interference pattern is a *stationary pattern*, i.e., it does not change in time. Interference can also take place if three or more sources produce coherent waves. Further analysis of interference can be found in **Math Note E.11.1** on **page 91** of *Appendix.pdf*.

In a typical interference experiment a wave is incident on a pair of holes (a *double slit*) which act as two sources for the other side. These sources are coherent, because they are created from the same "parent" wave. The interference pattern can be observed on a detecting screen, which could be a white screen in the case of light, or a detector (ears of a person in the case of sound) that can move and record the variation in intensity of the fringes.

Diffraction, another wave property.

Waves have another distinguishing property called **diffraction**. Simply stated, it is the bending of waves "around the corner." A wave approaching an obstacle with an aperture produces spherical (assuming that the aperture is circular) waves as it penetrates through the opening [Figure 11.8(a)], and a detector located at an angle from the direction of the motion of the original wave can detect the wave.

Huygens' principle.

The phenomenon of diffraction (and other wave properties) can be explained by **Huygens' principle.** which states that the motion of a wave can be determined by assuming that each wave front is composed of infinitely many point sources each producing spherical waves. Thus, starting with any wave front, we can construct the new front by drawing equal circles centered at every point of the old front (see Figure 11.9).

The alert reader may have noticed that the wave produced in the outgoing side of the

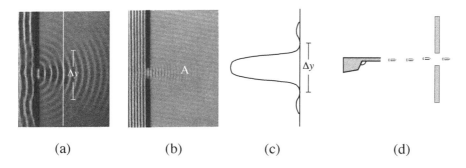

Figure 11.8: Only waves can produce diffraction (a), and only if the aperture diameter is comparable to the wavelength. If the aperture is much bigger than the wavelength, as in (b), diffraction pattern is considerably reduced. (c) The variation of the diffraction intensity with angle. (d) Particles, such as bullets, can go through the aperture only on a straight line.

aperture of Figure 11.8(a) is not composed of uniform circles, but rather of circles, along each of which the intensity varies. This is because the points of the wave at the aperture act as (infinitely) many little sources, which interfere as in the case of multiple slits, and produce a pattern analogous to the interference pattern. Figure 11.8(a) shows a vertical screen (the white line) that collects the wave and shows its intensity at various points on it. If we plot the intensity of the wave as a function of vertical distance on the screen, we get something similar to Figure 11.8(c). It is seen that there is a "central band," on both sides of which the intensity drops to zero. Beyond the central band the intensity is practically negligible, although one can detect some very weak "secondary bands." The amount of diffraction "bending" is thus related to the width of the central band, denoted by Δy. It turns out that Δy depends on the ratio of the wavelength to the size of the opening. If this ratio is of the order of unity the width of the band is noticeably large (as compared to the size of the opening). If, on the other hand, the ratio is too small, i.e., if the aperture is too large compared to the wavelength, then Δy becomes equal to the size of the opening, and the diffraction (bending) effect disappears.

A similar condition holds for interference: If the distance d between the two coherent sources of Figure 11.7 is too much bigger than the wavelength, the interference pattern will not be formed. The wave nature of light escaped detection for centuries because of this fact. Light's wavelength is much too small compared to any ordinary openings and distances. Thus, neither interference nor diffraction of light can be observed by passing it through ordinary holes and apertures. On the other hand, we witness the diffraction of sound everyday when we hear a speaker's voice around the corner of an open door. That is because the wavelength of a typical sound wave is comparable to the size of a typical door.

Interference and diffraction occur only if the appropriate size of the system is comparable to the wavelength.

Box 11.3.1. *For interference and diffraction to take place, the appropriate size (separation of the two holes for interference, and the diameter of the aperture for diffraction) must be comparable to the wavelength.*

It is important (and it becomes even more so when dealing with subatomic phenomena) to point out that

Box 11.3.2. *Only waves have interference and diffraction properties.*

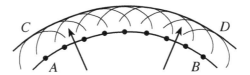

Source • S

Figure 11.9: Any point of a wave front can be considered a new point source. This is Huygens' principle.

If, instead of a wave, bullets are sent through the aperture of Figure 11.8, it is impossible to detect bullets at large angles from their original direction of motion [Figure 11.8 (d)]. This is the reason that for a long time light was assumed to be corpuscular. Every time light was sent through an aperture, it could be detected only directly opposite to it [point A in Figure 11.8 (b)]. The interference and diffraction phenomena have dramatic consequences in quantum theory of subatomic particles to which we shall return later in the book (see especially Section 23.2).

> **What do you know? 11.4.** In one experiment a wave with a frequency of 500 Hz approaches two holes whose separation is 50 cm. Does interference occur on the other side? The speed of the wave is 3 m/s. In another experiment the same wave approaches an aperture whose diameter is 1 cm. Does diffraction occur on the other side?

11.4   Doppler Effect

Everyone has experienced the change in the pitch (frequency) of the sound of a moving fire truck, ambulance, police car, or a train as they pass by. This is **Doppler effect** in action for sound waves. All waves exhibit this effect.

Consider a small source of waves at a point P, which produces spherical waves with P as their common center. As shown in Figure 11.10(a), the distance between consecutive spheres is simply the wavelength. A detector located at O detects the wavelength as λ and the frequency as $f = c/\lambda$ where c is the speed of the wave. Here we are adhering to the convention that, when the source or detector of wave is moving, c denotes the speed of the wave, leaving v for the speed of the source or detector.

Now suppose that the source starts to move to the right towards the detector. Between the time that the first wave front (sphere) is produced and the second is about to be produced,[3] the source has moved to the right. Since P is the center of the spherical waves, the center of the second sphere is displaced to the right [Figure 11.10(b)]. Similarly the third wave front (sphere) has a center displaced to the right of the second sphere's center, and so on. Such a displacement of centers squeezes the circles on the right while loosening them on the left. Thus, the effective wavelength (distance between wave fronts, i.e., spheres) will be shortened on the right and lengthened on the left. Since the velocity of the wave is not affected by the motion of the source,[4] the frequency, $f = c/\lambda$, increases on the right and decreases on the left.

Doppler effect when the source is moving.

---

[3]Recall that this time is, by definition, the period of oscillation of the source.

[4]The velocity depends only on the characteristics of the medium such as temperature, density, pressure,

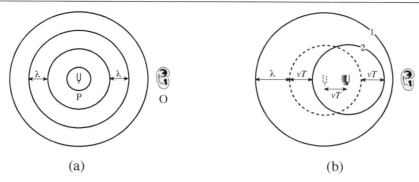

(a)   (b)

Figure 11.10: (a) The wavelength of the stationary source is the same in all directions. (b) As the source moves, the wave fronts are squeezed in front of the source and spread out in the back.

From the foregoing analysis we can conclude that

Doppler effect stated qualitatively.

> **Box 11.4.1.** *The frequency of the wave produced by an approaching source increases (the wavelength decreases) while that of a receding source decreases (the wavelength increases).*

This is the Doppler effect stated in a qualitative way. **Math Note E.11.2 on page 92** of *Appendix.pdf* quantifies the statement above, and obtains a formula connecting the wavelength (or frequency) at the source and at the detector, and **Example D.11.2 on page 33** of *Appendix.pdf* uses this formula to calculate the frequency change in the siren of a police car as it approaches and then recedes from you.

When the source moves the wavelength of the wave changes. What about the motion of the detector? Can it change the wavelength being detected? The first question we have to answer is: In the discussion above, with respect to which observer was the speed v of the source measured? The obvious answer—the detector—is not satisfactory, because the detector cannot be considered "at rest." Why? As we learned in Chapter 6, *motion is relative*. So, when we say that "the source is moving but the detector is not," we have a particular reference frame (RF) in mind.[5] This RF happens to be *the medium whose disturbance constitutes the wave* in question. For the wailing police car, it is the air—the medium of sound waves.

Having found an "absolute" reference frame, let us see what happens to the Doppler shift formula when the source is not moving relative to the medium, but the detector is. **Math Note E.11.3 on page 93** of *Appendix.pdf* finds the formula for this situation, and **Example D.11.3 on page 34** of *Appendix.pdf* uses the formula to find the frequency shift when the source is stationary while the detector is moving.

Doppler effect when the detector is moving.

Examples D.11.2 and D.11.3 seem to violate the principle of the relativity of motion: whether the source is moving or the detector, their *relative* speed is 170 m/s, yet in one case the frequency shifts from 450 Hz to 900 Hz, and in the other case, it shifts from 450 Hz to 675 Hz. This difference is caused by the assumption of the existence of an "absolute" reference frame, the medium in which the wave propagates. And this assumption violates the principle of the relativity of motion, because it grants privilege to a special reference frame, the RF of the medium.

---

etc. and not on the properties of the source

[5] For a discussion of reference frames, see Section 6.1.

For a long time it was thought that **ether** was the privileged medium for light waves, but the theory of relativity tells us that light—as well as all the other electromagnetic waves, including radar—has *no medium*! The correct investigation of the Doppler effect of electromagnetic waves requires the machinery of the special theory of relativity (see Part VI). However, as far as our present discussion is concerned, when the speed of the source or the detector is much smaller than the speed of light, nonrelativistic formulas obtained above apply to light waves as well. Math Notes E.11.3 and E.11.4 illustrate this conclusion. In fact, when the speed of the source or the detector is small compared to the speed of *any* kind of wave, the distinction between the two cases (source moving or the detector moving) is alleviated, as **Example D.11.4** on **page 34** of *Appendix.pdf* illustrates.

Doppler effect gives us a means by which to measure the relative speed of two moving objects S and A. Suppose that S is equipped with a transmitter and a receiver of radar. It sends a signal of wavelength λ to A, which reflects that signal back to S. The receiver at S picks up a signal whose wavelength has been Doppler-shifted from λ. This shift is the result of *two* Doppler effects: one on the transmitted wave and one on the reflected wave. As a result, we expect a Doppler shift that is twice the usual shift. The details can be found in **Math Note E.11.4** on **page 93** of *Appendix.pdf*.

Police cars equipped with a radar detector can measure the speed of cars using this double Doppler shift. The radar transmitter in the police car sends a signal; the car in the front receives this signal but with a different wavelength because the source (the police car) is moving relative to the car; the reflected wave, moving towards the police car has a new wavelength, which gets Doppler shifted again due to the motion of the detector (the police car). **Example D.11.5** on **page 34** of *Appendix.pdf* adds some numerical flavor to the discussion.

A more dramatic application of the Doppler effect than the impact it has on the pitch of a sound is in the study of the motion of stars and galaxies. By detecting the change in the wavelength of characteristic light of elements present in stars and galaxies, astronomers can easily detect whether a particular star or galaxy is moving away from us or is approaching us and with what speed. By careful measurements of such Doppler changes, Edwin Hubble, the American astronomer, was able to conclude, in the late 1920s, that all distant galaxies were receding from us, thus paving the way for the notion of an expanding universe. (See Chapter 29 for further detail.)

11.5   End-of-Chapter Material

11.5.1   Answers to "What do you know?"

11.1. Period T is the inverse of frequency. So, $T = 1/500$ or 0.002 s.

11.2. Distance between the two pulses is speed times time. So, multiplying 0.6 m/s by 0.5 s gives 0.3 m or 30 cm.

11.3. 10 constructive and 8 destructive.

11.4. Use $v = \lambda f$, to find $\lambda = 0.006$ m or 0.6 cm. The two holes of the interference are too far apart compared to the wavelength (50 cm is much much bigger than 0.6 cm) for interference to take place. The diameter of the aperture, on the other hand, *is* comparable with the wavelength. So, diffraction does occur.

11.5.2   Chapter Glossary

Amplitude The property of a wave corresponding to its strength. For water waves, for instance, it is the height of the wave.

Coherent Sources Sources of waves which oscillate in unison and maintain the relative motion of their oscillation.

Constructive Interference Regions of space where the waves of two coherent sources add to oscillate with twice the amplitude of the wave of each source.

Cycle The motion undergone by a simple harmonic oscillator during one period.

Destructive Interference Regions of space where the waves of two coherent sources cancel (destroy) each other.

Diffraction Interference of waves passing through a single aperture. When a wave diffracts, it bends as it passes through the aperture.

Doppler Effect The change in the frequency (or wavelength) of a wave when its source moves towards or away from the detector or the detector moves towards or away from the source.

Frequency The number of cycles a simple harmonic oscillator undergoes in one second. Frequency is the inverse of the period.

Hertz The unit of frequency.

Huygens' Principle states that the motion of a wave can be determined by assuming that each wave front is composed of infinitely many point sources each producing spherical waves.

Interference A property of waves in which two specially prepared sources (coherent sources) construct a pattern at some points of which the waves oscillate with double amplitude (constructive interference) and at other points the wave disappears (destructive interference).

Longitudinal Wave A wave for which the medium oscillates along the direction of wave motion.

Oscillation A motion that repeats itself.

Period The time required for a simple harmonic oscillator to return to its "original" position, which could be any position during the course of its motion.

Polarization A property of transverse waves whereby certain materials block the wave when held in a certain orientation in front of the wave, and allow the wave to pass when rotated 90 degrees from the blocking orientation.

Pulse A single disturbance that travels in a medium.

Simple Harmonic Motion (SHM) An oscillatory motion described mathematically in terms of trigonometric functions. A mass attached to one end of a spring while the other end is held fixed, describes a simple harmonic motion when the mass is displaced slightly and then released.

Simple Harmonic Oscillator (SHO) An object undergoing simple harmonic motion.

Simple Harmonic Wave (SHW) A wave produced in a medium whose source undergoes a simple harmonic motion.

Superposition The property of waves whereby two waves reaching a single point add to give the oscillation of the medium at that point.

Transverse Wave A wave for which the medium oscillates perpendicular to the direction of wave motion.

Wave A continuous succession of pulses traveling in a medium.

Wavelength The distance between two successive similar points of a wave. Denoted by λ, wavelength is measured in meters.

11.5.3   Review Questions

11.1. Who thought that light was composed of particles? Who thought that light was a wave?

11.2. What is a mechanical wave? How does it arise? Give an example of a mechanical wave.

11.3. What is oscillation? Give an example of a simple harmonic motion (SHM). What do we call a system which performs SHM?

11.4. Define period, cycle, and frequency and state what relation exists between period and frequency.

11.5. What is a pulse? What is a wave as defined in terms of pulses? What is a simple harmonic wave? State how you can produce a simple harmonic wave.

11.6. Define wavelength. How is it related to the period of a wave?

11.7. As you watch a wave on a rope, you notice a motion along the rope. Is it the rope that is moving? Explain!

11.8. What familiar property of sound is described by frequency? What familiar property of light is described by frequency?

11.9. What is the range of audible sound frequencies? What is the range of visible light wavelengths?

11.10. What physical property of a wave is associated with its energy (or intensity)?

11.11. What is a transverse wave? What is a longitudinal wave? Which one has polarization property?

11.12. What is constructive interference? What is destructive interference? Enumerate all conditions required of two sources to produce interference. Does *any* double-slit meet these conditions? Explain!

11.13. What is diffraction? Why do we expect diffraction based on multiple-slit interference? Enumerate all conditions required of an aperture to produce diffraction. Does *any* aperture meet these conditions? Explain!

11.14. What is Huygens' principle?

11.15. Why do we see diffraction of sound in our everyday experience, but not that of light?

11.16. What is Doppler effect? Explain why the wave fronts squash together in front of a moving source and separate behind it. What can you say about the frequency of the wave in front and behind the source?

11.17. Is there a Doppler effect when the source is stationary but the detector is moving? Explain why or why not.

11.18. Explain the principle behind a radar detector. How many Doppler shifts do you expect in the radar's operation?

11.5.4 Conceptual Exercises

11.1. If you double the frequency of a sound wave, what happens to its speed? To its wavelength? To the period of vibration?

11.2. Sound travels faster in solids than in air; however, its frequency does not change. Is the wavelength of sound longer or shorter in solids than in air?

11.3. Why does the amplitude of water waves produced in a pond decrease as the waves spread? Hint: Take the intensity of the wave into consideration.

11.4. Why do you see the lightning before you hear the thunder?

11.5. Point P is five wavelengths away from each of the two coherent sources which produce waves in phase (when one produces a crest, so does the other).
(a) Is P a point of constructive or destructive interference?
(b) One of the two sources is now moved so that P is 5.5 wavelengths away from it. Is P a point of constructive or destructive interference now?
(c) What if you further increase the distance of the displaced source so that it is now 6 wavelengths away from P?

11.6. Figure 11.11 shows two coherent sources producing waves. The distance between the sources, exaggerated for clarity, is comparable with the wavelength.
(a) Circle five points in the figure where constructive interference occurs.
(b) Circle five points in the figure where destructive interference occurs.

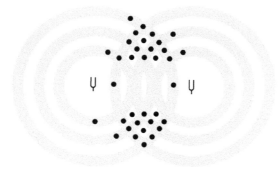

Figure 11.11: The two coherent sources producing waves. The distance between the sources is comparable with the wavelength.

11.7. A fire truck is blowing its siren while stuck behind a traffic jam. The traffic jam clears and the fire truck starts moving towards you. Does the frequency of the sound increase, decrease, or stay the same?

11.8. Using Doppler effect suggest a way of detecting whether Sun spins. Hint: Consider light coming from elements at opposite sides of Sun's edge.

11.5.5 Numerical Exercises

11.1. Audible sound frequencies range from 20 to 20,000 Hz. Visible light wavelengths range from 0.4 μm to 0.7 μm. Speed of sound is 340 m/s and that of light is 300,000 km/s.
(a) What is the range of audible sound *wavelengths*?
(b) What is the range of visible light *frequencies*?

11.2. The ISU public FM radio station WGLT operates at a frequency of 89.1 MHz. Radio waves travel at the speed of light $c = 300,000$ km/s. What is the wavelength of the radio wave broadcast by WGLT station?

11.3. The AM radio station WBBM in Chicago operates at a frequency of 780 kHz. Radio waves travel at the speed of light $c = 300,000$ km/s. What is the wavelength of the radio wave broadcast by WBBM station?

11.4. A wave whose speed is 50 m/s is approaching two interference holes 2 m apart. The frequency of the wave is 2 kHz.
(a) Will this wave produce an interference on the other side of the holes?
(b) The frequency is now reduced to 30 Hz. Is interference possible now?

11.5. A wave whose speed is 500 m/s is approaching a 15-cm aperture. The frequency of the wave is 5 kHz.
(a) Will this wave produce diffraction on the other side of the aperture?
(b) The frequency is now increased to 5 MHz. Is diffraction possible now?

11.6. A police car is moving at the rate of 50 mph while sounding its siren which produces a sound with a frequency of 400 Hz. Speed of sound is 340 m/s.
(a) What is the wavelength of the sound as measured by the driver?
(b) What is the wavelength of the sound as measured by an observer in front of the car?
(c) What is the frequency of this sound?
(d) What is the wavelength of the sound when the car recedes from the observer?
(e) What is the frequency of this sound?

11.7. The speedometer of a police car shows a speed of 95 mph as the policeman chases a speeder. He sends a radar wave with a wavelength of 4 m to the car and receives a signal whose wavelength has increased by 2×10^{-8} m.
(a) Is the police car approaching or receding from the speeder?
(b) What is the fractional change in the wavelength $(\Delta\lambda/\lambda)$?
(c) What is the speed of the police car *relative to the speeder*?
(d) How fast is the speeder going?

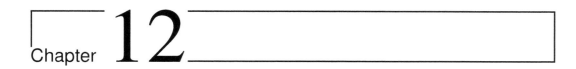

Chapter 12

Electricity

The evolution of electricity and magnetism constitutes one of the most fascinating chapters in the history of our civilization. What started as a childlike curiosity in the beginning of the seventeenth century turned into a major scientific enterprise in the beginning of the nineteenth century, developed into a dominant theoretical breakthrough in the middle of the nineteenth century, and put into some curious applications at the turn of the twentieth century. By the middle of the twentieth century, in combination with the quantum theory, electromagnetism became the cornerstone—no, the signature—of what we have come to call civilization. We can safely claim that electromagnetism is at the very foundation of every single modern technology.

The story of electricity, as many other intellectual developments, begins in a turbid antiquity. The first recorded observations of electric phenomena date back to the ancient Greeks who attributed the discovery of electricity (and magnetism) to **Thales of Miletus** (640?–546 BC). A contemporary of King Nebuchadnezzar of Babylon, Thales traveled extensively, receiving most of his education in physical sciences from the priests in Memphis and Thebes in Egypt. He founded the Ionian school of philosophy from which Socrates came. Thales appears to have been the first to observe that amber, when rubbed, was capable of attracting light objects.

Thales discovers electricity.

12.1 Modern Electrostatics

At about the same time that Galileo started the systematic study of motion in Italy, the wind of the Renaissance was fluttering the leaves of the tree of knowledge in other parts of Europe. The science of electricity and magnetism, the second major branch of physics after mechanics, started in sixteenth century England in the hands of William Gilbert. However, unlike the study of motion, which immediately demanded expertise and invention of a mathematical language, observation of electric and magnetic phenomena did not appear to require much fluency in mathematics at the beginning. Little wonder that the founder of modern electricity and magnetism was not a mathematician or a physicist, but a physician.

Gilbert's methodology of verifying the properties of electricity and magnetism by *experimenting* was a novel idea equaled only by his younger contemporary Galileo, who greatly admired Gilbert's book. However, this methodology was too radical for the "philosophers" and scholars of the day. And Gilbert showed no restraint in showering his contempt on them in his book, *De Magnete*.

Most of *De Magnete* deals with *lodestone* and its magnetic properties; however, Gilbert also touches on electricity. **Amber**, an ancient fossilized tree resin found in the Baltic region, and named by the Greeks "elektron," had a very special quality. When rubbed with

Gilbert discovers the attractive properties of other materials.

a cloth it could make chaff, bits of thread, and other light particles jump and stick to it.

This fascinating and almost supernatural property came under Gilbert's scrutiny. He soon discovered that it was not just amber that attracted light objects upon rubbing, but other substances such as rock crystals and some gems exhibited similar properties. He was the first to clearly differentiate between magnetic force, which acts only on iron and related elements, and the electric force which exists among many more substances. He is also credited with the word "electric," which he gave to all substances exerting "amber like" forces on light objects.

BIOGRAPHY

William Gilbert (1544–1603) was born in Colchester, England. After grammar school, William, in 1558, entered St. John's College, Cambridge, to study mathematics, and in 1564 he became an examiner in that subject for the Royal College of Surgeons. His interest then turned to medicine, and he graduated as a doctor in 1569. Upon receiving MD, he became a senior fellow of the college, where he held several offices. Gilbert set up a medical practice in London in the 1570s and became a member of the Royal College of Physicians (the body that regulated the practice of medicine in London and vicinity). He held a number of offices in the college and in 1600 was elected president. Gilbert, like Copernicus, did not limit himself to the practice of medicine, and soon found himself immersed in the mysteries and magic of electricity and magnetism. In 1600, he published the results of his findings in his great work *De Magnete*.

Otto von Güricke constructs the first electric machine.

For about sixty years, the findings of Gilbert on electricity lay dormant. Then in 1660, the German physicist and burgomaster of Magdeburg, Otto von Güricke (1602–1686), constructed the first electrical machine, in which he employed Gilbert's idea of "electrifying" substances by rubbing them. Using this machine, Güricke was capable of charging certain substances indefinitely and creating impressive sparks upon discharging them. Güricke's machine was pivotal in the study of static electricity in the eighteenth century.

Metals are discovered to be conductors of electricity.

In his attempts to electrify a wide variety of materials, Stephen Gray (1695–1736) discovered that metals cannot be electrified by rubbing. It then occurred to him that electricity could be transmitted to metals from other electrified objects such as glass. Soon he noticed that once a piece of metal is charged at one point, electricity appeared at other points of the metal. Gray's next challenge was to see how far he can transmit electricity. He chose longer and longer metallic rods, touched one end of them with electrified objects, and detected electricity at the other end. By 1734 Charles Du Fay (1698–1739) pushed the record electrical transmission distance to 1256 feet. Based on the observation that a charged body attracts an uncharged body, which after contact repels it, Du Fay also proposed that there are two kinds of electricity.

Electricity is proposed to be of two kinds.

> **What do you know? 12.1.** Why does a charged body attract an uncharged body and then repel it after contact?

The Leyden jar is invented.

One of the most important discoveries of this period was the **Leyden jar**, a bottle partly filled with water, containing a metal rod which projected through the neck. It was found that when the metal rod was connected to an electrical machine, the jar would "hold" the electricity. The news of this discovery spread rapidly throughout Europe, and within a short time many replicas of the jar were manufactured by various people. The Leyden jar, which is still in use, became a convenient device in which to store electricity.

Connection is made between lightning and electricity.

The word of the jar soon crossed the Atlantic Ocean and reached the shores of the American Continent. Benjamin Franklin (1706–1790), the American statesman and scientist, while toying with the Leyden jar, noticed a striking resemblance of the sparks and crackles of the jar to lightning and thunder. He thus came up with the conjecture that lightning is nothing but electrical discharges between the clouds and the Earth, and to prove his conjecture, he performed his famous experiment in which he flew a kite in stormy weather and noticed—quite painfully!—that electricity was carried from the tip of the kite to his hand.

As the end of the eighteenth century approached, enough experiments had been done on static electricity to render the question of the law of electrical forces relevant.

BIOGRAPHY

Charles Augustine de Coulomb (1736–1806) was born into an influential and wealthy family. After a solid training in language, literature, philosophy, mathematics, and science, Coulomb became interested in mathematics and astronomy. At the age of twenty, having had many arguments with his mother in Paris over his future career, Coulomb joined his father in Montpellier, where he became a member of the Society of Sciences. His interest in science drove him into a career in engineering in the French army, serving in the West Indies. When several bouts of fever forced his return to France, he began a lifetime of broad-ranging investigations in mechanics, electricity, and magnetism. As part of his investigations, he experimented with the twisting characteristics of various fibers under tension. This observation led Coulomb to the construction of a very sensitive instrument called the **torsion balance**, which he used to discover the mathematical law of the force of static electricity.

After a long series of experiments in electricity with the torsion balance, Coulomb, in 1785, discovered that the electrical force obeys a law similar to the gravitational force discovered over a century earlier by Isaac Newton [see Equation (9.1)]. The **Coulomb's law** states that the force F between two point charges of magnitudes q_1 and q_2 situated a distance r apart is given by

$$F = \frac{k_e q_1 q_2}{r^2},$$ (12.1)

Coulomb's law.

where k_e is a proportionality constant whose value, in the system of units we are using, is 8.988×10^9, which can be approximated as 9×10^9. The unit used for charges is **Coulomb** which can be defined as the amount of charge which exerts a force of 8.988×10^9 N on an equal charge when their distance is 1 m.

Unit of electric charge is **Coulomb** (C).

Although the laws of electrostatic and gravitational forces look alike, there is a vast difference between the two. First, gravitation is always attractive, while electrical force can be both attractive and repulsive: *like charges repel, unlike charges attract.* Secondly, the strength of the two forces are incomparably different as the following example shows.

> **What do you know? 12.2.** On a planet, a charged ball has a different acceleration than a neutral ball. What can be the reason?

Example 12.1.1. An electron has a mass of 9.1×10^{-31} kg and a negative electric charge of magnitude 1.6×10^{-19} Coulomb. Let us compare the strength of the gravitational and electrical forces between two electrons separated by 1 meter. The force of gravity is attractive, but the electrical force is repulsive. Do the two electrons attract or repel one another?

Recall that the gravitational force is given by $F_g = G m_1 m_2 / r^2$. Therefore, with $r = 1$, $G = 6.67 \times 10^{-11}$, m_1 and m_2 each equal to 9.1×10^{-31} kg, we get $F_g = 5.52 \times 10^{-71}$ N. On the other hand, with $k_e = 9 \times 10^9$, each charge equal to 1.6×10^{-19} C, and $r = 1$, the electrostatic force, $F_e = k_e q_1 q_2 / r^2$, yields $F_e = 2.3 \times 10^{-28}$ N. The ratio of F_e to F_g is 4.17×10^{42}.

Comparison of electric and gravitational forces.

Thus, the electric force is approximately 42 orders of magnitude larger than the gravitational force. To see the enormity of this ratio, suppose that we represent the gravitational force by an arrow as long as a *bacterium*. How long would the length of the electric arrow be? Much larger than the height of a man; much taller than the Empire State Building; much larger than the Earth itself; much larger than the solar system; and much larger than the Milky Way. In fact, the electric arrow would be over 10 billion times larger than the "size" of the visible universe! That is why, when electrical forces are present, gravitational forces can be completely ignored. ∎

If gravity is represented by a bacterium, how big is electricity?

12.2   Electric Field

Coulomb's law is an action-at-a-distance law possessing all the disadvantages of such laws. To remedy its shortcomings, let's introduce the concept of the electric field, which is very similar in concept to the gravitational field discussed in Section 9.2: If an electric charge (usually called a **test charge** in this context) feels an electric force (because of its charge) in a region of space, we say that an **electric field** is present in that region. The stronger the force, the more intense the electric field. More precisely, if $\mathbf{F}_e$ is the electric force acting on a charge q at some point, then the electric field $\mathbf{E}$ at that point is

Electric field defined.

$$\mathbf{E} = \frac{\mathbf{F}_e}{q} \quad \Rightarrow \quad \mathbf{F}_e = q\mathbf{E}. \tag{12.2}$$

This suggests that the unit of electric field is **Newton per Coulomb** or N/C. Electric field is thus simply the electric force per unit electric charge, and the connection between the field and the force concept is evident in this equation.

Box 12.2.1. *Equation (12.2) shows that the electric force $\mathbf{F}_e$ has the same direction as the electric field $\mathbf{E}$ when q is positive and the opposite direction when q is negative.*

The source of the electrostatic field is, of course, the electric charge, just as the source of the gravitational field is mass. The detection of an electric field implies the existence of an electric charge somewhere. The farther you are removed from the source of the electric field, the weaker the magnitude of that field. This is shown quantitatively in **Example D.12.1** on **page 35** of *Appendix.pdf*.

🛒 Food for Thought 🌿

Gilbert's pseudoscientific explanation of electricity and magnetism.

FFT 12.2.2. As valuable as Gilbert's observations were in promoting electricity and magnetism to the level of a science, his *explanations* of the phenomena bordered on the occult and mysticism. He speaks of electricity as *materia* and of magnetism as *forma*. He continues, "Electrical movements come from the *materia*, but magnetic from the prima *forma*; and these two differ widely from each other and become unlike, the one (magnetics) ennobled by many virtues, and prepotent; the other (electrics) lowly, of less potency, wherefore its nature has to be awakened by friction til the substance attains a moderated heat, and gives out an effluvium, and its surface is made to shine."

This semi-spiritual characterization of magnetics as "noble" and electrics as "lowly" is hardly scientific by today's standards. However, one has to realize that Gilbert's time was a time of transition from superstition to reason. Many great astronomers, including Copernicus and Kepler, believed in a connection between heaven and earthly happenings as today's psychics do. So Gilbert's "explanation" of the difference between magnetism and electricity should not be held against him. Moreover, as we shall see later, the electric and magnetic phenomena, with which Gilbert dealt, proved to be much more difficult to analyze scientifically and mathematically than motion, whose scientific analysis started at about the same time.

Today, we value Gilbert's work, and ennoble it by qualifying it as "scientific." Now imagine a hypothetical "scientist," who at about the same time, created the "science" of magnetozoology. This would be a man who would make many "systematic observations" of the magnetism "released" by hundreds of animals. He would write dozens of papers on the nature of animal magnetism, and publish several books on the subject. Why do we not call this man a scientist?

Every generation of physicists owes its success to the knowledge it inherits from the previous generation.

Although Gilbert's study of electricity and magnetism was not unlike our hypothetical magnetozoologist's study of animal magnetism, Gilbert's work was *followed up* by other physicists such as Güricke, who used Gilbert's discoveries to build his electricity machine; Gray, whose discovery of conductors was crucially based on the work of Gilbert and Güricke; Kleist, who combined Güricke's and Gray's discoveries to create a jar that was capable of storing electricity for a long time; and Coulomb, who used Kleist's jar in his monumental work that led to the mathematical expression of the electric force. And the process did not stop with Coulomb. Every generation of physicists owes its success to the knowledge it inherits from the previous generation. This is one characteristic of science that should be remembered at all times: *Science is a ladder whose upper rungs are crucially attached to its lower rungs*.

Four hundred years after Gilbert, we now know *why* the act of rubbing produces electricity: the elemental particles we call electrons get transferred from one object to the next, electrifying both. Gilbert was not aware of this, but his observations were the start of our modern understanding. On the other hand, our hypothetical magnetozoologist could not have laid the foundation of a new science, because animal magnetism was not (is not) the right material to study in a scientific investigation, and therefore, no scientist would have followed up on his investigations.[1]

The lesson to learn is that no amount of "systematic observation," no amount of emphasis by using the word "science," no amount of persuasion by gluing the suffix "-ology" or "-ics" to a discipline makes it a science. Science is no invention of one man or a group declaring the birth of a new "scientific discipline" after a conference. Science is as old as the human race itself. There is a continuous line that connects Einstein's general theory of relativity and the science of genetics to the discovery of fire 700,000 years ago. Beware of people who speak of "The New Science of . . . ." They are trying to break the line and artificially insert their discipline in the historical continuum.

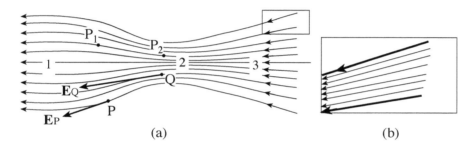

(a)   (b)

Figure 12.1: (a) Electric field lines, also called lines of force. The direction of the field is tangent to the field lines (curves). (b) The boxed region in (a) is magnified to illustrate that there are a lot more lines than meet the eye!

Electric (or any other) fields are depicted as streams of *directed* curves called **field lines** or **lines of force**. The direction of the electric field at any point through which a field line passes is along the *tangent* to the curve at that point. For example, the arrows $\mathbf{E}_P$ and $\mathbf{E}_Q$ in Figure 12.1 represent the electric fields at points P and Q. The field at other points can be drawn similarly. These arrows show the direction in which a *positive* test charge would experience a force when placed in the given electric field.

Electric field is represented as streams of directed curves.

In regions where the lines are more crowded together, the field is assumed to be stronger. Thus, the positive charge will experience a greater force to the left in Figure 12.1 when placed in region 2 than in region 1 or 3. That is why $\mathbf{E}_Q$ is drawn longer than $\mathbf{E}_P$. The sources producing the field are not shown in Figure 12.1, and most of the time they are not important.

Strength of electric field is related to the density of field lines.

What do you know? 12.3. Conductors have freely moving charges which easily rearrange themselves in such a way that the electric field inside the conductor is *always* zero. What happens if you force some charge inside a conductor?

Figure 12.2 shows some typical field configurations as well as the charges that produce them. In (a) the arrows point away from the positive charge because that is the direction

[1]Animal magnetism did become the subject of the *pseudo*scientific investigation of many people (see Section 43.4.3).

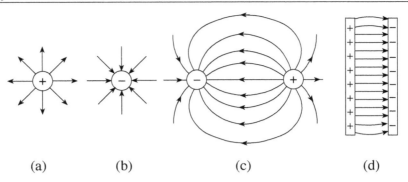

Figure 12.2: Some typical field lines with charges that produce them. (a) A positive point charge produces field lines that point radially outward. (b) Field lines of a negative point charge point inward. (c) Field lines of a dipole. (d) Field lines between a pair of oppositely charged plates.

of the force experienced by a positive test charge placed nearby. Notice that the lines are more concentrated near the charge indicating a stronger field there. The lines have changed direction in (b) because the source of the field is a negative charge causing attraction (arrows pointing toward a source) on a positive test charge. Figure 12.2 (c) shows an **electric dipole**. The field lines point away from the positive charge and toward the negative charge. The field lines in (d) are uniform except close to the edges of the charge distributions. This is what happens when two plates are charged oppositely and placed very close to each other ("close" meaning that the distance between the plates is considerably smaller than the dimensions of the plates).

Electric dipole.

The lines drawn in a typical representation of an electric field are only sample lines. Bear in mind that between any two neighboring lines there are (infinitely) many lines whose directions are intermediate between the two given lines. Figure 12.1(b) shows the magnified version of the region boxed in part (a) of that figure. Between the two lines in the upper right-hand corner of the figure in (a), there are (infinitely) many lines, only six of which are drawn in (b). Thus the reader should keep in mind that

Box 12.2.3. *In every region that there is an electric field, an infinite number of lines exist, of which only a finite sample is chosen to represent them pictorially.*

12.3   Electric Potential

Chapters 7 and 9 introduced the notion of the potential energy PE for the gravitational field of the Earth and other celestial bodies, and indicated that the higher an object is, the more PE it has. The natural tendency is for the object to move from a state of higher PE to that of a lower PE. The same considerations apply to electric fields. An **electric potential energy** is defined as the energy stored when an electric charge is moved from one point to another in an electric field.

Electric potential energy defined.

Consider Figure 12.1 again, and assume that a *positive* charge is moved from P_1 to P_2 along the field line connecting these two points. Since the direction of the field is from P_2 to P_1, the *electric* force will be in the same direction. To push the charge from P_1 to P_2, you have to provide it with a *mechanical* force directed from P_1 to P_2. Since the displacement and the mechanical force are in the same direction, the work done by this force is positive. This positive work is *stored* as the excess of electrical potential energy (EPE) at P_2 over that at P_1. We thus say that P_2 is at a higher electrical potential energy than P_1. A positive charge released at P_2 has a natural tendency to move towards P_1. Thus, in general, motion of positive charges take place from a region of high EPE to a region of low EPE.

The preceding discussion illustrates a specific instance of a general consideration. It turns out that P_1 and P_2 need not be on the same field line, and the work necessary to move the charge from P_1 to P_2 is independent of the path connecting the two points. Therefore, the EPE depends only on the points and not on how they are connected. If EPE_1 and EPE_2 denote the electric potential energies of a *positive* charge q at P_1 and P_2, respectively, then $EPE_2 - EPE_1$ is the work done by an external force to move q (slowly) from P_1 to P_2 irrespective of the path taken.

As in gravity (Section 9.2), we can introduce the notion of the potential. In fact, **electric potential**, denoted by V, is far more useful than its gravitational counterpart.[2] The electric potential difference between two points is the work required to move *one Coulomb* of electric charge from one point to the next. Equivalently, it is the difference between EPE at the two points divided by the displaced charge. Thus if V_2 is the electric potential at P_2 and V_1 that at P_1, then

Electric potential defined.

$$V_2 - V_1 = \frac{EPE_2 - EPE_1}{q} \quad \text{or} \quad EPE_2 - EPE_1 = q(V_2 - V_1). \qquad (12.3)$$

The concept of electric potential is useful because it does not depend on the charge. It only depends on the strength of the electric field between the two points. In fact **Math Note E.12.1** on **page 95** of *Appendix.pdf* shows that the electric potential difference between two (nearby) points is the electric field times the distance between the two points. The same Math Note also shows that the electric field points from a region of high electric potential to a region of low potential. Hence,

Box 12.3.1. *Positive charges move from high potential regions to low potential regions, while negative charges move in the opposite direction.*

What do you know? 12.4. Do you have to do positive or negative work to move a negative charge from a low potential region to a high potential region?

The unit of the electric potential is Joules per Coulomb, which is renamed **volt** to honor the Italian physicist Volta for his contribution to the development of electricity. In addition to N/C, electric field has a unit derived from Equation (E.14) of Math Note E.12.1. Since electric field can be written as potential difference divided by distance, its unit can also be **volt per meter**.

Volt defined.

By the definition of the potential difference, a positive charge q loses $q(V_2 - V_1)$ units of EPE when it goes from P_2, with higher potential V_2, to P_1, with lower potential V_1. Conservation of energy then implies that q acquires the same amount of kinetic energy. The same consideration applies to a negative charge moving from P_1 to P_2.

Example 12.3.2. A convenient unit of energy is **electron volt** (eV) defined as the kinetic energy gained by an electron when it moves in a potential difference of one volt. Electron itself has a negative charge $q_e = -1.6 \times 10^{-19}$. To define eV, consider a particle (such as proton) which has a charge of $+1.6 \times 10^{-19}$. Electron volt is then the KE gained by such a *positive* charge as its potential drops by one volt.

Electron volt defined.

Q: What is the kinetic energy in eV gained by $+1$ Coulomb of charge when it moves through a potential difference of one volt?

---

[2]The interested reader may refer to Math Note E.9.3 for a discussion of the gravitational analog of the present topic.

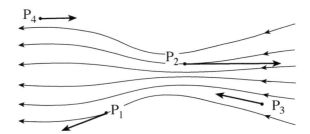

Figure 12.3: Electric fields and forces on some charges located in different regions of the field.

A: The charge of 1 Coulomb is $1/1.6 \times 10^{-19} = 6.25 \times 10^{18}$ times that of a proton. Thus, the energy gained by 1 Coulomb is 6.25×10^{18} times the energy gained by a proton. Since the latter is, by definition, one eV, we conclude that the kinetic energy gained by 1 Coulomb of charge in going through a potential difference of 1 Volt is 6.25×10^{18} eV.

On the other hand, Equation (12.3) indicates that the product of potential difference in volts and charge in Coulombs is simply energy in Joules. We thus have the conversion relations

$$1 \text{ Joule} = 6.25 \times 10^{18} \text{ eV} \quad \text{or} \quad 1 \text{ eV} = 1.6 \times 10^{-19} \text{ Joule}. \tag{12.4}$$

Electron volt, due to its smallness, is a convenient unit for atomic and subatomic phenomena. ∎

Example 12.3.3. In a certain region of space the electric field lines are as shown in Figure 12.3. If you introduce a charge of $+1$ Coulomb at point P_1, the force will be in the same direction as the field, i.e., tangent to the field lines and pointing in the direction determined by the arrows on the lines. Let us represent it as an arrow shown in the figure; it points approximately in the southwest direction. If the charge is released, it will accelerate in that direction *immediately after release*.[3] Now introduce a charge of -1 Coulomb at P_2. The electric field points to the left (west) there. Therefore, the force will point to the right (east). Why is the arrow longer at P_2? Because the field lines are more crowded there, indicating that the electric field is stronger.

What if you introduce a charge of $+1$ Coulomb at P_3? Is the force zero there because no electric field lines are shown to pass through P_3? No! Remember that we can afford to draw only a finite sample of the field lines. Between any two field lines at any point in space, there pass infinitely many other field lines, whose directions lie somewhere between the directions of the two original lines. So, the field line passing through P_3 must point a little north of west, and the force will be in the same direction as shown. ∎

What do you know?   12.5. A charge of -1 Coulomb at P_4 of Figure 12.3 will experience a force represented by the arrow in the figure. Explain.

12.4   Electric Charges in Motion

Galvani discovers the effect of electricity on frog's muscle and nerve tissues.

The history of the electric current begins with an accidental discovery by Galvani, who employed a variety of newly discovered electrical machines and the Leyden jar to produce and store electricity, and apply it to dissected frog legs—which were popular and convenient specimens for investigation—to study muscular and nerve activity.

Galvani and his laboratory assistants dissected and prepared a frog on a table on which there was an electrical machine. Quite accidentally, one of his assistants noticed that at the precise time that a spark was coming from the electrical machine, the muscles of the frog's limb were suddenly and violently convulsed. Galvani repeated the experiment many

---

[3]The subsequent motion cannot be determined simply from the initial direction of the force!

times, including some trials with the weather electricity, and each time the convulsions were inevitable. But it was another accidental observation that determined the future direction of research in electricity.

BIOGRAPHY

Luigi Galvani (1737–1798) wanted to study theology and enter a monastic order. His family, however, persuaded him to abandon that idea. Galvani eventually gained notoriety as a gifted scholar of medicine, and at age 25 became a Professor of Anatomy at the University of Bologna. As a skillful and diligent experimenter, and in keeping with the science of his time, he developed a keen interest in electricity and employed a variety of newly discovered electrical machines and the Leyden Jar to produce and store electricity. Galvani applied this electricity to dissected frog legs—which were popular and convenient specimens for investigation—to study muscular and nerve activity.

During his experimental trials with frog legs, Galvani noticed that frogs that were suspended by brass hooks through marrow, and which rested against an iron trellis, showed occasional convulsions with no active electrical machine around and no lightning in the atmosphere. In adjusting the specimens, he pressed the brass hook against the trellis by chance, and saw the familiar muscle jerk. He repeated the experiment and noticed that each time he completed the metallic contact between the brass hook and the iron trellis, the twitching of the muscle occurred.

In another similar experience, he noticed that when the specimen laid on an iron plate and he touched the brass hook in the spinal marrow against the plate, the convulsion took place. Galvani now recognized that a new principle was at work, and he decided to unravel the phenomena responsible for it. He replaced the iron plate with glass plate; the twitching did not occur. Using a silver plate restored the convulsion. He then joined equal lengths of two different metals and bent them into an arc. On touching the tips of the bimetallic arc to the frog specimen the reaction returned. He paired different metals into arcs and noted that the intensity of the twitching differed for different metallic pairs.

Galvani was now faced with the challenge of explaining the two kinds of convulsion he had observed in frog specimens: one coming from the sparks of electrical machines, and the other coming from the contact of dissimilar metals. The second kind of convulsion was especially puzzling: Did the electricity responsible for the convulsion reside in the anatomy of the specimen with the metals serving to release it, or was the effect produced by the bimetallic contacts, with the specimen serving as an indicator?

Being an anatomist, Galvani had a propensity toward living tissue, driving him to believe that the twitching of the muscle tissue was caused by "animal electricity." He took his discovery as one in physiology, and compared the body as a Leyden jar, in which the various tissues developed opposite electrical charges. In this, he was following the tradition of his day which ascribed the body motivation to a flow of "spirits" residing in the various body parts. This pseudoscientific "explanation" was soon to be undermined by another Italian, Alessandro Volta.

Volta is credited with inventing several important devices, including the **electrometer**, which is capable of measuring very minute electrical effects. Using these devices, he became a pioneer of the new science of electricity, and being an Italian, he was bound to come across the discovery of his fellow countryman, Galvani. Upon repeating and confirming Galvani's experiment, Volta also accepted "animal electricity" as the cause of the muscular convulsions. Both ascribed the activity of the muscle to an imbalance between the muscle and nerve electricity, which was restored to equilibrium when a metallic connection was made.

On continuing his investigations, however, Volta began to question the imbalance theory. He found in his experiments that the muscle would convulse when only the nerve was in the electrical circuit. Replicating the experiments of an earlier investigator, Volta placed a piece of tinfoil on the tip of his tongue and a silver coin at the rear of his tongue, and connecting the two with a piece of copper wire, he got a sour taste. Substituting a silver spoon for the coin, removing the copper wire, and connecting the spoon directly to the tinfoil, he got the same sensation. In another experiment, he connected his tongue to his forehead using dissimilar metals, and he got a sensation of light. All these effects would go away if the connections were made with a *single metal*. From these experiments, Volta concluded that

Box 12.4.1. *The dissimilar conductors did not act merely as agents for transferring electricity, but were actually the* source *of electricity.*

Volta does away with animal electricity.

By 1794 Volta had made a complete break from Galvani and the theory of animal electricity, and proposed the theory of "metallic electricity." To do away with the "animal electricity," Volta completely discarded the muscles and the nerves. In one of his experiments, he connected the two ends of the dissimilar metal to the two plates of his electrometer, and noticed the presence of electricity. No tissues were involved! Volta proceeded to test many metallic pairs and noted that the strength of the electricity varied considerably for different pairs. One important discovery along these lines was that the *moisture* of his hands considerably enhanced the strength of the electricity produced. Further investigation led Volta to add liquids such as brine and dilute acids in his conducting system.

BIOGRAPHY

Alessandro Volta (1745–1827) was very slow in learning as a child, so much so that his parents thought that he was mute. However, by age seven he turned into a bright student of exceptional promise. Within a few years, Volta had obtained an excellent classical education, and found himself under pressure from relatives to enter the priesthood. But, encouraged by an old friend who provided him with some experimental equipment, Volta decided to become a physicist. He quickly took an interest in the magical science of the time, electricity. His passion for electricity was so great that he wrote a poem in Latin on the subject. One of the greatest experimental physicists of all time, Volta invented electrophorous, studied the effect of sparks on the ignition of gases, discovered methane, and developed the voltaic pile, the precursor to modern batteries.

Volta invents the battery.

Volta discovered that in a circuit composed entirely of pairs of dissimilar metals, there was only a momentary movement of electricity. However, when he put two dissimilar metals in contact with a separator soaked with a saline or an acidified solution, he noticed a *steady* supply of electricity. Volta had accomplished the most decisive and practical discovery of his career: He had invented an electric **battery**.

What is the basic principle behind the operation of batteries? Charges, as all other objects, obey the laws of motion: They move in a wire because there are net forces acting on them. These forces are simply electric forces caused by various accumulation of charges in the batteries, producing electric fields in the wires, which in turn "push" the charges inside the wire, causing the flow of an electric current. If there were no forces to oppose the motion of the charges, not only would the charges move on their own, but they would even acquire *acceleration*, and there would be no need for a battery to "help" them along. However, just as a car moving on a highway is dragged by the ambient air, the charges moving in a wire are constantly dragged by microscopic objects. And just as the car needs a constant supply of force to keep it from slowing down and stopping, so do the charges in **Batteries and electric potential.** the wire need a constant supply of electric field to keep them from stopping. The presence of an electric field in a wire indicates an electric potential difference [see Equation (E.14) of Math Note E.12.1]. Therefore,

Box 12.4.2. *Batteries are devices which create electric field in—or potential difference between the two ends of—a wire.*

What do you know? 12.6. Earlier, we said that no electric field can exist in a conductor. How do you reconcile that statement with the one in Box 12.4.2?

A typical battery has two poles, positive and negative, and it can drive the charges only if these poles are connected by a conductor of electricity such as a metal. When we connect the two poles of a battery, we say that we have a **complete electrical circuit**. The carriers of electricity (negative electrons) are pushed away from the negative pole into the conducting wire, where under the influence of the field they are moved to the positive pole. To keep the electric current going, the battery now has to take the charge and put it on the negative pole. This is done internally by various chemical reactions. The ability of a battery to move the carriers from the attractive pole to the repulsive pole is measured by **electromotive force** (emf), which is denoted by $\mathcal{E}$, and has the same unit as electric potential, i.e., volt.[4] To remove the carrier of charge q from the attractive pole (where it wants to remain) to the repulsive pole, the battery does $q\mathcal{E}$ Joules of work. So, one can say that $\mathcal{E}$ is the potential difference between the two poles of a battery that a carrier must overcome in order to continue its periodic journey through the circuit.

Electromotive force or emf.

A battery is in many respects like a water pump. The pump takes water from underground, lifts it up to the surface by the expenditure of some energy (mechanical, electrical, or chemical), and releases it on the ground. The ground water then seeps through the ground layers and ends up underground, completing the water "circuit." Both electric and water flow need a source of energy (battery or pump).

Escher was a Dutch graphic artist whose interest in mathematics influenced his ingenious drawings of impossible constructions. One such construction (shown in the figure in the margin) defies energy conservation and the need for a pump to raise water to a height. Such a situation, of course, never occurs in nature. It, however, does occur in the greed of hoaxers and fraudulent entrepreneurs claiming the invention of machines that need no fuel.

The electric current is the effective motion of *positive* charges in a conductor from the higher potential at one end to the lower potential at the other end of the conductor. This requires some explanation. Historically, the founders of the science of electricity, when faced with the necessity to define concepts or make conventions, did so in terms of positive charges. Thus, for instance, electric current was defined in terms of the motion of positive charges. However, by the end of the nineteenth century it became clear that the particles whose motion gave rise to electric currents were *negatively* charged. Furthermore (as we noted earlier), positive charges move from high to low potential, and negative charges from low to high. This positive-negative and high-low potential asymmetry may appear as double jeopardy. In reality, however, they work to cancel each other as the following argument shows.

 Food for Thought 🌱

FFT 12.4.3. The use of batteries in modern society is as common as the consumption of food. Every appliance, large or small, uses batteries. Many inventors have become billionaires because of the existence of batteries. Yet, Volta, as staunch an *experimental* physicist as he was, showed no interest in his battery's entrepreneurial "potential,"[5] and that is what set him apart from an inventor.

---

[4]This is the voltage of the battery you buy at your local electronic store.

[5]No pun intended!

An inventor's main concern is the mass production of a product for human consumption. Therefore, his entire enterprise is focused on the ease of the utility of the product and the persuasion of the public to buy it. In fact, in modern consumer societies, the invention itself takes a back seat while its *marketing* becomes the center of attention. History is full of good inventions that never made it, because they were not advertised aggressively enough, and useless inventions that are found in every household because we see them on the screen every time we turn on our television set, and hear about them every time we listen to our radio.

While an inventor asks "What is the best way to make this gadget useful and popular?", an experimental physicist asks "What is the best way to understand how and why this gadget works?" The experimentalist pays absolutely no attention to the look and utility of his gadget, the two most important factors that drive an inventor. A successful inventor has to be a businessman, a politician, and a very clever salesman. A successful experimentalist has to be a devoted experimentalist [although, nowadays the procurement of research funds demands *some* element of political savvy in the experimentalist]. In short, an inventor is a *tech*nician, in the sense that he/she takes advantage of *tech*nology, while an experimentalist is a *scientist*.

> The same difference that exists between science and technology exists between an experimentalist and an inventor.

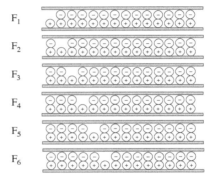

Figure 12.4: Negative charges moving in one direction is equivalent to positive charges moving in the opposite direction.

Figure 12.4 shows some fixed positive charges with respect to which some negative charges are moving to the left. Imagine that the negative charges proceed to the left by hopping from one site to the next. Thus in the first frame F_1 of the figure there is a positive vacant spot to the left. In F_2, this empty spot has been occupied by a negative charge and a vacant spot is created to the right of this negative charge. Every time a negative charge hops, it creates a vacant positive charge behind it. This positive vacant spot keeps moving to the right as the negative charge hops to the left. To an observer capable of measuring only the electric charge and not the motion of the individual particles, the process appears as a motion of the positive charge to the right. We, therefore, see that the two interpretations of the electric current—negative charges to the left or positive charges to the right—are completely equivalent. By convention, the latter interpretation is accepted although in reality the former is true.

What has been loosely called electric current has a precise meaning:

> Electric current and its unit **amp**.

Box 12.4.4. *Electric current is the amount of charge that passes through a given area (the cross section of a wire) in one second. In symbols, $i = q/t$, where i is the current and q is the charge passing through the cross section of the wire in time t.*

The unit of current is Coulomb per second, which is called **ampere**, or **amp** for short.

Batteries produce fairly steady electric currents, i.e., the amperage does not fluctuate appreciably. Such currents are called **direct currents** (DC for short). There are other sources of currents, discussed in the sequel, which produce **alternating currents** (AC for short). Such currents are time-dependent, and in most cases change directions regularly with a specified frequency, 60 Hz in the US and 50 Hz in Europe.

Direct current (DC) and alternating current (AC).

12.5   Electric Power Transmission

Batteries supply energy to the moving charges in a circuit continuously. It is therefore more natural to talk about *power*, which is energy supplied per unit time. **Math Note E.12.2** on **page 95** of *Appendix.pdf* shows that

> **Box 12.5.1.** *The power supplied by a battery (its wattage) is the product of the current in the circuit and the voltage of the battery.*

As a charge moves in a wire from a higher potential to a lower potential, it loses energy according to Equation (12.3). What happens to the energy lost by the carriers? If you bring your hand close to the wire you may note where the energy goes: it turns into heat. This heat is due to the drag forces operating inside the wire as mentioned earlier. The production of heat is inevitable whenever drag forces (e.g., friction) are present. The electrical drag in the wire is called **resistance** and is denoted by R. The German physicist, **George Simon Ohm**, showed in 1826 that there is a simple formula connecting the potential drop V, the current i, and the resistance R of a wire. This relation, which is now known as **Ohm's law** is

Resistance of a wire.

$$V = Ri. \tag{12.5}$$

Because of Ohm's contributions to electric circuitry, the unit of resistance is called **ohm** and denoted by Ω.

Example 12.5.2. A resistance of 100 Ω is connected to a 9-volt battery. What is the current in the circuit?

Because the two ends of the resistance are connected to the two poles of the battery, the voltage drop from one end to the other is the same as the voltage of the battery.[6] So, $V = 9$ volts. Equation (12.5) can now give the current: $9 = 100i$, or $i = 0.09$ amps.

With the current and the voltage at our disposal, we can use Equation (E.16) (which says that power is the product of current and voltage) to find the heat power loss in the resistance: $P = 0.09 \times 9 = 0.81$ Watt. ∎

The power loss in a resistor can be written in terms of its resistance and the current it carries. The result is that power loss increases as the *square* of the current. This has a very practical significance: If you cut the current in half, the power loss decreases by a factor of *four*. So, it pays tremendously in electricity transmission from power generators to consumers to reduce the current in the transmission lines as much as possible.

Power plants are "batteries" that have constant power (rather than constant voltage) outputs. The operators of the plant can vary the voltage (electromotive force, emf)—and with it, the current. Since power is the product of voltage and current, increasing the voltage decreases the current and vice versa. See **Example D.12.2 on page 35** of *Appendix.pdf* for a typical situation.

Example D.12.2 shows that reducing the current output of a plant can reduce the power loss in the transmission lines considerably. Of course, reducing the resistance of the wires

---

[6]Remember that potential difference is independent of the path. So whether you go from the positive pole to the negative pole inside the wire or inside the battery, you obtain the same potential difference.

can reduce the loss as well, and this is taken advantage of in electrical industry. However, there is a limit to the reduction of the resistance. The most common way of reducing resistance is by increasing the diameter of the transmission lines. But this increases the weight as well as the cost of the cables. Reduction of the current has its disadvantages as well. This reduction takes place at the cost of increasing the voltage to very high levels. Consumers, on the other hand, demand a much lower voltage for their appliances. Although there are ways of reducing the voltage to a manageable level at the consumption site, it does put certain limitations on how low the current can be.

12.6   End-of-Chapter Material

12.6.1   Answers to "What do you know?"

12.1. When you bring a charged body near an uncharged body, the *opposite* charges of the latter move towards the former. This proximity of the opposite charges results in an attraction. Upon contact, some of the charge of the charged body moves into the uncharged body, making the net charge of both bodies of the same sign; thus repulsion.

12.2. The planet has a net electric charge.

12.3. An isolated charge has electric field lines all around it. No matter what you do with the rest of the charges in a conductor, you cannot get rid of the electric field of an isolated charge. Since the conductor cannot tolerate any electric field inside, the charge cannot stay inside. It moves on the surface and rearranges itself in such a way as to make the field *inside* (but not outside) zero.

12.4. Electric field points from a high potential to a low potential. The force on the negative charge is opposite to that, i.e., it points from low potential to high potential. So, the charge wants to move from low to high potential on its own. The force *you* exert is opposite to this force. So, as you hold the charge, the displacement of your force is opposite to the direction of your force. So, you are doing negative work.

12.5. The electric field is to the left. So, the force must be to the right. The length of the arrow is determined by the strength of the field, which is smaller than the field at P_1.

12.6. The earlier statement was about electro*static* fields. No charges were moving, and if there was a motion, it was extremely rapid, lasting a tiny fraction of a second while the charges were rearranging themselves to ensure a zero field inside. Now the battery disrupts this rearrangement and keeps pushing the charges in the conductor.

12.6.2   Chapter Glossary

AC Alternating current.

Action at a Distance See the glossary at the end of Chapter 9.

Alternating Current (AC) A source of electricity producing a current that changes periodically with time.

Amber A fossilized tree resin found in the Baltic region. When rubbed with a cloth it could make chaff, bits of thread, and other light particles jump and stick to it.

Ampere (amp) The unit of the electric current named after the French physicist André–Marie Ampère. It is equal to one Coulomb of charge passing through a wire in one second.

Conductor A material in which electric charges are free to move.

Coulomb The unit of electric charge.

Coulomb's Law A mathematical statement, discovered by Charles Augustine de Coulomb, expressing the electrostatic force between two charges.

DC Direct current.

De Magnete The first scientific book on electricity and magnetism. It was written by William Gilbert.

Direct Current (DC) A source of electricity producing a current that does not change with time.

Electric Dipole Two opposite charges of equal magnitude separated by a small distance.

Electric Field The physical entity surrounding any (source) charge. A test charge introduced in this field experiences the electric force exerted by the source charge.

Electric Potential The difference between electric potential energy of a charge at two points divided by the charge.

Electric Potential Energy The energy stored in a charge when it is moved from one point to another in an electric field.

Electromotive Force (emf) The ability of a battery to move the carriers from the attractive pole to the repulsive pole. It is simply the voltage of the battery.

Electron Volt (eV) A very small unit of energy used in atomic interactions.

Elektron The Greek word for amber.

Field Lines Streams of directed curves whose direction at a point gives the direction of the field at that point, and whose density represents the strength of the field at that point.

Insulator A material which cannot conduct electricity.

Ohm The unit of electric resistance.

Resistance The property of a medium, in which electric current flows, which inhibits the flow of the charges.

Volt Unit of the electric potential.

12.6.3 Review Questions

12.1. Who was the first to discover electricity? How did he discover it? From where did he receive his education in sciences?

12.2. When did the experimental study of electricity start? Who started it? What was the difference between studying motion and studying electricity?

12.3. What is *De Magnete*? What is in it? What interesting property does amber have? Is this property exclusive to amber? What is the Greek word for amber? What word did Gilbert use to describe the "amber like" property of objects?

12.4. Who is Güricke? What did he invent? What use did he make of his invention? Who discovered the conducting property of metals?

12.5. Name the device that could "hold" electricity and Benjamin Franklin toyed with.

12.6. Who discovered the mathematical formula describing the force between two electric charges? What is the unit of charge? What are the similarities and differences between the force of electricity and the force of gravity?

12.7. How can you tell if there is an electric field in a region? What is the relation between the electric field and force?

12.8. How do you represent electric fields? From that representation, how can you determine the direction of the field? The strength of the field?

12.9. What is the relation between electric potential and electric potential energy? The electric field goes from P_1 to P_2. Which point is at a higher potential?

12.10. What is volt? Whom is it named after? What is an electron volt?

12.11. Who started the study of electric current? How did he start it? What was his profession? What was his theory concerning the twitching of the frog leg?

12.12. Describe some of the experiments Volta performed that got rid of "animal electricity." What was his conclusion about dissimilar metals? What was Volta's biggest invention?

12.13. What is emf? What device has emf?

12.14. Define electric current and say what unit it is measured in. What is AC? What is DC?

12.15. What is Ohm's law? What is the factor that connects voltage and current? What unit is used to measure this factor?

12.16. Explain the reason that power plants produce electricity at extremely high voltage at the plant only to reduce it at the consumer locations.

12.6.4 Conceptual Exercises

12.1. Figure 12.5 shows a sample of electric field lines.
(a) If a positive charge is introduced at A, which direction will the force on it point?
(b) If a negative charge is introduced at B, which direction will the force on it point?
(c) If a positive charge is introduced at C, which direction will it accelerate?
(d) If a negative charge is introduced at D, which direction will it accelerate?
(e) If a negative charge is introduced at E, which direction will it accelerate?
(f) Show the region of the weakest and the strongest electric field on the graph.

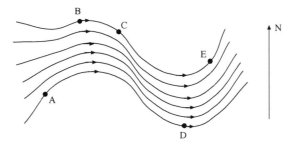

Figure 12.5: A sample of electric field lines.

12.2. Figure 12.6 shows a sample of electric field lines.
(a) If a positive charge is introduced at A, which direction will the force on it point?
(b) If a negative charge is introduced at B, which direction will the force on it point?
(c) If a positive charge is introduced at C, which direction will it accelerate?
(d) If a negative charge is introduced at D, which direction will it accelerate?
(e) If a negative charge is introduced at E, which direction will it accelerate?
(f) Show the region of the weakest and the strongest electric field on the graph.

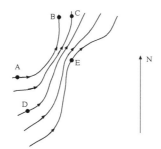

Figure 12.6: Another sample of electric field lines.

12.3. Suppose you wanted to float a negatively charged balloon in midair by placing it in a vertical electric field. Would you place it in a field that points up or down? Explain.

12.4. Figure 12.7 shows a sample of electric field lines on the left and two arrows representing the electric fields at A and C on the right.

Figure 12.7: Electric field lines and the electric field at A and C.

(a) Circle the arrow in the following that most closely represents the electric field at B.

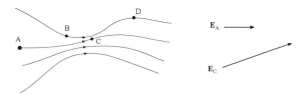

(b) Circle the arrow in the following that most closely represents the electric field at D.

12.5. Figure 12.8 shows some electric field lines and some labeled points.
(a) Which point has the strongest field?
(b) Which point has the weakest field?
(c) If a negative charge were introduced at B it would accelerate in which direction?
(d) If a positive charge were introduced at A it would accelerate in which direction?
(e) These field lines are produced by some charges. Circle the location of charges that produce these field lines. Determine the sign of these charges as well. Which charge is larger (in magnitude)?

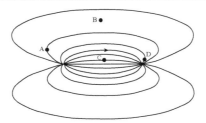

Figure 12.8: Some electric field lines.

12.6. Figure 12.9 shows some electric field lines and some labeled points.
(a) Which point has the strongest field?
(b) Which point has the weakest field?
(c) If a negative charge were introduced at B it would accelerate in which direction?
(d) If a positive charge were introduced at A it would accelerate in which direction?
(e) These field lines are produced by some charges. Circle the location of charges that produce these field lines. Determine the sign of these charges as well. Which charge is larger (in magnitude)?

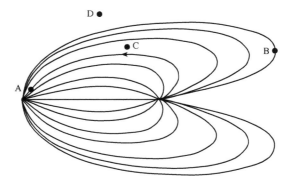

Figure 12.9: Some electric field lines.

12.6.5   Numerical Exercises

12.1. Mass of Earth is 6×10^{24} kg and that of Moon is 7.35×10^{22} kg. Suppose that you put equal amount of positive charges on Earth and Moon. How much do you have to put on each for the electric repulsion to neutralize the gravitational attraction? Note: You don't need the distance between Earth and Moon to answer this question!

12.2. A 10,000-kg sphere is to be suspended in midair 1 cm above the ground by charging it and a very tiny sphere placed right underneath it on the ground.
(a) What is the weight of the big sphere?
(b) If you put equal charges on the two spheres, how much charge is needed to suspend the big sphere?

Chapter 13

Magnetism and Electricity

Among the rather widespread and fairly abundant minerals found on Earth is a crystalline mineral ore of iron called *magnetite*, which, unlike any other iron ore, is magnetic. Occasionally, pieces of this ore, called **lodestones**, are found which are permanently magnetized. Some specimens have been known to have such a strong magnetization as to lift several times their weight. There is good reason to believe that the attractive power of lodestone has been known to mankind as early as 5000 years ago. It is said that Huang-ti, "an emperor of China in the year 2637 BC, had a chariot, upon which was mounted the figure of a woman, pivoted or suspended so that it was free to turn in any direction. The outstretched arm of the statue always pointed to the south under the influence of concealed lodestones" [Mey 71, p. 2].

The first person to study the magnetic properties of lodestones in the west appears to have been Thales, who was also the first person to note the electrical properties of amber. He experimented with lodestone and knew that it attracted iron. Thales did not write any books. What we know of him was transmitted orally until Aristotle recorded his teachings.

13.1   The Compass and the Lodestone

The era of modern magnetism, like other scientific endeavors, begins with the Renaissance. In 1269 Petrus Peregrinus, a soldier in the French army—and Roger Bacon's teacher—wrote a letter to one of his friends in which he described the details of the construction of a floating compass as well as a pivoted compass employing a steel needle.

The magnetic compass was the first practical device that arose from the study of magnetism. After the time of Petrus Peregrinus the compass soon came into general use, which in turn led to the serious study of magnetism. The first definitive work in which a detailed investigation of magnetism took place was *De Magnete* by William Gilbert, about which you have already learned in Section 12.1. By Gilbert's time the use of compass for navigation was much developed, and some of its basic properties were already known. For instance, the downward inclination of the needle of a compass in London led to the belief that there was a magnetic body inside the Earth.

Gilbert's investigation of lodestone led him to propose that the Earth itself was a huge magnet with north and south poles in close proximity of the opposite geographic poles (Figure 13.1). He showed by many experiments that all magnetic bodies were "regulated by the Earth." As a result, he concluded that lodestone was an iron ore directionally magnetized by the Earth when it lay along a meridian. Experimenting with a model of Earth he built himself, Gilbert explained why compass needles are parallel to the Earth surface at the equator, but bend downward in the northern hemisphere. And he attributed

Gilbert shows that Earth is a giant magnet.

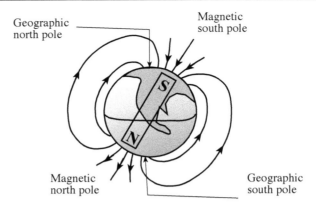

Figure 13.1: Earth is a giant magnet whose magnetic north pole is close to its geographic south pole and vice versa.

the motion and inclination of the compass needle entirely to the Earth magnetism.

Gilbert's diverse experiments with the lodestone led to some very useful discoveries such as the increase in intensity of the lodestone when capped by iron, and to some fundamental properties such as the fact that

Box 13.1.1. *Splitting a magnet repeatedly results in smaller magnets, each with its own north and south poles. Magnetic poles cannot be isolated.*

Gilbert proposed to catalog the declination and direction of the compass needle for various location as an aid to navigation, based on the belief that the two magnetic poles of the Earth are stationary. Hopes for this scheme faded with the discovery that, although the continents and the oceans are fixed on Earth, the *magnetic poles drift unpredictably from year to year*.

Drift of magnetic poles of Earth.

13.2   Magnetism from Electricity

Gilbert's painstaking experimentation of lodestone, although revealing in many respects, left one question unanswered: What is the source of magnetism? Unlike electrical forces which are caused by the electric charges, magnetic forces cannot be caused by magnetic charges, because—as demonstrated by Gilbert himself—magnetic charges do not exist in nature. The answer to this riddle had to wait for an "accidental" discovery.

The professor called on his class to gather around his laboratory table to show them the fascinating phenomena related to the electric current. In particular, he wanted them to see how the electric current from a battery could make a wire red hot, even melt it. He waved at his assistant signaling the connection of the battery. The assistant followed the motion, and all eyes were turned to the wire. All except the professor's. Something else caught his attention. Leaning forward, he gazed in amazement at the needle of a magnetic compass that happened to be close to the wire. The idle compass pointing lazily to the north had suddenly come to life, starting to swing strongly. To what was the compass responding?

Örsted discovers the connection between electricity and magnetism in the middle of a physics lecture at the University of Copenhagen.

At this point the professor completely forgot about the class, the lecture, and the students. He was on the verge of one of the most important discoveries of history. As the students watched their professor in bewilderment, he signaled to his assistant to disconnect the battery. To his delightful surprise, the needle went back to its original north-south position. He signaled the assistant to connect and disconnect the battery several times; and every time a current was established in the wire, the needle moved away from its normal position. The professor eventually turned to the class. They had witnessed their professor, Hans Christian Örsted (1777–1851), stumble on one of the oldest secrets of nature right on their campus at the University of Copenhagen: That *electric currents can produce magnetic forces.*

BIOGRAPHY

Hans Christian Örsted (1777–1851) was born in a small Danish town into a family that could not afford his education. It was the generosity of the townspeople that helped Hans and his brother Anders get an education. The boys were eager learners and made it a habit to read every book they could lay their hands on. Örsted followed his interest in science—an interest that was sparked when he helped his father, the village apothecary. He became especially interested in Volta's electric pile, and while still at the university, had resolved that experimentation with voltaic current would be a major field of interest. It was the pursuit of this interest that finally led to Örsted's discovery of electrical magnetism. Because of his background in his father's pharmacy, Örsted had an interest in chemistry as well. The production of aluminum is attributed to him.

On July 21, 1820 Örsted privately printed a 4-page pamphlet describing his discovery. Published in Latin, the universal language of science at the time, Örsted sent copies of the pamphlet to eminent scientists and learned societies of Europe. The announcement was so important that, within a few weeks, the paper was translated and reprinted in English, German, French, Italian, and Dutch.

July 21, 1820 marks the beginning of modern magnetism.

Although the discovery of the connection between electricity and magnetism *appears* to be accidental (and admittedly some unplanned incidents were crucial in the discovery), the hard work, diligence, and the relentless pursuit of knowledge of the discoverer cannot be neglected. We do not call the discovery of a gold mine by a gold digger who has spent all his life looking for gold "an accident." Granted that in the case of Örsted, luck played a more major role than in some other discoveries; but we should not forget that the element of luck—with varying degree of intensity—is present in almost all great achievements.

Luck versus perseverance.

Örsted's discovery set the course of physical research for the remainder of the nineteenth century. Its significance was clear from the start. Less than a year after its discovery, the French physicist Arago read Örsted's announcement of his experiments in the 11 September 1820 meeting of the French Academy of Sciences. Among the audience in that momentous meeting was **André-Marie Ampère** (1775–1836), a professor of mathematics at the École Polytechnique of Paris, and a distinguished physicist and chemist. Ampère immediately followed up on Örsted's discovery, and soon broke new grounds for the growth of the new science of electromagnetism. Within a week of Arago's announcement, Ampère had repeated Örsted's experiment, and on September 18, 1820 informed the Academy of his discovery that *electric currents in parallel wires produced a mutual magnetism that exerted a force between them.* The next few years saw Ampère producing a series of memoirs on the fundamentals of electromagnetism that led to the science of electrodynamics.

Building on Örsted's observation of the deflection of a compass needle due to the current in a wire, Ampère invented—and coined the name—**galvanometer** to measure the strength of the electric current precisely.

Ampere invents galvanometer and discovers the law of magnetic force on current-carrying wires.

From Örsted's experiments and his own, Ampère concluded that electric current is the source of all magnetic phenomena. In particular, inside every magnet must reside some form of electric current. He argued that since an electric current deflects a magnetic needle, and since the same needle is also affected by the Earth's magnetism, there must be electric currents circulating inside the Earth. He then deduced that, just as two magnets exert forces on one another, so must two electric currents, and successfully demonstrated this deduction and discovered the mathematical formula for the force between two current-carrying wires. **Example D.13.1** on **page 36** of *Appendix.pdf* gives this formula and calculates the force

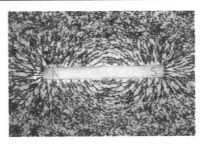

Figure 13.2: The regular arrangement of iron filings suggested to Faraday the existence of magnetic field lines.

between two sample wires.

BIOGRAPHY

André-Marie Ampère (1775–1836) had shown a precocity in mathematics by age four. Encouraged and tutored by his father, he developed such an intense scientific interest that at age 12 he was studying advanced treatises on analytical mechanics. At the age of 18, witnessing the execution of his father on the guillotine, Ampère was so devastated that his mind was temporarily deranged. In 1796 he met Julie Carron, and an attachment sprang up between them, leading to their marriage in 1799. From about 1796 Ampère gave private lessons at Lyons in mathematics, chemistry and languages; and in 1801 he moved to Bourg, as professor of physics and chemistry, leaving his ailing wife and infant son at Lyons. She died in 1804, and he never recovered from the blow. In the same year he was appointed professor of mathematics at the lycée of Lyons. From 1800 to 1814, Ampère devoted his time mainly to mathematics. And it was this strong background in mathematics that helped him later in his formulation of electrodynamics.

13.3   Faraday's Lines of Force

The next giant step in the development of electromagnetism was taken by Faraday, the English physicist/chemist. His superhuman physical intuition, grounded in his lack of mathematical ability, introduced the concept of field lines, paving the way for the electromagnetic revolution of the nineteenth century. The mere introduction of field lines into the consciousness of the physics community was a giant step. Their clever use in the establishment of the ultimate connection between electricity and magnetism was something that only Faraday could accomplish.

Take a bar magnet and place it on a cardboard. On top of the board sprinkle iron filings and tap on the board allowing the filings to align themselves. It is seen that the filing particles produce regular lines on the board similar to the ones shown in Figure 13.2. Faraday called these lines the **magnetic field lines** or **magnetic lines of force**, and suggested that the iron filing merely acts as a catalyst to make visible the invisible lines that are always there. This "visualization" of magnetism as invisible magnetic field lines was probably one of Faraday's most important contributions to physics. Ironically, Faraday's mathematical weakness forced him to invent the more intuitive concept of magnetic field lines, which later became one of the primary ingredients of the *mathematical* theory of electromagnetism in the hands of James C. Maxwell.

On these lines of force Faraday wrote:

> Now it appears to me that these lines may be employed with great advantage to represent the nature, condition, direction and comparative amount of the magnetic forces; and that in many cases they have, to the physical reasoner at least, a superiority over that method which represents the forces as concentrated in centers of action, such

Faraday introduces the magnetic lines of force.

as the poles of magnets or needles; or some other methods as, for instance, that which considers the north and south magnetisms as fluids diffused over the ends or the particles of a bar ...

I cannot refrain from again expressing my conviction of the truthfulness of the representation, which the idea of lines of force affords in regard to magnetic action .... In a straight wire, for instance, carrying an electric current, it is apparently impossible to represent the magnetic forces by centers of action, whereas the lines of force simply and truly represent them. The study of these lines have, at different times, been greatly influential in leading me to various results, which I think prove their utility as well as fertility.

With the notion of magnetic field lines, we can now summarize what we have learned about electromagnetism so far:

Box 13.3.1. *Electric current is the source of all magnetic fields. Moving electric charges not only create magnetic fields, but also experience magnetic forces when placed in a magnetic field.*

What do you know? 13.1. A current inside a wire creates a magnetic field outside. Now imagine riding along the wire with the same speed as the charges producing the current inside. The charges no longer move relative to you; i.e., you see stationary charges. And stationary charges produce only electric field. What's the catch?

History was on Faraday's side. James C. Maxwell not only found a precise mathematical meaning for these lines of force, but eventually turned them into *real physical* entities that made the radio talk, the television show, the microwave oven cook, and the telephone send a voice across the Atlantic and the Pacific oceans. But of these later.

Faraday next turned to the study of electrostatics. Instead of charges, he thought in terms of electric lines of force. To him the opposite charges were only the different ends of the lines of force. Since the field lines cause electrical forces on charges, and since charges move freely inside conductors, Faraday concluded that no electric field can exist inside a conductor—as long as no battery is attached to it. Because, if there were any electric field inside the conductor, it would move the charges, creating an electric current. With no battery attached to the conductor, such a spontaneous generation of electric current is impossible.[1]

To Faraday, opposite charges were only the different ends of the lines of force.

What happens if you add some charges to a conductor? Faraday concluded that it must reside *on the surface* of the conductor, because any charge inside would create lines of force around it. This is impossible by the argument above. To prove his point, he constructed a hollow cube twelve feet to a side. The outside of the cube was covered with a metal foil, and the cube was placed on insulating supports to prevent contact with the floor. With Faraday inside, the cube was charged from a powerful electrostatic machine. Here is how he reported on the experiment:

I went into the cube and lived in it, and using lighted candles, electrometers, and all other tests of electrical states, I could not find the least influence on them ... though all the time the outside of the cube was very powerfully charged, and large sparks and brushes were darting off from every part of its outer surface.

This experiment of Faraday, by the way, explains why it is quite safe to be inside a (metallic!) car when lightning strikes.

---

[1]Actually, when you put electric charges on a conductor, they initially *move* and rearrange themselves in such a way as to create a zero electric field inside. But this rearrangement takes an infinitesimally small time.

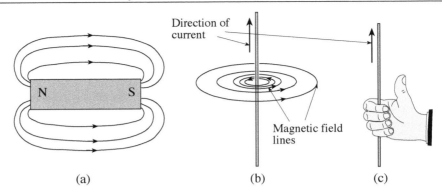

Figure 13.3: (a) A magnet has north and south poles. The magnetic field lines begin at north pole and end at south pole for a bar magnet, while (b) they circle around a current-carrying wire. (c) The RHR determines the direction of the magnetic field lines.

The electric lines of force emanate from positive and end on negative charges. What about the magnetic field lines? On a bar magnet, the field lines emanate from the north pole and terminate on the south pole [Figure 13.3(a)]. As in the case of electric field lines, the more concentrated the magnetic lines of force, the stronger the magnetic field. What about the magnetic field lines of a wire carrying an electric current? There is no north or south pole in the wire! It turns out that for a long current-carrying straight wire, the lines close on themselves in the form of concentric circles around the wire with the wire at their common center [Figure 13.3(b)]. A change in the direction of the current results in a change in the direction of the magnetic field. There is a rule, called the right-hand rule (RHR), similar to the one we encountered in Section 8.2, which relates the direction of the current to the direction of the magnetic field lines: If the thumb of the right hand points in the direction of the current, the fingers, when curled, show the direction of the magnetic field [see Figure 13.3(c)]. Actually, this rule applies to *any* electric current as long as we stay very close to the current.

Figure 13.4 shows the field lines for a single loop of wire, and a helical coil each carrying an electric current. Note the similarity between the outer magnetic field lines of the coil and a bar magnet. This similarity suggests that the magnetic field lines of a bar magnet may continue inside the magnet forming a *closed loop* just as in the case of a coil and a long straight wire. It is now an established fact that

Magnetic field lines of a bar magnet continue inside to form closed loops.

> **Box 13.3.2.** *The magnetic field lines always form a closed loop, and have no beginning and no end. Thus there are no isolated magnetic charges (monopoles).*

This means that it is impossible to have a north pole without a corresponding south pole attached to it. Even the atomic-sized magnets have a north pole and a south pole!

> **What do you know? 13.2.** Each iron atom is a microscopic magnet that aligns itself with the magnetic lines of force when placed in a magnetic field. If iron consists of such microscopic magnets, why is it not magnetic? Why does a piece of iron get attracted to either a north pole or a south pole?

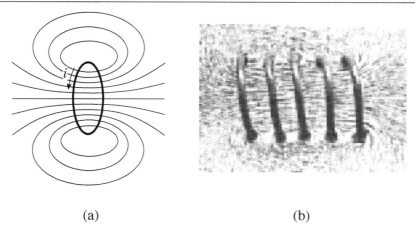

(a)   (b)

Figure 13.4: (a) A loop of current-carrying wire is a (very thin) magnet with a north pole and a south pole. (b) To thicken the (electro)magnet, use a solenoid. The magnetic field lines of a loop or a coil have no beginning and no end. This is true of all magnetic field lines.

13.3.1   Magnetic Fields of Currents

Only a month after Örsted's announcement was read in the September meeting of the French Academy of Sciences in 1820, two French physicists, **Jean-Baptiste Biot** and **Felix Savart**, in collaboration with Ampère, discovered how the magnetic "action" varied with distance from a long straight wire. The vague notion of "action" was later clarified and replaced by Faraday's concept of field. Biot and Savart discovered that, for a given current, the strength of the magnetic field of a straight long wire decreases in inverse proportion to the distance from the wire. **Example D.13.2** on **page 36** of *Appendix.pdf* gives the mathematical formula of this statement and uses it to show that it takes a gigantic current to produce any appreciable magnetic field.

BIOGRAPHY

The biography of **Michael Faraday** (1791–1867) is a paradigm of success through hard work, perseverance, the diligent pursuit of one's interest, and a little bit of luck. Being one of the ten children of a poor blacksmith, Faraday had to work as an apprentice to a bookbinder instead of going to school. Taking full advantage of this job he devoured all the scientific books he could lay his hands on. In 1810, he joined a group of teenagers interested in science, and attended their biweekly lectures regularly. Faraday's eagerness to learn was matched only by his hard work to overcome his lack of formal schooling. For instance, at the age of 22, he bore the humiliating task of running errands for the wife of Humphrey Davy (1778–1829), the English chemist, because, in return, he had the opportunity to travel with Davy all over Europe, attending his stimulating lectures, meeting great scientists of the time, and most of all being able to perform experiments in Davy's laboratory.

The unit of B in the formula of Example D.13.2 is **Tesla**. To get familiar with the size of tesla, note that the Earth's magnetic field is about 10^{-4} tesla. Another (much smaller) unit of magnetic field is **Gauss** which is 0.0001 tesla.

Tesla and **Gauss** are two common units of magnetic field.

A few years after Biot and Savart introduced their formula for the magnetic field of a long wire, Ampère, Biot, and Savart generalized the formula for wires of all shapes. Now they could not only recalculate the magnetic fields of a long wire—and get the same result as before—but also of a circular loop, a rectangular loop, etc. Of particular interest was the magnetic field of a **solenoid**, i.e., a wire wound in the form of a spring. For a (long) solenoid the magnetic field is concentrated mostly inside, where it is almost uniform. Outside the solenoid, the magnetic field is almost zero. **Example D.13.3** on **page 37** of *Appendix.pdf* calculates the magnetic field inside a solenoid and shows how hard it is to create sizable magnetic fields.

🗑 Food for Thought 🧠

FFT 13.3.3. Magnet "therapy?" Equation (D.12) shows that the magnetic field of a long (infinite) wire drops in inverse proportion to the distance. For shorter wires the field drops in proportion to distance squared. A permanent magnet is equivalent to a short wire; therefore *its* magnetic field drops in inverse proportion to distance squared. Magnetic sheets—of the type used on refrigerator doors—have magnetic fields that drop at a much faster rate. In fact, at a distance of only about one millimeter (about ten sheets of paper) the field of such magnetic sheets is almost zero.

With this in mind, one has to ask the question: How are the therapeutic magnets working? The claim is that because of the iron in the red blood cells, these cells get attracted to the magnet, and their arrival at the pain area relieves discomfort. There are two flaws in this argument:

- The iron in the red blood cells is in a chemical form that does not get attracted to magnets. In fact, it happens to be slightly *diamagnetic*—a form of iron that is *repelled* by a magnet. The arrival of red blood cells to a location beneath the skin causes a reddening of the skin. That is why the skin in the vicinity of a bruise gets red even though there is no cut there. However, placing a magnet on the skin, no matter how strong, will not cause a reddening of the skin. The readers can try this for themselves, and see the zero effect.

- The "therapeutic" magnets are usually placed in a velvet strap and tied to the skin. The thickness of the straps are at least a few millimeters, enough to stop the magnetic field from reaching even the outside of the skin!

Although simple physical and chemical arguments prove unequivocally that magnets have absolutely no effect on humans, millions of people resort to them for treatment of pain. If only they knew the very basics of physics and chemistry! And the fact that some bona fide doctors advocate such treatments does not validate their efficacy.

13.4 Electricity from Magnetism

Faraday's greatest discovery was the **electric induction**, which took place in the 1830s. He, like many other European scientists, was aware of Örsted's experiment. He was also aware that many scientists were tackling the question of the reverse process: If electric current can produce magnetism, is it possible to produce electricity from magnets? Earlier, in 1824, the French physicist François Arago mystified the scientific community with the apparatus shown in Figure 13.5. Arago demonstrated that by rotating the copper disk, one could "drag" the compass needle along. This mysterious effect occupied the attention of many great scientists including Ampère and Arago himself. But the puzzle was not solved until 1831, when Faraday made a breakthrough in one of the most brilliant experiments ever.

In his groundbreaking experiment on electromagnetism, Faraday based his procedure on Ampère's discovery of the forces between current-carrying wires. How would the activation of one circuit affect an adjacent circuit in terms of a possible transfer of magnetic induction? Would the creation of a magnetic force, such as the one created by an electromagnet, produce some "strain or vibration" in the adjacent wire that might give rise to an electric current? To test this, Faraday prepared two intertwined circuits wound around a wooden core with a battery and a switch in one and a galvanometer in the other. He noticed that, in his own words, "When the contact was made there was a sudden and very slight effect at the galvanometer, and there was also a similar slight effect at the galvanometer when the contact with the battery was broken."

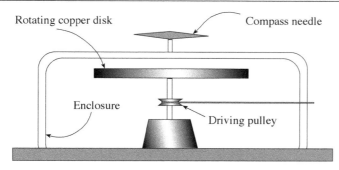

Figure 13.5: When the copper disk is rotated, the compass needle is "dragged" around with it.

It became clear to Faraday that the induction of electricity in the second circuit had to do with the making and breaking of electric current. To enhance the effect he used a (circular) iron core instead of wood, and noted that the deflection of the galvanometer was considerably more noticeable than before; but as previously, the effect was not permanent. The needle came to rest and was indifferent to the *steady* current. The needle responded in a similar way in a reversed direction when the battery was disconnected.

Faraday was aware of the fact that iron reinforced the *magnetism* of the current in the wire; he also realized that only when magnetism was changed did the galvanometer show a deflection. He then hit upon the idea of using permanent magnets to induce electric current in the coil connected to the galvanometer. He thrust a bar magnet into the coil; the needle responded. Quickly withdrawing the magnet reversed the swing of the needle. And with this experiment the age of electricity and electromagnetic communication became visible in the horizon.

Legend has it that the Prime Minister of England once asked Faraday what use could be made of his discoveries, to which Faraday replied "some day it might be possible to tax them." This was a remark made half jokingly, indicating the remote possibility of the use of a shear scientific discovery. However, this unintentional prophecy was to come true in a most dramatic way by the end of the nineteenth century when the foundations of a practical electric generator were laid.

13.4.1   Faraday's Law

Using the notion of the lines of force and pursuing his experiments further, Faraday showed that if the "number" of magnetic field lines cutting the area of a loop of wire changes, an electric current will be created in the loop. This current is the result of an induced electromotive force (emf). The faster the change, the larger the emf. The "number of magnetic field lines cutting the area of a loop" is an imprecise version of an exact mathematical quantity called the **magnetic flux**.

Magnetic flux.

Figure 13.6 illustrates the essential properties of the notion of the flux of a (magnetic, electric, or any other) field. The flux is proportional to the strength of the field. Thus, the loop at 1 has more flux than the loop at 2 although the latter is larger. It also depends on the size of the loop: More lines of force go through loop 3 than loop 2 because loop 3 is larger than loop 2. Finally, flux depends on the orientation of the loop relative to the field: Loop 4 is larger than loop 1, but the flux through loop 1 is larger because the lines are perpendicular to the surface of the loop.

With the notion of flux thus explained, **Faraday's law** can be stated as follows:

Faraday's law stated.

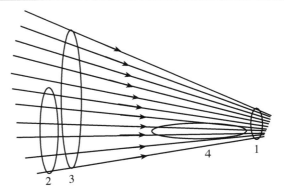

Figure 13.6: Flux depends on the size and orientation of the loop as well as the strength of the field.

Box 13.4.1. *The strength of the electromotive force induced in a loop of wire is proportional to the rate at which the magnetic flux through the area of the loop changes.*

Faraday's law does not give the *direction* of the induced current. This information is given by *Lenz's law*—named after the Russian physicist Heinrich Lenz (1804–1865)—which states that the current induced is in such a direction as to oppose the *change* in the magnetic flux.

There are a number of ways that one can change the magnetic flux:

- Change the orientation of the loop relative to the magnetic lines of force;

- Change the magnetic field itself (i.e., the number of field lines) by

  - moving a permanent magnet towards or away from the loop,
  - moving the loop towards or away from the permanent (or an electric) magnet,
  - changing the current in an electromagnet;

- Increase or decrease the area of the loop.

Any of the processes above results in the creation of an electric current in the loop.

What do you know? 13.3. How does the motion of the copper disk of Figure 13.5 drag the compass needle? Hint: Since motion is relative, assume a stationary disk and a moving compass needle; think of the disk as a lot of concentric loops; apply Faraday's law; and invoke the third law of motion.

Example 13.4.2. Let a wire loop lie flat on a surface (say this page). Bring the north pole of a magnet close to the loop. Since the magnetic field lines point away from the north pole, the lines cut the loop into the page. Since the magnet is brought closer to the loop, the number of field lines cutting the area of the loop (i.e., the magnetic flux) increases. According to Faraday's law, a current is established. According to Lenz's law, the direction of this current is such as to oppose the *change* in the flux. Since the flux increases *into the page*, the direction of the current is such as to produce a magnetic field out of the page. The right-hand rule suggests a counter-clockwise current.

If we move the north pole away from the loop, the magnetic flux *decreases*. To oppose this decrease, the current must have a clockwise direction to produce—by the RHR—a magnetic field into the page to compensate for the loss in the flux. The readers can convince themselves that if a south pole is chosen, the current in each case will be opposite the above currents.   ■

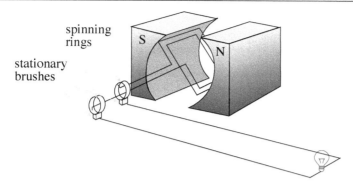

Figure 13.7: An electric generator. The rotation of the loop changes the magnetic flux through the area of the loop, creating an electric current in the loop, which can be drawn from the stationary brushes.

What do you know? 13.4. Loop number 2 in Figure 13.6 rotates three times as fast as loop number 1. Whose emf is bigger?

13.4.2 From Faraday's Law to Electric Power

Put a loop of wire in the space between the two poles of a (U-shaped) magnet (Figure 13.7). Depending on the angle between the loop and the magnetic field lines, the magnitude of the magnetic flux may be large or small. Now spin the loop. The magnetic flux changes and you can create an electric current in the loop. To make this kind of electricity production practical, connect each end of the loop to a conducting ring, which slides on a conducting brush, causing the transfer of charges from the loop to these brushes. To extract current from this device, connect the two leading wires of a circuit to these brushes as shown in Figure 13.7. This analysis illustrates the rudimentary principles of the operation of a **generator** of electricity. The rotation of the loop is done by mechanical means—steam-driven turbines in fuel-burning and nuclear power plants, or water-driven turbines in hydroelectric power plants.

Principle of generation of electricity.

The current and the voltage produced by a generator described above changes with time and is called **alternating current**. This current varies regularly with time, rising to a maximum, then dropping to zero and a minimum negative value before rising again. The shape of this variation is similar to a simple harmonic oscillation (see Figure 11.4). Such an oscillation leads to an *effective voltage*, which characterizes our appliances as 110- or 220-volt.

Alternating current.

What do you know? 13.5. Drill a small hole through the middle of a bar magnet. Place it on a pivot so that the magnet is horizontal and can be rotated about its pivot point. Place this assembly at the center of a vertical loop of wire. Rotate the bar magnet. Will there be an emf in the loop?

Transformers

Let an alternating current generator be attached to a solenoid. The current produces a magnetic field inside the coil as in Equation (D.13). Since the current is changing, the field in the coil is also changing. Now guide this alternating magnetic field to a second coil. This guidance can be accomplished by using an iron core that runs through both coils

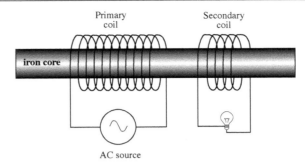

Figure 13.8: A transformer takes in a voltage at the primary coil and puts out a different voltage at the secondary coil. The iron core "guides" the magnetic field of the primary circuit to the secondary circuit.

How a transformer works.

(see Figure 13.8). The changing magnetic field in the second coil induces a changing flux, which, by Faraday's law, induces an emf (voltage) in the coil. Such an arrangement is called a **transformer**. The larger the number of windings in the second coil of a transformer, the larger the flux, and thus, the larger the emf. **Math Note E.13.1 on page 96** of *Appendix.pdf* finds a formula for the voltage in the second circuit in terms of the number of turns in each circuit and the voltage in the first circuit.

Step-up and step-down transformers.

Depending on the number of turns in the primary and secondary circuits, two kinds of transformers are in use: **step-up transformers** take in low voltages and deliver high voltages; **step-down transformers** convert a high voltage to a lower one. Small appliances (see **Example D.13.4 on page 37** of *Appendix.pdf*) that require very low voltages have a built-in step-down transformer.

Distribution of Electric Power

Generators of electricity are usually located in the country side, while most of consumption takes place in urban areas far from the power plants. Therefore, electricity has to be *transmitted* to consumers. Towards the end of the nineteenth century, when the practical utility of electricity was becoming evident, there were two schools of thought as to how to produce and deliver electric power to consumers. Thomas Edison (1847–1931), the American inventor, was a steadfast advocate of *direct current*, while Nikola Tesla (1856–1943),[2] a Croatian-American engineer and George Westinghouse (1846–1914) championed *alternating current*.

The biggest hurdle in transmitting electricity from the generator to the consumers was the loss of power in the transmission lines in the form of heat: $P = Ri^2$ [see Equation (E.17)]. It was therefore desirable to reduce both the resistance of the transmission lines and the current flowing in them, with the latter having much more dramatic effect than the former. However, the reduction of the current was limited by the appliances used by the consumers. Therefore, with direct current, there was no mechanism to reduce this heat beyond minimizing the resistance of the transmission cable. This meant locating the generators close to the consumption site. The generators of Edison Electric Light Company in New York City had to be centrally located. For buildings far from the generators, thick wires had to be used, and the farther the building, the thicker the wire.

Alternating currents, on the other hand, could go beyond the reduction of resistance and take advantage of transformers: increase the voltage at the generator using a step-up transformer thereby reducing the current fed into the transmission lines[3] (see Example D.12.2 and the discussion just before it), then decrease the voltage at the consumption site

---

[2]Yes, the same tesla as the unit of magnetic field.

[3]Recall that the product of current and voltage is the (predetermined) power output of the generator.

using a step-down transformer. The current in the transmission wires could be lowered, and the heat loss even more so. History sided with Tesla and Westinghouse. Today's generators, with the help of step-up transformers, deliver power at a voltage of a few hundred thousand volts and a relatively small current to *substations*, at which step-down transformers reduce the voltage and deliver the resulting power to smaller step-down transformers visible on the utility posts or backyards of residential buildings.

13.5   End-of-Chapter Material

13.5.1   Answers to "What do you know?"

13.1. There is no catch! Whether you see an electric field or a magnetic field (or both) depends on your state of motion. In fact, there is a deep connection between electromagnetism and relativity, among whose outcome is the transformation of electric to magnetic field and vice versa as the state of motion of observers changes.

13.2. Iron is not magnetic, because the microscopic magnets, of which it consists, are randomly oriented, so that the bulk iron does not have a north pole or a south pole. However, when placed in an external magnetic field, these microscopic magnets align themselves along the magnetic field lines and reinforce each other. If one pole of a magnet is closer to the bulk iron, the opposite poles of the majority of the microscopic magnets point to that pole, causing an attraction.

13.3. As the magnetic needle rotates, it changes the flux through the concentric loops, creating currents through those loops by Faraday's law. These currents are in the magnetic field of the needle. Therefore they experience a force by Box 13.3.1. So, the needle exerts a force on the copper disk. The third law of motion says that the disk should apply a force on the needle.

13.4. 11 lines pass through loop 1 and 6 through loop 2. Since loop 2 rotates three times as fast, its rate of change is $6 \times 3 = 18$ compared to 11. So, the bigger loop has more emf.

13.5. Rotating the bar magnet changes the direction and the magnitude of the magnetic field lines that pass through the loop. Therefore, an emf will be produced in the loop.

13.5.2   Chapter Glossary

Electric Induction The process of producing electricity from magnetism by changing the magnetic flux through a loop of wire.

Electromotive Force (emf) The voltage of the electricity produced in electric induction. It is also used for battery voltages: the emf of a battery is simply its voltage.

Faraday The English physicist who first discovered how to produce electricity from magnetism.

Galvanometer A sensitive device—built on Örsted's discovery—to measure small electric currents.

Gauss A unit of magnetic field. Magnetic field of Earth is about 1 Gauss.

Lodestone Pieces of magnetite which are permanently magnetized.

Magnetic Field The "imaginary" lines along which iron filings would line up when placed in the vicinity of a magnet. Invented by Faraday, these lines—and the corresponding magnetic field—became the cornerstone of Maxwell's mathematical theory of electromagnetism and led to the prediction of electromagnetic waves.

Magnetic Flux A mathematical quantity which is qualitatively described as the "number" of magnetic field lines crossing an area.

Magnetite A crystalline iron ore which has magnetic properties.

Örsted The Danish physicist who first discovered that electric currents produce magnetic fields.

Solenoid A long coil made to produce magnetic field by passing a current through it. An electromagnet.

Tesla A unit of magnetic field. Magnetic field of Earth is only 0.0001 Tesla.

Transformer An electric device which increases (step-up transformer) or decreases (step-down transformer) an AC voltage.

13.5.3 Review Questions

13.1. What are lodestones? How early were these known to humankind? How long ago were they used for navigation?

13.2. Who was the first person to study the magnetic properties of lodestones in the west?

13.3. Who invented the compass? Explain how it came to be believed that there was a magnetic body in the Earth.

13.4. What happens if you split a magnet in half? Can you isolate its poles?

13.5. Where is the Earth's magnetic south pole? Is it always at the same place?

13.6. Who was Hans Christian Örsted? What did he discover? Where did he discover it? What other big name is associated with magnetism? What was his major discovery? His major invention?

13.7. What is the source of magnetism? What is the source of magnetism of magnets?

13.8. Who was Faraday, and what important concept did he introduce in electricity and magnetism? How did Faraday interpret the electric charges in terms of his lines of force? How did he conclude that there should be no electric field inside a conductor?

13.9. What is the difference between the electric lines of force and magnetic lines of force? Which one has a beginning and an end? Which one is in the form of a loop?

13.10. How far does the magnetic field of a magnetic sheet used on refrigerator doors go? How do the therapeutic magnets work? Is the claim that they attract red blood cells valid? Why?

13.11. How do you produce electricity from magnetism? Who discovered the law of electric induction? Define magnetic flux. What is the relation between the magnetic flux and the electric current? What is electromotive force?

13.12. How does an electric generator work? Does it produce alternating current or direct current?

13.13. How does a transformer worK? What is a step-up transformer? What is a step-down transformer?

13.14. What is the advantage of AC over DC in the transmission of power? Why is the voltage high at the power plane? How is the voltage brought down at the consumption site?

13.5.4 Conceptual Exercises

13.1. Why was it plausible for Gilbert to assume that Earth was a giant magnet when he saw that magnetic needles always point to the north?

13.2. Why was it plausible for Ampère to assume that two current-carrying wires should exert forces on one another?

13.3. Explain why you can say that all magnets are electromagnets.

13.4. You let go of a ball; it falls to the ground. You say it is because a gravitational field acts on it. If you don't have a ball, can you say that there is no gravitational field? Is it safe to say that the gravitational field has an existence independent of the ball?

13.5. Does a solenoid give you a hint as to why the field lines of a bar magnet must loop? See Figure 13.4(b).

13.6. Compare the field lines of a long straight wire and those of a bar magnet or a solenoid and explain why the latter two can be thought of as having poles but the former not.

13.7. In a region of space, there is a uniform magnetic field; that is, its direction and strength does not change in that region. Hold a circular loop of wire perpendicular to the field, and move the loop parallel to itself.
(a) Will there be a current in the loop?
(b) What if you rotate the loop about an axis passing through the center of the loop and in the plane of the loop?
(c) About an axis passing through the center of the loop and perpendicular to the loop?

13.8. Is it possible to induce an electric current through a loop of wire without moving the wire or any of its parts?

13.9. Figure 13.9 shows a sample of magnetic field lines with two loops immersed in them. Each loop has an initial (i) position and a final (f) position.
(a) Which of the two loops carries more flux initially? If you were to assign numbers to fluxes, what would the ratio of the two fluxes be?

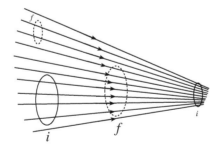

Figure 13.9: A sample magnetic field with fluxes through two loops.

(b) If the two loops were rotating with the same angular speed in their initial positions, which loop would carry more electric current?
(c) If the larger loop were rotating ten times faster than the smaller loop in their initial positions, which loop would carry more electric current?
(d) Suppose that the smaller loop moves to the left to its final position. Will there be an electric current produced in it while it is in motion?
(e) Suppose that the larger loop moves to the right to its final position. Will there be an electric current produced in it while it is in motion?

(f) How do you compare the *change* in the flux of the two loops?

(g) Suppose that the loops go from their initial to final positions in the same amount of time. In which loop do you expect a larger current to be induced?

(h) Suppose that the larger loop goes from its initial to its final position ten times faster than the smaller loop. In which loop do you expect a larger current to be induced?

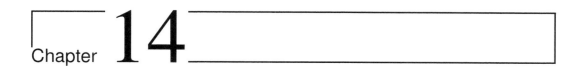

Chapter 14

Electromagnetic Waves

What makes a true champion? To a town or a village, it is a young athlete who, through devotion and perseverance, overcomes a physical handicap and goes on to bring an Olympic medal to the town or village; it is the honorable conduct of a daughter, a sister, or a mother who, in a popular war, rescued her comrades-in-arm at the risk of her own life. To a people it is a leader who rescues a nation from the tyranny and injustice of a corrupt and outmoded political system. Historical figures such as Spartacus, Joan of Arc, George Washington, and Martin Luther King are such national champions. Heroic as these figures were, their acts of heroism were confined to a local geography and a temporal slice of history.

There are more "quiet" champions who recognize no geographical borderlines, no temporal boundaries, and no national limits. They are champions of all people living in all places for all future time. Unaware of their heroism, they unselfishly pursue what is most important, not for a nation for some historical period, but for the entire human race for eternity. History is full of such global champions, most of them unknown to ordinary men and women who benefit from their acts of heroism on a daily basis. James Clerk Maxwell stands out tall among such champions.

14.1 Fields Are Primary

Faraday invented the concept of field lines as a means of visualizing the magnetic flux, thereby determining the electrical induction caused by the rate of change of this flux. The true power and the historic significance of fields did not emerge, however, until James Clerk Maxwell abstracted them and put them in their proper mathematical environment. And nature, in its most subtle way, awaited such sophisticated mental and mathematical achievement by humankind to guide them toward their most useful practical accomplishments.

Conventional physics was accustomed to "centers of force" acting at a distance, and when Faraday introduced the more intuitive concept of lines of force, it ran against the established belief. As Maxwell said, "Faraday, in his mind's eye, saw lines of force traversing all space. Where the mathematicians saw centers of forces acting at a distance, Faraday sought the seat of the phenomena in real actions going on in the medium."

Maxwell was the first strong supporter of the idea that Faraday's lines of force are indeed real physical and mathematical entities, and made them the cornerstone of his electromagnetic theory. In a paper presented to the Cambridge Philosophical Society in 1865, he said of Faraday's methodology:

> It was perhaps for the advantage of science that Faraday, though thoroughly conscious
> of the fundamental forms of space, time, and force, was not a professed mathematician.
> He was not tempted to enter into the many interesting researches in pure mathematics

which his discoveries would have suggested if they had been exhibited in mathematical form ....

It fell upon Maxwell himself "to enter into the many interesting researches in pure mathematics" which Faraday's field concepts suggested.

BIOGRAPHY

James Clerk Maxwell (1831–1879) came from a comfortable middle-class background. His mother died in 1839 from abdominal cancer, the very disease to which Maxwell was to succumb at exactly the same age. From an early age, he displayed a lively curiosity and a phenomenal memory. Maxwell's interests ranged far beyond the school syllabus, and he did not pay particular attention to examination performance. His first scientific paper on geometry was published at the age of 14. James attended Edinburgh Academy where he had the nickname "Daffy." His friend and biographer writes of Maxwell: "At school he was at first regarded as shy and rather dull. He made no friendships and spent his occasional holidays in reading old ballads, drawing curious diagrams and making rude mechanical models." This fascination with geometry and with mechanical models continued throughout his career and was of great help in his subsequent research. In 1850 he went to the University of Cambridge, where his exceptional powers began to be recognized. Maxwell is widely regarded as the nineteenth century scientist who had the greatest influence on twentieth century physics, making contributions to the fundamental models of nature. In 1931, on the centennial anniversary of Maxwell's birth, Einstein described Maxwell's work as the "most profound and the most fruitful that physics has experienced since the time of Newton."

14.1.1   Pre-Maxwellian Electromagnetism

The science of electromagnetism in the 1850s, when Maxwell started his work on the subject, consisted of four mathematical equations, which could be translated as the following four statements:

- Coulomb's law, which governed the (action-at-a-distance) force between two electric charges;

- Örsted–Faraday discovery that currents produce magnetism, especially the fact that magnetic fields have no beginnings and no ends;

- Faraday's law of magnetic induction of electricity in a wire;

- Ampère's law of production of magnetism from electric current.

One of Maxwell's great contributions to this science was his systematization of these equations, and their translation in the language of fields, a feat that history recognized by naming the field-equivalent of all the four statements above in his honor.

A thousand years from now, in the utopian world, in which every child is as fluent in mathematics as in his/her mother tongue; in which every adult appreciates the new kind of poetry, the poetry written in the language of mathematics; in which mathematical symbols of this poetry take on a new art form in the eye of ordinary men and women; **Maxwell's equations** will shine as a masterpiece among the poetic art pieces. We are far from this utopia, but as a reminder of this highest form of poetry, let us put each Maxwell equation together with a familiar art.

The following discussion explains, in nontechnical terms, the content of Maxwell's epoch-making work on electromagnetism. "Nontechnical explanation" is an oxymoron; but fortunately the richness of Maxwell's work makes it possible even for the non-mathematical—but motivated—person to appreciate the power and beauty of his creation. This creation is not unlike the creation of the most notable artists in history. If the genius of a great artist is to capture the beauty of nature, then Maxwell is one of the greatest artists of all time, because he was able to capture, on his mathematical canvas, the dazzling beauty of light itself.

🧺 Food for Thought 🍄

FFT 14.1.1. The word "field" has become a fashionable commodity among the practitioners of New Age medicine. When combined with another word taken from physics, "energy," it heralds a potent cure, a new state of mind and health, a higher level of consciousness. The purveyors of alternative medicine talk of the "interpersonal transfer" of the energy field by the laying-on-of-hand technique, the healing power of the magnetic fields of little pieces of magnets, the miracle of the energy field of the quantum healing techniques.

The only thing that is common between their use of the word "field" and the physicist's use of the word, as described in this chapter, is just that, the word. By their own admission, the field of alternative practitioners cannot be seen, cannot be touched, cannot be detected, or measured. In fact, they proudly claim that their "field" is nonmaterial.

The field of a physicist, on the other hand, is *entirely material*, and like any other material object under study in any branch of science, fields can be detected, measured, and demonstrably controlled. When a physicist speaks of a field, she can take you to her lab and show you where she produces it, how she guides it, and how she detects it at the receiver's end. Once she teaches you how all these are done, you can experiment with the fields yourself. Then you will know that when you hear a song on the radio, it is the *material* electromagnetic field that has turned into an electric current, which in turn has physically vibrated the diaphragm of your loudspeaker, which disturbs the air molecules around you, which *physically* vibrate the diaphragm of your ear, making you hear the song. When you watch your favorite program on television, it is again the *material* electromagnetic field that has created a tiny current, which has pushed some *material* electrons in a vacuum tube causing them to land on a coat of phosphorus material, which in turn emit a tiny amount of *visible* electromagnetic field, which your eye detects as TV images. There is no mysterious nonmaterial auras involved in a physicist's field.

There are two properties of the field concept that attract the alternative "doctors" to it: (a) fields are, for the most part, invisible to the human eye, and, (b) in principle, they spread over all space. These two properties fit very nicely with the (usually Eastern) spirituality of the practitioners of alternative medicine. After all, the "universal spirit," which is part of Eastern mysticism, is also invisible and spreads over all space. However, the word "spirit" is outdated and trivialized, but "field" carries the weight of the twentieth century physics. When you remove the cloak of modern verbiage from the "field" of the alternative medicine, you will discover the "spirit" of the ancient tribal shamans and witch doctors.

> "Field" of the alternative medicine and "spirit" of the witch doctors.

And neither the academic degree held by the practitioners of New Age medicine, nor their previous high rank in a hospital administration is relevant. The unawareness of the public—and the potential for turning this unawareness into personal wealth—is too luring to keep 100% of all medical doctors in their offices practicing honest and scientific medicine. It is very hard to resist the temptation of writing a national bestseller on modern spiritual healing, even if you hold a doctorate from an Ivy League Medical School.

It may not be money that attracts medical doctors and scientists to irrationalism. History gives us many examples of irrational inclination of (once-)rational people. Newton's fascination with alchemy is one example. Freud was a neurologist by training before he turned to the pseudoscience of psychoanalysis. Werner Heisenberg, the discoverer of the uncertainty principle in quantum theory, was lured to the Nazi ideology. Julian Schwinger, a Nobel Laureate and cofounder of the extremely successful quantum electrodynamics, was lured into the cult of cold fusion until his death in 1994. There are many other examples—both current and in history—of rational people turning to irrational doctrines. We don't know how or why such transitions take place. But we are certain of one thing: *Adherence of a well-known accredited scientist or doctor to irrational beliefs does not rationalize irrationality.*

> Adherence of a well-known accredited scientist or doctor to irrational beliefs does not rationalize irrationality.

Maxwell's First Three Equations

Maxwell's first equation, portrayed in Figure 14.1(a), also called **Gauss's law** after the great German mathematical physicist, states that the *divergence*[1] of the electric field is proportional to the concentration of the electric charge (ρ stands for the electric charge

---

[1]The symbol $\nabla \cdot \mathbf{E}$ stands for the divergence of the electric field.

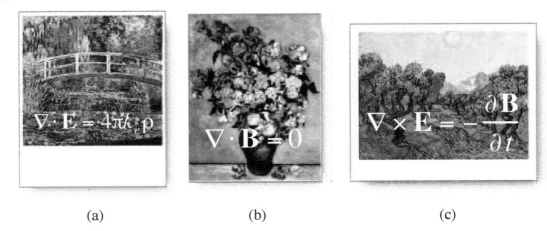

(a)   (b)   (c)

Figure 14.1: (a) Maxwell's first equation. The symbols $\mathbf{E}$, k_e, and ρ stand for the electric field, the constant appearing in Coulomb's law, and the electric charge (density), respectively. (b) Maxwell's second equation. The symbol $\mathbf{B}$, stands for the magnetic field. (c) Maxwell's third equation. The symbol $\nabla \times \mathbf{E}$ stands for the curl or the circulation of the electric field.

density). Maxwell's first equation is the reformulation of Coulomb's law [Equation (12.1)] in the language of fields.[2] **Math Note E.14.1** on **page 96** of *Appendix.pdf* takes you on a tour of this abstract symbolism. The essence of the first equation is that electric charge is the source of the electric field.

The second Maxwell equation, exhibited in Figure 14.1(b), states that the divergence of the magnetic field is zero. What does this mean? Math Note E.14.1 shows that divergence is closely related to the total flux through the bounding surface of a small cubic volume. The second equation is therefore saying that no matter where in space you pick your little cube, no net outward or inward flux will pierce its sides. Thus nowhere in space can you find a point from which magnetic field lines diverge, or to which they converge. Since such points indicate the presence of charges, the second Maxwell equation must be telling us that there are no magnetic charges in nature, or that magnetic field lines have no beginning or end.

The third Maxwell equation is displayed in Figure 14.1(c). This equation is the mathematical equivalent of the Faraday's law of induction expressed in the language of fields. **Math Note E.14.2** on **page 97** of *Appendix.pdf* shows the connection between the third equation and Faraday's law.

The Fourth Equation

Maxwell also inherited a fourth equation, the so-called *Ampere's law*, which is written in Figure 14.2(a). The LHS, as explained in Math Note E.14.2, is the circulation of the magnetic field around some (small) loop divided by the area of the loop. The RHS is (some constants times) the electric current divided by the area of the loop. Thus, the fourth equation states that the circulation of the magnetic field around a loop is proportional to the current piercing that loop. For example, if one calculates the circulation of the circular magnetic field of the long straight wire of Figure 13.3, one obtains $4\pi k_m$ times the current in the wire.

---

[2]As the alert reader may have noticed, the only surviving piece of Coulomb's law in this equation is the symbol k_e.

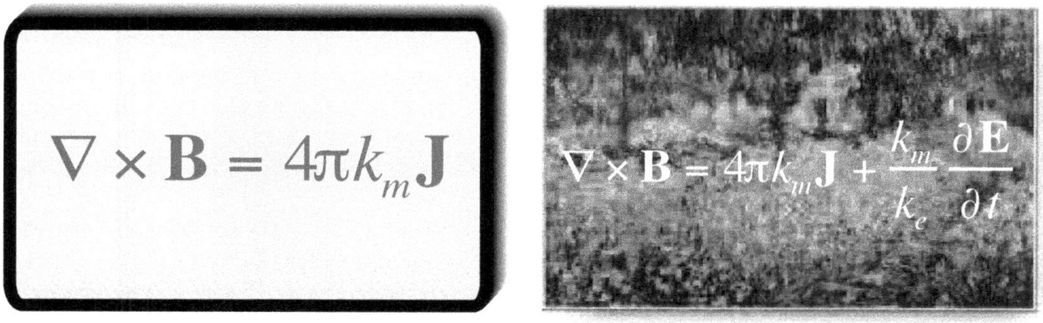

$$\nabla \times \mathbf{B} = 4\pi k_m \mathbf{J}$$

(a)

$$\nabla \times \mathbf{B} = 4\pi k_m \mathbf{J} + \frac{k_m}{k_e}\frac{\partial \mathbf{E}}{\partial t}$$

(b)

Figure 14.2: (a) The fourth equation as Maxwell inherited it. The symbol **J** stands for the electric current (per unit area). (b) Maxwell's fourth equation. Note that the term added by Maxwell is simply the time rate of change of the electric field.

14.2   Maxwell Corrects the Fourth Equation

Maxwell was one of the founders of vector calculus, the language in which electromagnetism is most fluently spoken. In fact, it was the need demanded by the electromagnetic theory—much like the need for calculus demanded by mechanics—that drove Maxwell in the direction of this mathematics. And although he did not use the abbreviated (and more sophisticated) symbols used in the artworks above, he was well trained in the subject to be able to analyze the four equations mathematically and logically. Such an analysis led Maxwell to an amazing discovery.

Electromagnetism forces new mathematics on mankind.

As early as 1855, the electromagnetic phenomena had occupied Maxwell's attention as well as many other prominent mathematicians and physicists. After a series of attempts at "constructing a mechanical model of electromagnetic action," Maxwell finally hit upon the key to such a dynamical theory. This key was the conservation of electric charge. Through a careful mathematical analysis he discovered that

*Here is the **true** power of the mind, not the "mind over body" of the alternative medical quackery!*

Box 14.2.1. *The fourth equation of electromagnetism was inconsistent with the conservation of electric charge.*

He also discovered the way to make the fourth equation consistent with charge conservation. This was a remarkable achievement, because it was entirely based on *the deductive power of the human mind*, Maxwell's mind. This remedial act of Maxwell immortalized his name, because today the four equations, each discovered by a different person, are justifiably called **Maxwell's equations**.

What do you know? **14.1.** Look at Figure 14.2 and explain why one equation is in a plain ugly box and the other on a masterpiece.

The only consistent way of changing the fourth equation was to add another term, the so-called **displacement current** to its RHS. The new equation, now called *Maxwell's fourth equation*, is displayed in Figure 14.2(b). The displacement current is simply the time rate of change of the electric field. In the absence of the electric current (i.e., when $\mathbf{J} = 0$), the new equation looks very similar to the third equation—except for the interchange of **E** and **B**.

If there are any instances in history in which the force of logic and impetus of mathematical consistency has changed the course of the human civilization, this is definitely one of them. The changing of the fourth equation, due to its momentous implication, is without a doubt one of the most significant events in the history of humankind, and definitely *the most important event of the nineteenth century.*

We have seen that for many years physicists had been looking for a connection between electricity and magnetism, and there were many experiments which showed such a connection. Maxwell's alteration of the fourth equation elevated the vague idea of a "connection" to a precise mathematical idea of *unity*. In Maxwell's theory, the electric and magnetic fields became inseparably intertwined and thus different manifestations of the same physical field, called the **electromagnetic field**. Maxwell's theory is the first true **unified field theory** which molds the electric and magnetic phenomena into a whole entity.

> The most significant event of the nineteenth century!

> The first true unified field theory.

14.3   Maxwell Predicts Electromagnetic Waves

Imagine a point in space where there is a changing electric field. According to Maxwell's theory, there must be a magnetic field close to that point. It turns out that this magnetic field is also changing. In reality, a changing electric field produces a changing magnetic field and vice versa. This changing magnetic field—again according to Maxwell's theory—produces a changing electric field in the neighboring points, which produces a changing magnetic field in the neighborhood of the neighboring points, ad infinitum! Such a progression of changing electric/magnetic fields is what is called an **electromagnetic wave**.

Maxwell showed, with mathematical rigor, that in regions of space where there are no electric charges and currents, his equations would lead to the equation displayed in Figure 14.3, *regardless of how the (distant) sources produce the electric and magnetic fields.* With proper identification of certain parameters, this is precisely the *wave equation* of Figure 11.1. He, thus, for the first time, predicted the existence of electromagnetic waves. And as **Math Note E.14.3** on **page 98** of *Appendix.pdf* shows

> **Box 14.3.1.** *The electromagnetic (EM) waves travel at the speed of light. Furthermore, this speed does not depend on how the waves are produced.*

Is this a mere coincidence? Is EM wave a completely new and different wave, which just *happens* to have the same velocity as light? Or, is light a form of EM wave? Maxwell boldly assumed the latter; and he was right. We now know that light is merely a form of EM wave which happens to be detectable by the human eye.

> **What do you know? 14.2.** As you ride a train going at 60 mph, throw a ball with a speed of 40 mph forward. The person outside will see the ball moving at 100 mph. This is called the law of addition of velocities. Now imagine getting on a spaceship going at half the speed of light and turn on your laser gun shooting laser light at the speed of light forward. What is the speed of this laser light relative to the observer outside? If you think it is 1.5 the speed of light, think again! And read the statement of Box 14.3.1, whose second statement precludes any dependence of the speed of light on the motion of its source.

Like other waves, EM waves satisfy $c = \lambda f$.[3] This brings up the question of: How large

---

[3] c is the universally accepted symbol for the speed of EM waves.

$$\nabla^2 \mathbf{E} = \frac{k_m}{k_e} \frac{\partial^2 \mathbf{E}}{\partial t^2}$$

$$\nabla^2 \mathbf{B} = \frac{k_m}{k_e} \frac{\partial^2 \mathbf{B}}{\partial t^2}$$

Figure 14.3: Two new equations that Maxwell derived from his set of four equations. These equations show that the electric field **E** and the magnetic field **B** are waves.

or how small can f and λ be? The answer: There is no limit to how large or how small they can be, as long as their product remains equal to c. We thus speak of the electromagnetic **spectrum** corresponding to the infinite range of all possible values of the wavelength or the frequency. Figure 14.4 shows the common range of EM wavelengths. Regions of the spectrum are given special names. These names become blurred at the overlaps between various regions. Thus, for example, the overlap between infrared region and the microwave region has no clear-cut name; similarly for the X-rays and the gamma rays, etc.

Electromagnetic spectrum.

> **What do you know?   14.3.** We claimed that Maxwell's correction of the fourth equation was the most significant event of the nineteenth century. Can *you* justify that? Hint: Look at the different categories of Figure 14.4.

The visible range consists of the rainbow colors. The longest visible wavelength is 0.7 μm (read *micrometer* which is 10^{-6} m) corresponding to red (R). The shortest visible wavelength is 0.4 μm corresponding to violet (V). In between these two extremes lie orange, yellow, green, blue, and indigo, making up a mnemonic acronym Roy G. Biv.

🧺 Food for Thought 🌱

FFT 14.3.2. What was it that motivated Maxwell? What application did he have in mind when working on electromagnetism? Was it the wealth that was involved upon patenting his discovery? Was he even interested in patenting his revolutionary findings? Perhaps it was the celebrity status that would surely come after his success? What *was* his motivation? Surely it wasn't just to solve an equation or unravel a mystery!

Indeed it was! Not a single practical consequence or invention resulting from EM theory could have been predicted by Maxwell. His only interest *was* to solve an equation, to unravel a mystery, to find the solution to a puzzle, to satisfy a personal curiosity. After all, this is the dominant characteristic of human species. As a species, our inquisitive mind, our curiosity and our desire to have answers to questions are our most effective weapons for survival. This is as true today as it was two million years ago. Look at a 4-year-old child whose mind is unfettered by any cultural or social bonds. The number of questions she asks per day runs into the hundreds! Yet, by the time she reaches adolescence, this number drops to single digits. There are many factors contributing to this disheartening apathy, but the dominant one, especially in the US, is the societal and parental emphasis on entertainment, sport, accumulating wealth, and having "fun," while stigmatizing children who like to exercise their intellect and curiosity as "nerds."

Fortunately there are exceptions. A small—and, sad to say, dwindling—fraction of the human species continues to ask questions, to think about problems, to have an ongoing desire to unravel the mysteries of nature. These are our scientists pursuing fundamental problems of our time. Problems that have no practical applications in sight, but have posed themselves in scientific communities as the most challenging. If we, as a species, are going to survive, we must support them unconditionally.

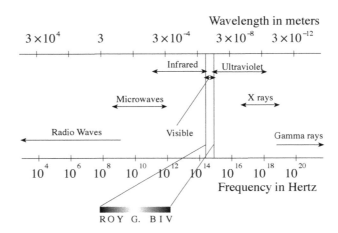

Figure 14.4: There is no limit to how large or how small a wavelength (or frequency) EM waves can have. The visible spectrum has the longest wavelength of 0.7 μm corresponding to red, and the shortest wavelength of 0.4 μm corresponding to violet.

Example 14.3.3. An AM (Amplitude Modulation) radio station transmits at 1000 kHz (kilohertz). What is the wavelength of the EM waves transmitted? Using the equation $c = \lambda f$ with $c = 3 \times 10^8$ and $f = 1,000,000 = 10^6$ Hz, we obtain $3 \times 10^8 = \lambda \times 10^6$ whose solution is $\lambda = 300$ m.

The radius of an atom is approximately 5 **Angstrom** (an Angstrom is 10^{-10} m or a tenth of a nanometer). What is the frequency of the EM wave having a wavelength this small, and what is the name of such a wave? Again using the formula $c = \lambda f$ with $\lambda = 5 \times 10^{-10}$ m, we obtain $3 \times 10^8 = 5 \times 10^{-10} f$, so that $f = 6 \times 10^{17}$ Hz. This is an X-ray. ∎

What do you know? 14.4. An EM wave has a wavelength of 250 nm. What category of EM spectrum does it belong to? What about 900 nm?

14.3.1   Production of EM Waves

Our discussion of EM waves has thus far assumed the prior existence of a changing electric (or magnetic) field which gives rise to a magnetic (or electric) field and so on, establishing a wave. But what causes the *original* changing field? We have seen that a charge is surrounded by an electric field, and a magnet by a magnetic field, whose strength at a given point depends on the distance of the point from the charge or the magnet. This suggests that if we move the charge or the magnet, the electric or the magnetic field will change and the wave will be produced. The mathematical analysis of Maxwell's equations shows that the suggestion is correct with a minor modification:

Accelerating charges produce EM waves.

Box 14.3.4. *Only an **accelerated** electric charge (or magnet) produces EM waves.*

Although a charge in uniform (constant velocity) motion has a changing electric field, it does not produce EM *waves*. To produce EM waves, the velocity of the electric charge must change.

There are many ways to produce EM waves, some practical and some not-so-practical:

- Throw a bottle of electric charge to the floor. The free fall and the stoppage of the charge create, respectively, an acceleration and a (very sudden) deceleration, causing EM wave production.

- Rub a piece of plastic against wool (this charges the plastic), and move the plastic up and down. This up-down motion of electric charge is an accelerated motion. Hence, EM waves are produced.

- Attach the above-mentioned electrically charged plastic to one end of a string, and, holding the other end, whirl the plastic above your head. This circular motion, like any other circular motion, has (centripetal) acceleration that can produce EM waves.

The three methods above are hardly the way EM waves are produced. The frequency in the second and third item is not even 10 Hz. Thus the wavelength is longer than 3×10^7 m or 30,000 km! A wave with such a large wavelength (the circumference of the Earth is about 40,000 km) is hardly called a wave. There are better ways of producing EM waves.

- An antenna carries an alternating current of high frequency. Because of the periodic change in the direction of the current, the charges responsible for the current undergo periodic acceleration producing EM waves.

- The charges in a wire carrying alternating current (AC) experience acceleration. Hence, they produce EM waves.

The difference between the last two methods is that in the first, the charges are given a well-defined frequency (the frequency of the radio station, for example), while in the second there is no well-defined frequency. That is why the EM signal from a wire is picked up as a noise by the receiver. The frequency of 60 Hz associated with the AC electricity is only the "average" frequency. There are many other "components" of higher and lower frequencies in the AC electricity.

What do you know? 14.5. In the classical picture of the hydrogen atom, the electron moves with a speed of 400,000 m/s on a circle of radius 0.1 nm. What kind of EM wave does it produce? Hint: Find the period and then the frequency.

The shape of an EM wave is very complicated in the vicinity of its source. However, if you look at the wave far away from the source, the shape is simple. Think of the wave front as a sphere that expands in all directions. If you are very far away, and have access to only a small portion of that sphere, that small portion appears as a flat plane (think of the surface of the spherical Earth). Because of this flatness, the wave detected at a large distance from its source is called a **plane wave**. The electric and magnetic field lines of plane EM waves lie in this planar front and are perpendicular to one another. Figure 14.5 shows a plane electromagnetic wave that moves along the z-axis. The electric field, which oscillates along the x-axis of the figure is perpendicular to the magnetic field which oscillates along the y-axis. Maxwell's equations show that for a plane EM wave, the electric and magnetic

Plane EM waves.

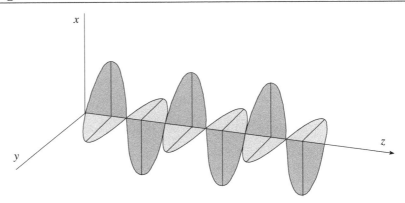

Figure 14.5: A plane electromagnetic wave has an electric field and a magnetic field that are perpendicular to one another. In this figure the electric field oscillates up and down (along the x-axis), and the magnetic field oscillates back and forth (along the y-axis). The wave is propagating in the positive z-direction.

fields are related at all times via $E = cB$ where c is the speed of light. Thus, the two fields oscillate in unison.

Like any other waves, EM waves carry energy, which is proportional to the square of their amplitudes (see Box 11.2.2). Since there are really two waves (electric and magnetic) associated with an EM wave, the energy is the sum of these two waves. This is the classical picture of an EM wave, and it is valid as long as we deal with low-frequency EM waves. One of the revolutionary discoveries at the turn of the twentieth century was that this classical assumption breaks down for high-frequency EM waves. The correct way of describing high-frequency (and in fact *all*) EM waves is the language of quantum theory. **Example D.14.1** on **page 37** of *Appendix.pdf* looks at the energy of the EM waves coming from the Sun and those coming from the power lines.

14.4   End-of-Chapter Material

14.4.1   Answers to "What do you know?"

14.1. The plain box contains the fourth equation that violated electric charge conservation, and is wrong. The masterpiece contains one of the masterpieces of mathematical physics: the fourth Maxwell's equation.

14.2. In fact, the speed of light is a universal constant independent of the motion of the source or receiver. This statement is at the heart of relativity theory, and leads to such weird phenomena as the relativity of time and length (see Chapter 25).

14.3. The correction of the fourth equation allowed the production of the entire spectrum of EM waves, which have penetrated every aspect of our modern civilization with dramatic and revolutionary consequences.

14.4. 250 nm is 0.25 μm or 2.5×10^{-7} m, which is slightly shorter than the violet wavelength of 4×10^{-7} m. So, it is ultraviolet. Similarly, 900 nm is infrared.

14.5. To find the period, you need the circumference of the orbit whose radius is given as 0.1 nm or 10^{-10} m. This gives $2\pi \times 10^{-10}$ or 6.28×10^{-10} m for the circumference. Divide the circumference by the speed to find the period. You should get 1.57×10^{-15} s. Take the inverse of this to get the frequency, 6.37×10^{14} Hz, corresponding to a wavelength of 0.47 μm, which is visible.

14.4.2 Chapter Glossary

Amplitude Modulation (AM) A technique of transmitting radio waves by modulating the amplitude of the carrier wave.

Displacement Current The term which Maxwell added to the fourth equation of electromagnetism, and which is the key to the prediction of the EM waves.

Divergence A technical term referring to the total flux through the bounding surface of a small volume divided by the volume.

Electromagnetic Spectrum The range of electromagnetic waves having different frequencies (or wavelength), all traveling at the speed of light.

Frequency Modulation (FM) A technique of transmitting radio waves by modulating the frequency of the carrier wave.

Wave Equation A mathematical equation obeyed by all waves. Maxwell showed that electric and magnetic fields obey this equation and therefore they can propagate as waves.

14.4.3 Review Questions

14.1. State the translation of the four equations that comprised electromagnetism in the 1850s. Which one of these was discovered by Maxwell?

14.2. Are there any similarities between the field as used by physicists and the field as used by the New Age doctors and nurses? What is the main difference between the two? Is it possible to detect the field of the physicists? Is it possible to detect the field of the New Agers? What properties of the field concept is attractive to the New Age doctors?

14.3. What is the essence of the first EM equation? What about the second equation? Whose name is associated with the third equation?

14.4. Who discovered the fourth equation of electromagnetism? What does the fourth equation say about a long wire carrying a current?

14.5. What is the significance of the fourth equation? What was wrong with it? How did Maxwell make it right?

14.6. What is the most important event of the nineteenth century? What physical entity was predicted as a consequence of this event? What was the connection between this physical entity and light?

14.7. Name the seven categories of the EM waves. Which category has the longest wavelength? The shortest wavelength? As you look at the EM spectrum, can you see why Maxwell's correction of the fourth equation is the most important event of the nineteenth century?

14.8. How can you produce EM waves? What is the most common way of producing them? Can you produce an EM wave by moving a charge up and down with your hand? What is a plane EM wave? Where, in relation to its source, do you see the wave as a plane wave?

14.4.4   Conceptual Exercises

14.1. Suppose you surround a point in space with a tiny box, and note that electric field lines are coming out of all sides of the box. From what you know about the field lines of electric charges, what do you think is inside the box?

14.2. Suppose you surround a point in space with a tiny box, and note that magnetic field lines are coming out of all sides of the box. You *have* to say "something is wrong." Why?

14.3. Take a loop of wire and place it in a magnetic field that changes with time. Make sure that the magnetic field is perpendicular to the area of the loop.
(a) Is the magnetic flux changing?
(b) Do you get a current in the loop?
(c) What kind of a field causes the charges in the loop to move (remember that magnetic field cannot accelerate charges)?
(d) Is it safe to say that the field which moves the charges has an existence of its own? That the motion of the charges is merely an indication of that existence just as the fall of an object is merely an indication of the existence of a gravitational field? You have just shown that a changing magnetic field produces an electric field that curls. This is the third Maxwell equation.

14.4. The equations in Figure 14.3 are wave equations. The ratio k_m/k_e is the inverse square of the wave speed. Look up the values of k_m and k_e and calculate the velocity of the EM waves. Do you see why light is an EM wave?

14.5. What category of EM waves has the longest wavelength? The shortest wavelength?

14.6. Is there any category of EM waves which has no practical application? Do you think that Maxwell was aware of these applications?

14.4.5   Numerical Exercises

14.1. An electric charge is oscillating with a frequency of 10 Hz.
(a) What is the wavelength of the EM wave produced? How does it compare with the circumference of the Earth (40,000 km)? Does it make sense to call such an EM wave a "wave"?
(b) The frequency is now increased to 5 kHz. What is the wavelength now? Does it make sense to call such an EM wave a "wave"?
(c) The frequency is further increased to 5 MHz. What is the wavelength now? Does it make sense to call such an EM wave a "wave"?

14.2. One of the fastest mechanical rotators—with an electric charge on it—spins at a rate of 50,000 rpm (revolutions per minute).
(a) What is the frequency of the rotation?
(b) What is the wavelength of the EM wave so produced?
(c) Does it make sense to call such an EM wave a "wave"?

Chapter 15

Epilogue: Mind and Matter

The development of electromagnetism took place very slowly compared to mechanics, although the two branches of physics started at about the same time. While only two decades after Galileo, Newton wrote down the laws of mechanics in their entirety, two *centuries* after Gilbert, there was still confusion about the laws of electricity and magnetism. The reason—aside from the fact that mechanics had a powerful mind such as Newton's on its side—was that electricity was intrinsically more complicated and required substantially more observations before it succumbed to a rigorous theoretical analysis.

15.1 Continuity

From Gilbert's first observations to Coulomb's mathematical statement about electrical forces, almost two centuries passed. The very control of electricity and its collection on various objects took a long time. All of this required patient and systematic observations by many generations of experimenters. There is no other place in the history of science than in experimental electricity where the *continuity* of science can be demonstrated more effectively. Every discovery became the *necessary* starting point for a new finding.

Örsted's discovery of the connection between electricity and magnetism in 1820 required a battery, which was invented by Volta in the 1790s. The invention of the battery would not have been possible without the earlier experiments by Galvani of the effects of electric currents on frog legs. These currents were produced using the Leyden jar, which was one of many devices that could store electricity. The construction of the Leyden jar relied on the conducting properties of metals, which was discovered by Gray at the beginning of the eighteenth century using Güricke's electric machine developed in the 1660s. And Güricke relied on the discoveries made by Gilbert, who started modern electricity and magnetism.

The discovery of electrical magnetism prompted one of the most intense research activities of human history. In less than two decades, all the observational and phenomenological pieces of the puzzle were in place, and there wasn't much of anything else physicists could do in the way of observation and data collection. The next step required the intervention of the human mind.

Not only were the experiments of electricity and magnetism connected like a chain that stretched from the 1830s all the way to 1600, but also the theoretical ideas introduced. For example, the mathematical formula for the force between two electrical charges [Equation (12.1)] uses the concept of force as introduced in the second law of motion by Newton in 1660s, and connects electricity to mechanics.

More importantly, all the intense experimental research in the magnetism of electric currents done in the decades of the 1820s and 1830s also used instruments that measured

Electricity, the epitome of the continuity of science.

mechanical forces introduced in the second law of motion. And all the intricate mathematical formulas of magnetism were based on such experiments and observations. One can say that the chain that connects over two hundred years of electromagnetism, branches off to the science of mechanics of the 1600s as well.

15.2   Specificity and Idealization

The entire subject of electromagnetism developed step by step by asking specific questions and studying specific objects. Gilbert, the father of modern electricity and magnetism, asked "Why does the *compass needle* point north?" "Why does *amber* attract pieces of paper?" "Can I *electrify* other objects?" Güricke, following Gilbert's footsteps and electrifying various substances asked the specific question "What is the best way to add more and more electricity to a substance?" And at the climax of electro*statics*, Coulomb asked the specific question "What is the electric force between two charged *metallic spheres*?"

Electromagnetism develops by asking specific questions.

After the discovery of electric conduction and batteries, the focus shifted to the magnetic effect of electric currents. But there was no shift in *specificity*. "How does an electric circuit produce a magnetic field?" No, more specifically, "How does a *straight* current-carrying wire produce a magnetic field?" No, that is too general! "How does an infinitesimal element of a current-carrying wire produce a magnetic field?" The more specific the questions became, the more widespread was their use. Once you know how an *infinitesimal* element of a current produces its magnetic field, the field of any other (large) current could be calculated, because any wire can be broken into a large number of infinitesimal pieces, and by adding the field of the infinitesimal pieces, one can (through the mathematical process of integration) calculate the field of the entire wire.

Idealization, another important step in the development of electromagnetism.

Every specific question or procedure accompanied some kind of *idealization*. When Coulomb measured the electric force between two charges, he used charged spheres. But when he wrote down the Coulomb force law [Equation (12.1)], he *idealized* the spheres to mathematical points. Furthermore, he ignored the air between the two charges and how it would affect the force. In other words, he assumed that the two charges were in a vacuum, although the air, with all the complex entities comprising it, was surrounding the charges. Was he justified in doing so? We now know that, although air is electrified by charges, and therefore alters the force between them, the effect is so negligible that one can safely ignore it, as did Coulomb.

The same idealization that Coulomb used to come up with his force law, was used in all aspects of magnetism of currents. Take an infinitesimal current. This is a mathematical concept just as a "point" is a mathematical concept. It is a point with a direction: take any point of a current-carrying wire and see which way the current is flowing *at that point*. The combination of the point and the arrow pointing in the direction of the current at that point is an infinitesimal current element. You cannot get any more ideal than this! Yet, the entire theory of magnetism is based on this idealization. Moreover, just as Coulomb ignored the air between the charges, so did the physicists measuring the magnetic forces and fields of currents.

15.3   Role of the Mind

Electromagnetism is sometimes considered the transition between classical and modern physics. It led directly to relativity; and in conjunction with thermodynamics and black body radiation, it also led to the quantum theory. However, this transition role did not come about as a result of the phenomenological developments of the 1820s and 1830s, but as a consequence of Maxwell's mathematical intervention of 1865.

The slow progress in the theoretical development of electromagnetism was due not only to the slow pace of its experimental development, but also to its mathematical complexity.

As a comparison take the gravitational (or even electro*static*) force between two point masses (or two stationary point charges). The only relevant direction is the line joining the two points. That is why both the gravitational force [Equation (9.1)] and the electrostatic force [Equation (12.1)] are relatively simple. When physicists tried to write down the expression for the magnetic forces between two (infinitesimal) currents, they had to take not only the direction connecting the two currents, but also the direction of the currents themselves. The intertwining of all these directions makes the mathematical formula of the magnetic force between two currents much more complicated.

As hard as the formulas were, by 1850s *all* the observed electromagnetic phenomena were condensed into a set of four equations. Every new experiment could be explained by these equations, and *as far as observation was concerned* there seemed to be no need for any modification of the four equations. However, two purely theoretical developments changed the course of history. Both came from James C. Maxwell's mind.

The first was Maxwell's brilliant recognition that electric and magnetic *fields* were of primary importance. The notion of fields was introduced by the mathematically illiterate Faraday first as an intuitive aid to understand magnetic forces, but then as a more serious quantity worthy of the attention of theoreticians. Maxwell, a supreme mathematical physicist, somewhat in isolation, took Faraday's advice and translated the four electromagnetic equations into the "language" of fields.

The second was nothing less than a miracle! It was this development that broke away from classical physics and paved the way for both relativity and the quantum theory. And this step was taken purely by the human mind, Maxwell's mind! Maxwell's mathematical mind! The details of this step are described in Chapter 14. Here I want to emphasize the power of the mind in uncovering the most hidden secrets of nature and the tool with which the mind digs through the surface, mathematics. This use of mathematics is different from its use prior to Maxwell. Pre-Maxwellian use of mathematics was in transcribing some experimental results in the proper language. Maxwell *intervened* in what was assumed to be the correct formulation of electromagnetism. And nature agreed with this purely mathematical intervention and rewarded Maxwell with the entire spectrum of the EM waves. *Real power of the mind.*

New Agers also talk about the "power of the mind," but their mind is a yogic spirit whose "power" comes from closing one's eyes and "concentrating" on the void. The power of the mind of the likes of Maxwell comes from years of strenuous training in the Olympian game of mental sport, mathematics, solving many unsolved riddles, and tackling the most difficult problem of the time. A toddler who throws her toy is not qualified to be called a "discus thrower." In the same vein, a yogi who closes his eyes and sits cross-legged is not qualified to be called a possessor of mental power, no matter how much he "concentrates" and how perfect his cross-leg position is. *New Agers' "power" of the mind!*

As we shall see in the sequel, Maxwell's "deductive" reasoning has become the norm of modern physics. In many cases, this deduction—without any observational necessity—leads to new branches of physics. For example, Einstein's special theory of relativity is entirely based on Maxwell's derivation of EM waves and not on any experiments or observations. Although almost all modern theories of physics are deductive theories, it does not mean that experiments play no role. On the contrary, crucial experiments such as the ones that led to the four equations of electromagnetism, set the *correct* stage on which the deductive theories act.

15.4 Materiality of Fields

The fields of Faraday and Maxwell have nothing in common with the "energy field" of the New Agers. The latter, borrowed from electromagnetism and modern physics, uses the word to give legitimacy to a nonmaterial, spiritual, voodoo, modern witchcraft. Electromagnetic fields, on the other hand, are as material as the electricity that turns into images coming from a TV, sound waves coming from a cell phone, and the multimedia of your favorite URL (see also Food for Thought 14.1.1).

The way one can test for the materiality of anything is to have it interact with a suitable form of matter and note that it affects the physical characteristic of that form of matter. Any physical quantity has this property; if it didn't we would not call it "physical." Electromagnetic fields or waves affect the motion of charged matter. When they reach the receiver of your radio, or TV, or cell phone, they set in *actual physical motion* the charges in the receiver and create an electric current that can get magnified and its energy transformed into audiovisual energy.

The New Agers' "field," or "energy," or "energy field" is, by their own admission, nonmaterial. These fields cannot be felt, cannot be detected by any physical instrument, do not have any visible or audible effect, and do not exist except in the minds of the New Agers or the guru whom they follow. Here is another difference between the physical fields and the New Agers' "field." Anybody can learn how to operate a device that produces EM fields, and if desired, learn the physical and electrical details of how the fields are produced; she can even be taught to construct a transmitter from wires, resistors, capacitors, etc. She can also learn how these fields are received by antennas, and if desired, learn the physical and electrical details of how the fields are received, and how to construct a receiver from wires, resistors, capacitors, etc. And if she fails, she is not told "You are not concentrating enough," or "You don't have enough faith in the EM fields." She is shown which wire or capacitor or another circuit element was connected incorrectly.

Difference between EM fields and New Agers' field.

Box 15.4.1. *The New Agers' field cannot be produced or received by any physical instrument. Only the "privileged few" who have faith or been anointed by their guru can "feel" the "energy." If I cannot feel the energy or otherwise detect it, I'll be accused of "not concentrating enough" or "not having enough faith."*

15.5   Support for Pure Research

Maxwell did not live to witness the materialization of his mathematical prediction. He died of stomach cancer at the relatively young age of 48 in 1879, the year Einstein was born. Had he lived out a normal life expectancy, he would have witnessed the production of EM waves by Heinrich Hertz in 1887, the invention of the radio by Marconi in 1901, and the evolution of his theory—in the hands of Albert Einstein—into one of the most significant discoveries of all time, the theory of relativity.

Today we can truly appreciate the indispensability of electromagnetism. Every radio or television set, every telephone, every computer, every large or small appliance in our homes, in short, every aspect of our contemporary civilization, one way or the other, has its roots in Maxwell's theory. As the American physicist Richard Feynman noted, "in the far future, when the historians look back at the 19th century and scrutinize the events that took place between 1800 and 1899, they will undoubtedly choose Maxwell's unification of electricity and magnetism as the landmark of that century." All the political, economic, and social upheavals such as the Napoleonic wars or the American Civil War will be dwarfed into insignificance next to the discovery of electromagnetism.

Maxwell's equations, the landmark of the nineteenth century.

A question often asked is "Why should we support research, especially if it is extremely expensive and has no immediate application?" Why should the tax payer's money go into fundamental research, such as high energy physics, requiring accelerators that cost millions (sometimes billions) of dollars? These researches aim at finding the ultimate structure of matter beyond molecules, atoms, and nuclei; researches whose results will definitely not be displayed on supermarket shelves in the near future! The subject of this part of the

Fundamental research lights our darkness.

book has a very good answer. If the founders of electromagnetism were concerned with its application and practicality, if Örsted had stopped his research on the effect of the electric current on magnets, Faraday had not diligently pursued his idea of electrical induction, and Maxwell had curbed his mathematical investigation of electromagnetism because they could not foresee any practical application of their research, we would still be burning candles to light our darkness.

Part IV

Thermodynamics

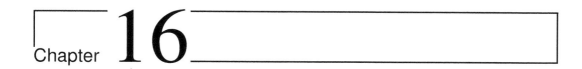

Chapter 16

Introduction to Probability

The third branch of post-Renaissance physics that started to develop at about the same time as the other two (mechanics and electromagnetism) is thermodynamics. It is amazing that, as soon as the chains of the Dark Ages were broken, the seeds of *all* areas of modern physics were planted almost simultaneously. Galileo, who gave birth to the science of mechanics, also launched the investigation of the properties of heat.

A serious investigation of thermodynamics, however, did not take place until the latter part of the eighteenth century, when thermometry became an art in which many scientists and entrepreneurs took part. The invention of the steam engine and the scientific examination of its operation had a profound effect on the development of thermodynamics. In fact, the second law of thermodynamics (which came before the first law!) was discovered in an attempt at improving steam engines. The laws as well as the methodology behind their discovery were completely phenomenological and, at times, mysterious.

By the middle of the nineteenth century it became increasingly plausible to assume that all substances were made up of particles which obey the laws of mechanics. No knowledge of the internal structure of these particles was either available or necessary for the study of the bulk matter. Any sample of matter was assumed to contain a very large number of these particles, and while an *exact* study of such a large collection of objects was impossible, this drawback became an opportunity to apply statistical methods to the sample, giving birth to **statistical mechanics**, which in turn shed light on the sources of the laws of thermodynamics and unveiled the mysteries of the phenomenological studies. To understand statistical mechanics, we have to know something about the fundamentals of probability theory, the subject of this chapter.

16.1   Basic Concepts

Probability is the mathematical investigation of **random events**. We have a good intuitive feeling of what a random event is, and this intuition is sufficient for our discussion. Examples of random events are the outcome of the throw of a die, of drawing a card from a deck of playing cards, and the toss of a coin. The latter is going to be the main focus of our attention in this chapter.

Let us start with a single coin. It is clear that the number of **possible outcomes** for a toss is 2: One head H and one tail T. We also say that the **frequency** of one H is 1 and that of one T (or zero H) is also 1. Probability is the ratio of the frequency to the total number of outcomes. Thus, the probability of 1 H is 1/2 and the probability of 0 H is also 1/2, and the sum of all probabilities is 1.

> **What do you know? 16.1.** What is the probability of drawing a red card from a deck of cards? Drawing a diamond? A king?

Now consider two identical coins. The possible outcomes are: (H,H), (H,T), (T,H), (T,T), where the first entry represents the outcome of the first coin and the second entry that of the second coin. Thus, the total number of outcomes is 4, the frequency for two H's is 1, for one H is 2, and for zero H is also 1. Therefore, the probability of zero H is 1/4, which is also the probability of 2 H, while the probability of 1 H is 2/4 or 1/2.

For 3 coins the possible outcomes are: (H,H,H), (H,H,T), (H,T,H), (T,H,H), (H,T,T), (T,H,T), (T,T,H), (T,T,T). Therefore, the total number of outcomes is 8, the frequency for three H's is 1, for two H's is 3, for one H is also 3, and for zero H is 1. With obvious notation, we therefore write $P(0) = 1/8$, $P(1) = 3/8$, $P(2) = 3/8$, and $P(3) = 1/8$. **Example D.16.1** on **page 38** of *Appendix.pdf* calculates the probabilities for four coins.

Can we generalize the arguments above to the case of many coins? In other words, can we derive a formula that finds the probability $P_n(m)$ of getting m heads in tossing n coins? The answer is yes, and **Math Note E.16.1** on **page 98** of *Appendix.pdf* shows how to find this general formula. That math note also shows the symmetry of the probability function about $n/2$: the probability of getting $n/2 + x$ heads in tossing n coins is the same as the probability of getting $n/2 - x$ heads for any positive integer x.

> **What do you know? 16.2.** What is $P_{10}(3)$, the probability of getting 3 heads in tossing 10 coins? What is $P_{10}(7)$?

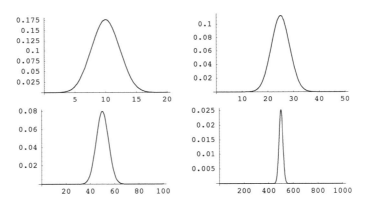

Figure 16.1: Plots of the probability distribution for the throw of 20, 50, 100, and 1000 coins. You can read off m_- and m_+ for each plot from the figure. Note how the graph becomes narrower as the number of coins increases. Note also how the maximum probability decreases as the number of coins increases.

It is convenient to plot the probability function versus the number of heads. Figure 16.1 shows the plot for 20, 50, 100, and 1000 coins. A few features of these plots are worthy of closer attention. First note that all plots have a common shape; in particular, they have a maximum at half the number of coins. This means that for any number of coins, the probability of obtaining half of them as heads is the greatest. Next note that the probability of getting a number of heads which is too much different from half the coins is so small that it cannot be plotted differently from zero. Denote by m_- and m_+ the values of m such that $P_n(m)$ is "negligible" when $m < m_-$ or $m > m_+$. The word "negligible" in the

m_- and m_+ defined.

present context means "indistinguishable from zero on a plot." In the throw of 20 coins the probability of getting 3 heads or less, or 17 heads or more is negligible as shown in Figure 16.1. Thus $m_- = 3$ and $m_+ = 17$ for $n = 20$. Similarly, the same figure shows that $m_- = 12$ and $m_+ = 38$ for $n = 50$; $m_- = 33$ and $m_+ = 67$ for $n = 100$; and $m_- = 440$ and $m_+ = 560$ for $n = 1000$.

16.2   Distribution Plots

The probability plots, also called **probability distributions**, can best be studied once we define some quantities. The **absolute width** Δ_n of the probability curve of n coins is the width of the plot on the horizontal axis, i.e., the difference between m_+ and m_-: $\Delta_n = m_+ - m_-$. The **relative width** δ_n of the probability curve, is the absolute width *Relative width of the* divided by the number of coins: $\delta_n \equiv \Delta_n/n$. Use the values of m_+ and m_- found above *probability curve.* and obtain $\delta_{20} = 0.7$, $\delta_{50} = 0.52$, $\delta_{100} = 0.34$, and $\delta_{1000} = 0.12$.

What is the significance of δ_n? Suppose we plot the probability distribution of n coins on a graph where the horizontal axis, whose range is 0 to n, is 10 cm long (slightly less than the width of the text you are reading right now). Then the width of the plot in centimeters—which we denote by x—is obtained by proportionality: $(10 \text{ cm})/(n \text{ coins}) = x/(\Delta_n \text{ coins})$. This yields $x = 10\Delta_n/n$ or $x = 10\delta_n$ cm. It is now clear that δ_n is proportional to the width of the graph of the probability distribution plot. This is clearly indicated in Figure 16.1.

Figure 16.2 shows the distributions for large numbers of coins. This set of plots differs very little from the set in Figure 16.1. What is not apparent from these figures is the sharp decrease in the quantity δ_n for larger n, because we have plotted points that are very close to the maximum to be able to read the values of m_+ and m_-. In other words, we have zoomed in on the middle of the plot. In Table 16.1 we record δ_n for Figures 16.1 and 16.2. This table indicates that as the number of coins increases, the width of the probability distribution curve—which is proportional to δ_n, as stated above—decreases.

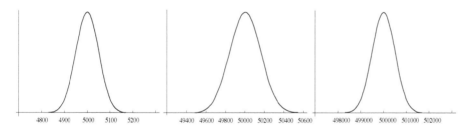

Figure 16.2: Plots of the probability distribution for the throw of 10,000 (left), 100,000 (middle), and 1,000,000 (right) coins. Note that the nonzero probabilities are concentrated in regions close to the maximum (half the number of coins).

The figures are more than just a pictorial representation of raw data. To appreciate them, let us consider 10 coins and turn the figure into a bar graph as shown in Figure 16.3(a). The area of each bar signifies the probability of getting the corresponding number of heads in the throw of 10 coins. If we are interested in the probability of getting between 3 and *Probability and area* 7 heads, we just add the areas of the bars. The shaded area in Figure 16.3(a) gives this *under the curve.* probability. The total area should add up to one. Thus to get the probabilities, we should really divide the area of interest by the total area. How does the area of interest compare with the corresponding area under the probability curve? For 10 coins the area of the bars between 3 and 7 heads is very different from the area under the curve between those values. However, as the number of coins increases, the difference between the area of the bars and the area under the curve decreases. For example, if one compares the area of the bars for

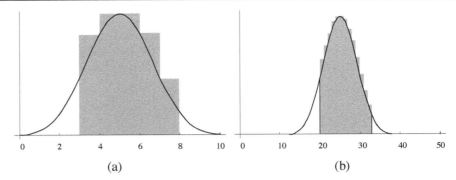

Figure 16.3: A bar graph shows the relation between probability and area. (a) The area is probability of finding between 3 and 8 heads in a throw of 10 coins. (b) The area is probability of finding between 20 and 33 heads in a throw of 50 coins.

the number of heads between 20 and 33 in a throw of 50 coins with the area under the probability curve, the horizontal axis, and the two vertical lines corresponding to $m = 20$ and $m = 33$, one finds that the two are almost equal [see Figure 16.3(b)]. This equality becomes more and more pronounced as the number of coins gets larger and larger.

Box 16.2.1. *For a large number of coins, the probability that the number of heads turning up is between m_1 and m_2 is the area under the probability curve between the two vertical lines corresponding to m_1 and m_2 divided by the total area.*

We defined m_- and m_+ as values of m, beyond which probability drops to (almost) zero on the distribution plot. This definition is inaccurate, because it relies on reading a value very close to zero on a plot. We make the definition more precise by saying that the probability is negligible if its value is less than one millionth of the maximum probability. In symbols, $P_n(m_\pm) = 10^{-6} P_n(n/2)$, because probability is maximum at $n/2$. Starting with this definition, you can obtain m_- and m_+ in terms of n. **Math Note E.16.2 on page 99** of *Appendix.pdf* provides the details.

By their very definition, m_- and m_+ bound a large fraction of the total probability. How large? The area between m_- and m_+ can be calculated numerically with great precision. It can be shown that in our case this area is 99.999985% of the total area! This means that the probability of getting a number of heads larger than m_+ or smaller than m_- is $1 - 0.99999985 = 0.00000015$ or 1.5 parts in 10 million! For $n = 1,000,000$ this means that the probability of getting more than 502,630 heads or fewer than 497,370 heads[1] is only 1.5×10^{-7}.

Probabilities are meaningful and practical only when we repeat the same experiment many times. When we say that the probability of a certain event E in an experiment is 0.001, we mean that if the experiment is repeated 1000 times, on the average, E occurs only once. Repeating the experiment 10,000 times results in about 10 E's, and in 1,000,000 experiments approximately 1000 E's will show up. In general,

---

[1] These numbers come from Equation (E.23) in Math Note E.16.2.

n	m_-	m_+	Δ_n	δ_n
20	3	17	14	14/20=0.7
50	12	38	26	26/50=0.52
100	33	67	34	34/100=0.34
1,000	440	560	120	120/1,000=0.12
10,000	4,840	5160	320	320/10,000=0.032
100,000	49,360	50640	1280	1,280/100,000=0.0128
1,000,000	498,000	502000	4,000	4,000/1,000,000=0.004

Table 16.1: n is the number of coins, m_- is the number of heads at which the plot is (almost) zero on the left of the maximum, m_+ is the corresponding zero on the right, Δ_n is the difference between m_+ and m_-, and δ_n is Δ_n/n. Note that δ_n gets smaller as n gets larger.

Box 16.2.2. *When we say that the probability of a certain event E in an experiment is p, we mean that if the experiment is repeated N times (where N is sufficiently large), on the average, E occurs pN times. We write this as*

$$\text{number of successes} = pN.$$

In particular, if we want to see E occur once, i.e., if we want one success, then N must equal $1/p$.

In the experiment of tossing a million coins, when we say that the probability of getting less than 498,000 or more than 502,000 H's is negligible (see Figure 16.2 or Table 16.1), we mean that if we toss 1,000,000 coins many many times, we will almost never see the number of heads to be smaller than 498,000 or larger than 502,000; equivalently, if we toss 1,000,000 coins many many times the number of heads almost always is between 498,000 and 502,000. Since both 498,000 and 502,000 are very close to 500,000, we can restate the above result as: *In many many tosses of* 1,000,000 *coins, the number of heads is (almost) always something extremely close to* 500,000. Math Note E.16.2 quantifies some of these discussions and concludes that for larger and larger number of coins:

- There is less and less chance of getting exactly half heads.

- Getting anything else is even less probable.

- If $m > m_+$ or $m < m_-$, then getting m heads is completely hopeless.

If the chance of getting $n/2$ (corresponding to the maximum probability) is negligible, and the chance of getting anything else is even more negligible, what *do* we get when we throw n coins? As n increases, so does the difference between m_- and m_+. In fact, from Math Note E.16.2, the number of heads lying between m_- and m_+—what we denoted by Δ_n—is simply $5.26\sqrt{n}$, which is clearly an increasing function of n. This increase in $m_+ - m_-$ compensates the decrease in the probabilities of the number of heads in this interval, so that the total probability is always 0.99999985. In short, although the probability of getting any individual number of heads between m_- and m_+ decreases with increasing n, there are so many probabilities between these two values that, when you add the little contribution of each, you get a significant total.

> **What do you know? 16.3.** How do you compare the probability of getting 500 heads in tossing 1000 coins with the probability of getting 5000 heads in tossing 10,000 coins? How do you compare the probability of getting 440 heads in tossing 1000 coins with the probability of getting 4400 heads in tossing 10,000 coins? Hint: For the second question, see Table 16.1.

16.3   The Law of Large Numbers

Let us now talk about an experiment which, although unwieldy, is in principle possible, and can shed some light on the subject. A typical physics experiment involves a very larger number of atoms participating in a large number of events. In *our* experiment, coins replace atoms and tosses replace events: We toss a trillion coins and count the number of heads that turn up in each toss. Expecting the result to be a number close to half the number of coins, i.e., 5×10^{11}, we continue until we get a number of heads that is "significantly" different from 5×10^{11}. How long before this happens?

> **What do you know? 16.4.** Compare Δ_n for 10 million and 100 million coins. Do the same with δ_n. Hint: See Table 16.1 for the trend.

First, let us estimate the time required to perform such an experiment. Assume that we ask all the working adult population of the US to participate in the counting as a substitute for their job, counting coins eight hours a day. Then, as **Math Note E.16.3** on **page 101** of *Appendix.pdf* calculates, for 10,000 trials, it will take over three years to perform the experiment, assuming that everybody can count at the rate of 3 coins per second! The reader may object that 10,000 trials is an overkill. However, it will become clear that well over 10,000 trials are needed to get a head count that is even slightly different from 5×10^{11}. Clearly, even ignoring the potential mistakes of bad counters, this version of the experiment is not plausible, and we have to look for a better version.

A slightly more practical alternative is to paint one side (say the tail) of each penny white and the other side black, load them on cargo planes, fly the planes to a high enough altitude, drop the coins in a flat desert, observe them from above, and check for "grayness." There are some "technical" difficulties, such as the enormity of the number of cargo planes (approximately 50), the size of the area used (about 40 square miles), the background color (color of the ground) which may interfere with grayness measurement, the fact that some pennies may be hiding behind others, and so on. We shall ignore such difficulties, because they are more manageable than the difficulties of the first method.

How does "grayness" measure the number of heads? Assume we are so far above the ground that the area covered with black and white pennies appears as only one square foot. Then a *perfect* gray square foot of area corresponds to half white and half black. In other words, if exactly half of the coins were head and half tail and they were arranged alternately side by side, we would see a shade of gray which we call *perfect*. If we shake the coins thoroughly before throwing them, chances are that white and black coins will approximately alternate once they land, and the odds of a large number of coins of the same color landing next to each other are very small.

The question asked at the end of the first paragraph of this section can now be rephrased: What are the chances of seeing anything *but* a perfect gray? An imperfect gray would have a shade of black or white as a result of an imbalance between the number of heads and tails. Let us take a shade of black, and ask the question: How many more black sides (than white sides) should turn up before the eye can distinguish this shade of black? The following example answers this question.

Example 16.3.1. Let us digress from coin tossing and get our hands dirty with paint mixing! The resolution of the eye in distinguishing various shades of gray varies from person to person. Assume that we can find somebody whose eyes are sharp enough to distinguish between perfect gray and the shade obtained when one drop of black paint is added to a gallon of perfect gray. Such an ability is very unlikely, but to be on the safe side, let us assume that such a person exists. However, we assume this is the limit. Thus by mixing less than one drop of black with a gallon of gray, we get a shade that is indistinguishable from perfect gray.

What is the ratio of black to white? There is half a gallon plus one drop of black and exactly half a gallon of white. There are approximately 100,000 drops in a gallon. Out of these, 50,000 are black and 50,000 are white when the paint is perfect gray. Adding one black drop changes the ratio to $50,001/50,000 = 1.00002$.

Going back to the coins, we conclude that, to be able to distinguish the shade of black, the ratio of the number of black coins to white coins must be the ratio obtained above, namely 1.00002. Since the total number of coins is one trillion, the actual number of black heads must be 500,005,000,000 leaving 499,995,000,000 white tails. The ratio of these two numbers is easily seen to be 1.00002 as desired.   ■

Food for Thought

FFT 16.3.2. We live by the rules of chance. Every time we step outside, there is a nonzero probability that we will get hit by a car, and another nonzero probability that once hit, we will die. Every time we go to the countryside, there is a chance that we will be struck by lightning. This chance increases in a thunderstorm, but even a sunny day can get cloudy quickly and provide a good condition for lightnings. As you[2] are sitting in your living room watching television, there is a nonzero probability that the small plane flying overhead will plummet into your house. There is about 0.03% chance that you will contract cancer this year, and once diagnosed, a 40% chance that you will die from it.

Unlike the tossing of coins, these probabilities cannot be calculated a priori, but a collection of data over many years can give a reliable magnitude for them. In fact, the figures of 0.03% and 40% were obtained from the data collected by the American Cancer Society over decades. These data consist of the number of new cases of cancer and the number of deaths caused by cancer every year. Using Box 16.2.2 in reverse—i.e., dividing the number of cases by the total population—can give you a very good estimate of the probability. For example, over the last few years, the number of new cases of cancer has been stable at about 1.2 million per year in the US. Similarly, the number of cancer deaths has been equally steady: about half a million per year. Of course, diet, exercise, and annual check-ups can *reduce* the risk (from 0.03% to 0.02% or even 0.01%), but can neither eliminate it nor contradict the fact that contracting cancer is probabilistic.

Not only is the occurrence of the disasters above totally probabilistic, but also surviving them. The probability that, with all proper treatments, a cancer patient will survive five years after being diagnosed is about 60%. This survival probability decreases as the amount of treatment is reduced and/or the period after diagnosis is increased, but it does not become exactly zero. Even at the extreme case of no treatment at all, the probability of surviving for a long time and the complete disappearance of cancer is not zero. The body itself has a mechanism of fighting the disease, and although most people's bodies lose the battle without outside help, the composition of the bodies of some patients is such that no outside help is necessary in the fight.

The survivor, of course, sees this completely differently. She looks at herself as one of the very few survivors of the disaster that kills hundreds of thousands of people. She thinks that she is one of the *chosen* few, whose survival must have a cosmic purpose. And if she has been seeing an alternative medical doctor or a faith healer (the chance of this is extremely high in cases of incurable diseases), she can be a priceless testimonial to the miraculous effectiveness of the alternative treatment or the prayer.

Probability and alternative medicine.

The same law of probability in Box 16.2.2 applies to extremely unusual and unexpected experiences such as the exact replay of a dream in real life. The images we see in our dreams are (sometimes distorted) copies of the most prominent scenes we see in our everyday lives. These images are, therefore, fairly limited in number, and they are shared by millions of people all around

Dreams (or nightmares) can come true!

---

[2] "You" is used as an indeterminate pronoun, and does not refer to you, the reader.

the world. So, a nightmare involving a car accident severely injuring a relative should not be uncommon. Even if the probability—p of Box 16.2.2—of the coincidence of the dream and the real event is very small, with the population of the Earth approaching seven billion, and the number of dreams a typical person experiences in a lifetime being approximately 100,000, the N of Box 16.2.2 approaches 7×10^{14}. Therefore, it should come as no surprise to hear *genuine* stories of such coincidences on a daily basis.

The lesson to learn from Box 16.2.2 is that, no matter how small the chance of the occurrence of some very unlikely event, as long as the population of the sample is large enough, there will be at least a few cases of that unlikely event. There is no need for introducing "miracles" and supernatural powers; the theory of probability and statistics can explain the unlikely event convincingly.

16.3.1 Law of Large Numbers in Numbers

The example above tells us that for an ideal sharp eye to see a shade of black which is minimally different from perfect gray, at least 500,005,000,000 heads must turn up in a toss of a trillion coins. Similarly, for the same eye to see a minimal shade of white, at most 499,995,000,000 heads must turn up. Putting these two statements together, we can say: *for the sharpest eye to see anything but perfect gray, the number of heads must be larger than* 500,005 *million or smaller than* 499,995 *million*.

What are the chances that in a random throw of a trillion coins, the number of heads is larger than 500,005,000,000 or smaller than 499,995,000,000? Using Equation (E.21) and a sufficiently accurate computer program, the probability can be calculated. The result is the staggeringly small value of 1.5×10^{-23}! Box 16.2.2 now tells us that, in order to see one occurrence ($N = 1$) of a non-perfect gray outcome, on the average, we have to throw the coins $1/1.5 \times 10^{-23}$ or 6.67×10^{22} times.

Let us rephrase this conclusion in a more comprehensible language: If we perform the same experiment (loading the 50 cargo planes, flying them to high altitudes, and dropping the black and white coins onto a desert) 6.67×10^{22} times, while the sharp-eyed observer is looking on, *only once* can he decide that the resulting color is not perfect gray. In the remaining times he will judge the color to be perfect gray!

Let's make this conclusion even more tangible. Imagine small coins. Very small coins! So small that a trillion of them can fit in an ordinary salt shaker. Put our black and white microscopic coins in a salt shaker, shake them, pour them, and spread them on a table top. Do all this in the presence of the sharp-eyed observer. Keep pouring and spreading the coins and ask him for the shade. His answer is almost always the same, perfect gray.

How long do you have to wait before he sees a shade of white or black? Well, it depends on how fast you can fill up the shaker, pour it, and spread the coins on the table top. Manual operation cannot be fast enough to give our sharp-eyed observer a chance to see a non-perfect gray. So, let us automate the entire process. After shopping around for a while, we find a machine that is really fast: it can fill the shaker, pour the content, and spread it on the table top 100,000 times a second! Remember that in order for our observer to see any shade of black or white, on the average we have to do the experiment 6.67×10^{22} times. This requires 6.67×10^{17} seconds. Since there are 3.15×10^7 seconds in a year, we would have to wait 2.1×10^{10} years, or 21 billion years. The age of the universe is 13.7 billion years!

Is there a supernatural power, an invisible hand, that flips the coins so we get perfect gray all the time?

This is a bafflingly counter-intuitive result. There are practically an infinite variety of shades of gray, from pure black to pure white. However, every time the experiment is performed, the result is perfect gray! Why is it that out of such an infinite variety of shades only the perfect gray shows up? Is there some kind of an "invisible hand" that purposefully arranges the coins so that the number of heads matches almost exactly with the number of tails? If we had not gone through the theory of probability, we would have suspected the intervention of an invisible hand. Such a suspicion could easily lead to superstition and the invention of the supernatural. However, as shown in this section, the persistent appearance of perfect gray is nothing but the manifestation of the blind force of mathematical probability. After all, each *individual* coin has a completely random 50-50 chance of showing up as a head or a tail.

16.3.2   Summary of the Law of Large Numbers

We have spent a great deal of time on probability theory of coins. Although intrinsically interesting and worthy of study, the theory will be useful only if we generalize it in two ways. First, the experimentation with coins is only a prototype of more general experiments studied in probability theory which deals with **distributions**. A distribution is a function or a rule which gives the probability of events. An example of distribution is the function $P_n(m)$ which gives the probability that m heads show up in a throw of n coins. We say that the **most probable** outcome for this distribution is $m = n/2$. Other terminologies used for "most probable" are **expectation value**, **average**, and **mean**. Using these terminologies, we generalize the important conclusion drawn from our long discussion of coins as follows:

Probability distribution.

> **Box 16.3.3.** (**Law of Large Numbers**) *For a sufficiently large number of random events, the probability of getting anything that is practically distinguishable from the expectation value is immeasurably small.*

This law is sometimes elegantly stated as *individual chaos leads to collective determinism.*

Secondly, in any physical experiment involving matter, the number of events is so large that the requirement imposed by the qualifier "large enough number" is very easily satisfied. There are approximately 10^{22} atoms in a liter of any gas at ordinary temperatures and pressures. This number is 10^{10} or ten billion times larger than the number of coins used in our hypothetical experiment. Such a large number makes the observation of any deviation from the expected average behavior of the gas absolutely impossible as the following argument reveals.

In our trillion-coin experiment, we noted that for the sharp-eyed observer to see a deviation from expected behavior, 5,000,000 extra heads (or tails) must turn up. This number was achieved by having in mind a device (the sharp-eyed observer) which could detect a deviation of one part in 100,000 (one drop in a gallon). Let us assume that for a liter of gas, we have a very sensitive instrument that can detect deviations of one part in a billion (a device that is at the forefront of our technology). For 10^{22} atoms the "extra" number required for the abnormal behavior of a liter of gas detectable by such an ultra-sensitive device is easily calculated to be 5×10^{12}. The probability of such an aberration is 3×10^{-2174}! This means that to see the aberration just once, you need to perform the experiments $1/3 \times 10^{-2174}$ or 3.33×10^{2173} times (see Box 16.2.2). Even if we could perform the experiment a billion times per second,[3] it would take 3.33×10^{2164} seconds, or 10^{2157} years to see a single occurrence of the aberration. This is 7.3×10^{2146} (73 with 2145 zeros in front of it) times the age of the universe! Hence, we can safely say that every time we observe a property of bulk matter, we expect nothing but its average behavior.

🧺 Food for Thought 🌱

FFT 16.3.4. The law of large numbers demands a great leap away from our intuition. Perfect gray all the time? (OK, not *all* the time, but a waiting period of 21 billion years for it not to happen!) How can it be? There are so many other shades of gray in a throw of a trillion black-and-white coins. Why are all these shades absent?

---

[3]There is a limit—imposed by relativity theory—to how fast physical processes can take place.

Questions like these are the manifestation of our desire to find a cause for any observable effect, a desire that can lead us astray. "When the outcome of every single experiment is consistently the same," we may argue, "there must be a reason for it." We have a fairly good intuition for probability as long as the number of random events is kept low. We are not surprised to see the outcome of a throw of ten coins clustered around five. However, we expect to see three, four, six, and seven as well, and indeed we do observe them. Not as frequently as five, but nevertheless they are frequent enough to be seen. But, when the same laws of mathematics—with the same degree of exactness and impartiality—tell us that, as the number of random events increases considerably, it is impossible to see anything that is ever so slightly different from the average, we start to question the laws.

This questioning is replaced by sheer disbelief when people unaware of the underlying laws encounter the same situation. "If there are such a large number of black and white coins," they say, "we should observe many different shades of gray. And if we see the same exact shade of perfect gray over and over," they continue, "then there must be some kind of an 'invisible hand' that is turning the coins to the same final outcome. There must be some kind of supernatural force. How else can so many other possibilities—shades of gray—disappear?"

Yet the same simple mathematical formulas of Math Note E.16.1, which predict the "acceptable" results of a small number of coins, also predict the seemingly unacceptable content of the law of large numbers. As seen before, when the number of coins gets larger and larger, the non-negligible probabilities cluster closer and closer to the average value. For an extremely large number of coins, this clustering becomes so close to the average that it renders any differentiation between the average (perfect gray) and all the neighboring values practically impossible. No invisible hand is needed to explain this. It is just the natural outcome of a large collection of random events.

There is a minor difference between probability distribution in matter and that of simple coin flips. In the latter case the number of possible outcomes is two while in the former this number is very large, in fact, infinite. For example, a molecule of oxygen moving through the room can have speeds ranging from zero to the maximum possible speed (speed of light). Thus the number of possible outcomes in a speed-measurement experiment is infinite. It turns out that this complication is not crucial in the large-number behavior of matter discussed above. However, it allows many possibilities for the average. In the case of the coins, we had only a single average outcome when their number grew indefinitely: The outcome was always gray when we tossed 10^{12} *identical* black-and-white coins.[4]

To allow for other shades of gray, keeping all elemental "units" identical, we have to change the units completely, say to dice. If we color 3 sides of dice white and 3 sides black, we get perfect gray as the expected color. Coloring 4 sides white and 2 sides black gives a whitish gray which we quantify as 2/3 white. If, instead, 5 sides are white and 1 side black, we get a 5/6-white gray. Similarly, we can get 2/3-black and 5/6-black shades of gray color. In all cases, all dice are identical.

It is now clear that if the number of possible outcomes for each elemental unit increases, the variety of the averages also increases. In particular, if this number becomes infinite, an infinite variety of averages will become possible. This means that an infinity of average-speed values will become possible for molecules in a gas. We shall see later that this average speed is related to the temperature of a gas. Therefore, the temperature of a gas can assume an infinite number of values. If there were only two possible outcomes in a speed-measurement experiment, corresponding to the two outcomes of a coin-flip, then the temperature of a gas would always be the same, corresponding to the single average of a trillion coins, perfect gray. The large number of possible outcomes for the elemental units is responsible for the infinite range of values that a typical bulk variable, such as temperature or pressure, can assume.

---

[4]The word "identical" is important. By coloring *both sides* of some coins black, we could obtain other shades of gray than perfect, but the coins would not be identical. In nature, all elementary components of matter are identical.

16.4 End-of-Chapter Material

16.4.1 Answers to "What do you know?"

16.1. There are 26 red cards in a deck of 52. So, assuming equal probability for all cards, the probability for drawing a red card is 26/52 or 0.5. There are 13 diamonds in a deck, so the probability of drawing a diamond is 13/52 or 0.25. The probability of drawing a king is 4/52 or 0.0769, because there are 4 kings in the deck.

16.2. Using the formula in Math Note E.16.1, we get $P_{10}(3) = 10!/(3!(10-3)!2^{10})$ or 0.117, which is also equal to the probability $P_{10}(7)$. The equality of these two probabilities is a special case of the symmetry of the probability distribution curve about half the number of coins: $P_{10}(5-2) = P_{10}(5+2)$.

16.3. The probability of getting exactly half the number of coins decreases with increasing n. Therefore, the probability of getting 500 heads in tossing 1000 coins is larger than the probability of getting 5000 heads in tossing 10,000 coins. Table 16.1 shows that 440 is the value of m_- for 1000 coins. However, 4400 is far to the left of 4840, the value of m_- for 10,000 coins. So, we expect the probability of getting 440 heads in tossing 1000 coins to be larger than the probability of getting 4400 heads in tossing 10,000 coins.

16.4. Table 16.1 shows that Δ_n increases for larger number of coins while δ_n decreases. So, we expect the Δ_n for 100 million coins to be larger than Δ_n for 10 million coins. On the other hand, δ_n for 100 million coins is smaller than δ_n for 10 million coins.

16.4.2 Chapter Glossary

Law of Large Numbers A very general law in probability theory stating that when the sample size in an experiment gets larger and larger, the chance of the outcome of the experiment being anything but the average (expected, mean, most probable value) becomes smaller and smaller.

Probability Distribution The plot of probability as a function of a random variable.

Random Event Any event whose outcome cannot be predicted by the laws of physics.

Statistical Mechanics The branch of physics which considers the bulk matter as an aggregate of microscopic particles obeying the laws of physics and applies these laws as well as statistical techniques to them to predict their bulk properties.

Thermometry The art and science of making and using thermometers.

16.4.3 Review Questions

16.1. What is the third branch of physics? When did it start and who started it? What was the role of steam engines in the development of this branch?

16.2. Is it practical to study the motion of particles in a sample of material exactly? How is this study accomplished? What is statistical mechanics and where does it get its name from?

16.3. Define random events, possible outcomes, frequency, and probability.

16.4. What are the possible outcomes of tossing two coins? What is the frequency of obtaining one head? What is the probability of obtaining one head?

16.5. Suppose you toss 20 coins. How do you compare the probability of getting 12 heads with that of 8 heads? How do you compare the probability of getting 14 heads with that of 6 heads?

16.6. Define, in words, the quantities m_-, m_+, Δ_n, and δ_n. What is the significance of δ_n? What property of the distribution plot does it measure?

16.7. On a probability distribution plot, how do you read the probability of obtaining a number of heads between two given numbers? For instance, in tossing a trillion coins, what is the probability of getting a number of heads lying between m_- and m_+? What is the probability of getting a number of heads smaller than m_- or larger than m_+?

16.8. Toss a trillion coins black on one side and white on the other. Spread them on a desert. Go to a high altitude and look at them. Do this 100,000 times a second. How long do you have to wait before you get anything but perfect gray? Explain how this result can lead to "invisible hands" and superstition.

16.9. State the law of large numbers. Does this law apply to 10 coins? 100 coins? A trillion coins?

16.4.4   Conceptual Exercises

16.1. Estimate the probability of getting a number of heads between 5000 and 5100 when tossing 10,000 coins (see Figure 16.2).

16.2. Using Figure 16.2, estimate the probability of getting a number of heads between 50,000 and 51,000 when tossing 100,000 coins

16.3. Suppose you toss two coins a million times.
(a) How many times do you expect to get two heads?
(b) In an actual experiment, do you get this exactly?
(c) How likely is it to get half of the number you expect?

16.4. The law of large numbers indeed requires *large* numbers. With 10 coins, the probability of getting 10 black sides is one in 2^{10} or about 1 in 1000.
(a) How many times do you have to toss 10 coins to see one occurrence of complete black?
(b) If the coins are microscopic and you use the machine that can toss the coins 100,000 times per second, how many occurrences of black do you see every second?

16.5. Figure 16.4 shows a plot of the probability curve for a certain random event which takes values from zero to infinity. Recall that the total probability is the area under the curve which is set equal to one. By "estimate" I mean a very qualitative description using words such as *unlikely, very likely, highly unlikely*, etc.
(a) Estimate the probability for the random event to have a value between 0.1 and 0.5.
(b) Estimate the probability for the random event to have a value between 0.05 and 0.08.
(c) Estimate the probability for the random event to have a value between 0 and 0.1.
(d) Estimate the probability for the random event to have a value between 0.1 and 2.
(e) Estimate the probability for the random event to have a value larger than 2.

16.6. Figure 16.5 shows a plot of the probability curve for the position of a subatomic particle as measured from a certain origin. Let O stand for the origin, A for the point 3 units away from the origin, and B for the point 6 units away from the origin. Recall that the total probability is the area under the curve which is set equal to one.
(a) How likely is it for the particle to be in the neighborhood of the origin?
(b) How likely is it for the particle to be in the neighborhood of the midpoint between the origin and A?
(c) How likely is it for the particle to be in the neighborhood of A?
(d) How do you compare the probability of finding the particle between O and A vs between A and B?

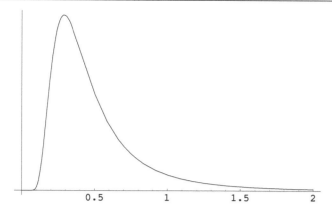

Figure 16.4: The probability curve for a certain random event which takes values from zero to infinity. Values between 0 and 2 are shown.

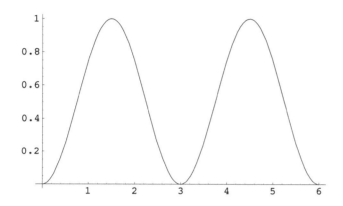

Figure 16.5: The probability curve for the position vector of a subatomic particle.

16.4.5   Numerical Exercises

16.1. For the toss of four coins:
(a) Enumerate the possible outcomes.
(b) Find the frequency for 4H, 3H1T, 2H2T,1H3T, and 4T.
(c) Find the probability for 4H, 3H1T, 2H2T,1H3T, and 4T.
(d) Use $P_n(m) = \dfrac{n!}{m!(n-m)!2^n}$ to find the probabilities.

16.2. In tossing 10 coins, use $P_n(m) = \dfrac{n!}{m!(n-m)!2^n}$ to
(a) find the probability of getting 5 heads; and
(b) to find the probability of getting 3 heads.
(c) How much more likely is it to get 5 heads than 3 heads?
(d) If you toss 10 coins 100,000 times, how many times do you see 3 heads? How many times 5 heads?

Chapter 17

Statistical Mechanics

The task of statistical mechanics is to "explain" thermodynamics by means of the basic assumption that all substances are made up of particles—atoms or molecules—obeying the laws of motion. As a model, we shall take an **ideal gas**, in which there are no internal forces among its constituent particles. This means that kinetic energy is the only kind of energy a typical particle has.

The ideal gas defined.

17.1 Ideal Gas Law

The simplest application of statistical mechanics is finding the pressure of a gas in terms of its other properties. The pressure a gas exerts on the walls of its container is the perpendicular force on the wall divided by the area on which the force is applied [Equation (8.6)]. From a microscopic viewpoint, the force is caused by the transfer of momentum of the particles to the wall per unit time (Box 7.2.1). This transfer occurs when the particles collide with the wall of the container. The following example should shed some light on the discussion.

Example 17.1.1. Consider a large rectangular sheet of rubber fastened at all sides, just like a trampoline. Suppose that we throw heavy balls at this sheet which we assume is held vertically [Figure 17.1(a)]. Let us throw the first ball. As soon as the ball hits the sheet, the sheet's side view will look like Figure 17.1(b). Because of its elasticity, the sheet will spring back to its original configuration, throwing the ball back, oscillating back and forth a few times, and finally coming to rest in position (c) in the figure. Now suppose we throw more and more balls more and more rapidly. Let us say that the balls always hit two points on the sheet designated A and B in Figure 17.1(d). If the balls arrive at A and B fast enough, the shape of the sheet will remain as (d) because the sheet will not get a chance to recuperate from its deformation.

Finally, assume that the balls come rapidly and at random so that all points of the sheet are exposed to the oncoming balls with equal probability. Then, the shape of the sheet will look like Figure 17.1(e), i.e., it is uniformly bulged (inflated) just like a tire or a balloon. The larger the speed of the balls, the deeper the bulge of the sheet. By throwing the balls rapidly enough, we can "explode" the sheet! ∎

Intuitively, the more particles we have in the gas, the more collisions take place, and the more pressure is exerted on the walls. Similarly, the faster the particles move the more frequently they collide with the walls, and the more pressure they create. Therefore, pressure should depend on both the number of particles and their speed. **Math Note E.17.1** on **page 101** of *Appendix.pdf* explains how this intuition can be made more precise.

Two results of Math Note E.17.1 are important enough to merit further discussion here.

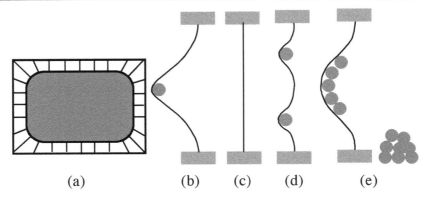

Figure 17.1: Pressure is caused by constant bombardment of the walls with molecules.

The first one is

$$\langle KE \rangle = \tfrac{3}{2} k_B T, \tag{17.1}$$

Boltzmann constant and Kelvin temperature scale.

$\langle Q \rangle$ is average of Q.

connecting the *average KE* of the gas particles to the temperature. In this equation, T is the temperature in K (for Kelvin, the scientific unit of temperature) and k_B is the **Boltzmann constant**, whose value is 1.3806×10^{-23} J/K. The **Kelvin temperature scale** is obtained from the Celsius scale by adding 273.16 to the Celsius reading. For example, water freezes at 273.16 K and boils at 373.16 K. In Equation (17.1) and in the sequel, the angle brackets flanking a quantity denote the average of that quantity.

> **What do you know? 17.1.** You raise the temperature of a gas from 150 °C to 300 °C. Does the average KE of the gas particles double?

Example 17.1.2. Equation (17.1) explains why a system cools down upon evaporation, as for example, the cool feeling we get when a breeze caresses our wet body at a beach. The breeze simply enhances evaporation. In any evaporation process molecules of the system are lost to the surrounding. Those molecules that are moving faster in the system have a better chance of escaping from the system. Thus, the faster molecules escape our skin upon evaporation leaving behind the more sluggish water molecules. The average kinetic energy of the remaining molecules is less than before, and therefore, the temperature of our skin drops. ∎

> **What do you know? 17.2.** When you put a hot object in contact with a cold object, the hot object cools down and the cold object warms up. Explain this from an atomic point of view and Equation (17.1).

Equation (17.1) and the definition of the kinetic energy can be used to calculate the average speed—also called the **root mean square** or *rms* speed—of the particles of a gas. **Example D.17.1 on page 39** of *Appendix.pdf* illustrates how to do that, and calculates the root mean square speed of helium to be over 1000 m/s.

Ideal gas law.

The second result of Math Note E.17.1 is the **ideal gas law**:

$$PV = N k_B T, \tag{17.2}$$

where P is the pressure of the gas in Pascal, V is its volume in m^3, and N is the number of particles in the gas. The other two quantities are as in Equation (17.1).

Historically, the ideal gas law was derived empirically, and the subsequent derivation of this law as outlined in Math Note E.17.1 was one of the triumphs of atomism and statistical mechanics. As early as 1787, through the work of people like Robert Boyle (1627–1691) and Jacque Alexandre Charles (1746–1823), it was shown empirically that for any given sample of gas the ratio PV/T was always a constant.

BIOGRAPHY

Ludwig Boltzmann (1844–1906), the father of statistical mechanics, was awarded a doctorate from the University of Vienna in 1866 for a thesis on the kinetic theory of gases. He taught at Graz, moved to Heidelberg and then to Berlin. Boltzmann, at least half jokingly, used to say that the reason he moved around so much was that he was born during the dying hours of a Mardi Gras ball. In 1873, he accepted the chair of mathematics at Vienna, but three years later he moved back to Graz. However, the following year Ernst Mach became the chair of history and philosophy of science at Vienna. Boltzmann had many scientific opponents but Mach was more than a scientific opponent as the two were on bad personal terms. Because of this, Boltzmann moved to Leipzig in 1900 but there he was close to one of his strongest scientific opponents, Wilhelm Ostwald. Despite their scientific differences Boltzmann and Ostwald were on good personal terms. However, his scientific arguments with Ostwald led to a severe depression in Boltzmann, to the point that he unsuccessfully attempted suicide during his time in Leipzig. In 1901 Mach retired from Vienna due to ill health, and this gave Boltzmann an opportunity to go back to Vienna. In 1902 he returned to Vienna to his chair of theoretical physics which had not been filled in the intervening period. Boltzmann had a soft-hearted personality, and his ideas of introducing probability into physics was not well-received. He was far ahead of his time. The enormous progress in today's physicists' understanding of thermodynamics is solely due to its foundation in statistical mechanics, which Boltzmann created almost singe-handedly.

> **What do you know? 17.3.** A mixture of He and Ne atoms has a temperature of 300 K. Which atoms have higher average KE? Which atoms move faster on the average?

In 1802 Joseph Louis Gay-Lussac (1778–1850) empirically showed that at a given pressure, say one atmosphere, the volumes of certain specific weights of gases increased by exactly the same amount per degree increase in temperature. For example, 1 gram of hydrogen, 16 grams of oxygen, or 44 grams of carbon dioxide all expand at exactly the same rate. In 1811 Amedeo Avogadro (1776–1856) came up with the brilliant idea that the reason for such a striking similarity between all gases is that the weights quoted above represent the *same number of molecules*. He noted that if one assumed that there were no attraction between molecules of a gas, and one ignored the size of the molecules, then as long as one had the same number of molecules in a given volume, one had to obtain identical results for all gases. So he concluded that 1 gram of hydrogen must contain as many molecules as 16 grams of oxygen and 44 grams of carbon dioxide. This number, now called **Avogadro's number**, turns out to be 6.02×10^{23}. A collection of 6.02×10^{23} molecules of any substance is called a **mole** of that substance. Avogadro's number was not determined until after the value of the Boltzmann constant was measured experimentally. Example **Example D.17.3** on **page 40** of *Appendix.pdf* illustrates how this is done.

*Avogadro's number and its relation to a **mole**.*

Equation (17.2) is a remarkably simple equation, yet it contains a mountain of information. It is one of the first windows through which humankind caught a glimpse of the otherwise invisible world of atoms and molecules. It contains a tiny number, the Boltzmann's constant, representing the world of the invisible, and a huge number N, typically of the order of the Avogadro's number, representing the visible world. This dichotomy is an essential feature of statistical mechanics, which bridges the macroworld of the observable to the microworld of the invisible.

Tying the observable to the invisible.

17.2   Most Probable Configuration

State of a system
defined.

Before applying *statistics* to a mechanical system, let's review how the system is described *exactly*. The discussion of Section 10.2 led us to conclude that if we know the position and velocity (or, equivalently, momentum) of every particle in a system at a particular time, we can predict its behavior for all future times (see Box 10.2.1). When the particles of a system take on specific values for their positions and velocities (momenta), we say the system is in a specific **state**. The conclusion of Section 10.2 can be reworded as follows: If we know the state of a system at one time, we can determine its state at any other time.

Picture the development of a system, say a gallon of air, as time passes. Imagine observing this system again and again, and each time taking a mental snapshot of the state of the system. You would then have a large collection of snapshots showing the locations of each air molecule and—since the picture is mental—its momentum (velocity) at various times. In principle, the laws of motion can predict the content of each snapshot precisely. In practice such a prediction is impossible (due to the large number of particles in the gas) and irrelevant (because the specific details of the system do not show up in its bulk properties). The task of statistical mechanics is to find the *probability* of each snapshot.

Use the knowledge of the theory of probability gained in Chapter 16 in the treatment of the present chapter, except for one caveat. In Chapter 16, we did not have to worry about any *physical constraints* that might be imposed on the system. The random events (the coin outcomes) could turn up in any possible way. Here, we have to impose certain restrictions. In the case of a gas, for example, the particles are confined within a volume, so their position cannot take on any arbitrary value. Similarly, the kinetic energy (hence, velocity) of particles cannot be completely arbitrary because the sum total of the kinetic energies of all particles must be a constant due to energy conservation. Because of such constraints we speak of accessible states rather than possible states: An **accessible state** of a system is any state of the system which respects the constraints imposed on it such as volume limitations and energy conservation. We denote the total number of accessible states by Ω.

Accessible state defined.

Example 17.2.1. To put some concrete flesh on the abstract bones of the preceding discussion, go back to the simple system consisting of, say 10 coins. However, instead of H and T, assign the values $+1$ and -1 to the sides of a coin, and find the total (algebraic) number showing up. This number is clearly between -10 (when all coins show up with -1) to $+10$ (when all show up with $+1$). It is clear that the most probable outcome is that for which we have equal numbers of $+1$ and -1, i.e., when the total is zero. This calculation assumes no constraint on energy, or any other physical quantity: There is an unlimited supply of energy that can be added to the system. This energy could be simply the effort of the tosser of the coins. Since there is no constraint, the total number of accessible states Ω is $2^{10} = 1024$.

How would the analysis change if energy conservation were taken into account? Assume that the numbers $+1$ and -1 represent the actual energy of each coin. A positive number represents a surplus of energy while a negative number indicates a deficiency of energy. For the sake of argument, assume that the value of energy is set at $+6$. The total energy of $+6$ corresponds to 8 positive and 2 negative coins. Using Equation (E.19) with $n = 10$ and $m = 8$ (or actually counting the possibilities) we find that $\Omega = 45$. Therefore, constraints reduce the number of accessible states. ∎

What do you know? 17.4. Instead of 10 coins of Example 17.2.1, say you have 5 coins with $+1$ on one side and -1 on the other. How many accessible states are there without constraint? What if the energy is to be $+1$? -2? -3?

17.2.1   Average Energy and Temperature

In our discussion thus far we have left one important element out: the interaction between two systems, i.e., one system coming in contact with another. Contact means *exchange of energy*. Of course, conservation of energy does not allow the total energy of the two systems to change. However, because of the exchange of energy, the system of interest has more accessible states available to it. We can now ask: In what states do we find our system when we allow it to interact with another system? Again, to understand the problem more fully, consider a concrete example.

Example 17.2.2. Assume that we have two systems A and B, composed, respectively, of 10 and 990 "energetic" coins with ± 1 unit of energy. As in Example 17.2.1, assume that the total energy of the combined system is $+6$. Because of the interaction with B, system A has a lot more possibilities open to it than in Example 17.2.1. For instance while the coins of A could not have been all $+1$ or all -1 before, now this possibility exists because the 10 coins can borrow energy from B and flip up or down. In fact, the only constraint is that the total number of positive coins outnumber the total number of negative coins by 6. This means that out of 1000 combined coins, we must have 503 positive coins and 497 negative coins. How these positive and negative coins distribute themselves in A and B is irrelevant. For example, one possibility is that no positive coin shows up in A, in which case all 503 positive coins are in B. Another possibility is that one positive coin shows up in A and 502 in B; or two positive coins in A and 501 in B, and so on. ■

An important question suggested by the example above is: *What is the most probable outcome for the two interacting systems?* In other words, out of all the possible configurations consistent with the constraint, which configuration occurs most frequently? The following example illustrates the significance of this question.

Example 17.2.3. Take the sample to be a little larger than the previous example. Let system A have 5000 coins, system B 6000 coins, and let the total energy be 1000. One possibility of getting a total energy of 1000 is, for example, for A to have all negative energies and B to have all positive. Clearly this configuration is very unlikely as there is only one way for it to happen. Another possibility is for A to have 4500 positive coins (and therefore 500 negative) and for B to have 1500 positive coins (and 4500 negative). This configuration is more likely, but still not the most probable. What *is* the most probable configuration?

Figure 17.2, whose plot uses Equation (E.35) of Math Note E.17.2, shows the probability as a function of the number of positive coins of A. It can be inferred from the figure that the peak occurs at approximately 2727. This is the most probable value of the number of positive coins of A respecting the constraint of the total energy of 1000. Since there are 11,000 coins altogether, to get $+1000$ total energy, we must have 6000 positive and 5000 negative. For the most probable configuration, out of the 6000 positive, 2727 are in A and the rest, i.e., 3273, must be in B. It is interesting to note that the ratio of the positive numbers in A to the total number of coins in A (or the *average* number of positive coins in A) is very nearly equal to the same ratio for B: $2727/5000 = 0.5454$ is very nearly equal to $3273/6000 = 0.5455$. This is not a coincidence, as the following discussion shows. ■

Example 17.2.3 calculated the most probable outcome for two specific systems in contact. But how does one find the most probable outcome in general? For systems consisting of small numbers of energetic coins, there is no simple formula. However, as **Math Note E.17.2** on **page 103** of *Appendix.pdf* shows, for sufficiently large numbers, a very interesting conclusion can be drawn:

Box 17.2.4. (Most probable configuration and average energy) *The most probable configuration of a system A in contact with a system B is that for which the **average energy** per coin for A and B are equal, and both are equal to the average energy of the combined system.*

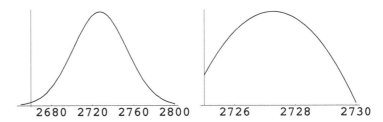

Figure 17.2: Plot of the probability as a function of the number of positive coins for A when A has 5000 coins, B has 6000 coins, and the total energy is 1000. The plot on the right is a blown-up version of the plot on the left.

This statement connects statistical mechanics with thermodynamics: We know from experience that when two systems are brought together, eventually they attain the same **temperature**. Thus, the balance of temperature is attained at the same time that the balance of average energies are achieved. It is, therefore, natural to conclude that

Temperature and average energy.

> **Box 17.2.5.** *The temperature of a system is proportional to the average energy of the particles of that system. When two systems, initially at two different temperatures, are brought in contact, they eventually reach the same final temperature. When this happens, we say that the two systems are in **thermal equilibrium**.*

For an ideal gas, energy is just the KE. Thus, Box 17.2.5 states that the average KE of the molecules of an ideal gas is proportional to the temperature. A little more (advanced) math will elevate this to Equation (17.1).

> **What do you know? 17.5.** Two systems have 10 and 15 "energetic" coins, respectively. The energy is $+5$. What is the most probable configuration when these two systems come in contact?

Math Note E.17.3 on **page 105** of *Appendix.pdf* shows how to find the final (equilibrium) temperature when a system A, with n particles and initial temperature T_A, is brought in contact with system B, with N particles and initial temperature T_B. If one of the systems is much larger than the other, the larger system is called a **reservoir**. Suppose B is the reservoir, then N is much larger than n, and in Equation (E.39), we can ignore n. Then $T_f \approx T_B$; i.e., a system in contact with a reservoir eventually attains the temperature of the reservoir. (If you are not convinced, look at **Example D.17.4** on **page 40** of *Appendix.pdf*.) The temperature of the reservoir does not change.

Reservoir and its unchanging temperature.

17.2.2   The Boltzmann Factor

For a collection of "energetic" coins, the calculation of probability and the decision concerning the most probable configuration involve specific formulas that apply to coins only [for instance, see Equation (E.35)]. A detailed analysis of a system in contact with a reservoir—i.e., a system with a well-defined temperature—yields a very general formula for the probability. Let the energy of the system be denoted by E and the temperature of the reservoir by T. Because of the exchange of energy between the system and the reservoir, E can take on many (in fact, often infinitely many) different values. The probability for the system to have *a specific* value E_s for energy is, as Ludwig Boltzmann discovered, proportional to $e^{-E_s/k_B T}$, which is now called the **Boltzmann factor** and is at the heart of statistical mechanics and thermodynamics.

In most cases, we have a system of many particles interacting among themselves, with the nth particle having energy E_n, which can assume many (even infinitely many) values. In such an important case, one can take a single particle to be the "system" and the rest of the particles, the reservoir. Then, the probability of a particle having energy E_n is proportional to $e^{-E_n/k_B T}$. In particular, the ratio of the probability $P(E_1)$ of a particle having energy E_1 to the probability $P(E_2)$ of the particle having energy E_2 is

$$\frac{P(E_1)}{P(E_2)} = \frac{e^{-E_1/k_B T}}{e^{-E_2/k_B T}} = e^{-(E_1-E_2)/k_B T}. \tag{17.3}$$

🧺 Food for Thought 🍶

FFT 17.2.6. Many people use statistics for the analysis of their data. Insurance companies use it to determine risk factors and set their premiums to maximize their profits. Marketing firms use it to decide which population group is suitable for a particular product. Pollsters use it to determine the political or economic trend of a group. The pharmaceutical companies use it to judge the efficacy of a drug. All the disciplines in the social sciences use it to study social trends of a community. And, of course, physicists use it as the title of this chapter indicates.

Partly due to the news coverage of the media, the word "statistical" has come to be synonymous with "scientific" to the layman; as if the use of charts, numbers, and a few statistical formulas are sufficient to render a discipline or an enterprise scientific. The general public, although lacking knowledge of its content, trusts science; and if a company, an enterprise, or an academic discipline can attach the word "scientific" to their products, it is likely that the public will accept that product without question.

But "statistical" is *not* synonymous with "scientific." Statistics is only a tool that can be used by many disciplines, just as a pen is a tool that can be used by many people: Einstein used it to write his theory of relativity, and his countryman, Hitler, used it to decree the murder of millions of Jews.[1] It is not "statistical" in *statistical mechanics* that makes a science; it is "mechanics." And that brings us to the crucial difference between the use of statistics in physics and in other disciplines.

"Statistical" is not the same as "scientific."

Whenever physics uses probability, it has a mathematical prescription for *predicting* that probability. In statistical mechanics, the mathematical prescription is called the *Boltzmann factor* discussed above. In quantum theory, to which we shall return later, it is the Schrödinger equation. No other discipline can claim this. The use of statistics in other disciplines involves collecting tons of data from the population by interviewing, sending questionnaire, conducting polls, etc. From these *statistical samples*, the trend is then predicted. A priori no clue, no theory, no formula, no predictive power for the outcome of the statistical samples exist. By contrast, physics has *a priori* universal formulas that it uses without resort to any interviews, questionnaire, or polls.

The difference between statistics in physics and in other disciplines.

17.3   Entropy

The number of accessible states is related to a thermodynamic quantity, which we want to explore in this section. This thermodynamic quantity has come to be associated with "messiness." What is the scientific definition of mess? A prime example of a mess maker is a toddler left in a room with a chest of drawers full of nicely folded socks, shirts, and underwear. Open all drawers and leave the child in the room, allowing her to do as she pleases for a couple of hours. Then go back in the room to find socks among the underwear, shirts among the socks, and underwear among the shirts, and all of these spread all over the floor! That's what you can call "a mess!" The content of the drawers—a small region

Scientific definition of mess.

---

[1] Also see Section 41.1.

of the room—has been transferred to the entire floor, a larger region of the room. This leads to the observation: *A messier room is that in which a larger area (or a larger number of "tiles") of the floor is covered by the contents of the drawer.* Let us look at some more examples of mess production.

Example 17.3.1. Fill one half of a salt shaker with salt and the other half with black pepper and start shaking it. After a sufficient amount of time the salt and pepper will be completely mixed. Is this a messier condition than the original one? Instead of floor tiles in the example of a toddler, we can imagine "tiling" the whole volume of the salt shaker with tiny cubes about the size of a grain of salt or pepper. Before shaking starts, the top half of the shaker is not available to the salt, and the bottom half is not available to the pepper. The process of shaking makes the unavailable cubes available. Once some of these available cubes are occupied, the system consisting of salt and pepper is said to be in a messier condition according to the definition of messiness used above.

When we open a bottle of perfume, the perfume particles will rush to occupy the available empty space of the room. Again according to the definition of messiness, the final condition (fragrant as it may be!) is in a messier state than the original state.

Pollution coming out of the smoke stacks of factories produces mess because it allows the smoke particles to occupy some of a large number of "tiles" available in the atmosphere. Similarly, pollution of rivers, lakes, and oceans by the waste of factories, and the resulting occupancy of the empty spaces by the waste particles, is another obvious example of an increase in the messiness of the system, the environment. ■

Entropy defined.   The state of messiness of a system is quantified by the concept of **entropy**, which is defined as the product of the Boltzmann constant and the natural logarithm of the number of accessible states: $S = k_B \ln \Omega$. The reason for taking the logarithm is to make the entropy an additive quantity. If one system has Ω_1 accessible states and another system Ω_2, then the total number of accessible states of the combined system is $\Omega = \Omega_1 \Omega_2$. For example, suppose that system 1 has 2 accessible states and system 2 has 3. If system 1 is in its first state, system 2 can be in state 1, 2, or 3, giving 3 possible states for the combined system. If system 1 is in its second state, system 2 can still be in state 1, 2, or 3, giving another 3 possible states. The total number of accessible states is 3+3 or 2 × 3. Using the fact that the logarithm of a product is the sum of the logarithm of the two factors, it is easy to show that the entropy of the combined system is $S = S_1 + S_2$.

17.3.1   The Law of Entropy Increase

Consider two separate systems with the number of accessible states Ω_1 and Ω_2. The initial total number of accessible states is $\Omega_i = \Omega_1 \Omega_2$, and the initial entropy is $S_i = S_1 + S_2$. When you bring the two systems together, allowing them to reach equilibrium, some of the states of either system that were not available to the other are now available to the combined system. Therefore, the final total number of accessible states Ω_f has increased, and the final entropy S_f is larger than S_i. This is an illustration of the **law of increase of the entropy**. **Example D.17.5** on **page 40** of *Appendix.pdf* illustrates this conclusion with numbers associated with energetic coins.

Example 17.3.2. A clamp confines a gas to one side of a container (see Figure 17.3). If the clamp is removed the gas gushes from left to right. The final equilibrium configuration is reached when the pressures on both sides are equal. This means that the gas on both sides is equally compressed. If the two half-volumes are equal, this means that there are (approximately) as many molecules on the right as there are on the left. The molecules are, of course, going back and forth between the two volumes, but at any given time, the number of molecules is almost the same.

From the entropy viewpoint, the initial configuration of the gas occupies a limited number of accessible states. Allowing the gas to freely occupy both halves increases the number of accessible states, and thus increases the entropy. It is also clear that the final state is messier—less organized, more chaotic—than the initial state. Any measurable excess of the number of molecules on one side than the other indicates that not all possible accessible states are included, and thus cannot correspond to an equilibrium state. Translated into the concept of entropy, this means that the

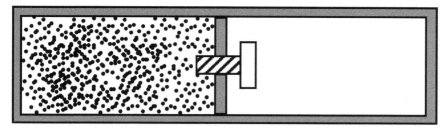

Figure 17.3: Once the clamp is removed, the gas will gush to the empty compartment.

configuration is not maximally chaotic and thus entropy has room to increase (by the transfer of molecules from the denser side to the rarer side). ∎

Example 17.3.3. Now imagine that the right half of the cylinder in Example 17.3.2 is filled with a different gas when the clamp is in place. As soon as the clamp is removed, the two gases will start to move in opposite directions. The maximum entropy is achieved when the two gases are completely mixed on both sides. ∎

Is there any situation in which the entropy of a system decreases? In principle, one can "undo the damage" by reversing the process that increased the entropy in the first place. In the case of the salt and pepper mixture, one can hire a very patient person and ask him/her to separate the salt and pepper grains and put them back in the shaker one by one. In the case of the perfume bottle, one can imagine a high-tech device that can attract individual perfume molecules and place them back in the bottle. Similarly, one can imagine microscopic devices that can reverse the entropy increasing processes. In all such instances, the system is no longer left alone. We say that the system is *open*, meaning that it is interacting with another system. For instance, the salt and pepper system is interacting with the "separator;" the perfume is interacting with the high-tech device, etc. The law of entropy increase applies only to *closed* systems.

Law of entropy increase is one form of the second law of thermodynamics.

Under very ideal conditions, it is possible to invent processes in which the entropy of a *closed system* does not change (but never decreases). Including these ideal cases, we can summarize all our observations by saying that

> **Box 17.3.4. (Law of entropy increase)** *The entropy of a **closed** system never decreases. In other words, no **closed** system tends to a less chaotic configuration over time. In mathematical symbols, $\Delta S \geq 0$.*

The word "closed" is significant. It is quite possible for the entropy of a system *that is in contact with another system* to decrease. But if you calculate (or measure) the change in the entropy of *both* systems, it is always positive (or zero, for no change), i.e., it always increases or remains the same. The statement of Box 17.3.4 is one form of the **second law of thermodynamics**.

🧺 Food for Thought 🧠

FFT 17.3.5. The law of entropy increase has been (mis)used by "scientific" creationists and the followers of the intelligent design "theory" (people who believe that the evolution of species could not have taken place without the intervention of an intelligent designer). They argue that the law of entropy increase prevents the assemblage of parts into a whole. The cells cannot form a complex animal or plant, unless an intelligent being actually puts them together. Otherwise, the more organized final product would indicate a violation of the law of entropy increase. They cite examples such as the making of a watch or a building, which cannot be done without the intervention of a watchmaker or a builder.

There are two flaws in this argument. Firstly, it assumes that somehow in processes involving human beings, the law of entropy increase *can* be violated. That this law does not apply to people! This is completely wrong. Any physical law that allows itself to be violated by humans, is not a law. Can you imagine a gravitational law that does not apply to humans? Or laws of motion that exclude human motions? Or an atomic theory that states that everything is made of atoms, except human beings? Physical laws apply to everything and every*body*. So, how do we reconcile the law of Box 17.3.4 with watch making and house building? By looking at the second flaw in the argument of scientific creationists and intelligent designers as follows.

> The entire universe is the only absolutely closed system.

The law of entropy increase applies to *closed* systems and only closed systems. To make a system closed, you have to include everything that is in contact with it. The building becomes *truly* closed only after we include the builder plus the workers that he hires, the equipments used in the construction, the fuel used in the equipments, the food taken in by the builder and the workers, the refinery in which the fuel is distilled, the farm on which the food grows, the Earth, the Sun, the solar system, the Milky Way, the entire universe! And the entropy of the universe increases regardless of what happens on Earth.

> The decrease in the entropy of an *open* system is not a violation of the second law of thermodynamics.

The universe is the only absolutely closed system. However, without extending every tiny system to the entire universe, we can *approximate* a closed system by including some of the more essential parts. For example, by including the construction machinery and the neighborhood in which the house is being built, we can already discover an increase in the entropy: Although the entropy of the house (an *open* system) *decreases*, it does so at the expense of polluting the air of the neighborhood with the exhaust of the fume given off by the machinery. If you add the entropy increase of the surrounding of the house (a positive quantity) to the entropy decrease of the house itself (a negative quantity), you'll find that the total is *positive*!

All cases of mess making accompanying the increase in entropy have at least one thing in common: *It is much easier to make a mess than to clean it up.* It requires little effort to throw the contents of a drawer all over the floor, or to mix salt and pepper in a shaker, and so forth. But it is not easy to put the scattered contents of a drawer back into it, and very difficult to separate the salt grains from the pepper grains. This has a practical implication in the construction of engines, devices that convert heat to useful work.

> Entropy and irreversible processes.

A process in which the entropy strictly increases is **irreversible**: it is impossible—highly (very highly) improbable—to see the uniformly distributed gas of Example 17.3.2 rearrange itself completely to the left; the gases of Example 17.3.3 to "unmix" and separate themselves to the two halves of the cylinder; the salt and pepper of Example 17.3.1 revert to their initial configuration; the perfume molecules collude to march into the bottle; or the two systems of Example D.17.5 to reach temperatures different from their equilibrium temperature. **Math Note E.17.4 on page 105** of *Appendix.pdf* derives a formula for the likelihood of two systems in thermal equilibrium to depart from that equilibrium, and gives 10^{4343} to one as the odds of the occurrence of a temperature difference of a millionth of a degree!

> An extremely small departure from equilibrium temperature has an even more extremely small odds of happening!

An important aspect of the entropy increase is that it is always accompanied by a loss of "useful work." This should be clear in light of the maximality of randomness. The energy of an organized state is concentrated and accessible. For example, the gas of Example 17.3.2 can turn a tiny wheel as it gushes through the hole on its way to the right half. Once it has settled to its equilibrium state of maximum entropy and equal pressure, the capability of turning a wheel will be lost.

17.3.2 Entropy and the Arrow of Time

Consider two different motion pictures. The first one shows a round bullet starting at an angle on the left of the screen, moving upward, reaching a maximum height and continuing downward, reaching the right side of the screen at an angle (Figure 17.4). The second motion picture shows smoke rising out of a pipe. The evolution of the process of the rise of smoke is depicted in Figure 17.5 with three representative "snapshots" at three times $t_1 < t_2 < t_3$.

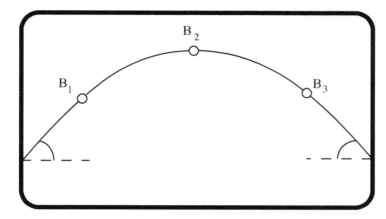

Figure 17.4: The trajectory of the round bullet as seen on the screen.

Now imagine running the projector backward. In the first case, we see a round bullet emerging at an angle on the right moving upward, reaching a maximum height, and continuing downward, reaching the left side of the screen at an angle. An observer watching this movie cannot tell whether the bullet actually started on the right, or it started on the left but the projector is running backwards. In other words, the two movies are completely equivalent. Such a situation is described by saying that the motion of the bullet is *invariant under time-reversal*. This time-reversal invariance, starring the bullet, is actually a consequence of the laws of motion, meaning that the time-reversed motion is also a possible motion.

Time-reversal invariance.

In the case of the second motion picture, when the projector runs backwards, the screen shows the smoke particles assembling themselves into smaller and smaller volume while at the same time moving towards the mouth of the pipe. An observer watching this sequence of events can easily conclude that the projector is running backwards. Therefore, the motion of smoke particles as a group is not time-reversal invariant. The time-symmetry breaks down for the smoke of the pipe.

What is the difference between the motion of a bullet and that of smoke? What ingredient is there in the smoke motion picture which allows the observer to unequivocally identify the backward motion of the film? The answer is *entropy*! Our discussion in this chapter has made it clear that entropy is a macroscopic quantity, i.e., it is well-defined only for a large number of particles (atoms, molecules). The motion of a single bullet—ignoring its irrelevant structure as a collection of more fundamental particles—cannot accommodate for entropy. On the other hand, smoke, consisting of many small particles, is a perfect arena on which entropy can unfold itself. Since for all natural macroscopic processes entropy increases, and since this increase accompanies a tendency towards what we identify as a more chaotic outcome, it is easy to detect the reversed motion of smoke. We can thus claim that entropy determines the direction of time:

Entropy is macroscopic.

Box 17.3.6. *Time flows in that direction as to make entropy grow.*

This growth of entropy can be detected only if we deal with macroscopic quantities. To appreciate this further, suppose that the projector, or the camera that took the motion

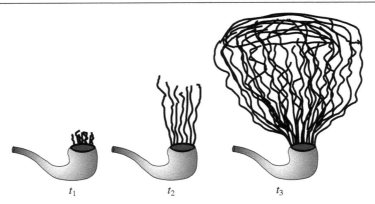

Figure 17.5: Three snapshots of the rise of smoke.

picture, has an incredibly high-powered zoom, so that we can magnify the view so much that individual particles of smoke appear on the screen. With a few smoke particles running at random on the screen, no viewer will be able to tell whether the projector is running forward or in reverse. As long as the screen is allowed to show only a few particles, the reversal of the motion of the projector will have no detectable consequences. It is true that the motion of all particles will be reversed, but the reversed motion is just as natural a possibility as the original motion. The slow "rise" of the particles is a minute effect completely overshadowed by their random (and very quick) back and forth motion.

The above analysis shows that as soon as we restrict ourselves to the microworld, the tendency to chaos, the law of the increase of entropy, fades away. In a sense the atoms, the molecules, and their constituents have no sense of time. To them, both directions of time are possible: they can travel back and forth in time.

17.4   End-of-Chapter Material

17.4.1   Answers to "What do you know?"

17.1. The average KE is proportional to *Kelvin* temperature, not Celsius. So, the average KE does not change by a factor of two. In fact, it changes by a factor of $(273.16 + 300)/(273.16 + 150)$ or 1.35.

17.2. Equation (17.1) says that the atoms of the hot object have more energy than those of the cold object. At the boundary of the two objects, the more energetic atoms of the hot object collide with the sluggish atoms of the cold object and make them move faster. This transfer of KE, reduces the KE of the hot atoms and increases the KE of the cold atoms at the boundary. The boundary atoms of the cold object share their extra energy with the other atoms of the cold object, making them more energetic. Similarly, the other atoms of the hot object share their energy with their less energetic partners at the boundary, making them more energetic. This process goes on, until the hot has no more energy to share with the cold, i.e., until both sides have the same temperature.

17.3. Average KE is proportional to the temperature. So, both kinds of atom have the same average KE. However, lighter atoms have larger average speed (to make the product of mass and velocity square equal). So, He atoms are moving faster on the average.

17.4. With no constraint, the number of accessible states is the same as the number of possible outcomes, which is $2^5 = 32$. If the energy is to be $+1$, we need 3 positive and 2 negative coins. By actually counting, or using Equation (E.19) with $n = 5$ and $m = 3$, you obtain the number of accessible states to be 10. There is no way to get a -2. For -3, you need 4 negatives and 1 positive. Again, either by actually counting, or using Equation (E.19) with $n = 5$ and $m = 4$, you obtain the number of accessible states to be 5.

17.5. When combined, you have 25 "energetic" coins. To get a net $+5$ for the total, you need 15 positives and 10 negatives. Let m_1 and m_2 be the number of positives in the first and second systems, respectively. Then, $m_1 + m_2 = 15$. Furthermore, Box 17.2.4 says the $m_1/10 = m_2/15$. These two relations have the unique solution of $m_1 = 6$ and $m_2 = 9$. So the most probable configuration has 6 positive coins in the first system and 9 positive coins in the second system.

17.4.2   Chapter Glossary

Avogadro's Number The number 6.02×10^{23}.

Entropy A thermodynamic quantity which measures the messiness of a system and how much useful energy is available when that system undergoes a process. The law of increase of entropy determines the arrow of time.

Ideal Gas A gas in which the constituent particles do not interact, so that their energy is only kinetic. Most gases behave like an ideal gas under normal conditions.

Irreversible Process A thermodynamic process which takes a system from an initial state to a final state, after which it is impossible to take the system from the final state to the initial state.

Kelvin The scientific temperature scale, equal to Celsius scale plus 273.16.

Mole A collection of 6.02×10^{23} (Avogadro's number of) atoms or molecules of a substance.

Reservoir A large body with which a (small) system can exchange energy. A reservoir is characterized by the fact that its temperature does not change due to the exchange of energy with a (small) system.

Reversible Process A thermodynamic process which takes a system from an initial state to a final state, after which it is possible to take the system from the final state to the initial state.

Root Mean Square (rms) The square root of the average of the square of a quantity.

Thermodynamics The branch of physics studying the bulk properties of matter.

17.4.3   Review Questions

17.1. What is the pressure of a gas from a microscopic point of view? How is it related to the number of particles in a gas? To the speed of the particles?

17.2. How is the average KE of the gas particles related to its temperature? How is the average speed of the gas particles related to its temperature?

17.3. In the ideal gas law, assume that the volume is fixed. From what you know about the microscopic viewpoint of pressure, does it make sense for the pressure to depend on the number of particles and temperature?

17.4. What is a mole, and how is it related to the Avogadro's number?

17.5. What physical quantities determine the state of a system? If you know the state of a system at one time, can you predict its behavior at a later time?

17.6. Define an accessible state of a system and its relation to the constraints on the system. What is the most common constraint imposed on a system?

17.7. What condition determines the most probable configuration of a system whose energy is constrained?

17.8. Explain why the increase in the number of accessible states is related to the increase in messiness. What is the relation between entropy and the number of accessible states?

17.9. State the law of entropy increase. Do *all* systems obey this law? Is it possible for the entropy of a system to decrease? If so, give an example of such a system.

17.10. What is an irreversible process, and how is it related to entropy? Give some examples of irreversible processes, and explain how these processes are probabilistic.

17.11. What is the relation between entropy increase and the availability of useful work?

17.12. Explain how entropy is related to the arrow of time. Is it possible for the smoke of a pipe to revert back into the pipe? If you see a movie that shows such a reversion, what would you conclude about the running of the movie? If you zoom in so much that you can see the motion of the individual smoke particles, can you determine whether the movie is running forward or backward?

17.4.4   Conceptual Exercises

17.1. You need three numbers to specify the location of a particle in space: one number tells us how much east (positive) or west (negative) it is; the second number tells us how much north (positive) or south (negative) it is; and the third number tells us how much up (positive) or down (negative) it is. Similarly, three numbers are required to specify the momentum (or velocity) of the particle: how fast is the particle moving in the east-west direction, etc. So, six numbers are needed to give the state of a single particle. How many numbers are needed to give the state of a two particles? Of 10 particles?

17.2. A gas has a temperature of 100°C. You double its temperature to 200°C. Does the average KE of its particles double? If not, to what Celsius temperature do you have to raise it for the average KE to double?

17.3. A mixture of helium and neon gas is held at a temperature of 30°C. Which gas has a larger average KE? Which gas has a larger average speed?

17.4. When you heat one part of a gas, at first that part is hotter. But eventually the temperature becomes uniform throughout the gas. Explain this from an atomic point of view and Equation (17.1).

17.5. What happens to the pressure of a gas if you heat it while keeping the volume constant? What happens to the volume of a gas if you heat it while keeping the pressure constant? What happens to the density of a gas if you heat it while keeping the pressure constant?

17.6. One mole of an ideal gas has a temperature of 600 K. You add another mole to it, lower its temperature to 300 K, and keep its volume the same. Does its pressure increase, decrease, or remain the same?

17.7. You have 4 "energetic" coins with +1 on one side and −1 on the other.
(a) How many accessible states are there without constraint?
(b) What if the energy is to be +1? −2? 0?

17.8. You have 6 "energetic" coins with +1 on one side and −1 on the other. How many accessible states are there without constraint? What if the energy is to be +1? −2? −4?

17.9. Explain (from a molecular viewpoint and accessible states) why the entropy of a system increases when you add heat to it.

17.10. Drop a ball. Just before it hits the ground, it has some KE. After it hits the ground it has no KE. What happened to its KE? From an atomic perspective, explain why there is an increase in entropy. Was useful work lost upon impact?

17.11. A human embryo, consisting of a few unorganized cells, turns into a highly organized baby. Is this a violation of the law of entropy increase?

17.4.5   Numerical Exercises

17.1. A mixture of neon and helium atoms is held at a temperature of 50 °C. Neon is five times heavier than helium.
(a) How many times larger is the average KE of the helium atoms than that of the neon atoms?
(b) How many times larger is the average speed (root mean square of velocity) of the helium atoms than that of the neon atoms?
(c) The temperature is now quadrupled to 200 °C. How do the answers to (a) and (b) change?
(d) Helium atomic mass is 6.64×10^{-27} kg. What is the average speed of the helium atoms when the temperature is 50 °C? When it is 200 °C? What are the average speeds of the neon atoms at these two temperatures?

17.2. What is the volume occupied by a *mole* of an ideal gas at room temperature (300 °K) and atmospheric pressure?

17.3. Two systems have 6 and 9 "energetic" coins, respectively. The energy is +5. What is the most probable configuration when these two systems come in contact? Hint: Look at the answer to **What do you know?** 17.5.

17.4. Two systems have 9 and 12 "energetic" coins, respectively. The energy is +7. What is the most probable configuration when these two systems come in contact?

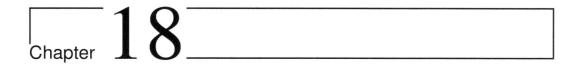

Chapter 18

Temperature and Heat

The machinery of statistical mechanics can help us examine thermodynamics with much more ease than the tools available to the physicists of the eighteenth and early nineteenth centuries, who were stymied by the complexity of the systems they had to study. Nevertheless, it is instructive to follow the historical development of concepts, and make connection with their statistical mechanical analogues. Let's start with the most basic quantity in thermodynamics.

18.1 Temperature

Temperature quantifies the notion of coldness or hotness. That the human sensation of hot and cold is inaccurate can be demonstrated by a simple experiment. Take three bowls labeled B_c, B_w, and B_h. Fill B_c with cold water, B_w with lukewarm water, and B_h with hot water. Place a finger of your left hand in B_c and one of your right hand in B_h. Wait a few minutes. Now place both fingers in B_w. Your left finger feels warm while your right finger feels cold.

Your tactile sense can fool you!

To measure the hotness of something one uses a **thermometer**. Thermometers have been in use for a long time. Galileo is credited with the invention of the first one using the expansion of air as its working principle. Most thermometers in use today employ the same basic principle, namely the expansion of substances such as alcohol and mercury (the two most common liquids used in thermometry) caused by heat.

Thermometers and how they work.

From the early seventeenth to the late eighteenth century, the art of thermometry developed considerably. By the middle of the eighteenth century it became known that the temperature at which ice melts, the so-called *melting point* of ice, does not change as the ice is heated up. Similarly, the *boiling point* of water is independent of how much the water is heated. Soon these two points became the standard fixed points of many thermometers. These fixed points were given arbitrary values by different thermometer makers, and the interval between these two values was divided into a number of equal parts. Daniel Fahrenheit (1686–1738) assigned 32 to the melting point of ice and 212 to the boiling point of water and divided the interval between these two points into $212 - 32 = 180$ equal parts. Anders Celsius (1701–1744), on the other hand, decided that a more appropriate pair of numbers is 0 for melting and 100 for boiling. Still other thermometrists chose other pairs of values for the fixed points.

> **What do you know? 18.1.** Why are the tubes of a thermometer, in which mercury or alcohol expand, so thin?

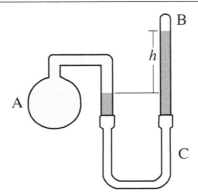

Figure 18.1: A constant-volume gas thermometer works on the empirical basis that, at constant volume, the pressure of a gas increases with temperature.

Clearly such a feudal state of affairs could not persist for long. Communication between users of different thermometers forced the adoption of a single *standard* thermometer. The problem was not so much in using different pairs of values for the fixed point (a simple rule can be used to convert from one scale to another), as in using different substances for thermometers. Suppose two people calibrate their thermometers according to Celsius scale; however, one uses alcohol and the other mercury. By their very construction these two thermometers read 0 degrees for the melting point of ice and 100 degrees for the boiling point of water. But, what guarantees that they will agree on the intermediate points and beyond? Such a guarantee would exist only if heat would expand alcohol and mercury at proportionate rates, which is not the case.

Constant-volume gas thermometers.

By the beginning of the nineteenth century it became clear that gas thermometers were the most accurate. One such gas thermometer is the so-called **constant-volume** gas thermometer shown in Figure 18.1. Put container A in contact with the material whose temperature is to be measured. The volume of the gas changes. Raise or lower the flexible tube C to restore the volume of the gas to its original value. This causes a change in the level of the liquid (usually mercury) in B proportional to the temperature.

Accurate as they were, gas thermometers still showed discrepancies among themselves. If one measured a single temperature, e.g., the boiling point of sulfur, with gas thermometers filled with oxygen, nitrogen, air, or hydrogen, the reading would vary from 444 °C to 446 °C, depending on the gas used. However, it was discovered later that, if one lowered the pressure of the gas *by decreasing the amount of gas* in the thermometer, the readings of thermometers were also lowered along straight lines. Figure 18.2(a), in which the temperature is measured in Celsius degrees, depicts this behavior. The figure shows that the disagreement among the thermometers is reduced for smaller pressures; the lower the pressure, the better the agreement. It was further seen that if one *extrapolated* the straight lines to zero pressure, all thermometers would agree exactly.

🎁 Food for Thought 🌶

FFT 18.1.1. It is appropriate at this point to remark briefly on the science and art of measurement. A thermometer is built to measure the temperature of an object. But it *uses* the assumption that water freezes and boils at constant temperatures! How do we know that the freezing (or boiling) point of water is fixed *prior* to the construction of our first thermometer?

Measurement and apparatus affect each other.

Measurement is an interactive process in which both the measuring device and the measured quantity participate. The very use of an apparatus presupposes a knowledge of what is being measured. To use a meter stick, we must have a fairly good idea of length. Clocks are useful because we have a fairly good idea of what time is. We rely on a balance when we weigh an object because we have a "hunch" that the two sides of the balance are pulled down with equal force. The inventor of the first balance knew nothing about the properties of the gravitational force that pulled the objects at the two ends of his balance. He did not know that the force of gravity was sensitive only to the mass of the objects (and not, e.g., on their color); that the gravitational acceleration did not vary appreciably from one end of the balance to the next; that this acceleration was the same for all objects. Yet the validity of all these properties of gravity is crucial for the effective operation of a balance.

Dissecting the operation of a balance.

As our understanding of physical quantities improves, our ability to build a better apparatus increases just as our need for such an apparatus becomes more and more urgent. While grandfather clocks were quite adequate for the eighteenth century, they are completely obsolete for modern-day scientists who have to measure the intricacies of the relativistic unification of space and time. On the other hand, an improvement in the measuring device entails a rectification of our knowledge of the quantity being measured. They go hand in hand, and one influences the other.

Thermometers and temperature are no exceptions to this rule. The first thermometer builders had a "hunch" that ice always melted at the same temperature. Their first crude thermometers verified this hunch. Improvements on these crude thermometers also pointed to the accuracy of the hunch. As our knowledge of the concept of temperature improved, our thermometers became more accurate, and such an accuracy allowed further progress in our understanding of temperature.

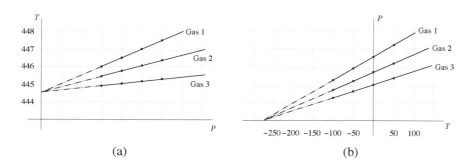

(a)   (b)

Figure 18.2: (a) There is a fair amount of disagreement among the constant-volume gas thermometers using different gases. However, as the pressure decreases, the disagreement is reduced. Extrapolation to zero pressure gives identical results. (b) When the amount of gas is fixed, P extrapolates to zero at $-273.16°$C.

Another observation that paved the way for the definition of a *universal temperature* was the variation of pressure with temperature *when the amount of gas was kept constant*, but small. It was noted that by making the gases colder and colder, their pressure became smaller and smaller, all maintaining a straight-line plot on a P vs T diagram [see Figure 18.2(b)]. When these lines were extrapolated to the left, they all crossed the T-axis at the same point. This "T intercept" turned out to be -273.16 °C.

Universal thermometer.

In 1848, the Scottish physicist, William Thomson (1824–1907), knighted as Lord Kelvin in 1892, suggested a new scale of temperature, known as **Kelvin temperature scale**, whose zero occurred at -273.16 °C, the **absolute zero**. Kelvin temperature scale, denoted by °K or simply K, is the scale used in scientific laboratories around the world. The vanishing of pressure at absolute zero agrees with Equation (17.2), whose temperature we have already identified as Kelvin.

Absolute zero is zero Kelvin.

18.2   The First Law of Thermodynamics

A simple system such as a gallon of gas (e.g., air) in a cylinder open at one end is a good tool to study thermodynamics. Put a piston at the open end of the cylinder. There are two ways to interact with the gas: pushing or pulling the piston and putting the walls of the cylinder in contact with a reservoir. This interaction transfers energy to the particles of the gas, changing its energy content. The energy associated with the motion of the piston is the **work** done on or by the gas. The energy transferred through the walls of the cylinder is *defined* to be the **heat** added or removed from the gas.

It is customary to denote the work done by the system by W so that the work done *on the system* will be $-W$. The reason for the minus sign is that in any process in which work is done and in which there are action/reaction forces, the work done by the reaction force is always negative of the work done by the action force. This is because the two forces are displaced by the same amount and in the same direction, while the directions of the two forces are exactly opposite by the third law. In the situation at hand, the outside agent and the gas constitute an action-reaction system.

First law of thermodynamics stated.

From energy conservation, we obtain

$$\Delta U = Q - W, \tag{18.1}$$

where ΔU is the change in the **internal energy**, Q is the heat added to the system, and W the work done *by the system*. Both Q and W can be positive or negative: $Q > 0$ means that heat is added to the system and $Q < 0$ means that the system gives off heat. Similarly, $W > 0$ means that the system does the work and $W < 0$ means that work is done on the system.

What do you know? 18.2. Recall that the particles of an ideal gas have only one form of energy: kinetic. How does the internal energy of an ideal gas consisting of N particles depend on the temperature of the gas? Hint: Equation (17.1) gives the average KE of each particle.

Equation (18.1) is the **first law of thermodynamics**. In words it says

---

Box 18.2.1. (First law of thermodynamics) *The amount of energy added to a system—for example, in the form of heat or work—is stored in the system as its internal energy.*

Math Note E.18.1 on **page 107** of *Appendix.pdf* applies the first law of thermodynamics to a very simple system and finds the heat transferred to this system in terms of the change in its temperature. In so doing, the math note also introduces **specific heat**. Specific heat is a quantity that is defined not only for a gas but also for solids and liquids. It is the amount of energy needed to raise the temperature of 1 kg of the system by 1 degree Celsius (or Kelvin). Table 18.1 shows the specific heat of some common substances.

What do you know? 18.3. Is it possible to add *only* heat to an ideal gas without changing its temperature?

18.2.1   Heat as a Form of Energy

Statistical mechanics derives Equations (18.1) and (E.48). However, statistical mechanics did not exist when people struggled with the concept of heat. The earliest theory of heat dates back to the 1770s when the French chemist Antoine Laurent Lavoisier (1743–1794) proposed the "caloric" theory. According to this theory, when a hot object comes in contact

Substance	Specific heat (in J/kg·°C)	Substance	Specific heat (in J/kg·°C)
Alcohol	2400	Lead	128
Aluminum	900	Mercury	140
Copper	387	Silicon	703
Germanium	322	Silver	234
Glass	837	Steam	2010
Gold	129	Water	4186
Ice	2090	Wood	1700
Iron	448		

Table 18.1: Specific heats of some common substances.

with a colder object, an imponderable fluid, called *calorie*, will flow from the hot to the cold object. There was some confusion with regards to the caloric theory at the beginning. For instance, some people believed that, just as there were positive and negative electricity, caloric came in two different forms: Hot objects have positive calorie while cold objects must have negative calorie. However, gradually it became clear that calorie existed in only one form and that it always flowed from hot to cold. By the late 1790s there were a number of scientists who questioned the caloric theory. One of these scientists was Benjamin Thompson (1753–1814), also known as Count Rumford.

🎁 Food for Thought 🌱

FFT 18.2.2. By the end of the eighteenth and the beginning of the nineteenth century, the three branches of physics, namely mechanics, electromagnetism, and thermodynamics, had progressed to various degrees. Mechanics was rigorously established on a firm mathematical foundation which itself had helped create. However, the other two branches were still in their infancy, and a lot of confusion was prevalent in them. Only in such a preponderance of confusion can a theory based on "imponderables" gain acceptance, albeit for a short while. The scientists today reject such imponderable entities as the "energy field" and the "Qi" of alternative medicine. But two centuries ago, it was not easy to do so. Even prominent scientists proposed mythical "theories."

These incipient sciences begged the attention of curious minds. And since the amount of knowledge accumulated in them was meager and unorganized, someone with little or no formal training could contribute to their further understanding. It is not surprising, therefore, to hear the names of Benjamin Franklin and Michael Faraday in the history of electromagnetism, and Benjamin Thompson in the history of thermodynamics.

Benjamin Thompson was born in Woburn, Massachusetts. Instead of a formal education, he became an apprentice in a store. When he was 19, he married a rich woman considerably older than him to get access to a comfortable life. However, the comfort was not to last for long, because when the Revolutionary War broke out, the 23-year-old Thompson sided with the British. His commitment to the British surpassed mere sympathy, and he became actively involved in spying for them.

After the Revolutionary War, he knew well that he would have to spend the rest of his life in exile. He first went to England where he was found taking bribes and selling war secrets to the French. Having made himself unwelcome in England, he left for Germany in 1783 and put himself at the service of Elector Karl Theodor of Bavaria, who, in recognition of Thompson's civilian and military services, granted him the title of a Count. Thompson chose the name Rumford because that was the town (now Concord) in New Hampshire where he used to live with his wife.

In 1799, after the death of the Elector, Count Rumford had to leave Germany because he had managed to accumulate many enemies during his stay there. He returned to England where, due to his scientific accomplishment (see below), he was admitted into the Royal Society, a prestigious scientific society established at the beginning of the seventeenth century in England. His second stay in England was not to be permanent either. He went to Paris in 1804 and married Lovoisier's

Benjamin Thompson's biography . . .

rich widow. However, their marriage was a disaster from the very beginning and lasted four bitter years. Thompson died in 1814 near Paris with his American daughter by his first wife at his side.

...contrasted with Michael Faraday's biography.

A possible explanation of why Faraday contributed to so many varied fields of science, but not to mechanics.

What a contrast with Faraday's biography! [For a concise biography of Faraday, see Section 13.3.] Faraday also had no formal education, but he was a man of strong character, a genius with an unprecedented devotion to science. Once he started his career as a scientist, he spent 100% of his time doing science. For him science was not just a diversion, it was his life. One can only speculate on the possible level of Faraday's scientific achievement had he had a formal training—especially in mathematics. It was a combination of his genius and the pristine nature of electromagnetism that led to his discovery of one of the most important pieces of the electromagnetic puzzle, the law of induction. To understand how critical the primitive nature of electromagnetism was to that contribution, one can merely ask why he did not contribute to the science of mechanics, although he advanced the other sciences of his time, including chemistry, considerably. The answer is simple: mechanics was by then almost 200 years old and had become highly mathematical, for which Faraday had no preparation.

Thompson and Franklin, on the other hand were not professional scientists. For them, science was only a diversion, a hobby, a pleasant curiosity; and if they stumbled on a discovery, so much the better. But, unlike Faraday, Thompson and Franklin did not—they could not, because there were other things occupying their time—pursue the discovery as a professional scientist would. And they were not alone in either their leisurely pursuit of knowledge or the importance of their discovery. In the following, we shall learn about Julius Meyer, a physician, and James Joule, a brewer, who contributed significantly to the development of the then pristine science of thermodynamics.

It is not our intention to undercut the discoveries made by such people. These discoveries were essential to the later development of science regardless of their discoverers. The point to be made is the difference between eighteenth century and contemporary physics. Today, without a total commitment, without an almost obsessive devotion, without a long and sophisticated training in physics, it is impossible to make any contributions to its development. Why? Because all branches of physics have become highly mathematical. Even Faraday, with all his genius and devotion, would be incapable of contributing to today's physics, because of his lack of formal training, just as he was incapable of contributing to mechanics during his distinguished career.

There are people who will say that this portrayal of physics discourages the youth from taking up physics and science as their career. But are we doing the youth service by giving them false hopes? Wouldn't it be a disservice to the youth if we tell them that they are learning arithmetic when they cannot do even a single-digit addition? Some educators advocate developing "positive attitude and self confidence" in our youth. That is good, but not at the expense of *replacing* the truth. We have to encourage our children that they *can* learn arithmetic. But if they have difficulty with the simplest addition, we should not hand them a calculator and teach them how to push buttons, and call that "learning arithmetic."

Similarly, we should not replace physics with the "fun" of computer animation and imagery[1] just to please the youth, because sooner or later the society will come to the realization that its youth lacks even the simplest understanding of physics. Part of the teaching of physics to the public ought to be to point out that becoming a physicist requires dedication and years of training. Then honest truth seekers will acknowledge the necessity of fully learning the existing theories—with all their mathematical, theoretical, and experimental intricacies—at the early stages of their career. Once this necessity is recognized, and they pursue it vigorously, honestly, and wholeheartedly, the society will benefit from their creativity.

What do you know? 18.4. Does adding heat to an ideal gas always accompany a rise in its temperature?

In 1798, when serving Elector Karl Theodor of Bavaria in the military, Thompson noticed that, in the process of boring cannons, the metal became so hot that it boiled the water used to keep it cool. The prevalent explanation was that, when the metal is broken to pieces, as into filing produced by boring, the caloric is liberated from the metal, giving rise to heat. However, Thompson realized that, even when filing is not produced, heat is given

---

[1] A practice that is sadly too common among educators.

off by mere friction. In fact, he showed that the amount of heat produced in the process of boring was so great that if it were poured back, it could melt the metal! In other words, more caloric could be released from the metal than it could possibly hold. Thompson's conclusion was that the heat was produced by the mechanical motion of the borer, and that the quantity of heat was equal to the motional energy of the borer. Therefore, he concluded, *heat is a form of energy.* Thompson even tried to measure how much heat was produced by a given amount of motion. He was thus the first to measure what is called the **mechanical equivalent of heat** (MEH). His figure of 5.57 Joules was, however, too high, and it was not until about 50 years later that the first acceptable value of 4.16 Joules was measured.

Heat is a form of energy.

Mechanical Equivalent of Heat (MEH).

Count Rumford's idea of the equivalence of heat and mechanical energy did not catch on immediately for two reasons. First, Lovoisier's fame and influence as the father of modern chemistry made it difficult to tackle the caloric theory which he had proposed earlier. Secondly, and more importantly, to convince the scientific community of the equivalence of heat and mechanical energy, a precise measure of such an equivalence was required. This was lacking up to 1842 when Julius Mayer (1814–1878), the German physician-turned-physicist, reported a rough estimate of the mechanical equivalent of heat.

The final and decisive blow to the caloric theory was imparted by James Prescott Joule (1818–1889), the English brewer. In 1847, Joule reported his measurement of the mechanical equivalent of heat. However, partly because of his profession (a brewer was not well-received in the academic circles of mid nineteenth century), and partly because of the fact that his results were based on measuring very small temperature differences, his work was not recognized at the beginning. He could not publish his result in any of the scientific journals of his time. With the help of his brother, a music critic, he managed to print his paper in a Manchester newspaper. After a few months he succeeded in reading his work to a hostile professional audience, among which was Lord Kelvin who commented favorably on Joule's work, and the interest in his work grew. Finally, in 1849, Joule read a revision of his paper to the Royal Society which accepted the results.

> **What do you know? 18.5.** Can the caloric theory explain the heat produced when you rub your hands together? Can the mechanical equivalence of heat?

Joule's work showed that each **calorie**,[2] the prevalent unit used to measure heat, contains 4.16 metric units of energy. Today, the accepted value is 4.186 metric units of energy per calorie. The metric unit of work and energy is named after Joule (see page 108) to honor his contribution to our understanding of the intricate concept of heat.

Joule finds the mechanical equivalent of heat.

To get an idea of how big a calorie is, we note that lifting a 1 kilogram object a distance of 1 meter requires approximately 9.8 Joules. Thus, such a lift is equivalent to $9.8/4.18 = 2.34$ calories of heat. This calorie should not be confused with the dietary calorie which is 1000 times bigger. Thus, to lose an equivalent of 2.34 dietary calories, one has to lift 1 kg a distance of 1 m for 1000 times, or to lift 50 kg a distance of 1 m 20 times! It is not easy to lose weight!

It is not easy to lose weight!

One of the methods Joule used to measure MEH was to drop a weight by a given distance. As the weight fell, it turned a paddle which stirred the water in a bucket. By measuring the distance the weight fell, Joule could calculate the work done. By measuring the temperature rise of the water, he could measure the heat produced. He then assumed that the work done was converted into heat. Equating the two, he could find the MEH. **Example D.18.1** on **page 41** of *Appendix.pdf* shows why a precise measurement of the mechanical equivalent of heat is so difficult.

Calculating the amount of heat transferred to an object requires knowing the value of the specific heat of that object. The determination of the specific heat of substances is made

---

[2]Calorie is defined to be the amount of heat needed to raise the temperature of one gram of water one degree Celsius. The reader is urged to verify this definition using Equation (E.48) and Table 18.1.

by heating a sample of the substance to a given temperature and subsequently submerging it in water at another temperature. By carefully insulating the whole apparatus so that no heat is lost to the surrounding, one can assume that heat given off by the hot object is equal to the heat absorbed by the cold object when the two come in *thermal equilibrium*, i.e., their final temperature reaches the same value. Equating these two quantities of heat, one can solve for the specific heat of the substance (see **Example D.18.2 on page 41** of *Appendix.pdf* for details of how this is done for copper).

We have casually used energy and its conservation, because it has become a standard part of physics literature, although the word "energy" is a fairly recent addition to the physics vocabulary. It was coined by Thomas Young (1773–1829) in a lecture delivered in 1801 to the newly established Royal Institution. Prior to that, there was a lot of confusion not only in using the word, but also in its precise physical meaning. Some physicists used the concepts of force, momentum, and energy interchangeably. Young clearly showed that momentum—known in those days as the quantity of motion—was quite different from what he called kinetic energy.

Precise physical concepts often start out as confusing and muddled.

It is ironic that the law of the conservation of energy, one of the most fundamental laws of physics, was developed and proposed by "outsiders." We have already mentioned Benjamin Thompson the diplomat, James Joule the brewer, and Julius Mayer the physician. And the person credited with the first clear statement of the law of energy conservation is Hermann Helmholtz (1821–1894) the physiologist. The law of energy conservation states that

Conservation of energy.

> **Box 18.2.3.** *There is a finite constant quantity, called energy, which can appear in various forms such as heat or mechanical work.*

Energy, in its broadest term, is the *capacity to do work*. Thus, this capacity could acquire many forms: mechanical, heat, electrical, nuclear, etc. Of course, nuclear energy had not been discovered in the 1840s. However, when it was discovered, it was shown to obey this all embracing principle.

18.2.2 First Law and Weight Loss

A useful application of the first law of thermodynamics is to the human metabolism, where Q is the calorie intake in the form of food, W is the work done in the form of muscular actions both internal (heart beat) and external (walking), and ΔU is the change in the internal energy of the human body. Unfortunately the ultimate form that this internal energy takes is fat.

Not all people have the same kind of metabolism. Certainly you have encountered people who eat a lot more than you do, exercise no more than you, and yet they seem to be a lot skinnier than you. This is not a violation of the first law of thermodynamics; it has to do with the "efficiency" with which human body uses the food intake. You are one of the unlucky people, if you have a very efficient digestive system, which processes a large portion of the food you eat. Your friend, with an inefficient digestive system can eat a lot, but a large portion of her intake is eventually disposed out of her body, leaving a small portion as Q. So although she eats a lot more than you, the Q of your friend may be actually smaller than yours. Furthermore, her *internal* muscular activity may be larger than yours, making her overall W larger, even though you may be doing more *external* exercise. In applying the first law of thermodynamics, Q should be regarded as the net calorie *intake* (not necessarily the calories in the food one eats), and W as the net work done (including the internal muscular activity, which varies from person to person).

Regardless of the variations among individuals mentioned above, if a person's metabolism is such that $Q > W$, he is eating more than he is exercising; thus $\Delta U > 0$, and the person will be accumulating fat. If $Q = W$, the person is eating exactly the same amount of energy as he gives off in the form of work; therefore, his weight does not change. If you think you are overweight, you have to make ΔU negative, i.e., make Q less than W. In other words: Eat less and/or exercise more!

Fat is *matter*. Both its accumulation and its loss require *material* physical processes obeying the first law of thermodynamics: To accumulate fat, ΔU must be positive; to lose it, ΔU must be negative. And to change ΔU, you need to change Q or W or both; i.e., you have to change your material intake of food, or your physical activity, or both. No amount of meditation, mental concentration, spiritual focus, or "positive attitude" will change the fat content of your body.

First law, human body, and gaining or losing weight.

Meditation and "positive thought" do not make you lose weight!

> **What do you know? 18.6.** Is it possible to lose weight by telling yourself that you *will* lose weight if you concentrate hard enough? By *only* exercising?

Next time you pass the local "health and beauty" store, advertising the "new aromatherapy" by means of which "mind speaks; body listens," think of the first law of thermodynamics. And remember that one cannot lose weight by soaking oneself in aromatic oils, closing one's eyes, and imagining a leisurely walk through a bed of flowers and herbs!

 Food for Thought

FFT 18.2.4. If there is one master universal result in physics, it ought to be the conservation of energy. It has been a guiding principle in many branches of physics. Heat was the first (macroscopically) non-mechanical energy to enter into the picture. At the beginning it was thought to be a nonmaterial "imponderable" quantity. Once the mechanical equivalence of heat was established, the idea of the conservation of energy won a strong foothold in the world of physics until the closing years of the nineteenth century.

In the 1890s, when radioactivity of certain elements was discovered, it seemed as if these elements had a supply of spontaneously generated energy, and the principle of the conservation of energy came under questioning by some physicists. In the meantime, a German mathematician by the name of Emmy Noether, pursuing a highly mathematical formulation of the laws of mechanics, discovered in 1918 one of the most beautiful and elegant results in mathematical physics. In what is now called the **Noether Theorem**, she showed that if time intervals are independent of their starting points [so that one hour is one hour whether it is measured from 1:00 pm to 2:00 pm, or from 9:00 am to 10:00 am], then energy must be conserved.

Noether Theorem and the connection between time and energy conservation.

With the exactness of a mathematical proof on its side, energy conservation seemed to be on a solid foundation, especially when quantum theory of 1925–26 appeared to be capable of solving the riddle of radioactivity. However, a closer look at these activities, especially what was later identified as the weak nuclear reaction, seemed to suggest a "missing energy." When physicists added the energies of all the particles produced in the weak nuclear decay and compared it with the energy before the decay, they found a depletion, a disappearance of energy! Some proposed non-conservation of energy in nuclear decay and suggested that somehow nuclear interactions are different from ordinary phenomena. Wolfgang Pauli, however, adhered to the principle of energy conservation, and proposed in 1930 the existence of a highly elusive particle, which Enrico Fermi later christened **neutrino**. Decades later neutrino was actually detected in ultrasensitive detectors.

Today, conservation of energy is one of the most firmly established principles of physics, and no violation of it has ever been observed. Indeed, by Noether's Theorem, such a violation is no less nonsensical than the statement that "the duration of one hour depends on when you start counting the minutes;" or that from 10:00 am to 11:00 am is one hour, but from 2:00 pm to 3:00 pm is only 58 minutes, or that Einstein lived for 76 years according to the Christian calendar, but 74 years according to the Jewish calendar!

Senselessness implied by the violation of energy conservation!

The next time you hear about claims of inventions of electric generators that consume no fuel, cars that run forever without gasoline or electricity, or any other *perpetual motion machine* (which by its very nature violates energy conservation), remember that, by Noether Theorem, the existence of these machines implies that time intervals depend on their starting moments! Therefore, energy machines and inexhaustible sources of energy sound as strange as the statement that Einstein lived for 76 years according to the Christian calendar, but 74 years according to the Jewish calendar.

18.2.3   Specific Heat

Equation (E.48) introduced the notion of the specific heat. It is appropriate to discuss the origin of the specific heat, especially its variation among different substances, from a statistical mechanical viewpoint. It will be more convenient to discuss **molar specific heat**. Recall that one mole contains 6.02×10^{23} molecules. Molar specific heat is the amount of heat needed to raise the temperature of one mole of a substance by one degree Celsius. We know that different substances have different specific heat. The question is, can we explain this difference by use of statistical mechanics? How can we explain the fact that some substances become hot quicker than other substances?

Consider two substances A and B. Assume that A has a simple molecular structure, while B's is more complex leading to more internal motions. For instance, different atoms in the molecule can vibrate relative to one another, or they can undergo rotational motion about each other. In the jargon of physicists, B is said to have vibrational and rotational **degrees of freedom**. Such internal motions either do not exist for A or they exist to a lesser degree. Now imagine adding the same amount of heat to the two systems which we assume have equal number of molecules, say one mole. This heat, or energy, will divide itself among all molecules. So each molecule of either system receives the same amount of energy on the average. This energy divides itself up into various forms that the molecule is capable of receiving, such as kinetic energy, vibrational energy, rotational energy, etc. Thus, molecule B, which has a lot of "internal" motions can absorb a large portion of the energy internally. Only a small part will be left to provide kinetic energy for the molecule. The situation is different for molecule A: a lot of energy goes into the kinetic "mode," and only a little in the internal modes.

BIOGRAPHY

Emmy Noether (1882–1935) studied German, English, French, and arithmetic in high school, and wanted to become a language teacher. However instead of finding a job as a language teacher, for which she became certified in 1900, Noether decided to study mathematics at university, but in those days, women were not allowed to officially register at universities. Each professor had to give a separate permission for his course. Noether sat in on courses at Erlangen and Göttingen, receiving her doctorate in 1907. She published many fundamental papers in algebra, but her monumental work is a theorem on conservation laws in mathematical physics. In 1933 the Nazis dismissed her from the University of Göttingen because she was Jewish. She accepted a visiting professorship at Bryn Mawr College in the USA. She died in 1935 due to post-surgery complications.

As discussed in Section 17.2, temperature is proportional to the average *kinetic* energy of molecules. We thus expect substance A's temperature to rise more than substance B's. Therefore A has a smaller molar specific heat than B, because less heat is required to raise its temperature. According to this argument, substances which have larger molecules must have larger (molar) specific heat. This is indeed the case.

> **What do you know? 18.7.** Two substances A and B both have rotational degrees of freedom. A requires less energy to excite its rotation than B. Which substance has a larger specific heat?

18.3   The Second Law of Thermodynamics

The first law of thermodynamics states the conservation of energy. It considers heat as a form of energy and incorporates it into the conservation law. It puts no restriction on the way various forms of energy can transform into one another. Thus, the first law is blind to the practicality of transforming work into heat or vice versa.

The process of the conversion of heat into work is of considerable practical importance, because that is precisely what an **engine** does. Engines were quite common in nineteenth century Europe, having been invented as early as 1698 by Papin (1647–1712), the French inventor, and improved by the Scottish engineer James Watt (1736–1819) in 1769. A common feature of all these engines was their extremely low efficiencies. The best of them had an efficiency of only 6%. This means that the engine would use, say 100 Joules of heat energy and produce only 6 Joules of work. The first law, on the other hand, did not restrict the 100% conversion of heat into work. Was it, in principle, possible to have an engine 100% efficient? It was the investigation of this question that led to the second law of thermodynamics.

Actually the second law was discovered before the first. In its earliest form it was formulated by the French physicist Nicolas Sadi Carnot (1796–1832) in 1824. Carnot was an army engineer who was interested in the efficiency of engines from a purely practical point. He considered the most efficient ideal engine, now known as **Carnot engine**, and showed that such an engine could not be 100% efficient. The ideality of Carnot engine makes it suitable for theoretical studies of thermodynamics. This situation should remind us of the first law of motion, which holds in the *ideal* circumstance, under which both friction and gravity are eliminated. The ideality of the first law of motion does not diminish its significance.

Carnot engine.

To understand the second law we must understand the basic properties of an engine. An engine, shown schematically in Figure 18.3, is a device which, in a *cyclic process*, extracts heat Q_h from a large hot reservoir at temperature T_h, converts some of it to work W, and dumps the rest of it Q_c in a cold reservoir at temperature T_c. Conservation of energy (18.1) implies that $\Delta U = Q_h - Q_c - W$, where both Q_h and Q_c are taken to be positive. The negative sign appears in front of Q_c, because Q_c is taken out of the system. Since the engine returns to its original state after one cycle, its internal energy will not change. So, $\Delta U = 0$, and we obtain $Q_h = Q_c + W$.

What is an engine?

The efficiency ϵ of the engine is the ratio of the work done by the engine to the amount of heat taken in (or the ratio of the useful work it produces to its energy intake):

$$\epsilon = \frac{W}{Q_h} = \frac{Q_h - Q_c}{Q_h} = 1 - \frac{Q_c}{Q_h}. \qquad (18.2)$$

It follows from this equation that a very efficient engine has a very low Q_c to Q_h ratio. How low can this ratio be? Can we make it zero and get an efficiency of 100%?

There are two obstacles to a very low Q_c/Q_h: practical and theoretical. The practical obstacles include such unwanted quantities as friction, viscosity, or any other dissipating forces, as well as loss of part of Q_h as it enters the engine. Such practical problems can in principle be minimized by lubrication, insulation, and other technological means. So, we can set aside the practical obstacles and concentrate on the theoretical ones.

Math Note E.18.2 on page 107 of *Appendix.pdf* shows how the theoretical limitation leads to the conclusion that for the most efficient (ideal, theoretical) engine, the change in the total entropy must be zero. When this happens, the engine is said to be **reversible**. The math note also proves:

Carnot's theorem.

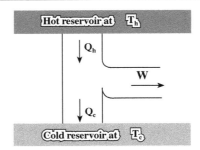

Figure 18.3: A schematic drawing of an engine. Every engine has a hot reservoir from which it extracts heat, and a cold reservoir into which it dumps heat.

Box 18.3.1. (Carnot's theorem) *No engine working between two heat reservoirs can be more efficient than a reversible (Carnot) engine working between the same two reservoirs. All Carnot engines working between two given reservoirs have the same efficiency of* $\epsilon = 1 - T_c/T_h$.

Sadi Carnot proved this in 1824 without any knowledge of entropy and its properties.

Carnot's theorem is one form of the **second law of thermodynamics** as applied to engines. There are other more "popular" versions of the second law. One is the law of the increase of the entropy encountered in Section 17.3.1, which Math Note E.18.2 used to prove Carnot's theorem. Another version is the so-called **Kelvin-Planck statement**: *It is impossible for an engine to have no other effect but to extract heat from a reservoir and to convert it completely into mechanical energy (work).* In other words, every engine must dump some of the heat it extracts. This is the heat that it cannot possibly convert into work. It can be shown that this statement is completely equivalent to Carnot's theorem.

Kelvin-Planck statement of the second law of thermodynamics.

What do you know? 18.8. Can you use Box 18.3.1 and the Kelvin-Planck statement of the second law to show that absolute zero is unattainable?

Refrigerator and heat pump.

A Carnot engine is, by definition, reversible. Therefore, it is possible to run it *backwards*. This means that all arrows in Figure 18.3 are to change directions. Such a device, with the help of some external work, takes away some heat Q_c from the cold reservoir and delivers $Q_h = W + Q_c$ to the hot reservoir. This is precisely what a **refrigerator** or a **heat pump** does. The notion of a refrigerator allows a third statement of the second law called the **Clausius statement** which can be shown to be completely equivalent to the other two statements: *It is impossible for a refrigerator to produce no other effect than the transfer of heat from a cold reservoir to a hotter reservoir.* A refrigerator, with the help of some work W, supplied by an electric motor, for example, removes heat from a cold reservoir (the freezer), and delivers heat to a hot reservoir (the kitchen). Clausius statement simply says that we cannot have a refrigerator without a motor (compressor).

Clausius statement of the second law of thermodynamics.

 Food for Thought

FFT 18.3.2. As if the first law of thermodynamics were not bad enough for the inventors of "free energy," there comes the second law, which ties their hands even more tightly.

Without the second law, one could imagine an engine which takes in a certain amount of energy and converts it completely into useful work. Now suppose that you take this work and feed it back *entirely* into the engine to do the work for you to feed into the engine to.... This process does not violate the principle of energy conservation. Of course, it is a useless machine because you cannot take any part of the work output without having to replace it. In fact, such a machine is undetectable, because to observe it, the machine has to give off *some* energy to the observer.

If the inventors of perpetual motion machines were claiming to have built simply a (undetectable) machine that runs forever, but does not yield any energy for human consumption, they would not be breaking the energy conservation law. However, their claim goes beyond that. They tell us that their machines can supply an infinite amount of energy *for us to use up*. This *does* violate the first law of thermodynamics.

The second law does not allow even an engine described above, because such a machine would convert the entire energy intake—what we called Q_h—into work W, violating the Kelvin-Planck statement of the second law of thermodynamics. The idea of machines that run on their own while performing *our* chores is nothing new. As early as 1618 a London physician named Dr. Fludd thought of running a mill by using the waterwheel not only to grind flour, but also to drive a pump, which pours water upstream in the millrace. Fludd's idea failed, of course, as well as all subsequent attempts. However, the pursuit of his idea led eventually to the principle of energy conservation.

Heat pumps have become popular as alternatives to other heaters. This is because a heat pump consumes much less energy than, say a space heater, to deliver the same amount of heat. It is, therefore, much cheaper to heat a room using a heat pump. Although more expensive than a space heater, heat pumps are more economical in the long run due to their low electric consumption. **Example D.18.3** on **page 42** of *Appendix.pdf* illustrates this discussion with some numbers.

18.4   End-of-Chapter Material

18.4.1   Answers to "What do you know?"

18.1. The expansion of material is an increase in *volume*. To make this increase noticeable, thermometer makers place the liquid in a thin tube, so that the increase gets translated almost entirely in a change in height.

18.2. If each particle has an average energy of $\frac{3}{2}k_BT$, and there are N particles, the total KE, which for ideal gas is the only energy, is $\frac{3}{2}Nk_BT$. So we write $U = \frac{3}{2}Nk_BT$ for an ideal gas.

18.3. If you add just heat (and no work) then ΔU is not zero. For an ideal gas U depends only on temperature. Therefore, a change in U means a change in the temperature of the ideal gas.

18.4. No! It is possible to add heat to a system and allow the system to do an equal amount of work. In such a process, the internal energy of the system does not change by the first law of thermodynamics. For an ideal gas, this means the temperature does not change either (see **What do you know?** 18.2).

18.5. No! The caloric theory stated that calorie is released only if you bore metals, i.e., only when you disintegrate objects. You are not disintegrating your hands when you rub them together. So, the caloric theory fails. However, the mechanical equivalence of heat easily explains it by allowing the mechanical energy of the motion of your hands to turn into heat.

18.6. No, period! No, not if you eat more than you exercise!

18.7. When you add the same amount of heat to A and B, the latter consumes more of it to excite its internal motion, leaving less for the KE of the molecules. Since KE determines the temperature, B will have a smaller rise in T than A. To have the same rise as A, B needs more heat. So, the specific heat of B is larger than A.

18.8. Maintain the cold reservoir at absolute zero, then the efficiency of Box 18.3.1 would be 1, violating the Kelvin-Planck statement of the second law.

18.4.2 Chapter Glossary

Absolute Zero The zero of the universal temperature scale.

Calorie The (old) unit of heat. It has been replaced with Joules, since heat is just another form of energy.

Constant-Volume Gas Thermometer A thermometer based on the physical principle that at constant volume, the pressure of a gas is proportional to its temperature. When the pressure of the working gas of the thermometer is reduced, the reading of the temperature becomes less and less dependent on the kind of gas used. This reading is the universal temperature of the substance being measured.

Engine A device which converts heat into mechanical energy.

Heat A form of energy associated with the overall mechanical energy of the particles making up a substance.

Heat Pump A device, which by using up some mechanical energy (work), extracts some heat from a cold reservoir (outside in winter and inside in the summer) and delivers some more heat to a hot reservoir (inside in winter and outside in summer).

Kelvin Temperature The scale of the universal thermometer, named after Lord Kelvin, who suggested the thermometer for scientific measurements.

Mechanical Equivalent of Heat (MEH) A conversion factor relating the old unit of heat (calorie) to the unit of energy, Joule.

Refrigerator A device, which by using up some mechanical energy (work), extracts some heat from a cold reservoir (inside of the refrigerator) and delivers some more heat to a hot reservoir (outside of the refrigerator).

Specific Heat The amount of heat needed to raise the temperature of one kilogram of a substance by one degree Celsius.

Universal Temperature The temperature scale derived from the constant-volume gas thermometer.

Universal Thermometer The constant-volume gas thermometer when the pressure of the working gas is extrapolated to zero.

18.4.3 Review Questions

18.1. What is the underlying property of substances used in constructing thermometers? Who invented the first thermometer, and what was used as it working substance?

18.2. What are the two "fixed points" used in designing thermometers? What numbers are given to these points in the design of thermometers using Celsius scale? Using Fahrenheit scale?

18.3. How does a constant-volume gas thermometer work? If you use two different gases in the construction of two constant-volume gas thermometers, do you get the same reading for the temperature of a single substance? Under what condition do you get the same reading?

18.4. What is the significance of extrapolating to zero pressure when using a constant-volume gas thermometer? Who suggested this extrapolated measurement of temperature as a universal scale? What is absolute zero?

18.5. In how many ways can you interact with a gas in a piston? State the first law of thermodynamics. Why should one introduce the notion of internal energy?

18.6. What is specific heat?

18.7. What is heat? What is the caloric theory of heat? What is the mechanical equivalent of heat? Who showed the equivalence of heat and energy most convincingly?

18.8. What does the first law of thermodynamics have to say about weight loss? Is it possible to lose weight by meditation?

18.9. What is the relation between energy conservation and time? What would a violation of energy conservation imply as far as time intervals are concerned? How weird would that be? You see how strong the principle of energy conservation is? Do you believe claims of generators that give you energy without burning any fuels?

18.10. Compare two substances: one, which is very simple and has very few internal structures; the other, complicated with a lot of internal structures. Which one has a larger specific heat?

18.11. What is an engine? What is the efficiency of an engine?

18.12. What is a Carnot engine? Can you find an engine that is more efficient than a Carnot engine? What is the efficiency of a Carnot engine? What does the second law of thermodynamics have to say about the efficiency of an engine? How did Kelvin and Planck state the second law of thermodynamics?

18.13. What is a refrigerator? A heat pump? How did Clausius state the second law of thermodynamics?

18.4.4 Conceptual Exercises

18.1. Keep you right hand in cold water for a while and then touch your own forehead. Does your right hand feel warm or cold?

18.2. You heat an iron ball until it is red hot, then place it in a glass of water. The water becomes hot (may even boil), while the iron ball cools down. What happens if you dip the iron ball in a swimming pool?

18.3. Which one has more internal energy, a small red hot iron ball or a swimming pool?

18.4. Suppose the internal energy of a system depends only on the temperature. If you perform some work on the system without any heat exchange, does the temperature of the system rise, fall, or remain the same?

18.5. Suppose the internal energy of a system depends only on the temperature. If you extract heat from the system without any work done on or by it, does the temperature of the system rise, fall, or remain the same?

18.6. What happens to the internal energy of a system if you add some heat to it, and at the same time let it do an equal amount of work for you?

18.7. Why does the presence of large bodies of water moderate the temperature of the land nearby? Hint: Look at Table 18.1 for specific heat of water.

18.8. Suppose you do absolutely no exercise. If you want your weight to stay the same, do you have to consume no food? Explain.

18.9. Show that the Kelvin-Planck statement of the second law of thermodynamics implies that no engine can be 100% efficient.

18.10. Why does a refrigerator need electric power to operate?

18.4.5   Numerical Exercises

18.1. What is 27 °C in Kelvin? What is human body temperature in K?

18.2. Each dietary calorie is 4186 J. A 60-kg person eats a cookie containing 100 calories. How high does he have to climb to burn up those calories? Does this sound reasonable? Are we missing something?

18.3. An engine burns 500 J of energy to give 100 J of useful mechanical work.
(a) What is the efficiency of the engine?
(b) How much energy does the engine exhaust to the environment?

18.4. An ideal engine operates between a cold reservoir at 30 °C and a hot reservoir at 500 °C.
(a) What is the efficiency of this engine?
(b) If the power intake of the engine is 2000 Watts, what is its power output? See Box 18.3.1 for help, and remember that T stands for Kelvin temperature.

Chapter 19

Epilogue: Whole and Parts

Thermodynamics is the only branch of physics motivated entirely by technology. Sadi Carnot, one of the founders of the subject, was a French army engineer interested in the efficiency of steam engines. His investigation led to the discovery of an ideal engine bearing his name, which is also connected to the second law of thermodynamics. However, this connection between physics and technology should not result in confusing the two. Ideas of science could start in a variety of ways. Astronomy was motivated by the religious desire of finding more about the gods in heaven. Archimedes started hydrostatics because of his interest in geometry of solids. Most ideas, however, come about simply because of the inner desire of the scientist to tackle an outstanding problem or an unanswered question.

19.1 Specificity and Idealization

Aside from the *art* of thermometry, which started with Galileo's invention of the first thermometer, the real science of thermodynamics began with the investigation of steam engines. In fact, for a long time, thermodynamics *was* the study of engines.

The second law of thermodynamics is a powerful universal law governing systems as small as a cup of water and as large as the universe itself. Yet it was discovered by examining a very *specific* system, an engine. Furthermore, this engine was not any real engine found in a mechanics shop; it was an *ideal* engine concocted only in the minds of physicists in the form of a schema such as the one shown in Figure 18.3.

Physicists have a natural instinct for idealization. Recall how Galileo idealized the blocks and inclines to discover the first law of motion; how Newton idealized the Earth-Moon system to discover the universal law of gravity; how Coulomb idealized two charged metallic spheres to discover the electrical force between them; and how the physicists working on the puzzle of magnetism idealized electric currents to an infinitesimal directional current element. Idealization is an inevitable prerequisite for discovering mathematical laws. And thermodynamics, with its emphasis on idealizing an engine, is no exception.

The very concept of **absolute zero** temperature is based on extrapolation and idealization. Although the process may cause a revulsion in an underdeterministic philosopher (see Section 5.3), the discovery of absolute zero was essentially based on connecting a few points and extrapolating the resulting line to regions beyond the reach of *any* experiments. In Figure 18.2(b), there are only 3 gases and 4 points per gas, making up 12 points altogether.[1] Yet these dozen points, when *extrapolated*, give us the absolute zero.

An underdeterminist would argue like this:

Idealization, the key to the discovery of absolute zero!

Highlights of the argument of an underdeterminist.

---

[1]In an actual experiment, you may have 4 or more gases and more than 4 points for each gas, but the total number of points will not be more than a few dozen.

Because there are only a finite number of points per gas (it does not matter whether you have 4 or 40,000 points per gas), the line connecting them is "underdetermined" and thus wrong! You cannot conclude that the line drawn through those points has any physical significance. How do you know what happens between any two adjacent points? Even if you fill the space between two adjacent points with thousands of other new points, how do you know what happens between two adjacent *new* points you just added? What is worse is the continuation of the lines into regions that *have* no points. This is doubly wrong! Physicists' conclusion about the absolute zero is wrong! The entire concept of absolute zero is wrong!

Physicists do not listen to underdeterminists, continue to use extrapolation and idealization, and discover universal laws, which make the human race a little wiser.

The first law of thermodynamics, possibly the most universal law of physics (when looked upon as the law of conservation of energy), was conceived by looking at specific cases such as the boring of the cannon or stirring water with a paddle. Joule, in his measurement of the mechanical equivalent of heat, used a specific substance (water), a specific method of heating it (paddles), and a specific form of mechanical energy (potential energy of a falling object). Yet he came up with the universal equivalence of heat (any kind of heat, not just the one produced by the paddles in a water bucket) and mechanical energy, and set the stage for the more universal concept of energy and its conservation.

Is the mechanical equivalence of heat wrong because Joule ignored some insignificant details?

Furthermore, Joule followed the other methodology of science: idealization. He was aware of the friction between the pulleys and their shafts and the heat produced by such a friction; yet he ignored it. He was aware of the air drag's effect on the motion of the pulleys, the rope attaching the pulleys to the paddle, and the paddle itself; yet he ignored it. He was aware that he was heating not only the water, but also the air surrounding it; yet he ignored it. Was he wrong in ignoring so many details? Some philosophers of science believe that exclusion of such details in a theory makes the theory wrong (see Section 20.2.2). But scientists not only think that Joule was right, but that science cannot be any other way! *Every* great discovery of science has taken place because certain inessential details were ignored. Scientists are quantifiers; they are interested in the *magnitude* of effects. Philosophers of science don't care about magnitudes; they care only about existence or nonexistence of effects.

Thermodynamic idealization does not end with the first and second laws. Many important *real* systems are treated as "ideal gases," and Equation (17.2) applies to them with great success. Of course, no real gas is ideal, just as no real surface was infinitely smooth in Galileo's derivation of the first law of motion (see Section 4.2). However, when the friction is small compared to other forces of interest, we treat a *real* surface as "infinitely smooth." Similarly, under certain conditions (conditions that were understood only after the development of statistical mechanics), a real gas can be treated as an "ideal gas."

19.2   Continuity

Carnot's derivation of his famous theorem (see Box 18.3.1) used the erroneous *caloric theory* of heat. However, Carnot's theorem itself is still valid and is a form of the second law of thermodynamics. After the discovery of the equivalence of heat and energy, Lord Kelvin *using Carnot's theorem* stated his more general statement (given on page 270), later refined by Planck.

The Kelvin-Planck statement of the second law envisioned an inevitable dissipation of useful energy. Rudolph Clausius picked up on the idea of the tendency toward energy dissipation and introduced the notion of entropy as the ratio of the heat absorbed (or given off) by a system divided by the system's absolute temperature. Using this definition of the entropy, Calusius showed that the dissipation of energy is equivalent to the increase in the entropy of a closed system.

We see, therefore, that the overarching statement of the law of the increase of entropy, which now appears to be completely detached from the specific (ideal) engine that Carnot studied, is linked step by step to that engine. What is even more remarkable is that the concept of entropy as a measure of the "messiness" of a thermodynamic system (a concept that could develop only through statistical mechanics) is *also* connected to Carnot's theorem.

19.3   A Case for "Fundamentalism"

The greatest impetus to the development of thermodynamics came when systems were assumed to consist of a large number of particles and the laws of motion and statistics were applied to it. Statistical mechanics has completely replaced thermodynamics and has made a fundamental understanding of the latter's concepts possible. For example, entropy which previously was conceived rather artificially by Clausius as the ratio of heat absorbed to the absolute temperature, was now defined as the (natural logarithm of the) number of accessible states. All processes involving various forms of entropy increase, including the transfer of heat and mixing, could now be explained with this single definition.

At the heart of statistical mechanics is the probability of individual constituents having certain physical properties. This probability was discovered by Boltzmann after whom it is called the *Boltzmann factor* (see Section 17.2.2). Once this probability is known, all the macroscopic (thermodynamic) properties of the system can be calculated. It is remarkable that the Boltzmann factor is given in terms of the physical quantities (mostly the energy) associated with the *individual* particles of the system. That is, the bulk property of the system is determined by the energy of a typical individual particle. For example, if the particles do not interact with one another, the only energy of a typical particle is its kinetic energy, and the system becomes an ideal gas. If there are inter-particle forces, then the system may show liquid and solid properties depending on its temperature. Even relativistic and quantum mechanical systems, with all their exotic behaviors, can be described by this simple Boltzmann factor.

> The physical properties of *individual* particles determine the bulk properties of the entire system.

The reduction of a thermodynamic system to its individual constituents may be philosophically unacceptable to those who think that "the whole is more than a collection of its parts." Nevertheless, statistical mechanics has shown that if we know the energy of each *individual* constituent of a thermodynamic system, we can predict the behavior of the entire system. This energy, of course, includes the *interaction energy* (the potential energy) between particles. But even the latter energy is given in terms of the properties (such as the location and the velocity) of a few (mostly two) particles. So although it is true that the parts *in isolation* cannot constitute the whole, once the interaction between pairs of particles is taken into account, the parts can determine the whole.

The deduction of the bulk properties from the properties of individual constituents is a proof that physics works best when physicists zoom their mathematical lens into the realm of the atoms and the subatomic particles. Nevertheless, some philosophers of science call this natural and rewarding process "fundamentalism," with all the negative connotations it carries. They chide those biologists who want to find the relation "between larger, richly endowed, complex [biological] systems, on the one hand, and fundamental laws of physics, on the other," and emphasize that they want to go beyond this: "Not only do I want to challenge the possibility of downward reduction, but also the possibility of 'cross-wise' reduction" [Pap 97, p. 316]. They ignore the fact that the tremendous progress in thermodynamics in the nineteenth century was the result of applying "fundamentalism" to its constituents. They are inattentive to all the progress in modern physics which is solely based on "fundamentalism." They brush aside the unprecedented advances in biology since the 1950s, when the "fundamental" unit of life, the DNA molecule, was unraveled.

> A philosopher of science attacks fundamental physics!

19.4 Thermodynamics and "Social Imagery"

The words in quotation marks in the title of this section are part of the title of an influential book written by a sociologist named David Bloor [Blo 83]. In it, Bloor proposes a "strong programme" designed to show that science, like any other human endeavor, is dependent on the social conditions under which it is developed. As a starting point, Bloor takes the undeniable fact that the popular ideas about the workings of the world vary from culture to culture and from time to time. For example, the cosmology to which Babylonians subscribed was different from Egyptian cosmology, and these two were different from the Greek and Roman cosmologies.

Bloor then takes the obvious fact that science has varied over time and developed by people of different cultural backgrounds, to *conclude* that science is *dependent* on the culture in which it has developed. That is, if relativity were discovered by a Japanese physicist, for example, it would take a different form and content than the one discovered by a German physicist! To see this, let's look at what Bloor is saying about mathematics on page 74 of his book:

> Everyone accepts that it is possible to have a relatively modest sociology of mathematics studying professional recruitment, career patterns and similar topics. This might justly be called the sociology of mathematicians rather than of mathematics. A more controversial question is whether sociology can touch the very heart of mathematical knowledge. ...By exhibiting the tactics that are adopted to achieve this end, I hope to convey the idea that there is nothing obvious, natural, or compelling about seeing mathematics as a special case which will forever defy the scrutiny of the social scientist. Indeed, I shall show that the opposite is the case.

At the beginning of his book, Bloor goes on to point out two examples relevant to thermodynamics, which we want to examine here. The first one is a fact that we have already discussed at the beginning of this chapter, namely that thermodynamics *started* with the investigation of steam engines. Bloor then equates the *motivation* for the discovery with the discovery itself, and ascribes the social conditions in which steam engines were developed to the science of thermodynamics. Thermodynamics becomes an enlightenment science! What Bloor fails to do is to mention that the *eventual* conclusion (what entropy is in terms of the number of accessible states of a thermodynamic system) is completely detached from its origin, the steam engines.

19.4.1 Kelvin's Mistake

The second example of Bloor's "strong programme" has to do with a mistake Lord Kelvin made in his calculation of the age of the Earth. It was not really a mistake, but insufficient knowledge of the workings of the interior of the Earth. Based on the two laws of thermodynamics and the theory of heat conduction, Kelvin estimated that it must have taken Earth 100 to 200 million years to cool down from an initial temperature of 4000–5000 °C to its present temperature. Furthermore, he estimated that the Sun could not have been illuminating the Earth for much more than a few hundred million years.

These calculations brought Kelvin face to face with Uniformitarian geologists, who believed that processes such as erosion had been continuing at the same rate for hundreds of millions of years. Kelvin's disagreement with these geologists was also an indirect disagreement with Darwin and the theory of evolution, because evolution required certain geological processes to have been going on for as long as 300 million years.

The debate between the physicists led by Kelvin and the geologists continued to the end of the nineteenth century. The puzzle of the *age of the Earth* created a bitter dispute between the physicists and the geologists, until the discovery of the nuclear radioactivity and the fact that the heat generated by the radioactive material in the core of the Earth

A sociologist claims that the content of science (even mathematics) is dependent on the social conditions under which it was developed.

is sufficient to compensate for the loss due to the conduction through the Earth's crust. Furthermore, nuclear radioactive processes gave the age of the Earth as over 3 billion years.

Kelvin, of course, was not the first physicist to make a mistake. The very process of discovery involves many trials and errors (mostly errors). Some of these errors may be more lasting than others. Classical physics lasted for over 200 years. And physicists believed in it, because it answered many questions, predicted many unexpected results, and seemed to work very well in all situations in which it was applied. The twentieth century brought about two realms in which classical theory failed: the realm of the small, and the realm of the fast. Physicists, faced with this failure, had to abandon the classical ideas and embrace the new ideas of quantum and relativity theories.

Scientists make mistakes before discovering the correct explanation!

As soon as radioactivity became a possible explanation for the source of heat generation, physicists abandoned Kelvin's estimate and agreed with the geologists that the Earth was much older than Kelvin had calculated. However, before the discovery of nuclear radioactivity, it was natural (and appropriate) for Kelvin and other physicists to hold on to theories that had been so deeply rooted in most phenomena of nature. It is essential that physicists do not give up their theories easily; that only on the face of hard observational evidence do they abandon their beliefs.[2] If physicists accepted any new "theory" simply because of the insistence of a certain group of people, then physics would no longer be a science.

19.4.2 Bloor's Opportunity

Kelvin's scientific blunder becomes a "social" opportunity for Bloor. Here is Bloor's assessment of this mistake:

> Kelvin's arguments caused dismay. Their authority was immense and in the 1860's they were unanswerable: they followed with convincing rigor from convincing physical premises. By the last decade of the century the geologists had plucked up courage to tell Kelvin that he must have made a mistake. This newfound courage was not because of any new dramatic discoveries, indeed there had been no real change in the evidence available. What had happened in the interim was a general consolidation in geology as a discipline with a mounting quantity of detailed observation of the fossil record. It was this growth which caused a variation in the assessments of probability and plausibility ... This example also serves to make another point. It deals with *social processes internal to science*, so there is no question of sociological consideration being confined to the operation of external influences. [Blo 83, p. 4] [Our emphasis.]

Notice the contradiction in Bloor's statement: "This newfound courage was not because of any new dramatic discoveries ... What had happened in the interim was a general consolidation in geology as a discipline with a mounting quantity of detailed observation of the fossil record." The "mounting quantity of detailed observation" is not to be considered as new discoveries!

Bloor is essentially saying that it was the *sociological* phenomenon of "courage" that prompted the geologists to (borrowing another sociological term) "stand up against" Kelvin. He ignores the very "consolidation in geology" as a result of "detailed *observation* of the fossil records" that he mentions in the passage. The reason that geologists "stood up" against Kelvin (if they ever did) was the only reason that scientists *can* stand up against a theory: *observation*. There is no doubt that social forces play a dominant role in the *development* of science and mathematics. The Dark Ages is a frightening example of how social structure can hamper the development of science for almost two millennia. But to say that social forces can alter science or mathematics *itself*—as Bloor is clearly implying about mathematics in the quote on page 278—is shocking.

---

[2]That the evidence presented by the geologists was not "hard" can be deduced from the fact that some supporters of Darwin's theory proposed that evolution did not depend on any particular timescale. Since there was no measure of the evolutionary rate, even 10 million years (rather than 300 million years) might be sufficient for the evolution to take place.

Part V

Twentieth Century Physics: Quantum Theory

Chapter 20

Birth of Quantum Theory

The beginning of the twentieth century seems to have been the converging point for the three main branches of physics: mechanics, electromagnetism, and thermodynamics. We have already seen how thermodynamics and mechanics unified to give birth to statistical mechanics, and how this unification not only explained mysterious quantities such as entropy, but also increased our understanding of the ultimate structure of matter.

The next convergence was that of thermodynamics and electromagnetism. After Maxwell showed that light was an electromagnetic wave, the interest in the study of the production of light grew rapidly. Intensifying this interest was the investigation of the spectra of stars, which seemed to provide important hints concerning stellar properties, such as their surface temperatures. Motivated largely by these two developments, physicists turned to a systematic exploration of the full spectrum of these waves, and discovered quite new forms of radiation.

20.1 Black Body Radiation

The interest in the connection between heat and radiation goes back to the very beginning of the nineteenth century when in 1800, William Herschel (1738–1822), the German-British astronomer, demonstrated that the Sun emits "invisible" radiation that is capable of heating objects. We now call this **infrared** radiation. To investigate the visible part of the spectrum, optical and photographic methods were used, with the latter being also useful for the ultraviolet part.

Infrared radiation discovered.

In 1896, Friedrich Paschen (1865–1947), extending earlier observations, showed that for objects emitting radiation in the infrared and visible regions of the spectrum, the wavelength corresponding to the maximally intense radiation decreased as the temperature of the radiating object increased. He also discovered an empirical formula which described the energy density distribution among the various wavelengths of the radiation emitted by a "perfectly black" object.

The notion of a perfectly black object or simply a **black body** was introduced by the German physicist Gustav R. Kirchhoff (1824–1887) who pioneered the systematic theoretical investigation of radiation due to heat in the late 1850s. At the end of 1859 Kirchhoff argued that at a given temperature and for a given wavelength, the ratio of the power of emission, denoted by e, to the power of absorption, denoted by a, is the same for all objects. A year later, Kirchhoff discussed the relation between e and a in great detail and he also introduced the concept of a completely black body which is capable of absorbing *all* radiation falling upon it. For such a black body, he concluded, the ratio e/a can depend only on the absolute temperature T, and the wavelength λ. He showed this dependence as $e/a = \Phi(\lambda, T)$. It was

Idea of black body introduced.

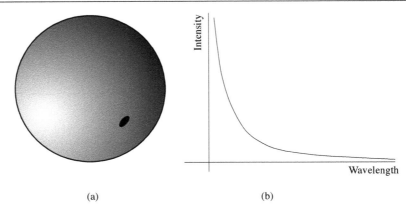

(a)   (b)

Figure 20.1: (a) The black body is the opening on the surface of the sphere. (b) Based on the classical EM theory, the intensity is suppressed for long wavelengths and enhanced indefinitely for short wavelengths.

for the function $\Phi(\lambda, T)$ that Paschen discovered an empirical formula in 1896. It was also this function that attracted the attention of most experimental and theoretical physicists of the latter part of the nineteenth century.

The physical interpretation of $\Phi(\lambda, T)$ is that if it is multiplied by the difference between two neighboring wavelengths, it gives the **energy flux** in that wavelength range. For this reason, $\Phi(\lambda, T)$ is called the **spectral energy flux**. *The flux of any physical quantity is the amount of that quantity crossing a square meter per second.* Thus, if we denote by $\Delta\lambda$ the difference between two wavelengths, one slightly larger than λ and one slightly smaller, then $\Phi(\lambda, T)\Delta\lambda$ gives the flux of the energy of all radiations whose wavelengths lie within the $\Delta\lambda$ range.

Opening in a cavity as black body.

A laboratory black body consists of a hollow metallic ball, called a **cavity**, with a small opening which allows electromagnetic waves to enter the cavity. This *opening is the black body* [see Figure 20.1(a)]. The EM waves incident on it will enter the cavity and get trapped there, because they keep getting reflected off the wall of the cavity with the chance to escape being proportional to the ratio of the area of the opening to the area of the entire cavity wall. The smaller the opening, the "blacker" the device. Almost all EM waves "shining" on the opening are absorbed, making it a very good approximation to the ideal black body. This same black body is also a radiator: Once the cavity is heated up, it gives off EM radiation. That is why an object that emits EM waves—including light—when heated up is also called a black body. It, therefore, should not come as a surprise to learn that the brightest object in the sky, Sun, is (almost) a "black" body. It turns out that

Sun approximates a "black" body.

Box 20.1.1. *Any object that emits EM waves solely due to heating is called a black body radiator. We say that a black body radiator is an object in which matter and EM waves are in thermal equilibrium.*

The most detailed study of a black body radiator involved the analysis of the *distribution* of the energy of the electromagnetic radiation among various wavelengths. It turned out that this analysis was completely equivalent to the determination of the function introduced by Kirchhoff. Since its understanding is crucial in appreciating the black body radiation, let us concentrate on the energy distribution of electromagnetic radiation.

What do you know? 20.1. Which one of the pair of spherical cavities on the right is a better approximation to a BBR? Which one on the left? Explain!

Suppose that a black body radiator, such as the Sun,[1] is held at a fixed temperature T. We can send this radiation through an ideal prism (as did Newton) to find all of its component "colors." In general, this color is not limited to the rainbow colors found by Newton. The ideal prism can also detect other colors, i.e., electromagnetic waves with wavelengths not included in the visible region of the spectrum. To get more information about the black body, we measure the intensity of various wavelengths: not only do we want to know what colors are emitted by the radiator at temperature T, but also, for example, how much brighter (or dimmer) is red compared to green or infrared or ultraviolet, etc.

How does classical EM theory explain the black body radiation? The size of the cavity in Figure 20.1 is sufficient to show the failure of the classical EM theory. The longer the wavelength of the wave, the harder it is to "fit" it in the cavity. If the size of the cavity is, for example, 50 cm, it does not accept wavelengths longer than 50 cm. On the other hand, any wavelength shorter than 50 cm should be easily accommodated, and the shorter the wavelength the more "room" for it in the cavity. So, long wavelengths are expected to be suppressed, and short wavelengths enhanced beyond limits in such a cavity [Figure 20.1(b)]. We shall see that this explanation contradicts the observations made at the end of the nineteenth century!

What do you know? 20.2. The spherical cavity on the right has a radius of 4 cm. The radius of the one on the left is 2 cm. Which one gives a more intense EM radiation with the wavelength of 1.7 cm? Explain!

20.1.1   Stefan-Boltzmann Law

In the last decade of the nineteenth century there was an intense search by many experimental and theoretical physicists to find an explanation of the intensity distribution of radiation among various wavelengths. The first theoretical attempts to derive the intensity distribution function were made by Ludwig Boltzmann and others. Boltzmann deserves particular attention because his method brought both electromagnetism and thermodynamics into the treatment of heat radiation. He showed in 1884 that electromagnetic radiation possesses pressure and, when combined with the second law of thermodynamics, it connects the energy flux J_e (energy radiated from each unit area of the radiator per second), sometimes called **brightness**, to the temperature by the formula $J_e = \sigma T^4$, where σ is a constant and T is the temperature of the radiator in Kelvin. σ has been measured accurately with a value of 5.67×10^{-8} in the scientific units we use in this book. So we write this equation as   Stefan-Boltzmann law.

$$J_e = 5.67 \times 10^{-8} T^4 \qquad (20.1)$$

and call it the **Stefan-Boltzmann law**, because the equation had been discovered empirically by Joseph Stefan in 1879. There is no wavelength in this equation because it gives the *total* energy flux with contributions coming from all wavelengths.

---

[1] As noted above, the Sun is not a "good" black body, but approximates it fairly well.

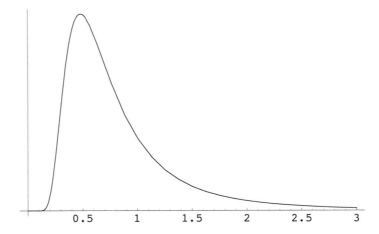

Figure 20.2: A typical black body radiation curve. The length of the wavelength at which the maximum occurs is inversely proportional to the temperature.

What do you know? 20.3. By what factor does the brightness of a BBR increase if you double its temperature?

20.1.2 Wien's Displacement Law

Black body radiation curve.

Wien's displacement law.

The next step was taken by Wilhelm Wien (1864–1928). Instead of the total energy density as given by the Stefan-Boltzmann law, Wien was interested in the distribution of energy among wavelengths. In a typical investigation, one plots the intensity of various wavelengths emitted by a black body as a function of wavelength. The resulting graph will look similar to the one plotted in Figure 20.2. This so-called **black body radiation curve** typically starts at zero, builds up to a maximum at a wavelength denoted as λ_{max}, and tails off to zero asymptotically for longer and longer wavelengths.

Wien's preliminary investigations and theoretical developments laid the foundation upon which Max Planck later erected the quantum theory of radiation. Working on the idea of the relation between the temperature, pressure, and the energy density of radiation as well as the Stefan-Boltzmann law, Wien was able to show that λ_{max} was inversely proportional to the Kelvin temperature of the black body radiator. His idea is written as

$$T\lambda_{max} = 0.0029, \tag{20.2}$$

and is called **Wien's displacement law**. It provides a method for determining the temperature of a (distant) black body such as a star. Plotting the curve of such a black body and reading off λ_{max} from the curve, you can determine the temperature of the distant radiator.

Example 20.1.2. Equations (20.1) and (20.2), discovered in terrestrial laboratories and offices of theoretical physicists, have a celestial ramification. Sun, radiating due to its hotness, is (almost) a black body radiator. Pass the sunlight through a good prism; measure the intensity of various wavelengths, and plot intensity as a function of wavelength to obtain a curve like Figure 20.2; read the wavelength at which the peak of the curve occurs; plug this in Equation (20.2) and find out that the temperature of the Sun's surface is about 6000 K.

Now use this in Equation (20.1) and find out that the surface of the Sun is as bright as 640,000 one-hundred-Watt light bulbs spread evenly on a table top! Look up the Sun's radius in an astronomy book and find out that it is 700,000 km. Use this to find the surface of the Sun in square meters; multiply by 640,000 to find how many one-hundred-Watt light bulbs the Sun's surface is equivalent to: about four trillion trillion!

Where does this power come from? What fuels this colossal release of energy? The answer to these questions is found (surprisingly) not in the cosmos, but deep within the atom: at its nucleus. Nuclear fusion converts mass to energy according to the famous $E = mc^2$. The mass depletion corresponding to the above power is over 4 million tons per second. Sun loses this much of its mass every *second* to shine. So, it will eventually die, but not when it loses *all* its mass. The combination of the nuclear and gravitational processes seal the fate of the Sun far ahead of its complete annihilation. Chapter 39 discusses these processes in some more detail. The interested reader may consult **Example D.20.1** on **page 42** of *Appendix.pdf* for some numerical embellishment. ∎

Sun loses over 4 million tons of its mass every second!

> **What do you know? 20.4.** Two BBR curves are plotted on the same graph. The wavelength at which the maximum intensity occurs for the hotter BBR is to the right of the colder BBR. True or false?

Wien's further theoretical studies of black body radiation resulted in separating the λ and the T dependence of the Kirchhoff function. By using statistical mechanics and applying the displacement law, he succeeded in expressing the Kirchhoff function as $\Phi(\lambda, T) = be^{-a/\lambda T}/\lambda^5$, where a and b are constants.

Wien's formula for Kirchhoff's function.

Food for Thought

FFT 20.1.3. "Are you telling me that you really believe that Sun's surface temperature is 5800 °K?," some people ask regarding the conclusion of Example D.20.1. "You can calculate numbers all you want. What convinces me," they say, "is the real physical measurement of this temperature."

The beauty and power of science is its interconnectedness. Seemingly unrelated phenomena may in fact be connected by the intertwining threads in the web of science. This interconnectedness not only holds the universal fabric of science together, but allows the scientists to constantly check the "strength" of the threads in the web. If one thread proves to be weak, the entire web may crumble. That is why the weak connections are constantly replaced by stronger ones. This is what we call "scientific progress."

We have already seen one instance of the interconnectedness of science in Example D.9.2: through a knowledge of the size of the Earth and two extreme possibilities for its constituent, we arrived at the upper and lower limits of the value of G, the universal gravitational constant. The precise measurement of G placed it somewhere in between. Now we offer another way of measuring Sun's surface temperature.

If Equation (20.1) [or Equation (20.2), since they are related] is correct and Sun's surface temperature is 6000 K, causing it to pour out the power equivalent of 4 trillion trillion one-hundred-Watt light bulbs, or 3.94×10^{26} Joules of energy every second in all directions, then any sphere centered at Sun should collect *all* this energy, and by dividing this energy by the area of that sphere (which is 4π times the square of its radius), we can calculate how much energy is collected by every square meter of that sphere. We, homosapiens, happen to be located on a sphere centered at the Sun with a radius of 1.5×10^{11} m having an area of 2.83×10^{23}. Therefore, the energy collected by each square meter of this sphere is $\frac{3.94 \times 10^{26}}{2.83 \times 10^{23}} = 1393$ Watts/m^2. Since Earth is located on this sphere, *it* should also collect this much energy per *its* square meter. The amount of solar energy hitting every square meter of the Earth's surface (above the atmosphere) has been *measured* and given the name **solar constant**. Its measured value is 1400 Watts/m^2! Of course, there are those, such as creationists and intelligent designers, who are not convinced by any evidence short of carrying a thermometer to the Sun and recording the reading on the thermometer!

20.2   Quanta Are Born

Max Planck enters the scene.

Max Planck (1858–1947) was aware of Wien's formula, and he knew how well it agreed with experiments. However, he did not like the way it was derived; he was after a new way and a new angle of treating the subject of radiation and thermodynamics and a more convincing derivation of that formula. Using the idea of *entropy*, which Planck thought to be fundamental, he re-derived Wien's equation and submitted his derivation to *Annalen der Physik* in November 1899. Whether his derivation was indeed more rigorous (as he claimed) or not is a matter of mathematical taste. Of more importance was that his method of derivation allowed the possibility for generalization to a new formula, whose discovery was indispensable for the creation of the quantum theory.

20.2.1   The Equation of a Curve

New experiments disagree with Wien's formula.

As the derivation of Planck was being printed, there was already experimental indications that Wien's formula was wrong. As early as February 1899, Otto Lummer (1860–1923) and Ernst Pringsheim (1859–1917), having invented a new method of accurate measurement, reported discrepancies between theory and their observations to the German Physical Society in Berlin. To eliminate experimental errors, they worked on the improvement of their methods in the following months. Later in the year, in the November meeting of the German Physical Society, Lummer again spoke of the discrepancies, this time with more confidence.

In the ensuing months there was a fervor of experimental and theoretical activities targeted at Wien's formula. Many theorists tried to generalize it so that it would agree with the recent results. To respond to such attacks on his formula, Wien scrutinized it more carefully and concluded, through thermodynamical arguments, that the *formula must be valid only for short wavelengths* and not necessarily for long wavelengths.

Reader, don't be alarmed by the appearance of Equations (20.3) and (20.5). We have introduced them as symbols not as mathematical equations necessary for understanding the discussion.

In the meantime, progress was being made at the other end of the spectrum both experimentally and theoretically. On October 7, 1900, Rubens, one of Planck's experimental colleagues, visited Planck at the latter's house and reported his recent findings concerning the long-wavelength behavior of black body radiation. He also mentioned the agreement between his observations with a formula derived by the British physicist Lord John W.S. Rayleigh (1842–1919) in June. Upon receiving this news, Planck immediately set out to try to connect the two ends of the black body spectrum. In this respect, he was at a great advantage over others, because, as mentioned earlier, his derivation of the Wien's law offered a much easier path to generalization. He could thus incorporate both Wien's formula and Rayleigh's formula in a single equation involving entropy. This equation gave the entropy equivalent of Wien's law when the product λT was small, and that of Rayleigh's law when λT was large. This entropy formula could now be translated into a formula for the spectral energy flux, $\Phi(\lambda, T)$. The historic result was

$$\Phi(\lambda, T) = \frac{b}{\lambda^5} \frac{1}{e^{a/\lambda T} - 1}, \tag{20.3}$$

Planck finds the fit to the black body radiation curve.

where a and b are constants similar to the ones in Wien's formula. One of Planck's students has reported on these historic events: "The same evening [Sunday, October 7, 1900, when Rubens visited Planck] he [Planck] reported this formula [Equation (20.3)] to Rubens on a postcard, which the latter received the following morning ... One or two days later Rubens again went to Planck, and was able to bring him the news that the new formula agreed perfectly with his [Ruben's] observations." Planck presented his new formula under the modest title *An Improvement of Wien's Spectral Law* to the German Physical Society on Friday, October 12, 1900. All the subsequent observations and precise measurements, up to the present, point irrefutably to the correctness of Planck's black body radiation formula. As great an achievement as it was in itself, the formula, to Planck, was only a small window to the esoteric world of radiation.

20.2.2 "Fundamentalism" Offers Quantum to Humanity

The pursuit of perfection and the emphasis on details are qualities found only in the best of us. It is also a driving force that pushes our civilization onto ever higher grounds. When Beethoven insists that a single note in a symphony of a few thousand notes should be C sharp and not D, when Van Gogh spends days on a single spot of his canvas, experimenting with many different hues of a particular color to which most ordinary eyes are insensitive, and when Dickens turns and tosses in a sleepless night trying to find the perfect word or phrase to put in his novel, in short, when the builders of our civilization go out of their ways to exalt their creation, they are instinctively pouring out the best of what sets humans apart from beasts.

In no other discipline is the effect of this pursuit more dramatic than in science. Any advancement in science opens a new door to new details and new questions. These questions in turn demand new experiments and technology, causing further advancement in our theoretical understanding of the subject, thus creating a new stage of the advancement, and a new cycle of evolution of scientific thought. With each new stage of scientific evolution, we get one step closer to the ultimate demystification of the universe, thus, one step farther in using our knowledge to better our lives.

One of the most momentous instances of this pursuit, upon which our great heritage of quantum physics is based, is Planck's attempt at unraveling the mysterious formula, Equation (20.3). The discovery of a formula that agrees perfectly with experiments is a Herculean feat for any human being. It requires a truly dedicated scientist, however, to go beyond the formula and uncover no less a riddle than the riddle of the quantum world. In deriving Wien's formula, Planck had used certain well-established thermodynamical principles. On the other hand, Wien's formula was based on the generalization—therefore, modification—of those very principles. Planck thus concluded that either those principles had to be changed or new assumptions must be made in the process of his derivation.

Food for Thought

FFT 20.2.1. The very insistence that leads to fundamental breakthroughs such as the discovery of quanta is attacked by some philosophers of science. Borrowing a detested political label, they call this pursuit of details "fundamentalism" and proclaim that "we [the philosophers] need to attack" it [Pap 97, p. 314]. Then they go on to assault all fundamental laws of physics, including Newtonian gravity. Their artillery is the following argument: Newtonian physics cannot predict where a 1000-dollar bill swept away by the wind will land; therefore, Newtonian theory is wrong. And by the same token *all* fundamental laws are wrong.

Quanta, "fundamentalism," and a philosophy of science.

Fundamentalism and idealization (see Food for Thought 4.2.2 and Section 5.2.4) come hand in hand: the fundamental laws of physics, as a rule, describe ideal situations. The first law of motion, as stated by Galileo, applies to *frictionless* surfaces. The second law of motion finds the acceleration of a *point* particle when the applied force is known. All the laws of electricity and magnetism apply to *ideal* point charges and straight *infinitely thin* currents. We cannot deny the first law simply because we cannot find a frictionless surface. And the second law and all the laws of electricity and magnetism hold even though we cannot create a point particle, a point charge, or an infinitely thin wire.

Newton's law of gravity applies to objects that are under the influence of *only* gravity. Given that, the law predicts remarkably well the motion of the planets around the Sun, the Moon around the Earth, a satellite, a projectile, and ballistic missiles (see Section 10.2.2 for some miraculous achievements of Newtonian physics). It can even predict the landing of the dollar bill: in the absence of the atmosphere, the dollar bill lands straight down in a uniformly accelerated motion with an acceleration of 9.8 m/s². We summarize this by saying "in the absence of the air, the dollar bill falls on a straight line under gravity."

Sun does not exist because its rays cannot penetrate clouds!

Now consider another physical process. Due to thermonuclear reactions, invisible light in the form of gamma ray is produced at the core of Sun, which travels to the surface losing energy and becoming visible. Then it leaves the surface of Sun and travels outward in all directions. In the absence of clouds, some of this light reaches the surface of the Earth and is caught by our eyes. We summarize this by saying "Sun shines." If we challenge the accuracy of the statement "the dollar bill falls on a straight line under gravity" because of the presence of the atmosphere on Earth, then we have to challenge the accuracy of the statement "Sun shines" because of the presence of of clouds in the sky!

This kind of philosophy is contaminated with the same kind of pragmatism that held back science for almost 2000 years under the Roman ideology. If we deny the laws of physics because the conditions for *their absolute validity* are never met *in real-life* situations; if we deny Newtonian physics because it cannot predict the random motion of a 1000-dollar bill swept away by the wind, then we are following the path of the ancient Romans who disregarded the beautiful Greek science developed over 300 years because they could not find a real-life situation in which it was directly applicable.

Planck had used purely thermodynamical reasoning to reach Wien's formula; and until the discrepancy with observation was discovered by Rubens, he was completely satisfied with this approach. However, there was another school of thought, led by Ludwig Boltzmann, which advocated the statistical-mechanical explanation of thermodynamics. Planck did not belong to this group, and, in fact, he was opposed to such a line of thinking. However, after the success of Equation (20.3), and the realization that his cherished thermodynamical principles might be at stake, he leaned more and more towards statistical mechanics. In his Nobel lecture in 1920, Planck recalled this proselytism: "This problem ... [did not lead] me automatically to a consideration of the connection between entropy and probability, that is, to Boltzmann's trends of ideas, until after some weeks of the most strenuous work of my life, light came into the darkness and a new, hitherto undreamt of perspective began to open up for me."

20.2.3   A Fundamental Discovery

Planck suggests quantization of EM radiation.

This "undreamt of perspective" turned out to be a perspective for all humankind, a radical departure from every scientifically "sacred" hypothesis, a toe-deep submersion into the ocean of the microworld. This perspective was the principle of the quantization of EM radiation. Planck reached this principle in mid-November of 1900, but presented it publicly at the meeting of the German Physical Society in Berlin on December 14, 1900. On this day the quantum theory was born.

BIOGRAPHY

Max Planck (1858–1947) was born into an academic family, his father, his grandfather, and his great-grandfather all being professors. He attended the famous Maximilian Gymnasium in Munich, where music was perhaps his best subject, but because of the encouragement of his physics teacher, he turned to physics and mathematics at the end. In July 1874, at the age of 16, Planck entered the University of Munich and took some courses in physics and mathematics. When he enquired about the prospects of research in physics, the professor of physics at Munich told him that physics was essentially a complete science with little prospect of further developments. However, 21 years after receiving his doctorate with a thesis on the second law of thermodynamics, Planck showed that physics was just beginning with his quantum theory of 1900.

Boltzmann's formulation of statistical mechanics showed that the higher the energy of a "particle" in a thermodynamic substance, the fewer the number of such particles. **Math Note E.20.2** on **page 109** of *Appendix.pdf* pursues this line of reasoning and concludes that

- a black body radiates EM waves in the form of "particles" or bundles or **quanta**; and

- the energy of each **quantum** is proportional to its frequency.

If, as is customary, we denote the constant of proportionality by h, the relation between energy of a quantum of EM radiation and its frequency (or wavelength) can be written as

Planck relation.

$$E = hf, \quad \text{or} \quad E = \frac{hc}{\lambda}, \tag{20.4}$$

where h is called the **Planck constant**, and has a value of $h = 6.626 \times 10^{-34}$ J·s.

Example 20.2.2. Let us look at the energies associated with some familiar EM waves. X-rays have a wavelength of about 10^{-10} m. So, the energy of their quanta is the product of the Planck constant and the speed of light divided by 10^{-10}. This gives 1.99×10^{-15} J. It is more common to express such small energies in terms of electron volt (see Example 12.3.2), which is 1.6×10^{-19} J. The energy of an X-ray is thus $1.99 \times 10^{-15}/1.6 \times 10^{-19}$ or 12437.5 eV or about 12 keV (kilo eV).

Similarly, violet light, with a wavelength of 4×10^{-7} m has an energy of 4.97×10^{-19} J or about 3 eV, while the energy of red light, with a wavelength of 7×10^{-7} m is 2.84×10^{-19} J or about 1.77 eV.

Going to the long-wavelength part of the EM spectrum, microwaves, with a wavelength of 1 cm have an energy of 1.99×10^{-23} J, or about 0.00012 eV, and radio waves with a wavelength of 1 m have an energy of 0.0000012 eV. ∎

With the assumption of the quantization of energy of EM radiation, one can derive (20.3) and express a and b in terms of more fundamental quantities. In fact, the spectral energy flux function of Equation (20.3) turns out to be of the form

$$\Phi(\lambda, T) = \frac{2\pi hc^2}{\lambda^5} \frac{1}{e^{hc/\lambda k_B T} - 1}, \tag{20.5}$$

where h, c, and k_B are, respectively, the Planck constant, the speed of light, and the Boltzmann constant. Planck's BBR formula (20.5) leads immediately to the Wien's displacement law and the Stefan-Boltzmann law. The interested reader can find the details in **Math Note E.20.1** on **page 108** of *Appendix.pdf*, which requires some understanding of calculus.

⚙ Food for Thought 🏺

FFT 20.2.3. The Carcinogenic EMF? The insatiable appetite of the public for sensationalism, and the equally insatiable appetite of the news media for attaining top ratings, have had a devastating effect on science and rational thinking. The case in point is the scare generated by some "investigative" reporters, who "discovered" that electromagnetic fields can cause cancer.

Nancy Wertheimer was an unemployed epidemiologist, who in 1979 drove around Denver looking for the "cause" of leukemia in children.[2] She noticed that many of the addresses of the victims of the disease were near power transformers. She sought the advice of a physicist named Ed Leeper, and together they reported that "high" magnetic fields were the cause of childhood leukemia.

Despite its flaws, the Wertheimer-Leeper report could not be dismissed. The magnitude of our exposure to EM waves is so enormous that even the slightest connection with cancer was cause for concern. Suddenly, there were reports of the growth of cancer rate among electrical workers; of the increase in miscarriages among women working at the computers; of farmers with power lines in their fields reporting that cows no longer gave milk, and chickens laid no eggs. Although none of these stories was substantiated by reliable statistical evidence, each new anecdote added fresh fuel to the inferno.

The current induced in our body is proportional to the frequency of the alternating magnetic field.

As mentioned in Example D.14.1, the energy of the EM wave is of no concern. However, the fact that the magnetic field is *alternating* may require some attention. It turns out that an alternating magnetic field can create a small electrical current in the body due to Faraday's law of induction (see Section 13.4.1). A consequence of this law is that the induced current is proportional to the frequency of the alternating magnetic field. Could these small currents somehow interfere with the body's defense mechanism against cancer?

---

[2]Most of this Food for Thought is adapted from *Voodoo Science*, by Robert Park [Par 02].

In June of 1989, the *New Yorker* ran a three-part series of articles on the hazards of power-line fields. The author of these articles was a veteran reporter named Paul Brodeur, who began his journalistic career exposing dark secrets of the CIA, and then switched to reporting on environmental and occupational hazards in 1968. His specialty, however, was microwaves.

A by-product of war technology,[3] microwaves were especially prone to suspicion by Brodeur as the "waves of terror," because most of the research on their biological effects had been supported by the Department of Defense, which concluded that microwaves were safe. When industry scientists reported similar findings, Brodeur saw it as a collusion of the electronics industry with the military. When scientists at universities dismissed any hazard associated with microwaves, he enlarged his umbrella of conspiracy to cover the academic scientists as well.

So, when the Wertheimer-Leeper report attracted the attention of the public to the hazards of the power-line "fields," there was no doubt in Bordeur's mind that the new hazards were of exactly the same nature as his evil microwaves regardless of the fact that microwave frequencies are about 100 million Hertz while the power-line fields had a frequency of only 50–60 Hertz!

The power-line panic soon spread among the public like an epidemic. In response to the environmental groups, many governments instigated thorough studies of the controversy. In 1994 a four-year study of 223,000 Canadian and French electrical workers reported that there was no increase in cancer risk associated with occupational exposure to EMF. A year later an even larger study of US electrical workers concluded that no increased risk of cancer was evident among the workers. In fact, both studies found the cancer rate among electrical workers to be *lower* than the general population. (But no scientist will conclude that EMF cures cancer; although one may ask why Wertheimer and Leeper don't!)

The bias inherent in the "study" of an investigative reporter guided by a mission can associate the effect with the cause of her choice. But an unbiased scientific investigation, in which the strength of the quantities in question are measured and the data are carefully analyzed, reveals a different outcome. As the chair of the review panel of the National Academy of Sciences summarized its exhaustive three-year study, "The question is what causes the association [with childhood leukemia].... We don't know," but he pointed out that neighborhoods with heavy concentration of power lines are usually poor, congested, and polluted—all of which are risk factors for cancer.

Partly due to such sensational journalism, there have been hundreds of lawsuits filed by cancer patients blaming EM radiation for their illness. And there have been thousands of "documentaries" aired by the sensationalist TV industry linking cancer to EM waves. In all these cases, there are of course "experts" who testify to the "scientific" validity of the claims, citing "researches" and "studies" to prove their points.

In a recent case, a Maryland neurologist claimed that his brain cancer was caused by cell phone use, and he asked 800 million dollars for damages. His claim rested on research by a Swedish oncologist, who published a study in the *European Journal of Cancer Prevention* that found long-term users of analog cell phones were at least 30% more likely to develop brain tumors than nonusers. This study was extensively reported by media. However, a review of epidemiological research on cell phone use, commissioned by the Swedish Radiation Protection Authority, described the study as "non-informative" and concluded that "there is no scientific evidence for a causal association between the use of cellular phones and cancer." This report was not mentioned by any popular media outlet, but was sufficient to dismiss the Maryland case.

It is well known that all cancer-causing agents act by breaking chemical bonds to produce mutant strands of DNA. Such chemical bonds require energies of about an electron volt. As we saw in Example 20.2.2, microwaves used in cell phones have an energy of 0.00012 eV, several hundred times weaker than the energy needed to break a DNA bond! This means that there is nearly as much chance to break a DNA bond with microwaves—and thus cause cancer—as to kill a man by throwing cotton balls at him!

The cost of the power-line scare is $25 billion!

The White House Science Office has estimated the total cost of the power-line scare, including relocating power lines and loss of property values. It is a staggering $25 billion! Was there not better venues of spending this money? Would it not have been better to fund the research aimed at *truly* fighting cancer? As long as the citizenry cannot differentiate between the rational cause for concern and the sensational panic howled by a biased reporter, such enormous plunders of resources are inevitable.

---

[3]Microwave ovens were introduced first as "radar ranges."

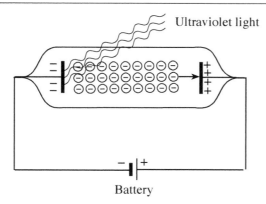

Figure 20.3: When UV light shines on the plate on the left, electrons are released, moving to the right and establishing an electric current.

20.3   Photoelectric Effect

Planck had much difficulty convincing (first himself, then) the physics community of the validity of his new revolutionary ideas. Many of the older generation of physicists resisted the idea of quantization of energy. But the emerging new generation embraced the idea and developed it to its fullest capacity. One member of this generation was a clerk at the Bern patent office. His name was Albert Einstein.

The year 1905 was the most productive year in the life of Albert Einstein, and possibly of any scientist in history. In that same year, he discovered the special theory of relativity, unraveled the Brownian motion (the random motion of objects under microscopes caused by its bombardment by the molecules of the solution), and explained the **photoelectric effect**.

When Heinrich Hertz was testing the success of Maxwell's prediction of EM waves, he was simultaneously paving the way for the antithesis of the wave theory, namely the particle theory of EM radiation. He realized that when the surfaces of the metallic balls used in producing sparks were polished, more sparks were produced. Later it was realized that the ultraviolet component of the spark light was capable of releasing negative charges (subsequently identified as electrons) from the clean surface of metals—thus the name, photoelectric effect.

The *existence* of the photoelectric effect could be explained by the classical wave theory of light. After all, it is conceivable that by shining some *intense* EM radiation on a metallic surface, one could provide enough energy to the electrons inside the metal to "jump" outside it. However, it was the *details* of this phenomenon that required a drastic shift in our view of the nature of light. The details that were subsequently discovered are as follows. Consider the *photoelectric cell* of Figure 20.3 consisting of two plates in a tube, from inside of which air has been pumped out (thus called a **vacuum tube**).

- Shine light of a given frequency on the left plate (also called the *cathode*), and observe that a current has been established, indicating the emission of electrons from the plate. If you increase the intensity of that light you'll observe an increase in the current, indicating the emission of a larger number of electrons from the plate. This is what is expected classically.

- The electrons ejected from the cathode have a kinetic energy (KE) right after ejection. This KE varies from electron to electron, but it is easy to measure the *maximum* energy of the electrons. If we reverse the polarity of the battery, then not all electrons will make it to the right plate. In fact, if the voltage of the reversed battery is too high, there will be so much negative charge on the right plate—and, therefore, so strong a repulsion on the incoming electrons—that no electrons will reach the right plate (and no current detected). So, if you start at a low voltage, and increase it slowly, you will reach a threshold voltage at which all the electrons will be stopped. This is called the **stopping potential**. Denote it by V_{stop}, and note that it is equal to the potential difference between the two plates. At this potential, the electrons with maximum KE will just barely make it to the right plate. Conservation of energy then implies that eV_{stop}, where e is the charge of the electron, must equal KE_{max}.

Classically, we expect that increasing the intensity of the incident light should increase KE_{max}, so that to completely stop the flow of electrons, the stopping potential should be increased as well. Not so! Observation shows that no matter how intense the incident light, the stopping voltage remains constant.

- How does V_{stop} depend on frequency? Classically, we expect no dependence on frequency, because the energy of a wave does not depend on its frequency. Experiments, however, indicate that V_{stop} is very much frequency-dependent, and that there is a minimum frequency f_0, below which—regardless of the intensity—no electrons are emitted, and no voltage is necessary to stop them (so that $V_{\text{stop}} = 0$ for this frequency). It is also found experimentally that f_0 varies from metal to metal.

- Because the relevant energy-related quantity for an EM wave is the energy flux [see Equation (D.14)], i.e., the amount of energy incident on a square meter of the metal per second, classically, we expect to impart more energy to the electrons the more intense the light and the longer we let it shine on the plate. Experiments completely contradict this conclusion. If the frequency of light is below f_0, no matter how intense the light and how long you wait, no electrons are ejected. On the other hand, once the frequency is increased above f_0, even the weakest light beam produces photoelectrons *instantaneously*.

All these puzzles were solved in a 1905 paper by Einstein entitled humbly, "On a Heuristic Point of View Concerning the Generation and Transformation of Light." In solving the riddle of photoelectric effect, Einstein assumed the following:

- Light (or more generally, any EM wave) consists of bundles or particles, which were later called **photons**. A photon of frequency f, or wavelength λ, carries an energy $E = hf = hc/\lambda$, as given by Planck.

- Electrons are bound to the photoelectric metal by an energy W, called the **work function**, which varies from metal to metal.

- If a photon hits one (and only one) electron, and imparts all its energy to that electron, then the electron acquires the maximum possible KE. Those electrons that receive less than the entire energy of a photon will have a smaller KE. Before being able to impart *any* KE to an electron, a photon must first overcome the binding energy with which the electron adheres to the metal.

Thus, basing his argument simply on the conservation of energy, Einstein came up with the photoelectric equations, which are given—along with the illustration of a couple of their uses—in **Example D.20.2** on **page 43** of *Appendix.pdf*.

What do you know? 20.5. Can you think of the vacuum tube of Figure 20.3 as a switch? Can you find an application for it?

20.4   End-of-Chapter Material

20.4.1   Answers to "What do you know?"

20.1. In the pair on the right, the cavity with a smaller hole is a better approximation to a BBR, because smaller hole gives less chance to the EM radiation inside to escape. In the pair on the left, the larger cavity is a better approximation to a BBR, because the trapped EM wave has more volume in which to remain trapped.

20.2. The wavelength of 1.7 cm is comparatively shorter for the bigger cavity than for the smaller one. Hence, the intensity of this wavelength from the bigger cavity is larger.

20.3. By a factor of 16.

20.4. False. The hotter BBR has a shorter $\lambda_{\max}$, and shorter means closer to the origin, i.e., to the left.

20.5. It *is* a switch; a light-sensitive switch: Shine light on it, it turns on; turn off the light, it turns off. It has a wide range of applications: all automatic doors and traffic lights use it.

20.4.2   Chapter Glossary

Black Body An idealized black object, one that absorbs 100% of the EM waves that fall on it. Being a perfect absorber, it is also a perfect radiator.

Brightness The flux of electromagnetic radiation energy.

Flux (of a physical quantity) The amount of a physical quantity crossing a square meter per second. Flux is a local quantity: Pick a point; pick a small loop perpendicular to the motion of the quantity; see how much of that quantity passes through the loop in one second; divide this by the area of the loop. That's how you find the flux of the quantity in question.

Photoelectric Effect An effect whereby electric current is emitted from the surface of certain metals when light shines on that surface.

Photon The particle of light.

Quantum The smallest unit of electromagnetic radiation carrying an amount of energy proportional to its frequency.

Wien's Displacement Law A law stating that the wavelength at which the maximum of a black body radiation curve occurs is inversely proportional to the temperature (in Kelvin).

20.4.3   Review Questions

20.1. Who discovered the infrared radiation? What did Friedrich Paschen discover?

20.2. What is a black body? Who introduced the notion of a black body? What did he discover concerning the ratio of the emission to absorption coefficients for a black body? What is the name of this ratio? What physical quantity does it measure?

20.3. Describe something that approximates a black body in the laboratory. Define a black body radiator.

20.4. What is the classical EM theory prediction for the behavior of a black body radiator? Based on the laboratory model of a BBR, can you anticipate such a prediction?

20.5. What is the sole physical quantity that the brightness of a BBR depends on? Who discovered this dependence?

20.6. What happens to the wavelength at which the maximum intensity of a BBR occurs when you increase the temperature of the BBR? What is the name of the law that governs this relation?

20.7. How, in principle, can you use Wien's displacement law to measure the surface temperature of a BBR?

20.8. What was the first step taken by Max Planck in the creation of the notion of quanta of the EM waves? Whose formula was he deriving? What did Planck do once he found out that Wien's law was not accurate? How well did Planck's new formula agree with observation? What was the title of Planck's talk at the German Physical Society that ushered us into the age of the quanta?

20.9. Why is "fundamentalism" so important in physics? Would Planck have come up with the quantum idea had he not gone after the "fundamental" reason behind the perfect agreement of his formula with observation? What was the role of statistical mechanics in reaching the quantum idea? Did Planck use statistical mechanics willingly?

20.10. State the fundamental quantum idea which Planck discovered. How are the EM waves treated now? What is the typical energy (in electron volts) of a visible quantum?

20.11. What were Einstein's three contributions to physics in 1905? Which contribution was related to Planck's idea of EM quanta? Can classical EM theory explain the photoelectric effect?

20.12. What is the stopping potential? How is it related to the maximum KE of the electrons emitted from the cathode?

20.13. From a classical point of view, do we expect the stopping potential to depend on the frequency? What does observation indicate?

20.14. Is it true that the classical theory predicts more energy for the emitted electrons the longer one shines light on the cathode? Explain!

20.15. How did Einstein solve the photoelectric puzzle? What was his basic assumption? What word did he use in place of quanta or bundle? What is work function?

20.4.4   Conceptual Exercises

20.1. The cube of side 2 cm in Figure 20.4 has a small hole that acts as a BBR. Consult the EM spectrum diagram in the textbook and answer the following questions *based on classical EM theory.*
(a) What is the longest wavelength category that can be fitted in the cube?
(b) Do you expect the radio waves in the BBR emitted by this cavity to be intense?
(c) Do you expect the intensity of UV waves in the BBR emitted by this cavity to be large or small?
(d) Which of the two waves do you expect to be more intense, IR or X-rays?
(e) Which of the two waves do you expect to be more intense, gamma rays or violet light?

20.2. Why are the pupils of our eyes black? Are our eyes colored black inside?

20.3. Which one is "blacker," a 1 cm hole on the surface of a 2 m spherical cavity, or a 2 cm hole on the surface of a 1 m spherical cavity? All numbers refer to radii.

20.4. As you cool a BBR more and more, the peak of its curve moves further and further to the right. True or false?

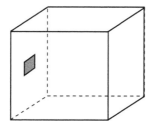

Figure 20.4: The hole in the cube can also act as a BBR.

20.5. Which EM wave has more energy, infrared or ultraviolet? Ultraviolet or gamma rays? Radio waves or visible light?

20.6. A hypothetical metal has a minimum frequency f_0 corresponding to green light. You shine orange, yellow, blue, and violet light on it. Which of these colors eject photoelectrons?

20.7. Suppose you shine orange, yellow, blue, and violet light on a metal whose minimum frequency f_0 corresponds to red light. Which of those colors eject photoelectrons?

20.8. If you shine orange, yellow, blue, and violet light on a metal whose minimum frequency f_0 corresponds to ultraviolet light, which of the colors eject photoelectrons?

20.4.5   Numerical Exercises

20.1. Figure 20.5 shows a black body radiation curve of a star in which the wavelength is given in units of μm.
(a) From the graph, read off the (approximate) wavelength corresponding to the maximum intensity.
(b) What is the surface temperature of the star?
(c) What is the brightness of the star?
(d) How much power (joules per second) is given off by the star if its radius is 1.5 million km?
(e) If the source of this power is conversion of mass into energy, how much mass is the star losing per second?

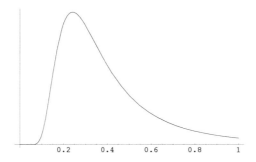

Figure 20.5: The BBR curve of a star. The values of the wavelength on the horizontal axis are in μm.

20.2. In its red giant phase, the Sun will have a surface temperature of about 4000 °K and a radius of about 100 million km.
(a) Calculate the brightness of the Sun.
(b) What is the total power output of the red giant Sun?

20.3. What is the surface temperature of a star whose BBR curve peaks at 0.29 μm? What is the brightness of this star?

20.4. An atom emits an EM wave of 4.5 eV.
(a) What is the frequency of this wave?
(b) What is its wavelength?
(c) Which category of EM spectrum does this belong to?

20.5. Red light with wavelength of 0.7 μm is incident on a hypothetical metal with a work function of 1.5 eV.
(a) What is the energy of the photon in eV?
(b) Will the photoelectrons be released?
(c) If so, what will their (maximum) kinetic energy be?

Chapter 21

The Atoms

The development of Planck's and Einstein's ideas of quanta eventually became tied to the understanding of the simplest atom, hydrogen. But the road to hydrogen was a long and bumpy road extending all the way back to the fifth century BC when the idea of "atom" acquired some serious credibility by the teachings of Democritus. Later, Epicurus, in trying to explain his primacy of senses, used the idea of an atom as the agent that stimulates sensation. Epicurean philosophy has survived in the writings of Lucretius, a Roman poet (see Section 42.1.1).

21.1 Modern Atomic Theory

The atomic theory of Democritus, Epicurus, and Lucretius lay dormant for over sixteen centuries until post-Renaissance "natural philosophers" adorned it with experimental testing and mathematical precision. Robert Hooke (1635–1703) used the atomic hypothesis to explain the "spring" of the air, which causes the restoration of, say a piston, when it is pushed in a cylinder full of air. He proposed that the spring of the air results entirely from the collision of atoms with the surface of their container.

Beginning of scientific atomism.

Daniel Bernoulli (1700–1782), another "modern" atomist, was a prodigious child born into a family with an extraordinary lineage of mathematicians and physicists. He applied Newton's laws of motion to a gas assuming that it was a collection of many particles. He showed that if a gas is imagined to consist of a large number of small particles, "driven hither and thither with a very rapid motion," their myriad collisions with one another and with the walls of their container would explain the phenomenon of pressure (see Section 17.1).

Dalton: father of scientific atomism.

With the development of experimental chemistry in the nineteenth century, atomic theory found its empirical domain in the works of John Dalton (1766–1844). Dalton was a self-educated Quaker who came across copies of Newton's *Principia* and Robert Boyle's *Works* in the library of a high school where he was a teacher. In *Principia*, Newton had speculated that the smallest indivisible bodies followed the laws of motion; that certain "hitherto unknown" forces were responsible for the adherence of these particles, or their recession from one another. Dalton decided to search for these indivisible particles and to include them in a quantitative scheme.

Having been involved with meteorology for some time, he chose air as his point of departure, and later generalized his investigation to all gases. The so-called "Law of Definite Proportions" proposed in 1788 by the French chemist, Joseph Louis Proust (1754–1826), was the first experimental clue for unraveling the mysteries of atoms. According to this law, a chemical compound can contain "elements" in the ratio of whole numbers. The law of definite proportions could easily be explained by supposing that each element was made up of indivisible particles. If the particle of element X weighed three times that of the element Y, and the compound was formed by combining one particle of each element, the ratio by weight would be 3 to 1 and never 3.5 to 1 or 2.5 to 1.

Law of definite
proportions: the
beginning of the atomic
theory.

Dalton noticed that sometimes elements might combine in different proportions to produce different compounds, and enunciated the law of multiple proportions in 1803. The explanation of this law, according to Dalton, was that a particle of a single element, such as carbon, might have a tendency to unite with various numbers of particles of a second element, such as oxygen, to produce different compounds. Thus, he explained, a particle of carbon may combine with one particle of oxygen to produce carbon monoxide, or it can combine with two particles of oxygen to produce carbon dioxide.

The more he examined the compounds the more evidence he found for the "indivisible particles." He did not have any difficulty finding a name for these indivisible particles. They had been named more than two millennia before by Democritus of Abdera: Dalton had finally "discovered" atoms! He published the results of his experiments in his epoch-making *TABLE: Of the Relative Weights of Ultimate Particles of Gaseous and Other Bodies*. Taking the weight of the hydrogen as 1, he itemized twenty-one elements in this Table.

Dalton finally discovers atoms.

Dalton only laid the foundation. The forthcoming generations of chemists and physicists would build a marvelously imposing edifice on this foundation. Amadeo Avogadro (1776–1856), the Italian chemist, showed that equal volumes of gases at the same temperature and pressure contained equal number of "particles." By simple weighing of these equal volumes, chemists could determine the relative weight of the particles of various gases.

Once the difference between an atom and a molecule was established, and the nature of the constituents of a gas—whether they were in atomic or molecular form—recognized, the notion of **atomic weight** was introduced. If the lightest element, hydrogen, is given the weight 1, other elements could be weighed accordingly. For example, oxygen gas, which consists of diatomic molecules, would be given the weight 32.

Atomic weight.

By mid nineteenth century, chemistry was developed to a point where certain patterns and similarities among elements were emerging. For example, the group of elements called *halogens* consisting of fluorine (F), chlorine (Cl), bromine (Br), and iodine (I)—although having very different physical properties and masses—have the strikingly similar property of having a strong tendency to combine with many metals to form white crystalline salts. Another group referred to as the *alkali metals*, includes lithium (Li), sodium (Na), potassium (K), rubidium (Rb), and cesium (Cs). These elements have a long list of identical chemical properties.

Such similarities among elements of completely different masses prompted Dmitri Mendeleev (1834–1907), the Russian chemist, to suggest that if elements were arranged in the order of increasing atomic weight, then groups of elements of similar characteristics would recur periodically. This **periodic table of elements** was a powerful tool for chemists in which to group many diverse elements and study them as a whole. Furthermore, careful placement of elements indicated some empty spots in the table, which presumably belonged to the as-yet-undiscovered elements. The discovery of such "missing" elements gave much credence to the table as well as Dalton's atomic hypothesis.

Periodic table of elements conceived.

The success of the atomic theory in chemistry rendered the theory so convincing that now *physicists* started to take it seriously. In 1860, James Clerk Maxwell began his mathematical investigation of the motion of molecules in a gas, and showed that these molecules have varying speeds: very few of them have very small or very large velocities while most of them have intermediate velocities. Independently, Ludwig Boltzmann came to the same conclusion in 1871 (see Chapter 17).

21.2   Dissecting Atoms

As the modern concept of the atom gained popularity among chemists and physicists, more and more attention was given to individual atoms. Since there were so many different atoms with varying physical and chemical properties, it became plausible to assume that they are made up of some simpler entity. Hydrogen, being the lightest atom, was a good candidate for the building block of all other elements, and much attention was given to it. However, it was soon realized that hydrogen itself was a composite entity.

We have already mentioned electrons in connection with the photoelectric effect. Prior to 1897, they were collectively called *cathode rays*. Many physicists, including Heinrich Hertz, who produced the first artificial EM wave and stumbled on the first indication of the photoelectric effect, were studying the rays that emanated from the cathode of the vacuum tube. The nature of these rays were quite a mystery, with many people believing that they were oscillation of ether, or some form of EM radiation.

Joseph John (J. J.) Thomson (1856–1940), the British physicist and director of the Cavendish Laboratory at Cambridge University, in his famous 1897 paper, "Cathode Rays," established quite convincingly that cathode rays were really a beam of negatively charged particles, whose charge-to-mass ratio (e/m ratio) he determined fairly accurately. And when the charge of these particles, now called **electrons**, was determined by the American physicist Robert A. Millikan (1868–1953), their mass was found to be orders of magnitude smaller than even the lightest atom, hydrogen. For the first time in the history of science, a particle was discovered that was lighter than the lightest atom. It was therefore natural to assume that hydrogen atom itself (as well as any other atom) must have structure, and that electron must be part of that structure.

Electrons are discovered in 1897, and atoms are no longer assumed indivisible.

BIOGRAPHY

Joseph J. Thomson (1856–1940) was born in 1856 in Cheetham Hill, Manchester in England, of Scottish parentage. In 1870 he studied engineering at University of Manchester, and moved on to Trinity College, Cambridge in 1876. In 1880, he obtained his BA in mathematics and MA in 1883. A year later he became Cavendish Professor of Physics. He was awarded the Nobel Prize in Physics in 1906, "in recognition of the great merits of his theoretical and experimental investigations on the conduction of electricity by gases." He was knighted in 1908 and appointed to the Order of Merit in 1912. One of Thomson's greatest contributions to modern science was in his role as a highly gifted teacher, as seven of his research assistants and his son, George Paget Thomson, won Nobel Prizes in physics. His son won the Nobel Prize in 1937 for proving the wavelike properties of electrons. In 1918 J. J. Thomson became Master of Trinity College, Cambridge, where he remained until his death. He died on 30 August 1940 and was buried in Westminster Abbey, close to Sir Isaac Newton.

If negative electrons are part of the atom, then an amount of positive charge equal to the charge of the electrons must reside inside the atoms, because atoms were known to be neutral. The question now was "How are the positive charges and the electrons distributed inside an atom?" The first suggestion came from Thomson himself, who imagined atoms as a sort of "raisin pudding." He considered the positive charge as being the background (the pudding) in which the electrons were embedded like raisins.

It is remarkable enough that, by the force of observation, a composition for the invisible atom had to be hypothesized. But it is even more remarkable that this hypothesis could actually be *tested*. Between 1909 and 1911, Ernest Rutherford (1871–1937)—who had come to Cambridge to work with J. J. Thomson—and two of Rutherford's coworkers, Hans Geiger (1882–1945) and Ernest Marsden (1889–1970), carried out a series of experiments that proved the Thomson model of the atom wrong.

Rutherford had performed some experiments involving the bombardment of heavy atoms with alpha particles, and asked Geiger to analyze the data and look for the scattering of the alpha particles. Geiger found no appreciable scattering (only of the order of one

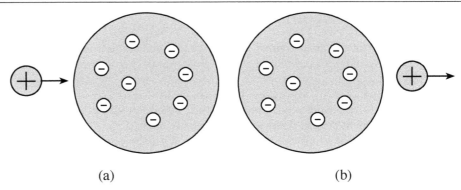

Figure 21.1: The raisin-pudding model of the atom predicts no scattering of the positive alpha particles. (a) The alpha particle approaches a Thomson atom. (b) The alpha particle comes out of the atom almost unaffected.

degree). Then in 1911, upon the recommendation of Geiger, Rutherford suggested to the young Marsden to repeat the earlier experiments and see if he could detect some *backward* scattering of alpha particles. What happened next was history. Rutherford later recalled:

*Rutherford and his coworkers discover the atomic **nucleus**.*

> Then I remember two or three days later Geiger coming to me in great excitement and saying, "We have been able to get some of the α-particles coming backwards...." It was quite the most incredible event that has ever happened to me in my life. It was almost as incredible as if you fired a 15-inch shell at a piece of tissue paper and it came back and hit you. On conclusion, I realized that this scattering backwards must be the result of a single collision, and when I made calculations I saw that it was impossible to get anything of that order of magnitude unless you took a system in which the greater part of the mass of the atom was concentrated in a minute nucleus. [Hol 01, p. 418]

To appreciate Rutherford's argument, consider a swarm of alpha particles impinging on a sheet of a heavy metal such as gold. If the gold atoms were as Thomson suggested, then we should see something like Figure 21.1, because the positive (and heavy) alpha particle encounters a more or less uniform distribution of positive charge and therefore a more or less uniform repulsion in all directions. The net result of repulsion is therefore almost zero. If the electrons are distributed almost uniformly throughout the atom, the same reasoning applies to give a net attraction of zero. On the other hand, if the distribution of electrons is not uniform, because of the fact that the electrons are so much lighter than the alpha particles, it will be the electrons that undergo large scattering not the alpha particles. It follows that the Thomson model of the atom is incapable of explaining large scatterings.

Thomson model of the atom contradicts the observed large scattering angles.

 Food for Thought

FFT 21.2.1. Prior to Rutherford, the study of atoms focused on their combination in chemical reactions, whose effects were visibly tangible: a gas combined with another gas to form a liquid; a gas and a liquid combined to form a solid; two colorless liquids combined to yield a bright red liquid. As long as the atomic *hypothesis* helped explain these visible phenomena, one could live with it.

But Rutherford, the skeptic philosophers would argue, was now telling us that he was "looking" inside the atom (to find its nucleus), and he was using abstract mathematics and physics as his tools. How can we be *sure* that atoms actually exist and are not some mental constructs that merely facilitate our understanding of certain phenomena? After all, we never have any direct sensory evidence for atoms. This view has persisted among a large group of the philosophers of science. If we extend this kind of argument to its logical limit, we have to deny what we see under a microscope or through a telescope. After all, we never have any direct visual sensation of the bacteria or moons of Jupiter!

Reality is not limited to our sensory perception. If it were, then we would have to deny the existence of invisible EM radiation and all of its manifestations including radios and televisions. People afflicted with color-blindness do not deny the existence of color simply because *they never have any direct sensory evidence for* color. They have learned about their sensory limitation, and rely on the experience of others concerning colorful matters. We, as human beings, have to learn and accept our biological limitations and—for our perception of reality—rely on the only truly solid collective knowledge of our race, science.

Human sensory perception and reality.

When Rutherford probed the interior of atoms by shooting alpha particles at them, he was relying on the repeatedly tested laws of nature: he relied on the fact that KE is given by $\frac{1}{2}mv^2$, that the force between two charged particles was given by $F = k_e q_1 q_2/r^2$, that the gravitational force between these particles is negligibly small, that the (electrical) potential energy of a system of two particles was $PE = k_e q_1 q_2/r$, that energy was conserved, that .... And if all of these proven laws lead to the definite conclusion that atoms have a very small positive nucleus, then we have to accept that conclusion. And this conclusion, though unperceived by any of our senses, becomes part of our knowledge to be used in the future to further our understanding of nature.

Now consider a different model, the one proposed by Rutherford, in which the positive charge is concentrated in a very small region of the atom. For the same reason as above, we can neglect the effect of the electrons. Figure 21.2 illustrates the essential points of Rutherford's argument. Because of the small size of the positive charge, most alpha particles will pass through the atom far away from the nucleus as in Figure 21.2(a) and will not be affected by the atom. Few alpha particles get relatively close to the nucleus and are scattered by a small angle as in Figure 21.2(b). On very rare occasions, an alpha particle will collide with the nucleus head-on and gets scattered backwards [Figure 21.2(c)]. This does not mean that the alpha particle actually touches the nucleus. The electric repulsion by the nucleus becomes infinitely strong when the alpha particle gets a chance to hit it. Since alpha particles are quite energetic, this impact could knock the nucleus out of the atom, *unless the nucleus happens to be extremely heavy.* Lack of observation of any nuclei being knocked out of atoms indicates that the nucleus of the atom is indeed very heavy.

Observational proof that atomic nuclei are extremely heavy.

Rutherford-Geiger-Marsden experiments therefore give us the following picture of an atom:

A new picture of the atom.

- An atom is generally neutral, with equal amount of positive and negative charges.

- The negative charge consists of the electrons.

- The positive charge is concentrated in the nucleus of the atom, which is about 1/100,000 the size of the atom itself. Therefore, an atom is mostly empty space.[1]

- The mass of the atom is almost entirely concentrated in the nucleus.

What do you know? 21.1. What would happen in Rutherford experiment if the nucleus of gold were light?

21.3 Bohr Model of the H-Atom

Rutherford's discovery of the nucleus drastically changed the "picture" of the atom. With an extremely heavy nucleus at the center, the atom appeared more like a mini solar system

---

[1]The electrons, if they have any size at all, must be orders of magnitude smaller than even the nucleus.

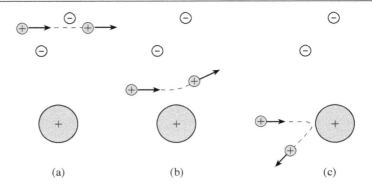

Figure 21.2: The Rutherford model of the atom predicts scattering of the positive alpha particles. (a) The alpha particle approaches a Rutherford atom far from nucleus, and comes out almost unaffected. (b) The alpha particle approaches the atom relatively close to the nucleus, and scatters through a small angle. (c) The alpha particle approaches the nucleus head on, and scatters backwards.

with nucleus playing the role of Sun and the electrons the role of the planets. However this **planetary model** of the atom did not last very long.

Recall from Chapter 14 that accelerating charges radiate EM waves, and since EM waves carry energy, the accelerating charge loses energy. **Math Note E.21.1** on **page 109** of *Appendix.pdf* calculates the energy of the electron of a hydrogen atom when it is located at a typical distance of 10^{-10} m from the nucleus, and finds the energy to be about -1.15×10^{-18} J.

> **What do you know? 21.2.** What is the significance of the fact that the energy of the electron is negative?

Math Note E.21.1 also shows that the electron rotates around the nucleus with a frequency of 2.53×10^{15} Hz, which is the frequency of the EM wave produced as a result of this rotation. The energy of such a wave is obtained from the Planck-Einstein relation, $E_{\text{rad}} = hf$, which, upon the substitution of the values of the Planck constant and electron's rotational frequency, yields $E_{\text{rad}} = 1.68 \times 10^{-18}$ J. It follows that the energy of the electron reduces by this amount after the emission of the radiation and becomes $E_{\text{new}} = -2.83 \times 10^{-18}$ J. Using Equation (E.57) of Math Note E.21.1, we can evaluate the

Failure of the planetary model of the atom.

new radius of the electron's orbit. It turns out to be 4.1×10^{-11} m. The electron has gotten closer to the nucleus. Since speed increases with decreasing radius [see Equation (E.54) of the Math Note], the electron is now moving faster, its frequency is therefore larger, and again by Planck-Einstein formula, it radiates more energy, and its radius decreases by a larger factor than before. This snowball effect continues until the electron collapses completely into the nucleus. One can calculate the loss of energy and the decrease of the radius classically (without resort to the Planck-Einstein relation), but the calculation is much more involved and beyond the scope of this book. The overall conclusion, however, is independent of the formula for energy. It can be shown that the entire process of collapse takes place in less than 10^{-8} s. Thus, this so-called **planetary model** of the atom fails.

Niels Bohr uses the quantum theory of Planck and Einstein to invent a stable model of the H-atom.

The discovery of the nucleus had created a seemingly insurmountable barrier on the path to understanding the structure of the atom. Many researchers in the Rutherford group were trying to come up with a model of the atom that did not collapse. Niels Bohr found a (partial) solution. Coming from the continent, he was at an advantage, because he was one of the few who knew the newly discovered quantum theory of Planck and Einstein. And he used it to invent a stable model of the (hydrogen) atom.

Bohr concentrated on the hydrogen atom (as did the generations coming after him) because of its simplicity: one light electron interacting with a heavy (and almost immovable) nucleus. That is why H-atom is sometimes called a "gift of nature." We shall not follow Bohr's original argument presented in 1913, which was extremely complicated. Instead, we use some simple algebra to arrive at his results. **Math Note E.21.2** on **page 110** of *Appendix.pdf* provides the details, of which we cite the following:

Hydrogen as a gift of nature.

Box 21.3.1. *The electron can be only on one of the infinitely many* **quantized** *orbits. The radius of the nth orbit is* $r_n = n^2 a_0$, *where n is a positive integer. When the electron is in the nth orbit, its energy is* $E_n = -13.6/n^2$ eV. *We also say that the atom is in its* **nth energy level**.

Here $a_0 = 5.3 \times 10^{-11}$ m is called the **Bohr radius**.

Bohr radius a_0.

Bohr further assumed that it was possible for an electron to make a transition from a higher orbit to a lower orbit. The difference in the energy of the two orbits is carried away by a photon. Denote the initial energy by E_i, the final energy by E_f, and the energy of the emitted photon by E_γ, then $E_\gamma = E_i - E_f$. The frequency and wavelength of this photon can then be determined by the Planck-Einstein relation $E_\gamma = hf$, or $E_\gamma = hc/\lambda$. **Example D.21.1** on **page 44** of *Appendix.pdf* gives a numerical application of this.

BIOGRAPHY

Niels Bohr (1885–1962) (right) attended Grammelholms school from 1891 to 1903. He did well at school without ever being brilliant. If he really excelled at a subject it was physical education. He and his brother Harald (left) were excellent soccer players. Harald, an excellent mathematician, even won a silver medal playing soccer for Denmark. Bohr entered the University of Copenhagen in 1903, receiving his doctorate in May 1911. With a travel grant from the Carlsberg Foundation, he went to the mecca of physics, Cavendish Laboratory in England to work with J. J. Thomson. After a few months he transferred to the Rutherford group where exciting new discoveries were being made. It was here that the foundation of the Bohr model of the atom was laid.

21.3.1   Spectral Lines

Transitions are usually depicted in **energy level diagrams** of the sort shown in Figure 21.3 for H-atom. An arrow connects the initial level to the final level. The transitions on the left (to $n = 1$) and on the right (to $n = 3$) are invisible, as explained below. Out of all the transitions to $n = 2$ only four are visible. These are shown in the middle of the diagram.

Energy level diagrams.

Although there are infinitely many of them, one can in principle, calculate the wavelengths of all the photons emitted by a hydrogen atom. Among these only a finite number are visible. We can identify them as follows. First we note that no transition to the $n = 1$ energy level creates a visible photon. This is because the lowest-energy photon for such transitions is the one emitted when the electron jumps from the $n = 2$ to $n = 1$ orbit. The energy of the photon, obtained from Figure 21.3, is $E_\gamma = -3.4 - (-13.6)$ or 10.2 eV. Example 20.2.2 calculated the energy range of the visible spectrum: it was from 1.77 eV for red to 3 eV for violet. So the 10.2-eV photon is not visible. All the other photons resulting from transitions from third orbits or higher will have even larger energy than 10.2 eV. So, they are all invisible.

Next consider transitions to $n = 2$ orbit. Example D.21.1 calculated the wavelength for $3 \to 2$ transition. Let us calculate the $4 \to 2$ wavelength. Again using Figure 21.3, we get the energy of the photon: $E_\gamma = -0.85 - (-3.4) = 2.55$ eV or 4.08×10^{-19} J. The wavelength of this photon is easily found to be 4.87×10^{-7} m or 487 nm (nanometer) corresponding to blue. Similar calculations show that the $5 \to 2$ transition yields a wavelength of 435 nm corresponding to indigo, and that the $6 \to 2$ transition yields a wavelength of 411 nm

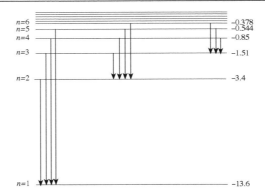

Figure 21.3: Energy level diagram for a hydrogen atom. Transitions to $n = 1$, $n = 2$, and $n = 3$ energy levels are shown. Only the middle four transitions are visible.

corresponding to violet. The $7 \rightarrow 2$ and and all the other transitions to $n = 2$ level result in ultraviolet (therefore invisible) photons. All transitions to $n = 3$ orbit are invisible because $E_3 = -1.51$ eV, which is already below the minimum energy of 1.77 eV for a visible photon.

What do you know? 21.3. What is the wavelength of the photon emitted when a hydrogen atom makes a transition from the fourth to the third level? Is it possible to have a photon of visible wavelength in a transition to the third level?

How do we observe these photons? Long before Bohr proposed his model of the atom, it had been known that light is emitted by gases and vapors when an electric spark is passed through them, or in some cases, when a continuous electric current is established through a thin tube filled with the gas. The light so emitted, when passed through a system of prisms (called a **spectroscope**), was found to have a completely different spectrum than sunlight. Whereas the latter's was a continuous spectrum consisting of all wavelengths between red and violet, the former's spectrum consisted of only thin lines of colored light. These lines came to be known as the **emission spectral lines**. It was further discovered that each element had its own unique set of spectral lines, a sort of fingerprint for the element.

Emission spectral lines are fingerprints for elements.

Food for Thought

FFT 21.3.2. Theories in physics are discovered in two different ways. Some theories are more or less a direct outgrowth of observation. Kepler's discovery of the elliptical orbit of Mars is one example of this process. Another example is the pre-Maxwellian electromagnetism. A third example is Planck's discovery of the formula describing the black body radiation curve. In all these cases, a rule or a mathematical formula is discovered, which summarizes the result of many observations. This process of theory making is called **induction** or **phenomenology**. Most of classical physics theories are inductive.

Modern physics is almost entirely deductive.

The other way of theory making uses a great deal of mental creativity. A prime example of this way of theory making, called **deduction**, is Maxwell's discovery of the electromagnetic waves. Modern physics is almost entirely deductive. Planck's fundamental quantum hypothesis and his relation connecting the frequency of the quantum of EM radiation to its energy is a good example of deduction. The relation $E = hf$ is not a direct (or even indirect) result of observation. It is a formula created solely by Planck's mind after months of thinking about ways to explain the black body radiation formula. Einstein's explanation of the photoelectric effect and his introduction of the photon idea in that explanation is another example. A third example, on which we have dwelt in this chapter is Bohr's idea of the hydrogen atom. The assumption that the angular momentum of an electron is an integer multiple of the Planck constant (see Math Note E.21.2) has no observational basis whatsoever; it was completely created by Bohr's mind. In the remaining chapters of this part, we shall see many more examples of deductive reasoning.

The deductive process has been so pervasive in modern physics that some theorists have expressed the belief that perhaps the human mind in and of itself can produce theories; that observation has a secondary, even an inessential role in theory making. That this belief is ill-founded can be demonstrated by the vast experimental work that went into the pre-Maxwellian electromagnetism, without which Maxwell could not have obtained his mathematical equations that led to the prediction of EM waves; by the intense observational activity of many experimental physicists, on which Planck relied heavily before writing down his black body radiation formula; and by many experiments on the photoelectric effect on which Einstein based his idea of a photon.

Can human mind in and of itself produce theories?

Perhaps the most convincing example of the crucial role that observation plays in a deductive theory is the Bohr model. Although Bohr's hypothesis was entirely a creation of his mind, it was a hypothesis *that applied to the Rutherford atom*. This atom had a heavy positively charged nucleus around which negative electrons revolved. And this picture of the atom was—quite surprisingly— discovered through, and only through, *experimental observation*. So, although deduction is central in making modern theories of physics, it cannot take place in a vacuum. Somewhere in the history of the development of every deductive theory there is a crucial observation, without which the theory could not exist. This observation may be in the immediate past of the theory as in the case of the Bohr model of the hydrogen atom, or in the distant past as in the case of Maxwellian electromagnetism, which is based on observations going back to Gilbert's in the late 1500s.

Important role of observation in the deduction process.

The exaggerated role of deduction advanced even by some physicists has become a base on which certain schools of social sciences erect their theoretical edifice. They propose "theories" upon "theories" based entirely on speculation and whim and argue as if those theories were irrefutable. Thus Latour *redefines* certain accepted concepts to give his version of the "theory" of relativity (see Food for Thought 27.5.3). And Bloor "theorizes" that beliefs change over time and are dependent on cultures in which they are believed; science also changes over time and has developed in many different cultures; therefore, science depends on culture (see Section 19.4)!

21.3.2 Shortcomings of the Bohr Model

When Bohr invented his model of the atom, there was an extensive collection of spectral lines data on which it could be tested. The success of the model in predicting the exact location of the spectral lines of the H-atom was an indication of the validity of the model. However there was an uneasy feeling among physicists about the model. For example, the very basic assumption concerning the angular momentum of the electron seemed to be very artificial. Why should the angular momentum be an integer multiple of $h/2\pi$ and not $h/3\pi$ or $h/2.3$, or any other odd combination? More seriously, what is the electron doing when it is in the first orbit? Is it sitting still? If so, then why is it not pulled into the nucleus by the electric attraction? Is it moving? Then why is it not radiating EM waves—as the laws of electromagnetism predict—and eventually collapsing into the nucleus due to loss of energy? When the electron is on a higher orbit, does it always jump to lower orbits? Which one? These, and many other unanswered and troubling questions, rendered the Bohr model of the atom unsatisfactory. The race was now on to find a new model that had none of these flaws.

21.4 End-of-Chapter Material

21.4.1 Answers to "What do you know?"

21.1. The alpha particles would be able to knock them out of the gold atoms, and one should be able to detect them outside the gold foil.

21.2. A negative energy indicates a bound system as we discovered in Example 9.2.5. This means that the electron cannot leave the atom on its own. You have to provide it with positive energy (such as the KE of a bombarding particle) to separate it from the nucleus.

21.3. The fourth level has an energy of $-13.6/4^2$ or -0.85 eV. The third level has an energy of $-13.6/3^2$ or -1.51 eV. The difference is 0.661 eV, which is the energy of the photon. Convert this energy to Joules, plug in the Planck relation (20.4), and find the wavelength to be 1.88 μm. This is infrared. The absolute maximum energy a photon can have in a transition to the third level is 1.51 eV. (Can you convince yourself why?) The wavelength of such a photon is 0..82 μm, which is infrared. Since visible light has shorter wavelength (and therefore higher energy), no transition to the third level is visible.

21.4.2   Chapter Glossary

Atomic Weight (really, atomic mass) The weight of a given collection of atoms or molecules of a substance. For hydrogen this is (approximately) one; all other substances are an (almost) integer multiple of the hydrogen weight.

Beta Rays A negatively charged ray discovered at the end of the nineteenth century, later identified as a beam of electrons.

Bohr Model A model of the atom which mixes the quantum ideas of Planck and Einstein with classical mechanics, leading to quantized orbits and quantized energy.

Cathode Rays Same as beta rays.

Planetary Model A model of the atom whereby the nucleus is considered a Sun and the electrons move around it as planets.

Raisin-Pudding Model A model of the atom whereby the negatively charged electrons (the "raisins") are embedded in a positively charged background (the "pudding").

Spectral Lines The lines appearing in the spectroscopy of gases of elements. Each element has its unique spectral lines, making them like a finger print of that element.

21.4.3   Review Questions

21.1. How did Robert Hooke explain the "spring" of the air? How did Daniel Bernoulli explain pressure?

21.2. What law of chemistry started modern atomic theory? How many elements did Dalton itemize in his book?

21.3. What was Avogadro's contribution to the atomic theory? How does weighing of equal volumes at the same temperature and pressure determine the relative mass of the particles of various gases?

21.4. How did the chemical similarities among elements lead to the periodic table? What were the heavier elements thought to be made up of?

21.5. What were cathode rays? What did they consist of? Were they positive, negative, or neutral? How heavy were they compared to hydrogen? Could physicists still assume that hydrogen was the building block of all other atoms? How did the first model of the hydrogen atom look?

21.6. When Rutherford bombarded atoms with alpha particles, what was he expecting? What did he actually see? What was the implication of his experiments?

21.7. What properties does the nucleus of an atom have? How large is it? How heavy is it? What charge does it carry?

21.8. With the nucleus being part of the atom, what model was naturally assumed concerning the configuration of electrons and the nucleus? Did this model guarantee the stability of the atom?

21.9. Why is hydrogen called a "gift of nature?" What theory did Bohr incorporate in modeling the H-atom? What are the essential features of the Bohr model of the H-atom?

21.10. If you pass the light emitted by atoms through a prism, do you see a continuous spectrum with all wavelengths present?

21.11. What are spectral lines? Why are they called the "fingerprint" of elements?

21.12. State some of the flaws of the Bohr model of the H-atom.

21.4.4 Conceptual Exercises

21.1. Why is the atomic hypothesis so natural for explaining the "spring" of the air?

21.2. What is the connection between the law of definite proportions and the atoms?

21.3. If the electrons happened to be more massive than a hydrogen atom, would there be reason to believe that atoms were divisible?

21.4. What does the scarcity of the backward (or large-angle) scattering of the alpha particles say about the size of the nucleus?

21.5. The energy of the electron in a hydrogen atom is a constant divided by the radius of the orbit of the electron. Can you deduce this from the information given in Box 21.3.1?

21.6. A hydrogen atom in its lowest energy state can absorb a photon and jump to a higher energy state. The energy difference between the lowest energy state ($n = 1$) and the next state ($n = 2$) is 10.2 eV. Is it sufficient for a photon to have 10.2 eV to be able to move an H-atom from $n = 1$ to $n = 2$? Consider both energy *and* momentum conservation.

21.7. If you pass the light from an incandescent bulb through a prism it comes out with a continuous spectrum (all the rainbow colors). If you send the same light through a hydrogen gas and then through a prism, some dark lines will appear in the spectrum. Why?

21.4.5 Numerical Exercises

21.1. A hydrogen atom makes a transition from the $n = 6$ to $n = 2$ state.
(a) What is the energy of the photon released?
(b) What is the photon's frequency?
(c) What is the photon's wavelength?
(d) Is it visible? If so, what color does it have? If not, which category of the EM spectrum does it belong to?

21.2. A hydrogen atom makes a transition from the n=20 to n=18 state.
(a) What is the energy of the photon released?
(b) What is the photon's frequency?
(c) What is the photon's wavelength?
(d) Is it visible? If so, what color does it have? If not, which category of the EM spectrum does it belong to?

Chapter **22**

Quantum Theory I

Bohr's model of the hydrogen atom was a hybrid between the classical Newtonian physics and the radical idea of the quantization of light. Newtonian physics by itself failed miserably when encountered with an atom that had a nucleus, and when Bohr mixed it with a dab of modern physics, it was improved; but only partially. It appeared that Newtonian physics needed a dramatic metamorphosis.

22.1 Electrons as Waves

Louis de Broglie (1892–1987), with a degree in history, had no intention of becoming a physicist, but when his older brother introduced him to the subject, he was so fascinated by it that he decided to make physics his profession. His interest in the fundamental principles led him to include in his doctoral dissertation of 1924 an idea that baffled his advisors and the physics community in general.

De Broglie argued that up until 1905, when Einstein—in his explanation of photoelectric effect—hinted at the existence of light particles, everybody thought that light was only a wave. Now everybody thinks that electrons are only particles. Is that really the case? What if we assume that they are also waves? Once he made this sweeping assumption, he sought to connect the wave and particle properties of the electron.

The most particle-like property of an object is its momentum, and it is the momentum that enters Newton's second law of motion, a law that describes the behavior of *particles*. So, for de Broglie, the question boiled down to this: Is there a relation between the wavelength (the definitive property of a wave) and momentum? He found the answer in photons. Planck and Einstein had shown that the energy of a photon was related to its wavelength via $E = hc/\lambda$. Einstein's theory of relativity connected the energy and momentum of the photon via $p = E/c$ [see Section 28.3.3, especially Equation (28.8)]. Therefore, concluded de Broglie, the momentum of a photon is related to its wavelength via $p = h/\lambda$. This, he argued, must be true for electrons as well, and came up with the following Nobel-prize winning relation:

$$p = \frac{h}{\lambda}, \quad \text{or} \quad \lambda = \frac{h}{p}, \quad \text{where} \quad p = mv. \tag{22.1}$$

de Broglie relation.

This is now called the **de Broglie relation**.

Example 22.1.1. Let us calculate the wavelength of a typical electron. The momentum of such an electron can be found from Math Note E.21.1, which gives the speed of the electron as 1.59×10^6 m/s. Multiplying the mass, 9.1×10^{-31} kg, of the electron and this speed, we get $p = 1.45 \times 10^{-24}$ kg·m/s. The wavelength of the electron is therefore $\lambda = 6.626 \times 10^{-34}/1.45 \times 10^{-24} = 4.57 \times 10^{-10}$ m, which is a typical size for atoms. ∎

The example above shows that if the electrons have wave properties as de Broglie suggested, then their typical wavelength is of the order of the atomic size. So, if we can find an aperture the size of an atom, we should be able to observe the diffraction of the electrons (see Box 11.3.1). This indeed took place in 1926 when two American physicists, Clinton Davisson (1881–1958) and Lester Germer (1896–1971), used the spacing between metallic atoms as their aperture to observe the electron diffraction. About a year later, across the Atlantic Ocean, the British physicist, George Thomson (1892–1975) using a different technique, proved the wave nature of the same *particle* which his father had discovered 31 years earlier.

Food for Thought

FFT 22.1.2. Science is an interconnected web of ideas, laws, theories, and experimental results. This interconnectedness solidifies science at the same time that it characterizes it. No other discipline enjoys such a strong connection among its parts. Political, social, economic, and historical theories are so loosely connected that two opposing theories (points of view, philosophies, or schools) could coexist.

The interconnectedness of science is especially pronounced in physics. It is very common to see ideas developed in one area of physics being used in other areas. Maxwell borrowed the mathematics of the motion of fluids in his description of the electromagnetic fields. The motion of a fluid is expressed in terms of *velocity fields*, whose variation is caused by "stress and strains" applied to the fluid. Even today people talk of "stress energy tensor" of electromagnetism, reminiscent of the fluid terminology borrowed by Maxwell over 150 years ago.

De Broglie's brilliant idea of applying to the electron the relation that exists between the wavelength and momentum of a photon is another significant example of the interconnectedness among areas of physics. In order to come up with the idea, de Broglie *had to know* relativity. In order for Maxwell to come up with his electromagnetic theory, he *had to know* the mechanics of fluids. Creative physicists must be well rounded, in command of the accumulated knowledge of their field, and in possession of a fertile imagination.

Each new connection is a new tie in the web that makes science ever more impermeable. It is impossible for a nonsensical theory or idea to penetrate this web. The internet, bookstores, and many libraries are filled with "unified theories" claiming to solve all the existing problems of physics. However, because of the detachment of these theories—often proposed by people with little or no background in physics—from the web of physics, they invariably make no sense, and fall flat on their face.

Other disciplines do not have such a tight web. One can even argue that they have no web at all. And if they do, the web's knots are so loose and the cracks so wide open that any "theory" can get through: Politicians can have any political ideology (or theory); philosophers of science can say anything they want about the nature and the working principles of science; educators can "educate" students based on constructivism, multiple intelligence theory, student-centered methodology, traditional approach, etc., many of which theories contradict each other; and sociologists can "voice" their opposing theories about society. In the "marketplace of ideas," a wrong theory can be right and a right theory wrong. Science does not have a "market place" for ideas.

"Marketplace of ideas," equal emphasis on right and wrong!

Later, when other sub-atomic particles were discovered, they too were shown to have wave properties. It was soon realized that Equation (22.1) described a much more universal property than was originally intended by it. Now we speak of the **wave-particle** duality of all objects. All objects? Are bullets, for example, waves as well? Can we see the diffraction of bullets? See what the following example tells you.

Wave-particle duality.

Example 22.1.3. A bullet has a mass of about 10 grams (0.01 kg) and moves with a speed of 40 m/s.

Q: What is the de Broglie wavelength of such a bullet?

A: First find the momentum of the bullet: $p = 0.01 \times 40 = 0.4$ kg·m/s. Now find the wavelength $\lambda = 6.626 \times 10^{-34}/0.4 = 1.66 \times 10^{-33}$ m, and conclude that to see the diffraction of a bullet, we need to pass them through an "aperture" whose diameter is not too much bigger than 1.66×10^{-33} m (see Section 11.3, especially Box 11.3.1). As a comparison, note that the size of the nucleus of an atom is about 10^{-15} m!

Clearly observing the diffraction of bullets is out of the question. What about considerably smaller objects, say a bacterium. The mass of a typical bacterium is 10^{-15} kg. If we want to see its diffraction, we should give it a very small speed, so that its wavelength would be large enough to be comparable with a reasonable-sized aperture. Let the speed be about 1 m/s. Its momentum is then $10^{-15} \times 1 = 10^{-15}$ kg·m/s. This gives a wavelength of 6.626×10^{-19} m. The diameter of a bacterium is about a micron or 10^{-6} m. Therefore, the required aperture through which the bacterium is to pass, must be 10 trillion times smaller than the bacterium itself! ■

What do you know? 22.1. A very small virus such as rhinovirus is approximately 30 nm in diameter. Its mass is about 10^{-23} kg, and it moves with a speed of about 1 cm/s. Can you see the diffraction of such viruses?

Aside from the experimental confirmation of the de Broglie hypothesis, there was another validation of the idea, which came from the Bohr model of the atom. It turns out that de Broglie's theory can explain the artificial assumption that was the starting point of Bohr's theory. The interested reader can find the details in **Math Note E.22.1** on **page 111** of *Appendix.pdf*.

22.2 Quantum Mechanics

The success of de Broglie's hypothesis drew the attention of the entire physics community to the theory of waves. Erwin Schrödinger (1887–1961) was somewhat of an outsider, as he lived in Austria, which, despite its proximity to the center of physics and mathematics, namely Germany, was rather isolated. Nevertheless, neither this isolation nor his engagement in active military duty prevented Schrödinger from contributing to physics. He was especially fascinated by de Broglie's revolutionary idea of the wave nature of the electron, and very quickly familiarized himself with it. On November 23, 1925, Schrödinger gave a seminar on de Broglie's work and a member of the audience suggested that there should be a wave equation. Within a few weeks Schrödinger had found the wave equation for the electron, which he published in January 1926.

22.2.1 An Equation for Everything

Schrödinger equation is depicted in Figure 22.1. A few features of this equation are worth mentioning. The most conspicuous feature is the appearance of the Planck constant $\hbar = h/2\pi$, confirming, along with Bohr and de Broglie, the importance of this constant in atomic physics. The second feature, just as important as the first, is the appearance of the imaginary number $i = \sqrt{-1}$. For the first time in the history of physics, imaginary numbers have entered an equation describing the behavior of a physical entity. The symbol V stands for the potential energy of the system, which in turn is determined by the forces of interaction. Finally, the symbol Ψ stands for the **wave function**, whose interpretation divided the physics community for a long time.

When applied to the hydrogen atom (one of the few systems which can be solved analytically), the Schrödinger equation gave exactly the same formula for its energy levels as the Bohr model. So, there seemed to be grounds for the credibility of this equation. In fact, the Schrödinger equation went far beyond explaining the hydrogen atom. One can say that any system of particles that move with speeds much smaller than light speed and interact via electromagnetic forces, is described correctly by the Schrödinger equation. This includes molecules (thus, the entire discipline of chemistry), solids, liquids, gases, plasmas,

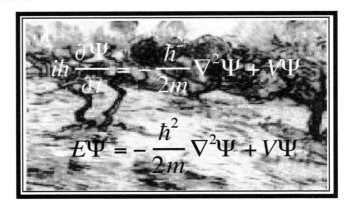

Figure 22.1: The Equation for Everything. Note the appearance of the Planck constant $\hbar$, the imaginary number $i = \sqrt{-1}$, and the wave function Ψ. The top equation involves time, the bottom does not.

and even life itself (molecular genetics). The Schrödinger equation appears to be indeed an "equation for everything."

BIOGRAPHY

Erwin Schrödinger (1887–1961) received lessons at home from a private tutor up to the age of ten. He then entered the Akademisches Gymnasium in the autumn of 1898. After graduation from the Gymnasium in 1906, Schrödinger entered the University of Vienna, and four years later, he was awarded a doctorate degree in physics. Soon he undertook voluntary military service in the fortress artillery. In 1914, with the outbreak of World War I, Schrödinger took up duty on the Italian border. A year later, he was transferred to duty in Hungary, where he submitted several research papers for publication. From 1917 onward he became interested in the quantum theory and in 1924 he read de Broglie's paper, the paper that changed the direction of his research, leading to the equation that rerouted physics in 1926.

22.2.2 Matrix Mechanics

Wave mechanics is Schrödinger's formulation of quantum theory.

In the spring of 1925, almost half a year prior to Schrödinger's formulation of what came to be known as **wave mechanics**, a young German physicist by the name of Werner Heisenberg, recovering from a bout of hay fever on the island of Helgoland in the North Sea, invented **matrix mechanics** which also explained the spectral lines of the hydrogen atom.

Matrices were highly abstract objects (consisting of rows and columns of numbers) introduced into the mathematical literature at the end of the eighteenth century. It is possible to define addition, multiplication, and division of matrices in a restrictive sense. The physicists of the early twentieth century were inconveniently unfamiliar with matrices. And when Heisenberg came across a set of numbers associated with the spectral lines of the hydrogen atom, it was a more mathematically inclined colleague who pointed out the true nature of those numbers. Soon, Heisenberg along with the rest of the physics community mastered the theory of matrices and solved some hitherto insurmountable problems.

The essence of matrix mechanics is that physical observables are represented not by ordinary numbers (or functions), but by matrices. The rows and columns of numbers in a matrix of an observable form the collection of the possible values the observable can assume in a measurement. One of the crucial properties of matrices is that, unlike ordinary numbers and functions, generally speaking, the operation of multiplication *is not commutative*: If **A** and **B** are matrices, then in general, **AB** $\neq$ **BA**. Two observables whose matrices do commute are called *compatible*; and if their matrices do not commute, they are *incompatible*.

Heisenberg discovered that if two physical observables are incompatible (their matrices do not commute), they cannot be measured with arbitrary certainty *at the same time*. There is no restriction on the accuracy with which any number of mutually compatible observables can be measured.

The state of a quantum mechanical system.

Box 22.2.1. *A complete description—or a **state**—of a quantum mechanical system is obtained when the values of the maximum set of mutually compatible observables are given for that system.*

Two fundamental incompatible observables are position **x** and momentum **p**. Heisenberg showed that if we denote the uncertainty in **x** by Δ**x**, and the uncertainty in **p** by Δ**p**, then

Heisenberg's uncertainty principle.

$$(\Delta\mathbf{x}) \cdot (\Delta\mathbf{p}) \geq \hbar, \tag{22.2}$$

where $\hbar = h/2\pi = 1.05 \times 10^{-34}$ J·s. Once again note the appearance of the Planck constant. Equation (22.2) is the celebrated **Heisenberg uncertainty principle**. Let us pause and examine this relation more carefully.

BIOGRAPHY

Werner Heisenberg (1901–1976) at the age of nine moved to Munich, where his father started to work as a professor of Greek. In 1914 World War I broke out and the Gymnasium in which Heisenberg was enrolled was occupied by troops. Students were left on their own, and he took the opportunity to concentrate on his favorite subjects: mathematics, physics, and religion. Upon graduation from the Gymnasium, he and his fellow student Wolfgang Pauli, began to study theoretical physics under Arnold Sommerfeld in October 1920. After gaining confidence, he switched from mathematics to theoretical physics. Heisenberg also began to plan to undertake research in relativity. However Pauli advised him against it, encouraging him to take up the atomic structure since theory and experiment did not agree.

Any measurement has an uncertainty—a margin of error. For example, if I ask you how tall you are, you may say 5 feet 11 inches. If I encourage you to be more accurate, you may come back with the answer 5 feet $10\frac{1}{2}$ inches after having gone to a drugstore and used a very accurate scale. If I persist in my demand for more accuracy, you may appease me by quoting a yet more accurate measurement: 5 feet $10\frac{5}{8}$ inches. Beyond this, you will probably not be able to give more accurate result. You will not be able to answer the question "Is your height 5 feet $10\frac{9}{16}$ or 5 feet $10\frac{11}{16}$? All you can say is that it is somewhere between the two. Your height measurement is uncertain to within $\frac{2}{16}$ of an inch, or the uncertainty in your height is $\frac{2}{16}$ inch. This uncertainty is, of course, the result of the crudeness of our measuring instruments and/or the bulkiness of what is to be measured. If the bulkiness of the object of measurement is reduced (What is the thickness of a 750-page book?), then the uncertainty in length can be reduced by using more accurate instruments.

The significance of Equation (22.2) is that it is *independent* of the measuring device. It says that regardless of the instrument and the level of our technology, the product of the uncertainty in position and momentum *in a simultaneous measurement* cannot be less that $\hbar$. In other words, it is impossible to build an apparatus that measures both the position and momentum of an object more accurately than is allowed by the uncertainty relation (22.2).

Heisenberg's uncertainty principle also applies to energy and time. If one tries to measure the energy E of a particle during a time interval t in which the particle has energy E, then it is impossible to determine both E and t more accurately than the following uncertainty relation:

$$(\Delta\mathbf{E}) \cdot (\Delta\mathbf{t}) \geq \hbar. \tag{22.3}$$

It is important to note that Heisenberg's uncertainty principle applies to the *simultaneous* measurement of incompatible observables. There is no limit set by the uncertainty principle on the accuracy of an isolated position measurement of a particle (or a simultaneous measurement of its position along with other observables *compatible* with position), as long as one does not try to determine its momentum at the same time. Similarly, one can determine the momentum of a particle as accurately as one pleases, as long as the position of the particle is not measured. The same statement applies to the time-energy pair.

What do you know? 22.2. A precision-tools company claims that it has invented a device that can measure the energy of an atom with an uncertainty of only 0.01 eV in a time interval of 10^{-14} second. Is the company's claim valid?

The appearance of the Planck constant in the uncertainty relations demands very small uncertainties in the measured quantities for the principle to be violated. This tells us that only in the subatomic world of the electrons and nuclei does the uncertainty principle put restrictions on measurements. The world of ordinary objects need not worry about violating the principle, because even for unrealistically accurate measurements of positions and momenta of ordinary objects, the uncertainty principle holds automatically. The following example helps clarify this point.

Example 22.2.2. Suppose that we can determine the position (of the center of) a tennis ball to within a hundredth of a millimeter (0.001 cm or 0.00001 m). This means that we can say with certainty that the ball is, for instance, 3.00001 m away from the center of the thickness of the net, and not 3.000015 m. This is a very (in fact, unusually) accurate measurement with a very small uncertainty of $\Delta \mathbf{x} = 0.00002$ m.[1] Similarly, suppose that we can determine the speed of the tennis ball to within 0.001 cm/s (0.00001 m/s), meaning that we can say for certain that the ball is moving with a speed of, say 50.00001 m/s and not 50.000015 m/s. This leads to an uncertainty of $\Delta \mathbf{v} = 0.00002$ m/s. If the mass of the ball is 200 grams (0.2 kg), then the uncertainty $\Delta \mathbf{p} = m\Delta \mathbf{v}$ in momentum will be 0.000004 kg·m/s. Therefore, the product of the uncertainty in position and momentum is $0.00002 \times 0.000004 = 8 \times 10^{-11}$, which is safely larger than $\hbar$. ∎

Schrödinger and Heisenberg pictures. We now have two theories—wave mechanics, also called the **Schrödinger picture** and matrix mechanics, also called the **Heisenberg picture**—that successfully describe the hydrogen atom. Both are highly mathematical, but use completely different mathematics in their treatment of the H-atom. It turns out that, just as a rose could be "described" by different words in different languages, wave mechanics and matrix mechanics describe the hydrogen atom in two different "languages." And just as one can go from one ordinary language to another by means of translation, one can also go from wave mechanics to matrix mechanics (and vice versa) by means of a process known as *transformation theory*, invented by one of the most brilliant mathematical physicists of all time, Paul Adrian Maurice Dirac, of whom we shall say more later.

What do you know?   22.3. What happens when Δp or ΔE of the Heisenberg uncertainty principle goes to zero?

22.3   The Meaning of Ψ

The Schrödinger equation contains the symbol Ψ, the "wave function," which is a function of the position $\mathbf{r}$ of the electron and time t, thus written as $\Psi(\mathbf{r}, t)$. The solution of the

---

[1]The uncertainty is twice the value quoted above, because we can either *overestimate* the position or *underestimate* it. Thus, we can only say that the position of the ball is greater than 2.99999 m and less than 3.00001 m. The difference between these two values is the uncertainty.

Schrödinger equation, for say the H-atom, yields not only the energy levels, but also a mathematical expression for $\Psi(\mathbf{r}, t)$. But what *is* Ψ? Schrödinger was the first to answer that question: Just as we have EM waves, sound waves, water waves, and waves on a rope or a membrane, so do we have "electron waves." However, there are at least three problems associated with this: (a) What exactly is waving? (b) If the electron is a wave, why do we always detect it as *localized* individual particles, contrary to our experience of waves being spread out? (c) Unlike all the other waves mentioned above, which are described by *real-valued* functions, Ψ is a *complex-valued* function due to the presence of $i = \sqrt{-1}$ in the Schrödinger equation. How can a real physical entity such as an electron be described by a complex-valued function?

22.3.1 Ψ and Probability

The revolutionary answer to these questions came from Max Born (1882–1970)—the German mathematical physicist and Heisenberg's mentor—in June 1926, six months after the Schrödinger equation was discovered. He stated quite radically that

Born introduces probability into the *foundation* of physics.

> **Box 22.3.1.** $\Psi(\mathbf{r}, t)$ *is a quantity whose absolute value squared times a small volume centered at* $\mathbf{r}$ *gives the probability of finding the particle in that volume at time t. If we denote this probability by* $P(\mathbf{r}, t)$ *and the small volume by* δV, *then* $P(\mathbf{r}, t) = |\Psi(\mathbf{r}, t)|^2 \, \delta V$.

$\Psi(\mathbf{r}, t)$ itself is called the **probability amplitude**.

The content of Box 22.3.1 is sometimes called the **Copenhagen interpretation**, because in the late 1920s, many young physicists gathered around Niels Bohr in Copenhagen to investigate the mysteries of the newly emerging quantum theory. At that time no one knew how to "interpret" the wave function Ψ coming out of the Schrödinger equation. However, after almost a century of harsh tests and experimental scrutiny, the word "interpretation" is no longer appropriate, because it invites the philosophization of physics, which in turn, gives the impression that there may be other (perhaps more mystical) interpretations of the quantum theory. Box 22.3.1 says how nature behaves in exact mathematical terms. It does not behave any other way, and there are no other "interpretations!"

Born's claim was more than most physicists could bear. The idea that physics, the most exact science, cannot predict the behavior of objects obeying its laws seemed to be undermining its very foundations. The science that could predict the location of planets to within fraction of a degree, that could foretell the existence of other unknown and unseen worlds in the solar system with miraculous determinacy, how could that science allow even the slightest hint of indeterminacy into its sacred shrine?

🧺 Food for Thought 🧠

FFT 22.3.2. Mathematics has always been inseparable from nature. Starting with the observation of the sky and the tilling of the land, humans have been forced to use mathematics whenever they have dealt with nature. The "experiments" of the Egyptians and Babylonians with geometry and arithmetic took the form of a most sophisticated discipline in the hands of the Greeks. By Plato's time, the need for mathematics in the study of nature was so compelling that he said, "Innocent light-minded men, who think that astronomy can be learnt by looking at the stars without knowledge of mathematics will, in the next life, be birds." And Archimedes naturally mingled the geometry of solids with hydrostatics to the point that he could not separate them, and his discoveries in one field led to the advancement of both.

After the Renaissance, Galileo, in his study of motion, had to invent a rudimentary form of calculus. And when Newton looked deep into the secret of motion, nature, in her most forceful way, compelled him to perfect calculus and beautifully apply it to the laws of motion and gravity.

Inseparability of mathematics and physics.

Later, the more physicists applied the mathematical laws of nature to new situations, the more new fields of mathematics opened up. Almost all the great mathematicians of the eighteenth and the first half of the nineteenth centuries were also great physicists.

There were some mathematicians, however, who concentrated on the inner abstract structure of the mathematics *that was earlier created by physics*. These abstract studies led to such pure and strictly mathematical ideas that many thought that the human race had finally created a form of mathematics that was completely separate from nature; that those branches of mathematics were so much the product of the human mind that nature would have nothing to do with them. These areas included imaginary numbers, infinite-dimensional spaces, non-Euclidean geometry, and group theory.

For a while, it seemed that the conjecture of the abstract mathematicians was correct. Physics seemed to have nothing to do with imaginary numbers, infinite-dimensional spaces, non-Euclidean geometry, and group theory. However, the twentieth-century physics changed all that. Einstein was forced to use the language of non-Euclidean geometry in his relativity theory; Schrödinger was forced to introduce imaginary numbers into his equation; Dirac showed that the only way to understand the new quantum theory was to incorporate the infinite-dimensional spaces into it. And group theory, which grew out of the attempts to solve algebraic equations (a truly and purely mathematical study), found its *indispensable* application in the study of the most fundamental constituents of matter.

The heart of matter speaks in mathematics, and only in mathematics. Just as the beauty of a poetry is best revealed in its original language, so does physics show its beauty and elegance in its language, mathematics. This, of course, does not mean that we have to be deprived of this beauty just because we are not well versed in mathematics. After all, Shakespeare can (and should) be appreciated by all humankind, not just English-speaking people. Göthe does not belong only to Germans, and Omar Khayyam only to Farsi-speaking people. Translations can convey—albeit to a limited extent—the essence of the beauty of the poetry from one language to another.

Impossibility of the translation from the nature's language to any human language.

However, while the translation from one human language to another can be as truthful as the talent and the ability of the translator (compare the Edward FitzGerald translation of the Rubaiyat of Omar Khayyam to other translations), a truthful translation from the nature's language to *any* human language is impossible. This is especially true of the mathematics of quantum theory, as it is highly abstract and formal. Nevertheless, it is an implied duty of physicists to make every effort to translate their language into human languages, so that nonprofessionals can appreciate the relevance of physics for two reasons: to support further development of physics (as the progress of physics *requires* public support), and to encourage the youth to become professional physicists. However, it is essential for both the physicists and the larger citizenry to keep the following wise remarks of Dirac in mind: "The amount of theoretical [mathematical] ground one has to cover before being able to solve problems of real practical value is rather large, but this circumstance is an inevitable consequence of the fundamental part played by transformation theory and is likely to become more pronounced in the theoretical physics of the future."

Schrödinger, Planck, and even Einstein criticize Born for introducing probability into physics.

Schrödinger's reaction was immediate and sharp. He remarked that had he known that his equation was to be interpreted as probabilistic, he would not have published it. Planck was on Schrödinger's side; and when Schrödinger took his place on his retirement in 1931, Planck praised him for bringing determinacy back into physics. These remarks by such well-known physicists were very unpleasant to Max Born, who was anxious to see how his idea would hold in the face of experiments and observations. But unpleasantness turned to disappointment when Einstein wrote to Born, "Quantum mechanics is certainly imposing. But an inner voice tells me it is not yet the real thing." Despite all these opposing remarks, Born prevailed.

Difference between probability used in statistical mechanics and in the quantum theory.

The notion of probability and statistics had already been applied to physics in the context of *statistical mechanics*, and both Planck and Einstein were masters of that application. But they were not concerned about the use of probability in statistical mechanics, because that usage was not at a *fundamental* level. The particles in a gas obeyed the *deterministic* laws of motion—and in principle one could keep track of the precise motion of each individual particle—but for a large number of molecules, it was more convenient and practical to use statistical methods. Max Born's probability, on the other hand, was used at a very

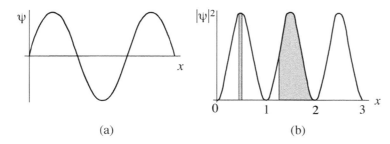

Figure 22.2: (a) The probability amplitude of an electron confined in a one-dimensional box. (b) The probability distribution of the same electron. The areas shown are the probabilities of finding the electron in the corresponding intervals.

fundamental level. One *could not* keep track of the motion of even a single electron. The best one could do was to determine the probability of finding an electron at a given location at a given time.

The infinitesimal size of the electron and other subatomic particles renders Max Born's idea applicably feasible. Even though the Schrödinger equation cannot predict the exact behavior of an *individual* subatomic particle, it predicts the behavior of *any macroscopic* collection of such particles with unprecedented accuracy. After all, even the smallest macroscopic volume such as the tip of a sharp needle contains over a million trillion atoms, and the law of large numbers (Section 16.3) makes the probabilistic prediction of the quantum theory all but deterministic.

Probability applied to a large number of subatomic particles gives deterministic results.

Figure 22.2 shows the probability amplitude and probability of an electron confined in a region along the x-axis (a one-dimensional box) of size 3 nm. The amplitude, as shown in Figure 22.2(a), can be positive or negative (in fact, in general, it can be a complex number involving $\sqrt{-1}$). The probability distribution (or probability density), given by $|\Psi|^2$, is always positive. To find the probability of finding the electron between two values of x, say x_1 and x_2, one calculates the **fractional area** under the graph of $|\Psi|^2$ bounded by those values. *Fractional area is the area as a fraction of the total area under the entire probability curve.* If x_2 is only slightly larger than x_1, so that $\delta x = x_2 - x_1$ is very small, this area can be approximated by $|\Psi|^2 \delta x$ as stated in Box 22.3.1 (here the infinitesimal "volume" is just δx). For example, the probability of finding the electron in a small interval of width δx at $x = 0.5$ nm is $|\Psi(0.5)|^2 \delta x$, where $\Psi(0.5)$ is the value of the wave function at $x = 0.5$ nm. This probability is shown as the narrow strip in Figure 22.2(b). If x_1 and x_2 are noticeably different, then one has to calculate the area either accurately by using integral calculus, or approximately by dividing the area between x_1 and x_2 into narrow strips of width δx and adding the areas of these strips, each one being $|\Psi|^2 \delta x$. One can also estimate the probability by just examining the graph, as the following example shows.

Probability distribution and area as probability.

Example 22.3.3. In Figure 22.2(b), we approximate each bump of the graph as a triangle. The base of each triangle is 1 nm. But what number do we assign to the height? It turns out that since we are interested in the *fractional area*, it doesn't matter, as long as we measure other heights in the graph relative to that number. So let us take the height to be 1. (Try another number and see that the final answer will not change.) Then the area of each triangle is $\frac{1}{2} \times 1 \times 1 = 0.5$. To get the area of the shaded region, subtract from this area the area of the right triangle in the middle hump on the left of the shaded area. This has a base of approximately 0.25 nm and an estimated height of 0.5. So its area is $\frac{1}{2} \times 0.25 \times 0.5 = 0.0625$. Therefore, the area of the shaded region is $0.5 - 0.0625 = 0.4375$. The probability of finding the electron between 1.25 nm and 2 nm is the ratio of this number to the total area (which is 3 times the area of each hump or $3 \times 0.5 = 1.5$). So, the probability of finding the electron between 1.25 nm and 2 nm is $0.4375/1.5$ or 0.2917. This calculation indicates that the electron spends about 29% of its time between 1.25 nm and 2 nm. ■

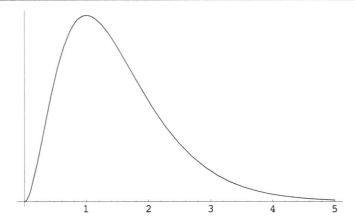

Figure 22.3: The probability distribution of the electron in a hydrogen atom which is in the $n = 1$ energy level. The horizontal axis is the distance from the nucleus in units of the Bohr radius a_0.

What do you know? 22.4. In Figure 22.2, how likely is it to find the electron between 1.99 nm and 2.01 nm?

22.3.2 Ψ for Hydrogen Atom

Let us now apply the Schrödinger-Born theory to the hydrogen atom. Recall that one result of this application was the quantization of the energy of the electron identical to the Bohr formula (see Box 21.3.1). What about the wave function Ψ? It turns out that the wave function—and therefore the probability—depends on the energy of the electron, which in turn depends on the integer n, called the **principal quantum number** in the context of the Schrödinger-Born theory. If $n = 1$, so that the energy is -13.6 eV, the probability distribution looks like the plot in Figure 22.3. The horizontal axis is the distance r from the nucleus in units of the Bohr radius a_0, equal to 5.3×10^{-11} m. It is clear from the graph that the maximum probability occurs when $r = a_0$. Hence, the electron is mostly found at a distance corresponding to Bohr's first orbit. But it can also be found at other distances as well, albeit with less probability.

Principal quantum number.

What do you know? 22.5. Assuming that a hydrogen atom is in $n = 1$ state most of the time, what is the radius of the hydrogen atom? Hint: Look at Figure 22.3.

Schrödinger-Born theory does not have the flaws of the Bohr theory.

The Schrödinger-Born theory corrects the flaws of the Bohr model of the hydrogen atom. It does not confine the electron to an orbit. Therefore, it does not lead to the difficulty of the emission of EM radiation due to its acceleration. As the figure shows, the theory actually allows the electron to be at all distances from the nucleus. However, it prohibits the electron from getting too close to the nucleus (and therefore collapsing into it) or too far away from the nucleus (and, therefore, escaping it altogether).

It turns out that for any value of the principal quantum number n, the probability function of the electron at the nucleus is zero, and drops to zero for very large values of r. However, the peak of the probability is no longer at $r = a_0$, but farther away from the nucleus. This is as expected, because, at least from the point of view of the Bohr model, the larger the n is, the farther the electron is expected to be from the nucleus.

The electron is not to be pictured as lying on orbits at fixed distances from the nucleus, but as "hovering" around certain orbits most of the time, with the possibility of being found very close and very far away from the nucleus. In fact, quantum theory defies any sort of mental picture for the behavior of the subatomic particles. However, if we were forced to imagine an electron in a hydrogen atom, we would have to replace the Bohr orbits with "fuzzy" orbits (sometimes called *electron clouds*) of the sort shown in Figure 22.4, where the stronger the shade of black the higher the probability of finding the electron there.

Figure 22.4: The first two "orbits" of the electron in an H-atom.

We now have a complete and satisfactory theory of the hydrogen atom. According to this theory, the electron can never be found at the nucleus, and therefore, the H-atom of this theory is stable. Furthermore, the quantum theory allows calculations which were completely out of reach in the Bohr model. For instance, it allows determining the probability that an electron makes a transition from a higher level to a lower level. It can even determine the "lifetime" of any level with a given principal quantum number. And while the Bohr model could not handle multi-electron atoms, the Schrödinger-Born theory is very well-suited for generalization to more complicated atoms, and even molecules. Although the analytical solutions of such generalizations are extremely complicated, powerful computers have been used very successfully in predicting the structure of complex atoms and molecules.

Schrödinger-Born theory can be applied to complex atoms and molecules.

22.4   End-of-Chapter Material

22.4.1   Answers to "What do you know?"

22.1.   Use de Broglie's relation to find the wavelength of the virus. First we need the momentum of the virus: $p = mv = 10^{-23} \times 0.01$. So, the momentum of the virus is 10^{-25} kg·m/s. Its wavelength is h/p or 6.626×10^{-9} m, or 6.626 nm. In order for the virus to pass through the hole, the hole must be at least 30 nm, which is less than 5 times the wavelength. So, if the hole is slightly more than 30 nm, the virus can get through it, and the hole is not too much bigger than the wavelength. So, diffraction can take place.
22.2.   The uncertainty in energy is 0.01 eV, or 1.6×10^{-21} J. The uncertainty in time is 10^{-14} s. The product of these two is 1.6×10^{-35} J·s. This is smaller than $\hbar$, and therefore, violates the uncertainty principle. The company is lying!
22.3.   As one factor in the Heisenberg uncertainty principle goes to zero, the other factor must go to infinity. Thus, precise knowledge of one precludes *any* knowledge of the other (an infinite uncertainty is equivalent to no knowledge at all).
22.4.   As Figure 22.2 shows, it is highly unlikely to find the electron between 1.99 nm and 2.01 nm, because the probability vanishes at 2 nm.

22.5. There is no such thing as the radius of an atom, because radius has a connotation of an exact distance, and we know that the electron's distance from the nucleus is only a probabilistic quantity. Nevertheless, we can ask where the electron is most likely to be found, and call that the "radius" of the atom. The most probable distance from the origin (nucleus) is a_0. That means that the electron is hovering around a_0 from the nucleus. Therefore, it is natural to call a_0 the radius of the H-atom.

22.4.2   Chapter Glossary

Compatible Observables Two or more observables which can be measured simultaneously without any restriction on the accuracy of their measurement.

de Broglie Relation A simple mathematical formula relating a particles momentum to its wavelength.

Incompatible Observables Two observables which cannot be measured simultaneously with unlimited accuracy.

Matrix Mechanics The study of microscopic systems employing Heisenberg's idea that all physical quantities are represented by matrices.

Observable Used in matrix mechanics to describe any physical quantity.

Probability Amplitude The wave function Ψ; the solution of the Schrödinger equation.

Schrödinger Equation A fundamental equation describing the behavior of subatomic particles under the influence of (usually electrical) forces. This equation is based crucially on the assumption that particles have wave properties.

Uncertainty Principle A physical principle, discovered by Werner Heisenberg, stating that two incompatible observables cannot be measured simultaneously with unlimited accuracy.

Wave Mechanics The study of microscopic systems employing the Schrödinger equation.

Wave-Particle Duality The idea that all particles have wave property and all waves have particle property.

22.4.3   Review Questions

22.1. What analogy did de Broglie use to endow electrons with wave properties? Which two theories did he incorporate? What physical quantity describes waves? What describes a particle? How are these two properties related in the context of de Broglie's hypothesis?

22.2. In principle, do all particles have wave properties? Can we detect such properties for ordinary particles? Why or why not?

22.3. State some of the features of the Schrödinger equation. What two quantities did the Schrödinger equation yield when applied to the H-atom?

22.4. State the basic assumption of matrix mechanics. Do matrices always commute when multiplied together? Define compatible and incompatible observables. What is a complete description of a quantum mechanical system?

22.5. State Heisenberg's uncertainty principle. What pair of observables does it apply to? Can you measure a single physical quantity with arbitrary accuracy? Can you measure a pair of physical quantities with arbitrary accuracy, if they are incompatible? If they are compatible?

22.6. Name two pairs of incompatible observables. Can you simultaneously measure time and position of a particle with arbitrary accuracy? What about time and energy? What about momentum and position?

22.7. What is the correct "interpretation" of Ψ, the wave function? Why is "interpretation" flanked by quotation marks?

22.8. What is the difference between the probability used in statistical mechanics and that used in quantum mechanics? Is the Schrödinger equation useless simply because it only predicts the probability of the behavior of the individual subatomic particles? How does the law of large numbers come to our rescue?

22.9. When applied to H-atom with the principal quantum number $n = 1$, what predictions does the Schrödinger-Born theory have? Does it give orbits for the electron? Is it possible for the electron to be extremely close to the nucleus, closer than Bohr's first orbit? Is it very likely? Is it possible for the electron to be very far the nucleus? Is it very likely?

22.10. If you were to define "orbits" for the electron, would they be sharp or fuzzy? What are such orbits usually called?

22.11. How complete is the Schrödinger-Born theory? What other phenomena does it predict? Can it be applied to complex atoms? How successful is this application?

22.4.4 Conceptual Exercises

22.1. Electron A moves twice as fast as electron B. Which one has a shorter wavelength? A proton and an electron move with the same speed. Which one has a longer wavelength?

22.2. To "see" an object with a wave (such as light), the wavelength of the wave should be comparable to the size of the object. How small an object can you see with an optical microscope? With an electron microscope?

22.3. Complex numbers did not exist prior to the sixteenth century! So, the Schrödinger equation could not have been discovered earlier than the sixteenth century. Think about this, and the connection between mathematics and physics.

22.4. If you multiply the position matrix and the momentum matrix with position first, do you get the same thing if momentum came first in the product?

22.5. Can you measure the position of an electron with unlimited accuracy? Can you measure the momentum of an electron with unlimited accuracy?

22.6. Could probability amplitude be negative? Could probability be negative?

22.7. At a certain location, the probability amplitude of an electron oscillates between a maximum of 5 and a minimum of -5. Between what values does the probability (density) oscillate at that location?

22.8. Is it correct to say that whenever the probability amplitude is minimum, so is the probability (density)? Explain.

22.9. Referring to Figure 22.2, what is the probability for the electron to be between zero and 1 nm from the left side of the box? Between 2 nm and 3 nm from the left side of the box?

22.10. Would the Schrödinger equation be a good equation for the H-atom if it predicted a nonzero probability at the origin? What would be the consequences?

22.11. Would the Schrödinger equation be a good equation for the H-atom if the probability did not drop to zero for large values of r? What would be the consequences?

22.4.5   Numerical Exercises

22.1. A bacterium has a mass of 5×10^{-18} kg, a diameter of 1 μm, and moves with a speed of 0.5 m/s.
(a) What is the wavelength of this bacterium?
(b) Is it possible to see the diffraction of such bacteria?

22.2. How slow should a 100-gram bullet move so that it can exhibit diffraction? Take the diameter of the bullet to be 1 cm.

22.3. A bacterium of length 10^{-6} m has a mass of 10^{-15} kg and moves with a speed that can be measured to within 0.001 cm/s. What is the minimum uncertainty in its position?

22.4. Referring to Figure 22.2, approximate the three humps as triangles with unit height. What is the probability for the electron to be between 0.45 nm and 0.55 nm from the left side of the box? Hint: Take the strip to be a rectangle.

22.5. Referring to Figure 22.2, approximate the three humps as triangles with unit height. What is the probability for the electron to be between 2.5 nm and 3 nm from the left side of the box?

Chapter 23

Quantum Theory II

The highly mathematical nature of the quantum theory, combined with the probability inherent in it, makes it one of the "weirdest" theories ever discovered by humankind. This weirdness comes partly because of our attempt at "translating" the consequences of the quantum theory into our own language. Such a "philosophization" of the quantum theory, a theory which works smoothly and perfectly in the language of mathematics, has led to its different "interpretations." In this chapter, we summarize some consequences of the quantum theory.

23.1 Quantum Tunneling

When we store candies in a jar, we are assured that when we reach out in our cabinet and take the jar out, the candies will be there, even after a long time. In the absence of a mischievous child (or a weak-willed adult), it would be impossible to find a candy outside the jar. Nevertheless, in the strange world of the quanta, this is precisely what happens!

Suppose that we, as well as everything around us, shrink to subatomic sizes. Then, the behavior of everyday objects in our world is governed by the laws of quantum mechanics. Now, let us concentrate on the motion of a single (quantum) candy in a (quantum) jar [Figure 23.1(a)]. The candy obeys the Schrödinger equation, and the jar is described by a mathematical formula representing the potential energy of the candy. When the Schrödinger equation is solved for this jar-candy system, two basic results follow: (a) there is a principal quantum number quantizing the energy of the system as in the hydrogen atom, and (b) each quantized energy has a wave function Ψ.

Figure 23.1(b) shows the probability distribution $|\Psi|^2$ for the lowest energy level (or lowest principal quantum number). The two vertical lines show the boundaries of the jar. The shaded region gives the probability of finding the candy in the jar. Note that the shaded area does not cover the entire area under the probability curve. A small fraction of the curve lies outside the boundary of the jar, indicating a nonzero probability for finding the candy outside! This phenomenon is called **quantum tunneling**, which is, unfortunately, a misnomer, because no tunnel connects the inside of the jar to the outside.

Quantum tunneling.

If you put a quantum candy in a quantum jar, set it on a table and do not disturb it, chances are that you will eventually see the candy outside. And this is not because the candy is small and the jar may have a hole—a "tunnel"—in it! You can make the jar as solid as you wish.

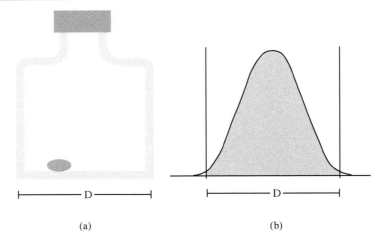

Figure 23.1: (a) A quantum jar with a quantum candy inside. (b) The probability distribution of the candy in the jar.

Box 23.1.1. *The probability of finding the candy outside the jar is nonzero. This probability depends on how tall the jar is; it decreases as the height of the jar increases. For an infinitely tall jar, the probability is zero.*

A classical candy can be found outside only by lifting it and putting it there.

To understand this weird behavior better, let us go back to our real world of classical physics. The height of the jar can be represented mathematically by a potential-energy function with a nonzero value, say 1 J (one Joule), in the thickness of the jar wall. If the total energy E of the candy is less than 1 J, say $E = 0.5$ J, then the candy cannot penetrate the wall of the jar, because as soon as it arrives in the thickness of the wall, the potential energy becomes larger than the total energy, indicating that the kinetic energy is negative, which can never happen! Therefore, a candy whose total energy is less that the potential energy of the height of the jar is permanently confined inside the jar. The only way that the candy can be found outside is to give it enough extra energy so that its total energy becomes larger than the potential energy of the wall. This means, in effect, lifting it from the bottom of the jar and putting it outside.

Tunneling is easily explained in the language of mathematics, but impossible to explain in any human language.

How does the quantum candy end up outside without any lifting? There is really no answer in the "human language" to this question. The nonhuman (mathematical) answer is "The solution to Schrödinger equation gives a nonzero probability. Therefore, there is a chance that the candy is found outside." A more "convincing" answer may be as follows. Heisenberg's uncertainty principle in time and energy [Equation (22.3)] allows the candy to have a huge uncertainty in its energy if the time spent in the wall is infinitesimally small. So, suppose we keep track of the candy and note that one moment it is inside and the next moment outside. The uncertainty in time is very small, making the uncertainty in energy big. What is the consequence of this? Well, if we are hugely uncertain about the energy, then we cannot say whether its conservation has been violated or not. Therefore, it is possible for the candy to violate the (classical) energy conservation, "create" extra energy to overcome the potential barrier of the wall, and land outside.

The foregoing "explanation" of the weirdness of the quantum tunneling is really cheating our way out. We invoked the uncertainty principle to explain the tunneling, and in the process, we shifted from one form of weirdness to another. Uncertainty principle is just as weird as the tunneling effect; and there is no way that one can "explain" the uncertainty principle. It is a *mathematical* consequence of the *mathematical* matrix mechanics.

The probabilistic nature of tunneling requires a large sample of jar-candy system to test the prediction of the theory. For example, if the theory predicts a probability of 10^{-15} for the candy to be seen outside, then we need a sample considerably larger than 10^{15} jar-candy systems to be able to see any candies outside. A sample of 10^{20} systems, for instance, would, on the average, result in (see Box 16.2.2) $10^{20} \times 10^{-15} = 10^5$ or 100,000 candies outside, a large enough number to be easily detectable.

The weird quantum tunneling can actually be observed!

What do you know? 23.1. Suppose that the tunneling probability for a jar-candy system is a billionth. How many jar-candy systems do you need to see 1000 candies out?

As weird as quantum tunneling is, nature abounds in it.

- A special form of nuclear **radioactivity** is a prime example of quantum tunneling. Some heavy nuclei act as jars that hold quantum candies called the *alpha particles*.[1] From the knowledge of the size of the nucleus and some rudimentary understanding of nuclear forces, it is possible to construct a jar-like model of the nucleus with a nuclear potential representing the wall of the jar. One can then solve the Schrödinger equation for such a model and predict the probability of seeing an alpha particle outside. With even a small sample of nuclear material having a huge number of "jars," the prediction can be easily tested against observation. Such tests have been performed and the predictions confirmed.

- **Nuclear fusion**, the process that generates energy for stars to shine and give off heat, is another example of quantum tunneling. Two ("heavy") hydrogen atoms can fuse together to form a helium atom. This process of fusion releases a tremendous amount of energy, which we see as the light and heat from a star (our Sun). For fusion to take place, the nuclei of the two hydrogen atoms *must get immeasurably close* to each other. But since the two nuclei are positively charged, they repel each other; and the closer they get, the stronger this repulsion. The *potential barrier* corresponding to this repulsion is so high that, if classical physics were to apply, it would require an infinite amount of energy to overcome the barrier. Fortunately, due to quantum tunneling, the particles can "tunnel" through the potential barrier and fuse. Fusion is the inverse of the alpha emission. In the latter case, the particle tunnels "out," while in the former case the two particles tunnel "in."

Nuclear fusion is impossible without quantum tunneling.

- Aside from the two *natural* examples of quantum tunneling above, there are many *artificial* examples, which have far reaching technological ramifications:

  - **Tunnel diodes** are semiconductor devices consisting of two oppositely charged regions separated by a gap (the potential barrier). This potential barrier can be controlled by the voltage between the two regions. By varying this voltage, one can vary the number of electrons tunneling through the barrier.

  - **Quantum dots** are artificial quantum "jars" constructed by quantum technologists, in which a single (or multiple) electrons can be trapped. There is a tremendous amount of effort devoted to the construction and use of these dots in electronics.

  - **Josephson junction** consists of two superconductors separated by a thin insulating layer. Under certain conditions, it is possible to tunnel a pair of electrons from one superconductor to the other through the insulating barrier.

---

[1] The same α-particles that Rutherford used to discover the atomic nucleus.

– **Scanning tunneling microscope** is probably the masterpiece of technology using the quantum tunneling effect. It is discussed in some detail in Section 23.6.

What do you know? 23.2. When two positive light nuclei approach infinitesimally close to each other, the electric repulsion becomes infinitely large. So, the penetration "wall" is infinitely tall. According to Box 23.1.1, the probability of tunneling is zero. How can you have fusion then?

23.2   The Double-Slit Experiment

In Chapter 11, we talked about the interference of waves originating from two coherent sources, which are usually two slits in a solid barrier. From de Broglie and Schrödinger we have learned that electrons have wave properties. Therefore, we should be able to see the interference of the electrons. However, the electrons are always detected as particles. How does this dual nature of the electron play itself out in a two slit experiment? To appreciate the consequences of this wave-particle duality, let us perform three separate double-slit experiments: with bullets, with waves, and finally with electrons.

23.2.1   Bullets and Double-Slits

A gun fires towards a distant vertical barrier with two holes, which allow bullets to go to the other side, being subsequently collected on a second (wooden) barrier (see Figure 23.2). We perform three different experiments with this gun-barrier set up.

In the first experiment, we cover the lower hole [see Figure 23.2(a)], and fire a large number, say 1000, bullets. After all the bullets are fired, we go to the other side and count the number of bullets collected by the wooden obstacle. Let us say that 50 bullets made it to the other side. It then follows that the probability—call it P_1—for the bullets fired from that particular gun to pass through the upper hole is 50/1000 or 0.05 or 5%. The 50 bullets that passed through the upper hole will not all be landing at exactly the same point on the wooden obstacle. There will be a lot landing at A_1 directly opposite the center of the hole, but there will also be some landing on points close to A_1. This indicates that P_1 has a distribution that looks approximately like the curve shown at the bottom of Figure 23.2(a).

In the second experiment, we cover the upper hole [see Figure 23.2(b)], and fire 1000 bullets as in the first experiment. Since the two holes are assumed to be identical, we expect that around 50 bullets will make it to the other side. Thus, the probability—call it P_2 this time—for the bullets fired from that particular gun to pass through the lower hole is also 0.05 or 5%. P_2 is expected to have a distribution that looks very much like P_1 with maximum at A_2 directly opposite the center of the lower hole. The curve for P_2 is shown at the bottom of Figure 23.2(b).

Finally, in the third experiment, we leave both holes open [see Figure 23.2(c)], and fire 1000 bullets as in the first two experiments. If we count the number of bullets collected by the wooden obstacle, we will see that around 100 bullets will have made it to the other side. Thus, the probability—call it P_{12}—for the bullets fired from the gun to pass through both holes is 0.1 or 10%. P_{12} is expected to have a distribution of the shape shown at the bottom of Figure 23.2(c). And indeed that is what one finds in an actual experiment, because each bullet arriving at the first barrier can go either through the upper hole, in which case it will be part of the P_1 distribution, or through the lower hole, in which case it will be part of the P_2 distribution. If we want to see which bullet goes through which hole, we place a sufficiently intense light source[2] behind the first barrier and watch the bullets one by one

---

[2]This light source could be the Sun if the experiment is performed outside.

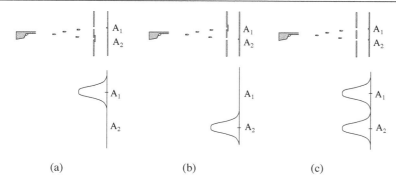

(a)   (b)   (c)

Figure 23.2: The gun on the left fires bullets toward a distant barrier with two holes in it. (a) Lower hole is covered. (b) Upper hole is covered. (c) Both holes are open. The three curves at the bottom of the set-ups show the probability distribution for the bullets to land on the wooden screen. The subscripted letter A designates the point directly opposite the center of each hole. Note that the distance $\overline{A_1 A_2}$ in the plots has been enlarged for clarity.

as they come out the back side.

A consequence of this experiment is that probabilities add. The probability for the bullets to land on the wooden screen when both holes are open is the sum of the probabilities for each hole when the other is blocked. In terms of the symbols defined above, $P_{12} = P_1 + P_2$. In particular—when the holes are far apart—the probability that a bullet lands opposite the midpoint of the two holes is negligible (almost zero).

What do you know? 23.3. When the two holes are apart, the probability of finding a bullet at the midpoint of the two holes is zero [see Figure 23.2(c)]. As you bring the holes closer, it is no longer zero. Can you make this probability equal to P_1? Larger than P_1?

23.2.2 Waves and Double-Slits

A wave approaches an obstacle—from the left—with two apertures whose separation is not too much larger than the wavelength of the wave (see Figure 23.3). The wave is then stopped at a second barrier—with an array of detectors, if necessary—which measures the intensity of the arriving wave at various (vertical) points of the barrier. As for the bullets, we perform three experiments. In the first experiment, we close the lower aperture as shown in Figure 23.3(a). Assuming that the size of the aperture is not exceedingly large, the wave will *diffract* through the upper hole. The diffracted wave will have a (wide) central band and two faint secondary bands on both sides of it [see Section 11.3, especially Figure 11.8]. The intensity curve of the diffracted wave is shown in Figure 23.3(a). For all practical purposes, the secondary bands can be neglected, giving an intensity curve that is similar to that of the probability curve of the bullets of the previous subsection. Let us call this intensity I_1. We could also interpret I_1 as the probability of finding "a wave" on points of the detecting barrier.

In our second experiment, we close the upper aperture. Now the wave will *diffract* through the lower hole, and the intensity curve of the diffracted wave is shown in Figure 23.3(b). Let us call this intensity (or probability) I_2.

The third experiment leaves both apertures open. The approaching wave now *splits up*. Part of it goes through the upper hole and part through the lower hole [Figure 23.3(c)]. The two holes act as coherent sources, which emit waves on the right that will interfere and

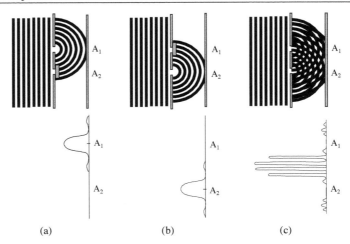

Figure 23.3: The wave approaches a barrier with two holes in it. (a) Lower hole is covered. (b) Upper hole is covered. (c) Both holes are open. The three curves show the intensity (probability) distribution for the wave to land on the detecting screen. The extra bumps on either side of the central band are due to diffraction. The subscripted letter A designates the point directly opposite the center of each hole. Note that the distance $\overline{A_1 A_2}$ in the plots has been enlarged for clarity.

produce the typical fringes discussed in Chapter 11. The intensity curve will look like that shown in Figure 23.3(c). Let us call this intensity I_{12}. It is clear that $I_{12} \neq I_1 + I_2$; and we expect that, because, for waves, we add the *amplitudes* not the intensities.

Let us denote the amplitude of the wave coming through the upper hole by a_1, and that through the lower hole by a_2. The total amplitude—call it a_{12}—will then be $a_{12} = a_1 + a_2$. Since intensity is proportional to the square of the amplitude, we have $I_{12} = (a_{12})^2 = (a_1 + a_2)^2$, which gives

$$I_{12} = I_1 + I_2 + 2a_1 a_2, \tag{23.1}$$

where I_1 is the intensity of the upper hole and I_2 that of the lower hole. This shows that the total intensity is not equal to the sum of the two intensities. If the two holes produce waves of equal amplitude, and two crests or two troughs meet at some point, then we have $a_1 = a_2$, where the amplitudes can be positive (crest) or negative (trough). Substituting a_1 for a_2 in Equation (23.1), we obtain $(a_{12})^2 = (a_1 + a_1)^2 = (2a_1)^2 = 4(a_1)^2$, or $I_{12} = 4I_1$. This is constructive interference. If one crest and one trough meet at some point, then we have $a_1 = -a_2$ (it does not matter which one is crest and which one trough). Substituting $-a_1$ for a_2 in Equation (23.1), we obtain $(a_{12})^2 = (a_1 - a_1)^2 = 0$, or $I_{12} = 0$, which is destructive interference.

23.2.3   Electrons and Double-Slits

Now we come to the crucial double-slit experiment with electrons. Electrons are particles that can be fired from an "electron gun."[3] We set up three experiments similar to the ones for bullets. When the upper hole is open, we get what we expect: a probability distribution P_1 very similar to the one shown in Figure 23.2(a). When the lower hole is open, we get a probability distribution P_2 of Figure 23.2(b).

Now we open both holes, expecting to see a probability distribution P_{12} of Figure 23.2(c). But that is not what we get! Instead, we observe a probability distribution very similar to Figure 23.3(c). How can that be? The answer is simple: according to de Broglie, electrons have wave properties, and when you send them through a double slit, they show the typical

---

[3]Such a device actually exists.

interference fringes expected of waves. In fact, if the holes are small enough, even a single slit can exhibit diffraction, so that the probability distributions P_1 and P_2 will look more like the intensity curves in Figures 23.3(a) and (b). But we ignore these diffraction effects.

Is the answer really that simple? After all the electrons are particles, and we can, if we want to, fire them one at a time. And when each electron reaches the double-slit, it cannot "split up" as a wave does.

> **Box 23.2.1.** *The electron has to go either through the upper hole or the lower hole.*

If it goes through the upper hole, it belongs to the distribution P_1; if it goes through the lower hole, it belongs to the distribution P_2. There is, therefore, no way for the electron to end up at the midpoint of $\overline{A_1 A_2}$, where both P_1 and P_2 are almost zero. But it does! How can it be? We answer this question in two different ways: In the language of nature, mathematics, and in a human language, English.

First mathematics. In this language, we have to forget the statement of Box 23.2.1. It is a statement that comes from picturing the electrons as bullets. Although the electrons are *detected* as particles, they are not *entirely* particles, as humans have experienced particles such as bullets. In the language of mathematics, the electrons that go through the upper hole are described by a wave function (or probability amplitude), say Ψ_1 at the detecting screen. Similarly, the electrons that go through the lower hole are described by another wave function Ψ_2. The total probability *amplitude*—call it Ψ_{12}—for the electrons detected at the screen is simply the sum of the two amplitudes: $\Psi_{12} = \Psi_1 + \Psi_2$. Exactly like waves! To find the total probability, we must square the total amplitude. Now, in general, Ψ_1 and Ψ_2 are complex numbers, adding a slight complication in the algebra. To avoid this inessential complication, let us assume that Ψ_1 and Ψ_2 are real, as they are in many actual situations. Then, squaring Ψ_{12} as in Equation (23.1), we obtain the total probability P_{12}:

When spoken in its native language (mathematics), quantum theory is easy to understand.

$$P_{12} = P_1 + P_2 + 2\Psi_1\Psi_2. \tag{23.2}$$

This shows that the total probability distribution has an interference pattern, exactly as in waves. Thus, in the language in which nature likes to speak, there is no difficulty explaining the behavior of the electrons.

We humans, however, have no "feel" for such mathematical explanation, and insist that we should explain the electrons' behavior in our own language, in which the statement in Box 23.2.1 is quite legitimate. Let us see how far we can go.

READER: The electron that reaches the double-slit is a particle. Therefore, we have no choice but to assume that it either goes through the upper hole or the lower one.

When spoken in a foreign (human) language, quantum theory is incomprehensible.

AUTHOR: But contrary to the bullet, an electron is not easily visible. So, you can't tell whether it went through the upper hole or the lower one. We should not make any assumptions unless we can verify it.

R: Okay. So, tell me, is it possible to put a light bulb behind the first barrier to actually *see* the electrons? If so, then I have proved my point. Just illuminate the path of the electrons behind the slits as shown in Figure 23.4, and you will see for yourself that the electrons are either going through the upper hole or the lower hole.

A: This much I know: In the so called *gedanken* experiments, anything that does not violate any known laws of physics is allowed. So, I don't see any reason why we can't put a light bulb behind the barrier.

R: Okay. Then put a really bright light bulb behind the slits to make sure that we can see the electrons.

A: I am sorry, but there is a problem with putting a light bulb behind the slits. It will interfere with the motion of the electrons.

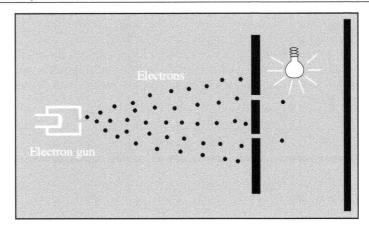

Figure 23.4: A light bulb behind the slits will help locate which hole each electron goes through.

R: Wait a minute! You are contradicting yourself. If you have forgotten, let me refer you to your own discussion of bullets and double-slits in which you said, and I quote, "If we want to see which bullet goes through which hole, we place a sufficiently intense light source behind the first barrier and watch the bullets one by one as they come out the back side."

A: That is a very good point. However, it is the difference in the size of a bullet and an electron that forces me to be careful when analyzing their motion. Any time you shine light, you are sending a large number of photons, each of which carries some energy. (Remember the Einstein-Planck formula $E = hf$?) However large the number of these photons is, and however large the amount of energy each one carries, their effect upon impact on a bullet is infinitesimal. It is like hitting a bowling ball with tiny grains of salt. But the same photons, even a single one of them, if it carries enough energy, can affect the motion of the electron considerably.

R: I get it now. But still, I don't see any reason why we can't go on with our experiment. Let the photons interfere with the electrons. As long as we can see which hole each one passes through, it shouldn't matter.

A: It *does* matter, because when the light interacts with the electrons, it randomizes their path and smears out the interference fringes on the detecting screen. In other words, the electrons will behave like bullets, and their wave nature disappears.

R: Okay. If the interaction of light with electrons destroys the wave property, let's minimize this interaction. Use a dimmer light bulb.

A: Good idea! But it does not completely obliterate the problem. You see, when you dim the light, you are just reducing the *number* of photons not the energy of each one. And when you send fewer photons to the slits, they can interact with fewer electrons.

R: You mean that, only for some electrons can we tell which hole they went through. That is fine as long as we can track them to the detecting screen. Maybe we will finally be able to see the interference fringes formed by these electrons. Is there a way of differentiating the electrons that we see at the slit from those that we miss?

A: (*After a long, thoughtful pause*) I think I have found a way. The electron detectors are counters that click when an electron arrives at them. Since we are sending the electrons one by one, we can keep track of them by a sequence of flashes and clicks. Here is how. If there is a flash at the upper hole and then a click at the screen, we know the electron came from the upper hole. Same for the lower hole. Finally if there is no flash at the holes but there is a click at the screen, we know the electron was missed.

R: That is great! So, now we *can* track the electrons. Do we see an interference?

A: We do and we don't! The electrons that we detected at the holes form a bullet-like pattern. *Their* potential fringe is smeared out again due to the randomization by interacting with photons.

R: Let me guess. The missed electrons will form an interference fringe.

A: Exactly! As long as you don't interfere with electrons, they exhibit wave properties. As soon as you disturb them, they exhibit only their particle properties.

R: That is frustrating. There must be a way of seeing *all* the electrons without disturbing them too much. You said that reducing the intensity of light changed only the *number* of photons, but not their energy. What if we reduce the *energy* of the photons rather than their number? Hey, we can do that, can't we? If I remember correctly, the energy of the photon is given by $E = hc/\lambda$. So, let's make the wavelength of the photons as large as possible. Let's replace the light bulb with a source of long EM waves.

A: That is very ingenious! There are microwave sources producing photons with wavelengths a few centimeters long, or about 100,000 times longer than the photons of a light bulb.

R: Good! Let's use a microwave source.

A: We certainly can; and when we do, we indeed see the interference of the electrons on the detecting screen.

R: So, that's it! We have accomplished what we wanted.

A: I hate to disappoint you, but with a long-wavelength photon, we encounter a new obstacle.

R: What *now?*

A: A long wavelength cannot resolve the two slits, because the separation between the slits is much much smaller than the wavelength. It is like trying to hit *only one* of a pair of billiard balls separated by just a few millimeters with a "stick" that is as thick as a football field! If the wave cannot tell the two slits apart, it is certainly incapable of determining which hole a given electron passes through.

R: So, are you saying that we are stuck?

A: I'm afraid so. In fact, our inability to determine the hole through which each electron passes, and at the same time observe the interference fringes is related to the uncertainty principle. When we determine the hole through which the electron passes, we are localizing it, i.e., we are making the uncertainty in position small. The uncertainty principle then enlarges the uncertainty in momentum, which is related to the wavelength via de Broglie relation. A large uncertainty in wavelength translates into the absence of the wave nature, i.e., interference. On the other hand, when we do observe the wave nature of the electrons by using long-wavelength photons, we lose information about their location at the slits.

R: Wait a minute! Maybe we are not stuck after all. Since it was the small size of the slit separation relative to the photon wavelength that prevented us from determining which hole the electron went through, we can increase the separation until it is much larger than the (long) wavelength of the photons. That way we do not disturb the electrons—because our photons have very long wavelengths—and we can tell which holes each one goes through— because the photon wavelength is small compared to the slit separation.

A: But you are forgetting something. If we increase the separation between the two slits to the point that it is much larger than a few centimeters, we won't see interference, because that kind of separation is also much much larger than the electron wavelength. See Section 11.3, if you don't remember the condition for interference.

R: I don't like quantum theory. It just doesn't make sense.

A: It doesn't make sense because we are trying to adapt it to *our* way of thinking using *our* language. If you express it in its own language of mathematics, it makes perfectly good sense. The double-slit experiment is only one example of the "weirdness" of quantum theory. There are other examples. I already mentioned tunneling and the spooky way that a quantum candy can "get out of" a quantum jar. Another example is quantum angular momentum.

> **What do you know? 23.4.** How do you reconcile the statement of Box 23.2.1 with the non-bullet-like behavior of the electrons?

23.3   Angular Momentum and Spin

Quantum physics is strongly bound to classical physics.

At first glance there seems to be no relation between the quantum theory and the classical Newtonian theory that it replaced. After all, in the classical theory there is no quantization of energy, no spooky tunneling, and no probability. On closer examination, however, a deep relation between the two theories emerges. For example the fact that the concept of energy—developed laboriously in Newtonian physics—can be used in the quantum theory speaks to such a relation. Many other concepts such as momentum, potential energy, kinetic energy, electric and magnetic fields, etc. can also be "translated" into the quantum language. It was indeed the translation—sometimes called the *first quantization*—of the conservation of mechanical energy that gave rise to the Schrödinger equation.

23.3.1   Angular Momentum

A classical quantity that gives truly astonishing results upon "quantization" is angular momentum. Let us recall from Section 8.2 that classical angular momentum is a vector quantity **J** related to rotational motion. For the simplest case of rotation about an axis, the direction of this vector is given by the right-hand rule: curl the fingers of your right hand in the direction of rotation, your thumb points in the direction of **J**. It therefore points along the axis of rotation. And for an even simpler case of a single particle of mass m moving on a circle of radius r with speed v, one can give the *magnitude* of **J**: it is rmv or rp, with p the (linear) momentum of the particle.

> **What do you know? 23.5.** What is the direction of the Earth's angular momentum?

Angular momentum, like any other vector, is determined completely by its components.

By varying the speed or the orientation of the axis of rotation, we can vary the magnitude and the direction of the classical angular momentum at will: holding a single bicycle wheel by its shaft, we can rotate it as fast as we please, and can orient the shaft in any direction we desire. There is no theoretical restriction on the direction and magnitude of the classical angular momentum. Another way of stating this is to say that classical angular momentum can have any (precise) value for its projections along three mutually perpendicular axes. Figure 23.5 shows the vector **J** and its projections J_x, J_y, and J_z along the three mutually perpendicular axes, x, y, and z. The three projections (also called *components*) uniquely determine—and are uniquely determined by—**J**. There is no restriction on the values of the components J_x, J_y, and J_z. There is also no restriction on how accurately we can measure these components *simultaneously*. For any given **J**, of course, the values of the components are restricted to lie between $|\mathbf{J}|$ (when **J** lies along the positive direction of the corresponding axis) and $-|\mathbf{J}|$ (when **J** lies along the negative direction of the corresponding axis). Thus, if **J** lies along the positive z-axis, then $J_z = |\mathbf{J}|$ (and $J_x = 0$, $J_y = 0$); and if **J** lies along the negative z-axis, then $J_z = -|\mathbf{J}|$ (and $J_x = 0$, $J_y = 0$).

In quantum theory, angular momentum, like other physical quantities, is represented by a matrix; in fact, by three matrices corresponding to the three components J_x, J_y, and

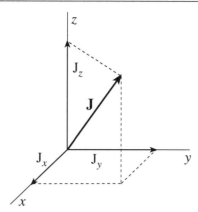

Figure 23.5: Any vector, including **J**, has three projections (or components) along three mutually perpendicular axes.

J_z. The mathematics of matrices (see Section 22.2.2) then yields the result that no two components of a given angular momentum commute. Hence, by matrix mechanics,

Box 23.3.1. *No two components of angular momentum can be measured simultaneously.*

It turns out, nevertheless, that the matrix representing the magnitude[4] of **J**—denoted by |**J**| or J—commutes with all three components. Therefore, this magnitude and one (and only one) of the components—usually taken to be the z-component, J_z—can be used to determine an *angular momentum state* of a system (see Box 22.2.1).

What do you know? 23.6. Can you simultaneously measure J_z and the *projection* of **J** onto the xy-plane (not J_x or J_y individually)?

When the elegant machinery of matrix theory is applied to quantum angular momentum some surprising results emerge. The first result is that the *magnitude of the angular momentum is quantized*, and the theory gives precisely how it is quantized. It turns out that $|\mathbf{J}| = \sqrt{j(j+1)}\,\hbar$, where j is a nonnegative integer and $\hbar$ is the Planck constant divided by 2π. Thus, we can have angular momenta of magnitudes 0 (corresponding to $j = 0$), $\sqrt{2}\,\hbar$ (corresponding to $j = 1$), $\sqrt{6}\,\hbar$ (corresponding to $j = 2$), etc. This is in complete contrast to the classical angular momentum. While classically, any nonnegative number was a possible value for the magnitude of angular momentum, in quantum theory any magnitude that cannot be written as $\sqrt{j(j+1)}\,\hbar$ with j a nonnegative integer (e.g., a magnitude of $\sqrt{\frac{21}{16}}\,\hbar$ corresponding to $j = \frac{3}{4}$) is forbidden.

*The magnitude of angular momentum |**J**| is quantized.*

What do you know? 23.7. A physicist claims that in an experiment, he measured the magnitude of angular momentum of a particle and obtained $\sqrt{306}\,\hbar$. Is he talking nonsense?

The second result is that J_z—the other quantity describing the quantum state of the

Components of angular momentum are also quantized, and they are further restricted.

---

[4]The square of the magnitude of any vector is, by Pythagoras' theorem, the sum of squares of its components. If you are interested in more details, see Appendix C.

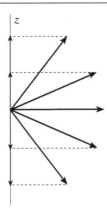

Figure 23.6: The quantum angular momentum of magnitude $\sqrt{6}\,\hbar$, corresponding to $j = 2$. Note that the maximum projection along the z-axis is not $\sqrt{6}\,\hbar$ but $2\hbar$. Altogether there are 5 allowed values for J_z.

angular momentum—is restricted to a value *not* between $-|\mathbf{J}|$ and $|\mathbf{J}|$, but between $-j\hbar$ and $j\hbar$. Furthermore, it cannot take just any value between those two numbers, but only values obtained from them in steps of 1. So, (in units of $\hbar$) the smallest value for J_z is $-j$; the next value is $-j + 1$, the next one is $-j + 2$, etc., all the way to $j - 1$, and finally j. Figure 23.6 illustrates this for $j = 2$. The length of the angular momentum vector is $\sqrt{6}\,\hbar$. The maximum and the minimum values for J_z are not $\sqrt{6}\,\hbar$ and $-\sqrt{6}\,\hbar$, but $2\hbar$ and $-2\hbar$. The other allowed values for J_z are $\hbar$, 0 and $-\hbar$. The figure shows that only 5 angles are allowed for the orientation of $\mathbf{J}$ relative to the z-axis.

We chose J_z as the second observable to label the quantum angular momentum state. We could have chosen J_x or J_y. Moreover, the orientation of a coordinate system is completely arbitrary. We can choose any direction to be our positive z-direction (the x- and y-directions are of course not arbitrary once we choose a z-axis). Therefore, the quantization of projections applies to any direction. We summarize our findings in the following Box. However, as we shall see shortly, our conclusions apply only to orbital angular momentum, i.e., the angular momentum resulting from the orbital motion of a particle around a center.

Box 23.3.2. (Quantization of orbital angular momentum) *In quantum theory the (magnitude of the) orbital angular momentum can take on only values given by* $|\boldsymbol{J}| = \sqrt{j(j+1)}\,\hbar$, *where j is a nonnegative **integer**. The projection of the orbital angular momentum along any line can take on values between $-j\hbar$ and $j\hbar$ incremented by $\hbar$ from the lowest to highest values. The number of these projections is $2j + 1$.*

Quantization of direction?

Sometimes the quantization of the projection of angular momentum is interpreted as "quantization of direction." This is not entirely correct, because there is no restriction in quantum theory on our choice of direction for the axes of our coordinate system. We can choose any direction we want along which to project the angular momentum vector. However, once this (arbitrary) projection axis is chosen, the angular momentum vector can have "quantized" direction (angle) with respect to it.

What do you know? 23.8. What is $|\mathbf{J}|$ for $j = 3$? What are the values for its projection along an arbitrary line? How many directions does angular momentum have?

23.3.2 Spin

The pace at which the quantum weirdness popped at the early twentieth century physicists was phenomenal. No sooner had one weirdness been "resolved" than another jumped at them. People were just beginning to get used to the weird idea of quantized angular momentum when they were hit with the weirder notion of spin.

Angular momentum is closely related to a quantity called *magnetic moment*. In fact, it can be shown that (for electrically charged rotating objects) magnetic moment is always proportional to angular momentum. Classically, a magnetic moment can be thought of as a (usually microscopic) magnet with north and south poles. A magnetic moment has a direction: it is the arrow that connects its south pole to its north pole. When you place a magnetic moment in a magnetic field it aligns itself with the field so that its arrow is in the same direction as the field.

Angular momentum and magnetic moment.

In the absence of a magnetic field, a collection of classical magnetic moments will be randomly oriented: their arrows are equally likely to point in any direction. When you place these magnetic moments in a magnetic field, many (but not all) of them will align with the field. Although the tendency is for all moments to align themselves, collision and exchange of energy among the moments will reorient many moments in other directions. Thus, the arrows of the moments can point in all directions, although the direction of the field is slightly preferred.

Now suppose that the magnetic field is not homogeneous, but varies in strength in a particular direction. Then, the physics of magnetism says that there is a force on each moment, and this force depends on the angle between the moment and the field. Let's say that the magnet is arranged so that the moments that are aligned with the field (the angle between the moment and the field is 0) are pushed up by this force while those aligned opposite to the field (the angle between the moment and the field is 180 degrees) are pushed down. All the other moments (with angles lying between 0 and 180 degrees) are pushed somewhere between these two extremes.

A magnetic moment in an inhomogeneous magnetic field experiences a force.

Figure 23.7(a) shows an apparatus that demonstrates the effect of the force of an inhomogeneous magnetic field on the magnetic moments. A source produces magnetic moments randomly distributed in all directions. These moments travel horizontally through an inhomogeneous magnetic field, causing them to pick up a transverse component of speed, so that when they come out of the field, they will be diverted slightly up or down from their original direction. Moments that are aligned with the field will deflect farthest upward and those anti-aligned farthest downward. The other moments lie between these two extremes.

If the magnetic moments behaved classically, the deflection would be continuous and the deposit on the collector screen would be a continuous blob shown at the top of Figure 23.7(b). Of course, that is not what we expect, because microscopic magnetic moments behave quantum mechanically. A quantum magnetic moment—being proportional to the quantum angular momentum—has a discreet number of projections along any axis. In fact, this number is given in Box 23.3.2 as $2j + 1$, which is an odd number for any integer j. Therefore, regardless of the magnitude of the magnetic moment, the number of angles it makes with *any* direction is *odd*, and correspondingly, the number of blobs on the collector screen has to be odd. For example, if the angular momentum to which the magnetic moment is proportional happens to have $j = 1$, then 3 blobs are expected, as shown in the middle of Figure 23.7(b).

When Otto Stern and Walther Gerlach, two German physicists, performed an actual experiment of the type described above using silver atoms, they observed a truly weird outcome: The number of blobs turned out to be 2. It was hard enough to get accustomed to the quantization of angular momentum with j being an integer. Now Nature was telling us that j was not necessarily an integer. The *orbital* angular momentum, the kind that results from the motion of the electron around the nucleus, could not be responsible for this double-valuedness; its mathematical theory was air tight, and did not allow any noninteger

Discovery of spin.

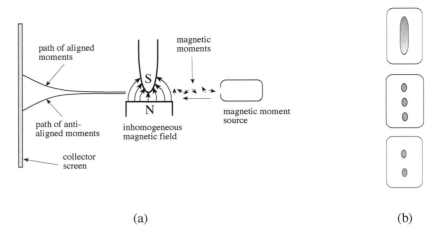

Figure 23.7: (a) The magnetic moments spread out on the collector screen after they pass through the inhomogeneous magnetic field. (b) Classical moments yield the single long perpendicular blob on the top. Quantum theory predicts an odd number of blobs, as shown in the middle. Some magnetic moments yield the lower blobs on the collector screen.

value for j. It was therefore suggested that the electron has not only an orbital angular momentum, but an intrinsic **spin**. The name "spin" is a misnomer, suggesting an actual intrinsic rotation of the electron. There is no such spin just as there is no "orbital" angular momentum, because there is no "orbit."[5] The correct way of dealing with spin is to consider it as another intrinsic property of a particle. A particle is identified by specifying its various physical characteristics such as mass, charge, speed, etc. Spin is one more such characteristic.

Since $2j + 1$ determines the number of blobs on the screen of Figure 23.7 and since for spin this number is 2, we conclude that $j = \frac{1}{2}$ for the intrinsic spin of some particles (including the electron). It turns out that there are other possibilities for j. In fact, the following box summarizes our understanding of a total angular momentum, consisting of orbital and spin angular momenta:

Box 23.3.3. (Quantization of total angular momentum) *In quantum theory the (magnitude of the) total angular momentum can take on only values given by* $|\boldsymbol{J}| = \sqrt{j(j+1)}\,\hbar$, *where j is a positive integer or half-integer (an odd integer divided by 2). The projection of the angular momentum along any line can take on values between $-j\hbar$ and $j\hbar$ incremented by $\hbar$ from the lowest to highest values. The number of these projections is $2j + 1$.*

Orbital angular momentum, whose value of j is usually denoted by l and referred to as *orbital angular momentum quantum number*, is always an integer (see Box 23.3.2). Thus, any half-odd-integer valuedness is associated with the spin. Instead of j one uses s for spin.

A particle whose spin is $\sqrt{s(s+1)}\,\hbar$ is called a **spin-s** particle. Thus, the electron is a *spin-half* (or *spin-$\frac{1}{2}$*) particle; its projection with $s = \frac{1}{2}$ is called **spin up**, and with $s = -\frac{1}{2}$, **spin down**. Like the total angular momentum, s can assume only integer or half-odd-integer values. Particles with integer spin are called **bosons**; those with half-odd-integer spins are called **fermions**. If the particle is massive, there are $2s + 1$ projections of

Distinction between **bosons** and **fermions**; projection of spin for massive and massless particles.

---

[5] Orbits were denounced in Section 22.3.2.

spin along any given direction starting at the minimum of $-s\hbar$ and incrementing by $\hbar$ to the maximum of $s\hbar$. Massless particles have only two projections: $-s\hbar$ and $s\hbar$.

What do you know? 23.9. In a Stern-Gerlach type experiment, 6 blobs are formed. What is the spin? What are the possible values for s_z if the particle is massive? If massless?

The concept of spin was the missing link in explaining the periodic table of elements. Recall that the energy levels of a hydrogen atom are determined by the principal quantum number n. Matrix mechanics states that a complete description of any quantum system is given by a list of possible values of a complete set of compatible observables. For the H-atom this list consists of n, l (the orbital angular momentum quantum number), l_z—often denoted by m—(the projection of the orbital angular momentum along the z-axis), and s_z (the projection of the electron spin along the z-axis). Sometimes we write this complete set of compatible observables as $|n, l, m, s_z\rangle$. A rule that emerges out of solving the Schrödinger equation for the H-atom is that l is restricted to nonnegative integer values that are less than n. Thus, when $n = 1$ (i.e., when the H-atom has its lowest possible energy), $l = 0$; and when $n = 2$, $l = 1$ or $l = 0$, and so on. For each l, there are $2l + 1$ values for m, and the value of s_z is restricted to $\pm\frac{1}{2}$. Any set of possible values of these numbers determines a quantum mechanical state of the H-atom.

What do you know? 23.10. A physicist claims to have discovered a particle with a spin magnitude $1.5\sqrt{7}\,\hbar$. Has he discovered a boson, a fermion, or is he talking nonsense?

To build more complicated atoms, one adds an appropriate number of electrons to the H-atom (and a corresponding number of positive charges to the nucleus), assuming that the states of the resulting atom are also described by $|n, l, m, s_z\rangle$. The question is: As we add more and more electrons to build more and more complicated atoms, in what states do they place themselves? One is interested in the lowest possible energy state also called the **ground state**. One might think that keeping $n = 1$, the lowest possible energy state, and adding all the electron in this state might be a possibility. However, the resulting "table of elements" will not be periodic at all. In fact, it will consist of a single column in which all elements lie.

Ground state is the state of lowest energy of a quantum system.

In order to reconstruct the *periodic* table of elements, Wolfgang Pauli (1900–1958), an Austrian physicist, proposed the

Pauli exclusion principle.

Box 23.3.4. (Pauli Exclusion Principle:) *No two identical fermions (such as two electrons) can be placed in the same quantum state.*

This principle turned out to have far-reaching ramifications involving the role of relativity in the formulation of quantum theory. At this point, we simply accept the principle and apply it to the construction of the periodic table.

Exclusion principle explains periodic table of elements.

For $n = 1$ there are only two states: $|1, 0, 0, \frac{1}{2}\rangle$ and $|1, 0, 0, -\frac{1}{2}\rangle$. Therefore, we can place only two electrons in the $n = 1$ state. This is sometimes restated as *two electrons are needed to fill the first* **shell**. The (neutral) atom that has two electrons is helium. Therefore, both electrons in the ground state of the helium atom have $n = 1$, and helium has a *filled* (first) shell. The next electron cannot go in the first shell; it must start filling up the $n = 2$

states—the second shell. There are 8 such states, whose $|n, l, m, s_z\rangle$ values are given below:

$$l = 0 \text{ for these:} \quad \left|2, 0, 0, \tfrac{1}{2}\right\rangle, \left|2, 0, 0, -\tfrac{1}{2}\right\rangle \qquad m = 0$$

$$l = 1 \text{ for these:} \quad \begin{cases} \left|2, 1, -1, \tfrac{1}{2}\right\rangle, \left|2, 1, -1, -\tfrac{1}{2}\right\rangle & m = -1 \\ \left|2, 1, 0, \tfrac{1}{2}\right\rangle, \left|2, 1, 0, -\tfrac{1}{2}\right\rangle & m = 0 \\ \left|2, 1, 1, \tfrac{1}{2}\right\rangle, \left|2, 1, 1, -\tfrac{1}{2}\right\rangle & m = 1 \end{cases}$$

The ground state of lithium—the element that has 3 electrons—has 2 electrons in the first shell and 1 electron in the second shell. The ground state of beryllium—the element with 4 electrons—has 2 electrons in the first shell and 2 electrons in the second shell. The ground state of carbon—the element with 6 electrons—has 2 electrons in the first shell and 4 electrons in the second shell. The ground state of neon—the element with 10 electrons—has 2 electrons in the first shell and 8 electrons in the second shell. The next electron can no longer end up in an $n = 2$ state: this shell gets filled with 8 electrons. Therefore, sodium, which has 11 electrons, starts filling up the $n = 3$ states (the third shell).

What do you know? 23.11. Fluorine has 9 electrons. With what elements does it combine most strongly?

Helium and neon are the first two of the so-called "inert gases." These are elements that have filled (or *closed*) shells, and are chemically inactive, because they cannot "share"[6] electrons with other elements. These electrons are rather tightly bound to the rest of the atom. The inert gases appear in the same column of the periodic table of elements. The element that comes right after an inert gas is chemically very active, because it has an extra electron outside a closed shell that is loosely bound to the rest of the atom and can be easily shared by other atoms. Hydrogen, lithium, and sodium are examples of such elements, and they appear in the same column of the periodic table.

The other entries of the periodic table can also be explained using the exclusion principle. However, as the number of electrons increases in the atom other considerations have to be taken into account, complicating the details of the theory. For example, the third inert gas, argon, does not have the entire third shell closed. Only the $l = 0$ and $l = 1$ *subshells* of the third shell are filled. This is because for multielectron atoms it is possible for the lower values of n to have higher energies. These complications render the construction of the periodic table of elements beyond the scope of this book.

The discovery of spin was important not only because it explained the periodic table, but also because it was a part of a larger quantity that we have called *total angular momentum*. The significance of this quantity is that

---

Box 23.3.5. *The total angular momentum of a quantum system is conserved. The system may consist of different parts each having its own spin and orbital angular momentum. These parts may be isolated or interacting among themselves.*

There is a deep connection between this conservation law and the fact that physical phenomena are independent of the orientation of the objects from which they emanate. For example, when we measure the gravitational force between two masses at a given distance, we don't have to take into consideration whether the two masses lie along a line extended from left to right, or up to down, or front to back, or anything in between. In general,

---

[6]Electron sharing—whereby one atom shares one or more of its electrons with another atom—is at the root of all chemical reactions.

neither each individual spin (even their total), nor each individual orbital angular momentum (even their total) is conserved. It is only the total angular momentum that does not change. Of course if the system has only spins, or only orbital angular momenta, then the total of that quantity is conserved.

23.4 Quantum Measurement

It is appropriate at this point to discuss an important consequence of the quantum theory: measurement. Any act of measurement involves a *system*, some of whose properties are being measured, and an *apparatus*, which does the measuring. In classical physics it is assumed that the act of measurement does not influence the property being measured. This is so ingrained in our thinking that we never question the impact of the apparatus on the system. And there is good reasons for this: in almost all cases, the measuring apparatus is much "smaller" than the system. A telescope can hardly influence the motion of a galaxy, a star, or a planet; watching a race car as we time its travel from one point to another in a speed measurement has no impact on the car whatsoever; the influence of ammeters and voltmeters on the currents and voltages *for which they are designed* is minimal.

A quantum mechanical measurement tells a completely different tale. Two important factors enter a quantum measurement:

- The apparatus is much larger that the quantum system being measured. Therefore, its influence on the system can certainly not be neglected: the process of measurement becomes an *interaction* between the apparatus and the quantum system.

- This interaction demands a chain of events, the beginning link of which is a quantum interaction, and the end link of which is a *macroscopic* signal that we can detect.

We now focus on the theory of measurement in quantum physics.

Any (quantum) measurement is made on a system in some quantum state. The latter is defined in terms of—or, as is more commonly stated, **labeled** by—a *complete set of compatible observables*. For example, states of a hydrogen atom (or its electron) are labeled by the energy (with quantum number n), the orbital angular momentum (with quantum number l), the projection of the orbital angular momentum along some axis, say the z-axis (with quantum number m), and the projection of the electron spin along the same axis (with quantum number s_z). We have denoted such states by $|n, l, m, s_z\rangle$. For example, $\left|1, 0, 0, \frac{1}{2}\right\rangle$ describes the hydrogenic electron in its lowest energy (-13.6 eV), with (necessarily) $l = 0$ and $m = 0$, whose spin projection along the z-axis is $\frac{1}{2}\hbar$.

Quantum states and how they are labeled.

A general state of the hydrogenic electron is given in terms of a **superposition** of labeled states. This means that Ψ for the electron is a sum of different $|n, l, m, s_z\rangle$'s each multiplied by a number (coefficient). The square of each coefficient represents the probability of the electron to be in that particular state. For instance,

$$\Psi_1 = \frac{1}{\sqrt{2}}\left|1, 0, 0, \tfrac{1}{2}\right\rangle + \frac{1}{\sqrt{2}}\left|1, 0, 0, -\tfrac{1}{2}\right\rangle$$

describes a hydrogenic electron that is in the lowest energy state with equal probability of having spin up or down. On the other hand,

$$\Psi_2 = 0.6\left|1, 0, 0, \tfrac{1}{2}\right\rangle + 0.8\left|2, 1, 0, -\tfrac{1}{2}\right\rangle \tag{23.3}$$

is an electron that has 36% chance of being in the lowest energy state with spin up and 64% chance of being in the second energy state with an orbital angular momentum 1, a 0 projection of orbital angular momentum, and a spin down. Other sums with more (even infinitely many) terms are also possible.

Measurement results can
be **sharp** or **blurred**.

Suppose that a hydrogen atom is "prepared" in state Ψ_1, and we measure its energy. What is the result of the measurement? It is clear that although Ψ_1 is a superposition of two states, the result of the measurement will be -13.6 eV (corresponding to $n = 1$), because both states have the same energy. Similarly, a measurement of orbital angular momentum or its projection yields 0. We say that energy, angular momentum, and its projection are **sharp** in Ψ_1. On the other hand, measurement of the spin projection does not yield a definite result: it gives $\frac{1}{2}\hbar$ half the time and $-\frac{1}{2}\hbar$ the other half. We say that spin projection is **blurred** in Ψ_1.

Thus far, we have measured only compatible observables. What about the measurement of incompatible observables? It should be clear from the uncertainty principle that such a measurement is *always* blurred. In the case of hydrogen, for example, position is not one of the compatible observables. Hence, any position measurement on a state $|n, l, m, s_z\rangle$ (or superposition of such states) is always blurred. This is the reason behind Figures 22.3 and 22.4. We, therefore, conclude that

---

Box 23.4.1. *A measurement of one of the compatible observables made on a system labeled by those observables can be either sharp (if the corresponding probability is 1) or blurred (if the corresponding probability is less than 1). Any measurement of an incompatible observable is always blurred.*

Example 23.4.2. What is the probability that a measurement of the spin projection on Ψ_2 of Equation (23.3) yields $-\frac{1}{2}\hbar$? Since the coefficient of the term with $s_z = -\frac{1}{2}$ is 0.8, the probability is $(0.8)^2$ or 0.64, indicating that the spin projection is blurred.

What is the probability that a measurement of the orbital angular momentum projection on Ψ_2 above yields 0? Since both terms have $m = 0$, the probability is 1. Stated differently, the probability is the sum of the squares of the coefficients of all terms with $m = 0$. In the case of Ψ_2 we have $(0.6)^2 + (0.8)^2$, which is 1. Thus, m has a sharp value in Ψ_2.

What is the probability that a simultaneous measurement of energy and spin projection yields -13.6 eV and $-\frac{1}{2}\hbar$, respectively? Since no labeled state exists in Ψ_2 with $n = 1$ and $s_z = -\frac{1}{2}$, i.e., since the coefficient of such a labeled state is 0 in Ψ_2, the probability is zero. ∎

What do you know? 23.12. Referring to Equation (23.3), what is the probability for orbital angular momentum to be 1? To be 2? To be zero? What is the probability for the energy to be -13.6 eV?

Measurement sends a
quantum system into a
state labeled by the
value of the measured
observable.

A measurement of an observable property of *any* state of a quantum system results in one of the possible values of that observable. Since this value is necessarily sharp immediately *after* the measurement, we must have "channeled" the system into a state with a sharp value for that particular observable. Thus the act of measurement sends a quantum system into a state labeled by one of the values of the measured observable. This is the essence of a quantum measurement.

23.5   Quantum Entanglement

The spookiest outcome of the quantum theory is **quantum entanglement**, a phrase coined by Schrödinger himself. It started with Einstein's struggle to show that the quantum theory proposed by Max Born, in which probability played a fundamental role, was incomplete. In a number of conferences attended by the prominent physicists of the time, including Einstein and Bohr, Einstein would come up with some ingenious *gedanken* (or thought) experiments, pointing, at first glance, to the incompleteness of the quantum theory. Bohr,

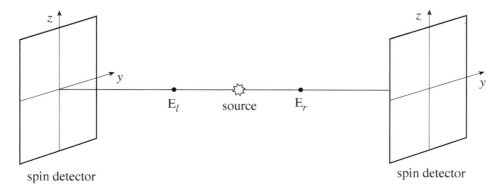

Figure 23.8: Two electrons fly off in opposite directions. The total spin of the two electrons is zero.

after a few hours of thought, would discover the flaw in Einstein's arguments and report it to him. One such argument seemed to have the potential of undermining the quantum theory.

Einstein and two of his collaborators, Boris Podolsky and Nathan Rosen, wrote a paper in 1935 that appeared to have given a dizzying jolt to (the completeness of) the quantum theory. In this paper, the authors suggest an experiment, now known as the *EPR experiment*, which demonstrates an ostensibly "telepathic communication" between two distant particles, violating the quantum theory. Instead of the original experiment, we shall examine a modification of it, proposed by David Bohm, that involves the spin of an electron.

A pair of electrons are produced at a source, flying off in opposite directions to two distantly separated spin detectors as shown in Figure 23.8. The source prepares the two-electron system in a state in which each orbital angular momentum is zero and the total spin is also zero. Since the total angular momentum is just the sum of the spins in this case, and since this total angular momentum is conserved, we conclude that—in the absence of any disturbance—the total spin must remain zero. In particular, just before they enter the detectors, the electrons must have opposite spins: if E_l is up (its s_z is $+\frac{1}{2}$), then E_r is down (its s_z is $-\frac{1}{2}$), and vice versa.

Now suppose we measure the projection of the spin of E_l along the z-direction in the left detector—which could be located in New York—without measuring or otherwise disturbing E_r. If the result of our measurement turns out to be "spin up," we conclude that E_r, which by now has reached the right detector—which could be located in London—must be in a spin-down state, and vice versa. In other words, the two electrons are *entangled*.

The mere fact that a measurement could influence the outcome of another performed far away is not a quantum mechanical spook. It happens classically as well. Take a penny and cut it thin in the middle so that one half bears the head and the other the tail; put each half in a locked box so no one can see it; give one box to Jack heading toward New York, and the other box to Jill heading toward London. If Jack opens his box, he will know exactly what Jill will find in *her* box when she opens it.

Einstein, Podolsky, and Rosen—and later, Bohm—were trying to show that *quantum mechanics* was wrong or incomplete. They argued that because the determination of the spin projection of E_l instantaneously specifies the spin projection of E_r, and because nothing can move faster than light, E_r must have had the specified value all along. In other words, the s_z-value of E_r must have been *sharp*. Now suppose we take our projection line to be the y-axis and measure s_y of E_l. Again, we may get either $+\frac{1}{2}$ or $-\frac{1}{2}$ for E_l, and the opposite for E_r. From this measurement we have to conclude that the s_y-value of E_r must have been *sharp*. Therefore, the quantum mechanical state of E_r must have been labeled by the sharp values of s_y *and* s_z. This violates the uncertainty principle, because, as Box 23.3.1

EPR paradox stated.

indicates, s_y and s_z are not compatible! How does quantum theory respond to this?

The best way to analyze the response is to speak in the language in which quantum theory is most comfortable: mathematics. The source prepares a *two-electron system* with total spin zero. The only way it can do this is by placing the left electron in the up position and the right one in the down position or vice versa. So the state Ψ prepared by the source is a *superposition* of two states, which we designate as $|u\ d\rangle$ and $|d\ u\rangle$. Let us first explain these states. u stands for "up" and d for "down." The words "up" and "down" have nothing to do with vertical direction. The symbol $|u\ d\rangle$ means that if we choose *any* line, the left electron E_l will have some projection on this line which we *arbitrarily* call "up"—thus the placement of the letter u on the *left* in the brackets. E_r will then have an opposite projection, which had better be called "down." A similar meaning is attached to the symbol $|d\ u\rangle$. Thus the state of the two-electron system can be written as

$$\Psi = \frac{1}{\sqrt{2}}\,|u\ d\rangle + \frac{1}{\sqrt{2}}\,|d\ u\rangle\,.$$

Note that Ψ does not have any sharp labels: no measurement of Ψ gives a sharp value for *any* relevant observable.

When we make a measurement on the system, we channel it into a particular labeled state; and this measurement *need not involve both electrons*. We have encountered this kind of situation before: if we measure the energy of the hydrogen atom and get -13.6 eV (corresponding to $n = 1$), we conclude, *without any measurement*, that the orbital angular momentum is zero (see the discussion following Box 23.3.3). Thus, a measurement of the spin projection of E_l alone *is* a measurement of the system, regardless of the distance between its parts. Furthermore, *before* such a measurement, Ψ has only blurred values for both spins. Only *after* our measurement along the z-axis does s_z—and *only* s_z—become sharp. And if we choose the y-axis, then s_y becomes sharp, *but never both s_y and s_z*. We therefore conclude that the EPR argument against the uncertainty principle is not valid.

It should be evident from the discussion above (as well as the double-slit experiment with electrons) that all the "weirdness" of the quantum theory stems from the *superposition principle*. What Schrödinger called "entanglement" was nothing but this principle, which from a mathematical point of view is so natural and so obvious that no attention or discussion is (or need be) devoted to it. Superposition is so elementary that we encountered it in disguise at the beginning of the book: the rule of the addition of vectors (see Appendix C) is nothing but the superposition principle. As long as we recognize the essential role this principle plays in the (necessarily mathematical formulation of the) quantum theory, we will encounter no "weirdness." It is only when we step outside the natural habitat of the theory, and enter the realm of "interpretation," that the quantum weirdness emerges.

23.6   Quantum Technology

The legitimacy of science lies in its ability to explain the mysteries of nature and to predict unforeseen phenomena. No other area of science has been able to accomplish this better than quantum physics. The drive for explanation is the sole motivation for all great scientists; and the founders of the quantum theory were no exceptions. In their attempt at explaining the hydrogen atom they stumbled on one of the most abstract and "strange" theories ever. They did not foresee the awe inspiring impact it came to have on every aspect of industry of the second half of the twentieth century and beyond.

Unlike some science educators, we will not "legitimize" science in general, physics in particular, and especially *quantum* physics by devoting pages upon pages of this book on the praise of their pragmatic utility. To us, when science predicts unseen elements that are found later, when it describes the behavior of the simplest element, hydrogen, with sophisticated mathematical precision, when it predicts the bending of light in gravity and measures this bending with utmost ingenuity, when it predicts the mechanism of the creation of the universe and its expansion, when it foretells the existence of a background radiation in our universe and measures the properties of this radiation with a disagreement of less than one percent with the theoretical prediction, then science has proven its legitimacy far beyond any of its application in a rocketship, a sports utility vehicle, a wide screen television set, or a dazzling special effect of a Hollywood movie.

However, quantum physics has given so much to our civilization that a short list of the devices that are the backbone of modern technology would be informative. We want to emphasize that none of the following inventions would have been possible without the highly abstract, highly weird, and highly counter-intuitive theory behind them. In fact, most of the inventors were physicists who (necessarily) had a good grasp of this theory.

> None of the following inventions would have been possible without the highly mathematical and abstract quantum theory.

- **Transistor**, invented in 1947 by three American physicists at the Bell Telephone Laboratories, proved to be a viable alternative to the vacuum tube. This invention was based crucially on the developments in the quantum theory of 1923–1930. By the late 1950s transistors supplanted vacuum tubes in many applications, and played a pivotal role in the advancement of electronics. Their small size, low heat generation, high reliability, and relatively small power requirements made possible the miniaturization of complex circuitry such as the ones required by computers. During the decades of the 1960s and 1970s individual transistors were superseded by integrated circuits in which a multitude of transistors and other components (e.g., diodes and resistors) were put together on a single tiny wafer of semiconducting material.

- **Laser.** Although Albert Einstein, using the newly discovered quanta, recognized the existence of stimulated emission in 1917, no methods were found to use it in devices until the 1950s. Two American and two Soviet *physicists* independently showed that it was possible to construct such a device using optical light. The first laser, constructed in 1960, used a rod of ruby. Since then many types of lasers have been built.

- **Electron microscope.** After the quantum theoretical assumption—and its later experimental verification—that electrons have wave properties, it was demonstrated in 1926 that magnetic or electrostatic fields could serve as lenses for electrons or other charged particles, and this discovery initiated the study of electron optics. In 1935 the construction of the first commercially built electron microscope was begun in England, and it was soon followed by the production of microscopes in Germany and the United States.

- **Liquid crystal** is a substance that flows as a liquid but maintains some of the ordered structure characteristic of a crystal. Certain organic substances when heated will not melt directly but will turn from a crystalline solid to a liquid-crystal state. Although liquid crystals were discovered in the 1920s, understanding of their properties and their commercial applications (as in wristwatches and laptop computers) were boosted only after the discovery and mastery of quantum mechanics.

- **Microprocessors** are the byproducts of the transistor technology. Early integrated circuits (ICs) consisted of about 10 individual components on a silicon chip 3 mm square. The development of large-scale integration (LSI) during the early 1970s made it possible to pack thousands of transistors and other components on a chip of roughly the same size. This technology gave rise to the *microprocessor*, an IC that contains all the arithmetic, logic, and control circuitry needed to carry out the functions of a digital computer's central processing unit. Very large-scale integration (VLSI), developed during the 1980s, has vastly increased the circuit density of microprocessors (as well as of memory and support chips). This technology has yielded microprocessors containing more than 20,000,000 transistors on a chip with an area of less than 2 cm^2.

- **Scanning Tunneling Microscope (STM)** appeared in 1981 when *physicists* at the IBM Zürich Research Laboratory set out to build a tool for studying the local conductivity of surfaces. Its principle of operation is based on the *quantum mechanical* phenomenon of tunneling discussed at the beginning of this chapter. The probability of the tunneling of electrons into a needle-like probe electronically attached to a monitor decreases exponentially as the distance of the probe from the sample surface increases. The scanning tunneling microscope makes use of this extreme sensitivity to distance to map the bumps and grooves of a sample onto the monitor.

  The magnification of STMs are so large that individual atoms can be seen! And whenever humans "see" something, they want to "touch" it. After "seeing" the Moon, it took hundreds of thousands of years before humans were able to touch its surface in 1969. We have also been able to touch the surfaces of Mars and Jupiter, and have just sent probes to touch the surface of Saturn and its moons. Only a few years after seeing the atoms, the scientists at IBM found a way of not only touching the atoms, but *moving* them! Figure 23.9 shows how by *moving* the atoms of iron in a background of the atoms of copper, IBM scientists have been able to spell "atoms" in Japanese.

Figure 23.9: The Kanji characters representing the word "atom." The literal translation is approximately "original child."

- **Nanotechnology.** Products are manufactured by casting, grinding, and milling. From an atomic point of view, it is like trying to lay bricks using cranes and forklifts. What if you could rearrange molecules and atoms individually? Nanotechnology is the emerging technology based on the possibility of manipulating individual atoms and molecules.

  Nowadays physicists are talking about molecular robotics, i.e., robots that are molecular both in their size and precision. These robots could be built to perform specific tasks; for example, attacking the invader in the body of a sick person. As Feynman said in a classic talk in 1959: "The principles of physics, as far as I can see, do not speak against the possibility of maneuvering things atom by atom."

All the considerations above should teach us—particularly those responsible for drafting society's science policies—that

> **Box 23.6.1.** *It is not technology that drives science, but science that drives technology. The most dramatic revolutions in our technology have come about as a result of finding answers to (entrepreneurially inconsequential) childlike curiosities.*

23.7   End-of-Chapter Material

23.7.1   Answers to "What do you know?"

23.1. A billionth is 10^{-9}. In the language of Box 16.2.2, the number of successes is 1000, and the probability is 10^{-9}. So, $1000 = 10^{-9}N$, and N can be easily found to be 10^{12}, or a trillion.

23.2. At short distances, the *attractive* nuclear force takes over, and overcomes the infinite electric repulsion. The wall is no longer infinitely tall, but has a finite height, allowing the tunneling of the nuclei.

23.3. Remember that Figure 23.2(c) is simply the sum of Figure 23.2(a) and (b). Imagine bringing the two curves of Figure 23.2(c) closer together. When they overlap, the sum can be large. In particular if the two peaks get close enough to each other, the value between them could equal or exceed each peak value.

23.4. Box 23.2.1 assumes that the electrons behave like bullets. It "pictures" the electrons as classical particles. Both the assumption and the "picture" are wrong!

23.5. The right-hand rule gives the direction as a directed line joining the south pole to the north pole.

23.6. No! If you know the value of the projection, you can get J_x and J_y from it. Then you would know the values of J_z and J_x simultaneously, and this is not allowed by the uncertainty principle.

23.7. No, because $306 = 17 \times 18$, and he has found a particle with $j = 17$.

23.8. From Box 23.3.2, $|\mathbf{J}| = \sqrt{3(4)}\,\hbar$ or $|\mathbf{J}| = \sqrt{12}\,\hbar$. The values of the projection along any path are $-3\hbar$, $-2\hbar$, $-\hbar$, 0, $\hbar$, $2\hbar$, and $3\hbar$. There are 7 directions.

23.9. The number of blobs is $2j + 1$; and if $6 = 2j + 1$, then j must be $\frac{5}{2}$. The projection values are, therefore, $-\frac{5}{2}\hbar$, $-\frac{3}{2}\hbar$, $-\frac{1}{2}\hbar$, $\frac{1}{2}\hbar$, $\frac{3}{2}\hbar$, and $\frac{5}{2}\hbar$ if the particle is massive. If massless, then the particle has only two projections: $-\frac{5}{2}\hbar$, and $\frac{5}{2}\hbar$.

23.10. Note that $1.5\sqrt{7} = \sqrt{15.75}$, and that $15.75 = 3.5 \times 4.5$. So, the particle has a spin of $3.5 = \frac{7}{2}$, making it a fermion.

23.11. Two of the 9 electrons fill the first shell. The remaining 7 end up in the second shell. Since 8 electrons are needed to fill the second shell, fluorine combines best with elements that have an electron by itself in a shell. Hydrogen, lithium, and sodium have such an electron.

23.12. The coefficient of the state which has $l = 1$ is 0.8. Hence, the probability is 0.8^2 or 0.64. There is no $l = 2$ state in Ψ_2; therefore, the probability is zero. The coefficient of the state which has $l = 0$ is 0.6. Hence, the probability is 0.6^2 or 0.36. Energy of -13.6 eV corresponds to $n = 1$, and the coefficient of the state with $n = 1$ is 0.6. Hence, the probability is 0.6^2 or 0.36.

23.7.2   Chapter Glossary

Blurred Quantity A physical observable not appearing in the labeling of a quantum state. A measurement of this quantity does not give an exact result, but several results with different probabilities.

Boson A particle whose spin is a multiple of $\hbar$.

Electron Microscope A microscope using the wave property of electrons and the fact that electric and magnetic fields could act as lenses for the electrons.

Exclusion Principle A principle stating that two electrons (more generally, fermions) cannot occupy the same quantum state.

Fermion A particle whose spin is an odd multiple of $\hbar/2$.

Josephson Junction Two superconductors separated by a thin insulating layer. Under certain conditions, it is possible to tunnel a pair of electrons from one superconductor to the other through the insulating barrier.

Laser (Light Amplification by the Stimulated Emission of Radiation) A purely quantum mechanical invention, allowing the production of highly intense and monochromatic EM waves.

Liquid Crystal A quantum mechanical substance that flows as a liquid but maintains some of the ordered structure characteristic of a crystal.

Magnetic Moment A physical quantity associated with some particles proportional to their angular momenta. A magnetic moment placed in a magnetic field aligns itself with that field.

Microprocessor Another device, used frequently in computer technology, whose invention relied heavily on the quantum theory.

Orbital Angular Momentum The angular momentum resulting from the motion of a particle around a center of force.

Quantum Angular Momentum The quantum analogue of classical angular momentum. Unlike its classical counterpart, no two components of the quantum angular momentum can be measured simultaneously.

Quantum Dot A quantum box in which one can trap an electron.

Quantum Entanglement A quantum phenomenon involving two subatomic particles in which the measurement of a property of one subatomic particle influences such measurement of the other even though the two particles may be completely separated.

Quantum Tunneling A quantum phenomenon by which the probability of finding a particle in a classically forbidden region is nonzero.

Scanning Tunneling Microscope (STM) A microscope, whose construction depends on the quantum phenomenon of tunneling. STM's magnification is so large that with them, one can see atoms.

Sharp Quantity One of the physical observables appearing in the labeling of a quantum state. A measurement of this quantity yields an exact result.

Spin The angular momentum intrinsic to a particle. It is named so, because it was (erroneously) thought that particles have an intrinsic rotational motion similar to the Earth spinning about its axis.

Transistor A device whose invention depended crucially on quantum mechanics. It replaced vacuum tubes which were used in the construction of many electronic gadgets.

Tunnel Diode A semiconductor device consisting of two oppositely charged regions separated by a gap. Electrons can tunnel through this gap, and their numbers can be controlled by varying the potential between the two regions.

23.7.3   Review Questions

23.1. What are the two basic results obtained when you apply the Schrödinger equation to a quantum jar-candy system?

23.2. What is quantum tunneling? Is there a tunnel? Can you follow the motion of a quantum candy as it goes out of its jar?

23.3. Does the probability of finding the quantum candy outside its jar depend on the height of the wall of the jar? How?

23.4. What is the best language in which to talk about tunneling? Can a human "explain" the phenomenon?

23.5. Is it possible to observe the prediction of the Schrödinger equation as applied to jar-candy systems? What kind of a sample do you need? Is this kind of sample easily obtainable?

23.6. Give some examples of tunneling actually occurring in nature. And some man-made examples.

23.7. When you send bullets through a single hole, what do you expect to see where bullets are collected? When you send bullets through two holes, what do you expect to see? How does the probability through two holes relate to each of the probabilities through single holes?

23.8. When you send a wave through a single hole, what do you expect to see where the wave is detected on the other side? When you send a wave through two holes, what do you expect to see? How does the probability (intensity) through two holes relate to each of the probabilities through single holes?

23.9. When you send electrons through a single hole, what do you expect to see where the electrons are detected on the other side? When you send electrons through two holes, what do you expect to see in light of the fact that electrons are particles? How does the probability through two holes relate to each of the probabilities through single holes?

23.10. Since the electron is a particle, does it not have to go through either the first hole or the second hole? What kind of probability would you expect for those that go through the first hole? For those that go through the second hole? So, shouldn't the total probability be the sum of these two probabilities?

23.11. How does mathematics answer all these questions about the electrons? Does it give you what is expected of the behavior of the electron?

23.12. How does a human language answer them? What is one way of seeing which hole an electron goes through? What is wrong with that?

23.13. What is the difference between seeing a bullet and seeing an electron? What is the effect of light on either one? What happens to the outcome when the light used to see the electrons is dimmed? What kind of pattern do you see for the detected electrons?

23.14. Dimming the light source only decreases the number of photons, not their energy. What if you decrease photon energy by looking at the electrons with long wavelength photons? Do these photons affect the motion of electrons? What other problem does this pose? Can you tell which hole the electron went through? Do you see interference?

23.15. What happens if you move the two holes farther apart? Do you see interference for such a large separation?

23.16. Is quantum theory detached from classical physics, or is it related to it? Explain either way, and support your explanation with examples.

23.17. Explain how, in classical physics, angular momentum can have arbitrary magnitude and direction. Can you give an example of this arbitrariness? What does this imply regarding the three components of the angular momentum vector?

23.18. Can you simultaneously measure the three components of the angular momentum vector in quantum mechanics? Which two quantities related to angular momentum are usually used to define an angular momentum state of a system?

23.19. What are the possible values of $|\mathbf{J}|$? How does this compare with the classical angular momentum? What are the possible values for J_z?

23.20. Is there anything special about the z-axis? If instead of the z-axis we choose the x-axis, what would the possible values for J_x be? If instead of the z-axis or x-axis, we choose any arbitrarily directed line, what would be the possible values for the component of $\mathbf{J}$ along that line?

23.21. What is the relation between angular momentum and magnetic moment? What happens if you place a magnetic moment in a homogeneous magnetic field? What happens if you place it in an inhomogeneous magnetic field?

23.22. What do you expect classically, if you send magnetic moments through an inhomogeneous magnetic field? What do you expect quantum mechanically? How many "bobs" do you expect to see?

23.23. What is the j value for spin? What are the possible values of $|\mathbf{J}|$ in light of spin? What is a boson? A fermion?

23.24. How many numbers do you need to specify a quantum mechanical state of the H-atom? What are these numbers?

23.25. State Pauli's exclusion principle and explain how it results in the periodic table of elements. How many electrons are needed to fill the first shell? Which states do they fill? How many electrons are needed to fill the second shell? Which states do they fill?

23.26. What is the difference between a classical and a quantum mechanical measurement?

23.27. How do you label a quantum mechanical state of a hydrogen atom? What is a superposition of labeled states? What is the significance of the numbers multiplying each labeled state in a superposition?

23.28. What is a sharp and blurred measurement result? The measurement of which quantity, if any, of Ψ_2 of Equation (23.3) yields a sharp result?

23.29. What does the act of measurement do to a quantum mechanical system?

23.30. Describe David Bohm's version of the Einstein-Podolsky-Rosen experiment. How does quantum theory resolve the paradox implied by the EPR experiment?

23.7.4 Conceptual Exercises

23.1. Can you follow the motion of a quantum candy as it creeps out to the outside? What would you get for the KE of the candy when it is in the thickness of the wall of the jar where the potential energy is larger than the total energy?

23.2. Why is radioactivity a probabilistic process? Do you expect nuclear fusion to be a probabilistic process as well?

23.3. In the bullet experiment of the double-slit, each bullet can either go through hole 1 or hole 2. Is it true that the probability of going through hole 1 plus the probability of going through hole 2 should add up to 1? Explain.

23.4. Is it possible for the probability of bullets landing on the wooden barrier to have a maximum at midway between the images of the two holes [midway between A_1 and A_2 in Figure 23.2(c)]?

23.5. Show directly from Equation (23.1) that, for (unequal) waves, two crests or two troughs give larger intensity than the sum of the two intensities, and for a crest and a trough, less than the sum of the two intensities.

23.6. What is wrong with the statement in Box 23.2.1? What does it imply regarding the electron probability when both holes are open? Is it plausible to think of electrons as bullet-like particles?

23.7. Compare the energy of a photon (a few electron volts) with the energy of a bullet (a few Joules). Do you see why the photons hitting the bullet have absolutely no effect on its motion?

23.8. If electron were a boson, would we have a periodic table of elements? Would we have a filled shell? If yes, which one would be filled?

23.9. Suppose you make an energy measurement of a hydrogen atom and obtain -3.4 eV. (a) What principal quantum number did H-atom have? (b) What value for the energy do you expect to get if you make another measurement immediately after the first?

23.10. Does quantum entanglement violate causality, i.e., the transmission of information with a speed faster than light?

23.7.5 Numerical Exercises

23.1. In a quantum jar-candy experiment, the probability of finding the candy outside is 10^{-9}. A typical sample contains 10^{22} jar-candy systems. How many candies do you expect to see outside?

23.2. Compare the energy of a photon (a few electron volts) with the kinetic energy of an electron whose mass is 9.1×10^{-31} kg and moves with a speed of 10^6 m/s. Do you see why the photons hitting the electron have such a dramatic effect on its motion?

23.3. A physicist claims that in an experiment, he measured the magnitude of angular momentum of a particle and obtained $\sqrt{210}\,\hbar$. Is he talking nonsense? If not, how many projections does this angular momentum have along a given line?

23.4. A hydrogen atom is described by Equation (23.3). In an energy measurement, what is the probability that you get an energy of -13.6 eV? Suppose you make that measurement and do get $-13,6$ eV. Immediately you make an orbital angular momentum measurement. What is the probability that you get $l = 1$? What is the probability that you get $l = 0$?

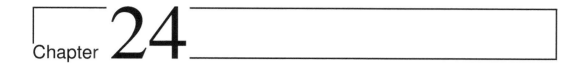

Chapter 24

Epilogue: No "Interpretation"

Quantum mechanics has given scientists valuable lessons in how science works. Unfortunately, it has also been misinterpreted and misrepresented by pseudoscientists and portrayed as an antipode of classical physics. As a result, the majority of the educated public is confused about both the quantum theory and classical Newtonian physics and the relation between the two. We postpone the examination of the misrepresentation of the quantum theory in pseudoscience until Chapter 35. This chapter concentrates, in part, on the characteristics of the theory and how it is similar to (and different from) the classical theory.

24.1   Continuity

Every final chapter of each part of the book so far has shown the continuity of physical ideas. If those parts only illustrated the chain-like attachment of each new idea to one or more old ideas, quantum theory by itself is a paradigm of the continuity of ideas. The very concept of quantum or bundle of energy occurred to Max Planck only *after* he accepted—somewhat reluctantly—the tenet of statistical mechanics. There is no doubt that without statistical mechanics the birth of the quantum theory would have been next to impossible.

The real progress in quantum theory started when Niels Bohr used Planck constant in the theory of the hydrogen atom. His theory, although utterly artificial and incomplete, paved the road to a thorough understanding of the atomic hydrogen. The first step on this road was taken by Louis de Broglie, who used Einstein's relativistic and quantum mechanical ideas of a photon to come up with the wave theory of the electron. De Broglie's electron waves could "explain" some of the artificiality of Bohr's theory. One can therefore say that the wave nature of the electron is fundamentally tied to both relativity theory and the quantization of the EM waves.

De Broglie's idea fascinated Erwin Schrödinger, whose equation is a manifestation of his commitment to the importance of the wave aspect of the electron. Furthermore, all the quantities and concepts that comprise the Schrödinger equation—such as momentum, potential energy, kinetic energy, and the conservation of mechanical energy—were taken from the classical physics of Newton and Galileo. Thus, not only are the new ideas of the quantum theory connected *together* like the links of a chain, but there is a huge link that connects the quantum mechanical chain to the classical chain.

24.2   Role of "Fundamentalism"

Many philosophers of science, especially those with an affection for Eastern mysticism, abhor fundamentalism and reductionism. Yet every significant progress in every scientific discipline has been precisely the result of digging deeper into the domain of the "fundamentals" for the answer. The entire subject of quantum theory is a paradigm of reductionism in action.

Reductionism led to the notion of the **quantum**.

A Planck with a holistic philosophy would have stopped once he had discovered the correct formula [Equation (20.3)] for the black body radiation curve. He would have probably thought to himself "My philosophy tells me not to fathom any further. Not to try to find a fundamental reason for why the equation I have discovered works. That would be a reductionist approach, which is the opposite of my holistic philosophy." But Max Planck, being a true scientist, saw in the formula an opportunity to discover a fundamental secret of nature, a fundamental constant h that is as important as the gravitational constant G and the speed of light c.

The pursuit of fundamentals is one of the hallmarks of modern science, and every development of the quantum theory attests to this fact. Niels Bohr correctly identified the fundamental nature of a hydrogen atom, simply a nucleus with one electron attracted towards it electrically. Louis de Broglie zoomed his investigation entirely on the nature of the most fundamental particle, electron. Only by isolating this fundamental entity from a holistic and crowded surrounding was he able to discover the wave nature of the electron, paving the way for Erwin Schrödinger to find the master equation of quantum physics. And it was the application of this equation to the most fundamental interacting entity, a hydrogen atom, that showed the power and promise of the quantum theory.

24.3   Specificity

Hydrogen atom is sometimes called the "gift of Nature." Its simplicity (consisting of a nucleus and a single electron) purifies its theoretical study considerably. Chemists were attracted to hydrogen because it seemed to be the "building block" of all other elements. However, with the discovery of the electrons and the realization that even hydrogen atom has internal structure, physicists, like all other scientists, considered hydrogen the key to the understanding of *all* atoms.

Specific study of the H-atom led to the **universal** quantum theory.

Niels Bohr was the first to apply the newly discovered quantum ideas to atoms. And he was intuitively aware of the fact that it would be a waste of time to adhere to a "holistic" principle and tackle a collection of atoms or even a complex atom. In his fertile imagination, Bohr isolated a single hydrogen atom from the rest of the universe and applied his inventions to that single atom.

After Bohr's 1913 paper on the quantization of the hydrogen atom, the eyes of the minds of all physicists turned to one single hydrogen atom. Both Heisenberg and Schrödinger concentrated their effort on the solution of the hydrogen puzzle. Their discovery of the all-encompassing quantum theory was the fruit of such a concentration on that *specific* system. Just as Kepler's discovery of the planetary motion came out of his study of Mars, and Newton's discovery of the *universal* law of gravity came out of his concentration on an apple and the Moon, and Maxwell's discovery of the electromagnetic waves came out of his focusing on the mathematics of the electric and magnetic fields, so did the universal quantum theory come about when Bohr, Heisenberg, Schrödinger and others focused their attention on the simplest atom.

24.4 Induction versus Deduction

Scientific progress depends crucially on the interaction between observation and theory. Observation provides the necessary framework in which theories can be built. The degree to which observations influence the theories varies. In some cases the theory is strongly influenced by one or more observation (or experiments). In others, there seems to be no connection between a theory and *any* observation. The former procedure is called **induction**, the latter, **deduction** (see also Food for Thought 21.3.2).

Classical physics, for the most part, was an inductive science. The discovered laws were by and large direct results of observations. A good example of induction at work is Kepler's discovery of the laws of the planetary motion. These laws were chiefly Kepler's translation into the mathematical language of the patterns that he discovered in the data collected by Tycho Brahe. In a similar vein, Galileo, Newton, Coulomb, Ampère, and all the others that studied electricity and magnetism "induced" physical laws by translating experimental results into mathematics. This mathematics, of course, could be the starting point for the discovery of other important *formal* developments leading to new *mathematical* breakthroughs. However, these formal accomplishments, important as they were, did not lead to significant *physical* discoveries.

Maxwell's prediction of the electromagnetic waves was, on the other hand, one of the first applications of *deductive reasoning* in the development of physics. He did not merely translate the results of some experiments or observation into mathematical formulas. Maxwell *changed* the formulas so obtained to account for the more reliable charge conservation.

Physics, starting with the works of Planck and Einstein, has become a deductive science. Granted that Planck used the results of the black body observations to come up with the mathematical formula for the black body radiation curve, but his real breakthrough, the idea that electromagnetic waves were composed of "bundles of energy," was a result of purely deductive reasoning. Similarly, the special theory of relativity (STR) was a natural *deductive* extension of the Maxwellian electromagnetic theory as we shall see in the next part.

Possibly because of the intangibility of the objects under study by modern physics (subatomic particles that we cannot see and feel, and motion with speeds approaching the speed of light), physicists must rely more on the deduction of the mind than the induction of the senses. When Niels Bohr took the first step toward applying the quantum ideas to the hydrogen atom, he did so purely from his imagination. When de Broglie proposed the wave nature of the electron, it was simply a conjecture fermenting in his mind. Schrödinger, Heisenberg, Dirac, and other founders of the quantum theory all used the extraordinary imaginative power of their minds to develop their ideas.

The influence of deduction is so prevalent in modern physics that some physicists have come to believe that only deductive reasoning matters; that laws can be discovered by purely mathematical thoughts. The development of quantum theory clearly appears to attest to such a belief. Bohr did not use any experimental result to discover the theoretical foundation of the quantization of the hydrogen atom; de Broglie did not resort to any experiments to introduce the wave property of an electron; Schrödinger wrote down his equation only on the basis of the wave nature of the electron.

So why bother with experiment and observation? The answer is that in any and all deductive reasonings *there is* one or more experimental result that play a crucial role. The experiments may not be the most recent ones, and may not have any *direct* bearing on the theories; but without them the theories could not come into existence. While Maxwell's prediction of the electromagnetic waves was a deductive process, it could not have occurred were it not for the experiments of the previous generation of physicists on how electric currents produced magnetic phenomena. Similarly, Bohr's application of the quantum idea to the hydrogen atom gave results only because *he used Rutherford's experimental finding* that atoms consist of a small heavy positively charged nucleus as well as the negatively

Modern physics demands much deduction; but it cannot exist without observation and experimentation.

charged electrons. Had Bohr used the raisin-pudding model of the atoms, he would not have gotten anywhere.

24.5   No Interpretation for Quantum Theory

"Mathematics is the language of Nature." This statement, coined by Galileo and implied by as early a thinker as Plato, is at the heart of the constitution of science. Nowhere else does the statement apply more appropriately than to modern physics. Due to its abstract character, quantum physics can be spoken *exclusively* in the language of mathematics. A sign of this exclusiveness is the highly deductive nature of quantum physics. The appearance of $\sqrt{-1}$ in the Schrödinger equation, the emergence of matrices in the Heisenberg treatment of the hydrogen atom, the infinite-dimensional space that naturally connects the Heisenberg and Schrödinger picture, and the association of the Ψ function with probability all point to both the kinship of quantum physics to mathematics and its alienation from ordinary human experience.

Quantum theory speaks **very clearly** in the language of mathematics. It cannot tolerate any "interpretation."

The abstract nature of quantum theory, along with the intangibility of the objects of its study (subatomic particles, atoms, and molecules), has brought about the uneasy possibility of "interpretation" (and therefore, misinterpretation). The practitioners of quantum physics have learned to apply it to many situations using the only language in which the practitioners can converse with the theory, mathematics. The paradigm of this inseparability of mathematics and the quantum theory is the advice Richard Feynman, an American physicist, gave in response to a question about how to interpret the theory:[1] "Shut up and calculate!" This quotation is by no means to be interpreted as the decree of an authority to his subordinate. It simply embodies the fact that quantum physics is best spoken in the language of mathematics, and that a translation into any human language risks the loss of the theory's message.

Unfortunately, many "scholars" insist on interpreting the quantum theory in the context of human experience. We shall talk about the fallacy of requiring any translation or interpretation of the quantum theory in some length in Chapter 35. This fallacy is very easy to understand:

Box 24.5.1. *While a translation of poetry or any masterpiece of literature from one human language to another is fruitful but imperfect, the translation of quantum physics from mathematics into* any *human language is next to impossible.*

That is not to say that physicists ought to reject communicating the implications of the quantum theory with the public. On the contrary, it is the *responsibility* of the physicists to inform the public of the grandeur of the theory and to encourage the participation of the youth in the noble pursuit of discovery. However, in so informing the public, it is also the responsibility of the physicists to warn the layman of the limitation of the information and the necessity of learning mathematics for a full understanding of the theory.

---

[1] *Physics Today*, Vol. 57, May 2004, pp. 10–11.

Part VI

Twentieth Century Physics: Relativity Theory

Chapter 25

Birth of Relativity

The twentieth century is only five years old and a multitude of events which are later to become the hallmark of this century have already taken place: massive demonstrations are being reported in many Russian cities, the biggest one taking place in St. Petersburg and being brutally crushed by the czarist police; George Bernard Shaw has just published his "Major Barbara." A year earlier, the Rolls Royce company was founded and Picasso completed his famous painting, "The Two Sisters." Two years before, Orville and Wilbur Wright successfully flew a powered airplane and Jack London finished his well-known novel "The Call of the Wild." Four years before, the first year of the twentieth century, as if ushering us into the new century, Marconi transmitted telegraphic radio messages from Cornwall to Newfoundland; the first motor driven bicycle was invented; the first Mercedes car was constructed, the first Nobel Prize was awarded; and, as if officially closing the gates on the nineteenth century, the prominent figure of that era who ruled England for 64 years, Queen Victoria, died.

Most important of all these events were probably *thoughts* brewing in the mind of a quiet young clerk at the Bern patent office in Switzerland. The young man was Albert Einstein (1879–1955) and the thoughts, later published in *Annalen der Physik*, were the special theory of relativity (STR), the photoelectric effect, and the theory of Brownian motion, each one a revolution in scientific thinking and a cornerstone upon which to build most of future physics. Such an outburst of creativity in the span of only a few months was unprecedented in the history of science. Although Einstein won the Nobel prize for the photoelectric effect, his work on relativity—both the special theory of 1905 and the general theory of 1916—is by far one of the most pivotal works in the history of science.

25.1 Law of Addition of Velocities

Chapter 20 showed how the marriage of thermodynamics and electromagnetic radiation led Planck to the concept of quantum, a radical departure from classical Newtonian physics. In this chapter we see how the conflict between mechanics and electromagnetism led Einstein to his epoch-making discovery of relativity.

This conflict is between one of the simplest conclusions of Newtonian mechanics, the **law of addition of velocities**, and the electromagnetic theory. To understand the law of addition of velocities, we use the notion of a *reference frame* (RF) introduced in Section 6.1. It is very common to replace an RF with one of the observers who reside in it. The word "observer" is not to be construed necessarily as a human being; and if it *is*, all of his/her human traits are to be ignored (see Food for Thought 27.5.3 for this seemingly unnecessary precaution).

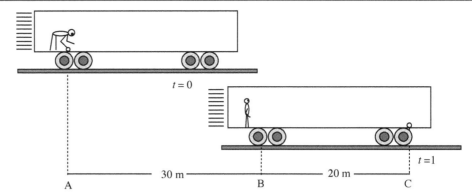

Figure 25.1: Illustration of the law of addition of velocities. Initially ($t = 0$) the ball and the person are at A. One second later the ball is at C while the person has moved to B. Relative to the ground, therefore, the ball has moved from A to C in one second.

Consider a train moving to the right with a passenger (observer) holding a ball in his hand ready to throw it in the forward direction. All this is taking place while you, the ground observer, are watching. For definiteness, let us assume that the train moves at the speed of 30 m/s and the ball is thrown at 20 m/s relative to the thrower. Intuitively, we expect the speed of the ball relative to you—relative to the ground observer—to be $30 + 20 = 50$ m/s. Figure 25.1 gives a "proof" of this intuition. We can summarize this as

> **Box 25.1.1. (Law of addition of velocities)** *The velocity of an object (the ball) relative to observer G (the ground observer) is equal to its velocity relative to observer P (passenger) plus the velocity of P relative to G.*

Law of addition of velocities.

Denoting the velocity of the ball relative to G by $\mathbf{v}_{bg}$, that of the ball relative to P by $\mathbf{v}_{bp}$, and that of P relative to G by $\mathbf{v}_{pg}$, we can write the above statement as

$$\mathbf{v}_{bg} = \mathbf{v}_{bp} + \mathbf{v}_{pg}. \tag{25.1}$$

This is the law of addition of velocities.

> **What do you know? 25.1.** Al Capone and his gang are moving on the streets of Chicago at 60 mph firing bullets with a speed of 200 mph. What is the speed of the bullets relative to the innocent bystanders if the bullets are fired in the forward direction? In the backward direction?

The motion of a boat crossing a river (Figure 25.2) is a good example of the validity—and the vector nature of—Equation (25.1). In this case, we speak of velocities of RFs relative to one another, thus eliminating the "object" which may appear to have a different role than the observers. With the introduction of RFs, objects and observers are treated equally. Thus, in the analysis of the motion of the boat, we can refer to the water RF (w), the boat RF (b), and the ground RF (g). While the paddler paddles the boat perpendicular to the flow of the water (in the direction of $\mathbf{v}_{bw}$ in Figure 25.2), the actual motion of the boat is in the direction of $\mathbf{v}_{bg}$, because the water gives the boat an additional motion to the right. The law of addition of velocities is so ingrained in our mind that we cannot accept anything that appears to violate it. One instance of the violation of this law is Maxwell's electromagnetic theory.

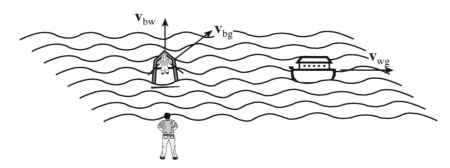

Figure 25.2: A boat moves sideways relative to Karl's RF standing at the bank of the river. Emmy, who is paddling the boat, is in the boat's RF, and the abandoned boat following the current of a river is in the river's RF.

What do you know? 25.2. Car number 1 moves at 70 mph on a highway. Car number 2 moves at 60 mph in the same direction on the same highway. Let 1, 2, and g denote the reference frames of car 1, car 2, and the ground, respectively. Use Equation (25.1) (being careful about the order of the subscripts) to find v_{12}, the velocity of car 1 relative to car 2.

Recall from Chapter 14 (and Math Note E.14.3) that the equations of electricity and magnetism—in particular the constants k_e and k_m—lead to electromagnetic (EM) waves, which travel with a speed $c = 300,000$ km/s in vacuum. Maxwell's derivation of the EM wave equation and his prediction of their speed *had nothing to do with their source* (see Box 14.3.1). This means that

Box 25.1.2. *Every time you detect an electromagnetic wave, it is moving at the rate of 300,000 km per second in vacuum.*

Now suppose that Emmy, riding on a train, performs all her experimental and theoretical investigations on electricity and magnetism on the train. Then she discovers the values of the two constants k_e and k_m precisely as was done on the ground (after all, why should physics be different on a train than outside it). And if she has a Maxwell on board, he will discover the existence of electromagnetic waves and that their speed is $300,000$ km/s in vacuum regardless (of the motion) of their source. Not only that, if Emmy sends these EM waves out of the train and Karl, who is standing outside, detects them (not knowing that they came from the train), he measures their speed to be 300,000 km per second.

The law of the addition of velocities was so firmly established in the nineteenth century, that the notion of light traveling with the same speed relative to two different RFs had no right to exist despite the above discussion involving Emmy and Karl! So, what was the alternative? One possibility was to consider the Earth as the special RF in which Maxwell's equations hold. But this was immediately dismissed because Copernicus, three centuries earlier, had already cautioned everybody about the dangers of bestowing upon the Earth a special position in the universe. Moreover, the orbital motion of the Earth, with its semi-annual change of direction, would have to have distorting effects on the light of stars and planets which would be easily detectable. Lack of such distortions immediately rules out the Earth as a privileged frame. If the Earth is not the privileged RF, then, for the same reason, no other planet, nor the Sun nor any other star or galaxy can be privileged by Maxwell's equation.

The rise and fall of ether theory.

The only remaining alternative is the "medium" surrounding the celestial bodies and filling the space between them. This medium was called **ether**. In fact, the idea of ether was invented and used by Huygens as early as 1690, but it became the subject of intense scrutiny in the nineteenth century after Maxwell's prediction of the EM waves. Ether theory demanded so many strange and complicated notions that by the last quarter of the nineteenth century, physicists started to doubt its existence. The final blow to ether came through the works of Albert Michelson (1852–1931) and Edward Morley (1838–1923), one of whose crucial experiments showed that there was no detectable motion of Earth relative to ether, thereby undermining the theory and the concept of ether.

25.2   Principles of STR

When he was only 16 in 1895, Einstein wondered what the world would look like if one could catch up with light; in particular, what would light itself look like. The intriguing nature of this question can be appreciated by an analogy.

How would water waves look if you caught up with them?

Water waves are easily made and seen to travel as concentric circles. Imagine that we are quietly hovering over these circles at exactly the same speed as their expansion rate. How will the waves appear to us? Concentrating only on the local waves—clearly we cannot move with all the waves as they go in different directions—and assuming that we ride just above a crest, we see the crest right underneath us, and since we move with it at exactly the same speed, it will remain underneath. The other crests and troughs in the vicinity also move at the same speed, and will also remain static. Thus, locally at least, the wave ceases to exist! It will appear as a static and permanent deformation of water surface with no motion and no oscillation.

How would light look if you caught up with it?

If we could catch up with light, and if ether (the medium whose undulation was thought to create light waves just as the undulation of the water in a pond creates water waves) really existed, the ether wave (i.e., light) would also cease to exist! What is even more intriguing is that, since ether was supposedly invisible, there would be no trace of light left! Although in the case of water waves, a static deformation of water, the medium of transmission, was left to remind us of the once traveling wave, no such reminder will remain in the case of light. As soon as we catch up with it, light will be completely gone! Einstein reckoned that light must be moving at a special speed.

In the intervening years between 1895 and 1905, the question of the behavior of light and motion was constantly on Einstein's mind. As his scientific and mathematical abilities matured, the question took on the more definite form of how light (or electromagnetic waves) behaved in moving frames. He even came up with a version of the Michelson-Morley experiment but, being theoretically inclined, never performed it. Once he became acquainted with the works of Lorentz (who solved the problem of electrodynamics of moving frames assuming that the speed of this motion was negligible compared to that of light), he concluded that the speed of light must be the same in the moving frame and outside. This, however, contradicted the addition law of velocities. These thoughts occupied Einstein's mind for months. In his own words,

Einstein explains how he created STR.

> Why do these two concepts contradict each other? I realized that this difficulty was really hard to resolve. I spent almost a year in vain trying to modify the idea of Lorentz in the hope of resolving this problem.
>
> By chance a friend of mine in Bern (Michele Besso) helped me out. It was a beautiful day when I visited him with this problem. I started the conversation with him in the following way: "Recently I have been working on a difficult problem. Today I come here to battle against that problem with you." We discussed every aspect of this problem. Then suddenly I understood where the key to this problem lay. Next day I came back to him again and said to him, without even saying hello, "Thank you. I've completely solved the problem." An analysis of the concept of time was my solution.

Time cannot be absolutely defined, and there is an inseparable relation between time and signal velocity. With this new concept, I could resolve all the difficulties completely for the first time.

Within five weeks the special theory of relativity was completed. I did not doubt that the new theory was reasonable from a philosophical point of view.... This is the way the special theory of relativity was created.

Einstein based his entire theory of relativity on a few simple and appealing principles. His STR follows from two principles which we state immediately and discuss in detail later:

Principles of STR.

Box 25.2.1. (The two principles of relativity:)

1. *The laws of physics are the same for all inertial observers. Therefore, it is impossible to detect the (absolute) uniform motion of an RF through measurements done entirely in that RF.*

2. *The speed of light in empty space is the same for all observers irrespective of their motion or the motion of the source.*

The word "uniform" describing the motion in the first principle is important. If the motion becomes nonuniform (accelerated), then it *is* possible to detect it through measurements done entirely in the RF. An airplane that is cruising smoothly with constant velocity does not "feel" to be moving, i.e., its motion cannot be detected internally (without looking outside to see the clouds move underneath). However, as soon as the plane enters a region of rough weather, its uniform motion turns nonuniform, and the passengers can "feel" the motion of the plane without looking outside: acceleration can be detected, uniform velocity cannot.

Acceleration can be detected, uniform velocity cannot.

What do you know? 25.3. You are in a spaceship in empty space away from all stars and planets. (a) Can you tell if you are moving or not? (b) What if you look out your window and see something pass by your spaceship; can you *now* tell if you are moving?

The second principle is the direct result of Maxwell's electromagnetic theory. In choosing between the law of addition of velocity and EM theory, Einstein picked the latter and came up with his second principle of relativity. He based his choice on theoretical grounds only. At the time that he postulated the STR he was not familiar with the Michelson-Morley experiment. The theoretical reasoning—at least in *his* mind—was strong enough to warrant the second postulate. Let us examine some of the most immediate consequences first, leaving the more dramatic ones for later sections.

Relativity is a direct result of Maxwell's electromagnetic theory.

The first principle is really a generalization of Galileo's observation concerning the laws of mechanics. In his *Dialogue Concerning the Two Chief World Systems-Ptolemaic and Copernican*, Galileo discusses many aspects of motion through three main characters, Salviati, Sagredo, and Simplicio. Salviati expresses Galileo's own views, namely the Copernican viewpoint; Sagredo defends Aristotle and Ptolemy; and Simplicio is a neutral listener. In one of the chapters of the book, called the *Second Day*, the three are discussing the motion of a projectile and how it would appear if thrown on a ship. They also compare the flight of a projectile and that of a bird. In an attempt to explain the differences and similarities of all such motions, Galileo lays down the **Galilean Principle of Relativity.**. Listen!

Galilean Principle of Relativity.

Shut yourself up with some friend in the main cabin below decks on some large ship, and have with you there some flies, butterflies, and other small flying animals. Have a large bowl of water with some fish in it; hang up a bottle that empties drop by drop into a wide vessel beneath it. With the ship standing still, observe carefully how the little animals fly with equal speed to all sides of the cabin. The fish swim indifferently in all directions; the drops fall into the vessel beneath; and, in throwing something to your friend, you need throw it no more strongly in one direction than another, the distances being equal; jumping with your feet together, you pass equal spaces in every direction. When you have observed all these things carefully (though there is no doubt that when the ship is standing still everything must happen in this way), have the ship proceed with any speed you like, so long as the motion is uniform and not fluctuating this way and that. You will discover not the least change in all the effects named, nor could you tell from any of them whether the ship was moving or standing still. In jumping, you will pass on the floor the same spaces as before, nor will you make larger jumps toward the stern than toward the bow even though the ship is moving quite rapidly, despite the fact that during the time that you are in the air the floor under you will be going in a direction opposite to your jump. In throwing something to your companion, you will need no more force to get it to him whether he is in the direction of the bow or the stern, with yourself situated opposite. The droplets will fall as before into the vessel beneath without dropping toward the stern, although while the drops are in the air the ship runs many spans. The fish in their water will swim toward the front of their bowl with no more effort than toward the back, and will go with equal ease to bait placed anywhere around the edges of the bowl. Finally the butterflies and flies will continue their flights indifferently toward every side, nor will it ever happen that they are concentrated toward the stern, as if tired out from keeping up with the course of the ship, from which they will have been separated during long intervals by keeping themselves in the air. And if smoke is made by burning some incense, it will be seen going up in the form of a little cloud, remaining still and moving no more toward one side than the other. The cause of all these correspondences of effects is the fact that the ship's motion is common to all the things contained in it, and to the air also ...

In Galileo's time physics consisted of rudiments of mechanics. His conclusions applied only to motion of birds, flies, cannon balls, etc. Einstein, on the other hand, possessed Newton's mechanics and Maxwell's electromagnetism. Therefore, his first principle of relativity has more far-reaching consequences than Galileo's. In fact, the first principle of relativity suggests the second! To see this, imagine performing all the important electromagnetic experiments and drawing all the important conclusions, including Maxwell's equations, in another reference frame, a moving laboratory (such as one on a train). The scientists in this laboratory, inspired by Maxwell's theory, will produce electromagnetic waves and conclude that, in empty space, they travel at 300,000 km/s. The first principle now implies that the production of electromagnetic waves, and thus their speed, must be independent of the reference frame. Therefore, the speed of these waves, including light, must be frame-independent. This is the second principle of relativity.

25.2.1   Inertial Frames

The word "inertial" in the first principle needs some clarification. Imagine sitting in the cafeteria of a moving train sipping on your cup of coffee. All of a sudden you notice a "jerk" pushing you in your chair and shaking your coffee cup in its saucer. You know the reason behind this sudden jerk: the train track must have had some bumps, causing the "smooth" motion of the train to turn jolty. In the language of physics we say that the train initially had a constant velocity but the bumps on the track caused a change in this constant velocity. In other words, the bumps caused an *acceleration* of the train, and this acceleration was "transferred" to you and your coffee cup.

The second law of motion states that the acceleration of an object is caused by forces acting on it. Are there any forces acting on you and the coffee cup when the train moves over a bump on its track? You may say that the bump pushes the train up (or down), and this force is transferred to the chair, on which you are sitting and the table, on which your coffee cup is standing. Then the chair and the table, *being in contact* with you and your coffee cup, physically exert forces on you and your coffee cup. It appears that there are indeed physically real forces involved.

BIOGRAPHY

Albert Einstein (1879–1955) began his schooling in Munich. At age 15 he entered the Luitpold Gymnasium. He studied mathematics, in particular the calculus, beginning around 1891. In 1895 Einstein failed the entrance examination to Eidgenössische Technische Hochschule in Zurich, where he planned to study electrical engineering. As a preparation for the ETH in Zurich, he attended secondary school at Aarau. This helped Einstein get into ETH, from which he graduated in 1900 as a teacher of mathematics and physics. Three of Einstein's fellow students, including Marcel Grossmann, a noted mathematician who later introduced Einstein to differential geometry (an essential tool for general theory of relativity), were appointed assistants at ETH in Zurich but not Einstein. By mid 1901 he had a temporary job as a high school teacher. In 1902 Grossmann's father landed Einstein a job as a clerk at the patent office in Bern, where he worked until 1909. It was here that Einstein, in his spare time, completed his papers on relativity, photoelectric effect, and the Brownian motion, without the benefit of close contact with the scientific community.

Now suppose you have just boarded the train, and as you are trying to find your seat, the train starts to accelerate. In order to prevent yourself from falling you grab the bar on top of your seat. Again, you feel a force, and trying to find the source of this force, you note that the train is experiencing a force (traction) due to the track, and this force is transferred to you because your feet are in contact with the floor of the train: it is like somebody is grabbing your feet and is pulling them, causing you to go off balance.

So far, you can understand the jerk and the fall from the second law of motion and the actual forces acting on you. Now suppose you get on the train early and go directly to the "game room." You are standing by one of the air-puck tables, and notice a strange phenomenon: as the train starts to accelerate (and you start to fall) the puck on the table starts to move *in the direction opposite to* the train's motion. How do you explain this motion? There is no friction (or traction)[1] between the puck and the table! *There is no real force causing the backward acceleration of the puck.* That is why such a force is called a **fictitious force**. Whenever an object is in an *accelerated* RF, it experiences a fictitious force.

Fictitious force in accelerating frames.

> **What do you know? 25.4.** In some amusement parks, you get on a ride which is a cylindrical enclosure. You stand by the wall. The cylinder starts rotating faster and faster. You stick to the cylindrical wall so hard that when the floor is removed, you don't fall. How can you explain this?

We can now define an inertial RF: A reference frame is inertial if no fictitious forces exist in it. In such an RF, an isolated object will not change its motion unless a (real) force acts on it, i.e., the first law of motion holds in this RF. We, therefore, define an inertial frame as follows:

Inertial frame and the first law of motion.

> **Box 25.2.2.** *An inertial frame is an RF in which Newton's first law holds: Isolated objects will continue their state of rest, or their motion on a straight line with constant speed, indefinitely.*

---

[1]There *is* a little friction, but, following Galileo's footsteps in his discovery of the first law, we can imagine reducing this friction further and further, until it is completely absent (in an ideal setting).

Let us call two RFs *relatively inertial* if one of them moves with constant velocity relative to the other. It is not hard to convince oneself (see **Example D.25.1** on **page 44** of *Appendix.pdf*) that if RF_1 is an inertial frame and RF_2 is a second frame inertial relative to RF_1, then RF_2 is also inertial, i.e., Newton's first law holds in RF_2. Therefore, if we can find one fiducial inertial frame, we can construct unlimited numbers of them by setting in uniform motion cars, trains, spaceships, etc., relative to the fiducial frame. Whether such a fiducial frame exists or not is not answered by the special theory of relativity which is concerned with the consequences of the principles of relativity assuming such a frame exists. We shall come back to this question when we take up the general theory of relativity.

25.2.2   Law of Addition of Velocities and Second Principle

The second principle is most intriguing. To grasp how extraordinary this principle is, we compare it with a familiar experience. Consider a train moving with a speed of 60 mph to the right. Inside the train stands a passenger holding a gun which fires a bullet with a speed of 180 mph. This means that the passenger measures the speed of the bullet relative to the gun—and himself—to be 180 mph. If the gun is aimed in the direction of motion of the train what do we (standing on the ground) measure the speed of the bullet to be? Let us look at the situation a minute after the bullet is fired. Since the train is moving at 60 mph, the passenger moves a distance of 1 mile. Since the bullet is moving at 180 mph relative to the passenger, its distance from him will be 3 miles a minute later. So we, on the ground, see the bullet move a distance of 4 miles in one minute. We thus conclude that the speed of the bullet must be 240 mph, the sum of the two speeds. This is the Newtonian law of addition of velocities which we also discussed in Section 25.1.

If we denote the velocity of the bullet relative to the passenger by $\mathbf{v}_{bp}$, that of the passenger relative to the ground by $\mathbf{v}_{pg}$, and the velocity of the bullet relative to the ground by $\mathbf{v}_{bg}$, then we get a relation identical to Equation (25.1). And, in general, for any three objects a, b, and c in motion, we have (note the order of indices)

$$\mathbf{v}_{ca} = \mathbf{v}_{cb} + \mathbf{v}_{ba}. \tag{25.2}$$

Example 25.2.3. Car A passes us at 65 mph. Immediately, we get in our car B and chase it at 60 mph.

Q: What is the speed of car A relative to our car?

A: Both values of speed given above are relative to the ground G. So, we have $v_{AG} = 65$ mph and $v_{BG} = 60$ mph. Equation (25.2) now yields $\mathbf{v}_{AG} = \mathbf{v}_{AB} + \mathbf{v}_{BG}$, or $65 = v_{AB} + 60$, yielding $v_{AB} = 5$ mph. Thus, the speed of the car has been cut considerably relative to us. This is, of course, because we have (partially) caught up with the car.

Q: What happens if our car reaches 65 mph?

A: In this case, we have $65 = v_{AB} + 65$, or $v_{AB} = 0$. This means that A appears motionless relative to us. If two cars go at the same speed, the distance between them remains unchanged so the passengers in either car will detect no relative motion.

If our car goes at 70 mph, then $65 = v_{AB} + 70$, or $v_{AB} = -5$ mph. The negative sign means that the velocity of car A relative to car B is in the opposite direction to the other velocities. Thus, car B will be approaching car A. This is again what we expect: If we go faster than the car in front of us, that car appears to be approaching us backwards. ∎

> **What do you know? 25.5.** Suppose you travel at the speed of light. You look at a light beam (or a spot on it). The beam should appear stationary to you, as the second question of Example 25.2.3 indicates. So what happens to the second postulate?

The example above and the discussion preceding it illustrate the law of addition of velocities in Newtonian physics which perfectly agrees with common sense. Now, let us analyze the second principle of relativity in this context. Replace the train with a super fast spaceship and the gun with a flashlight. Assume that the spaceship travels at 200,000 km/s. What is the speed of the light emitted by the flashlight relative to us, on the ground? Newtonian physics, and common sense, says that it should be $200,000 + 300,000 = 500,000$ km/s. The second postulate says that it is 300,000 km/s!

To tickle our common sense a little more, let's assume that we see a light beam passing us at 300,000 km/s. Immediately, we get on our super fast spaceship and chase the beam with a speed of 299,000 km/s. Newtonian physics (and common sense) tells us that the speed of the light beam relative to us must be $300,000 - 299,000 = 1000$ km/s. The second postulate says that even though we are chasing the light beam with a speed of 299,000 km/s, as we measure the speed of the light beam just ahead of us, we find it to be 300,000 km/s! Incredible, strange, impossible, you say. Incredible? Yes. Strange and impossible? No. There is nothing less strange and impossible than a proven law of nature. In fact, it is we that are strange, judging a perfectly normal nature (after all, what is more normal than nature itself?) by our crude senses and declaring the imperfections we perceive through them as "normal" and "possible".

Strange consequences of second principle.

Nature is not strange. We are!

What do you know? 25.6. A beam of light L and a super fast spaceship S start at he same time according to observer B. S travels at 299,000 km/s, and L at 300,000 km/s. After one second, B measures the distance of L from her, and finds that it is 300,000 km. She measures the distance of S from her, and finds that it is 299,000 km. She concludes that L is only 1000 km ahead of S. How can S say that L is moving at 300,000 km/s?

There have been many direct and indirect verifications of the second principle starting with the discovery of the aberration in the seventeenth century, passing through the null experiments of Michelson and Morley, and ending with the most recent astronomical observations. There is absolutely no evidence to the contrary. Light always travels with the same speed in vacuum irrespective of the motion of its source and detector. No matter how "strange" and "impossible" this may seem, we have to accept it and make it our starting point, our pillar upon which to build all our theories. And if we find "strange" phenomena in these theories, such as the relativity of time and space, slowing down of time in moving RFs, equivalence of mass and energy, etc., we must accept them and let only objective and precise experimentation be the ultimate judge of these strange phenomena.

25.3 Relativity of Simultaneity

Before discussing the *relativity* of simultaneous "events" we need an operational definition of simultaneity. **Events** are the most essential building blocks of relativity. Although it is hard to define an event without circular logic, it is an intuitively obvious notion. For our purposes, an event is any abrupt physical phenomenon that can be recorded. In particular, one can talk about the position of an event and its time of occurrence. An event is relativity's equivalent of a point in geometry. Examples of events are explosion of a firecracker, take-off of a space shuttle, collision of two particles, emission of a light signal, turning on a switch, birth of a child, and death of a salesman. In our discussion we always deal with ideal events, i.e., those which do not have any extension either in time or in space. Thus, the explosion of a supernova, as seen from the Earth, is a good approximation to an ideal event, but as seen from a nearby star it is not ideal because the exploding star goes through some preliminary stages before preparing itself for the final act of explosion. The most convenient definition of an ideal event is the following:

Events as building blocks of STR.

Event defined.

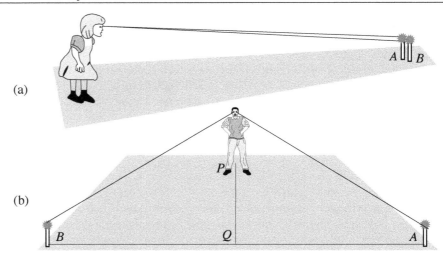

(a)

(b)

Figure 25.3: (a) Two adjacent firecrackers are said to have exploded at the same time if we receive their signals at the same time. (b) Two widely separated firecrackers are said to have exploded at the same time if we receive their signals at the same time while being located on the perpendicular bisector of the line segment joining the firecrackers.

Box 25.3.1. *An event is described by a single point in space and a single instant in time.*

We note that all observers agree on a single ideal event. In other words, if a physical phenomenon is stamped as an event by one observer, then it is stamped as an event by all observers. As a prototype of events, we consider the explosion of a firecracker in the sequel.

Simultaneity of two events occurring at the same point. Our next consideration centers around the notion of **simultaneity**. How do we determine whether two events are simultaneous or not? Firecrackers A and B explode somewhere in space and we receive the signals of their explosions. Question: Did the explosions occur at the same time? If A and B were at the same point when they exploded, it is easy to answer this question: If we receive their light signals at the same time, they must have occurred at the same time [Figure 25.3(a)]. Note that because of the special property of light, this conclusion is independent of the state of motion of the firecrackers: A and B could be moving at very high speeds in arbitrary directions at the time they reach the common point in the figure.

Simultaneity of two events occurring at different points. What if A and B are not located at the same point? Figure 25.3(b) shows the line segment $\overline{AB}$ and its perpendicular bisector $\overline{PQ}$.[2] If the observer happens to be on $\overline{PQ}$ and receives the two light signals at the same time, he can conclude that they must have occurred at the same time. Again, due to the invariance of light speed, this definition of simultaneity is independent of the state of motion of A, B, and the observer. Thus, to conclude whether two events separated in space are simultaneous, we must locate ourselves on their perpendicular bisector and see if we receive the light—really electromagnetic—signals at the same time. If we are off perpendicular bisector, we must allow for the difference in the travel time of light signals.

Example 25.3.2. Assume that the line segment $\overline{AB}$ of Figure 25.4 is 300,000 km and $\overline{AO}$ is 150,000 km.

---

[2]Recall that the perpendicular bisector of a line segment has the property that all its points are equidistant from the two end points of the line segment.

Q: Under what conditions can we conclude that events A and B are simultaneous?

A: Clearly, if O receives the two signals at the same time, she cannot conclude that A and B are simultaneous, because the light signal from B must have traveled an extra 300,000 km before reaching A and accompanying the signal from A. So, if O receives the two signals at the same time, she must conclude that B occurred before A. If the two signals did occur at the same time, then O must receive the signal from A first and then—precisely one second later—the signal from B. ∎

The example above illustrates that the simultaneous *reception* of signals from two events does not imply that the two events occurred at the same time. To facilitate the understanding of simultaneity, let us assume that, for each pair of events, every reference frame has a designated observer O_{des}, who happens to be equidistant from that given pair. Then, we can define simultaneity as follows:

> **Box 25.3.3.** *A pair of events A and B are* **simultaneous** *in a reference frame (RF) if the designated observer of that RF receives the light signal from A and B at the same time.*

Designated observer of a reference frame.

The significance of light signals used in the definition of simultaneity can be appreciated by looking at events A and B which occur in *moving* reference frames. Suppose we use bullets instead of light signals.

Karl and Emmy are standing on two trains parked next to each other at point C a distance of 300 m from James. They fire bullets—moving at 100 m/s toward James—from their rifles simultaneously. Three seconds later, James sees the bullets pass by him together. He concludes that the two events of firing the bullets must have occurred simultaneously because he knows that the two rifles were practically at the same location and the speed of the bullets were equal.

Now suppose that Karl is moving away from James at 50 m/s, while Emmy is moving towards James at 50 m/s. When they reach C at the same time, they fire bullets towards James. Emmy's bullet has a speed of 150 m/s while Karl's bullet speed is 50 m/s. So, although fired simultaneously, Emmy's bullet reaches James in 2 seconds while Karl's reaches him in 6 seconds, and he cannot conclude that the two events occurred simultaneously. The motion of the rifles has affected the speed of the bullets relative to James. Light does not have this problem, because the speed of light is not affected by the motion of its source. If Karl and Emmy were moving at half the speed of light in opposite directions, the light signals they emit at C will reach James *at exactly the same time*. That is why the very notion of simultaneity is so intricately tied to the speed of light.

> **What do you know? 25.7.** In the discussion above, suppose that Karl and Emmy move at 200 m/s (in opposite directions, of course). How long does it take Emmy's bullet to reach James? How about Karl's bullet?

25.3.1 Simultaneity without Relative Motion

Having defined the notion of simultaneity, we can now ask: If the designated observer O_{des} sees events A and B as simultaneous, does another observer O perceive them as simultaneous? If O_{des} and O do not move relative to one another, i.e., if both are in the same RF, the answer is yes, and here is why. Being aware that he is not equidistant from the two events, O knows that he cannot rely on the simultaneous reception of the signals. He, therefore, must turn to the designated observer for help.

Suppose O_{des} has a (very accurate) watch that emits a light signal every second. Any other observer, including O, *at a fixed distance from* O_{des}, will receive those signals every

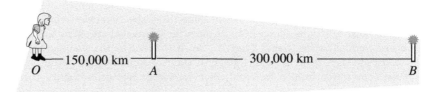

Figure 25.4: Determination of simultaneity when firecrackers are located at different distances from the observer.

second, because each signal has exactly the same distance to travel. Therefore, O's clock can be compared and made to run at exactly the same rate as the clock held by O_{des}, and vice versa. In particular, O_{des} can send a coded message to O at the exact "second" that she receives the two simultaneous signals from events A and B, informing O that those two events occurred at the same time.

All clocks of an RF can be made to run at the same rate and be synchronized.

Furthermore, all clocks of the RF can be completely synchronized. Here is how to do it. Let us call clock C the reference clock. This is the most accurate clock available. In fact, we assume it is so accurate that we can call it an *ideal clock*.[3] We can then synchronize all other clocks with C. How do we do this, if other clocks are located far, very far—light years—away from C? By light signals! An operator at C can synchronize any clock C′ as follows. First she determines the distance between C and C′ by measuring the time interval between sending a light signal to C′ and receiving the reflected echo at C. Denoting this time interval by T, she determines the distance between C and C′ to be $Tc/2$. For instance, if T is 2 seconds, the distance is 300,000 km.

Once the distance between C′ and C is determined, the operator can send a code at 12:00 containing the sentence "As soon as you receive this signal, set your time to 12:00 plus $T/2$!" Then the operator at C′ can follow the order, setting his time to be exactly the same as C. This is what takes place at the Earth's RF.[4] All clocks are set to Greenwich Mean Time. However, since it would be fatuous for the residents of Tokyo to be forced to call it 12:00 noon when everywhere is dark, the synchronization of the Earth clocks follows a more democratic rule by incorporating the geographic location in the process.

It is crucial to note that the synchronization procedure is applicable because all clocks are stationary relative to one another. As we shall see, relative motion between two clocks affects their operation. Theoretically, one can synchronize all the ideal clocks of a single RF to tick in perfect harmony. We summarize the foregoing discussion as

Box 25.3.4. *All observers in a single RF keep the same time and agree on the notion of simultaneity.*

25.3.2 Simultaneity with Relative Motion

What happens to simultaneity when two RFs move relative to one another? To investigate this question, let us get back on our train and consider two firecrackers at its two ends. Karl is on the ground and Emmy stands at the midpoint of the train. Suppose that the

---

[3]An ideal clock never runs fast or slow. Such a clock does not exist, of course, but *atomic clocks* approximate this ideal clock to a very high degree of accuracy.

[4]Strictly speaking, not all residents of Earth have the same RF because of the Earth's spin: People in China and the US move in opposite directions! But such a difference in speed is so much smaller than the speed of light that we can ignore it.

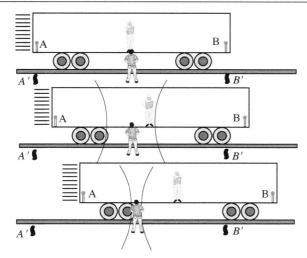

Figure 25.5: Different observers disagree on the simultaneity of two events. The ground observer (Karl) sees the two events as simultaneous, while the train observer (Emmy) does not. The marks labeled A' and B' are left on the ground by the explosion of the firecrackers A and B, respectively. These marks are as seen by Karl. To see how these same marks appear to Emmy, refer to Figure 25.7(a).

firecrackers produce smoke that can leave permanent marks on the platform. Suppose also that Karl finds himself right in the middle of these two marks. Finally, assume that he receives the signals from A and B at the same time. Figure 25.5 shows the situation as seen by Karl. The top figure is the moment the explosions occur.[5] The middle figure shows the wave fronts moving away from their sources. Note that Emmy has received the signal from B, and that the two wave fronts are equidistant from Karl and from the positions of the firecrackers *at the time of explosions* (i.e., positions of A and B in the top figure). Finally, in the bottom figure the two wave fronts have reached Karl while Emmy is still waiting for the signal from A.

After receiving the two signals simultaneously, Karl walks to A' counting his steps, walks to B' counting his steps, notices that the two distances are equal, and concludes that events A and B were simultaneous. What about Emmy? After receiving the two signals, she measures her distance from A and B and verifies that she was halfway between them. She therefore concludes that B must definitely have occurred before A.

This discussion brings out the essence of the relativity of simultaneity. What Newtonian mechanics, and common sense, thought obvious and self understood, is dramatically undermined by the scrutiny of relativity principles. There is no escape from this! Once we accept the principles of relativity, in particular the invariance of the speed of light, we have to face the (initial) discomfort of the realization that simultaneity is not universal.

Although simultaneity of two events is not universal, the fact that Karl sees A and B as simultaneous, is. Not only does he say "A and B are simultaneous for me," but also Emmy and all other observers in the universe say "A and B are simultaneous for Karl" despite the fact that they themselves do not see A and B at the same time. This notion can be further clarified by assuming that Karl is equipped with a special firecracker C that he can trigger when the two signals reach him at the same time. Thus, the explosion of C heralds the simultaneity of A and B as seen by Karl, and all observers keeping an eye on C can conclude that Karl saw A and B at the same time (if they see the single explosion of C) or not (if they don't see the explosion of C).

The preceding discussion may have given the reader the impression that Karl is being

---

[5]As *calculated* by Karl after he receives the signals.

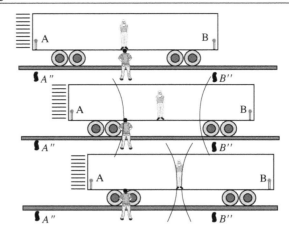

Figure 25.6: Here Emmy sees the two events as simultaneous. Karl sees A before B. The marks on the platform are as seen by Emmy. To see how these same marks appear to Karl, refer to Figure 25.7(b).

treated as special because he is the one who sees the two events simultaneously. To erase any impression of such special treatment, let us give Emmy the "honor" of observing the two events at the same time. Figure 25.6 shows the situation as seen by Emmy. The top figure is the moment the explosions occur.[6] The middle figure shows the wave fronts moving away from their sources. Note that Karl has received the signal from A, and that the two wave fronts are equidistant from Emmy and from the positions of the firecrackers. Finally, in the bottom figure the two wave fronts have reached Emmy while Karl is still waiting for the signal from B. As before, after receiving the two signals, Karl and Emmy measure their distances from A and B (or their marks on the platform) and verify that they were both halfway between them. Emmy concludes that the two events occurred simultaneously while Karl decides that A must have occurred before B.

> **Box 25.3.5.** *Simultaneity is a relative phenomenon. If one observer sees two events as simultaneous, other observers moving relative to the first, in general, do not. However, the fact that a given observer sees two events as simultaneous (or not) is universal.*

25.4 Relativity of Length

The relativity of simultaneity has an immediate consequence, another counter-intuitive result. The discussion of the last section led to an apparent paradox. It is clear that A and B are located at the two ends of the train in Figure 25.5, with Emmy in the middle. Karl can see the two ends of the train coincide with the firecracker marks on the ground. So, he is puzzled: Why doesn't Emmy receive the signals at the same time? After all, each signal has to travel half the length of the train to reach Emmy. And surely, Emmy doesn't see a different length for the train than Karl! Or does she?

Before scrutinizing the effect of motion on length, we need to know how to measure the length of an object when it moves. We could jump on the object and use a tape measure. But this would be the same as measuring the length of the object when it *does not move*, because motion is relative and when you are on the object, the object is not moving relative to you. You can't use a tape measure while the object is moving: put the head of the tape

How to measure the length of an object in motion.

---

[6]As *calculated* by Emmy after she receives the signals.

measure at the front of the object when it reaches you; but by the time you get to the end of the object, it has moved from where it was when you started the measurement. This should tell you that, somehow you must put the head and the tail of the tape measure *simultaneously* at the two ends of the object in motion. One way to do this is to have two events occur simultaneously at the two ends of the object, and make sure that those events leave permanent marks in your reference frame. The distance between these two marks, which you can measure at your leisure, is the length of the object.

25.4.1 Contraction of Length along Motion Path

Look at the events of Figure 25.5 from Emmy's point of view. Further assume that we are *repeating* the exact same experiment. So, there are already black marks on the ground from the previous experiment, and we perform the experiment with blue smoke. Emmy is standing in the middle of the train, anticipating the occurrence of B. She eventually receives the signal from B and concludes that a fraction of a second earlier the front end of the train must have coincided with the black mark at B'. Immediately she looks back, and since the signal from A has not arrived yet, she concludes that the rear end of the train has not reached the mark on the ground yet. She, therefore, perceives a picture shown in Figure 25.7(a), and concludes that the *train is longer* than the distance between the ground marks! This conclusion is an inescapable consequence of the relativity of simultaneity which in turn is a result of the second principle.

It seems that Karl and Emmy are in disagreement concerning the length of the train. Karl's measure of the length of the train is shorter than Emmy's. Could it be that their contradicting conclusions are due to the fact that Karl saw the two events simultaneously? That if Emmy sees the two events at the same time the conclusions will be reversed? To see this, let us conduct the (double) experiment in which Emmy sees the explosion of the firecrackers simultaneously, as shown in Figure 25.6. Karl is standing in the middle of the two marks on the platform, anticipating the occurrence of A. He eventually receives the signal from A and concludes that a fraction of a second earlier the rear end of the train must have coincided with the black mark at A''. Immediately he looks to his right, and since the signal from B has not arrived yet, he concludes that the front end of the train has not reached the mark on the ground yet. He, therefore, perceives a picture shown in Figure 25.7(b), and decides that the *train is shorter* than the distance between the ground marks! It does not matter who sees the two events at the same time; *the train is shorter for Karl.*

Is there anything special about the train? Is it because it is moving and the platform is not? But wait! It can't be, because motion is relative; to Emmy it is the platform that is moving, and if relativity is correct, lengths on the platform should appear shorter to Emmy. Let's see if this is indeed the case.

Place the firecrackers on the platform instead of the train and let Emmy and Karl measure the distance between them. Since our preceding discussions have shown that the distance measurement (whether it is shorter or longer) is independent of who sees the explosions simultaneously, let us assume that Emmy, riding on the train, sees the firecrackers explode simultaneously. Karl, on the other hand, receives the signal from A before that from B. Figure 25.8 explains this situation.

As usual, assume that an identical experiment has been done before, and it is now being repeated. So, there are already black marks on the train from the previous explosions indicating the distance between the firecrackers *as measured by Emmy*. Karl is standing in the middle of the two firecrackers on the platform, anticipating the occurrence of A. He eventually receives the signal from A and concludes that a fraction of a second earlier the black mark closer to the rear end of the train must have coincided with A. Immediately he looks to the front, and since the signal from B has not arrived yet, he concludes that the mark closer to the front end of the train has not reached B yet. He, therefore, decides that

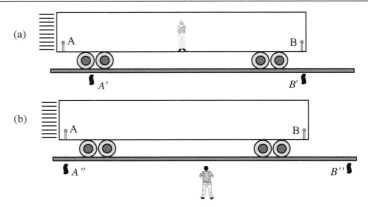

Figure 25.7: (a) The distance between the firecracker marks is the length of the train as seen by Karl. Emmy is seeing the train longer. (b) The distance between the firecracker marks is the length of the train as seen by Emmy. Karl is seeing this distance to be longer than the length of the train he measures.

the distance between the two firecrackers on the platform is larger than the marks on the train. Therefore, *Emmy measures the distance between A and B to be less than what Karl measures.*

Note the symmetry between the two observers. Neither is any more special than the other. Both are claiming that lengths measured in their RF are longer than the moving lengths.[7] In other words, they both conclude that *moving lengths shrink.*

Moving lengths shrink.

And the shrinkage is not due to some kind of a mechanical compression of the train or the platform (or cars, trucks, planes, or meter sticks). The two ends of a stick merely indicate two points in space. It is this distance that is shrinking. Thus, *motion affects the space itself.* For instance, these two points could be the locations of two stars. As seen from Earth, these two stars, say the Sun and Alpha Centauri, which are almost fixed relative to the Earth's RF, are seen to be about 4 light years away from one another. The crew of a spaceship, traveling at very high speed towards Alpha Centauri, sees this distance shrunk because, to them, the length between the Sun and alpha centauri is in motion. We shall see some striking examples of such space travels in Chapter 26 where we quantify the shrinkage of lengths.

25.4.2   Length Transverse to Motion Path

All the contractions treated so far occur for lengths that are along the path of motion. Would the same effect occur if the length moved perpendicular to itself? We ask Karl and Emmy to help us figure this out. Karl gets into a spaceship while Emmy climbs to the top of a mountain. We give each a long stick at the end of which are two lights, which are off for now. Karl's lights are red; Emmy's green. We instruct each to hold his/her rod firmly in the middle[8] and make sure that it remains perpendicular to the path of motion, and we arrange the altitude of the spaceship in such a way that the midpoints of the rods coincide at the moment that Karl passes Emmy [see Figure 25.9(a)]. The rods are equipped with a trigger device that turns the lights on and off right at the moment that they pass each other. We now use the first principle of relativity and simple logic to conclude that neither rod shrinks.

Suppose that Emmy saw a shrinkage of Karl's rod. Then the red lights would be in between the green lights; and since her rod is covering Karl's rod, she would not see the red

---

[7]Remember that it is *the platform* that is moving relative to Emmy!

[8]To be precise, the middle of the rod should be at the level of his/her eyes (detectors of the light flashes).

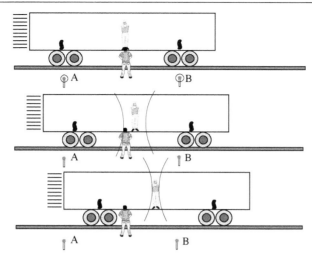

Figure 25.8: Emmy sees the two events as simultaneous and decides that the length $\overline{AB}$ is the distance between the two marks on the train. Karl sees A before B and decides that the distance between the marks is shorter than $\overline{AB}$.

flashes. But she would not be alone in seeing this situation. Since the two rods coincide instantaneously as they pass each other, everybody in the universe that happens to (be behind Emmy and) see the event, would see the ephemeral green flashes. Thus, Emmy concludes that everybody in the universe (at least, everybody behind her) should see only two green flashes as shown in Figure 25.9(b).

So far so good! But now we invoke the first principle of relativity: if Emmy concludes that moving lengths perpendicular to the direction of motion shrink, Karl has to come to the same conclusion. So, he sees Emmy's rod moving, and expects it to shrink. His experience will be that the green flashes will occur between the red flashes, and if his experience were to materialize, he would conclude that everybody in the universe behind Emmy should see two green flashes between two red flashes as shown in Figure 25.9(c). But this is impossible. The same universal observers cannot reach two different conclusions for the same experiment. We have to conclude that neither rod changes its length due to its motion. Let's summarize our observation in the last two subsections as:

> Only lengths parallel to motion shrink.

Box 25.4.1. *All moving lengths parallel to the direction of motion shrink. The lengths perpendicular to the direction of motion are not affected by the motion.*

Example 25.4.2. A circle in motion will appear as an ellipse, because the diameter parallel to the direction of motion will shrink while the diameter perpendicular to the direction of motion will remain unchanged. Similarly, a square moving parallel to one of its sides appears as a rectangle. If the square moves along one of its diagonals, it will appear as a rhombus. Don't expect to see a change in the shape of a circle or a square by holding them in your hand and whizzing them in front of your face! The change in the shape is noticeable *only* if the circle and the square move with speeds close to light speed (see Example 26.2.1 in Chapter 26). ∎

> A circle becomes an ellipse; a square becomes a rectangle (or a rhombus) as they move close to light speed.

It is instructive to analyze the effect of motion on a length parallel to the motion path with a length perpendicular to the path in terms of simultaneity. In the former case, the arrival time of the light signal coming from different sources depends on the observer. If one observer receives the two signals at the same time the other observer does not; and the reason for this asymmetry is that the motion of one of the observers is towards one

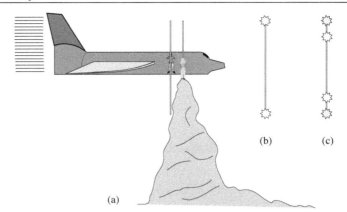

Figure 25.9: Analysis of the effect of motion on length perpendicular to the direction of motion. (a) Karl is approaching Emmy, both holding their rods. (b) Emmy's conclusion of what people see behind her. (c) Karl's conclusion of what people see behind emmy.

source and away from the other (see Figures 25.5 and 25.6). However, in the perpendicular case, both observers (really, their eyes) are equidistant from the two sources *at all times*, regardless of their motion or the time at which the flashes are sent. In other words, since both observers *always* remain on the perpendicular bisector of *both* rods, they *both* receive the signals of the two events simultaneously. Therefore, they have to conclude that the two lengths are equal.

25.5   End-of-Chapter Material

25.5.1   Answers to "What do you know?"

25.1. In the forward direction, the two speeds are in the same direction. Therefore they add to give 260 mph for the speed of the bullets relative to the bystanders. In the backward direction, the two speeds are in opposite directions. Therefore they subtract to give 140 mph for the speed of the bullets relative to the bystanders.

25.2. Write Equation (25.1) as $v_{1g} = v_{12} + v_{2g}$ and note that $v_{1g} = 70$ mph and $v_{2g} = 60$ mph to get $v_{12} = 10$ mph. This means that car 1 appears to be moving at 10 mph relative to car 2.

25.3. (a) No. No matter where you look, you see darkness in empty space with flickering starlight in the background. (b) Still, no! All you can say now is that something is moving *relative* to you, or you relative to it.

25.4. The cylinder has a centripetal acceleration. So there must be a fictitious force in the *opposite* direction, i.e., towards the cylindrical wall. This force causes you to stick to the wall and not fall down.

25.5. Since the second postulate seems to be violated, it must be impossible to move at the speed of light! Later, when we consider relativistic energies in Chapter 28, you'll see that it requires *infinite* amount of energy (or fuel) to accelerate an object to the speed of light.

25.6. As you will discover later in this and in the next chapter, both time and distance are relative concepts. One second for B is not one second for S, and 1000 km for B is not 1000 km for S. When S measures the distance and the travel time of L, he finds that L is moving at 300,000 km/s according to him!

25.7. Since Emmy is moving towards James, the bullet's speed is the sum of the two speeds, i.e., 300 m/s. So, the bullet reaches James in one second. The speed of Karl's bullet is the difference between the two speeds, i.e., 100 m/s *away* from James. So, Karl's bullet never reaches James!

25.5.2   Chapter Glossary

Ether A hypothetical medium whose undulation was believed to manifest itself as light. Relativity theory did away with ether altogether.

Event Something that happens at a point in space and at a single instant in time. A point in the four-dimensional spacetime. A point with four coordinates, one of which is time.

Fictitious Force A force in a reference frame created solely due to the acceleration of the RF.

Galilean Relativity A restricted version of relativity which applies only to mechanics and motion. On a ship moving smoothly, all physical experiments look identical to those done on land.

Inertial Frame A reference frame in which the first law of motion holds. Usually it is an RF with no acceleration.

Length Contraction A relativistic effect whereby a length shrinks in the direction of its motion.

25.5.3   Review Questions

25.1. What is the law of addition of velocities? When crossing a river flowing eastward on a boat, what direction does the boat move if you paddle northward?

25.2. What does electromagnetic theory say about the speed of EM waves? How does the speed of the waves depend on the speed of the source?

25.3. What was ether and why was it hypothesized? Which experiment showed that it did not exist? What would happen if you could catch up with light? Explain in terms of ether as the medium of light wave.

25.4. State the two principles of relativity. Can observers detect their uniform motion by doing experiments entirely in their RFs? Can observers detect their acceleration by doing experiments entirely in their RFs? What theory yields the second principle of relativity?

25.5. What is the difference between Galilean and Einsteinian relativity?

25.6. What is an inertial RF? What is a fictitious force? How do you define an inertial frame in terms of the first law of motion?

25.7. Can you apply the law of addition of velocities to light? If you chase a light signal faster and faster, can you see it slow down?

25.8. What is an event? How do you define two simultaneous events? What is the role of light in this definition?

25.9. If two observers are stationary relative to one another, and one observer concludes that two events are simultaneous, does the second observer conclude the same thing? If the first observer receives the two light signals at the same time, does the second observer as well?

25.10. Explain in detail how two clocks in the same RF can be synchronized, even though they may be millions of kilometers apart.

25.11. If two observers are moving relative to one another, and one observer concludes that two events are simultaneous, does the second observer conclude the same thing?

25.12. How do you measure the length of an object in motion? Explain how the relativity of simultaneity implies the relativity of length. What happens to the length of an object when it moves along its length? Is the change due to some mechanical compression? What happens to the length of an object when it moves perpendicular to its length?

25.13. What happens to a circle when it moves close to light speed? What happens to a square when it moves close to light speed along one of its sides? Along one of its diagonals?

25.5.4   Conceptual Exercises

25.1. An airplane is to go straight north in a region where an easterly wind is blowing. Which direction should its wind velocity be?

25.2. A train is moving at 80 mph. A passenger throws a ball towards the front of the train at 60 mph. How fast is the ball moving relative to a ground observer?

25.3. A train is moving at 80 mph. A passenger throws a ball towards the back of the train at 80 mph. How fast is the ball moving relative to a ground observer?

25.4. A spaceship is moving at 250,000 km/s. A passenger sends a beam of green laser light towards the front of the spaceship at 300,000 km/s. How fast is the laser light moving relative to a ground observer?

25.5. A spaceship is moving at 250,000 km/s. A passenger sends a beam of green laser light towards the back of the spaceship at 300,000 km/s. How fast is the laser light moving relative to a ground observer?

25.6. You are in a train moving at 65 mph. You hold a ball in your hand, go to a window, stick your hand out, and throw the ball up. Ignore air resistance. Where does the ball land on its way down, towards the front of the train, towards the back of the train, or into your hand?

25.7. At what point is an airplane closest to an inertial frame, when taking off, just after landing, or when cruising at high altitiudes?

25.8. Karl is on a super fast spaceship moving at 250,000 km/s. He sends a beam of light perpendicualr to his direction of motion. A second later he sees the beam at a distance of 300,000 km/s up. Emmy watches Karl. She sees Karl sending the light beam, and notices that a second later Karl has moved 250,000 km horizontally and the light beam 300,000 km vertically. So, to her the light path is the hypotenuse of a right triangle of sides 250,000 km and 300,000 km. Is the light beam, which covers this hypotenuse in one second, going faster than 300,000 km/s according to Emmy? Hint: Read Einstein's account of the discovery of relativity on page 362. Also see Figure E.18.

25.9. You receive the light signal from two events at the same time. Are these two events simultaneous? Explain.

25.10. Emmy and Karl are not moving relative to one another. Karl is in the middle of the location of two events, receives the signals from the two events at the same time, and concludes that the two events were simultaneous. Can Emmy say that they were simultaneous as well?

25.11. Figure 25.5 shows that when the relative motion of two observers is along the line joining two events, then if the two events are simultaneous according to one observer, they cannot be simultaneous according to the other. Is it possible for two events to be simultaneous according to two observers moving relative to one another? Hint: Consider the perpendicular bisector of the line segment joining the two events.

25.12. Event A is simultaneous with event B, and event B is simultaneous with event C. Is it true that event A is simultaneous with event C? If so, can you find a single designated observer who receives the three signals at the same time assuming that the three events form a triangle? Can you find such an observer if the three events lie along the same line?

25.13. You want to measure the length of a train moving close to light speed. Put two smoking firecrackers at its two ends. After the train passes you by, you note that the light signals from the two firecrackers arrive at you simultaneously. Do you conclude that the marks they left on the ground represent the length of the train in motion? Explain.

25.14. Is it true that the distance between the marks made by the two ends of a moving object is the length of the object if the two marks occurred at the same time?

25.15. Karl and Emmy each hold a meter stick parallel to the direction of their relative motion. Which statement is correct? (a) Karl's meter stick is shorter for Emmy. (b) Emmy's meter stick is shorter for Karl. (c) Both (a) and (b).

Chapter 26

Relativity of Time and Space

The qualitative discussions of the last chapter made it clear that motion affects time and space. This chapter investigates these effects quantitatively and determines the extent to which motion affects the working of clocks and rulers, and addresses the question of whether we have to correct our watches or adjust our height every time we hop in and out of a car.

26.1 Time Dilation

As Einstein said "Time is what is measured by clocks." So, let us look at the effect of motion on clocks. And by the "effect of motion" we do not mean the effects of the bumps and puddles of a rough road which may actually damage a clock. We have in mind the smoothest possible ride such as the sailing of a spaceship in outer space. Any effect on clocks must be interpreted as an effect on time itself, although we may be using a particular clock to detect the effect.

The clock best suited for this investigation is the "arm" of the Michelson-Morley apparatus shown in Figure 26.1. It consists of a source S of light, or electromagnetic waves, and a mirror M. The distance between S and M is L which we can conveniently take to be, say 1.5 m. Then it takes light $1.5/3 \times 10^8$, i.e., 5×10^{-9} seconds or 5 ns (nanosecond, as nano means 10^{-9}) to go from S to M, and 10 ns to go to M and come back to S. If we place a light sensitive "ticker" at S, it will tick every 10 ns. We call such a clock a Michelson-Morley clock, or **MM clock** for short. **Math Note E.26.1** on **page 111** of *Appendix.pdf* compares two such clocks which are moving relative to one another and connects the time intervals measured by the two clocks via a simple mathematical relation, which we reproduce below:

The Michelson-Morley clock.

$$\Delta t = \frac{\Delta \tau}{\sqrt{1 - (v/c)^2}} = \gamma \Delta \tau, \qquad \gamma = \frac{1}{\sqrt{1 - (v/c)^2}}, \tag{26.1}$$

Proper time is measured by a clock present at the two events.

where $\Delta \tau$, the **proper time** is the time *interval* between two events measured by a clock that is present at the two events. If the clock, and therefore the observer, is present at both events, the events must occur at the same spatial point for that observer. For example, to the captain of a spaceship who looks at the take-off from Earth and landing on another planet from the window of the spaceship, these two events are taking place at the same spatial point: the window. We can thus define proper time in two different ways:

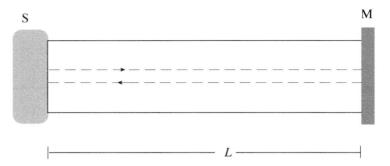

Figure 26.1: A Michelson-Morley clock. A "tick" of this clock occurs when the light signal makes a round trip along the length L.

> **Box 26.1.1.** *The proper time interval between two events is the time measured by an observer who is present at both events, or for whom the two events occur at the same spatial point.*

The time interval Δt, called the **coordinate time**,[1] is the time measured by another observer O', moving relative to O with speed v, for whom the two events occur at two different spatial points. Equation (26.1) applies to situations in which the relative velocity of the two observers remain constant. If there is an abrupt change in the speed or the direction of motion, we have to apply (26.1) *separately* to those portions of the motion in which the velocity does not change.

Gamma factor introduced.

The expression $1/\sqrt{1-(v/c)^2}$, to which we assigned the symbol γ, is sometimes called the **gamma factor**. It determines whether relativistic considerations are essential or not. If the speed v of an object is very small compared to c, then γ is very nearly 1 and relativistic effects can be ignored. On the other hand, when γ is much larger than 1, classical mechanics fails and relativity theory must be invoked.

Before discussing (26.1) any further, notice the factor $\sqrt{1-(v/c)^2}$. A physical quantity cannot be the square root of a negative number. Therefore, the expression under the radical sign must be nonnegative: $1-(v/c)^2 \geq 0$, i.e., $v/c \leq 1$. We, therefore, conclude that nothing can travel faster than light. Since objects can carry information and since information can be interpreted as signals, we obtain the important result:

Never faster than light!

> **Box 26.1.2.** *No signal can be transmitted with a speed greater than the speed of light.*

Light, of course, travels at the speed of light. Is it possible to attain this speed for anything else? STR predicts that for any *massive* object, no matter how small the mass, the answer is no. In fact, the daily activity of the accelerators (atom smashers) around the world is a testimony to this impossibility. By providing more and more energy to an elementary particle, these machines can accelerate the particle to higher and higher speeds, just as the provision of fuel to the engine accelerates a car. The amount of energy delivered to particles has grown steadily several hundred thousand times over the last few decades. However, in no instance has a single particle been observed to have a speed greater than, or even equal to, the speed of light. In 2000, the Large Electron Positron Collider (LEP) at CERN was able to accelerate the electrons to a speed of $0.999999999997c$, corresponding

No massive object can move at the speed of light!

---

[1]The strange naming of Δt will make more sense when t becomes a coordinate in Chapter 27.

to a γ of 409000. The Large Hadron Collider (LHC), also at CERN, is due to accelerate protons to a speed of $0.999999991c$ corresponding to a γ of 7461.[2]

"But light, which Einstein himself discovered to be composed of *particles* (photons), does travel at the speed of light," you may remark, "How do you explain the speed of photons?" In the last paragraph, we were careful to emphasize the word "massive." It turns out that photons are massless, and a full account of this and other energy-related topics can be found in Chapter 28.

Equation (26.1) is at the heart of the relativity theory, and as you will see in the sequel, almost all relativistic results are obtained from it. Before discussing the implications of this formula, let's look at some numerical examples.

Example 26.1.3. The captain of the Spaceship Enterprise cruising at the speed of 1.8×10^8 m/s (60% speed of light, or $0.6c$) wakes up at time zero (event E_1), goes about doing his chores for 16 hours—according to his clock, of course—and goes to bed (event E_2).

Q: How long does this appear to the Earth observers?

A: Note that $\Delta\tau = 16$ hours is the proper time, because the two events occur at the same point in the spaceship, the bed. Stated equivalently, this time interval is measured by a clock that is present at both events. To calculate Δt, the time interval between E_1 and E_2 according to the Earth clock, we need v/c. But we are given that $v = 0.6c$. Thus, $v/c = 0.6$ and $\Delta t = 16/\sqrt{1-(0.6)^2} = 20$ hrs. The Earth observers, therefore, conclude that the captain has a long 20-hour work day! ∎

> **What do you know?   26.1.** The gamma (or dilation) factor is 1.25 for Example 26.1.3. What is the gamma factor if the speed is 90% of light speed? How many hours would a work day be for the captain?

Space travel beyond the closest stars may seem physically impossible. The ambition of exploring stars hundreds of light years away appears to be futile. After all, it takes light, the fastest object in the universe, 100 years to travel from a star that is 100 light years away to Earth. How can the crew of a space probe, even moving infinitesimally close to the speed of light, hope to survive the journey? All members of the crew will be dead by the time the probe reaches the star. **Example D.26.1 on page 44** of *Appendix.pdf* shows that, due to the relativity of time, such space travels are possible if we can achieve speeds close to light speed.

This relativity of time intervals is called **time dilation** and is one of the most difficult concepts to grasp in relativity theory, because it leads to what has incorrectly come to be known as *time warp* or *time travel*.

Time dilation.

Captain Kirk, of the Spaceship Enterprise is 30 years old when he starts on a mission that takes him and his crew to a distant star, leaving his wife and his newborn daughter behind. After spending a few months on one of the planets of the star, he and the crew of the Enterprise head home and reach Earth 15 years after they took off. When he asks for his wife, he is told that she died of old age decades ago; and as he looks in the crowd welcoming the crew, he finds a 60-year-old lady who resembles his wife. He is shocked to find out that the old lady is his daughter! If you don't believe the story, look at **Example D.26.2 on page 45** of *Appendix.pdf*: Numbers don't lie!

Daughters older than fathers!

> **What do you know?   26.2.** Captain Kirk's spaceship moves at $0.99c$. He is 30 years old and has had a baby just before he leaves on his mission to a distant planet. It takes him 2 years to get to the planet. How old is his daughter when he lands on the planet?

---

[2]CERN is not retrogressing! The key factor in progress is the *energy* of the particle. The speed of the electrons corresponds to an energy of 200 GeV (200 billion eV) while the speed of the protons corresponds to an energy of 7000 GeV.

26.2   Length Contraction

Example D.26.2 may seem to suggest a violation of the supremacy of the speed light. It takes light 24.63 years to travel the Earth-planet distance, yet Enterprise covers it in 5 years. Is Enterprise going faster (almost five times faster) than light? Of course not! The key to this puzzle is in the notion of length contraction encountered before, for which **Math Note E.26.2** on **page 113** of *Appendix.pdf* derives the following formula:

Length contraction formula.

$$L = L_0 \sqrt{1 - \frac{v^2}{c^2}} = \frac{L_0}{\gamma}, \tag{26.2}$$

where γ was introduced in Equation (26.1). L_0 is length measured by an observer stationary relative to the two ends of the length; we call L_0 the **rest length**. L is length measured by an observer moving relative to the two ends of the length; we call L the **moving length**. We see that a moving length is shrunk by a factor of $\sqrt{1 - (v/c)^2}$ to an observer with respect to whom the length moves with a speed v.

Length contraction can "explain" why space travel to distant stars is possible, even though light itself may take hundreds of years to travel such distances. The point is that, as a spaceship travels at a speed close to light speed, the distance between Earth and the destination star becomes very short. So, the crew does not see the star 100 light years away, but only 5 light years (if the spaceship is moving fast enough). **Example D.26.3** on **page 46** of *Appendix.pdf* provides a sample numerical detail.

Example 26.2.1. Having learned about the length contraction and having read Example 25.4.2 in Chapter 25, Emmy wants to try to make an ellipse out of a circle by moving it. First she compares lengths differing slightly from 1 meter to see how much a meter stick should shrink before she can actually notice the difference by merely eyeballing it. She decides that she can tell the difference between 100 cm and 95 cm by merely looking at them, i.e., if somebody showed her a meter stick (100 cm long), and a little later a 95-cm stick, she could "remember" that the first stick was longer.

Next, she tries to see how fast a circle with a diameter of 100 cm should move horizontally so she could see it as a vertical ellipse. Clearly, it has to move at such a speed that the horizontal diameter shrinks at least to 95 cm. She uses Equation (26.2) with $L = 95$ cm and $L_0 = 100$ cm: $95 = 100\sqrt{1 - (v/c)^2}$, whose solution gives $v/c = 0.312$, or $v = 0.312c$. Plugging in 3×10^8 m/s for c, she gets $v = 9.4 \times 10^7$ m/s, or 209 million mph. Little wonder we do not see people and objects shrink as they move by us! ∎

> **What do you know? 26.3.** In the previous **What do you know?**, the spaceship is moving at almost the speed of light and it takes them 2 years to get to the planet. So, the distance from Earth to the planet is very nearly 2 light years. What is the distance according to Earth people?

26.3   The Twin Paradox

You may very well have a feeling of uneasiness regarding time dilation, especially when the aging process comes into consideration. Let us dwell on this topic for a while. Karl and Emmy are twins. As soon as they are born, Karl is put on the spaceship Marinarus heading to the star Epsilon Eridani about 11 light years away from Earth to explore its planets. The speed of Marinarus is $0.9c$. Both the ground control and the crew of the spaceship agree to send "happy birthday" signals to the twins on each anniversary of their birth.

On the first anniversary, the ground control sends the first "happy birthday" signal. They wait for a month, then 2 months, then the whole second year to receive Emmy's "happy birthday" message, all in vain. They send Karl his second happy birthday signal and wait another year. Still no signal! On the third anniversary, they repeat their annual message, and they urge the crew to respond. Still no response! Although it has given up hope by now, the ground control nevertheless sends a "Happy Fourth, Karl!" knowing that the message is probably aimless. Marinarus is now considered "lost in space," but the communication channel is left open just in case. Exactly 131 days after the fourth message is sent the ground crew receives the signal saying "Happy first, Emmy!" Puzzled by this strange time warp, the ground control consults its theoretical physicist for an explanation. His crude response is as follows: There are two kinds of time delays involved. One is the relativistic time dilation which stretches the year in the spaceship to something more than a year. The second is the time delay due to the travel time of the radio signal from the spaceship to the Earth. His exact response is explained in **Math Note E.26.3** on **page 113** of *Appendix.pdf*.

How do we separate the delay due to the signal travel from the relativistic effect of time dilation? In other words, given Δt, the *total* elapsed time measured by the Earth clock, how can we separate from it $\gamma \Delta \tau$, the dilated time? This is a relevant question because in an actual experiment, it is the total time that is measured by the Earth clocks, and one may be interested in the relativistic effect only. The *qualitative* answer is discussed below. The quantitative answer can be found in **Math Note E.26.4** on **page 114** of *Appendix.pdf*.

Let us start the qualitative explanation by recalling (see Box 25.3.4) that *each reference frame has a unique time.* Therefore, instead of communicating directly to the ground control, the Marinarus crew can relay their messages to outposts located at strategic distances from the main control room. For instance, the theoretical physicist, foreseeing the relativistic time dilation, can suggest outposts located approximately 2.065 light years apart. These correspond to locations which Marinarus reaches just at the time that the crew sends their annual signals. In this case, the signal will be received immediately after it leaves the spaceship. How did the physicist come up with 2.065 light years? See the following example.

Example 26.3.1. Where should the first outpost be located so that the signal is received immediately after it leaves the ship?

According to the Earth clock, the crew of the spaceship send their first signal $\Delta t_1 = \gamma \Delta \tau_1 = 1/\sqrt{1 - (0.9)^2} = 2.2942$ years after they leave. During this time, the Earth observers measure the distance covered by the spaceship to be

$$\text{distance} = (\text{velocity})(\text{time}) = (0.9c)(2.2942 \text{ years})$$
$$= 2.06478c \text{ years} = 2.06478 \text{ light years} \approx 2.065 \text{ light years}.$$

Thus, an outpost located approximately at 2.065 light years, receives the signal as soon as it is sent. ∎

The usefulness of these outposts comes from their ability to record the information immediately. Thus, when the crew of Marinarus sends the code "Happy first, Emmy!" the inhabitants of the outpost can immediately look at their clock and see that 2.29 years have passed, not one year!

Is there something magical about the spaceship that keeps people young? Do we start staying young as soon as we jump on a spaceship headed into outer space? This seems to be contradictory to the principle of relativity which gives equal rights to all RFs. To understand the question, let us look at the same situation this time from the point of view of the spaceship RF. Marinarus crew see Earth receding from them at the speed of $0.9c$ and after one year they send their signal and wait for the corresponding signal from the Earth. In fact, without going any further, one can switch Earth and the spaceship, and all their corresponding devices and people everywhere in the foregoing discussion of the trip, and come up with the spaceship version of the journey. Thus, for instance, the crew of Marinarus will receive the "Happy first, Karl!" 4 years and 131 days after departure. *In each RF, the moving twin ages less.*

object	v (mph)	v (m/s)	v/c	$\frac{1}{2}(v/c)^2$	NY to LA
car	65	29	9.69×10^{-8}	4.7×10^{-15}	46 hrs
race car	230	103	3.43×10^{-7}	5.9×10^{-14}	13 hrs
jet plane	600	268	8.94×10^{-7}	4×10^{-13}	5 hrs
super-sonic jet	2240	1000	3.34×10^{-6}	5.5×10^{-12}	1.34 hrs
satellite	17,900	8,000	2.67×10^{-5}	3.5×10^{-10}	10 min

Table 26.1: The relativisticity of familiar moving objects.

How can this be? How can both twins see the other one grow less rapidly? What if we introduce a TV set that continually monitors the other twin? Then Emmy sees Karl as a 1 year old when she is more than 4 years old. On her 13th birthday she sees her brother as only 3. When she is married and has a child of her own at the age of 26, the monitor shows Karl as only 6. Finally, at the age of 91, in a retirement house where Emmy is spending the few remaining years of her life, she hardly recognizes her brother's face on the monitor appearing as a young 21-year-old man.

Such a state of affairs is hard to accept. What is worse, we have to accept the fact that Karl sees on *his* TV screen the youthfulness of his twin sister! Yet this is precisely what the special theory of relativity predicts. Is there no way that we can determine once and for all who ages less? As long as the two are moving at constant speed relative to each other, no. Both RFs conclude exactly the same thing because time is a relative physical quantity. We are used to an absolute time because all the speeds we deal with are so small that we never experience the dilation of time. Thus, a watch in a car, a clock in a ship, and a timer in a plane all are treated the same as the watch on our wrist. "Wait a minute," you may say. "There is a way of determining whether Karl stays younger or Emmy. Bring them face to face! Reverse the course of Marinarus on Karl's first birthday and bring it back to Earth. Then, when Karl is 2 years old, he will face his twin sister. Will she be younger, as the Marinarus crew may suspect, due to the motion of Earth clocks relative to them, or older as the ground control may calculate due to time dilation?" Stay tuned until Chapter 27 for the answer.

26.4   Relativisticity

All the foregoing discussions dealt with speeds approaching light speed. Now take some ordinary vehicles and see how important relativistic effects would be for them. Values of γ corresponding to some typical speeds are given in Table 26.1, which makes it clear that for all humanly possible speeds—even for the highest attained—the relativistic effects are extremely small. The measure of "relativisticity" is $\gamma - 1$ which, for the fastest speed attainable, is less than four parts in 10 billion! Such a small effect is not ordinarily detectable. **Math Note E.26.5** on **page 115** of *Appendix.pdf* shows how to approximate $\gamma - 1$ when the speed of the moving object is much smaller than light speed.

γ − 1 measures relativisticity.

> **What do you know? 26.4.** What speed is needed to give a gamma factor of 1.0001?

Example 26.2.1 showed how hard (actually impossible) it was to shrink a meter stick by a mere 5 cm. Take the fastest and largest man-made object, a 100-meter-long satellite moving at 18,000 mph. How much does it shrink as it passes by? By only the length of a molecule! It is impossible to see the difference between 100 m and another length that differs from it by the size of a molecule. What about time dilation? By how much does a clock slow down when placed on a jet plane and kept in motion for hours? Only a few nanoseconds! See **Example D.26.4** on **page 47** of *Appendix.pdf* for details.

Since we have been encountering the gamma factor γ on so many occasions, it is convenient to plot it as a function of the fractional speed v/c. Figure 26.2 shows such a plot. It should be clear from the graph that to get a γ that is even slightly different from 1, you need extremely high speeds, although such speeds may be negligible compared to light speed.

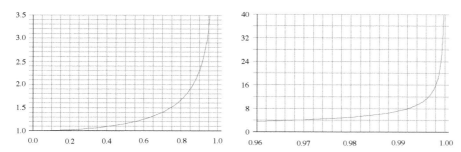

Figure 26.2: The gamma factor γ plotted as a function of the fractional speed v/c. The left figure plots the smaller values of γ. The right figure plots the larger values of γ.

While at ultra-relativistic speeds, strange phenomena, such as a daughter getting older than her father, are possible, even the fastest and the farthest round trip achievable today— an Apollo mission to the Moon, for example—yields only minute fractions of a second as a time-dilation effect (see **Example D.26.5** on **page 47** of *Appendix.pdf* for details).

Examples D.26.4 and D.26.5 show that even for humanly extraordinary speeds, the relativistic effects are immeasurably small. The dramatic relativistic effects of time dilation and length contraction demonstrated in Examples D.26.2 and D.26.3 require enormous vehicular speeds, speeds unavailable to our race now or in the near future. Stated differently, ordinary phenomena, in which relevant speeds are much smaller than light speed, obey classical ideas. It does not mean that relativity gives wrong results when applied to ordinary phenomena, it just means that for ordinary phenomena the relativistic effects are so small as to be unobservable by crude measuring devices such as ordinary watches and meter sticks. In fact, all the formulas of relativity turn into classical formulas when speeds are much smaller than light speed. For example, if v/c is much smaller than 1, then $\sqrt{1 - v^2/c^2} \approx 1$, and Equations (26.1) and (26.2) yield $\Delta\tau \approx \Delta t$ and $L \approx L_0$, respectively, as expected in classical physics. The correspondence between the relativity theory and Newtonian physics is stated as follows:

Box 26.4.1. *All the formulas and concepts of relativity reduce to the corresponding Newtonian formulas and concepts when all relevant speeds become much smaller than light speed.*

26.5 End-of-Chapter Material

26.5.1 Answers to "What do you know?"

26.1. $\gamma = 1/\sqrt{1 - 0.9^2} = 2.29$ and $\Delta t = 16\gamma = 36.7$ hrs.

26.2. Here $\gamma = 1/\sqrt{1 - 0.999^2} = 22.37$ and $\Delta t = 2\gamma = 44.7$ years. So the captain is 32 years old, while his daughter is 44.7 years old!

26.3. According to Earth, the spaceship is moving at almost the speed of light and it takes the spaceship 44.7 years to get to the planet. So, the distance from Earth to the planet is very nearly 44.7 light years.

26.4. From definition of the gamma factor get $1.0001 = 1/\sqrt{1 - (v/c)^2}$. So, $\sqrt{1 - (v/c)^2} = 1/1.0001 = 0.9999$. Square both sides, find $(v/c)^2$, take the square root of the result and get $v/c = 0.014$. So, v is 0.014 times the speed of light or 4.2 million m/s. This is equivalent to 9.5 million mph! It takes a lot of speed to get anywhere near light speed!

26.5.2   Chapter Glossary

Coordinate Time Time kept by a clock that is moving relative to the clock that keeps the proper time.

Gamma Factor A quantity defined for a moving object. If the object has speed v, its gamma factor is $1/\sqrt{1 - (v/c)^2}$.

Length Contraction The shrinkage of length to an observer with respect to whom the length moves.

Proper Time Proper time of an observer is the time kept by his/her clock.

Time Dilation The slowing down of clocks (including biological clocks such as aging) in motion.

26.5.3   Review Questions

26.1. What is proper time interval? What is coordinate time interval? Which one is shorter?

26.2. What is the ultimate speed for sending information? Is it possible for a massive object to move at the speed of light? What does this say about the mass of a photon?

26.3. What is time dilation? Is it possible for a father to be as old as his child? Can he be older? Can these happen without traveling?

26.4. What is the rest length? What is the moving length? Which one is longer? Is it possible to see length contraction of ordinarily moving objects?

26.5. What is the twin paradox? Why is it called a paradox?

26.6. How much does the fastest and largest man-made object, a 100-meter-long satellite moving at 18,000 mph shrink as it passes by? (See the beginning of Section 26.4.)

26.7. What is the relationship between relativity and Newtonian physics? What happend to relativity theory when the speed of objects under investigation is much much smaller than light speed?

26.5.4   Conceptual Exercises

26.1. One clock is on a spaceship that goes from Earth to a distant planet. Another clock is at the space center on Earth. For the time interval between the two events, take-off from Earth and landing on the planet, which clock, if any, measures the proper time?

26.2. One clock is on a spaceship that goes from Earth to a distant planet, turns around immediately and comes back to Earth. Another clock is at the space center on Earth. For the time interval between the two events, take-off from Earth and landing on Earth, which clock, if any, measures the proper time?

26.3. Light travels from a star to Earth in 60 years. The captain of Enterprise gets only 5 years older when he goes from Earth to that star. Is there something wrong with these statements? Has he traveled faster than light?

26.4. Is it, in principle, possible for the proper time to be 1 second and the coordinate time to be 1 year? How big is the gamma factor?

26.5. Alpha Centauri is a star that is 4 light years[3] away from Earth. Is it, in principle, possible for the crew of a spaceship to measure the distance between Earth and Alpha Centauri to be 4 km? How big is the gamma factor?

26.6. Box 26.1.2 says that no signal can travel faster than light, yet we have seen examples where the crew of a spaceship can travel a distance of tens of light years and get only a few years older. Are they violating the statement in Box 26.1.2?

26.7. Use Figure 26.2 to estimate the Earth-Alpha Centauri distance as seen by the crew of a spaceship moving at $0.997c$ towards Alpha Centauri.

26.8. A rectangle is twice as long as it is wide. How fast, and along which side, should it be moving for it to appear as a square? Figure 26.2 can be helpful.

26.9. Emmy's spaceship is 100 m long. How fast should it be moving relative to Karl so its length is 77% of its actual length as measured by Karl? Figure 26.2 can be helpful.

26.10. Karl and Emmy are in two different spaceships in outer space, moving at $0.94c$. Karl watches a movie which lasts 1.5 hours. How long does the movie appear to Emmy? Emmy watches the same movie. How long does the movie that Emmy watches in her spaceship last for? How long does this movie appear to Karl? Figure 26.2 can be helpful.

26.11. Karl and Emmy are cosmonauts assigned to two missions. Karl is to travel to planet Neemaz, while Emmy goes to planet Zohal in opposite direction to Karl's motion with the same speed as Karl's relative to Earth. The planets are equidistant from Earth. The relative speed of the spaceships is $0.995c$. Refer to Figure 26.2. It takes Karl 1 year to get to Neemaz.
(a) How long does it take Emmy to get to Zohal?
(b) How long does it take Karl for Emmy to go from Earth to Zohal?
(c) How long does it take Emmy for Karl to go from Earth to Neemaz?

26.12. Karl and Emmy are newly born twins. Karl is put on a spaceship traveling at $0.99875c$. When Karl is 1 year old, how old is Emmy? When Emmy is 1 year old, how old is Karl? Refer to Figure 26.2, and in each case determine carefully who is measuring proper time.

26.13. Karl and Emmy are newly born twins. Karl is put on a spaceship traveling at $0.997c$ to a planet 13 light years away. Refer to Figure 26.2.
(a) What is the Earth-planet distance according to Karl?
(b) How old is he when he lands on the planet?
(c) How old is Emmy when Karl lands on the planet?

26.5.5 Numerical Exercises

26.1. The crew of a spaceship goes to a distant planet with a speed of $0.999c$. It spends a year exploring the planet, and comes back with the same speed. The captain of the spaceship is 35 years old and has just had a baby when he leaves on the mission. The whole

---

[3]A light year is not a year with half the calories! Neither is it a unit of time. It is the *distance* light travels in 1 year, which is about 9.47×10^{12} km.

trip takes 5 years for the crew.
(a) How many years does the captain spend on the way from Earth to the planet?
(b) How many years does it take the "baby" for her father to land on the planet?
(c) How old is the "baby" when the captain returns? How old is the captain?
(d) Is it likely for the crew to be able to see any of their friends and relatives?
(e) How far is the planet from the Earth?
(f) What is the Earth-planet distance according to the crew?

26.2. Alpha Centauri is 4 light years away from us. How fast (in terms of a fraction of light speed) should a spaceship move so the distance between Earth and Alpha Centauri becomes 1 light day to its crew?

26.3. How fast should a meter stick move for its length to shrink to 99.9 cm? Do you see how hard it is to observe relativistic effects?

26.4. The crew of a spaceship goes to a distant planet 35.36 light years away with a speed of $0.9999c$. It spends a year exploring the planet, and comes back with the same speed. The captain of the spaceship is 30 years old and has just had a baby when he leaves on the mission.
(a) What is the distance between Earth and the planet according to the crew?
(b) How many years after take-off does it take the crew to land on the planet?
(c) How many years does it take the "baby" for her father to land on the planet? Find the answer in two different ways!
(d) How old is the "baby" when the captain returns? How old is the captain?

Chapter 27

Spacetime Geometry

Relativity theory abolishes the notion of absolute time and space. An event, specified by a location in space and an instant in time, is described differently by different RFs. For example, an observer on a train may say "Firecracker A, located in front of the train exploded at noon on 1, January 2003." The same event may be described by the observer on the platform as "Firecracker A, exploded at 1:00 pm on 1, January 2003 while passing the edge of the platform." The task of this chapter is to give geometrical meaning to events and relate the coordinates of the same event as measured by different observers.

27.1 Space + Time = Spacetime

The most elegant way of relating an event's space and time properties assigned by two observers is to use geometry. But before getting into the geometry of relativity, let's review the ordinary (Euclidean) geometry.

A position can be "coordinatized" by three numbers corresponding to three axes. For example, to specify an office in the Sears Tower, we first give the location of the building by specifying the north-south and east-west coordinates (two numbers),[1] then give the height (floor number). These three numbers are usually denoted by (x, y, z), where x is the east-west coordinate (positive for east, negative for west), y the north-south coordinate (positive for north, negative for south), and z is the up-down coordinate (positive for up, negative for down or basement).

Review of Euclidean geometry.

Abstracting the idea, we can study *solid* geometry by assigning three numbers to each point. If a certain collection of points happen to lie on a geometric figure, then those points are related via a mathematical formula. For instance, a surface is described by an equation giving one coordinate of a general point in terms of the other two. All points in space whose coordinates satisfy the equation lie on that surface. All curves and surfaces are determined by equations. Thus, solid geometry turns into coordinatized or (*analytic*) geometry.

Many of the features of solid geometry can be derived by studying **plane geometry** as is done in a typical elementary geometry course; and we shall do the same. From now on we shall confine ourselves to studying objects in a plane. Thus, as shown in Figure 27.1(a), a point is specified by only a pair of numbers (x, y). Lines and curves and other geometric figures are objects for which x and y are related via equations. For instance, any point whose coordinates satisfy the equation $y = 2x + 3$ lies on a line that has a slope of 2 and an intercept of 3. Similarly points whose coordinates satisfy the equation $x^2 + y^2 = 1$ lie on a circle of unit radius centered at the origin.

Solid geometry is hard; simplify to plane geometry.

---

[1]These two numbers are usually replaced by a word (the name of the street) and a number (address on the street).

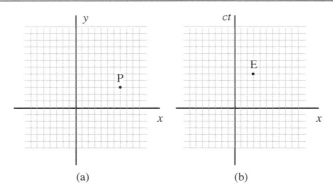

Figure 27.1: (a) A point in the plane is specified by a pair of numbers or coordinates (x, y); for instance P has coordinates $(7, 3)$. (b) An event in the plane spacetime is specified by a pair of numbers or coordinates (x, ct); for instance E has coordinates $(3, 5)$.

A job interview at 10:00 a.m. on the 53rd floor of the Sears Tower can also be "coordinatized" by adding another number to (x, y, z). The interview is an event that is a union of a spatial point and an instant in time; it is described by a set of four numbers (x, y, z, t). Geometrically, we have added the extra "dimension" of time to the three-dimensional space to create the four-dimensional **spacetime**. If it was only *hard* to picture three-dimensional geometric figures, it is *impossible* to picture four-dimensional objects. In fact, the only way to work with spacetime objects *is* through coordinates and mathematical relations between them.

Spacetime introduced.

Fortunately, just as plane geometry helped us understand many of the features of solid geometry, *plane spacetime*, shown in Figure 27.1, can help us understand the subtle features of the full 4D spacetime. Now, in the case of Euclidean geometry, it did not matter which two coordinates we chose to represent the plane. We chose (x, y); but any other pair would have been equivalent: they would have amounted to holding our plane vertical.[2] For spacetime, we *must* include time as one of the coordinates, because time is not equivalent to other (space) coordinates.[3] As the other coordinate, we take x, although y and z are just as good. Thus, our plane spacetime geometry consists of a set of events described by pairs like (x, t).

> **What do you know? 27.1.** How do you describe events that lie on a line parallel to the ct-axis? Parallel to the x-axis?

It is convenient to change this pair slightly. The first coordinate is a space coordinate measured in meters; the second coordinate is measured in seconds. We want a pair both of whose members are measured in the same units. To achieve this, we can either divide x by a speed or multiply t by a speed. This speed has to be a universal speed so that different observers do not use different speeds at random. The *only* speed that is universal is the speed of light c. So, we choose to *multiply* the time coordinate by c and describe an event by (x, ct). Figure 27.1(b) shows an event E, whose x is 3 m and whose ct is 5 m.

Example 27.1.1. At what time did the event E of Figure 27.1 occur? Since we are given $ct = 5$ m, we can easily find t by dividing by the speed of light *in meters per second*:

$$t = \frac{ct}{c} = \frac{5}{3 \times 10^8} = 1.67 \times 10^{-8} \text{ s}.$$

---

[2]The plane described by (x, z) is a plane, one side of which runs east-west, the other up-down. Similarly (y, z) is a plane, one side of which runs north-south, the other up-down.

[3]This is the reason that Δt was called the *coordinate time* in Chapter 26.

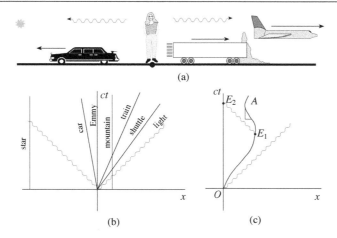

Figure 27.2: (a) Some moving objects in Emmy's RF. (b) The worldline of those objects. (c) The worldline of an object A whose velocity changes.

Thus, the event occurred 3 meters away from the origin and 1.67×10^{-8} seconds after the time origin. ■

27.2   Events and Worldlines

The building blocks of spacetime geometry are *events* just as the building blocks of Euclidean geometry are *points*. Events of spacetime geometry and points of Euclidean geometry exist independently of the observers plotting them on their coordinatized sheets of paper. However, for purposes of calculation, it is convenient (in fact necessary) to construct a coordinate system in which to assign numbers as coordinates of points or events.

In Euclidean geometry one picks a point of the plane, calls it the origin, erects his axes from there, and measures the coordinates of all points relative to this coordinate system. In plane spacetime geometry one picks an *event*, calls it the origin, erects her axes from it, and measures the coordinates of all events relative to that coordinate system. These origins and axis orientations are quite arbitrary and are chosen to make calculations as simple as possible.

A collection of points arranged and connected in a certain succession constitutes a curve in Euclidean plane. What is the analogue of a curve in spacetime plane? Although it is possible to "connect" a random collection of events in a spacetime plane, it is more appropriate to think of the collection as the *location* of a moving object at different *times*. The resulting "curve" is known as the **worldline** of that object.

Worldlines are curves in spacetime.

Figure 27.2(a) shows Emmy, standing at the origin of her coordinate system, as she watches some objects, each assumed to be moving with a constant velocity that is comparable to light speed. In Figure 27.2(b) are drawn the worldlines of these objects. All the worldlines are straight lines, because all speeds are assumed constant. The slope[4] of an object moving with speed v_b relative to Emmy[5] is $c\Delta t/\Delta x$ or $1/(\Delta x/c\Delta t)$ or $1/(v_b/c)$. Introducing the symbol $\beta_b = v_b/c$, we write the slope as $1/\beta_b$. For light, this is equal to 1; therefore, on a spacetime coordinate system, light has a worldline that makes a 45° angle with the positive x-axis. We often draw the light worldline as a wavy line. Since $\beta_b < 1$

The symbol β is used for the ratio of speed to the speed of light.

---

[4]As usual, slope is measured relative to the horizontal axis. However, later we will find it convenient to use slopes relative to the vertical axis as well.

[5]The subscript "b" reflects our choice of a *b*ullet or a *b*all as a typical moving object. You will see this subscript later as well.

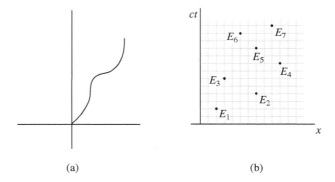

Figure 27.3: (a) A worldline. (b) Some events.

for all other objects, their slope must be *larger* than 1, i.e., the angle that they make with the x-axis must be larger than 45°. This goes not only for objects that move with constant speed as in Figure 27.2(b), but also for objects such as A in Figure 27.2(c), whose speed is changing. In the latter case, the restriction applies to *instantaneous* speed: at every moment, any infinitesimal segment of the worldline must have a slope that is larger than 1. Figure 27.2(c) shows a typical segment of the worldline and the rise and run for that segment. We see that the rise is larger than the run. This must hold at every point (event) of the worldline.

What do you know? 27.2. What is wrong with the worldline in Figure 27.3(a)?

The worldline of any observer is his/her time axis.

An object that is stationary relative to Emmy has a vertical worldline, because its distance (x-coordinate) does not change. The mountain to her right and the star to her left are such objects [their positions in her spacetime plane in part (b) of the figure are not to scale]. Emmy herself has a worldline that is vertical (her time axis), because she is not moving (and her x-coordinate is always zero). This is true for all observers: *the worldline of any observer is his/her time axis.*

What do you know? 27.3. Which pairs of events in Figure 27.3(b) are causally connected? For those that are causally connected, determine if a massive probe can be present at both events; if so, what the fractional speed of that probe should be.

When an observer O' is present at two events, his worldline, which may not be a straight line in the coordinate system of another (inertial) observer O, passes through (or connects) those two events. The clock of O', therefore, measures the proper time between those events (see Box 26.1.1). If the events are very close together on the worldline of O', the arc of this worldline connecting those events is almost a straight line and can be identified as the *time axis* of O', because the two events have the *same spatial coordinates*. If O' is not moving with constant speed relative to O, this time axis keeps changing, but if O' has constant speed relative to O, the worldline of O' is his time axis.

In general, many observers can be present at any given two events, but only one of them has constant speed: it is the observer whose worldline is a *straight line* passing through the two events. This straight line must have a slope larger than 1. When an observer can be present at two events, we say that those events are **causally connected**. In fact,

Causally connected events.

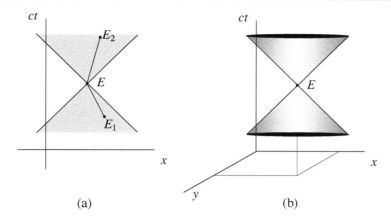

Figure 27.4: (a) Events E_1 in the past and E_2 in the future are causally connected to E. (b) One more axis of the space is added to show why the shaded region is called the light "cone." (c) The curved worldline of O' relative to O.

causal connection is generalized to also include the case when the worldline of a *light signal* connects the two events.

Let us summarize our observation so far:

> **Box 27.2.1.** *Relative to an observer O, every object moves on a worldline whose instantaneous slope at every event of the worldline is larger than 1, except for light, whose slope is 1. The instantaneous time axis of an observer O' is the tangent to his/her worldline. If O' moves with constant speed relative to O, the worldline of O' is his/her time axis. Two events that can be connected by a straight worldline whose slope is larger than (or equal to) 1 are called* **causally connected,** *otherwise they are* **causally disconnected.**

Given any event E, we want to identify the collection of all events that are causally connected to E. By definition, it is all the events that can be connected to E by a straight line whose slope is larger than or equal to 1. Some of these, such as E_1 in Figure 27.4(a), occurred in the past of E; some, such as E_2, will occur in the future. The shaded region collects all events that are causally connected to E. The region is called the **light cone**, because it is the projection of a four-dimensional region onto our two-dimensional spacetime plane. We cannot show all the three space *and* the time axes at the same time on a two-dimensional page. However, we are used to seeing three-dimensional objects drawn on a two-dimensional paper. So we add only one more space axis (the y-axis) and redraw the shaded region in Figure 27.4(b) to explain the word "cone" in *light cone*.

Causally disconnected events, i.e., those lying outside the light cone, are inaccessible to E. It is impossible for E to communicate with those events, because the signal of communication would have to travel faster than light. On the other hand, it *is* possible for E to send a "probe" to any event that lies within the light cone. The events that lie on the surface of the cone require a light probe (signal); those inside require a probe traveling slower than light.

Light cone as the collection of all events causally connected to a given event.

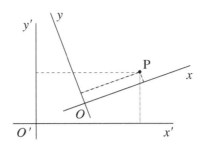

Figure 27.5: The same point P has different pairs of coordinates in different coordinate systems.

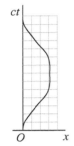

> **What do you know? 27.4.** Describe the motion of O', with the curved worldline in the accompanying figure, relative to O. The units are in light years. In particular, answer the following questions raised by O and according to O:
> (a) How far away does O' go from O?
> (b) How long does it take him to get to his destination?
> (c) Approximately how many years does he stay at his destination?
> (d) How many years does it take for O' to go and come back?

The intersection of the worldlines of two objects is an *event* (just as the intersection of two lines is a *point* in Euclidean geometry) at which both objects are present. If these objects happen to be observers, then the *observers* are present at that event. For example, observers O and A in Figure 27.2(c) are both present at the event with coordinates $(0, 0)$ in the RF of O. For the same reason, both A and a light signal are present at E_1, with the light signal initiating on the worldline of A and moving away from it. We call E_1 the *emission* of a light signal by observer A. Similarly, E_2 is the *reception* of the same light signal by O.

27.3   Space Transformation

Every spacetime plane has an origin, which is both a space origin and a *time origin*. You are familiar with the space origin from your algebra course. But what exactly is the origin of time? It is really analogous to the origin of space (coordinates). When we draw our x- and y-axes in a plane, we are doing it quite arbitrarily, and some random point in the plane of the sheet of paper—the point at which the two axes cross—receives the "honor" of being the *origin*. The next time we draw the axes, another random point becomes the origin. As I am writing these lines, I am 2003 years, 15 days, 10 hours, and 8 minutes away from the *arbitrary* origin of time chosen by a certain population of the Earth. This same instant (specified to within a minute) of time is described by another population—for example, the Moslems or the Chinese—differently. Thus, the origin of time is an arbitrary instant with respect to which we measure the times of the occurrence of all events.

Not only is the origin of a coordinate system (CS) arbitrary, but so is the *orientation* of the axes. In the case of plane geometry, this causes the *same* point to have *different* pairs of coordinates in different CSs. Figure 27.5 shows two CSs O and O', whose origins and orientations are different. It is seen that the same point P is represented by two different pairs of numbers in the two CSs.

Coordinate axes need not be horizontal and vertical.

What do you know? 27.5. In the xy coordinate system a point P has coordinates $(1,1)$. What are the coordinates of P relative to $x'y'$ whose origin coincides with that of xy, but the x'-axis makes an angle of 45 degrees counterclockwise with the x-axis?

The freedom to choose the origin and axis-orientation entails the necessity of communication among observers—using different coordinate systems—the way in which one set of coordinates transforms into another. In other words, any observer should have the ability to transform his/her pair of coordinates of a given point to the pair used by any other observer. This means that there should be a mathematical recipe for finding the pair (x', y') that observer O' assigns to a point, from the pair (x, y) that observer O assigns to the same point. **Math Note E.27.1** on **page 116** of *Appendix.pdf* provides this recipe.

Coordinate transformation.

Transformation of single points is important in studying coordinated geometry. What is more important, however, is properties of objects consisting of a collection of points; objects such as lines and curves. Since any curve (including lines) can be built up from very small (infinitesimal) line segments,[6] understanding the nature of a line segment becomes an essential undertaking.

Math Note E.27.2 on **page 118** of *Appendix.pdf* calculates the length of a line segment as measured in two different coordinate systems and shows that the value for the length is the same in both CSs. This is of course obvious, because we don't use coordinates to measure the distance between two points, and if one person measures the distance and gets a certain value, everybody else will get the same value. However, the connection between distance and coordinates is important because, as we shall see, in relativity we have no choice but to use coordinates.

Because we are dealing with Euclidean geometry, let us call $\overline{P_1 P_2}$ (or Δr) the **Euclidean distance**, and summarize the finding in Math Note E.27.2 this way:

Euclidean distance and its invariance.

Box 27.3.1. *The **Euclidean distance** between any two points in a plane is independent of the coordinates used to describe those points. We say that Euclidean distance is an **invariant** geometric quantity.*

Since Euclidean geometry was discovered before the notion of coordinates was introduced in mathematics, the idea of the distance being independent of the coordinate system used is deeply ingrained in our mind. We never question the invariance of the Euclidean distance, and when we transform the coordinates of points from a system O to another system O', we expect that the distance between any two points remain the same. That this is indeed the case is shown in **Math Note E.27.3** on **page 119** of *Appendix.pdf*.

In deriving the coordinate transformations of Math Note E.27.1 we used Euclidean *geometry*, in which the statement of Box 27.3.1 holds. Can we derive the rules *algebraically*? In other words, can we start with the algebraic equation for distance between two points [Equation (E.82)] and derive from it the transformation rules using only algebra? The reason for such a derivation is that we have no "pictorial" conception of an *invariant* distance in the spacetime plane, and the only way of deriving the transformation rules in the spacetime plane is the algebraic way. **Math Note E.27.4** on **page 119** of *Appendix.pdf* derives the Euclidean transformation rules using algebra.

---

[6]In the age of personal computers and graphic software, it is not hard for the reader to see this: draw any "smooth" curve on your computer screen using a drawing program; magnify the image sufficiently and you will see the line segments (even the jagged points) that make up the curve.

27.4   Spacetime Distance

Our knowledge of spacetime is restricted to the coordinates; we don't have a visual image of the geometry of spacetime. On the contrary, Euclidean geometry was developed over centuries by studying *visual* (and visualizable) images. This does not mean that we cannot draw pictures in spacetime; it means that the pictures we draw have to be interpreted differently than the familiar Euclidean pictures.

The first thing we need to do is to find an expression for the *invariant* distance between two events. For this, we use the Euclidean analogy with the caveat that in plane geometry we look at points in a *real* plane, in which we can draw real lines two of which we can pick as our axes. In spacetime plane only one of the axes is real; the time axis is not real. Why? Because, we can easily move back and forth along a space axis at will without leaving our reference frame, but we cannot move backward along the time axis, and the passage of time in the forward direction is out of our control.[7]

Math Note E.27.5 on **page 121** of *Appendix.pdf* derives the expression for the invariant **spacetime distance** (also called **spacetime interval**), which we reproduce in the following

Spacetime distance and its invariance.

> **Box 27.4.1.** *The **spacetime distance (interval)** Δs between two events E_1 and E_2, with respective coordinates (x_1, ct_1) and (x_2, ct_2), is $\Delta s = c\Delta\tau = \sqrt{(c\Delta t)^2 - (\Delta x)^2}$, where $\Delta\tau$ is the proper time interval between the two events, $\Delta t = t_2 - t_1$, and $\Delta x = x_2 - x_1$. This spacetime distance is independent of the observer.*

The negative sign under the radical has a dramatic physical consequence: It allows for the distance between two *distinct* events to be zero! **Example D.27.1** on **page 48** of *Appendix.pdf* shows how this can happen, and concludes that

Spacetime distance for light.

> **Box 27.4.2.** *For any light signal $\Delta s = 0$. In other words, if two events are connected by a light signal (i.e., if a light signal is present at the two events), then the spacetime distance of those events is zero.*

In Euclidean geometry the distance between two points is zero only if the two points coincide, i.e., if they are the same point. Not so in spacetime geometry! Two *different* events have a zero spacetime distance if they can be connected by a light signal.

> **What do you know? 27.6.** Does it make sense to talk about the distance between two causally disconnected events?

> **What do you know? 27.7.** Is it possible for a massive object to be present at two events for which $\Delta s = 0$?

> **What do you know? 27.8.** Suppose two events are causally disconnected for one observer. Is it possible to find an observer for whom the events are causally connected?

---

[7] As we saw in Chapter 26, it is possible to move forward in time (as in the case of a captain of a spaceship who remained younger than his daughter) in a controlled way, but you have to change your reference frame.

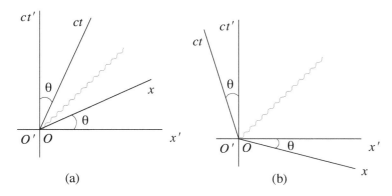

Figure 27.6: (a) O is moving in the positive direction of O'. (b) O is moving in the negative direction of O'. The relative speed is larger in (a) than in (b). Note how the light worldline (represented by the wavy line) makes equal angles with both axes of O and O'.

27.5   Rules of Spacetime Geometry

Geometry is the most elegant and powerful way of studying relativity. Because of the difference between the formula for the distance in the Euclidean and spacetime planes, distortions take place, not unlike distortions resulting from mapping the spherical geometry of the globe onto a flat piece of paper. (See for example, Box 27.6.1 below.) However, we can understand and manipulate these distortions by resorting to a set of rules derived from the invariant distance formula of Box 27.4.1. These rules are summarized in Boxes F.0.2 and F.0.3 in *Appendix.pdf*.

Based on these rules, one can draw geometric figures and extract algebraic relations governing events and worldlines. We shall not exploit the full extent of these rules and how to get to the algebraic results from them, because such discussions are slightly technical. Nevertheless, we urge the reader who is mathematically inclined (at the level of high school geometry and algebra) to consult Appendix F in *Appendix.pdf*.

For the purpose of the simplest application of these rules, we state their most qualitative features:

The word "stationary" in this box means that the reader is in the RF of O'.

Box 27.5.1. *Suppose that O moves relative to O'. The axes of O', the stationary observer, are drawn perpendicular. The axes of O form an acute angle if O moves in the positive direction, and an obtuse angle if O moves in the negative direction of O'. Light worldline makes equal angles with the two axes of all observers.*

Figure 27.6 shows the axes of two observers moving relative to one another. In Figure 27.6(a), O is moving in the positive direction of O', while in Figure 27.6(b) O is moving in the negative direction of O' with a smaller speed (corresponding to a smaller θ). The light worldline, represented by a wavy line, always makes equal angles with any pair of axes.

What do you know? 27.9. How would you draw the axes of Figure 27.6 from the perspective of O (i.e., assuming that O is stationary)?

Simultaneity Revisited

The diagrammatic approach to relativity can elucidate some of the notions we discussed earlier. Take, for instance, the relativity of simultaneity, which was one of the first topics we encountered. Figure 25.5 showed a picture in which Karl (observer O') detects a simultaneous explosion of two firecrackers A and B. To Emmy, on the other hand, B happens before A. Spacetime diagrams can further unravel the succession of these events, as discussed in Section F.2.

If, for simplicity, we assume that the explosions occur—according to Karl—at exactly the same time that Emmy passes him, then the diagrammatic analysis shows that for Emmy B occurred *before* she reached Karl, but its signal reaches her *after* she passes Karl, consistent with the middle picture of Figure 25.5.

The discussion above, although qualitative, sheds some light on the notion of simultaneity as perceived by two different observers. If you know Emmy's speed relative to Karl and Karl's measurement of the length of the train, you can calculate the time difference between the explosions, the separation between the two firecrackers, and the time of the reception of the two signals all according to Emmy. The interested reader can find the details in Example D.27.5.

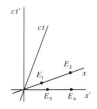

> **What do you know? 27.10.** The accompanying figure shows observers O with axes x and ct and O' with axes x' and ct'. Are E_1 and E_2 simultaneous according to O or O'? What about E_3 and E_4? In each case, if they are not simultaneous, say which occurs first.

The Train and the Tunnel

In the early days of relativity, there appeared to be a "paradox" having to do with the contraction of length; it was called the *pole-and-the-barn paradox*. We consider a "modern" version of it and call it the *train-and-the-tunnel paradox*. Emmy's train moves close to light speed as it approaches a tunnel. Karl measures the contracted length of the train and concludes that it should nicely fit the tunnel he is standing by. Emmy, on the other hand, sees her train to be *longer* than what Karl takes to be the length of the train and the tunnel to be *shorter*[8] than what Karl takes to be its length. So she concludes that there is no way she can fit her train in that tunnel. What is going on?

The spacetime geometric analysis of the problem in Section F.3 of *Appendix.pdf* shows that while the two ends of the train coincide with the two ends of the tunnel *according to Karl* (and therefore he concludes that the two have the same length), Emmy sees the rear end of the train outside while its front is just emerging from the tunnel. The paradox is therefore related to simultaneity: what is simultaneous to Karl need not be simultaneous to Emmy. **Example D.27.2 on page 48** of *Appendix.pdf* adds some numerical spice to the discussion.

The Doppler Effect

Emmy is moving with speed v away from Karl along his positive direction. She sends a light signal of wavelength λ to him, which is Doppler shifted to λ'. What is the relation between λ, λ', and v?

Figure 27.7(a) shows Karl's (the primed axes) and Emmy's (the unprimed axes) RFs, whose origins are assumed to have coincided at $t = 0$ and $t' = 0$. Event E_1 is the emission of a crest of the light wave by Emmy's light source. After T seconds (T is the period of

---

[8]Because the tunnel is moving relative to Emmy; so, its length should shrink for her.

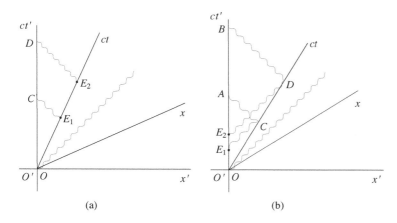

Figure 27.7: (a) The spacetime diagram of the Doppler effect. (b) The spacetime diagram of the Doppler effect including reflection from a moving object.

the light wave related to its frequency via $f = 1/T$), the next crest is emitted at event E_2. These two crests travel along the two wavy worldlines (making a 45° angle with the axes as they should), and are received by Karl at C and D. Therefore, $\overline{CD} = cT'$, where T' is the period of the light signal as measured by Karl. Since $\lambda = cT$ and $\lambda' = cT'$, what is left to do to obtain the Doppler formula is to find $\overline{CD}$ in terms of v and T. **Math Note E.27.8** on **page 123** of *Appendix.pdf* shows how to accomplish this task.

Spacetime diagrams can also be used to derive the Doppler shift of an electromagnetic wave reflected from a moving object. Figure 27.7(b) shows Emmy moving relative to Karl. Karl sends a radar signal to Emmy and measures the wavelength of its reflection, whereby determining Emmy's speed. How does he do it? Let E_1 be the sending of a wave crest, and E_2 that of the next crest, so that $\overline{E_1E_2}$ is cT, where T is the period of the radar. These waves intersect Emmy's worldline (are received by her) at C and D, whereupon they get reflected and are received by Karl at A and B. Note that the radars emitted at E_1 and E_2 and their reflections $\overline{CA}$ and $\overline{DB}$ travel on worldlines making 45° angle with Karl's axes. Our task is to find $\overline{AB}$ in terms of $\overline{E_1E_2}$. The derivation can be found in **Math Note E.27.9** on **page 123** of *Appendix.pdf*, which also shows us precisely where the factor of 2 in Equation (E.13) comes from.

Time Travel?

Humankind have been traveling in three dimensions for millions of years. The two-dimensional aspect of this travel was magnified when man invented the wheel, horse and buggy, and particularly the automobile. Travel in the third dimension became excitingly frequent with the invention of the airplane. Now we have learned that relativity makes time a "fourth dimension." So, a natural question to ask is: Can we travel in time? A typical situation is the occurrence of an event E in the past of an observer O' as shown in Figure 27.8(a). Since O' has no "time machine," he exploits the variation of time as measured by others, and finds an observer O for whom E occurs now. O must lie on the x'-axis (because O' wants to do the "time traveling" now, i.e., at $t = 0$), and E on the x-axis (because E must be taking place now for O). This determines the x-axis as the line EO. By drawing a light worldline at O (which makes a 45° angle with O' axes) and reading the angle it makes with the x-axis, you can draw the ct-axis as well.

Giordano Bruno was a fiercely independent Renaissance thinker whose ideas of Copernican astronomy and infinite worlds with infinite (perhaps similar) histories were blasphemy in

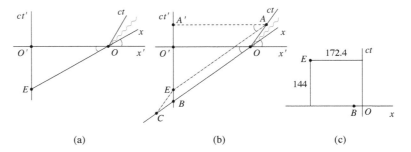

Figure 27.8: (a) Spacetime diagram for rescuing Bruno by finding a reference frame (RF) whose present time is Bruno's execution. (b) The coordinates of E in the nonperpendicular coordinate system O. (c) The events B and E as seen by O.

Can we stop the execution of Giordano Bruno?

the eyes of the sixteenth-century church (see page 38). After a long period of imprisonment and defiance, the church decided to silence Bruno for good. On February 17, 1600 Giordano was burned at the stakes. Now that we are familiar with the geometry of spacetime, can we stop the execution of this free thinker? **Example D.27.3 on page 48** of *Appendix.pdf* finds an RF for which the execution occurs *now*. Can the crew of this RF save Bruno? No, because they are not present at the *location* of the event.

What if we find an RF for which 10 *Earth* years *before* Bruno's execution is *now* [as in Figure 27.8(b)]? For this RF Bruno's execution actually occurs 144 years from now! So the crew of this RF have 144 years to plan to save Bruno. Can they? To save him, they have to send somebody whose worldline connects O and E in Figure 27.8(c). Such a worldline lies outside the light cone, i.e., the person must travel faster than light!

Bruno's execution took place in the past. What about time-traveling in the opposite direction? Can we stop *future* mishaps from happening? **Example D.27.4 on page 51** of *Appendix.pdf* has yet another disappointing answer! All these examples have shown us that *using the numbers given in those examples*, it is impossible to travel back and forth in time. Can we generalize this? Is it impossible to prevent *any* past or future event? The following general argument shows that the past is indeed out of our reach. A similar argument shows that the future is also inaccessible.

Proof of impossibility of going back in time.

The reference frame O that we are seeking must lie on the Earth's x'-axis (see Figure 27.9). It could be any point of this axis with one condition: the resulting coordinate system for O must have its *time* axis in the light cone at O.[9] It follows that the x-axis must be to the right of the cone. But the x-axis is the line OB, because B is happening NOW (at the same time as the origin of time) for O. So we must choose O on the horizontal line in such a way that the line BO is outside the light cone. This makes B causally disconnected from O. The line $O'O$ is also outside the line cone. Since the line EO lies between $O'O$ and BO, it too must lie outside the light cone. Therefore, E is causally disconnected from O, and no probe (even light) can reach E from our RF.

Lorentz Transformations

The rules of spacetime geometry, as outlined in Boxes F.0.2 and F.0.3 have enabled us to connect the coordinates and intervals measured by one observer to those measured by another. In each example that we encountered, we calculated the coordinates and intervals based on the particular numbers given in that example. It is now worthwhile to introduce an *algebraic* recipe for relating the coordinates and intervals of two observers. Section F.4 finds this recipe, which is at the heart of relativity theory and is called the **Lorentz**

---

[9]Because the time axis of a RF is its worldline and no worldline can be outside the light cone.

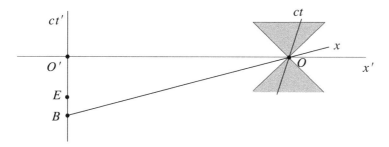

Figure 27.9: We want to get to E, but we go to B to "wait."

transformation.

Lorentz transformations are sometimes used to simplify the analysis of physical processes. A process may be easier to analyze in some particular reference frame. Then one analyzes the process in that RF and Lorentz transforms it to any other desired RF. A corollary of this procedure is the following:

Box 27.5.2. *If a physical process is impossible in one reference frame, then it is impossible in **all** reference frames.*

Lorentz transformation quantifies all the qualitative discussions of relativistic effects. For instance, the discussion of simultaneity above and in Section F.2 concentrated on the qualitative aspect of simultaneity in the context of spacetime geometry. Using Lorentz transformation, we can calculate the time difference between two events according to Karl, when Emmy, moving at relativistic speed sees them simultaneously. **Example D.27.5** on **page 51** of *Appendix.pdf* finds such a time difference.

Other examples of the Lorentz transformations are given in Section D.27. One is **Example D.27.6** on **page 52** of *Appendix.pdf*, which shows that Lorentz transformations imply both time dilation and length contraction. Another is **Example D.27.7** on **page 52** of *Appendix.pdf*, which looks at time travel—in this case, to John F. Kennedy's assassination in 1963—using Lorentz transformations. The third is **Example D.27.8** on **page 55** of *Appendix.pdf*, which resolves the following seemingly peculiar paradox.

🧺 Food for Thought 🌱

FFT 27.5.3. Latour is a well known social scholar who thinks that scientists should not be allowed to express their opinions about their trades. Such opinions, he argues, belong only in the domain of the social sciences. In the last few decades these social scholars have been playing with the idea that science is not independent of the social environment in which it has developed (see Section 19.4); that *literally* we would have a different kind of physics if the social conditions under which the physical ideas have been developed had been different. For example, the solar system would obey a different kind of gravitational law had Newton been raised in a society other than the seventeenth-century England. In order to buttress this claim, Latour wrote an article entitled "A Relativistic Account of Einstein's Relativity."

Latour uses two techniques to make his point: he *redefines the "usual concepts"* of the social sciences to fit his needs, and he makes his own strange interpretation of relativity by reading a single book, the book that Einstein wrote for the layman in 1917. To show the reader that the first technique as stated above is not tongue in cheek, we let Latour describe it in his own words:

> Instead of extending the social sciences' usual concepts to the natural sciences, I want
> to redefine these very social concepts in order to be able to make them explain the

more formal sciences. The task at hand is to keep the same strong programme, but to doubt what social sciences have to say about society. [Lat 88, p. 4]

In other words, if the social sciences cannot prove that natural sciences are social, then change the social sciences so much as to be able to prove that the natural sciences *are* social. This socialization of sciences is, by the way, the task of what Latour calls the "strong programme."

The word "observer" is used frequently in physics literature, especially the literature intended for the layman, and we have used it frequently in our treatment of relativity thus far. We have also seen its use in the quantum theory, where it has been (ab)used to infer an "observer-created reality." Latour's interpretation of the relativity theory relies heavily on this word.[10] In popular accounts of relativity theory observer is used instead of "reference frame," because the latter is an unfamiliar word to most readers. This usage is definitely not intended to "humanize" the theory. Yet that is precisely what Latour does: observer for Latour is only a "delegate."

One of the cornerstones of relativity is *Lorentz transformations*, the subject of this chapter. As the reader has noticed, Lorentz transformations are mathematical statements connecting the locations and times of the same event in two different reference frames. In most treatments of relativity however, these transformations are assumed to connect the observers living in the corresponding reference frames. Latour's reading of the transformations is the semiotic concepts of "shifting in" and "shifting out," whereby an author changes the personae of a narrative from third person, say to first person and back again. More precisely, Latour identifies the two "observers" as two imaginary "delegates" between whom the narrator "shifts in and out," and the mathematical idea behind the Lorentz transformations becomes merely a "shifting in" and "shifting out" of the narration. Here is Latour's interpretation of relativity in his own words:

> The peculiarity of Einstein's narration is not that it puts to use shifting in and out, since every narration does the same, but that it focuses the reader's attention upon these very operations.
>
> He [Einstein] is interested only in the way in which we send any actor to any other frame of reference. Instead of describing laws of nature, he sets out to describe how any description is possible....
>
> Technically his book is about delegation ...and is a book of meta-linguistics or of semiotics, one which tries to understand how any narration is constructed. [Hut 98, p. 185]

Thus Latour reduces the entire theory of relativity, on which hundreds of scientific books and articles are written, to a single expository book *written for nontechnical readers*, and then reduces *that* book to a trivial narration. To him the mathematically elegant and physically all-encompassing theory of relativity becomes "Einstein's narration;" the ground-breaking Lorentz transformations become the "peculiarity of Einstein's narration;" and Einstein is blamed for being interested in Lorentz transformations, which to Latour are "the way in which we send any actor to any other frame of reference." The last sentence of the quotation gives an excellent summary of the extent of Latour's understanding of relativity and its disfiguration brought about by his injection of "meta-linguistic" and "semiotic" potion into the theory.

An peculiar paradox!   We started our quantitative discussion of relativity with time dilation, in which we held a Michelson-Morley clock *vertical* and moved it horizontally with high speed (see Section 26.1). Then using the Pythagorean theorem, we derived the formula (26.1) connecting proper time and the coordinate time. In the meantime, we have discovered the phenomenon of length contraction. This creates a dilemma.

Suppose Emmy and Karl are on a spaceship moving at 86.6% light speed, and each carries a Michelson-Morley clock 3 meters long. Emmy holds her clock vertically while Karl keeps it horizontal. As they travel from Earth to a distant planet, they both age 5 years. We know how to calculate Earth time interval from Emmy's clock: using Equation (26.1), we deduce that the Earth inhabitants age

$$\Delta t = \frac{\Delta\tau}{\sqrt{1-(v/c)^2}} = \frac{5}{\sqrt{1-(0.866)^2}} = 10 \text{ years.}$$

---

[10]For a thorough critique of Latour's article see [Hut 98, pp. 181–192].

Karl, on the other hand, holding his clock along the direction of motion, causes a shrinkage of its length for the Earth people. Are they to conclude that Karl's ticks are shorter than Emmy's, and therefore, that the Earth people must age more than 10 years? Which one is it? Have they aged 10 years or more? Example D.27.8 resolves this strange double timing.

Relativistic Law of Addition of Velocities

Relativity prohibits any object to go faster than light speed. What happens, then, when a supertrain passenger fires a supergun, shooting bullets at almost light speed while the train is also moving at almost light speed? Would a platform observer see the bullet moving faster than light? Our intuition says that if the train moves at $0.9c$ and the bullet is fired in the forward direction with a speed of $0.9c$ with respect to the train, then the platform observer should see the bullet move at $1.8c$. *Our intuition is wrong, of course!* Lorentz transformations are right and will give us the right answer, as **Example D.27.9 on page 56** of *Appendix.pdf* illustrates.

Example D.27.9 shows that even though two speeds may be very close to the speed of light, they do not "add up" to a speed larger than light speed. The Lorentz transformations played a crucial role in this limitation, and its general form can be used to obtain a formula for adding velocities *relativistically*. Suppose that a bullet is moving in O with speed v_b and O is moving relative to O' with speed v. What is the speed v'_b of the bullet relative to O'? **Math Note E.27.13 on page 128** of *Appendix.pdf* derives Equation (E.106), a mathematical formula giving v'_b in terms of v_b and v.

This law of addition of velocities never violates relativity theory as the reader may verify by inserting some large values (but smaller than c, of course) for v and v_b and noting that v'_b comes out smaller than c as well. In fact, Math Note E.27.13 shows this quite generally.

What do you know? 27.11. Observer O sees spaceship A move at $0.99c$ to the right and spaceship B move at $0.99c$ to the left. So, in 1 second he sees the distance between A and B increase to almost 600,000 km. Is the speed of A relative to B almost twice the speed of light?

Equation (E.106) also agrees with the second principle of relativity. If instead of a bullet, O sends a light beam so that $v_b = c$,[11] (E.106) gives $v'_b = (c + v)(1 + vc/c^2)$, which is easily shown to be equal to c. Hence, O' also measures the speed of light to be c.

27.6 Curved Worldlines

The straight worldlines that we have considered so far do not describe the most general type of motion. Objects and observers (i.e., reference frames) often change their speed. In fact, starting from rest, to attain a final constant speed, one has to accelerate. The worldline of an accelerating RF is a curve in the spacetime plane such as the one shown in Figure 27.10(a), where the slope changes from event to event. If the curve represents a real worldline, then the tangent[12] to the curve at every point must have a slope that is larger than 1, because the inverse of this slope is v/c, with v the instantaneous speed of the accelerating RF.

Although straight worldlines are restrictive, they are fundamental in the same sense that straight lines are fundamental in Euclidean geometry: *any* worldline can be built up from small straight worldline segments. For example, Figure 27.10(b) shows a worldline

A curved worldline is the union of a lot of small straight worldlines.

---

[11]We still use v_b, although a light signal is the object that is moving, not a bullet.
[12]Recall that the tangent line to a curve at a point P is a straight line that touches the curve only at P in the vicinity of P.

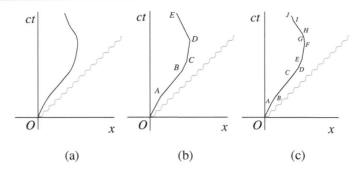

Figure 27.10: (a) The worldline of an accelerated frame. (b) Five *straight* worldlines approximating the curve. (c) Ten *straight* worldlines approximating the curve.

consisting of five straight segments ($\overline{OA}$ through $\overline{DE}$). This collection of five straight worldlines approximates the curve of Figure 27.10(a) fairly well, but if we want a better approximation, we can increase the number of straight worldlines. Figure 27.10(c) shows a worldline consisting of ten straight segments ($\overline{OA}$ through $\overline{IJ}$), which obviously approximate the curve more accurately. For even better accuracy we can further increase the number of worldline segments. All these straight segments must have slopes that are larger than 45°.

27.6.1   The Spacetime Triangle Inequality

Given any two points in Euclidean geometry, there are infinitely many curves that connect those points. These curves have different lengths, and only one—the one we call *straight*—has the shortest length. At the heart of this property lies the *triangle inequality*, which states that the sum of the lengths of any two sides of a triangle is greater than the length of the third side. If triangle inequality holds, then one can show that indeed a straight line is the shortest path. Here is how.

Consider two points P_1 and P_2 in a Euclidean plane. Draw a straight line and a curve through the points as shown in Figure 27.11. Pick a (strategic!) point A on the curve and form the triangle P_1P_2A. By triangle inequality, $\overline{P_1A} + \overline{AP_2} > \overline{P_1P_2}$. Now choose a point B on the curve between P_1 and A and another point C between A and P_2. Invoking the triangle inequality again, we see that $\overline{P_1B} + \overline{BA} > \overline{P_1A}$ and $\overline{AC} + \overline{CP_2} > \overline{AP_2}$. Substituting these inequalities in the previous one, we get $\overline{P_1B} + \overline{BA} + \overline{AC} + \overline{CP_2} > \overline{P_1P_2}$. Continuing this process ad infinitum, we conclude that the length of the curve is larger than the length of the straight line.

Is there a triangle inequality in spacetime geometry? A **spacetime triangle** consists of three mutually causally connected events connected by straight worldlines (the slopes of these lines are therefore larger than 1). The spacetime distance between two events at the two ends of a side of the triangle is the **spacetime length** of that side. **Math Note**

A weird inequality! **E.27.14** on **page 129** of *Appendix.pdf* shows that

> **Box 27.6.1.** *The sum of the spacetime length of any two sides of a spacetime triangle is **less than** the spacetime length of the third side.*

This is a surprising result, but by now we must be used to surprises in relativity! The inequality, which is quite the opposite of what we are accustomed to, is the direct result of the formula for the spacetime distance, especially the negative sign in the equation of Box 27.4.1.

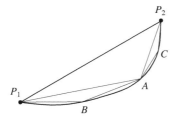

Figure 27.11: The triangle inequality in Euclidean geometry proves that a straight line is the shortest distance between two points.

Given any two causally connected events in spacetime geometry, there are infinitely many worldlines that connect those events. Using the triangle inequality of Box 27.6.1 and an argument similar to the one used in the Euclidean case of Figure 27.11, we can show that out of all these worldlines the straight one has the *longest* spacetime length. These worldlines represent observers who travel with different speeds and accelerations, and the spacetime length becomes essentially the proper time of those observers.[13] Since straight worldlines represent inertial observers, we have the important result:

Box 27.6.2. *The longest proper time between two events is measured by the inertial observer present at those events. Stated differently, the inertial observer ages more than any accelerated observer between two events.*

27.6.2 The Twin Paradox Revisited

Section 26.3 discussed the twin paradox, and it emphasized the *symmetry* between the twin observers: each twin sees the other age slower. We may accept this reluctantly as long as the twins are far apart and one is moving relative to the other. But what if we bring them face to face? Which one will be younger?

Suppose Emmy's rocketship takes off (event E_1), immediately reaches a speed close to light speed, and heads to a distant planet. As soon as she reaches the planet, she abruptly turns around and heads back to Earth to land at event E_2. Karl, her twin brother is waiting for her at the landing site. From our experience with the example of the captain and his daughter (e.g., see Example D.26.2), we know that Emmy will be younger, but why?

Karl's worldline in Figure 27.12(a) is the vertical axis labeled ct'; Emmy's worldline is the broken line E_1PE_2; the planet's worldline is the vertical line passing through P. The spacetime triangle inequality of Box 27.6.1 implies that Emmy's spacetime distance between E_1 and E_2 is *shorter* than Karl's, and therefore, she will be younger.

What do you know? 27.12. In Figure 27.12(b), the worldline of O' is the ct'-axis and that of O is the curved path. Which one measures the proper time interval between E_1 and E_2?

"But," the reader may say "we haven't really resolved the paradox. There is still a symmetry between the two twins: Emmy looks at her brother and sees that he is moving away from her as she takes off toward the planet. On her way home, she sees Karl approaching her, exactly as Karl sees his twin sister. So, there is a complete symmetry. How do you

---

[13]Recall from Box 27.4.1 that the spacetime distance is c times the proper time.

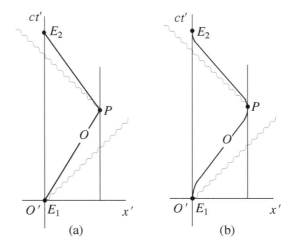

Figure 27.12: The twins' worldlines. (a) The accelerations are abrupt (infinite). (b) The accelerations are smooth.

explain the fact that only Emmy stays younger?" The answer is that there is *no symmetry* between the two! Emmy is *accelerating* for part of her journey; Karl never experiences an acceleration. In fact, Emmy experiences four kinds of acceleration: her take-off requires an acceleration until she reaches her steady speed; once she approaches the planet, she has to slow down to a stop; then she has to accelerate toward Earth until she reaches her steady speed for return; finally, she has to slow down for landing. Emmy's true worldline is shown in Figure 27.12(b), where the curvatures at the beginning, in the middle, and at the end of the journey indicate the acceleration and noninertiality of Emmy's RF. And, as Box 27.6.2 indicates, accelerated RFs experience a shorter passage of time.

27.7   End-of-Chapter Material

27.7.1   Answers to "What do you know?"

27.1. All these events have the same projection on the x-axis. Therefore, they have the same x values; i.e., they are at the same location. Similarly, all events lying on a line parallel to the x-axis have the same projection on the ct-axis. Therefore, they have the same ct values; i.e., they occur at the same time—they are simultaneous events.

27.2. In the middle of the worldline the slope is less than 1.

27.3. Draw the light cone (past and future) of each event and see which of the remaining events lie in that light cone. You should get the following results: E_1 is causally connected to E_3, E_5, E_6, and E_7. E_2 is causally connected to E_4, E_5, E_6, and E_7. E_3 is causally connected to E_1, E_5, E_6, and E_7. E_4 is causally connected to E_2 and E_7. E_5 is causally connected to all but E_4. E_6 is causally connected to E_1, E_2, E_3, and E_5. E_7 is causally connected to all but E_6.

27.4. O' starts from rest relative to O accelerates to very high speed. Then slows down until he stops relative to O. After a while O' starts coming back, first accelerating, then slowing down and finally stopping.

(a) A little over 3 light years.

(b) Approximately 6 years.

(c) About 2 years.

(d) About 13.5 years.

27.5. The x'-axis goes through P. So, $y' = 0$, and P has coordinates $(\sqrt{2}, 0)$ in the $x'y'$ coordinate system.

27.6. If you define the distance as $\sqrt{(\Delta s)^2}$, no, because $(\Delta s)^2$ is negative for two events that are causally disconnected. But if you consider $(\Delta s)^2$ itself as an invariant quantity, then you can say that for two causally disconnected events, $(\Delta s)^2$ is negative.

27.7. No. Because $\Delta s = 0$ only if $c\Delta t = \Delta x$; or $\Delta x/\Delta t = c$, implying that the object must be traveling at light speed, which is impossible.

27.8. No. When the first observer calculates $(\Delta s)^2$, she gets a negative number. Since $(\Delta s)^2$ is *invariant* (i.e., the same for *all* observers), every observer obtains the same result. In particular, every observer obtains (the same) negative number for $(\Delta s)^2$, and concludes that the two events are causally disconnected.

27.9. From the perspective of O, ct and x are perpendicular. Then ct' and x' axes make an obtuse angle in Figure 27.6(a) and an acute angle in Figure 27.6(b).

27.10. To find the ct coordinate of an event, draw a line through the event parallel to the x-axis. Where it intersects the ct-axis is the time coordinate of the event. To find the x coordinate of an event, draw a line through the event parallel to the ct-axis. Where it intersects the x-axis is the space coordinate of the event. Thus, E_1 and E_2 are simultaneous according to O but not according to O'. Similarly, E_3 and E_4 are simultaneous according to O' but not according to O. E_1 occurs before E_2 according to O', and E_4 occurs before E_3 according to O.

27.11. Of course not! From B's perspective, neither the time is 1 second nor the distance 600,000 km. These quantities change in such a way as to keep A's speed relative to B less than light speed.

27.12. Both! Because both observers (and their clocks) are present at both events.

27.7.2   Chapter Glossary

Causally Connected Refers to two events. If an observer or a light signal can be present at both events, then the events are causally connected.

Euclidean Distance The distance with which the reader is familiar. It is the notion that describes how far one house is from another or one city is from another.

Light Cone (of an event) is the collection of all events causally connected to that event.

Lorentz Transformation A mathematical rule that gives the coordinates of an event as measured by one observer in terms of those measured by a second observer.

Spacetime Distance A property associated with two events that is independent of the coordinates of the two events.

Worldline A curve in a spacetime coordinate system.

27.7.3   Review Questions

27.1. What is spacetime? How many coordinates do you need to specify a spacetime point?

27.2. What is an event in the context of spacetime geometry? What is a worldline? What is the restriction on the shape of the worldline of a physical object?

27.3. Given an event E, what is the light cone at E? Is E causally connected to the events inside its light cone? Outside its light cone? On its light cone?

27.4. What is coordinate transformation? Are the coordinates of a point the same in all coordinate systems? If they are not the same, can you find one set in terms of the other?

27.5. What is Euclidean distance? What does it mean to say that Euclidean distance is an invariant quantity?

27.6. What is spacetime distance between two events? How is it related to the proper time interval between those events? Is it possible for the spacetime distance between two distinct events to be zero?

27.7. State the train and the tunnel paradox. Resolve the paradox in terms of the notion of simultaneity.

27.8. What is time travel? Is the daughter getting older than her father considered a time travel? Does relativity allow time travel?

27.9. Does a curved worldline represent an accelerated RF? How does the spacetime triangle inequality compare with ordinary triangle inequality? How does this resolve the twin paradox?

27.7.4   Conceptual Exercises

27.1. You are given two points P_1 and P_2 and two coordinate systems O and O'.
(a) Are the coordinates of P_1 the same in O and O'?
(b) Are the *difference* in the coordinates of P_1 and P_2 the same in O and O'?
(c) Are the sum of the squares of the difference in the coordinates of P_1 and P_2 the same in O and O'?

27.2. You are given two events E_1 and E_2 and two plane spacetime coordinate systems O and O'.
(a) Are the coordinates of E_1 the same in O and O'?
(b) Are the *difference* in the coordinates of E_1 and E_2 the same in O and O'?
(c) Are the difference between the squares of the difference in the coordinates of E_1 and E_2 the same in O and O'?

27.3. Suppose you are given the spacetime coordinates of two events for observer O. Can you find the spacetime distance between the two events for another observer O'?

27.4. Is it true that if the spacetime distance between two events is zero, then those two events are necessarily connected by a light signal?

27.5. The spacetime axes of observer O are obtuse in the spacetime coordinate system of O', whose axes are necessarily drawn perpendicular to each other. Is O' moving in the positive or negative direction of O?

27.6. Two events E_1 and E_2 occur simultaneously according to observer O, with E_2 farther away from the origin than E_1. Observer O' is moving in the negative direction of O and shares the same origin with it. Are E_1 and E_2 simultaneous to O'? If not, which event occurs first?

27.7. A moves relative to B at $0.9c$. B moves relative to C at $0.95c$. Using the formula for the relativistic law of addition of velocities, you can find the speed of C relative to A (you don't have to for this question). Is this speed (a) larger than the speed of light? (b) Larger than $0.95c$ but smaller than c? (c) Between $0.9c$ and $0.95c$?

27.7.5   Numerical Exercises

For qualitative spacetime questions, graphs are very convenient. However, application of quantitative rules of spacetime geometry rules, given in Appendix F, are harder. These rules could be used to answer the following questions. But use of the Lorentz transformation of Box F.4.1 simplifies the calculations, and we encourage the reader to use them in the following exercises.

27.1. Emmy (observer O) is in the middle of a 200-meter train car moving at 99.9% the speed of light. Karl (observer O') is standing on a platform seeing Emmy pass by in his positive direction. Emmy passes Karl at time zero for both observers. At the moment that she passes him, Karl observes that firecrackers A (at the rear end of the train) and B (at the front end of the train) explode simultaneously. With Emmy's spacetime axes as perpendicular, sketch the following:
(a) Karl's axes, and the location of the firecrackers (label the points A and B).
(b) The space coordinates for the two explosions as seen by Karl (label these x'_A and x'_B).
(c) The actual value of x'_A and x'_B in meters.
(d) The time coordinates for the two explosions as seen by Emmy (label these ct_A and ct_B).
(e) The actual value of ct_A and ct_B in meters.
(f) The actual value of t_A and t_B in μs.
(g) The time coordinate at which Karl receives the two light signals from A and B (label it T'_{AB}).
(h) The actual value of cT'_{AB} in meters.
(i) The actual value of T'_{AB} in μs.
(j) The time coordinates at which Emmy receives the two light signals from A and B (label them cT_A, cT_B).
(k) The actual value of cT_A and cT_B in meters.
(l) The actual value of T_A and T_B in μs.

27.2. (See Example D.27.7 for help.) Suppose we are in the distant future when speeds have reached close to light speed. On the 10th anniversary of her mother's tragic death in a car crash Karl tries to prevent the event from happening. So he plans to find a spaceship, for which 5 years earlier than the accident is NOW. That way, he would have some time to prepare for the prevention of the accident. He finds the spaceship Diracus, which is 16 ly away. Diracus happens to be just passing an outpost there, so the plan can be immediately communicated to Diracus.
(a) Draw the Earth's coordinate axes with origin O' and axes x' and ct' and place the origin of Diracus (event O), Karl's mother fatal crash (event E), and 5 years earlier (event B) in the Earth's spacetime plane.
(b) Draw the Diracus axes x and ct?
(c) How fast is Diracus moving (i.e., what is β)? What is γ?
(d) What is the space separation between O and B in the Earth's RF?
(e) What is the time separation between the same two events in the Earth's RF?
(f) What is the time separation between O and B in the Diracus RF?
(g) How far is the site of the event B from Diracus according to the Diracus crew?
(h) When is E happening according to the Diracus crew?
(i) What is the x-coordinate of E according to the Diracus crew?
(j) Draw E and B in the rest frame of O, i.e., in a coordinate system in which ct-axis is perpendicular to the x-axis.
(k) Is it possible to save Karl's mother?

27.3. The year is 2139 and the Intergalactic Space Federation (ISF) is trying to go back to the year 1939 to stop the World War II by capturing Hitler and taking him to outer space. It finds the spaceship Diracus, which is 201 ly away and for which the event of 1939 is NOW. Diracus happens to be just passing an outpost there, so the plan can be immediately communicated to Diracus.
(a) Draw the Earth's coordinate axes and place both the origin of Diracus (event E_1) and the event of Hitler's capture (event E_2) in the Earth's spacetime plane.
(b) What is the space separation between the two events in the Earth's RF?
(c) What is the time separation between the two events in the Earth's RF?
(d) What is the time separation between the two events in the Diracus RF?
(e) How fast is Diracus moving (i.e., what is β)? What is γ?

(f) How far is Hitler's residence from Diracus according to the Diracus crew?

(g) Can the crew prevent WWII?

27.4. A particular spectral line of the hydrogen atom coming from a distant galaxy is seen to have a wavelength that is 3 times longer than the same line seen on Earth. How fast is the galaxy moving and in what direction?

27.5. A particular spectral line of the hydrogen atom coming from a galaxy is seen to have a wavelength that is 10% shorter than the same line seen on Earth. How fast is the galaxy moving and in what direction?

27.6. Emmy is moving in the positive direction of Karl at 0.95c. She fires a bullet (event E_1) in the positive direction and a second later (event E_2) finds it at 270,000 km away from where she fired it.

(a) What are Δx and Δt according to Emmy?

(b) What are $\Delta x'$ and $\Delta t'$ according to Karl? What is the speed of the bullet according to Karl?

27.7. Jill is on a spaceship that travels to a planet of a star system 20 l.y. away on a world line shown in Figure 27.13 as seen by observer O, Jack. All units are in light years.

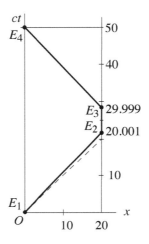

Figure 27.13: The heavy path is the worldline of the spaceship. All units are in light years.

(a) How long is the time interval between take-off from Earth (E_1) and landing on the planet (E_2) according to Jack?

(b) How long is the time interval between landing (E_2) and departure (E_3) from the planet according to Jack?

(c) How long is the time interval between departure (E_3) and landing on Earth (E_4) according to Jack?

(d) How long does the entire trip take according to Jack?

(e) From the figure determine what Δs is for the two events E_1 and E_2.

(f) From the figure determine what Δs is for the two events E_2 and E_3.

(g) From the figure determine what Δs is for the two events E_3 and E_4.

(h) What is Δs for the entire trip? How long does this trip take according to Jill?

(i) Who measures the proper time interval between E_1 and E_4, Jack or Jill (or both)?

(j) What is the speed of the spaceship in m/s between E_1 and E_2? Between E_2 and E_3? Between E_3 and E_4?

Chapter 28

E = mc²

Chapter 27 introduced the spacetime distance and the relevant transformations (the Lorentz transformations) of coordinates that leave this distance unchanged: Given two events the spacetime distance between those events is the same for *all observers*. We also saw that, except for a factor of c, the spacetime distance is simply the proper time interval between the two events, where as the reader may recall, the proper time is the time measured by a clock that is present at both events. In this chapter, we use the proper time to define some important relativistic quantities.

28.1   Coordinate Velocity

The first such quantity we wish to introduce is the generalization of ordinary velocity, and in order to understand the generalization, we have to look at the ordinary velocity from a new perspective.

Terry travels from Bloomington, Illinois to Chicago on an almost straight path as shown in Figure 28.1(a). The distance between the two cities is about 130 miles, and if Terry covers this distance in 2 hours, his average speed will be 65 mph. This is the way we normally think of car travels. But there is another—more precise—way. As Terry moves 130 miles northeast from Bloomington to Chicago, his eastward displacement is about 82 miles and his northward displacement 101 miles. We can say that Terry moved 82 miles east and 101 miles north simultaneously. There is more information in the second description of motion than in the first. Saying Terry moved 130 miles northeast in 2 hours does not tell us how much in each direction he traveled; but saying he traveled 82 miles eastward and 101 miles northward specifies the direction (*as well as the distance*) exactly.[1]

Velocity can be broken up (resolved) into its **components**.

We can go even further and calculate his speed in each direction. Since he covered 82 miles in the east direction in 2 hours, we say his *speed east* was 41 mph. Similarly, we say that his *speed north* was 50.5 mph. These speeds form a right triangle and if we use the Pythagorean theorem, we get $\sqrt{41^2 + 50.5^2} = 65$ mph, which is his "real" speed. We can denote these speeds by v_{east} and v_{north}.

Instead of east and north, we can use the more general x and y directions, and denote these **coordinate** velocities by v_x and v_y, where the motion is considered to take place between two points, say P_1 and P_2. Furthermore, since the "eastward distance" is nothing but the increment in x and the "northward distance" the increment in y, we can write the x- and the y-component of the velocity: $v_x = \Delta x/\Delta t$, $v_y = \Delta y/\Delta t$, where Δt is the time it takes the object to go from P_1 to P_2 [see Figure 28.1(b)].

---

[1] For the distance, simply apply the Pythagorean theorem to the triangle in Figure 28.1(a).

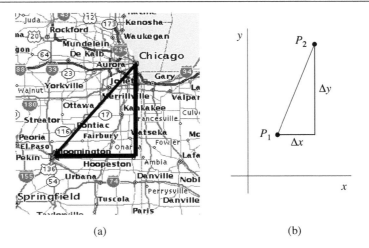

(a)   (b)

Figure 28.1: (a) A northeast path can be "decomposed" into an east path and a north path. (b) A general path can be "resolved" into an x path and a y path.

What do you know? 28.1. Emmy moves at 30 mph eastward and 40 mph southward. How fast is she moving in the southeast direction?

28.2 Spacetime Velocity

How do we apply the discussion above to relativity? Imagine an object moving in space while carrying a clock with it. The points of space through which the object passes, plus the instants at which the object happens to pass through those points constitute a series of events. The object is of course present at all those events, and its clock measures the proper time. Not all observers have access to this clock, and therefore, they cannot read the proper time directly. However, each observer can measure the distance the object travels in a particular time interval in his own RF, and using Box 27.4.1, can calculate the proper time interval for that distance. The important point to emphasize is that once the calculation is done, all observers get the same number. Thus,

---

Box 28.2.1. *Every moving object has a unique and universal time attached to it, its proper time.*

It is therefore natural to define the physical quantities associated with that object in terms of this universal time.[2] One such quantity is the **spacetime velocity**.

Box 28.2.2. Stop! *The remainder of this chapter gets slightly technical. The only parts of this chapter you would need are Equation (28.6) (which you already know). Also look at Example 28.3.2, Box 28.3.3, and, of course, the Food For Thoughts!*

---

[2]Let us immediately emphasize that although each object has a universal time attached to it, this time is different for different objects.

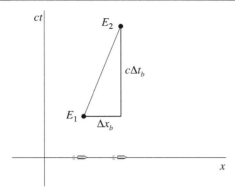

Figure 28.2: The displacement and flight time of the bullet according to Emmy. Note that the bullet is moving along the x-axis only.

Emmy looks at a bullet moving with a speed v_b *relative to her*; measures its displacement Δx_b—between two events E_1 and E_2—and the time Δt_b it takes the bullet to cover the distance Δx_b (see Figure 28.2). These are the analogues of Δx and Δy of the coordinate velocity of Figure 28.1(b). However, instead of dividing these by Δt, which is universal only in nonrelativistic situations, we divide by the next best thing: the bullet's proper time. To do this, Emmy calculates the spacetime distance Δs_b—from Δt_b and Δx_b—of the bullet and from that $\Delta \tau_b$, the proper time between E_1 and E_2. This yields two components: the space component $\Delta x_b/\Delta \tau_b$ and the time component $c\Delta t_b/\Delta \tau_b$. We denote the space part by u_{bx} and the time part by u_{bt} (b for "bullet," x for space, and t for time). Then **Math Note E.28.1** on **page 131** of *Appendix.pdf* gives

$$u_{bx} = \gamma_b v_b, \quad u_{bt} = \gamma_b c, \quad u_{bt}^2 - u_{bx}^2 = c^2, \tag{28.1}$$

where $\gamma_b = 1/\sqrt{1 - (v_b/c)^2}$. The "bullet" represents *any* moving object of interest, including a light signal, for example. The last equation, which holds in all RFs, can be interpreted as the **invariant length** of the spacetime velocity in the two-dimensional spacetime geometry.

Karl watches Emmy and the bullet move relative to him. Anxious to see what he gets for the spacetime velocity, he concentrates on the bullet, moving with a speed v_b' *relative to him*. He measures its displacement $\Delta x_b'$—between the same two events E_1 and E_2 that Emmy measured—and the time $\Delta t_b'$ it takes the bullet to cover the distance $\Delta x_b'$. He then calculates the spacetime distance Δs_b of the bullet and from that $\Delta \tau_b$, the proper time between E_1 and E_2, and he gets exactly what Emmy gets, of course. The spacetime velocity that Karl measures for the bullet has the space part $\Delta x_b'/\Delta \tau_b$ and the time part $c\Delta t_b'/\Delta \tau_b$. We denote Karl's space part by u_{bx}' and the time part by u_{bt}'. Then, using Equation (26.1) and noting that this time the speed is that of the bullet as measured by Karl, we have $u_{bx}' = \gamma_b' v_b'$ and $u_{bt}' = \gamma_b' c$.

The spacetime velocity that Karl measures is of course different from what Emmy measures. As Equation (28.1) and the corresponding one for Karl show, only the *space* component of the spacetime velocity is related to the ordinary velocity of the bullet. Indeed, when the bullet moves with speeds considerably smaller than light speed, γ_b will be indistinguishable from 1 and u_{bx} from v_b, the classical velocity. The time component has no classical analogue.

How are the components of the spacetime velocity of the bullet as measured by Karl related to those measured by Emmy? From the definition of the components of the spacetime velocity it should be evident that they are related via Lorentz transformations. In fact, the same Lorentz transformations that connect $(\Delta x_b', c\Delta t_b')$ to $(\Delta x_b, c\Delta t_b)$ connect (u_{bx}', u_{bt}') to

Spacetime velocity.

Invariant length of spacetime velocity.

(u_{bx}, u_{bt}). More specifically, if we divide both sides of the Δ-equations of Box F.4.1 by $\Delta\tau_b$, we get the primed spacetime velocity on the left-hand side and the unprimed spacetime velocity on the right-hand side. Thus, we can write

$$u'_{bx} = \gamma(u_{bx} + \beta u_{bt}), \quad u'_{bt} = \gamma(\beta u_{bx} + u_{bt}). \tag{28.2}$$

> **What do you know? 28.2.** Emmy sees a relativistic bullet move at $v_b = 0.9c$. What are the values of u_{bx} and u_{bt}, the components of the spacetime velocity of the bullet relative to Emmy? Now suppose that Emmy moves at $0.6c$ relative to Karl. What are the values of u'_{bx} and u'_{bt}, the components of the spacetime velocity of the bullet relative to Karl? Verify that the invariant length of the spacetime velocity is the same for Emmy and Karl. Hint: You need to consult Math Note E.28.1.

The first equation in (28.2) gives the space component of the bullet's spacetime velocity as measured by Karl in terms of the space and time components of the bullet's spacetime velocity as measured by Emmy. Since both components of the spacetime velocity are given in terms of the ordinary velocity, this equation is a relation between the ordinary velocity of the bullet as measured by Karl and the ordinary velocity of the bullet as measured by Emmy (as well as the relative speed of the two observers). It follows that the relativistic law of addition of velocities should be somehow hidden in Equation (28.2). **Math Note E.28.2** on **page 131** of *Appendix.pdf* shows you how to uncover the relativistic law of addition of velocities from Equation (28.2).

28.3   Spacetime Momentum

Spacetime momentum.

Spacetime velocities are important in their own right. However, their true significance lies in their use for building other quantities. One such quantity is the **Spacetime momentum.**, which is naturally defined as the product of mass and the spacetime velocity. Since the spacetime velocity has two components, so does the spacetime momentum. Thus, we define the space component (denoted by p_{bx}) and the time component (denoted by p_{bt}) of the spacetime momentum (of a bullet) as

$$p_{bx} = mu_{bx} = \gamma_b mv_b = \frac{mv_b}{\sqrt{1 - (v_b/c)^2}} = \frac{m\beta_b c}{\sqrt{1 - \beta_b^2}},$$
$$p_{bt} = mu_{bt} = \gamma_b mc = \frac{mc}{\sqrt{1 - (v_b/c)^2}} = \frac{mc}{\sqrt{1 - \beta_b^2}}, \tag{28.3}$$

where m is the mass of the bullet.

How are these two components related to the classical properties of the bullet? To answer this question we go to the limit of low velocities and compare the results to the corresponding Newtonian quantities (see Box 26.4.1). The first equation in (28.3) becomes **Relativistic momentum.** $p_{bx} \approx mv_b$ when v_b/c is very close to zero (and thus γ_b very close to 1). Therefore, it is natural to identify $\gamma_b mv_b$ with the **relativistic momentum** of the bullet.

28.3.1   Relativistic Energy

The identification of p_{bt} is a little harder, because its comparison with the corresponding Newtonian quantity requires some algebra, the details of which can be found in **Math Note E.28.3** on **page 132** of *Appendix.pdf*. It turns out that if we multiply p_{bt} by c we get the **Relativistic energy.** **relativistic energy** of the bullet E_b:

$$E_b = p_{bt}c = mcu_{bt} = \gamma_b mc^2 = \frac{mc^2}{\sqrt{1 - (v_b/c)^2}} = \frac{mc^2}{\sqrt{1 - \beta_b^2}}. \tag{28.4}$$

What do you know? 28.3. How much energy is required to accelerate a 1000-kg car from rest to $0.999c$ according to (the wrong) classical physics? According to (the correct) relativistic physics?

The spacetime momenta of an object as measured by two observers are related by Lorentz transformations. In fact, if we multiply both sides of (28.2) by m, we immediately obtain the Lorentz transformations for spacetime momentum. However, we want to change this result slightly. The first change is a matter of notation: we use Equation (28.4) to replace p_{bt} with E_b and change p_{bx} to p_b, because, with p_{bt} removed, the x subscript is superfluous. So, multiplying both sides of 28.2 by m and incorporating these changes, we get

$$p_b' = \gamma(p_b + \beta E_b/c), \quad E_b' = \gamma(\beta p_b c + E_b). \tag{28.5}$$

These equations connect the energy and momentum of an object as measured by an observer O to those measured by O', relative to whom O is moving with speed v.

Two special cases of Equation (28.4) require our attention: when $v_b = 0$ and when $v_b = c$. In the first case the denominator is 1 and we get $E_b = mc^2$, from which we remove the subscript b and write the most famous equation in physics:

$$E = mc^2, \tag{28.6}$$

The most famous equation in physics.

showing the equivalence of mass and energy. It implies that it is in principle possible to convert mass into energy and vice versa. Because of the enormity of c, a tiny amount of mass can turn into a tremendous amount of energy. In a nuclear power plant uranium mass is turned into energy through the nuclear process of **fission**,[3] in which the large uranium nucleus fragments into two smaller **daughter** nuclei whose total mass is *smaller* than the uranium mass. The remaining mass of the uranium is converted into the kinetic energy of the daughter nuclei, which is ultimately converted into electrical energy.

🧺 Food for Thought 🌱

FFT 28.3.1. Example 28.3.2 illustrated the awesome power hidden in a small amount of mass. A nuclear power plant has mechanisms in place to control this awesome power in a way that useful electricity comes out of the fission process. An uncontrolled nuclear fission reaction is a nuclear bomb.

Both the controlled and the uncontrolled nuclear fission reactions, as used in military and in industry, are *technological* enterprises. The *science* behind this technology—the main part of which is $E = mc^2$—is only a small corollary of the much more encompassing relativity theory.

The development of the first two atomic bombs in the US and their subsequent deployment on Hiroshima and Nagasaki constitute one of the darkest moments in human history. Many scientists were dragged into the construction of the bomb either by persuasion or by the whirlwind of the political events; most of them later regretted to have been involved in it. Einstein, for example, who wrote several letters to President Roosevelt urging him to support the construction of the bomb, regretted such actions tremendously in his later years. "Had I known that the Germans would not succeed in producing an atomic bomb, I would not have lifted a finger," he said later.[4]

The image of science that some critics portray to the public is that of Hiroshima. Yet the bomb has as much to do with science (of relativity) as the electric shocks used in the torture chambers of dictatorships have to do with electricity. The reader has hopefully seen what an insignificant part of relativity $E = mc^2$ is. All the conclusions about the relativity of time, the shrinkage of space, the slowing down of aging, the possibility of space travel to the worlds far away, and of course the equivalence of mass and energy should not be reduced to the construction of the atomic bomb,

---

[3] See Section 31.2.2 for a discussion of fission.
[4] *Subtle is the Lord,* A. Pais, Oxford University Press, 1982, p. 454.

which employs a *technology* that makes use of the equivalence of mass and energy. There are other enterprises that use the same equivalence for peaceful means—as nuclear medicine and electric production do—or for furthering our understanding of the universe around us—as the physics of fundamental interaction does.

Example 28.3.2. To see the immensity of the energy hidden in even a small mass, suppose that one kilogram of uranium is used in a nuclear power plant to produce electricity. If all the mass is converted into usable electric output, this amount of mass produces

$$E = mc^2 = 1 \times (3 \times 10^8)^2 = 9 \times 10^{16} \text{ J}$$

of energy.

A large power plant feeding a small city produces about one *gigawatt* of power or one billion (10^9) Joules per second. If this power plant had only one kilogram of uranium that it could convert entirely into energy, the production of electricity could continue for

$$\frac{9 \times 10^{16} \text{ J}}{10^9 \text{ J/s}} = 9 \times 10^7 \text{ s.}$$

Since there are about 30 million seconds in a year, this is equivalent to about three years!   ∎

What do you know? 28.4. Spaceship Enterprise uses matter antimatter annihilation to produce energy. The power consumed in Enterprise is 10 megawatts (10 million Joules per second). If half a kilogram of matter annihilates half a kilogram of antimatter, how much energy is produced, and how long does it take Enterprise to use up this energy?

28.3.2   No Mass at Light Speed

The second special case of Equation (28.4), $v_b = c$, makes the denominator of Equation (28.4) equal to zero, giving rise to an infinite energy. This is unacceptable *unless the numerator also vanishes*, i.e., $m = 0$. The reverse argument is also true: if $m = 0$ (and the particle has some energy), then the denominator must also vanish, i.e., $v = c$. Since photons (see Section 20.3), being particles of light, travel at the speed of light, they must be massless. We therefore conclude that

---

Box 28.3.3. *Only massless particles such as photons can travel at the speed of light. Moreover, if a particle moves at light speed, it is necessarily massless. No massive particle can attain light speed.*

It gets progressively harder to increase the speed of an object as this speed gets closer and closer to light speed. **Example D.28.2** on **page 60** of *Appendix.pdf* shows why.

28.3.3   Invariant Length of Spacetime Momentum

Spacetime momentum is defined directly in terms of the spacetime velocity. Since the latter satisfies the invariant "length" equation [the last equation in (28.1)], the former—being just m times the spacetime velocity—also satisfies a similar equation: $p_{bt}^2 - p_{bx}^2 = m^2 c^2$. In terms of energy (and removing the subscript x), this becomes $(E_b/c)^2 - p_b^2 = m^2 c^2$. Multiplying both sides of this equation by c^2 and removing the subscript b yields

Relation between energy momentum and mass.

$$E^2 - p^2 c^2 = m^2 c^4. \tag{28.7}$$

Once again, this holds in all RFs, and one can think of it as the **invariant length** of the spacetime momentum.

An interesting consequence of Equation (28.7) is that it allows the possibility of setting $m = 0$. For massless particles such as photon, we have

$$E^2 - p^2 c^2 = 0 \quad \text{or} \quad E^2 = p^2 c^2 \quad \text{or} \quad E = |p|c. \qquad (28.8)$$

The absolute-value sign around p is necessary because E is always positive, but p could be positive or negative. However, often we just write $E = pc$, and introduce a negative sign if p points in the negative direction. De Broglie used the last equation of (28.8) to derive his famous formula $\lambda = h/p$ for electrons. **Example D.28.3** on **page 60** of *Appendix.pdf* and **Math Note E.28.4** on **page 132** of *Appendix.pdf* investigate an interesting consequence of combining the energy-momentum Lorentz transformations (28.5) and the relation between energy and momentum of a photon (28.8). The outcome of the investigation is the relativistic Doppler formula.

28.3.4   No Relativistic Mass

Many authors write Equation (28.4) as $E_b = Mc^2$, identify the result with Equation (28.6), let M depend on velocity, and call it the **relativistic mass**! There is a historical justification for it.[5] The notion of the dependence of mass on velocity was introduced by Lorentz (of Lorentz transformation fame) in 1899 and then developed by him and others in the years preceding Einstein's formulation of special relativity in 1905, as well as in later years.

If you apply Newton's second law of motion $\mathbf{F} = \Delta \mathbf{p}/\Delta t$, which holds in relativity theory as well, and *define* relativistic mass as the coefficient of acceleration in $\mathbf{F} = M\mathbf{a}$, then M will have different values when the velocity is perpendicular to the force (the transverse mass) than when it is in the direction of the force (longitudinal mass). This already should be an indication of the trouble with the notion of "relativistic mass." Nevertheless, these are the very expressions with which Lorentz introduced the two masses. Together with the "relativistic mass" in the relation $\mathbf{p} = m_r \mathbf{v}$, these masses formed the basis of the language physicists used at the beginning of the twentieth century. Making the trouble even more lasting, it was decided to call the "relativistic mass" simply "mass" and to denote it by m, while the normal mass m was nicknamed "rest mass" and denoted by m_0.

 **Food for Thought**

FFT 28.3.4. The masslessness of photons has created a philosophical dilemma for some people. How can a particle exist if it does not have any mass? Some go so far as to interpret photons as "nonmaterial" and draw mystical conclusions from such an interpretation of physics. [See Section 35.3.3 for a "medical" application of this interpretation.] At the heart of this question is the definition of a particle.

Why do we call a bullet a particle, but do not call the wind a particle? The answer: a bullet is *localized*, but the wind is not. Extending this notion to the subatomic particles, we call an electron a particle because it is localized. But how do we know this? By the interaction of an electron with a detector. When the detector of the electrons goes "click," we know that a *localized* electron has arrived.

But localization alone is not the main characteristic of a particle. A point in the vacuum is also localized, but it is not a particle. A particle is usually moving, and in its motion carries energy and momentum. How can we tell if a particle is carrying energy and momentum? By observing how it transfers some of them to other particles. When the particle in question hits another particle (large or small) it transfers part of its energy and momentum to the target particle. A bullet hitting a stationary block of wood imparts some of its momentum and energy to the wood, setting it in motion (bulk energy and momentum) and heating it up (internal energy and momentum). An

Massless does not mean nonmaterial!

---

[5]For a detailed discussion of the notion of the relativistic mass see [Oku 89].

electron hitting a stationary molecule imparts some of its momentum and energy to the molecule, setting it in motion (bulk energy and momentum) and causing its part to move as well (internal energy and momentum).

What about a photon? Well, it also moves; it also carries energy and momentum; and when it hits another particle (large or small) it transfers part of its energy and momentum to the target particle. Just as a bullet is not "more of a particle" than an electron simply because it is more massive than an electron, so is an electron not "more of a particle" than a photon simply because it is more massive than a photon. Mass is only one of a myriad attributes characterizing particles. It is by no means the determining attribute. A photon is as "material" as any other particle, be it an electron or a bullet.

The real trouble with the notion of relativistic mass comes in when we associate a mass of E/c^2 to photons when they experience a gravitational force. When the correct theory of gravity, the general theory of relativity, is applied to photons, it is found that a photon would have different masses for different inclinations relative to the gravitating body. If it falls directly toward the center, its mass is E/c^2, but if it moves perpendicular to the radial direction, its mass is $2E/c^2$.

Equation (28.7) clearly identifies mass (on the right-hand side of the equation) as an *invariant* quantity, which is the same for *all observers*. Only if the particle is at rest relative to an observer (so that $\mathbf{p} = 0$) does E, which is now called the rest energy and usually denoted by E_0, equal mc^2. Einstein, who at the beginning of relativity theory talked about a "relativistic mass," in a letter dated June 19, 1948, writes, "It is not good to introduce the concept of the mass $M = m/\sqrt{1 - v^2/c^2}$ of a moving body for which no clear definition can be given. It is better to introduce no other mass concept than the 'rest mass' m. Instead of introducing M it is better to mention the expression for the momentum and energy of a body in motion."

28.4   Conservation of Momentum

One of the distinct features of momentum is its conservation, which is the content of the first law of motion as stated in Box 7.1.2 in Chapter 7. While momentum conservation was discovered early on alongside the laws of motion, energy conservation required more time because of the variety of forms in which it shows itself. It was not until the middle of the nineteenth century that its formulation appeared as the first law of thermodynamics.

As **Food for Thought 18.2.4** illustrated, conservation of energy goes a lot deeper than the first laws of motion and thermodynamics. In her celebrated paper of 1918, Emmy Noether, the German mathematician showed that conservation of momentum and energy is related to the invariance, respectively, of space and time under *translation*. Invariance under space translation means that the space interval of one meter does not depend on its initial point: A meter stick always measures a meter no matter where you put its zero. Similarly, invariance under time translation means that an hour is an hour whether it is measured from 1:00 to 2:00 or from 4:12 to 5:12, and Jane is 20 years old whether she uses the Christian, Jewish, Moslem, or Chinese calendar. Noether's theorem does not allow any violation of the conservation of energy or momentum.

In the context of relativity theory, the discussion above means that energy and momentum conservation holds for *all* observers. This universality has some consequences that we investigate now. Conservation laws are best studied in collisions. To appreciate the relativistic conservation laws and their consequences, we first turn our attention to the simpler case of the classical laws. All collisions, both classical and relativistic, are assumed to take place in one dimension.

Conservation of energy and momentum and its relation to the invariance under spacetime translations.

Figure 28.3: The conservation of momentum in a collision.

28.4.1   The Classical Case

Two masses m_1 and m_2 are moving with velocities v_1 and v_2, respectively as shown in Figure 28.3. Since $v_1 > v_2$ a collision takes place and two new masses M_1 and M_2 are produced, which move with velocities V_1 and V_2. All of this takes place on a table top in Emmy's laboratory in the reference frame O. Conservation of momentum tells her that

$$m_1 v_1 + m_2 v_2 = M_1 V_1 + M_2 V_2.$$

The same process is observed by Karl who sees Emmy and her lab move in the positive direction with velocity v. Therefore everything that moves in Emmy's lab appears to move faster by v, for example m_1 moves with a speed $v_1 + v$ relative to Karl.[6] For him, the conservation of momentum becomes

$$m_1(v_1 + v) + m_2(v_2 + v) = M_1(V_1 + v) + M_2(V_2 + v),$$

or

$$m_1 v_1 + m_2 v_2 + (m_1 + m_2)v = M_1 V_1 + M_2 V_2 + (M_1 + M_2)v.$$

Emmy's conservation law equates the first two terms on the left-hand side to the first two terms on the right-hand side. It follows that

$$(m_1 + m_2)v = (M_1 + M_2)v \quad \text{or} \quad m_1 + m_2 = M_1 + M_2,$$

i.e., the total mass does not change in a collision. We thus have the following important result

> **Box 28.4.1.** *In any classical process (be it a collision or a chemical reaction) in which masses of the participating objects change, in addition to the conservation of classical momentum, the initial total mass must equal the final total mass. This is restated by saying that **the total mass is conserved** in a classical reaction.*

28.4.2   The Relativistic Case

Now look at the process of Figure 28.3 from a relativistic viewpoint. We can go through the same argument as above, but use the *relativistic* law of addition of velocities to come up with a relativistic version of Box 28.4.1. This turns out to be very complicated. The easier way is to use the Lorentz transformations for momentum and energy as given in Equation (28.5). **Math Note E.28.6** on **page 134** of *Appendix.pdf* shows that

> **Box 28.4.2.** *In any relativistic process (be it a collision or a nuclear reaction) in which masses of the participating objects change, in addition to the conservation of relativistic momentum, the initial total energy must equal the final total energy; i.e., **the total energy is conserved** in a relativistic reaction.*

---

[6]The law of addition of velocities hold in a *classical* process.

This does not say anything about the conservation of mass. In fact, **Example D.28.4** on **page 61** of *Appendix.pdf* shows that *mass is not conserved in a relativistic reaction*:

What do you know? 28.5. Two "particles," each having a mass of 1 kg, move at $0.9c$ towards each other. They collide and form a new particle. What is the mass of the new particle? This is an example of the conversion of energy into matter.

28.5   End-of-Chapter Material

28.5.1   Answers to "What do you know?"

28.1. Use Pythagorean theorem to get $\sqrt{30^2 + 40^2} = 50$ mph.

28.2. Equation (E.108) shows that $u_{bx} = \gamma_b v_b$ and $u_{bt} = \gamma_b c$, where $\gamma_b = 1/\sqrt{1 - 0.9^2}$ or $\gamma_b = 2.29416$. Thus, $u_{bx} = 2.29416 \times 0.9c$, or $u_{bx} = 2.0647c$, and $u_{bt} = 2.29416c$. The components of the spacetime velocity of the bullet relative to Karl can be obtained from Equation (28.2), with γ and β being related to the *relative speed* of Emmy and Karl. With this relative speed being $0.6c$, β is 0.6 and γ turns out to be 1.25. So,

$$u'_{bx} = 1.25(2.0647c + 0.6 \times 2.29416c) = 4.3015c,$$
$$u'_{bt} = 1.25(0.6 \times 2.0647c + 2.29416c) = 4.4162c.$$

The invariant length according to Emmy is $(2.29416c)^2 - (2.0647c)^2$, which is very nearly c^2. Similarly, the invariant length according to Karl is $(4.4162c)^2 - (4.3015c)^2$, which is also very nearly c^2.

28.3. Classically, the required energy, ΔE, is just the difference between the initial and final KE. Since the car starts from rest, this is just the final KE, which, with $m = 1000$ kg and $v = 2.997 \times 10^8$, gives $\Delta E = 4.49 \times 10^{19}$ J. Relativistically, the required energy, ΔE, is the difference between the initial and final energy. However, the initial energy is not zero, it is the rest energy mc^2 or 9×10^{19} J. The final energy is given by Equation (28.4) with γ_b corresponding to a speed of $0.999c$, or $\gamma_b = 22.366$. This gives a final energy of 2.013×10^{21} J. So, $\Delta E = 1.9 \times 10^{21}$, considerably larger than the classical value.

28.4. Altogether a mass of 1 kg is converted into energy. So, $E = mc^2$ gives 9×10^{16} J. Since each second Enterprise uses 10^7 J of energy, 9×10^{16} J will last 9×10^9 seconds, or (noting that there are approximately 31.5 million seconds in a year) approximately 285 years.

28.5. The two particles have equal and opposite momenta. So, their total momentum is zero. Therefore, the final particle will be at rest. The initial energy is twice the energy of each particle, or twice $\gamma_b mc^2$, with γ_b corresponding to a speed of $0.9c$, or $\gamma_b = 2.29$. This gives a total energy of 4.13×10^{17} J, which is the rest energy of the final particle. Dividing this energy by c^2 gives the mass M of the final particle: $M = 4.59$ kg.

28.5.2   Chapter Glossary

Coordinate Velocity The rate at which the coordinates of a moving object change with time.

Spacetime Momentum Mass times spacetime velocity.

Spacetime Velocity The rate at which the spacetime coordinates of a moving object changes with its proper time.

28.5.3   Review Questions

28.1. Does coordinate velocity have more information about the motion or speed plus the direction?

28.2. What is spacetime velocity? Is the spacetime velocity of a bullet the same for all observers? What about the invariant length of the bullet's spacetime velocity? What is this invariant length in terms of the components of the spacetime velocity?

28.3. With what classical quantity do you identify the space component of the spacetime momentum? With what classical quantity do you identify the time component of the spacetime momentum?

28.4. Is it possible to accelerate massive objects to speed very close to light speed? Is it possible to accelerate massive objects to light speed? How does the dependence of energy on speed affect this?

28.5. If a particle moves at light speed, what can you say about its mass? If a massless particle has energy, what can you say about its speed?

28.6. Is it correct to say that in a nonrelativistic reaction, the initial total mass is the same as the final total mass? Is it correct to say the same thing about relativistic reactions?

28.5.4   Conceptual Exercises

28.1. If you travel 60 miles west and 80 miles south simultaneously in 2 hours,
(a) what is your speed in the southwest direction?
(b) What is your westward speed?
(c) Your southward speed?
(d) Can you get (a) from (b) and (c)?

28.2. You travel 141 miles at 45 degrees north of east in 2 hours. What is your eastward speed? What is your northward speed?

28.3. Can observer O "read" the clock of observer O' even though the latter is moving very fast very far? If yes, explain how. If observer O'' also reads the clock of observer O', do O and O'' agree on the reading? Does O' agree with either O or O'' or both?

28.4. A relativistic bullet is moving at $0.968c$ according to Emmy. Using Figure 26.2 and Equation (28.1) find the two components of the bullet's spacetime velocity according to Emmy.

28.5. A proton is moving at $0.995c$. Using Figure 26.2 and Equation (28.1) find the two components of the proton's spacetime velocity. Does it have an invariant spacetime velocity length of c^2, as it should?

28.6. When multiplied by mass, spacetime velocity becomes spacetime momentum, whose time component is γmc. When speed is small compared to c, this becomes mc. Does this mean that mass m is moving at the speed of light to give a momentum of mc?

28.7. A proton is moving at $0.9995c$. Using Figure 26.2 and Equation (28.4) find the energy of the proton in GeV (giga electron volt, with giga meaning 10^9). Mass of proton times c^2 is 938.27 MeV (million electron volt).

28.8. How fast should a neutron be moving to have an energy of 1 GeV (giga electron volt, with giga meaning 10^9)? Mass of neutron times c^2 is 939.57 MeV (million electron volt). Hint: Use Equation (28.4) to find γ; then use Figure 26.2 to read off speed from the γ you found.

28.9. How much energy is needed to accelerate a proton from rest to $0.866c$? Hint: Use Figure 26.2 to read off γ; then use Equation (28.4) to find the final energy.

28.10. If Newtonian physics were correct and we wanted a proton to have a kinetic energy of 2 GeV (giga electron volt, with giga meaning 10^9), how fast should it be moving? Mass of proton is 1.67×10^{-27} kg. How fast should it *really* be moving (i.e., based on relativity)?

28.11. Show that Box 28.4.1 implies that the number of atoms of a particular species does not change in any chemical reaction.

28.12. A particle of mass 1 kg is moving at $0.986c$. It collides with an identical particle at rest and sticks to it. What is the momentum of the moving particle? What is the energy of the moving particle? What is the momentum of the stationary particle? What is the energy of the stationary particle? What is the momentum of the final product? What is the energy of the final product? What is the mass of the final product? Hint: Use Figure 26.2 to find γ of the moving particle. Use Equation (28.7) to find either mass or energy when the other (plus momentum) is given.

28.5.5   Numerical Exercises

28.1. An electron is moving at $0.9921567c$ according to Emmy.
(a) What are the two components of the electron's spacetime velocity according to Emmy [see Equation (28.1)]?
(b) Karl is moving in the negative direction relative to Emmy with a speed of $0.998045c$. What are the two components of the electron's spacetime velocity according to Karl [see Equation (28.2)]?
(c) In each case, verify that the invariant length of the spacetime velocity is c^2.

28.2. How much energy is needed to accelerate a proton from rest to $0.99c$? To $0.999999c$?

28.3. A very compact car has a mass of 200 kg. It is desired to accelerate this car to speeds very close to the speed of light.
(a) Calculate the energy needed to accelerate the car from rest to a speed of $0.99c$.
(b) How much energy does it take to accelerate the car further from $0.99c$ to $0.9999c$?
(c) How much energy does it take to accelerate the car further from $0.9999c$ to $0.999999c$?
(d) How much energy does it take to accelerate the car further from $0.999999c$ to $0.99999999c$?
(e) How much energy does it take to accelerate the car further from $0.9999999999c$ to $0.999999999999c$? Each gallon of gasoline stores approximately 10^9 J of energy. How many gallons of gasoline are needed to accomplish this acceleration?
(f) Compare this with the nonrelativistic solution.

28.4. The spaceship Enterprise in Star Trek is fueled by matter–antimatter annihilation into energy. Suppose 1 kg of matter annihilates 1 kg of antimatter.
(a) How many Joules of energy is released?
(b) If the power consumption of Enterprise is 10 million Watts, how many seconds would this energy last? How many years?

28.5. A 0.5 kg ball moving at 20 m/s hits a stationary 1.5 kg ball and sticks to it, together forming a new ball.
(a) What is the mass of the new ball? Is it the same as the total mass of the original balls?
(b) What is the momentum of the new ball? Is it the same as the total momentum of the original balls?
(c) What is the speed of the new ball?
(d) What is the kinetic energy of the new ball? Is it the same as the total kinetic energy of the original balls?

28.6. A 0.5 kg ball moving at $0.995c$ hits a stationary 1.5 kg ball and sticks to it, together forming a new ball.

(a) What is the momentum of the new ball? Is it the same as the total momentum of the original balls?

(b) What is the energy of the new ball? Is it the same as the total energy of the original balls?

(c) What is the mass of the new ball? Is it the same as the total mass of the original balls?

(d) What is the speed of the new ball?

Chapter 29

General Theory of Relativity

The special theory of relativity (STR), as we presented it in the previous chapters, deals mostly with inertial reference frames. Only Chapter 27 briefly considered noninertial RFs, but did not talk about how the laws of physics would look like in such RFs. Furthermore, although the entire discussion of STR was based on inertial RFs, it did not give any instructions as to how to build one. It is the task of the general theory of relativity (GTR) to do so.

29.1 The Equivalence Principle

In describing inertial frames, Section 25.2.1 outlined how the acceleration of an RF relative to the ground translated into the acceleration of objects in the RF relative to the RF itself. What effect does the acceleration of an RF have on the motion of objects inside the RF? Emmy decides to find out. To avoid complications arising from gravity and contact forces, she takes the RF (a rocket) to outer space, far from any planets and stars, and places an apple inside the RF [see Figure 29.1(a)]. The rocket, the apple, and Emmy (not shown) all float in space, and no force exists to pull or push the apple. The apple floats inside the rocket, and as Emmy gives it a jolt, it starts to move with constant velocity until it hits the wall of the rocket. Emmy, invoking Box 25.2.2, concludes that the first law of motion holds in the rocket frame; it is an inertial RF.

To experience the effect of the acceleration of the rocket on the motion of the apple, Emmy triggers the boosters. Three snapshots of the subsequent motion [Figure 29.1(b)] explain what Emmy sees in the rocket. The apple, unaware of the thrust of the boosters, stays put in its location in space. The rocket accelerates "upward" with its bottom (floor) catching up with the apple, first slowly, later faster, until it hits the apple. Inside, things look differently. The apple, floating in midair before the boost, appears to be moving toward the floor of the rocket; first slowly, later faster, until *it* hits the floor of the rocket. *It drops to the floor*, as if there were gravity in the rocket!

But wait! We cannot call any apparent acceleration "gravity." To qualify for this special label, the acceleration must be independent of the object. Only if an apple and an orange experience exactly the same acceleration can it be said that they are being gravitated. Place an orange next to the apple in the first snapshot of Figure 29.1(b); it will remain next to the apple in the other two snapshots and all the untaken snapshots in between. The rocket will hit both at the same time. Inside, both the apple and the orange will hit the floor at the same time, reminiscent of Galileo's famous leaning-tower-of-Pisa experiment. Emmy will, of course, also experience the same acceleration. If she is initially floating in the middle of the rocket in a squatting position, she will fall to the floor, unfolding herself to land on her

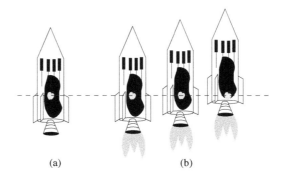

Figure 29.1: (a) The rocket RF in outer space with an apple inside it. (b) As the rocket accelerates, its bottom catches up with the apple.

How to create artificial gravity. feet; and if she is hovering above her chair, she will fall into it and feel its pressure on her back, in exact analogy to a gravitational acceleration. The acceleration inside the rocket has all the signs and signatures of gravity.

> **Box 29.1.1.** *To create a (artificial) gravitational acceleration (or force) in a given direction in a reference frame, accelerate the frame in the opposite direction.*

Now take the rocket to a region of space where a gravitational field is operating, say close to a planet. Shut off the boosters to allow the rocket RF to fall freely. Discover that all traces of the gravitational field have disappeared! Box 29.1.1 gives us an explanation: when the RF falls freely, it is accelerating in the direction of the gravitational field, creating a new (artificial) gravitational acceleration in the opposite direction; this acceleration cancels the existing gravitational acceleration of the planet, so that no acceleration remains inside the RF. Another equivalent explanation is what we have already encountered in Section 9.3: that gravity pulls all objects with the same acceleration, causing the RF as well as everything inside it to fall at exactly the same rate, and giving the appearance of the absence of any force.[1]

> **What do you know? 29.1.** A space station is in the shape of a giant bicycle wheel as shown, where people live in the tube of the wheel. To create an artificial gravity, it is set to rotate. Which direction does the artificial gravity point?

Emmy, being inside a freely falling RF and noticing that objects float in midair, tries a little experiment: she jerks the apple and notices that it moves on a straight line with constant speed all the way to the wall of the rocket. The first law of motion holds in her RF; hers seems to be an *inertial* reference frame. Earlier, when she was in outer space away from the influence of planets and stars, she experienced exactly the same thing. She cannot find any difference between her freely falling RF and her RF when she was far into space. **The equivalence principle.** Emmy is experiencing the **equivalence principle**:

---

[1]Gravity is the only force with this special property. No other force in nature pulls all objects at the same rate. For example, the electric force on an object depends on the electric charge the object carries. Even if two objects have the same charge, their electric acceleration depends on their masses: the heavier object acquires a smaller acceleration.

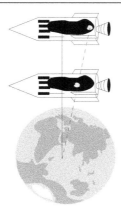

Figure 29.2: The apple moves toward the center of the rocket as both fall toward the center of the planet.

Box 29.1.2. (The equivalence principle) *All* small *freely falling reference frames are inertial. A small accelerated RF is indistinguishable from an RF, in which there exists a gravitational field.*

The first part of the equivalence principle answers the question which the special theory of relativity could not answer. The second part summarizes the fact that you can create gravity in an RF by accelerating it, and that you can eliminate gravity by letting the RF fall freely.

The word *small* has a significance worth exploring. The large RF of Figure 29.2 falls freely toward a planet. The apple, being pulled to the center of the planet along the dashed line, moves toward the center of the RF without any push or pull as the RF falls. The RF appears to have a *fictitious* force, and it is not qualified for an inertial frame. The source of the problem is just the *size* of the RF. If we shrink the RF to a very small size (compared to the planet's), everything will (almost) fall along the solid line and no object will move *relative* to the RF; the RF becomes inertial. For this reason, the equivalence principle is *local*: it applies only to local (i.e., small) RFs as opposed to global (i.e., large or infinite) RFs.

What do you know? 29.2. How long should the rocket be so that the angle between the dashed line and the solid line in Figure 29.2 is one degree? Assume that the rocket is two Earth radii from the Earth center. Hint: The ratio of the side opposite the angle to the side adjacent to the angle gives the angle in radian, where each radian is 57.3 degrees.

29.2   Consequences of Equivalence Principle

The general theory of relativity starts with the equivalence principle, but goes far beyond it. Nevertheless, the equivalence of gravity and acceleration of RFs makes it possible to learn qualitatively some of the outcomes of GTR without a detailed (and necessarily mathematical) understanding of the latter. This section introduces some of these outcomes.

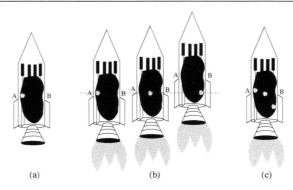

Figure 29.3: (a) The apple, thrown at A towards B, moves on a straight line to B. (b) The apple is thrown towards B, and it moves on a straight line *in space*, but rocket's acceleration moves the floor of the rocket closer to the apple. (c) The path of the apple as it appears inside the rocket.

29.2.1   Bending of Light

Throw an apple in outer space; it moves on a straight line with constant velocity until it hits something. Throw it inside a rocket floating in outer space; it again moves on a straight line with constant velocity until it hits the side of the rocket. And if the rocket is *not* in outer space, but *is* freely falling, the outcome is the same, because both RFs are inertial by the equivalence principle. This is illustrated in Figure 29.3(a).

Now start the booster engines and throw the apple from one side of the rocket to the other. Figure 29.3(b) shows three snapshots of the subsequent motion of the apple and the rocket. Although the apple moves on a straight line in space (along the dashed line in the figure), the rocket's acceleration moves the floor of the rocket closer to the apple. Instead of hitting point B on the other side of the rocket, the apple hits the wall at a point below B. To an observer inside the rocket the path of the ball is that of Figure 29.3(c): the acceleration of the RF has caused a bending of the path of the apple. Invoking the equivalence principle, we conclude that gravity bends the path of the apple. But this is a roundabout way of discovering a truth that every child experiences on Earth! What was the purpose of this exercise?

Replace the apple with a beam of light. In the absence of any acceleration, light travels from A to B on a straight line exactly like the apple (albeit much faster). After starting the boosters, light travels along the straight dashed line *in space* as the apple did in the previous experiment. In the (very small) time it takes light to go from A to B, the floor of the rocket gets nearer the dashed line, causing the beam to hit the other side of the rocket somewhat below B, just as in the apple case (although not as far from B as the apple did). The acceleration of the RF has caused a bending of the path of the light beam. Invoking the equivalence principle, we conclude that

Box 29.2.1. *Gravity bends the path of light.*

And this is definitely not a commonplace conclusion!

In retrospect, one should expect the bending of light from Newtonian gravity. Although the Newtonian gravitational *force* on light (a photon) may be zero because its mass is zero, a photon can be expected to accelerate in a gravitational field because this acceleration is independent of mass: If all objects accelerate identically in a gravitational field regardless of their masses, it is natural to assume that a photon accelerates the same way.

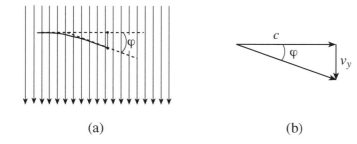

(a)   (b)

Figure 29.4: (a) A light beam bends in a gravitational field. The length of the line segment joining the two dots in the figure is what we calculated in Example 29.2.2. (b) The deflection angle is the angle between initial and final velocity vectors. The angle is exaggerated a great deal; it is very small for ordinary gravitational fields.

Example 29.2.2. The width of a rocket having an acceleration of 9.8 m/s^2 is 10 m. We throw an apple from A (see Figure 29.3) with a speed of 25 m/s (about 56 mph).

Q1: How long does it take the apple to get to B?

A1: Time is distance divided by speed. So, it takes the apple $10/25 = 0.4$ s to get to the other side.

Q2: How far below B does the apple hit the other side?

A2: Recall that the (vertical) distance an object travels when it has a (vertical) acceleration a is $\frac{1}{2}at^2$. Since the apple's vertical acceleration *relative to the rocket* is 9.8 m/s^2 and it covers the vertical distance in exactly the same time that it covers the horizontal distance $\overline{AB}$, it falls below B by $\frac{1}{2} \times 9.8(0.4)^2$ m, i.e., 0.784 m.

Now we send a beam of light across the rocket from A to B.

Q3: How long does it take the light to get to B?

A3: Light is moving a lot faster than the apple, but the time of its travel is given as before: distance divided by speed. This gives $10/3 \times 10^8 = 3.33 \times 10^{-8}$ s.

Q4: How far below B does the beam of light hit the other side?

A4: The vertical distance is still $\frac{1}{2}at^2$. In the case of light this is $\frac{1}{2} \times 9.8(3.33 \times 10^{-8})^2$ m, or 5.4×10^{-15} m. For comparison, the size of the atom is about 10^{-10} m! So this "bending of light" is completely unobservable. ∎

The example above illustrates how small the bending of light is for ordinary accelerations (or gravitational fields) and distances. To enlarge this bending one has to pass light over a very long distance in a very strong gravitational field. A way of measuring the bending is to find the angle between the original direction of the light signal and its direction where it is detected. Figure 29.4(a) shows the bending of a light beam in a uniform gravitational field. The distance between the two dots is what was calculated in Example 29.2.2. What we really want is the *angle* φ between the initial and final directions of motion. **Math Note E.29.1** on **page 134** of *Appendix.pdf* finds a formula for this angle and calculates it for the numbers of Example 29.2.2: it is less than a billionth of an arc second! It also calculates the deflection angle of light on the surface of a white dwarf, with a gravitational acceleration of a million m/s^2, when the light travels a distance of 1000 km: it is 2.3 arc seconds. So, to see a noticeable bending of light, you need to go on something like a white dwarf with an enormous gravitational acceleration, and throw a light beam across a fairly long distance.

Isn't it possible to see the light deflection for an ordinary object like the Sun with an acceleration of about 270 m/s^2? **Math Note E.29.2** on **page 135** of *Appendix.pdf* shows that the angle of deflection, i.e., the angle between the path of light before approaching a spherical celestial body and the path after receding from it is

$$\varphi = \frac{2GM}{Rc^2}, \tag{29.1}$$

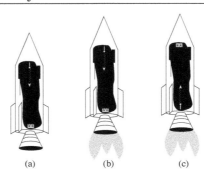

Figure 29.5: (a) As the rocket cruises in space, no Doppler shift is detected. (b) When the acceleration is opposite to the direction of the light beam, the wavelength shortens. (c) When the acceleration is in the direction of the light beam, the wavelength lengthens.

where G is the universal gravitational constant, M is the mass, and R is the radius of the body. For the Sun, $M_\odot = 2 \times 10^{30}$ kg and $R = 7 \times 10^8$ m, giving $\varphi = 0.874$ arc second.

29.2.2   Gravitational Doppler Shift

A rocket is cruising in space with its boosters turned off. A flashlight at its top sends a beam of light, which gets detected by a device at the bottom [see Figure 29.5(a)]. The wavelengths at the source and at the detector are of course the same, because neither the source nor the detector change their velocity: there is no Doppler shift.

Karl, the rocket pilot, turns on the boosters, and the rocket starts to accelerate [Figure 29.5(b)]. The flashlight emits a light beam. A fraction of a second later the beam is detected at the bottom of the rocket. But during this time the velocity of the rocket has changed. The beam leaves the source when the speed of the entire rocket (including the source) is, say v; it arrives at the detector when the speed of the entire rocket (including the detector) is, say $v + \Delta v$. Thus, as far as light is concerned, there is a difference of Δv between the source speed and the detector speed, and this difference is opposite to the direction of light motion, i.e., toward the source. Therefore, light "thinks" that the detector "approaches" the source, and the detected wavelength must be shorter than the original wavelength. We say that the light beam is **blue shifted**, because blue is at the short-wavelength end of the visible spectrum (see Figure 14.4 in Chapter 14).

> Blue shifted means wavelength has decreased.

Figure 29.5(c) shows the case where the source is "chasing" the detector. Once again, let the speed of the source at the time of emission be v and that of the detector at the time of detection $v + \Delta v$. Thus, as far as light is concerned, there is a difference of Δv between the source speed and the detector speed, this time *in the direction of* light travel, i.e., away from the source. Therefore, the detected wavelength is longer than the original light wavelength. We say that the light beam is **red shifted**, because red is at the long-wavelength end of the visible spectrum.[2]

> Red shifted means wavelength has increased.

The last two paragraphs tell us that light (or any EM wave) moving in an accelerated RF undergoes a Doppler shift: EM waves traveling opposite to the acceleration direction are blue-shifted; those traveling in the direction of acceleration are red-shifted. But our real goal is to see the effect of *gravity* on the wavelength of EM waves. Invoking the equivalence principle and recalling that the gravitational field is in the opposite direction to the acceleration to which it is equivalent, we arrive at

---

[2] A more appropriate nomenclature would be *gamma-shifted* and *radio-shifted*, as gamma rays and radio waves have, respectively, the shortest and the longest wavelengths in the entire EM spectrum.

> **Box 29.2.3.** *Electromagnetic waves are blue shifted when moving in the direction of a gravitational field, and red shifted when moving opposite to it.*

It is interesting to note that quantum physics, which is totally unrelated to GTR, predicts the same qualitative result. When a photon (a particle of EM waves) "falls" in a gravitational field it gains energy (just like a ball falling), i.e., its energy at the bottom of its fall is larger than at the top. The energy of a photon is inversely proportional to its wavelength. Therefore, a photon must have a shorter wavelength at the bottom than at the top. And this applies to all the photons in an EM wave, i.e., to the EM wave itself. Similarly, when one "throws" an EM wave "uphill," i.e., opposite to the gravitational field, it loses energy, and its wavelength increases.

Box 29.2.3 tells of the existence of the Doppler effect when EM waves move in a gravitational field. Does a red light "dropped" from a mountain top appear blue at the foot of the mountain? How strong is the Doppler shift? **Math Note E.29.3** on **page 135** of *Appendix.pdf* develops a formula and **Example D.29.1** on **page 62** of *Appendix.pdf* shows how tiny the effect is on Earth.

29.2.3 Gravitational Time Dilation

The gravitational Doppler effect discussed above leads directly to another intriguing result. The change in the wavelength must be accompanied by a change in the frequency (or, equivalently, the period), because the speed of the EM wave is constant. It follows therefore, that gravity changes the period of the EM waves. But the period of an EM wave is simply a length of time.[3] The conclusion is that time is affected by gravity. There is another route to discovering the effect of gravity on time. Section 27.6 showed how time slows down for accelerated RFs. It follows from the equivalence principle that

> **Box 29.2.4.** *Gravity slows down clocks. Clocks in stronger gravitational fields run slower than those in weaker fields.*

Math Note E.29.4 on **page 136** of *Appendix.pdf* derives the formula for the gravitational time dilation and shows how it depends on height or distance from the center of the gravitating object. When you are in space, the notion of height (and the related concept of the gravitational potential) acquires new meaning: If the field lines point from P_1 to P_2, then by definition, height and potential of P_1 is larger than height and potential of P_2.

How do you define "higher" in outer space?

Example D.29.2 on **page 62** of *Appendix.pdf* shows that the effect of (ordinary) gravity on time is very small: after a year, a clock on top of Mount Everest runs only 3×10^{-5} s faster than a clock at the sea level! This is so small that even the most accurate clocks on the shelves of your favorite electronic store fail to measure it. *Any* two such clocks will differ by a lot more than a few 10^{-5} s in as little as <u>one day</u>, *even if they are ticking next to each other*! **Atomic clocks** are, however, *super* accurate: they can keep time with an accuracy of a few seconds in 10 million years, or a fraction of a microsecond (10^{-6} second) in a year. Such clocks are indeed sufficiently accurate to measure time intervals of the order of 10^{-5} s (10 microseconds).

---

[3]In fact, a second is now defined to be 9,192,631,770 times the period of vibration of radiation from the cesium-133 atom.

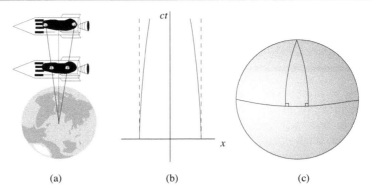

(a)   (b)   (c)

Figure 29.6: (a) The two apples move toward each other although they are in an "inertial" frame and no one is pushing or pulling them. (b) The expected worldlines of the apples (dashed lines) in the coordinate system of the rocket, and their actual worldlines (solid curves). (c) The two *straight* lines approach each other.

29.3   Einstein's General Theory of Relativity

The equivalence principle equates gravity with acceleration, but does not tell us anything about how either of them is produced. In his attempt to reconcile Newtonian gravity with the special theory of relativity, Einstein created the **general theory of relativity** (GTR), one of the most elegantly encompassing theories produced by humankind. His fundamental idea was to focus on the behavior of the path of freely falling objects. Why? Because Einstein believed that the independence of the gravitational motion of objects from their size and structure, had a geometric characterization, a property that was part of the fabric of spacetime.

29.3.1   Geometrization of Gravity

To appreciate the geometric nature of gravity refer to Figure 29.6(a). According to the first law of motion objects should remain at rest in an inertial frame if they are not moving to begin with and no forces are acting on them. Equivalence principle designates the rocket of the figure as an inertial frame because it is in free fall. Yet the two apples that are initially at the two ends of the rocket tend to move towards each other for no apparent reason!

From the point of view of spacetime geometry, the two apples are at rest and no forces are acting on them; therefore, they should remain at rest and their worldlines in the coordinate system of the rocket (with origin at the center of the rocket) should look like the dashed lines of Figure 29.6(b). However, their actual worldlines look more like the solid curves of that figure. The size of the rocket is, of course, the reason for this *observably large* deviation. Reducing the size of the rocket, although diminishing the effect, will not completely erase it. How can one explain this "deviation" of inertial worldlines?

To understand the deviation of inertial worldlines, confine yourself to a two-dimensional surface such as a perfectly smooth table top. Place two small disks on the table and note that they don't move. Push them slightly *in parallel* directions; each will move on a *straight* line, and these straight lines will keep their distance from one another. No deviation of inertial paths!

Einstein argued that in a *curved* "tabletop" such as the one shown in Figure 29.6(c), the straightest possible paths will deviate after a while even though you start them parallel. Thus if you could build a spherical tabletop and you could confine the disks to this table, the disks would move along the great circles, the straightest possible paths, the inertial paths; but these paths get closer together (they deviate). This deviation is, of course,

Figure 29.7: Einstein equation of general relativity, another masterpiece of physics.

due to the inherent curvature of the two-dimensional space of the tabletop. Einstein then made the bold and sweeping assumption that the deviation of the apples' worldlines in Figure 29.6(b) and the deviation of the straight lines in Figure 29.6(c) are exactly of the same kind. *Gravity is related to the curvature of spacetime.* He then resorted to the classical gravity and noted that the source of gravity is mass. Since mass and energy are equivalent in relativity, Einstein generalized the source of gravity to energy (which includes mass via $E = mc^2$). He then wrote an elegant mathematical equation, the **Einstein equation**, which we reproduce in Figure 29.7.

Einstein equation.

The left-hand side of Einstein equation describes curvature and the way distances are defined in spacetime (R is the symbol for curvature and g for a recipe of how to find the shortest distance between two spacetime points); the right-hand side is energy (mass included).

Unlike classical gravity, in which space is a predefined entity with matter moving in it under the influence of the gravitational forces, Einstein equation ties the matter (the right-hand side of the equation) and the spacetime (the left-hand side of the equation) together: matter determines the structure of the spacetime, and the spacetime determines how matter moves in that structure. In short, Einstein equation says:

> **Box 29.3.1.** *Gravity is a manifestation of the curvature of spacetime. Matter tells spacetime how to curve; spacetime tells matter how to move.*

29.3.2   Experimental Tests of GTR

The simplicity of the idea on which the general theory of relativity is based, and the elegant geometrical soil, in which the theory is planted, make GTR one of the most encompassing theories of all time. It was this simplicity and elegance that prompted Einstein, normally a modest and down-to-earth person, to refer to GTR as so magnificent that he would have been "sorry for dear God" had any experimental result contradicted it. The ultimate test of any theory is, of course, observation. But elegance and simplicity are great indicators of the validity of a theory: if a theory is based on simple assumptions and is also expressed in terms of "elegant mathematics," chances are that it is correct. General theory of relativity enjoys both of these criteria. Many rigorous experimental tests of the GTR have been conducted, and every time the theory has emerged triumphant. Here is a partial list.

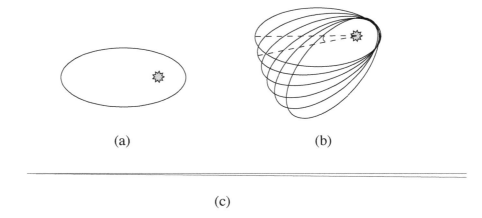

(a)   (b)

(c)

Figure 29.8: (a) Mercury's Newtonian orbit when it is the only planet in the solar system. (b) Precession of the perihelion of Mercury. (c) A one-degree angle.

Precession of the Perihelion of Mercury

Newton's theory of gravity predicts a perfectly elliptical orbit for a single planet going around the Sun (see Math Note E.10.2 for details). If the solar system consisted only of the Sun and the planet Mercury, the latter's orbit would—according to Newtonian theory of gravity—be an ellipse on which Mercury would be moving over and over again for eternity [Figure 29.8(a)].

In the real solar system, the other planets perturb the perfect ellipticity of Mercury's orbit. This perturbation had been calculated by the middle of the nineteenth century, and it had been shown that the overall effect of the perturbation was the **precession of the perihelion** of Mercury. This means that every time Mercury completes its revolution, its orbit tilts ever so slightly. Figure 29.8(b) shows a hugely exaggerated version of this tilt. Suppose that in some revolution, Mercury's perihelion (the point of its orbit nearest the Sun) is horizontal. The second time Mercury comes around, its perihelion will be shifted slightly, so that the line joining the Sun to the new location of the perihelion makes a small angle with the horizontal (the previous perihelion). Figure 29.8(b) shows the angle for the farthest point of orbit for clarity (this angle is equal to the angle that perihelion makes with the horizontal).

As Mercury repeatedly revolves around the Sun, the perturbation of the other planets causes its perihelion to change further and further, each time making a slightly larger angle with the horizontal. After a very long time, the perihelion will have advanced to a large angle. How long and how large? In 1859, the French astronomer Leverrier calculated this advance to be 531 seconds of arc per century! A second of arc is 1/3600 degree; and for comparison, an angle of one degree has been drawn in Figure 29.8(c). Thus, over 100 years of revolution, Mercury's perihelion advances by about 1/7 degree, or 1/7 of the angle shown in part (c) of the figure. A very small effect indeed! But not small enough to go undetected. In fact, not only did Leverrier calculate this small number, but he was able to compare his result with the observations of the previous 300 years and detect a discrepancy. Those observations led to a value of 574 seconds of arc per century. This disagreement of 43 seconds of arc was a puzzle for astronomers and physicists until 1915 when Einstein introduced the GTR and calculated the orbit of Mercury in the absence of other planets. GTR predicted that if Mercury were the only planet going around the Sun, it would not move on a perfect ellipse, but its perihelion should advance approximately 43 seconds of arc per century!

Bending of Light

The equivalence principle predicts the bending of a beam of light in a gravitational field, and Example 29.2.2 calculates the magnitude of this effect when light bends in a gravitational field (acceleration) of 9.8 m/s^2 over a distance of 10 m. To be measurable, the bending of light must occur over a larger distance in a stronger gravitational field. However, the prediction of the equivalence principle is not expected to match the experimental measurement, even at the large distance and strong gravity limit. The reason is that the equivalence principle does not incorporate the curvature of spacetime. The Einstein equation, on the other hand, predicts a bending of light that is twice that predicted by the equivalence principle, and its verification by astronomical observation is one of the most exciting chapters in the history of science.

Einstein had been working on the effect of gravity on light since 1911 using mainly the equivalence principle. He was also aware of the possibility of measuring the effect. By 1915, in which year he discovered the general theory of relativity, he had calculated the amount of bending when a star light grazed the Sun and was detected on Earth. In Figures 29.9(a) and 29.9(b), the light rays from two stars A and B are shown to be received by Earth at a certain location on its orbit around the Sun. The angular separation α between the two stars can be measured and recorded. Six months later, the Earth moves to a location "behind" the Sun. The light rays coming from A to this new location of the Earth graze the Sun and bend. The light rays coming from B stay farther from the Sun and bend much less. Figure 29.9(c) shows the bending of these light rays. The amount of bending has been exaggerated in the figure so you can see the details of the effect.

The observer or the instrument receiving the rays from A and B records the angular separation as β, because this is the angle that the two rays make with one another on Earth. It, therefore, appears as if A's location has been shifted to A′. A comparison of the photographic images of the two stars taken at the Earth's two locations should reveal this relativistic bending of light, and with calibration, it can quantitatively measure this effect for verifying the theory.

There is, however, a catch. While taking a picture of the positions of the stars A and B in the night sky of Figure 29.9(a) poses no problem, recording the positions of the stars in the sky of Figure 29.9(c) is impossible: no stars are visible when the Sun is in the sky! Well, not quite impossible. If we could block the sunlight so much so that—at least momentarily—a darkness as black as the night could fall on Earth, we might be able to detect the bent light rays. Such blockades of the Sun's light occur naturally and periodically; they are called eclipses of the Sun!

Upon the urging of Einstein himself in 1916, astronomers developed a keen interest in verifying the GTR's 1915 prediction of the bending of starlights. By early 1919 attention was focused on the total eclipse of the Sun predicted to occur in May 1919 in the southern hemisphere. Arthur Eddington, a British astronomer set out on two expeditions that took him and his team to Sobral, Brazil, and to Principe, an island off the coast of West Africa. At both these sites, the British astronomers were able to photograph the separation of two stars—the rays of one of which grazed the Sun—and compare the result with the separation of the same stars 6 months earlier. Their result verified Einstein's general relativistic calculations!

As mentioned before, the equivalence principle (EP) is only the first step towards the general theory of relativity. Applied to the bending of light, EP, which does not go beyond the Newtonian mechanics, yields Equation (29.1), and for the Sun, an angle of 0.874 arcsecond. Einstein's GTR, on the other hand, predicts a value that is twice that given in Equation (29.1), and a bending of light due to the Sun's gravity equal to 1.75 arcseconds. This factor of two comes from the curvature of spacetime that is naturally accommodated in the GTR. Thus, the 1919 measurement did not simply verify the bending of light, it provided evidence for the curvature of spacetime as proposed by Einstein in his GTR.

The 1919 observation of light bending verified that spacetime is curved.

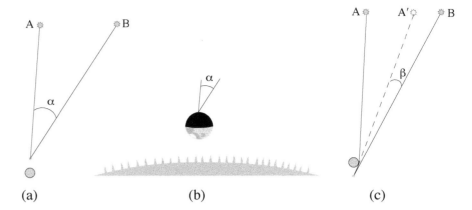

Figure 29.9: (a) In the night sky of Earth, stars A and B have an angular separation α. (b) The blown-up version of (a). (c) In the day sky of Earth, rays coming from A and reaching the Earth have to bend toward the Sun, while those from B are more or less unaffected, because they don't get close to the Sun. The result is a reduction of the angular separation to β. The amount of bending has been exaggerated to show the details.

Modern technology has enabled physicists to test the bending of electromagnetic waves (of which light is an example) as they graze the Sun. In one experiment, a radar signal was sent to a planet such as Venus in two different ways: just before Venus went behind the Sun as seen from Earth and when the Sun was far away from the radar path. It takes the radar slightly longer to traverse the bent path than the straight path. The experiment measured the difference in the round-trip travel times of these two cases and compared the result with the GTR's prediction. Within small experimental errors, GTR's prediction was confirmed.

One of the most intriguing confirmations of GTR's prediction of the bending of light took place in 1979. As early as 1937, Einstein had predicted a substantial bending of light in a region of very strong gravitational field, much like the bending of light when it passes through an optical lens (Figure 29.10). He showed that if a large concentration of mass existed between a light source and Earth, then the extraordinary gravitational field of this mass would bend the light, causing the formation of a double image of the source. This so-called **gravitational lensing** was observed in 1979 for the first time.

Gravitational lensing.

In Figure 29.10 a **quasar**—a very distant source of very strong radio waves and light, also known as a *quasi stellar radio source*—labeled A sends EM signals in all directions. An extremely massive object M (a giant black hole, for example) blocks the signals that would have come directly to Earth. However, the strong gravitational field of M can bend some other rays in such a way that a telescope on Earth could detect them. The image would appear as two quasars A$'$ and A$''$, which look almost identical. The figure in the margin shows a real gravitational lens.

> **What do you know? 29.3.** Suppose that M in Figure 29.10 is a perfect sphere. What would the image of A look like on Earth?

Gravitational Waves

Einstein's GTR equation is a mathematical statement about the effect of energy on the curvature of spacetime (and vice versa). As long as the source of the energy is static, the curvature of spacetime remains unchanged. Thus, the orbits of the planets and the bent

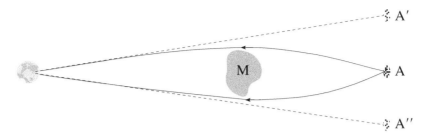

Figure 29.10: The quasar A sends EM signal in all directions. Those that head directly to Earth are blocked by a very massive object M. However, the strong gravitational field of M bends other rays toward Earth. These rays appear to be coming from A′ and A″.

path of light are embedded in a static (small) curvature in the vicinity of the Sun. The universe, however, is brimming with exotic objects; objects with extraordinary accelerations and gravitational fields. Such combination of acceleration and gravity makes these objects a *dynamic* source of spacetime curvature, a curvature that changes with time, a curvature undulation, a **gravitational wave**.

The prediction of the gravitational waves, ripples in the very fabric of spacetime, is one of the most unexpected and exotic triumphs of GTR. It is as triumphant as the prediction of the electromagnetic waves in the middle of the nineteenth century. It took about 20 years before a direct confirmation of the EM waves was established in the laboratory. The experimental confirmation of the gravitational waves has been coming much slower.

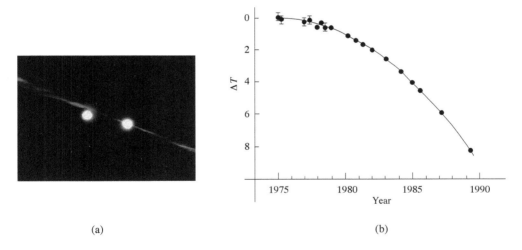

(a)   (b)

Figure 29.11: (a) A pair of binary pulsars. (b) As the binary pulsars lose energy to the gravitational waves, their period decreases. Here the change in the period of the binary pulsar $1913 + 16$ is measured over a span of 14 years. The solid line is the curve predicted by the GTR; the dots are observation points.

The first indication of the gravitational waves was discovered in the observation of binary pulsars. A pulsar is a collapsed star, whose small size endows it with an extraordinarily fast rotation around its axis. The collapse also causes a phenomenal increase in the magnetic field of the star. This combination (strong magnetic field and incredibly fast rotation) makes the star pulsate radio signals with extreme regularity and accuracy. The first pulsar discovered "beeped" every 1.33733 seconds ... exactly! This accuracy led some people to believe that the beeps were signals sent by some alien civilization.

Figure 29.12: An aerial view of the L-shaped structure of LIGO in Hanford, WA.

Binary pulsars give indirect evidence for gravitational waves.

A **binary pulsar** is a pair of pulsars (see Section 39.5.2) that rotate about one another—each acting as a "planet" of the other [see Figure 29.11(a)]. The accuracy of their pulsation helps astronomers keep track of the motion of the pulsars. In particular, it is possible to measure the period of their rotation precisely. The interest in the binary pulsars arose because of general relativity's prediction that they should emit gravitational waves due to their mutual rotation. Based on this theory, the energy carried by the waves should cause a decrease in the period, very much like the decrease in the period of the planetary model of the atom discussed at the beginning of Section 21.3. The change in the period is very small for short intervals of time, but over many years this change can accumulate to measurable quantities.

Figure 29.11(b) shows the decrease in the period of the binary pulsar 1913 + 16 over a span of 14 years.[4] A comparison of the GTR's prediction (the solid curve) and observation (the dots) reveals the exceptional accuracy of GTR. Figure 29.11 is also a testament to the phenomenal advances made in instrumentation: to be able to measure the period of two stars as they pirouette around one another to within a few seconds is indeed remarkable.

Although the *indirect* evidence of the diminishing period of a binary pulsar as described above is fairly convincing, almost a century after their prediction, a *direct* evidence for the gravitational waves is still lacking, due entirely to the expected weakness of the signal to be detected. Nevertheless, a global effort is underway to catch the first glimpse of a gravitational wave signal coming from the explosion of a supernova or the collision of two black holes.

LIGO is designed to detect the gravitational waves directly.

The *Laser Interferometer Gravitational-Wave Observatory* (LIGO) is a facility built to detect the extraordinarily dim signal received on Earth thousands of years after a cataclysmic event took place at a cosmic distance. LIGO employs an L-shaped 4-foot diameter vacuum pipe with 4-kilometer (2.5-mile) arms (Figure 29.12). Unlike EM waves, gravitational waves penetrate even the thickest material unimpeded. Therefore, LIGO detectors need not be "looking" at the sky. In fact, they are entirely shielded in a concrete cover. At the vertex of the L, and at the end of each of its arms, are mirror surfaces, which are the sensors of gravitational waves.

When the gravitational waves pass through LIGO's L-shaped detector they will cause such a ripple in the space (and the time) surrounding the detector as to decrease the distance between the mirrors in one arm of the L, while increasing it in the other. These changes are minute: just 10^{-16} centimeters (or a hundred million times smaller than the diameter of a hydrogen atom) over the 4-kilometer length of the arm. As small as this difference is, it is sufficient to cause a detectable interference between the two laser beams that recombine at

---

[4]This plot is adapted from a paper by J. Taylor and J. Weisberg published in *Astrophysical Journal* in 1989.

the vertex of the L after traversing along its two arms.

At least two detectors located at widely separated sites are required for the unequivocal detection of gravitational waves. Local phenomena such as micro-earthquakes, acoustic noise, and laser fluctuations can cause disturbances that can pass for a gravitational wave event. This may happen locally at one site, but a simultaneous occurrence of such disturbances is extremely rare at two widely separated sites. The two sites for the LIGO observatories are located near Livingston, Louisiana, and Hanford, Washington.

Although the idea of LIGO was conceived in the late 1970s, only recently have scientists been able to make their first trial runs of the facility; and in a few years, the news of the arrival of the first gravitational wave is expected to make headlines across the globe.

Global Positioning System

One of humankind's distinguishing traits is exploration. It spurred our most distant ancestors out of Africa, the primitive tribes into initial conflicts and final mergers, the early civilizations into war and peace, and the medieval sailors to unfamiliar and unknown territories. The earliest explorations were conducted at random. Later, they demanded a guiding mechanism to assure the explorer's correct choice of path. At the beginning, the stellar and planetary positions in the sky were the only guides known to humankind. After the discovery of magnets, compasses became the most effective navigational tools available to the explorers. And until very recently, they were the only tools used in navigation.

With the advent of satellites a new kind of navigation became possible. The **Global Positioning System** (GPS) uses a network of clocks mounted on satellites to pinpoint the location of a person (called *receiver*, as he/she is presumed to be carrying a hand-held GPS "receiver") on, or above, the Earth surface. The position and time of a receiver are specified by four quantities (three spatial coordinates and time). Therefore, if four satellites, situated at four *known* locations, send light signals at *known* times to a receiver, the latter can calculate his/her coordinates and time by invoking the second postulate of the special theory of relativity, namely that the distance between the receiver and a satellite is the speed of light times the time interval between the transmission of the signal from the satellite and its reception by the receiver.

Precision is the key to the proper working of the GPS. In particular, the clocks must keep a very accurate time. If one of the clocks deviates by a microsecond (10^{-6} s), the coordinates measured will be off by the speed of light times a microsecond or about 300 meters. To avoid an error larger than a meter, the clocks must keep the same time to within $1/(3 \times 10^8)$ or 3.3 nanoseconds (a nanosecond is 10^{-9} s). Such an accuracy must be maintained in the time intervals between periodic synchronizations. This leads to a fractional time stability, $\Delta t/t$, of 10^{-13}. Only atomic clocks can achieve this.

Relativistic effects are much larger than 10^{-13}, and therefore, *they must be accounted for* in the design and operation of GPS. For example, a satellite located at 2.5×10^7 m from the Earth center moves with a speed of about 4000 m/s. A clock carried by this satellite will run slower than the Earth clock by a factor of $\frac{1}{2}(v/c)^2$ [see Equation (E.72)]. For $v = 4000$ m/s, this gives 8.9×10^{-11} which is 890 times larger than the fractional time stability of 10^{-13}. Gravitational effects are even larger. In fact, Example D.29.2 shows that for a satellite on this orbit, the gravitational time dilation is 5.1×10^{-10}, or 5100 times larger than the fractional time stability.

Suppose that the designers of GPS were unaware of the importance of relativity, and to tame the spread of error over long periods of time, they wanted to build a system so precise that it could locate a receiver to within a centimeter (0.01 m). This requires a clock accuracy of $0.01/(3 \times 10^8)$ second or 3.3×10^{-11} second. For simplicity, consider only the gravitational time dilation, for which $\Delta t/t \approx 5 \times 10^{-10}$. Then in only $t = 0.066$ s, relativistic effects already reach the temporal tolerance level of $\Delta t = 3.3 \times 10^{-11}$ second. Stated equivalently, in only 0.066 second, the relativistic effects achieve the spatial tolerance level of one centimeter, so that by the passage of every 0.066 second another indeterminacy of one centimeter is added to the spatial determination of the receiver; and every second $1/0.066 = 15$ cm gets added to the uncertainty in the location of the receiver! Therefore, if relativistic effects were not accounted for, satellite clock errors building up in just one hour would cause navigational errors of 3600×15 centimeters, or about 540 m; and in a single day this error builds up to over 12 km, rendering the system useless. All this tells us that

> **Box 29.3.2.** *The proper operation of GPS depends crucially on the theory of relativity (both special and general).*

29.4   GTR and the Universe

In physics, simple becomes complex and complex becomes simple!

The biggest surprise of GTR was its prediction of how the entire universe behaved. How can *any* theory predict an entity as complicated as the universe in its entirety? The universe, with countless galaxies, within each, billions of stars, around each star, numerous planets, and each planet a complicated world, full of complex material including life! The answer lies in the spirit of physics itself. Physics can take a simple structure such as a hydrogen atom and make it so complicated that an entire generation of physicists could spent their scientific life studying it and still not get the complete answer. It can also take a complex structure such as the entire universe and simplify it so much that the laws of physics become applicable. In the first picture, the atom becomes a universe, in the second picture the universe becomes a collection of tiny particles, each particle a galaxy!

29.4.1   Expansion of the Universe

It was this simplifying assumption that permitted the Russian physicist Aleksandr Aleksandrovich Friedmann (1888–1925) to apply the general theory of relativity to the entire universe. What exactly was his simplifying assumption? Actually there were three underlying assumptions that went into **Friedmann universe**. He thought of the universe as a dust cloud, each dust particle being a galaxy. He also assumed that the universe is **isotropic** and **homogeneous**.

Isotropy explained.

Isotropy is the property of the universe that does not allow a preferred direction. Looking along different directions in an isotropic universe you should not be able to detect any features distinguishing one direction from another. "Wait a minute!" you may say. "I happen to drive eastward to my work every morning, and I clearly see a Sun in front of me. But I am pretty sure that there are no suns to my left or right or back! Forget about the Sun. I am standing on the floor of my room. I see a huge round sphere beneath me that I have learned to call 'Earth,' but I can't find any Earth in any other direction. How can you tell me that the universe does not recognize any preferred direction?"

The kind of isotropy we are talking about operates on a very large scale. Remember that galaxies are "dust particles" of the universe. So the isotropy that goes into the Friedmann model of the universe is of the kind that operates at intergalactic distances. In fact, galaxies that are "close," i.e., that are closer than millions of light years are to be lumped into a single "dust particle." Only at separations of hundreds of millions of light years do we begin to see the isotropy property. Standing behind a giant telescope and looking at the distribution of "distant" galaxies, you cannot detect any variation of this distribution at different angles of observation. The pattern of Figure 29.13(a) is isotropic at the center, because all directions are identical. However, it is not isotropic at any other point. Figure 29.13(b) shows a pattern that is not isotropic at any point, while Figure 29.13(c) shows a pattern that is isotropic at all points.

Homogeneity explained.

Homogeneity is the property that precludes any special or preferred *point* in the universe.

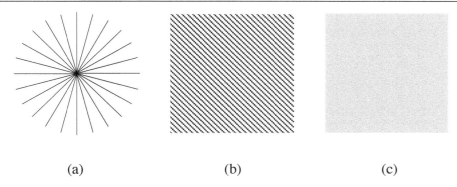

(a)   (b)   (c)

Figure 29.13: (a) A pattern that is isotropic at the center, but not at any other point. (b) A pattern that is homogeneous at all points but not isotropic at any point. (c) A pattern that is both isotropic and homogeneous at all points.

Again, this property operates only at distances larger than hundreds of millions of light years. This means that the universe looks the same whether you observe it from the Milky Way[5] or any other *distant* galaxy. It *does not* look the same if you observe it from Milky Way as opposed to the Andromeda galaxy, a mere 2 million light years away! Of the three patterns in Figure 29.13, the first is not homogeneous while the second and third are. If you want to "picture" the Friedmann universe as a pattern, then Figure 29.13(c) is the closest pattern representing it.

What do you know? 29.4. In a huge forest, trees are in rows parallel to each other. Is the forest homogeneous or isotropic or both or neither?

When, in 1922, Friedmann solved Einstein's GTR equation for a homogeneous and isotropic universe, in which (distant) galaxies were treated as dust particles, he discovered that *the universe must be expanding*! Einstein's reaction to Friedmann's work was that of disbelief. He thought that Friedmann had made a mistake; and when he realized that Friedmann's conclusion was correct, he decided to change the GTR equation, adding a **cosmological constant** to it to make the universe static, because no sign of an expanding universe existed at that time. But when indications of an expanding universe surfaced in the observations of the American astronomer Edwin Hubble in the late 1920s, Einstein regretted the addition of the cosmological constant and referred to this mutilation of his equation as "the biggest blunder of my life."

Einstein modifies his beautiful equation.

What exactly did Hubble observe? Galaxies are made up of some of the same elements that terrestrial objects are composed of. Each of these elements has its own unique spectral lines, a finger print. The motion of a galaxy changes the wavelength of these spectral lines (measured on Earth) due to the Doppler effect. By measuring the change in the (wavelength of the) spectral lines of a given element, one can determine the (radial) speed of a galaxy relative to us as well as the direction of the speed:[6] an increase in the wavelength, also called a **red shift**, indicates that the galaxy is moving away; a decrease in the wavelength, called a **blue shift**, means that the galaxy is approaching. Using the Doppler shift in the spectral lines of known atoms in many galaxies, Hubble discovered that *all **distant** galaxies move away from us*.

---

[5]Disregarding the distinctive *local* features of the Milky Way such as the presence of other "nearby" galaxies.

[6]See Equation (E.10).

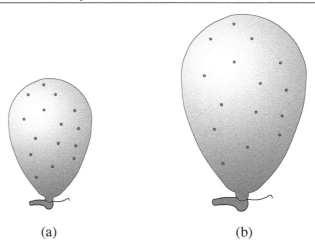

(a)   (b)

Figure 29.14: (a) A balloon with some round stickers. (b) As the balloon expands, the distance between *any* two stickers increases.

Are we at a privileged position in the universe, from which (and only from which) the rest of the universe recedes? Are we at the "center" of the universe? The thought crossed some minds, even the minds of some physicists. But Copernicus and Kepler showed us how the old astronomy went astray by trying to place Earth at the center of the universe. Earth, as Bruno taught us, is only one planet (among many) that circles one star among billions of others, which moves around the center of a galaxy among billions. Homogeneity of the universe tells us that, from the perspective of *any* galaxy, the distant galaxies should move away. How can this be? How can all galaxies see all the other distant galaxies recede? The following analogy should be helpful.

On a balloon, attach some tiny round stickers (Figure 29.14). Blow into the balloon and see the distance between *any* two stickers increase. If there were some two-dimensional creatures living on one of these stickers, they would see all stickers moving away from them; and since they cannot dig into the balloon (this would require a sense of a third dimension, which our creatures lack), they are incapable of seeing the balloon expand. First they would think they are at the center of the expansion. However, if they are intelligent enough to think of their "universe" as isotropic and homogeneous, then they would have to conclude that the only way that *all* stickers can recede from one another is for the entire balloon to expand. One extra bonus of this balloon analogy is that the stickers themselves *do not expand.*

We live in a three-dimensional "expanding balloon." Our galaxy plus all the other local group of galaxies, including our two backyard neighbors, the two **Magelanic clouds**[7] and the nearby Andromeda galaxy[8] constitute our tiny sticker. No distance at this tiny scale increases. In fact, we know that Andromeda is approaching the Milky Way. However, once we go beyond the local structures and observe the *distant* galaxies, they all move away from us. And since the universe is homogeneous and isotropic, any distant galaxy behaves the same way, like the stickers on a balloon. We, therefore, have to conclude that *the universe is expanding.*

The universe has no boundary.

"But if the universe is expanding," you may say, "it must be expanding into something." Not at all! Remember that the universe is homogeneous; so no point (on a large scale) can be different from any other point. This prevents the universe from having any edge or

---

[7]At a tiny distance of only 300,000 light years!
[8]At a close distance of 2,000,000 light years!

boundary. In order for it to expand into "something" it must have a boundary. Since it does not, it cannot expand into any "empty space." The balloon of the two-dimensional creatures has the same property. It does not have a boundary: if an adventurous two-dimensional creature starts on a cosmic journey in search of the "edge" of their universe, he will fail; he'll be just going round and round the balloon. To us, their entire balloon is an "edge" separating the inside from the outside. But the two-dimensional creatures have no sense of a third dimension; their entire world is the surface of the balloon, and this world gets bigger and bigger, i.e., their two-dimensional *space* is constantly being *created*! And that is precisely what is happening to *our* universe: as it expands, *our universe creates new space*. Let us summarize our findings so far:

> **Box 29.4.1.** *On a large scale (of hundreds of millions of light years), our universe is expanding. This expansion manifests itself as the recession of all distant galaxies from one another. The universe has no boundary; therefore, its expansion entails the creation of new space.*

29.4.2   Hubble Law

Homogeneity and isotropy put a severe restriction on the relative speed of two galaxies. **Math Note E.29.5** on **page 137** of *Appendix.pdf* shows that the speed of one galaxy relative to another must be proportional to the separation between the two. This so-called **Hubble law** can be written as

*Homogeneity and isotropy lead to **Hubble Law**.*

$$v = Hd, \qquad (29.2)$$

where H is the **Hubble parameter**, whose most recent value is observationally found to be between 21 and 23 km/s per Mly (million light year). This means that for every million light years increase in distance, the speed of the galaxies increases by 21–23 km/s. The Hubble law is a consequence of the application of GTR to a homogeneous and isotropic universe. In fact, GTR yields the result that the speed with which *any two points* of a homogeneous and isotropic universe recede from one another is proportional to their separation. That we have to look at *distant* galaxies to observe this is a result of the breakdown of homogeneity and isotropy at small distances.

> **What do you know? 29.5.** A galaxy is 500 Mly away from Earth. How fast is it moving away from us? Take the Hubble parameter to be 22 km/s per Mly.

The actual value of H is determined only through observation. By looking at the Doppler shift in the spectral lines (see Section 21.3.1) of elements emanating from galaxies, Edwin Hubble (1889–1953) discovered that, except for the galaxies neighboring our Milky Way, the spectral lines coming from *all* distant galaxies were shifted towards *longer* wavelengths. Since this kind of **red shift** meant a recession (see **Example D.29.3** on **page 63** of *Appendix.pdf*), he concluded that all distant galaxies were moving away from us, and the farther they were, the faster they moved, obeying the simple relation of Equation (29.2).

Hubble himself overestimated the constant to a value of about 150 km/s per Mly. Since then, the observational refinements have been able to narrow the range, very slowly. Even as late as the late 90s, the value was reported to lie between 12 and 31 km/s per Mly. However, with the advent of the Hubble Space Telescope, the gap between the upper and lower limits has been constantly diminishing. The range of values of H quoted above is the result of the most recent observations (as of July 2008).

29.4.3   Age of the Universe

One of the immediate consequences of the expansion of the universe is its contraction in the past. If we retrace the size of the universe backward in time, it must get smaller and smaller as we go further and further in the past.

> **Box 29.4.2.** *GTR predicts that at a* finite time *in the past, the universe must have had an infinitesimally small size, indicating the time of the inception of the universe, or what has come to be known as the* **big bang**.

In fact, the simple Hubble law can give us an approximate age of the universe. If the present speed of the recession of distant galaxies persisted all the way to the moment of creation,[9] then we can obtain the time in the past, when two given galaxies were right "on top of each other."

Consider the Milky Way and another distant galaxy separated by some large distance. How long did it take for the galaxy to reach this distance starting at the same point as the Milky Way? If the galaxy is moving (and has been moving all this time) with the same speed, the time it took to reach the current separation is just distance divided by speed. **Example D.29.4** on **page 63** of *Appendix.pdf* follows this line of argument quantitatively and reaches the conclusion that the time is just the inverse of the Hubble constant: $t = 1/H$, independent of the distance. This means that all galaxies have taken this much time to reach their present position starting at the same point as the Milky Way. It follows that the inverse of the Hubble constant gives the time in the past at which *all* galaxies were on top of each other. Example D.29.4 calculates this time to be between 13 and 14 billion years for the range of values of the Hubble constant given above.

Recently the age of the universe has been measured very accurately.

> **Box 29.4.3.** *The universe is 13.7 billion years old. In other words, the big bang occurred 13.7 billion years ago.*

Considering the crude assumptions we made in reaching our estimates, it is remarkable that we obtained values that do not disagree considerably with the observed age! This agreement is deceiving. As we shall see in Section 37.2, a more precise calculation of the age of the universe actually *widens the disagreement*!

29.5   End-of-Chapter Material

29.5.1   Answers to "What do you know?"

29.1. The acceleration is centripetal (towards the center). Therefore, the artificial gravity is centrifugal (away from the center). So, people will be walking inside the tube with their feet sticking to the outside rim of the tube from inside.

29.2. Denoting the length of the rocket by L and the radius of Earth by R, we want the ratio of $L/2$ to $2R$ to be $1/57.3$ radian, or $\frac{L}{4R} = \frac{1}{57.3}$. This gives $L = 4R/57.3$. With the radius of Earth being 6400 km, we get $L = 447$ km or 279 miles!

29.3. A ring.

29.4. As you move in the forest, you see the same view. That is homogeneous! If you turn and look at the forest at a different angle, you see different things. That is not isotropic.

---

[9]We know that this assumption cannot be true, because the speeds were a lot larger than the present speeds near the moment of creation.

29.5. Speed of the galaxy in km/s is the product of the Hubble parameter and the distance. So, multiply 22 by 500 and get 11,000 km/s.

29.5.2   Chapter Glossary

Atomic Clock A type of clock that uses an atomic resonance frequency standard as its timekeeping element. They are the most accurate clocks in existence.

Blue Shift A decrease in the wavelength due to the Doppler effect.

Equivalence Principle The principle that identifies the acceleration of an RF with the gravity felt in that RF.

Gravitational Lens is formed when the light from a very distant, bright source is bent around a massive object between the source and the observer.

Homogeneity The property of being the same at all points.

Isotropy The property of being the same in all directions.

Precession of Perihelion A slight rotation of the major (or minor) axis of the elliptical orbit of a planet each time it completes its revolution around the Sun.

Pulsar A collapsed star, whose small size endows it with an extraordinarily fast rotation around its axis.

Quasar (*Quasi-stellar radio source*) A powerfully energetic and distant galaxy with an active galactic nucleus. Quasars were first identified as being high redshift sources of electromagnetic energy, including radio waves and visible light, that were point-like, similar to stars, rather than extended sources similar to galaxies.

Red Shift An increase in the wavelength due to the Doppler effect.

29.5.3   Review Questions

29.1. How can you create an artificial gravity in a reference frame? How can you eliminate an existing gravity in an RF?

29.2. State the equivalence principle and analyze the role of the size of the RF in your statement.

29.3. What are some of the immediate consequences of the equivalence principle? How noticeable are these consequences?

29.4. Explain how the equivalence principle leads to the bending of light in gravity. How does it imply the gravitational Doppler shift? How does Doppler shift lead to the gravitational time dilation? Can ordinary clocks measure time dilations induced by Earth's gravity? Can atomic clocks?

29.5. What property of spacetime is gravity related to according to Einstein's GTR? Why is this called the geometrization of gravity?

29.6. If Mercury were the only planet of the Sun, what would its orbit look like according to Newtonian gravity? According to GTR? How large is the precession of the perihelion of Mercury?

29.7. The equivalence principle and GTR both predict the bending of light. Do they predict the same amount? Which one agrees with observation? What is gravitational lensing?

29.8. What is a pulsar? What is a binary pulsar? What is an indirect evidence for gravitational waves? Has gravitational wave been observed directly? Is there hope for their direct observation? What experiment is designed to observe them directly?

29.9. Name a device in whose construction both STR and GTR play a dominant role.

29.10. What happens when you apply GTR to the entire universe? What assumptions do you make about the universe? What do galaxies turn into based on these assumptions?

29.11. What is a homogeneous universe? What is an isotropic universe? Is our universe homogeneous? Is it isotropic? On what scale?

29.12. Who predicted the expansion of the universe? Who observed this expansion? In what form did he observe the expansion?

29.13. What is the implication of expansion of the universe when going backward in time? What is GTR's prediction about the size of the universe in a sufficiently remote past?

29.14. What is big bang? How big was the universe at the moment of the big bang? When did big bang occur?

29.5.4   Conceptual Exercises

29.1. You are confined to a spaceship which has no windows to let you see the outside world. You are wearing magnetic shoes which hold you to a magnetic floor. Now let go of a few things and note that they all fall. Should you conclude that there is a gravitational field, or that the spaceship is accelerating?

29.2. You are confined to a spaceship which has no windows to let you see the outside world. You are wearing magnetic shoes which hold you to a magnetic floor. Now let go of a few things and note that they all float in midair. Should you conclude that you are in outer space away from all stars and planets, or that you are in free fall near a gravitating body?

29.3. You are confined to a spaceship which has no windows to let you see the outside world. You are wearing magnetic shoes which hold you to a magnetic "floor" of a room. Now let go of a few things and note that they all fall to the "ceiling." What should you conclude? (a) There is a gravitational field directed from your foot to your head. (b) The spaceship is accelerating, and your feet point in the direction of acceleration. (c) Either of the above.

29.4. You are confined to a spaceship which has no windows to let you see the outside world. You are wearing magnetic shoes which hold you to a magnetic floor. Now throw a few things and note that they all move on a straight line with constant speed. Should you conclude that you are in outer space away from all stars and planets, or that you are in free fall near a gravitating body?

29.5. In discussing the bending of apples and light in a gravitational field (see Figure 29.3), one property of light made its bending immeasurably small. What property was it?

29.6. The deflection angle (29.1) of light when it grazes the Sun is 0.874 arc second. What would it be for a star which is 100 times heavier and 100 times smaller than the Sun?

29.7. You are in outer space in a region where there is a very strong gravitational field. You take out your flashlight and send a beam of orange light along the gravitational field to your space buddy, who is stationary relative to you. Is the signal that he receives red or yellow?

29.8. You are in outer space in a region where there is a very strong gravitational field. You take out your flashlight and send a beam of green light opposite to the direction of the gravitational field to your space buddy, who is stationary relative to you. Is the signal that he receives yellow or blue?

29.9. Which atomic clock runs the fastest, the one on Earth, the one on Mount Everest, or the one at the surface of the Moon? Which one runs the slowest?

29.10. Why can gravity be geometrized but electricity cannot? Do all particles, regardless of their physical properties, behave the same under gravity? Under electricity?

29.11. Why do we talk about the precession of the perihelion of Mercury and not of other planets? Do they have any precession? How do the precessions compare with that of Mercury? What is the reason for the difference?

29.12. Does a satellite with an elliptical orbit, in principle, have a precession of its perihelion as it goes around Earth? If so, is it measurable?

29.13. Is the prediction of the equivalence principle for the bending of light correct? What property of space is missing in the equivalence principle?

29.14. Explain how the shape of the mass M in Figure 29.10 determines the nature of the images of A received on Earth. For example, what kind of image would we receive on Earth if M were very long in the up-down direction of the figure? Or in the into-out-of-the-page direction of the figure?

29.15. Why is the reduction in the period T of a binary pulsar an indirect indication of gravitational waves (see Figure 29.11)? Hint: Use Kepler's third law to deduce that a reduction in T indicates a reduction in orbit radius r. Now use Equation (E.57) to convince yourself that a reduction in energy E entails a reduction in r.[10]

29.16. When you see that all distant galaxies move away from you, and that the farther they are the faster they move, do you conclude that you are at the center of the universe? What is a reasonable conclusion?

29.17. Put some helium in a cylindrical tank, keep it at a uniform temperature and pressure, and ignore the walls of the tank. Is this gas isotropic and homogeneous? Now bring one end of the cylinder in contact with a hot reservoir and the other with a cold reservoir. Is the gas still isotropic and homogeneous?

29.18. In the balloon analogy of the expansion of the universe, why is the fact that the size of the stickers does not increase important? What does the size of a sticker correspond to in the actual universe?

29.19. If a galaxy is 5 Mly far, does it move away from us? What if it is 600 Mly far? Are there any galaxies that are 20 billion light years away from us? If so, how fast are they moving? Is such a speed possible?

29.5.5   Numerical Exercises

29.1. A space station in the shape of a wheel has a radius of 100 m. It is desired to create an artificial gravity in the station mimicking Earth's by rotating the wheel about its shaft. What should the period of the revolution be? Hint: Find the speed at the rim which gives rise to a centripetal acceleration equal to Earth's gravity. From the speed and the circumference of the wheel find the period.

---

[10]Although Equation (E.57) was derived for an atom, a similar formula is obtained for gravitational energy because gravitational and electrostatic forces have identical dependence on r.

29.2. A beam of light is sent across a distance of 100 km on a white dwarf whose gravitational acceleration is 3.2×10^6 m/s^2.
(a) How long does it take light to go from one end of the field to the other?
(b) How far does this light beam falls during this time?
(c) What is the deflection angle of light, i.e., the angle between its original direction of motion and its direction at the other end of the field?

29.3. A beam of starlight grazes a neutron star of mass 9×10^{30} kg and radius 25 km.
(a) What is the deflection angle for this light beam in radians?
(b) What is the deflection angle in degrees?

29.4. A satellite 5000 km above the surface of a white dwarf carries an atomic clock synchronized with a similar clock on the white dwarf. The mass of the white dwarf is 8×10^{29} kg and its radius in 5000 km. The satellite is kept in orbit for 1 month. Consider only the GTR effects.
(a) Which clock will be running faster?
(b) How many seconds will the faster clock be ahead of the slower one after 1 month?

29.5. A distant galaxy is 750 Mly away. Assume that the Hubble parameter is 23 km/s per Mly.
(a) What is the speed of the galaxy in m/s?
(b) How long (in seconds) did it take this galaxy to move this distance?
(c) How many *years* ago was this galaxy on top of the Milky Way?

29.6. What is the maximum distance a galaxy can be away from us? Hint: What is the maximum speed?

Chapter 30

Epilogue: No "Marketplace"

Quantum physics and relativity are the two pillars of modern physics. Often in this context, the word "modern" is misused as an antonym of "classical." The abstract and mathematical nature of both quantum and relativity theories makes them easy target for abuse in the hands of the purveyors of mysticism. We shall see some detailed examples of the abuse of quantum theory in Chapter 35. It will be shown there how New Age authors misrepresent quantum physics to their readers and use some twisted interpretation of the theory to advocate a parallel between quantum physics and Eastern mysticism.

Classical physics is more immune from such misinterpretation because (a) it deals with everyday tangible objects, and (b) the mathematics used in classical physics is less abstract. The easier accessibility of classical physics plus the negative publicity of technology (atomic bombs, chemical warfare, nuclear proliferation, etc.) attributed wrongly to science, has been used by mystics to put a hideous mask on classical physics, making it evil, destructive, and "mechanistic." On the other hand, the incomprehensible modern physics can be put forth to the nonexperts as a gentle, "organic," and "holistic" science. Chapter 24 showed, and Chapter 35 will expand on, how fallacious this separation is and how quantum physics is vitally connected to classical physics. This chapter presents—among other things—the connection and similarity between relativity and classical physics.

> There could be no quantum theory without classical physics.

30.1 Continuity

The title of Einstein's first paper on relativity was "On the Electrodynamics of Moving Bodies." It sounds like a dull paper on electricity and magnetism. Dull, it wasn't! But it *was* a paper on Maxwell's electromagnetism, and how a moving observer perceives them. And that is how relativity began.

Relativity is an offspring of Maxwell's electromagnetism, and as such its umbilical cord is attached to the womb of classical physics. Any claim that relativity and classical physics have nothing in common is equivalent to the claim that an infant has no connection with its mother. The essential ingredient in any scientific discovery is the mastery of the existing body of knowledge in a given field. If Einstein had not mastered *classical* mechanics, *classical* electromagnetism, *classical* wave theory, and the mathematics of *classical* physics, he would not have been able to discover relativity.

> There could be no relativity without classical physics.

Electromagnetism is not the only "classical" subject that helped the development of relativity. Many of the concepts defined in classical physics carried over to relativity. For example, relativistic momentum and energy are generalizations of the corresponding classical concepts (see Section 28.3). Furthermore, every time a new formula was discovered in relativity, classical physics was consulted to unravel the true meaning of that formula.

Section 28.3.1 and Math Note E.28.3 are prime examples of how classical physics was crucial in identifying the "time" component of relativistic momentum as the *relativistic energy*.

In his general theory of relativity (GTR), Einstein relied heavily on the Newtonian gravity. In fact, as we mentioned in Chapter 29, Einstein experimented with just the equivalence principle—a direct result of Newtonian gravity—for many years before he stumbled on the full GTR. More importantly, when he did discover GTR, Newtonian gravity was vital in writing the Einstein equation (the equation in Figure 29.7).

30.2   Specificity

Special theory of relativity (STR) embodies a universal statement about space and time. Every physical theory dealing with the extremely fast *has to* use this universal relativistic language of space and time. Einstein himself used it in his formulation of gravity, and discovered the general theory of relativity. Dirac used it in his description of a fast electron, and discovered the existence of antimatter. Weinberg, Salam, and Glashow used it in their description of the decay of fundamental particles, and, as a result, unified the weak and electromagnetic interactions and anticipated the existence and properties of the massive weak gauge bosons. Gross, Wilczek, and Politzer used it in their description of the strong interaction among quarks, and discovered quantum chromodynamics and quark confinement.

Yet all this universality came about because Einstein looked for an answer to a *specific* question: "How do Maxwell's equations look in moving bodies?" Our treatment of the STR started by looking at a *specific* clock and how motion affected its ticking (see Section 26.1). From that, all the *universal* conclusions about length contraction and time dilation, Lorentz transformations, relativistic energy, even $E = mc^2$ followed.

The general theory of relativity was born out of Einstein's *specific* question: "How do I incorporate STR into gravity theory?" Nevertheless, this child of specificity predicted the universal expansion. Who would have imagined that the answer to such a specific question could make such a universal statement as "the universe is expanding," whose logical (backward) extension resulted in no less a conclusion than the creation of the universe? Of course, this is not the first time we have encountered this. Every universal statement in physics (and in all branches of science) stems from a specific question or a specific analysis. No holistic approach has ever resulted in a discovery, because by its very nature it has to encompass so many variables and parameters that no scientist can handle.

30.3   Use/Creation of Mathematics

Three years after the discovery of STR, the Russian-German mathematician, Hermann Minkowski delivered an address at the 80th Assembly of German Natural Scientists and Physicians held at Cologne on September 21, 1908. Minkowski began his opening remarks with the following statement:

> The views of space and time which I wish to lay before you have sprung from the soil of experimental physics, and therein lies their strength. They are radical. Henceforth space by itself, and time by itself, are doomed to fade away into mere shadows, and only a kind of union of the two will preserve an independent reality. [Ein 52, p. 75]

Minkowski then went on to consider time as a fourth dimension and show that in such a four-dimensional continuum, a distance could be defined which was intrinsic to the space, and independent of the coordinates used. In short, he, for the first time, derived what we have been calling the *spacetime distance*.

Until 1908, the notion of distance between two points had been strictly Euclidean, which, in terms of the coordinates of the two points, was given by the square root of the sum of the squares of the difference between corresponding coordinates [Equation (E.82) gives a two-dimensional version of this distance]. The special theory of relativity suggested—to Hermann Minkowski for the first time—the invention of a new kind of geometry with a new definition of distance, whose two-dimensional version is given in Box 27.4.1.

The interplay between mathematics and physics is one of the most fruitful enterprises of humankind. The study of motion and gravity led to the invention of calculus and differential equations. The application of the laws of motion to the fluids gave rise to vector calculus and tensor analysis, whose generalization to the electric and magnetic fields gave us the EM waves. STR gave birth to a new notion of distance and a new kind of geometry. On the other hand, a systematic study of the differential equations led to the abstract Lie group theory, which has been indispensable in understanding subatomic particles and forces.[1] Moreover, abstract differential geometry developed by nineteenth century mathematicians, in combination with Minkowski's notion of distance, found a natural dwelling in the general theory of relativity.

30.4   Induction versus Deduction

In his later years, Einstein focused more and more on formalism of relativity theory and less and less on its experimental success. And this, for good reason: experiments played no (direct) part in either special, or general theory of relativity. The former was built on the constancy of the speed of light, which in turn, was a consequence of the electromagnetic *theory*. The latter was based on the equivalence *principle*, which was derived from Newtonian gravitational *theory*.

Einstein was, therefore, led to believe that physics was purely deductive; that the future theories can be built entirely from existing *theories*. Although this may be true in the majority of cases (STR and GTR being prime examples), we cannot ignore the vital role of experiments in other cases. Maxwell's electromagnetism is a good example. Maxwell's *deductive* reasoning created the correct set of equations for electromagnetism (see Sections 14.2 and 14.3). Nevertheless, the raw material with which he started was the result of over three decades of observation and experimentation, and *induction* thereof. Quantum theory is another equally good example. Schrödinger's wave mechanics and Heisenberg's matrix mechanics were created as the two engaged in unraveling the secrets of the hydrogen atom, with some help from Bohr's theory. But we should never forget that the structure of the H-atom—upon which all three theories were based—was purely an experimental outcome: if Rutherford had not experimented with the gold atoms, we would not have known about the *nucleus* of the H-atom, a crucial ingredient in the quantum recipe.

The failure of Einstein in unifying electromagnetic and gravitational forces—an endeavor that preoccupied him many of his later years—is a testimony to the importance of experiments. In the 1930s and 40s, when Einstein was thinking about such a unification, the other two fundamental (the strong and weak nuclear) forces were neither understood nor taken seriously. Yet their knowledge is crucial in any attempt at unifying forces, as we shall see in the next part of the book (see Section 34.2).

A (harmful) side effect of ignoring the inductive process of physics,[2] is the Platonic overemphasis on the mind and the subsequent derailing of the scientific process. Plato thought that all knowledge came from human mind and the human mind alone. Although he was a staunch supporter of mathematics and geometry, he chided those who *experimented* with geometrical shapes. It is, therefore, not surprising that Plato's philosophy includes the supernatural elements of Form, Idea, and the Universal Soul.

---

[1]See Sections 33.1, 33.2.1, and 34.1 for details.

[2]The process that creates phenomenological theories directly out of experimental and observational data, much like the electromagnetic equations that were discovered as a result of numerous experiments (see also Chapter 15).

Harmfulness of the idea
of the "marketplace of
ideas."

Another (even more harmful) side effect is the lift received by certain social scholars, who advocate the "marketplace of ideas," in which any and all competing theories are sold. In this marketplace, socialization of scientific content (Food for Thought 27.5.3), quantization of Ayurveda (Section 35.3.3), parallelism of modern physics and Eastern mysticism (Section 35.3), and the "theory" of the dependence of mathematical content on social conditions (Section 19.4) compete with the theory of relativity, quantum theory, and the electroweak unification. In this marketplace, observation has no kiosk, and objectivism gets "deconstructed" by a Derridean literary critique.

Part VII

Twentieth Century Physics: Microcosmology

On the Experimental Front

Chapter 29 revealed that at a finite time in the past (about 13.7 billion years ago) the universe was infinitesimally small. A compression of the material of the universe into an infinitesimal volume will surely result in extreme conditions of heat, pressure, density, and energy. At a mere 20 million °K in the core of a star, already novel physical processes such as nuclear fusion take place. The early universe with temperatures running into hundreds of billions of degrees (and more!) must have been the arena of some of the most spectacular physical phenomena. To understand such breath-taking happenings, we need to go back to the early twentieth century and follow two developments seemingly unrelated to the early universe, one in the experimental investigations of the nucleus, the other, in the theoretical attempts at combining relativity and the quantum theory. This chapter is devoted to the experimental front.

31.1 The Nucleon

The discovery of the atomic nucleus by Rutherford and his collaborators was a milestone in physics. But it was only the tip of the iceberg. As the future generations of physicists were to discover, the nucleus housed many of the fundamental constituents of the universe.

The scattering of alpha particles from nuclei of atoms revealed not only the existence of the atomic nucleus, but also its structure. The very particles (α particles) that were being used to probe the structure of atoms were coming from the nuclei of radioactive material. So, it was natural to assume that they were the building blocks of all nuclei. Soon, however, it was found that the lightest atom, hydrogen, could not be made up of such particles. Rutherford had already shown in 1903 that the α particles were *doubly* charged. If the hydrogen nucleus were to be made up of α particles, it had to carry *at least* two units of positive charge. However, all attempts at observing a doubly ionized hydrogen atom failed. All ionized hydrogen atoms carried only one positive unit of charge. Furthermore, the mass of hydrogen was measured to be smaller than that of an α particle. Rutherford called the nucleus of the hydrogen atom **proton**.

*Rutherford proposes the existence of the **proton**.*

What do you know? 31.1. What is the significance of not observing a doubly ionized hydrogen atom, and why does this indicate that its nucleus consists of only one proton?

Other elements had already been placed in the periodic chart according to what had been vaguely called the **atomic number**. The connection between the atomic number and the number of positive charges in the nucleus was proved by a set of experiments that a

*Moseley makes precise the notion of the **atomic number**.*

student of Rutherford, H. G. J. Moseley (1887–1915), carried out on various elements using X-rays. In fact, it was through Moseley's work that the concept of atomic number was made precise and the location of missing elements in the periodic table accurately identified.

BIOGRAPHY

Ernest Rutherford (1871–1937) was born near Nelson, New Zealand, studied at Nelson College and won a scholarship to study at Canterbury College, University of New Zealand. In 1895 Rutherford traveled to England for postgraduate study at the Cavendish Laboratory, University of Cambridge. In 1898 Rutherford was appointed to the chair of physics at McGill University where he demonstrated that radioactivity was the spontaneous disintegration of atoms, the work which gained him the 1908 Nobel Prize in Chemistry. This is ironic given his famous remark "In science there is only physics; all the rest is stamp collecting." In 1907 he took the chair of physics at the University of Manchester, where he discovered the nuclear structure within atoms and was the world's first successful "alchemist": he converted nitrogen into oxygen. In 1917 he returned to the Cavendish as Director. Under him, Nobel Prizes were awarded to Chadwick for discovering the neutron, Cockcroft and Walton for splitting the atom using a particle accelerator, and Appleton for demonstrating the existence of the ionosphere.

It was now established that the nucleus of an atom of an element contains a number of protons equal to the atomic number assigned to that element. This number is denoted by Z. A neutral atom, therefore, contains an equal number of electrons. A hydrogen atom has thus a proton as its nucleus with one electron at a "large" distance (recall—from Section 21.2—that the nucleus has a very small size compared to the entire atom) from it. Many subsequent nuclear experiments confirmed the existence of protons. During one such experiment, in which Rutherford discovered what he called "An Anomalous Effect in Nitrogen," he showed that when α particles collide with the nitrogen nucleus, a singly ionized hydrogen atom is produced, establishing further proof of the reality of proton.

> The existence of the **neutron** is suspected.

The structure of the nucleus was beginning to take shape except for one piece of the puzzle. In a 1920 paper entitled "Nuclear Constitution of Atoms," Rutherford proposed a second constituent of the nucleus, based on the results of the decisive experiments of 1911–1913. As soon as the atomic nucleus was discovered, and its charge determined, it was found that the *mass* of the nucleus (in units of hydrogen mass) was *almost twice its atomic number*. This naturally led to the assumption that other particles must be present in nuclei beside protons; and that these particles must be electrically neutral. This neutrality prompted Rutherford to suggest that these particles must consist of a proton and a "nuclear electron."

While charged particles could be detected by the visible path they left behind in a detector (a cloud chamber), the **neutron**—as the neutral particle was to be called later—was not easily detectable. Many experiments designed specifically to find this elusive particle failed, but in 1930 it was reported that the collision of α particles with light elements such as lithium could create a radiation so penetrating that it was considered to be very high energy gamma rays. However, no gamma rays of this energy had been known to exist. Then in 1932 the French physicists Irène Curie (Marie Curie's daughter) and her husband Frédéric Joliot published a paper, in which they showed that this radiation, when passed through materials such as paraffin wax, which contained hydrogen, could eject protons of very high speed.

> The word **nucleon** refers to either a proton or a neutron.

Immediately after the publication of this paper, James Chadwick (1891–1974), one of Rutherford's assistants, published a paper entitled "Possible Existence of a Neutron," in which he proved that the experiment of Joliot and Curie could not involve gamma radiation, but could best be explained by a neutral particle that was almost as heavy as the proton. In the same year, Werner Heisenberg proposed that all nuclei are made up of protons and neutrons, and that neutrons are not to be regarded as made up of an electron and a proton, but rather as a "fundamental" particle just like a proton. The word **nucleon** now refers to either a proton or a neutron.

It is common to designate a nucleus by placing the atomic number Z as a subscript and the total number of nucleons (protons plus neutrons) as a superscript to the left of the chemical symbol of that element. The total number of nucleons is denoted by A and is usually called the **atomic mass number**, but a more appropriate name would be the **nucleon number**.[1] If N denotes the number of neutrons, then $A = Z + N$, and if X represents the chemical symbol, then ^A_ZX is the nucleus of X atom with Z protons and $A - Z$ neutrons. Thus, $^{14}_6$C is carbon with 8 neutrons, and $^{17}_8$O is oxygen with 9 neutrons.

Symbolic representation of nuclei.

> **What do you know? 31.2.** How many protons and neutrons are there in each of the following nuclei? $^{70}_{32}$Ge, $^{148}_{62}$Sm, $^{113}_{47}$Ag, $^{58}_{26}$Fe.

31.1.1   The Nuclear Force

The breakdown of a nucleus in terms of its constituents, proton and neutron, introduces a dilemma: If the atomic nucleus consists of *positive* protons and *neutral* neutrons, why does it not burst asunder due to the repulsive electrical forces? Are there stronger *attractive* forces at work? The answer is "yes." In fact, as early as 1921, Chadwick had shown that the angular distribution of the scattered protons in the collision of hydrogen atoms with α particles indicated that forces other than electricity must be at work when α particles interact with protons *at very short distances*. This **short-ranged** nuclear force operates only when nuclear particles get extremely close to one another (so close that the separation between them is of the order of a nuclear diameter); unlike the electric or gravitational forces that operate at *all* separations. Another evidence for the short range of the nuclear force is the very existence of molecules: if the nuclei of the atoms in a molecule felt the strong attractive nuclear forces, they would collapse into a single nucleus, and the molecule into a single atom.

Evidence for a very short-ranged and very strong nuclear force.

The strong nuclear force is responsible for holding (or **binding**) the nucleons of a given nucleus together, just as the gravitational force is responsible for binding satellites to the Earth, moons to their planets, and planets to their stars (see Example 9.2.5); and just as the electromagnetic force is responsible for binding the electrons (see Math Note E.21.1) and the nucleus of an atom together. However, while in the case of electromagnetic force the binding energy is of the order of a few eV (see Section 21.3 and Math Notes E.21.1 and E.21.2), in the nuclear case the binding energy is typically millions of eV (a few MeV).

> **What do you know? 31.3.** What would happen to the world if the strong nuclear force were long-ranged?

The stability of a nucleus is determined by how tightly its nucleons are bound, i.e., by the binding energy each nucleon experiences. Because of the enormity of the (total) binding energy, and because of the equivalence of mass and energy, the total *binding energy contributes to the mass of the nucleus*. When we speak of the binding energy, we ignore the fact that *it is negative*. However, in calculating its contribution to the mass of the nucleus, we must incorporate this negative sign. Let e_b stand for the binding energy *per nucleon* of a nucleus that has Z protons and N neutrons. Then the mass M of such a nucleus is given by

Mass of nucleus in terms of proton mass, neutron mass, and binding energy per nucleon.

$$M = Zm_p + Nm_n - (Z+N)m_b \quad \text{or} \quad Mc^2 = Zm_pc^2 + Nm_nc^2 - BE, \qquad (31.1)$$

---

[1] The writing of the atomic number is really redundant once the chemical symbol is used. However, because it is hard to memorize the atomic number of all entries of the periodic table, it is helpful to have it next to the chemical symbol.

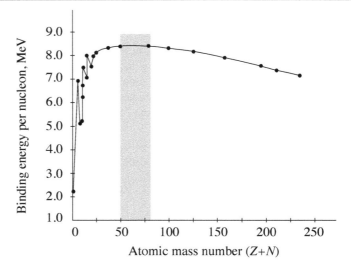

Figure 31.1: The stability curve for nuclei. The shaded region includes the most stable nuclei.

where m_p and m_n are the masses of the proton and neutron, respectively, BE is the binding energy of the nucleus, and $m_b \equiv e_b/c^2$ is the mass-equivalent of e_b.

Unified atomic mass unit.

A convenient unit of mass for nuclear interactions is the **unified atomic mass unit** denoted by u and equal to $1.6605388 \times 10^{-27}$ kg, which we designate as m_u sometimes. The equivalent energy, e_u, of 1 u is obtained by multiplying it by c^2. This gives a value of 931.494 MeV for e_u when the accurate value of c is used. The mass of the proton in atomic mass unit and its equivalent energy in MeV are 1.0072764669 u and 938.272 MeV, respectively. The corresponding values for neutron are 1.008664916 u and 939.565 MeV.

> **What do you know? 31.4.** The energy equivalent of the mass of a deuteron (the nucleus consisting of a proton and a neutron) is 1875.6 MeV. What is the total binding energy of the deuteron?

Mass excess defined.

In some tables of nuclear data, instead of the binding energy of nuclei, a quantity known as **mass excess**—denoted by Δ—is recorded. Mass excess is, by definition, the difference between the actual mass of the nucleus and the number of nucleons times the unified atomic mass unit. **Example D.31.1 on page 64** of *Appendix.pdf* gives the formula that connects Δ to e_b. For the majority of nuclei, Δ is a negative number: the smaller the mass excess, the larger the binding energy, and the more stable the nucleus. **Example D.31.2 on page 64** of *Appendix.pdf* calculates the mass excess for iron.

Iron is one of the most stable nuclei.

Iron, with a binding energy of 8.55 MeV per nucleon is one of the most stable nuclei. Much heavier and much lighter nuclei than iron have a value of e_b smaller than 8.55 MeV; therefore, they are less stable than iron. Figure 31.1 shows the plot of the binding energy per nucleon as a function of the atomic mass number (the total number of nucleons in the nucleus). It should be clear from the figure that the most stable nuclei have atomic mass numbers approximately between 50 and 80.

31.2   Nuclear Reactions

The strong attractive nuclear force between the nucleons holds the nucleus of the atom together. But for how long? Certain nuclei are stable, i.e., they remain unchanged forever.

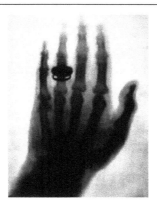

Figure 31.2: One of the first X-ray pictures. This is the left hand of Röntgen's wife showing her wedding ring as well.

Other nuclei tend to "transmute" over time entirely on their own. Certain other nuclei can transmute once they are triggered by some external stimulus. In fact, one can provide energy to any two nuclei and let them collide. The result of the collision could be other nuclei plus other particles as well as some released energy which shows up as the kinetic energy of the final products. A typical nuclear reaction is

$$B + C \rightarrow G + H + \text{other particles} + \text{energy},$$

where B and C are the *incoming* nuclei and G and H the *outgoing* nuclei. This looks like a chemical reaction. However, there is a difference! In a chemical reaction involving molecules, the atoms at the beginning are equal to the atoms at the end.

Box 31.2.1. *In a nuclear reaction, the number of incoming nucleons is equal to the number of outgoing nucleons.*

Typically, one of the incoming nuclei is light: a neutron, a proton, or a helium nucleus. For example,

$$_2^4\text{He} + _7^{14}\text{N} \rightarrow _8^{17}\text{O} + \text{p}.$$

The three most important types of nuclear reactions are radioactivity, fission, and fusion, which we shall discuss in some detail.

31.2.1   Nuclear Radioactivity

Röntgen's discovery of X-rays in 1895 created a great deal of excitement in the scientific community. The historic photo of his wife's hand showing the bones of her fingers, including the wedding ring (Figure 31.2), which appeared in Röntgen's second article on X-rays, opened the door to modern medicine (and Nobel Prize). No wonder many physicists sought novel ways of producing these mysterious rays!

In 1896, Henri Becquerel (1852–1908), the French physicist, found a chemical compound— potassium uranyl sulfate—that (he thought) spontaneously emitted X-rays without any stimulus. However, the Polish-French physicist, Marie Curie, and her husband Pierre Curie realized that Becquerel's radiation was not X-ray, but a completely new physical process. In search of the source of the radiation, she worked painstakingly to chemically isolate the element mixed in minute amounts in uranium ores. She discovered *radium*.

Rutherford identifies α and β rays. Becquerel and Villard identify γ rays.

Once radium was isolated, physicists could study the nature of its radiation. In 1899 Rutherford identified the α and the β components of the radiation, and a year later Becquerel and Villard established the existence of the third component, the γ rays. α rays consist of two protons and two neutrons (the helium nucleus); β rays are beams of electrons or positrons, the antiparticle of the electron; and γ rays are very energetic EM waves. These rays are the by-products of any natural nuclear radioactivity. As discovered by Rutherford in 1902, because they carry electric charge, the emission of α and β rays changes the atomic number Z of the nucleus, and therefore, entails a **transmutation** of one element into another. γ rays, on the other hand, do not change the identity of a nucleus; they are the nuclear—and thus much more energetic—version of the atomic spectral lines (Section 21.3).

What is an isotope?

The chemical properties of an element, which *determine* the element itself, are specified by the number of electrons—and, therefore, by the atomic number Z—of that element. In fact, Z pins down the element itself: $Z = 1$ means hydrogen, $Z = 2$ means helium, $Z = 6$ means carbon, etc. An element with its specific Z can have different number N of neutrons in its nucleus. For example, carbon's nucleus is usually found to have 6 protons and 6 neutrons. However, a small percentage of carbon comes with 7 or 8 neutrons in its nucleus. Each variety of an element with a given number of neutrons in its nucleus is called an **isotope**. Thus, the isotopes of an element all have the same Z but different N.

BIOGRAPHY

Marie Sklodowska Curie (1867–1934) was a physicist and chemist of Polish upbringing and, subsequently, French citizenship.[a] She was a pioneer in the field of radioactivity, the first person honored with two Nobel Prizes, and the first female professor at the University of Paris. She was born in Warsaw and lived there until she was 24. In 1891 she followed her elder sister Bronislawa to study in Paris, where she obtained her higher degrees and conducted her subsequent scientific work. She founded the Curie Institutes in Paris and Warsaw. Her husband Pierre Curie was a Nobel co-laureate of hers, and her daughter Irène Joliot-Curie and son-in-law Frédéric Joliot-Curie also received Nobel prizes. Her achievements include the creation of a theory of radioactivity (a term coined by her), techniques for isolating radioactive isotopes, and the discovery of two new elements, polonium and radium. It was also under her personal direction that the world's first studies were conducted into the treatment of neoplasms ("cancers"), using radioactive isotopes. While an actively loyal French citizen, she never lost her sense of Polish identity. She named the first new chemical element that she discovered (1898) "polonium" and in 1932 she founded a Radium Institute (now the Maria Sklodowska-Curie Institute of Oncology) in her home town Warsaw, headed by her physician-sister Bronislawa.

[a]This biography was taken from Wikipedia.org.

Radioisotopes.

The radioactivity of an element depends on which isotope is involved. We speak of *radioactive isotopes* or **radioisotopes** of an element. For example, while $^{12}_{6}\text{C}$ is stable (i.e., it is not radioactive), $^{14}_{6}\text{C}$ is a β emitter. Beta radiation is caused by the transformation of one of the neutrons of the nucleus into a proton. For $^{14}_{6}\text{C}$, one of the 8 neutrons becomes a proton, turning carbon into a nitrogen. This nuclear reaction, or **nuclear decay** as it is often called, is represented by

$$^{14}_{6}\text{C} \rightarrow {}^{14}_{7}\text{N} + \text{e}^{-}.$$

A nuclear decay can take place only within the context of relativity theory: the mass of the decaying nucleus transforms partially into the kinetic energy of the end products.

Half-life of a nuclear decay.

Being a quantum interaction, nuclear decay can only be predicted with a probability. The probabilistic character of a nuclear decay translates into **half-life**. This is the time in which half of a radioactive sample decays. Starting with an initial number of nuclei N_0, after one half-life $N_0/2$ nuclei remain, after two half-lives $N_0/4$ survive, after three half-lives $N_0/8$ survive, etc. In general, if the half-life is denoted by t_{half}, then after a time t, the number $N(t)$ of nuclei surviving is given by

$$N(t) = \frac{N_0}{2^{t/t_{\text{half}}}}.$$ (31.2)

Note that t need not be an integer multiple of t_{half}; it can be any (positive) number.

What do you know? 31.5. An isotope of Ni has a half life of 20 s. Starting with 1 kg of this isotope, how much of it is left after 40 s? After 1 minute? After 2 minutes?

Example D.31.3 on **page 65** of *Appendix.pdf* shows that if we know the initial population of a radioactive isotope and its half-life, we can predict the surviving population at any later time. It also shows how crucial the magnitude of the half-life is in this survival. For times long compared to the half-life, the surviving population becomes so small as to be hard to measure.

Carbon Dating

The calculation of Example D.31.3 could be turned around for a much more useful application: If we can measure the ratio $N(t)/N_0$ of a radioactive substance now, and if we know its half-life, then (31.2) will give us t, the time in the past when $N(t)$ was equal to N_0. For this kind of *dating* to work, our radioisotope must have an appropriate half-life. For instance, if we are trying to find the age of a sample that is centuries old, $^{19}_{8}\text{O}$ with a half-life of 27 seconds is utterly useless, because even after a few *hours*, the fraction becomes immeasurably small.

Carbon 14, with a half-life of 5730 years, is most suitable for archaeological dating. This isotope of carbon is produced in the upper atmosphere by the nuclear reactions triggered by cosmic rays. The ratio of the abundance of $^{14}_{6}\text{C}$ to $^{12}_{6}\text{C}$ is about 1.3×10^{-12}. The live organisms, interacting with the atmosphere, inherit the same ratio of the two isotopes of carbon in the carbon dioxide of their bodies. So, as long as the organism is alive this ratio is retained. Once an organism dies, the ratio decreases due to the ^{14}C decay. From the $^{14}\text{C}/^{12}\text{C}$ ratio of an archaeological sample one can determine the age of the find.

What do you know? 31.6. The $^{14}\text{C}/^{12}\text{C}$ ratio of an archeological sample is 6.5×10^{-13}. How old is the sample?

Example 31.2.2. The Dead Sea Scrolls were found by a shepherd in 1947. Because of their religious importance and the implications they might have on Christianity and Judaism, they were kept out of reach of the scientists and archaeologists for almost half a century by a self-appointed priesthood of less than a dozen scholars who claimed exclusive rights to the document. It was only in 1991 that scroll fragments began surfacing in the antiquities market, and the content of these Scrolls revealed.

A Roman historian had written (in AD 79) about a deeply religious tribe named Essenes, which lived in the same area that the Scrolls were found. Although there were no women in the tribe, there were sufficient number of recruits from outside to keep the tribe alive for many generations. Living in caves, the Essenes wrote and collected hundreds of documents ranging from personal matters such as divorce papers to the collective belief of the tribe.

Dead Sea Scrolls and their age.

In many of these Scrolls, there were mentions of a Messiah that shared an ancestry with King David and was put to death. The oldest of these "Messianic" scrolls was almost 2250 years old, indicating that it must have been written around 250 BC. **Example D.31.4** on **page 65** of *Appendix.pdf* shows how this might have been calculated. ∎

Age of Earth

Radioactivity shows that
Earth is 4 to 5 billion
years old.

All of the nuclear material (with the exception of hydrogen and helium) was made in the interior of stars by nuclear reactions. These reactions produce not only stable nuclei, but radioactive ones as well. When this nuclear soup condenses to form new stars and their planets, the radioactive elements are carried along. If a radioactive element has a half-life that is much shorter than the age of the planet, the element will disappear, and only the stable by-product of its decay will remain.

Natural radioactivity.

A few of the radioactive elements present at the time of Earth formation have half-lives that are comparable to the age of the Earth, and so are still present and still radioactive. These elements constitute some of the **natural radioactivity** around us. From the present ratio of the number N_d of the stable by-product (the daughter nucleus) to the number N_p of the naturally radioactive nucleus (the parent nucleus) in a rock sample, it is possible to estimate the age of the Earth, as **Example D.31.5** on **page 65** of *Appendix.pdf* shows.

Example D.31.5 gives the simplest method of applying nuclear radioactivity to the determination of the age of Earth. In real situations, rocks don't have only one daughter nucleus, and not all of it may have been the result of the decay of the parent nucleus. Nevertheless, alternate (and more complicated) procedures involving more than one daughter nucleus have been devised to calculate the age of the Earth. All these procedures lead to an Earth that is between 4 and 5 billion years old.

31.2.2   Nuclear Fission

The electrical neutrality of the neutron makes it a good candidate for probing the atomic nucleus. Unfettered by the charges of the electrons surrounding the nucleus, neutrons move straight to the nucleus where they can interact directly with their nuclear kindred. In one such process a (very slow) neutron penetrates a heavy nucleus (the **parent nucleus**), makes that nucleus unstable, causing it to split into two lighter nuclei (the **daughter nuclei**) plus a few neutrons. This process is called a **nuclear fission**.

Transformation of mass
to energy in a nuclear
fission reaction.

The parent nucleus is typically $^{235}_{92}\text{U}$, and a typical fission looks like

$$^{1}_{0}\text{n} + {}^{235}_{92}\text{U} \rightarrow \text{X} + \text{Y} + \text{neutrons},$$

where X and Y are any of the 90 or so possible daughter nuclei. If you add the mass of the neutron to that of $^{235}_{92}\text{U}$ on the left you get something that is larger than the total mass of the daughter nuclei plus the neutrons produced on the right. This **mass defect** becomes the source[2] of the kinetic energy of the fission products on the right via $E = mc^2$. Since the total number of protons and neutrons does not change in the reaction, the mass defect arises because of the difference in the binding energies of the daughter nuclei and the uranium, as **Math Note E.31.1** on **page 138** of *Appendix.pdf* shows rigorously.

What do you know? 31.7. (a) What are the missing numbers in each of the following fission reactions? (b) How many neutrons are produced?

$$^{1}_{0}\text{n} + {}^{235}_{92}\text{U} \rightarrow {}^{140}_{55}\text{X} + {}^{92}_{??}\text{Y} + \text{neutrons}$$

$$^{1}_{0}\text{n} + {}^{235}_{92}\text{U} \rightarrow {}^{142}_{??}\text{X} + {}^{92}_{36}\text{Y} + \text{neutrons}$$

$$^{1}_{0}\text{n} + {}^{239}_{94}\text{Pu} \rightarrow {}^{141}_{??}\text{X} + {}^{94}_{37}\text{Y} + \text{neutrons}$$

$$^{1}_{0}\text{n} + {}^{239}_{94}\text{Pu} \rightarrow {}^{142}_{58}\text{X} + {}^{96}_{??}\text{Y} + \text{neutrons}$$

---

[2]The general statement is that the initial energy must equal the final energy. But since the neutron and uranium are moving very slowly, their total energy is just the sum of their masses (times c^2).

Example 31.2.3. A possible fission reaction is

$$\,_0^1\mathrm{n} + \,_{92}^{235}\mathrm{U} \ \rightarrow \ \,_{56}^{141}\mathrm{Ba} + \,_{36}^{92}\mathrm{Kr} + 3(\text{neutrons}).$$

Q: How much energy is released in this reaction?

A: The energy released is the energy equivalent of the mass defect. The mass defect is simply the difference in the total binding energies of $\,_{56}^{141}\mathrm{Ba}$ plus $\,_{36}^{92}\mathrm{Kr}$ and the binding energy of $\,_{92}^{235}\mathrm{U}$. To find this difference, we refer to Figure 31.1, and note that at the atomic mass number 141 the binding energy per nucleon is about 8 MeV and at 92 it is about 8.3 MeV. For uranium-235 the binding energy per nucleon is approximately 7.3. Therefore, the energy-equivalent of the mass defect is

$$141 \times 8 + 92 \times 8.3 - 235 \times 7.3 = 176.1 \text{ MeV}.$$

This is over a hundred million times larger than the energy released in the chemical reaction of a typical combustible molecule! ∎

When a uranium nucleus fissions, a few (say 3) neutrons are produced which carry some of the energy released. These neutrons can fission 3 new uranium nuclei causing the release of 9 neutron, which in turn can fission 9 uranium nuclei releasing 27 neutrons, etc. If not controlled, this **chain reaction** could spread very quickly, releasing a tremendous amount of energy in a very short time. Such an uninhibited chain reaction is a nuclear bomb. On the other hand, if the release of energy is controlled, it is possible to turn it into other forms of energy such as electricity, as is done in a nuclear power plant.

Nuclear chain reaction.

 # Food for Thought

FFT 31.2.4. The investigation of nuclear fission, like all other scientific investigations, was spurred by the natural curiosity of a few physicists, who were trying to understand the newly discovered nucleons and how they made up the nucleus of an atom. The precursor of fission was artificial radioactivity discovered by Irène Curie and Frédéric Joliot, who bombarded certain light nuclei with α particles and noted that many of the by-products of nuclear radioactivity were produced *even after the α particle source was removed.*

The new world of the nucleus attracted the attention of many young physicists around the globe. Many puzzling questions in this new world demanded an answer, and the curiosity of these physicists became the playing ground of the questions. It was this curiosity that prompted Enrico Fermi (1901–1954) to redo Curie-Joliot experiments using uranium instead of the light elements and neutron instead of α particles. The plan of his group in Italy was to make artificial elements heavier than uranium.

In 1938 Otto Hahn (1879–1968), a German chemist, and Fritz Strassmann (1902–1980), a German physicist, showed that one of the elements that Fermi thought was transuranium, was actually an isotope of barium ($Z = 56$). Then Lise Meitner (1878–1968), another German physicist, and her nephew Otto Frisch, proposed the theory that the neutron caused the uranium to split into two lighter elements releasing a gigantic amount of energy in the process, and they named this reaction *fission*.

It was an unfortunate twist of history that this source of enormous power caught the attention of politicians and military men at a time when a global war was looming. Two weeks after the outbreak of World War II in 1939, some British government officials, threatened by Adolf Hitler's speeches, in which he warned of some "secret weapons,"[3] asked some British scientists to look into the possibility of using nuclear energy in the production of weapons. This possibility seemed so remote[4] to the British (and the United States) government that the immigrant physicists were given the task of working on the project while being kept out of the more "promising" radar.

It was not until 1942 that the United States government took the development of the *atomic bomb* seriously enough to set up the *Manhattan Project*. Subsequently, the British terminated their program and sent many of their scientists to the US to help the Americans produce the bomb. And Fermi, who had moved to the US from Italy, led a group at the University of Chicago to investigate

---

[3] It turned out that Hitler's "secret weapons" was his technologically advanced Air Force.

[4] For example, the initial budget that the US government allocated to this research was a mere \$6000.

the feasibility of a nuclear chain reaction, the first step towards the construction of a nuclear bomb or a nuclear power plant. This group succeeded in achieving the first controlled chain reaction on December 2, 1942.

This turn of events has been used by some sociopolitical zealots to spearhead such a public hysteria that any mention of the word "nuclear" is equated with destruction. For example, the procedure of obtaining three-dimensional images of human tissues in medicine, which has saved thousands of lives, uses "nuclear" magnetic resonance (NMR). However, the word "nuclear" had to be removed from the procedure to ease public concern, and it was renamed *magnetic resonance imaging* or MRI.

The nuclear bomb did not affect just nuclear physics; the very science of physics has been under attack by many social critics of science. But is it really fair—not only to physics and physicists, but to the entire humanity—to blame nuclear power for what the political and military establishments have used it for? If so, then we have to blame another tremendous source of power unleashed many many years ago. This source has also had a double role for mankind: it has been used as a source of energy and warmth, and as a source of destruction in wars and acts of vengeance. It is the source that has provided light at night, digestible food, and heating for cold winter days. It has also destroyed armies of men, houses, villages, and cities for centuries. This powerful source of destruction and construction was discovered over 500,000 years ago. It is called *fire*!

31.2.3   Nuclear Fusion

The peak of stability in Figure 31.1 occurs at the atomic mass numbers between 50 and 80. The nuclei on the far right and the far left of this region are less stable. The lower stability of heavy nuclei was the raison d'être for fission. The lower stability of light nuclei also has an important consequence. If two light nuclei could fuse together to form a nucleus closer to the stability region, then the sum of the original masses will be more than the sum of the final masses due to the higher (negative) final binding energies. This *missing mass* can turn into the (enormous) kinetic energy of the by-products of the nuclear **fusion reaction**.

> **What do you know? 31.8.** What would happen if the plot in Figure 31.1 were a flat horizontal line?

The proton-proton cycle.   An example of nuclear fusion is the **proton-proton cycle** whereby hydrogen nuclei (protons) fuse to form helium nuclei. This takes place in three steps in the interior of stars where the temperature is 15–20 million °K. In the first step, a **deuteron**—heavy hydrogen nucleus—is formed:

$$p + p \rightarrow {}^2_1\mathrm{D} + \mathrm{e}^+ + \nu,$$

where e^+ stands for *positron*, the antiparticle of the electron, and ν stands for *neutrino*, of both of which we shall say more later. In the second step, a proton fuses with a deuteron to form an isotope of helium, releasing energy in the form of gamma ray:

$$p + {}^2_1\mathrm{D} \rightarrow {}^3_2\mathrm{He} + \gamma.$$

There are actually two possibilities for the third (and final) step of the cycle. ${}^3_2\mathrm{He}$ can fuse with a proton:

$$p + {}^3_2\mathrm{He} \rightarrow {}^4_2\mathrm{He} + \mathrm{e}^+ + \nu,$$

or two ${}^3_2\mathrm{He}$ nuclei can fuse:

$${}^3_2\mathrm{He} + {}^3_2\mathrm{He} \rightarrow {}^4_2\mathrm{He} + p + p.$$

Example 31.2.5. Let us calculate the energy released in the proton-proton cycle. The net effect of the first possibility is the fusion of four protons into ${}^4_2\mathrm{He}$ and the release of two positrons, two neutrinos, and a massless gamma particle[5] (see **Example D.31.6** on **page 66** of *Appendix.pdf*

---

[5]Gamma particles are very energetic photons; and photons, as discussed in Section 28.3.2, are massless.

for a "proof" of this and the result of the second possible reaction). Thus, neglecting the neutrino masses,[6] the mass defect is $4m_p - m_{He} - 2m_e$. In atomic mass units, this gives

$$4 \times 1.0072764 - 4.002602 - 2 \times 0.0005486 = 0.0254 \text{ u},$$

which is equivalent to 23.7 MeV.

The second possibility involves two ^3_2He nuclei. Thus, we have to multiply both sides of the first two steps by 2. The net effect of the first two steps would then be six protons fusing to give two ^3_2He nuclei, two positrons, two neutrinos, and two gamma particles. Then in the last step the two ^3_2He nuclei combine to give one ^4_2He and two protons. The overall net reaction is thus fusing four protons to give one ^4_2He, two positrons, two neutrinos, and two gamma particles. The difference between this and the first possibility is the production of one extra γ, which carries no mass. Therefore, the mass defect is the same as before.   ■

The fusion of light nuclei requires bringing them close enough for the attractive nuclear forces to take over. For this to happen, the light nuclei should have at least as much kinetic energy as the electric repulsive potential energy between them. For example, each proton of the first step of the proton-proton cycle should have at least 5 keV of kinetic energy[7] to overcome the repulsive potential energy of the two protons. This may seem a very small number, but when translated into the temperature (as discussed in Section 17.1 and Box 17.2.5) required for a **thermonuclear fusion reaction**, it becomes tens of millions of degrees Kelvin, as the following example shows.

Thermonuclear fusion reaction.

Example 31.2.6. Estimate the temperature of a hydrogen gas that can undergo fusion by assuming that the average KE of the hydrogen atoms is 5 keV. Equation (17.1) gives the average KE, $\langle KE \rangle$, as $\frac{3}{2}k_B T$ where k_B is the Boltzmann constant and T the temperature in Kelvin. The average KE is 5 keV or 5000 eV or 8×10^{-16} J. So, from $\langle KE \rangle = \frac{3}{2}k_B T$, we get

$$8 \times 10^{-16} = \tfrac{3}{2} \times (1.38 \times 10^{-23})T \quad \text{or} \quad T = \frac{8 \times 10^{-16}}{2.07 \times 10^{-23}} = 3.86 \times 10^7 \text{ °K},$$

or about 39 million degrees. However, because of the probabilistic nature of the relation between temperature and KE, even at temperatures of 15–20 million degrees, there are sufficient number of "fast" protons for fusion to take place.   ■

Food for Thought

FFT 31.2.7. March 23, 1989 is a particularly significant day in the modern history of science. On that day, two electrochemists, Stanley Pons from the University of Utah, and Martin Fleischmann from Southampton University, reported a major breakthrough in nuclear fusion research. There were a few indicators that pointed to suspicion about this report. (See [Hui 92] for details.)

- All genuine major breakthroughs in science have been reported through professional journals, whose editors and referees put the submitted articles under severe scrutiny. The prestige of such a professional journal rests on the quality and the reputation of its editors and referees. The discovery of Pons and Fleischmann was announced in a *news conference* intended for commercial networks! All the experts in the field of fusion learned about this "cold fusion" on the radio, or television or in the commercial newspapers and magazines.

- The process of fusion, as noted above, takes place only under extremely high temperatures and pressures such as those inside stars. Pons and Fleischmann claimed that their "fusion" took place at room temperature.

---

[6]Neutrinos were believed to be massless until the so-called "neutrino oscillations" were discovered, giving a very small mass to these particles.

[7]So, strictly speaking, we should add this to the mass energy of the two protons when we calculate the mass defect. But these are so small compared to the proton mass, or even the electron mass, that we can neglect them.

- Although fusion requires extreme temperature and pressure, it does not mean that terrestrial fusion has not taken place. In fact, many laboratories in the US, Japan, Russia, and Western Europe had been (and still are) working on the same problem for decades. However, in all conventional cases, the process takes place for a fraction of a second, and requires huge equipments and millions of dollars. Pons and Fleischmann claimed that their "fusion" required table-top chemistry equipment much like those used in high school laboratories.

- As our discussion has shown, the fusion processes are accompanied by the production of lethal levels of radioactive material [such as the (positive) beta particles and the highly lethal gamma rays in the proton-proton cycle discussed before]. The healthy presence of Pons and Fleischmann in the news conference was proof for the absence of such processes.

- Both Pons and Fleischmann were *electrochemists* and all their publications prior to their March 23 announcement were in that field. Fusion requires expertise in the field of nuclear physics or nuclear chemistry, of which neither knew enough to warrant this potentially Nobel-Prize caliber discovery.

- Finally, unlike all scientific discoveries, the announcement came after a long nervous period of secrecy. Many experts, upon hearing about the discovery in the news media, contacted their colleagues in the physics department at the University of Utah to learn more about the Pons-Fleischmann experiment. To their surprise, they all realized that no one in that department knew of the experiment! A legitimate discoverer involves his/her professional colleagues and coworkers in the discovery. And this is driven not only by the involuntary desire to share the excitement with all people nearby, but also by a need to check the accuracy of the process and the result of discovery.

In the nearly two decades since its announcement, cold fusion has been tried over and over again by many physicists and chemists around the world with no single legitimate success.

Plasma confinement.

The enormous energy released in the nuclear fusion (as illustrated in Example 31.2.5) has prompted many physicists and engineers to look into the possibility of commercial production of fusion energy, and there are many facilities around the world devoted to this challenge. The plan is to use deuterium (heavy hydrogen found abundantly in ocean water) as a fusion fuel. There are many practical obstacles to this plan, not the least of which is **plasma confinement**.[8] This refers to confining the deuterons at high enough temperature for a long enough time so that the fusion (and the release of energy) can be sustained. Despite great progress in the "fusion industry," no commercial production of fusion energy has yet been possible.

31.3   The Emergence of Particle Physics

The experiments of Rutherford and his colleagues in 1911–1913 opened the gates to heaven ... literally. These experiments, and especially the ingenious techniques invented as a result of them, eventually led the physicists to a scientific solution to no less an enigma than the creation of the universe itself. The steps that led to that solution were taken quite haphazardly and independently. The connection between the subnuclear phenomena and cosmology came much later.

31.3.1   Neutrino

Of the three radioactive rays, beta rays turned out to be the most interesting. It can be easily explained as the decay of a neutron into a proton and an electron; and for a long time people thought that only these three particles were involved. Thus, when a carbon-14 decays, one of the eight neutrons turns into a proton, resulting in a nitrogen nucleus

---

[8]Because of the extremely high temperature, the heavy hydrogen atom is separated into a positive deuteron and a negative electron; the resulting "soup" is a plasma—a collection of positive and negative charges.

with seven protons and seven neutrons, plus an electron which flies off with great speed. However, the experimental analysis of the energy of the electron revealed that it did not have a unique value. Why should the value of the electron energy be unique?

For the sake of simplicity, concentrate on the decay of a single neutron:

$$\mathrm{n} \rightarrow \mathrm{p}^+ + \mathrm{e}^-.$$

Look at this process in the neutron's reference frame (RF), where n is necessarily stationary. Since the momentum of n is zero, the proton and electron must have equal and opposite momenta. The equality of momenta and the conservation of energy determines the energy of the electron uniquely as shown in **Math Note E.31.2** on **page 138** of *Appendix.pdf*. This value happens to be 1.289 MeV. If instead of a neutron, a *nucleus* β-decays, the flying electron will still have a single energy value, albeit one different from 1.289 MeV.

What do you know? 31.9. Why can a stationary neutron decay into a proton and other things, but a stationary proton cannot decay into a neutron and other things?

Observations, however, indicated that the electron energy in the β decay is not unique. And this created a turmoil in the physics community. Some suggested that energy was not conserved in nuclear reactions. Others took a less dramatic approach and suggested that different nuclei *of the same sample* decay differently. The correct explanation came from Wolfgang Pauli, who suggested that the end product of a neutron β decay is not just an electron and a proton, but also a third—extremely elusive—particle, which was later called **neutrino** (little neutron) by Fermi. Neutrino carries some of the available energy in a β decay, leaving the remaining energy to be shared by the electron and the proton; and the available energy can be divided among the three particles *in anyway possible*, as long as the momenta of the proton, the electron, and the neutrino add up to the initial momentum of the neutron, and their energies to the initial energy of the neutron.

31.3.2 Muons and Pions

In the first 35 years after the discovery of the first elementary particle (the electron), the number of what seemed to be "fundamental" particles was a mere five: electron, photon, proton, neutron, and neutrino.[9] In the second 35 years a deluge of fundamental particles caught the physicists off guard, and brought about such a state of confusion that only very radical ideas could resolve it.

🧺 Food for Thought 🌿

FFT 31.3.1. The introduction of the elusive neutrino in the explanation of the β decay brings out a very sensitive issue concerning the validity of scientific claims. The *elusion* of neutrino, a psychoanalyst may argue, allows the introduction of other *elusive* phenomena. If physics is allowed to cope with a particle that cannot be detected, then psychoanalysis should be allowed to introduce the undetectable and nonmaterial id, ego, and superego. And if neutrino can explain the variation in the energy of the electrons in β decay, why can id, ego, and superego not explain the disorders in personality? There are two major differences between a neutrino and an id (or ego or superego).

First, although neutrino was not physically detected, physicists could *measure* many of its material properties. They could determine that neutrino was either massless or had a very small mass; that it had an intrinsic spin (see Section 23.3) equal to that of the electron (and proton and neutron); that in every β decay one could assign unambiguously a momentum and energy to it

---

[9] A particle was also discovered in 1932, which was identical to an electron except for its *positive* electric charge. This particle was later identified as *positron*, the antiparticle of the electron. We are not counting the positron as an independent particle.

consistent with the energy and momentum conservation. No such determination has been made for id. In fact, no one (including the psychoanalyst) knows what id is. Is it a "subatomic particle" of psychoanalysis? Is it an "atom" of psychoanalysis? A collection of such "atoms?" Is id a wave? If so, what is its frequency? Its amplitude? Its wavelength? The fact is that the building blocks of psychoanalysis are as much void of any physical properties as phantoms and spirits.

Second, as elusive as neutrinos were at the beginning, they came eventually under control. About four decades after they were "hypothesized," real neutrino beams were produced and scattered off electrons, as real as billiard balls scattering off each other! The reality of neutrinos was finally established. Almost a hundred years have passed since Freud introduced the id-ego-superego trinity, and still there is not even a smidgen of evidence for their physical reality. The elusive id is still as elusive as a phantom!

Yukawa predicts the existence of the meson.

The sixth "fundamental" particle was predicted by the Japanese physicist Hideki Yukawa (1907–1981) in 1935. Following up on a suggestion by Werner Heisenberg concerning the quantization of the strong nuclear force, Yukawa argued that just as the electromagnetic field has a particle (the photon), the "nuclear field" must also have a particle, called the **meson**. However, photon, moving at the ultimate speed, can carry the electromagnetic force to large distances. That is why the electromagnetic force is *long-ranged*. Since nuclear forces are short-ranged, the meson must be massive.[10] Knowing the range of the nuclear forces (essentially the size of a nucleus), Yukawa could predict the mass of the meson to lie between the mass of an electron and that of a proton.

Two years after Yukawa's prediction of the meson, several teams of physicists reported the discovery in cosmic rays of a new particle with a mass between the electronic and protonic masses. At first, everyone thought that this was the Yukawa particle. But a number of experimental and theoretical analyses proved otherwise. If this newly discovered particle was the Yukawa meson, then—being the "carrier" of the strong nuclear force—it had to interact strongly with (the nucleus of) matter. That wasn't the case! Furthermore, theoretical calculations concluded a very short half-life for the Yukawa meson. The new particle was over a hundred times longer-living! By 1947 it was becoming evident that the particle found by the cosmic ray experimenters was not the Yukawa particle, but a "daughter" of its *decay*. Thus, the seventh particle was discovered before the predicted sixth! This cosmic-ray particle was called the μ-*meson* at that time, but the name was later changed to **muon**.

Muons came in two varieties, positive and negative, and each had a half life of about a microsecond, in which time it decayed into an electron plus at least two other particles. How did physicists know that the electron was accompanied by at least two other particles? Our discussion of the beta decay in Section 31.3.1 implies that if a particle decays into only *two* lighter particles, the end products have well-defined energies. Lack of a well-defined energy for the electrons implied the existence of neutrinos. In a muon decay precisely the same phenomenon occurs: the electrons do not have a well-defined energy, indicating that at least three particles are produced in a muon decay. The elusive nature of the other particles hinted at the production of two neutrinos beside the electron.

Pions are discovered.

What of the Yukawa meson? Just as the muon was being proposed as the daughter of the decay of the Yukawa particle, several groups of experimenters discovered a particle that interacted strongly with matter as expected of the Yukawa meson. The *pi-meson* (as this particle was called at the beginning) came in three kinds: negative (discovered first), positive, and neutral. Now we call these particles **pions**.

Cosmic rays seemed to be the ultimate source of new particles, and novel techniques of detection[11] paved the way for discovering and identifying many exotic particles. In the

---

[10]For a more rigorous discussion of the relation between the range of a force and the mass of the particle responsible for it, see Example 34.1.4.

[11]Such as the Wilson cloud chamber, in which charged particles create bubbles along their path, allowing physicists to identify them by studying the shape of the track they leave behind.

same year that the pion was discovered, two British physicists found evidence for a neutral particle, whose mass lay between a pion mass and a proton mass. This particle came to be known as **K-meson** or **kaon**. And by 1953, in which an international conference of cosmic-ray physicists took place, several other particles were identified which had masses larger than the nucleon. These massive particles were collectively called the **hyperons**.

Kaons and hyperons are discovered.

31.3.3   Cyclotron and the Particle Explosion

Cosmic rays provided the physicists with a noticeable collection of new particles on which to ponder. But the **elementary particle physics** would not have been born were it not for the invention of the **particle accelerators**. Accelerators brought the study of the newly discovered particles from mountaintops to laboratories, where they could be produced in high quantities under controlled environments.

Elementary particle physics is born!

The idea of using electromagnetic fields to accelerate charged particles to high energies goes back to the heydays of nuclear physics. In 1932 John Cockroft and Ernest Walton, in their attempt at bombarding the lithium nucleus, built the first **linear accelerator** to speed up protons to several hundred keV. The grandchild of Cockroft-Walton accelerator is the one at Stanford Linear Accelerator Center (SLAC), which is two miles long and can accelerate particles to several *giga* electron volts (GeV)!

The higher the energy desired of a charged particle, the greater the length of the linear accelerator. This puts some restrictions on the facility, not the least of which is its mere linear size. Could we somehow "bend" the linear tube to reduce its effective size? Could we build a *circular* accelerator?

When an electrically charged particle enters a magnetic field, it bends and moves on a circular path, whose plane is perpendicular to the direction of the magnetic field and whose radius is proportional to the momentum of the particle.[12] Furthermore, *the period of the revolution of the particle on the circle is independent of the size of the circle*. Because the force induced by the magnetic field is always perpendicular to the motion of the particle, the magnetic force cannot accelerate a charged particle. To accelerate particles, we need an *electric* field.

What do you know? 31.10. Why can we not use just a magnetic field to accelerate particles?

Ernest Lawrence, an American physicist, came up with the idea of *periodically* accelerating the particles with an electric field while sending them through a magnetic field to "circulate." Figure 31.3 shows two semicircles called the "dees," placed between the two poles of a magnet (not shown) so that a magnetic field is established in the dees perpendicular to the plane of the paper. The dees are separated by a gap, in which an *alternating* electric field can accelerate charged particles. If the period of oscillation of the electric field is equal to the (constant) period of the circulation of the particle in the dees, the field will always *accelerate* the particle in the gap. The device is called a **cyclotron**.

Place a positive particle near the center of the cyclotron when the electric field points to the left. The electric field will accelerate the particle in the gap and sends it into the left dee with a small velocity. As the particle enters the magnetic field of the dee, it circulates and comes out of the left dee (with the same speed as it went in). The electric field, whose frequency is chosen so that now it points to the right, will accelerate the particle *further* and sends it into the right dee. Having a slightly larger speed, the particle moves on a slightly larger circle, but—because the period of revolution does not depend on the size of the circle—the time spent in the right dee is the same as that spent in the left dee. So when

---

[12]This property is used to determine the momentum of a particle in a cloud chamber.

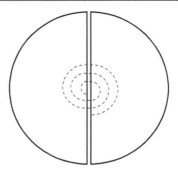

Figure 31.3: The "dees" of a cyclotron. The dotted line is the path of the accelerated particle.

the particle comes out of the right dee, the electric field has changed direction again and can accelerate the particle further. This process can spiral the particle to very high speeds.

The first cyclotron, built in 1931 by Lawrence's student M. Stanley Livingston, had a diameter of 2.5 inches and accelerated protons to 80 keV. It was soon followed by a 9-inch cyclotron capable of accelerating protons to one MeV (10^6 eV). By 1945, the largest cyclotron (built also by Lawrence's group) had a diameter of 184 inches and was capable of achieving energies up to 190 MeV. Because of such high energies, relativity played an important role in the design of the new generation of particle accelerators, including the 184-inch cyclotron. Theory of relativity predicts that the period of revolution of a particle in a magnetic field is not constant. One way to incorporate this relativistic effect was to synchronize the frequency of the electric field with the speed of the particle entering the gap. Another, more cost effective way, was to synchronize both the electric and the magnetic fields of the cyclotron. Such a machine is called **synchrotron**.

> Cyclotron evolves into synchrotron.

Synchrotrons mushroomed in the 1950s and 60s all over the world, especially in the US. The high energies achieved in a *controlled* environment paved the way for unraveling some of the most protected secrets of the universe. Through a merger of experiments, theories, and mathematics, a very successful model of fundamental particles emerged by the end of 1970s. To understand this model, we have to move on to the theoretical front.

31.4   End-of-Chapter Material

31.4.1   Answers to "What do you know?"

31.1. A neutral atom is said to be ionized when one of its electrons is pushed out of the atom. If one (negative) electron leaves the neutral atom, the atom ends up with a single positive charge. It is *singly* ionized. When two or more electrons are pushed out, the atom is doubly or multiply ionized. When you cannot see *any* doubly ionized hydrogen, it must mean that neutral hydrogen has no "second" electron to give out. It consists of one positive proton and one negative electron.

31.2. $^{70}_{32}$Ge has 32 protons and $70 - 32 = 38$ neutrons; $^{148}_{62}$Sm has 62 protons and $148 - 62 = 86$ neutrons; $^{113}_{47}$Ag has 47 protons and $113 - 47 = 66$ neutrons; $^{58}_{26}$Fe has 26 protons and $58 - 26 = 32$ neutrons.

31.3. All the nuclei in the universe would coalesce into a single gigantic nucleus (more likely into a multitude of gigantic nuclei) with the electrons running around them.

31.4. In Equation (31.1) with $Z = 1$ and $N = 1$, use 1875.613 MeV for M on the left, 938.272 MeV for m_p and 939.565 MeV for m_n on the right. Then the total binding energy—which is $(Z + N)m_b$—would be 2.25 MeV.

31.5. 40 s is two half lives; so $1/2^2 = 1/4$ or 250 grams of the sample will be left after

40 s. One minute is 3 half lives; so $1/2^3 = 1/8$ or 125 grams of the sample will remain after 1 minute. Two minutes is 6 half lives; so $1/2^6 = 1/64$ or 15.625 grams will be left after 2 minutes.

31.6. The $^{14}C/^{12}C$ ratio for live organisms is 1.3×10^{-12}. The given ratio is one half of this. So, one half life must have gone by since the sample was living. Therefore, the sample must be 5730 years old.

31.7. (a) 37; 56; 57; 36. (b) 4; 2; 5; 2.

31.8. Neither fission nor fusion would take place, because there would be no difference between binding energy per nucleons. All nuclei would be equally stable, and there would be no preference to form new nuclei.

31.9. Neutron is more massive than proton. The extra mass can turn into very light particles and the kinetic energy associated with their motion. Although a stationary proton *by itself* cannot produce a neutron, if it is *bombarded* with something like an electron, a neutron *can* be produced.

31.10. When a force such as the magnetic force is perpendicular to the direction of motion of a particle, it does no work on that particle (see Box 8.1.1 and Figure 8.1). If work is zero, the change in the KE is also zero by Equation (8.1), and if the KE does not change, speed will not change.

31.4.2   Chapter Glossary

Accelerator Machines designed to accelerate particles to very high energies for the purpose of colliding them with target nuclei and other particles to probe what is inside the targets.

Alpha Ray A by-product of nuclear radioactivity. It carries two units of positive charge, and was later identified as the nucleus of the helium atom consisting of two protons and two neutrons.

Atomic Number The number of protons in a nucleus. This determines what element the nucleus belongs to.

Beta Ray A by-product of nuclear radioactivity. It carries one unit of charge, and was later identified as electron if the charge is negative and positron (anti-electron) if the charge is positive.

Binding Energy The energy that holds a system together due to some attractive force. The value of a binding energy is given in positive numbers, but the energy itself is negative.

Chain Reaction A nuclear process in which the newly formed neutrons are absorbed by heavy nuclei to produce lighter nuclei *and some neutrons*, which are absorbed by more heavy nuclei to produce even more neutrons to be absorbed by even more heavy nuclei, etc.

Cyclotron One of the first particle accelerators in which an alternating electric field accelerated charged particles, while a magnetic field kept them in circular orbits.

Daughter Nucleus One of the lighter nuclei produced in a nuclear (usually fission) reaction.

Fission A nuclear interaction in which a heavy (parent) nucleus absorbs a slow neutron and disintegrates into lighter (daughter) nuclei plus some extra neutrons.

Fusion A nuclear reaction in which lighter nuclei fuse together to form heavier nuclei and in the process release energy.

Gamma Ray Another by-product of nuclear radioactivity. It is a very energetic photon, carrying an energy of the order of a few MeV or more.

Half-Life Describes radioactivity. It is the time in which half of the initial (large) number of radioactive nuclei disintegrate.

Isotope A variation of the nucleus of an element determined by its neutron number. While the nucleus of an element is *determined* by the number of protons it contains, the number of neutrons is not. Different isotopes of the nucleus of an element have different number of neutrons.

Mass Excess The difference between the mass of a nucleus and the number of nucleons times the unified atomic mass unit.

Meson The name given initially to the particle responsible for the short-ranged nuclear force.

MeV Million electron volt, is a convenient unit of energy for nuclear interactions. It is also the unit in terms of which the masses (times c^2) of some particles are given.

MRI Magnetic resonance imaging. See *nuclear magnetic resonance.*

Natural Radioactivity Radioactivity that occurs naturally (as opposed to artificial radioactivity).

Neutrino A "little" neutron. A very weakly interacting particle produced in some nuclear processes.

Neutron The neutral particle found in all nuclei (except hydrogen). All nuclei are made up of protons and neutrons.

NMR See *nuclear magnetic resonance.*

Nuclear Decay A process in which a nucleus disintegrates into other nuclei. Same as nuclear radioactivity.

Nuclear Magnetic Resonance The nuclear process used in medicine to image parts of the body and diagnose diseases. Replaced by "magnetic resonance imaging" due to the public fear of the word "nuclear."

Nucleon The constituent of all nuclei. Nucleon refers to either a proton or a neutron.

Parent Nucleus The heavy nucleus that undergoes a nuclear (usually fission) reaction and disintegrates into some lighter (daughter) nuclei.

Plasma Confinement A process referred to fusion reactors in which the deuterons are confined for a long enough period so that the fusion process can take place.

Positron The anti-particle of the electron.

Proton The nucleus of hydrogen atom. It carries one unit of positive charge, and together with neutron they constitute all nuclei. An element is determined by the number of protons in its nucleus.

Proton-Proton Cycle A fusion reaction taking place in the core of stars to convert hydrogen to helium and in the process create energy necessary for the sustenance of the star.

Radioactivity A process in which a heavier nucleus disintegrates into lighter nuclei, and in the process, emits one of three radioactive decay products: alpha, beta, or gamma radiation.

Radioisotope An isotope of an element which is radioactive.

Short-Ranged Force A force that operates only at short distances (i.e., distances smaller than the nuclear size).

Synchrotron The second-generation cyclotrons which incorporated the effects of relativity, which became important when the accelerated particles achieved a speed close to light speed.

Thermonuclear Fusion Nuclear fusion taking place in the core of stars due to the extremely high temperature.

Unified Atomic Mass Unit Denoted by u, is a convenient unit in which to measure nuclear masses. It is $1.6605388 \times 10^{-27}$ kg, very nearly equal to a nucleon mass.

31.4.3 Review Questions

31.1. Why is it important to understand nuclear physics for the study of the early universe?

31.2. What prompted physicists to abandon the idea that alpha particles were the building blocks of all nuclei? What role did the ionization of hydrogen atom play in this endeavor?

31.3. Who proposed the existence of the proton?

31.4. What is the atomic number, and how is it related to the number of electrons in a neutral atom?

31.5. How did the mass of the nucleus indicate the existence of another particle beside proton in the nucleus? What could the charge of this particle be? What is this particle called, and who discovered it?

31.6. What is a nucleon? How many nucleons are there in a nucleus consisting of 13 protons and 14 neutrons?

31.7. What are some pieces of evidence for the existence of *short-ranged* nuclear force? Is this force attractive or repulsive? Is there a nuclear force between protons and protons? Between protons and neutrons? Between neutrons and neutrons?

31.8. What is binding energy? How do you compare the binding energy that holds an atom together with the binding energy that holds a nucleus together? Does the binding energy of an atom contribute substantially to its mass? Does the binding energy of a nucleus contribute substantially to its mass?

31.9. What is the unified atomic mass unit? If the unified atomic mass unit is "mass," how can it be given in MeV—as is customary—which is the unit of energy?

31.10. Who discovered radium? How many components are there in the radiation of radium? What are these components? Who identified these components?

31.11. What is alpha ray? What is beta ray? What is gamma ray? Which of these are electrically charged? How are they related to various particles?

31.12. What is an isotope? What distinguishes among the isotopes of an element? Give an example of an element with different isotopes.

31.13. What is a radioisotope? What is the half-life of an isotope?

31.14. What radioisotope is used in archeological dating? Why? What radioisotope is used in determining the age of Earth? Why? Can you reverse the role of these two isotopes? Why, or why not?

31.15. What is fission? What role does the stability curve of nuclei (Figure 31.1) play in making fission possible? What is the purpose of fission? What is a "parent nucleus?" What is a "daughter nucleus?" Does a daughter nucleus have more or less binding energy per nucleon than a parent nucleus?

31.16. What is a chain reaction? How is a nuclear bomb different from a nuclear reactor in a power plant?

31.17. Who discovered fission? What was the motivation behind their discovery?

31.18. What does NMR stand for? What does MRI stand for? What is the relation between the two? Why was one changed in favor of the other?

31.19. What is fusion? What role does the stability curve of nuclei (Figure 31.1) play in making fusion possible?

31.20. What is the proton-proton cycle? Where does it take place? Does it require any energy input? Does it give off any energy? How does the output energy compare with the input energy?

31.21. What is cold fusion? Who were the two scientists who claimed they observed it? When was it claimed to have taken place? What are some of the suspicious indicators of the report of cold fusion? Could any other scientist repeat the experiment?

31.22. What is plasma confinement? Why is it necessary for the production of energy in a fusion reactor?

31.23. What is neutrino? How is it related to the energy considerations in the beta decay of a neutron or a nucleus? Who proposed the existence of neutrino?

31.24. What is the meson and how is it related to the short-ranged nuclear force? Is μ-meson or pi-meson responsible for the nuclear force?

31.25. What is a particle accelerator? What are the linear accelerator and the cyclotron? Why can you not accelerate particles using only a magnetic field? What is the difference between a cyclotron and a synchrotron?

31.4.4   Conceptual Exercises

31.1. The number of neutrons in a *light* nucleus is equal to the number of protons, but is larger for heavier nuclei. How does this increase contribute to the stability of heavier nuclei? Hint: Is it more likely for two or three protons to fly apart due to electric repulsion?

31.2. Why is neutron a better nuclear probe than a proton?

31.3. Does the binding energy of an atom contribute significantly to its mass?
(a) What is the mass of a hydrogen atom in MeV if you neglect the binding energy between the proton (with a mass of 938.27 MeV) and the electron (with a mass of 0.51 MeV)?
(b) How does this change if you include the hydrogen binding energy of approximately 14 eV?

31.4. Can you say that the larger the *total* binding energy of a nucleus the more stable the nucleus? Why, or why not?

31.5. Can you say that the larger the binding energy per nucleon of a nucleus the more stable the nucleus? Why, or why not?

31.6. A neutron is captured by the nucleus of a hydrogen atom. Is it still a hydrogen nucleus or the nucleus of a different element?

31.7. Is it true that the number of protons is conserved in a nuclear reaction due to the conservation of the electric charge?

31.8. $^{4}_{2}$He is the most abundant isotope of helium nucleus. But there are other isotopes such as $^{3}_{2}$He, $^{5}_{2}$He, $^{6}_{2}$He, and $^{7}_{2}$He. Why isn't there a $^{2}_{2}$He isotope?

31.9. A very slow neutron can be captured by a heavy nucleus. But a proton requires a much higher speed to be captured by the same nucleus. Why?

31.10. Identify x in $^{226}_{88}$Ra $\rightarrow$ $^{222}_{86}$Rn $+ x$. Can you name the element?

31.11. Identify x in $^{226}_{88}$Ra $\rightarrow$ $^{212}_{82}$Pb $+ x$. Can you name the element?

31.12. Why do alpha and beta radioactivity change the chemical character of the decaying nucleus but gamma radiation does not?

31.13. Both heavy nuclei, with nucleon number above 100, and light nuclei, with nucleon number below 30 can undergo beta and gamma radioactivity, but only heavy nuclei can undergo alpha radioactivity. Why?

31.14. What happens to the energy released in a nuclear radioactivity process? Can you explain why a sample of a radioisotope is warmer than its surrounding?

31.15. In a beta radioactivity, in which an electron is produced, a neutron is assumed to decay into a proton and an electron (plus a neutrino ν). Since neutron is heavier than proton one can say that the extra mass of neutron transforms into an electron (plus a neutrino) and some kinetic energy. How can you explain

$$^{40}_{19}\text{K} \rightarrow {}^{40}_{18}\text{Ar} + \text{e}^{+} + \nu,$$

in which a proton transforms into a neutron, a positron (plus a neutrino), and some kinetic energy? Is it possible for a *free* proton (one that is not bound to a nucleus) to undergo the same process?

31.16. Is it possible to use mass instead of number of nuclei in Equation (31.2)? How would you write the new formula?

31.17. Start with 4 radioactive nuclei with a half life of 1 minute. Can you say for sure that two nuclei will be left after 1 minute? Why, or why not?

31.18. Radioactivity is a statistical process, yet some tables of radioactive isotopes list half-lives to 3 or 4 significant figures. How can this be? Hint: See Section 16.3.

31.19. Some transuranium elements (elements with more protons than uranium) undergo "spontaneous fission." Why does spontaneous fission occur only in very heavy nuclei? Why can't light nuclei such as helium and lithium spontaneously fission?

31.20. What happens to the energy that is released in a fission reaction?

31.21. With just a little bit of energy you can split a heavy nucleus and get a tremendous amount of energy. That is what fission is. Is it possible to do the same thing with light nuclei? Explain.

31.22. It is possible to split a deuteron into its constituents, a proton and a neutron, by bombarding it with a gamma particle. Can you get more energy from such a splitting than what you put into it? Explain!

31.23. Suppose that in a fission process, a uranium-235 split into two nuclei of the same element and two neutrons. What element would the final products be? How much energy is released? Use Figure 31.1 to estimate the binding energies.

31.24. Suppose that in a fission process, a uranium-235 split into three nuclei with equal nucleon numbers and two neutrons. How much energy is released? Use Figure 31.1 to estimate the binding energies.

31.25. Suppose that in a fission process, a uranium-235 split into four equal nuclei, and no neutrons. What element would the final products be? How much energy is released? Use Figure 31.1 to estimate the binding energies.

31.26. Explain why uranium-235 cannot split into 20 (almost) equal nuclei. See Figure 31.1.

31.27. In the core of a star, light elements fuse to produce heavier elements in various stages until the only element left is iron. Explain why this process stops at iron.

31.28. Is it possible to fuse two elements, whose nucleon numbers are greater than 50 to produce a net amount of energy? Hint: See Figure 31.1.

31.4.5   Numerical Exercises

31.1. $^{55}_{25}$Mn has an atomic mass of 54.938049 u.
(a) What is the total binding energy of $^{55}_{25}$Mn in MeV?
(b) What is its binding energy per nucleon?

31.2. The mass excess for aluminum 27 is -17.2 MeV.
(a) What is the total binding energy of aluminum 27 in MeV?
(b) What is its binding energy per nucleon?
(c) What is its mass in MeV?
(d) What is its mass in unified atomic mass unit?

31.3. An isotope of iodine has a half-life of 60.1 days.
(a) What fraction of a sample of this isotope remains after 1 year?
(b) How long do we have to wait for a sample of this isotope to reduce to 10% of its original population?

31.4. A hypothetical fission reaction is given by

$$n + ^{255}_{94}X \rightarrow ^{140}_{50}Y + ^{109}_{?}Z + ?n$$

The binding energy per nucleon of X, Y, and Z are 7.4 MeV, 8 MeV, and 8.2 MeV, respectively.
(a) What is the number of protons in Z?
(b) How many neutrons are produced?
(c) What is the energy released in the reaction?

31.5. In the carbon cycle 3 helium nuclei turn into one carbon nucleus. The masses of carbon and helium are 12 u and 4.0015 u, respectively. What is the energy released in the reaction?

Chapter 32

On the Theoretical Front

With the discovery of the quantum theory (describing the very small) and the special theory of relativity (describing the very fast), it was natural to seek a theory that described very fast subatomic particles. Such a theory had to combine STR and the quantum theory, resulting in an equation that would replace the Schrödinger equation. Early attempts yielded results, which, although deemed accurate later on, were considered nonsensical at the time. For example, one formulation appeared to yield negative probability.[1] The negative sign of the probability can be traced to Equation (28.7), which contains the square of the energy, making it possible for the energy to be negative (keeping its square positive). This negative energy led to what was interpreted as negative probability.

32.1 Mathematical Prediction of Antimatter

To circumvent the problem of negative probability, Paul Dirac (1902–1984) invented an equation in 1928, now bearing his name. Most great theoretical discoveries in physics are based on some simple principle. Maxwell's equations were based on the conservation of the electric charge; the special theory of relativity was based on the constancy of the speed of light with respect to all observers; the general theory of relativity was based on the equivalence principle; and the Schrödinger equation was based on the fact that electrons are waves. In the case of the **Dirac equation**, the principle was invariance under the Lorentz transformations. What came out of this basic principle—after some elegant mathematical manipulation—was truly revolutionary. Dirac proved that if STR is to hold and negative probability is to be absent, then the Ψ function of the Schrödinger equation cannot be just a single function, but a collection of four functions. Two of these were associated to the motion of the electron. Why two? It turned out that the Dirac equation included the *spin* of the electron *automatically*.[2] Thus, the mysterious "spin" was explained by a *relativistic quantum theory*.

> Dirac finds a relativistic quantum equation for the electron.

What about the other two components? Employing some mathematical techniques, which the quantum theory itself helped develop, it was shown that the other two components described a particle with *positive* electric charge. It was therefore natural to ascribe them to the proton. However, soon (using the same mathematical techniques) it was discovered that the mass of the assigned particle must be identical to the mass of the electron. Proton with a mass almost 2000 times the electron mass could not be the right candidate. And

---

[1]The negative sign was later attributed to the electric charge rather than probability, rendering the formulation plausible.

[2]Recall from Section 23.3.2 that a spin-$\frac{1}{2}$ particle such as an electron is described by two components.

when Carl Anderson discovered a positively charged particle having the mass of the electron in 1932, it was not hard to associate it with the Dirac equation.

BIOGRAPHY

Paul Adrien Maurice Dirac (1902–1984) was born in Bristol, England, of a Swiss, French-speaking father and an English mother. Dirac went to Merchant Venturer's School, the public school where his father taught French, and while there displayed great mathematical abilities. After graduating from Bristol University at the age of 19, and not being able to find a job as an electrical engineer, he ended up at Cambridge as a research student in 1923. His appointment as university lecturer came in 1929. He assumed the Lucasian professorship following Joseph Larmor in 1932 and retired from it in 1969. Two years later he accepted a position at Florida State University where he lived out his remaining years. In response to a question posed to him in Moscow in 1955 about his philosophy of physics, Dirac wrote on the blackboard "Physical Laws should have mathematical beauty." This writing is still preserved today.

The first antiparticle is discovered in 1932.

This positive electron, or **positron**, as it has come to be called, has some interesting properties. For instance, Dirac theory predicts that, by merely putting an electron and a positron next to each other, it is possible for them to completely annihilate one another into pure (electromagnetic) energy! This can happen only because the two particles have *exactly* the same mass. Because of this process of annihilation, positron has also been designated as the **antimatter** of the electron.

Food for Thought

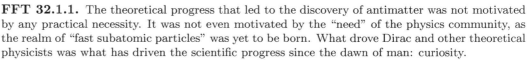

FFT 32.1.1. The theoretical progress that led to the discovery of antimatter was not motivated by any practical necessity. It was not even motivated by the "need" of the physics community, as the realm of "fast subatomic particles" was yet to be born. What drove Dirac and other theoretical physicists was what has driven the scientific progress since the dawn of man: curiosity.

Special theory of relativity (STR) was born in 1905, and as soon as it was born it bedazzled everybody by its simplicity, beauty, and power. Quantum theory was born in 1925/26, and immediately it was recognized as one of the most comprehensive theories ever. Dirac was familiar with both theories. In fact, he contributed substantially to the quantum theory by introducing a method of combining the Heisenberg's matrix mechanics and Schrödinger's wave mechanics. And because STR had shown itself as a more complete theory than Newtonian mechanics, a relativistic description of the subatomic world seemed to be the correct course of action. As a matter of fact, Schrödinger himself started with a formulation of *relativistic* quantum mechanics, and only after he failed, he resorted to "quantize" the Newtonian mechanics.

Practical application: the last thing on a great physicist's mind!

Dirac's attempt at inventing a relativistic quantum mechanics was not unique, although the outcome of his attempt was. Scientists in general, and physicists in particular, have a childlike curiosity. Just as a child likes to play with letter cubes to make words, physicists like to "play" with theories to make new ones. And just as a child is oblivious to the "practical application" or the potential "utility" of her game, so are the theoretical physicists who work on the most fundamental riddles of our universe.

And the successful work on the most fundamental riddles often accompanies surprises; delightful surprises. Dirac's work not only clarified the mysterious property called "spin" and predicted the existence of antimatter, but also created a new branch of elegant mathematics called *spin geometry*, affirming the prophesy of the great mathematician of the nineteenth century, Joseph Fourier, who said: "The profound study of nature is the most fruitful source of mathematical discoveries."

"The profound study of nature is the most fruitful source of mathematical discoveries."

Joseph Fourier

The history of physics and mathematics is full of examples of great discoveries made as a result of "the profound study of nature." Newton's discovery of the laws of motion, universal gravitation, and the development of calculus and differential equation, Maxwell's prediction of the electromagnetic waves, Einstein's discovery of the theory of relativity, and Dirac's prediction of antimatter and the creation of spin geometry are but a few instances of the fruits of curiosity.

Right after the identification of the positron as the antiparticle of the electron, physicists started hunting for other antiparticles. Proton, being the second particle discovered,[3] was the natural target. With the development of synchrotrons (see the end of Chapter 31), physicists sought a machine that could produce an *antiproton*. The plan was to accelerate a proton to a sufficiently high energy that upon its impact with a stationary proton, the relativistic conversion of the energy of the impinging proton could result in an antiproton. Simple relativistic calculation, reproduced in **Math Note E.32.1** on **page 139** of *Appendix.pdf*, revealed that the machine had to accelerate protons to about 6 billion eV. Such a machine, called the *Bevatron*[4] was developed at the University of California at Berkley, and put in operation in 1955. In the same year, the antiproton was discovered.

Anti-proton is discovered in 1955.

It is now known that every elementary particle in existence has an antiparticle with identical mass. Any property of the particle that has an opposite is shared by the antiparticle with the opposite sign. The only property of a particle familiar to the reader and having an opposite is the electric charge. Proton has a positive charge, antiproton negative; electron has a negative charge, positron positive. There are, however, other "charges," of which we shall say more as we learn more about other particles. Furthermore,

Box 32.1.2. *Any process (such as decays) involving particles and antiparticles has an identical partner in which all particles are turned into antiparticles and vice versa.*

For example, if neutron decays into a proton, an electron, and an antineutrino, then an antineutron decays into an antiproton, a positron, and a neutrino.

What do you know? 32.1. A charged particle clearly cannot be its own antiparticle. Could neutron be its own antiparticle?

32.2 Quantum Electrodynamics (QED)

At about the same time that Dirac was occupied with the relativistic quantum mechanics, other physicists were taking up another fundamental problem: EM waves are made up of photons, as Einstein demonstrated in 1905; they are also described by Maxwell's equations. So how do you get photons out of these equations? How do you *quantize the EM field*? How do you *quantize Maxwell's equations*? Pursuit of questions like these created one of the most successful mathematical theories of all time, **quantum electrodynamics** (or **QED**, for short). This was also the first **quantum field theory** and the first **gauge theory**, of which we shall say more later.

Quantum electrodynamics studies the interaction of electromagnetic field with electrically charged elementary particles. The only elementary particles, with which we are familiar at this point are electrons and positrons. Protons and other seemingly "elementary" particles are—as we shall see in the sequel—in reality composite.

How do electrons and positrons (and other truly elementary particles that we shall encounter later) interact electromagnetically? The old idea of action-at-a-distance, whereby one particle influences the other *instantaneously*, violates causality and relativity. The more realistic idea, that the *field* of one charge at the location of the other charge influences the latter *locally*, while correct, treats the electromagnetic field classically. The complete

Electromagnetic force is a result of the exchange of photons.

---

[3]Photon was "discovered" before proton, but photon is massless, and as we shall see later, it has no antiparticle (or, equivalently, it is its own antiparticle).

[4]A billion eV was called a BeV at the time of the operation of the Bevatron. However, to avoid the confusing use of "billion" for a million million (10^{12}) in Britain, and for a thousand million (10^9) in the US, the international community of physicists later decided to use "giga" to mean 10^9, and GeV to mean 10^9 eV.

and correct description quantizes the electromagnetic field, identifies Einstein's photon as the quantum of the EM field, and treats the electromagnetic interaction among charged elementary particles as the **exchange of photons**. According to this picture, the reason that an electron feels the electromagnetic force of another electron is that the two send photons to each other. So everything, including the force, becomes a (quantized) particle.

But something is missing from this picture that is related to the quintessential dichotomy of the quantum theory: particles are waves are particles! The field-photon duality meets this dichotomy criterion: After all, it is the oscillation of the EM *field* that creates EM waves, which in turn consist of photons. What about the electrons and positrons (and other elementary particles)? Shouldn't *they* also have a dual nature? Shouldn't electrons and positrons be fields as well?

> **Box 32.2.1.** *Quantum electrodynamics and other quantized field theories assume that all elementary particles and all fundamental interactions (forces) are represented by fields, whose quanta are the particles they represent. Matter particles exert forces on one another through the exchange of force particles.*

According to this picture, electrons and positrons are quanta of a field that obeys the Dirac equation, just as photons are quanta of the electromagnetic field obeying Maxwell's equations. In the jargon of theoretical physics, *quantization of Maxwell's equations yields photons and quantization of the Dirac equation yields electrons and positrons*. We shall use Box 32.2.1 to study some of the processes involving electrons, positrons, and photons. At the heart of this study is the **Feynman diagrams**, which are pictorial representations of a highly mathematical procedure for calculating the quantitative aspects of the processes. Because of their intuitive appeal, we shall consider these diagrams in some detail.

BIOGRAPHY

Richard Feynman (1918–1988) was born into a Jewish emigrant family from Russia and Poland. His father, Melville, was a business man whose real interest was in science, but he never had the opportunity to make a career out of it. Melville wanted his first child to be a son and he also wanted him to become a scientist. Richard was Melville's first son, and it was clear what lay ahead for him. Feynman really enjoyed mathematics competitions and was a real star in his school. Upon graduation from high school, he applied to several universities, but was rejected by most; partly because of his poor grades in literature and art, but mostly because he was a Jew. Nevertheless, he was accepted by the Massachusetts Institute of Technology, which he attended in 1935. Feynman was very much at ease with physics, taking graduate courses as a sophomore, and reading the new subject of quantum theory on his own. Upon graduation in 1939, he went to Princeton University and worked under the supervision of John Wheeler. It was during this time that he invented a new, more intuitive, formulation of the quantum theory. He later used this formulation to quantize the electromagnetic field leading to the development of quantum electrodynamics. As a by-product, Feynman invented the intuitive graphical representation—now called the *Feynman diagrams*—of the mathematical formulas of QED. He won the Nobel Prize in physics in 1965 for his contribution to QED.

32.2.1   Feynman Diagrams

The exchange of a photon between two electrons requires its emission by one electron and its subsequent absorption by the other. Since emission and absorption are events, and since such processes are inherently relativistic, they are prime candidates for actors on the stage of spacetime diagrams.

The most elementary Feynman diagram is shown in Figure 32.1. A stationary[5] electron, represented by the directed line, emits a photon—represented by the wavy line—at event

---

[5]Relative to some observer who uses the coordinate system shown in (a).

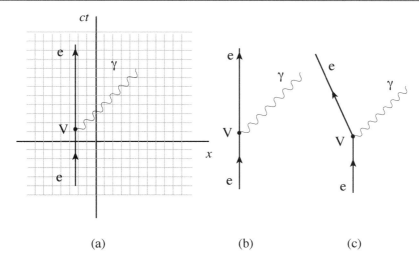

Figure 32.1: (a) The spacetime diagram of a stationary electron emitting a photon and remaining stationary. (b) The same diagram without the coordinate grid. (c) The spacetime diagram of a stationary electron emitting a photon and moving in opposite direction to the photon. (a) and (b) violate both momentum and energy conservation; (c) violates energy conservation.

V, the **vertex**. The photon moves away while the electron remains at rest. Since we normally are not interested in where and when the photon was emitted, we shall do away with the grid and the axes and represent the same process by Figure 32.1(b), in which it is understood that time runs vertically upward. Anything below the vertex is labeled **initial** state, and above the vertex **final** state. Thus, Figure 32.1 describes a process in which the initial state is an electron at rest, and the final state is an electron at rest plus a traveling photon.

The process of Figure 32.1(a and b) is not physical and cannot take place, because it violates the momentum conservation. Here is why. The momentum of an electron at rest is zero, making the initial momentum zero. The final momentum is not zero, because the electron is not moving, but the photon carries momentum. The process also violates the energy conservation, because initially the total energy is the energy of the stationary electron ($E = m_e c^2$), but finally it is that plus the energy of the photon. One may think that something like Figure 32.1(c) can make the process go; however, while momentum conservation may be saved by having the electron move in the direction opposite to the photon, the energy conservation is still violated, because now even the final electron *alone* has more energy than the initial electron. Setting the *initial* electron in motion will not save the process either, because we can always observe the process in the rest frame of the initial electron.

So what is the use of the process of Figure 32.1 if it cannot happen? Well, it turns out that by combining two such *impossible* processes, we can make a process that is *possible*. After all, it is the *exchange* of photons that creates the force between charged elementary particles. Figure 32.2(a) shows two electrons at rest. The left electron emits a photon at V_i, which is subsequently absorbed by the right electron at V_f. Anything below the earlier vertex is called the **initial state**, and above the later vertex the **final state**. Only the initial and final states represent actual real particles; particles connecting vertices (such as the photon in the figure) are called **virtual**, because they are not observable (they are not present either in the initial or in the final state).

Because the photon connecting V_i and V_f is not real, we don't know when it was emitted and when absorbed. In fact, we don't know whether it was emitted by the left or the right

Vertex of a Feynman diagram.

Violation of energy and momentum conservation in a "single-vertex" Feynman diagram.

Initial and final states of a Feynman diagram.

Virtual particles in a Feynman diagram.

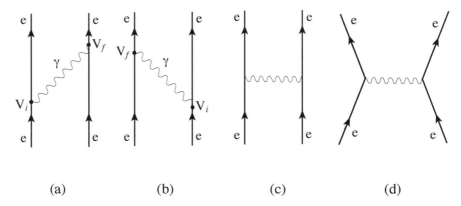

Figure 32.2: (a) The photon emitted at V_i by the left electron is absorbed at V_f by the right electron. (b) The photon emitted at V_i by the right electron is absorbed at V_f by the left electron. (c) The photon drawn horizontally and the dots of vertices removed. (d) The motion of the initial and final electrons are also of no consequence. So sometimes we draw (c) as (d) and vice versa.

electron and consequently absorbed by the other. Figure 32.2(b) describes exactly the same process as Figure 32.2(a). Therefore, it is customary to detach any significance from vertices, remove the dots (representing events), and draw the photon *horizontally*, treating emission and absorption equally [Figure 32.2(c)]. This, of course, puts the photon *outside* the light cone; but since the photon is virtual and unobservable, it does not violate the theory of relativity. Furthermore, since it is the exchange of photons that carries significance (and not the state of motion of the initial and final electrons), sometimes we draw Figure 32.2(c) as Figure 32.2(d), where the initial electrons appear to be moving towards, and the final electrons away from, one another. Because of this, both Figure 32.2(c) and Figure 32.2(d) are labeled *electron-electron scattering* or simply **e-e scattering**.

Electron-electron (or e-e) scattering.

We have not as yet talked about the positron's role in QED interactions. The only difference between a positron and an electron is that the former is represented by a line *with its arrow pointing down*. With this addition, we can now state the QED rules for Feynman diagrams.

> **Box 32.2.2. (Feynman Rules for QED)** *Processes in quantum electrodynamics can be represented diagrammatically using the **Feynman diagrams**, which obey the following rules:*
> *1. Charged particles are represented by up-going directed solid lines; antiparticles' arrows point down.*
> *2. A photon, represented by a wavy line, can be attached to any point of a directed line at a **vertex**, designating an emission or an absorption.*
> *3. The region of the diagram below all vertices is called the **initial state**, and that above all vertices the **final state**.*
> *4. All lines having no end point in either the initial or the final state represent virtual particles and cannot be observed.*
> *5. Any process obeying these rules, whose final state has as much energy-momentum as its initial state is a possible physical process.*

Although stated for QED, most of the Feynman rules given here apply to other theories as well.

Electron-positron (or e^+-e^-) scattering.

Example 32.2.3. The initial state of the Feynman diagram of Figure 32.3(a) consists of a

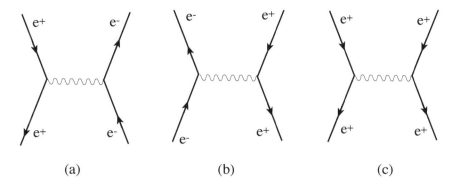

(a)   (b)   (c)

Figure 32.3: (a) Electron-positron scattering. (b) Also electron-positron scattering. (c) Positron-positron scattering.

positron on the left and an electron on the right. A photon is exchanged between the two, indicating an electromagnetic interaction. The final state is the same as the initial state. The process is an electron-positron scattering, or an e^+-e^- scattering. Figure 32.3(b) switches the positron and the electron, but it represents an identical process, i.e., the same e^+-e^- scattering.

In Figure 32.3(c) the initial and final states of the Feynman diagram consist of two positrons, which exchange a photon at an intermediate time. The diagram represents a positron-positron, or an e^+-e^+ scattering. ∎

Positron-positron (or e^+-e^+) scattering.

The rules of Box 32.2.2 allow other less obvious processes. For example, the diagram of Figure 32.4(a) has a photon and an electron in the initial state and the same thing in the final state: it is a photon-electron scattering, which is also called **Compton scattering**, named after Arthur Compton (1892–1962), the American physicist, who performed this scattering experiment in 1923 and showed that light indeed consisted of photons. For aesthetic reasons, one "bends" the lines at vertices and redraws the diagram of Figure 32.4(a) as Figure 32.4(b). Note that the electron between the two vertices is virtual and cannot be observed.

Compton scattering.

What do you know? 32.2. The accompanying figure shows some Feynman diagrams of QED. Determine which ones are correct and which ones are wrong.

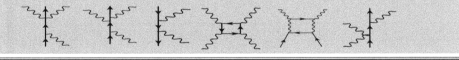

There is another rule that also applies to Feynman diagrams, which can give rise to new processes out of old ones:

Box 32.2.4. *One can twist and bend the lines joined at a vertex of a Feynman diagram or rotate an entire Feynman diagram as long as one does not detach any line. If the resulting diagram does not violate energy-momentum conservation, then it represents a legitimate physical process.*

We have already used this rule in going from (a) to (b) in Figure 32.4. Figure 32.4(c) is obtained from Figure 32.4(b) by a counterclockwise rotation of 90 degrees. The new diagram has an electron and a positron in the initial state and two photons in the final

Electron-positron (or e^+-e^-) annihilation.

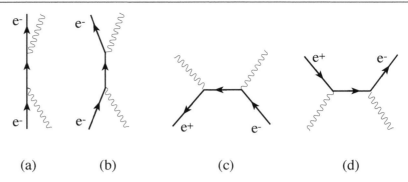

(a)   (b)   (c)   (d)

Figure 32.4: (a) and (b) Electron-photon (or Compton) scattering. (c) Electron-positron annihilation. (d) Electron-positron pair production.

state. It is an e^+-e^- **annihilation.** Conservation of energy does not allow annihilation into absolute "nothingness." The energy of the e^+-e^- pair (the rest energy, for example) has to go somewhere. In Figure 32.4(c), it is carried away by the two photons. The process is the annihilation of matter and antimatter into EM energy. Rotating Figure 32.4(b) by 90 degrees clockwise [or rotation of Figure 32.4(c) by 180 degrees] yields Figure 32.4(d), which has two photons in the initial state and an electron-positron pair in the final state. It

e^+-e^- pair production. represents the reverse process of Figure 32.4(c); it describes the process of e^+-e^- creation out of EM energy, which is called e^+-e^- **pair production.**

Rule 2 of Box 32.2.2 precludes a process in which an odd number of directed lines end at a vertex. Such a process would violate conservation of electric charge. For example,

Charge conservation forbids certain vertices. diagram (a) in Figure 32.5 has an electron in its initial state and two photons in the final. As it describes a process in which the negative charge of the electron disappears, it is forbidden. The conservation of charge is a lot stricter than the conservation of energy and momentum; it forbids diagrams that violate it even at the single-vertex level. On the other hand, while the diagrams in Figure 32.5(b) by themselves violate momentum conservation,[6]

e^+-e^- scattering. they can combine to form a diagram that conserves energy-momentum and can indeed take place. Figure 32.5(c) shows such a two-vertex diagram describing another version of an e^+-e^- scattering [see Figure 32.3(a) for the earlier version]. Moreover, Figure 32.5(c) can be obtained from other legitimate Feynman diagrams by a 90-degree rotation. In fact, any one of the diagrams of Figure 32.3 yields Figure 32.5(c) when rotated by 90 degrees.

> **What do you know? 32.3.** Is it possible to have photon-photon scattering in QED? That is, a scattering with only two photons in the initial state and two in the final state?

A vertex that violates electric charge conservation cannot stand either alone or in conjunction with other vertices. One of the reasons for the "soft" energy-momentum conservation is the uncertainty principle. The uncertainty in energy allows the violation of its conservation for very short time intervals [see Equation (22.3)]; short, but long enough for the virtual energy-violating particle to go from one vertex to the other. No uncertainty relation exists for the electric charge, and no violation of its conservation is allowed under *any* circumstances.

---

[6]To see this, go to a reference frame in which e^+ and e^- move with equal speed in opposite directions. The total momentum is zero, but the initial or final photon cannot have a zero momentum. Since the process is impossible in this particular RF, Box 27.5.2 implies that it is impossible, period!

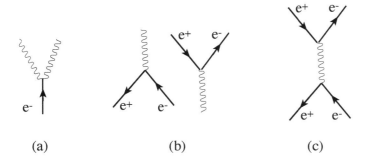

Figure 32.5: (a) This vertex violates charge conservation. (b) These two are good vertices but they violate energy conservation. (c) Putting the two vertices of (b) yields another diagram describing electron-positron scattering.

32.2.2   Predictions of QED

The simple diagrams of the previous subsection are only a small fraction of the complicated QED industry which was developed after WWII. All the real processes discussed in Section 32.2.1 have only two vertices. However, for a given physical process, Feynman diagrams with more vertices exist, whose contribution to the process must be taken into account. For example, the e^+-e^- scattering of Figure 32.2 can also be described by any one of the diagrams in Figure 32.6. These are all of **second order** because they include two (virtual) photons. Figure 32.6(a) describes a process in which two photons are exchanged between the two scattering electrons. In the four diagrams of Figure 32.6(b) one of the electrons emits a photon and subsequently absorbs it. These diagrams represent the self-interaction of the electrons or their **self-energy**. The two diagrams of Figure 32.6(c) alter the vertices of the scattering and are therefore called **vertex corrections**. Figure 32.6(d) represents a "self-interaction" of the exchanged photon. It is also called **vacuum polarization**.

Higher-order Feynman diagrams.

> **What do you know? 32.4.** How can the photons of Figure 32.6(b) bend back to be recaptured? Are they being accelerated?

All these diagrams are legitimate and have to be included in any mathematical calculation of the e^+-e^- scattering. Furthermore, "higher-order corrections" involving six vertices and more should also take part in e^+-e^- scattering. There are two opposing factors to take into account when including higher-order diagrams. First, the vertex, where a photon line is attached to a charge line, has a numerical value representing the strength of the interaction. In QED this numerical value is about 0.0854 per vertex. The strength associated with a diagram—which is roughly the numerical value that the diagram contributes to the process—is the *product* of the strength of all its vertices. Thus, the contribution of the diagram of Figure 32.2 (recall that all diagrams in that figure are identical) is $0.0854 \times 0.0854 = 0.00729$, because there are two vertices in that diagram. The contribution of the diagrams of Figure 32.6 is $0.0854^4 = 0.0000532$, because there are four vertices in those diagrams. We conclude that the diagrams of Figure 32.6 contribute much less to the process than those of Figure 32.2.

Second, when calculating the diagrams *that have loops*—such as all diagrams of Figure 32.6—physicists encountered mathematical expressions that made no sense! These "loop" diagrams yielded *infinite* contributions! Through the long and winded process of **renormalization**, physicists were able to "tame" the infinities, and make sense out of the higher-order diagrams. Once the renormalization process was carried out, QED made

Loops, infinities, and renormalization.

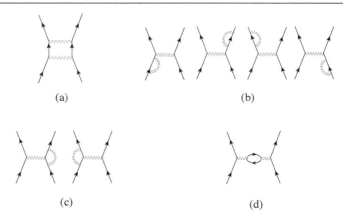

Figure 32.6: Second-order Feynman diagrams contributing to the $e^- - e^-$ scattering. Note how all these diagrams have a "loop" in them. (a) Two-photon exchange. (b) Electron self-interaction or self-energy. (c) Vertex corrections. (d) Photon self-interaction or vacuum polarization.

some remarkable predictions with unprecedented accuracy, making it the most precise theory ever. We briefly discuss two of these predictions: the Lamb shift and the anomalous magnetic moment of the electron.

The vacuum polarization diagram of Figure 32.6(d), with the electron-positron loop in its middle, affects the nature of the exchange of photons between charged particles. This in turn affects the electromagnetic force exerted by charged elementary particles on *Vacuum polarization* one another. QED, through the process of renormalization, can calculate the effect of the *changes the Coulomb* vacuum polarization on the force. In essence it changes the Coulomb law of Equation *law.* (12.1) ever so slightly. This slight change affects the potential energy of an electron in the atom, and consequently its total energy. What this translates into is that the energy levels of say hydrogen atoms are not $-13.6/n^2$ as predicted by both the Bohr model and the Schrödinger theory, because both these theories assume that the electric force obeys the Coulomb law. The inclusion of vacuum polarization effect changes the $n = 2$ energy state of the hydrogen atom from -3.4 eV to -3.40000001122 eV. This is part of what is *Lamb shift.* called the **Lamb shift**, which, although tiny, is measurable; and the measurement agrees remarkably with the theoretical calculation.

The other prediction that we intend to discuss, the **anomalous magnetic moment** of the electron, has an interesting history worth telling. When an electric charge rotates about a fixed point (e.g., the center of its orbit if the orbit is circular) it acquires two properties, angular momentum L and magnetic moment μ. We are already familiar with the former (see Section 8.2.2). The latter arises due to the *electric charge* of the revolving object. A charge in motion is an electric current; this current produces a magnetic field. When the charge moves on a loop, the magnetic field can be associated with a *magnetic moment*,[7] *Relation between* which is related to the angular momentum of the electric charge. **Math Note E.32.2** on *angular momentum and* **page 140** of *Appendix.pdf* shows how to obtain this relation for a circular motion: for a *magnetic moment.* charge q with mass m, the result is $\mu = (q/2m)L$.

In the early days of quantum mechanics, when angular momentum was quantized and the spin of the electron was discovered, it became apparent that the magnetic moment μ should also be quantized. Moreover, since electron spin (its intrinsic angular momentum) is $\frac{1}{2}\hbar$, it was concluded that the magnetic moment of the electron should be $\mu_e = q\hbar/(4m)$. Early experiments, on the other hand, were measuring μ_e to be $q\hbar/(2m)$ exactly *twice* the expected value! This was partly the reason to realize that "spin" had nothing to do with

---

[7]Magnetic moment is a property of a magnetic dipole; the latter being a (small) bar magnetic.

the "angular rotation."

One of the early successes of the Dirac equation (see Section 32.1) was the prediction that μ_e should be $q\hbar/(2m)$. Thus, Dirac's marriage of relativity and quantum theory not only gave us the antimatter, but also clarified the mysterious notion of spin. As experimental techniques improved, physicists could measure μ_e more and more accurately, and in the process realized that it was slightly larger than that predicted by the Dirac theory. This slight difference came to be known as the *anomaly* of μ_e.

Anomaly in the magnetic moment of the electron.

It was quickly realized that the anomaly could be explained by going beyond the simple Dirac theory and using the quantized version of the Dirac equation, i.e., employing the full power of QED. When diagrams such as the two shown in Figure 32.6(c) and higher-order Feynman diagrams are calculated and renormalized, QED predicts a very precise value for the **anomalous magnetic moment** of the electron. In the following equation, we compare the predicted theoretical value and the measured experimental value of the anomalous magnetic moment of the electron:

$$\mu_e^{\text{theory}} = \frac{e\hbar}{4m}\left(2.0023193048 \pm 8 \times 10^{-10}\right),$$

$$\mu_e^{\text{exprmnt}} = \frac{e\hbar}{4m}\left(2.0023193048 \pm 4 \times 10^{-10}\right). \tag{32.1}$$

The 2 comes from the Dirac equation, the rest from QED corrections to it. That we have a theory that can predict a physical quantity to such a high precision is remarkable. It is even more remarkable that we have *instruments* that can measure that quantity with equal precision. And it is miraculous that the two agree!

🧺 Food for Thought 🌱

FFT 32.2.5. Equation (32.1) greatly deserves our awe and admiration. It is the culmination of a game; the kind of scientific game that dwarfs any major-league baseball event, any football bowl, any basketball, wrestling, or hockey extravaganza. It is a 400-year-old game that Galileo started with his inclined planes, a game between experimentalists and theorists that draws in the utmost intellectual prowess of our race.

Not a single "practical application" of the precise determination of the electronic magnetic moment anomaly has yet been found. Nevertheless Equation (32.1) remains an invaluable jewel of humanity. Hidden from the eyes and minds of ordinary men and women, it shines with a light that only the mind-eye of a few can detect.

Pause and ponder for a moment! Equation (32.1) contains a physical property of one of the most fundamental constituents of nature, a particle that exists in everything we see, hear, touch, and smell (as well as many other things that we don't see, hear, touch, or smell); a microscopic part of a fragrant rose and an unseeable galaxy. Equation (32.1) also contains the result of a calculation performed by the human mind using mathematical symbols invented by the human mind. Now take one of the countless number of the electrons in the rose, or an unreachable one in the unseeable galaxy; take it to any laboratory that can measure its magnetic moment accurately; read the measurement on the computer interface; then consult any book on quantum field theory and learn the mathematical techniques developed there; take a piece of paper (a large one!) and calculate the precise value of the magnetic moment of the electron; and feel the miracle of Equation (32.1)!

How is it that the human mind can produce (predict) a number that happens to be hidden in the heart of the matter at the heart of an invisible galaxy? Isn't it spectacular that nature has developed an organ that, through abstract manipulation, can understand its own origin and constituents? Isn't it unfortunate that only a small fraction of our race can appreciate this? And isn't it a disaster that this fraction is dwindling year in year out?

32.3   End-of-Chapter Material

32.3.1   Answers to "What do you know?"

32.1. No! If it were, then Box 32.1.2 would allow a neutron of the nucleus of a radioisotope to decay into an antiproton, an electron, and a neutrino. The antiproton then would annihilate a proton of the nucleus. This would increase the ratio of neutrons to protons, making the nucleus less stable, causing more neutron decays, annihilating more protons, etc., until the entire nucleus is annihilated.

32.2. Starting from left: First diagram is wrong because it has two photons attached to a single vertex; second diagram is okay; third diagram is okay, it is positron-photon scattering; fourth diagram is okay, it is photon-photon scattering; fifth is wrong (you tell me why!); last diagram is also wrong.

32.3. Yes! But the Feynman diagram is of a "higher order!" That is, you need four vertices to get an appropriate Feynman diagram. The diagram does the job!

32.4. These photons are not physical. They are virtual photons and can violate certain physical principles. After all, the photons of Figure 32.3 are horizontal, but it makes no sense to say that they are going faster than light, because they are virtual (they are not part of the initial or final state).

32.3.2   Chapter Glossary

Anomalous Magnetic Moment A tiny correction to the magnetic moment of an electron as predicted by quantum electrodynamics.

Final State The configuration of real particles after all interactions among them has taken place. It is used in the context of Feynman diagrams.

Initial State The configuration of real particles before any interactions among them takes place. It is used in the context of Feynman diagrams.

Lamb Shift A tiny correction to the Coulomb's law due to the self-interaction of a charged particle as it occurs in quantum electrodynamics.

Pair Production A process in which a particle and its antiparticle are created out of pure energy (usually the energy of a pair of photons).

Photon The quantum of the electromagnetic field (see quantum field theory).

Positron The antiparticle of the electron predicted by the relativistic quantum mechanics and Dirac equation.

Quantum Electrodynamics A quantum field theory describing the interaction of charged fundamental particles with photons.

Quantum Field Theory A mathematical theory in which every interaction and particle is considered as a field, which is quantized in the sense that it represents a specific particle. The matter particles interact via the exchange of particles representing the force.

Renormalization A mathematical technique which tames the nonsensical infinities that show up in the calculations associated with higher order Feynman diagrams (diagrams with more than two vertices).

Scattering A process in which two particles interact (usually with very high kinetic energy) to produce two or more particles. It often involves the conversion of the initial KE to mass via $E = mc^2$.

Vertex Used in conjunction with Feynman diagrams. It is a point in the diagram where three or more particles meet. Vertex in the most fundamental part of a Feynman diagram and represents a particular kind of interaction.

32.3.3   Review Questions

32.1. What was the motivation behind combining special theory of relativity with quantum theory? What was wrong with the theory that came out of the combination?

32.2. Who successfully combined STR and quantum theory? What simple principle guided him? What was the conclusion concerning the wave function Ψ? How was this conclusion related to spin?

32.3. What was the significance of the multicomponent nature of Ψ? What particles were associated with the components of Ψ? What is the name of the particle predicted by the Dirac equation? What is its relation to electron?

32.4. Does proton have an antiparticle? What kind of electric charge does it have?

32.5. What does quantum electrodynamics study? How do charged particles interact according to quantum electrodynamics? What is the particle of the quantized electromagnetic field?

32.6. What is a Feynman diagram? What is a vertex of a Feynman diagram? How do you describe the initial and final states of a Feynman diagram? Is there a real physical process in QED represented by a Feynman diagram having only one vertex?

32.7. What is a virtual particle? What is the significance of Compton scattering? When an electron annihilates a positron, what is the end product? What is pair production?

32.8. How many vertices are there in a second order Feynman diagram? What is a self-interaction Feynman diagram? What is a vertex correction? What is vacuum polarization?

32.9. What happens when higher order Feynman diagrams are calculated? What is the name of the procedure used to tame the bad mathematical behavior of these diagrams? Once this procedure is used, what kind of predictions does QED provide? Give two examples of the application of QED and the results obtained from these applications.

32.10. What is magnetic moment? How is it related to angular momentum? Is spin related to magnetic moment in the same way that classical angular momentum is related to magnetic moment?

32.3.4   Conceptual Exercises

32.1. Why does negative probability not make sense? Hint: Consider two random events, one with positive and the other with negative probability of equal magnitude.

32.2. To create an antiproton, a proton had to be accelerated to a KE of almost 6 billion eV, while the rest energy (mc^2) of an antiproton is just under one billion eV. Why can't the KE of a proton with KE of one billion eV turn into an antiproton? Hint: Look at Math Note E.32.1.

32.3. What physical process do you get if you rotate Figure 32.3(a) by 90 degrees clockwise? Counterclockwise? What if you do the same thing to Figure 32.3(b) and (c)?

32.4. Two charged particles p and q (and their antiparticles $\bar{p}$ and $\bar{q}$) have Feynman vertices identical to the electron vertex discussed in the book. These vertices, shown in Figure 32.7(a), can be combined to give description of physical experiments, much like those discussed in the book for electrons and positrons. With two particle, new diagrams are possible, one of which is shown in Figure 32.7(b), and which we call "p-q scattering into p-q." Draw all the Feynman diagrams similar to Figure 32.7(b) involving particles as well as antiparticles.

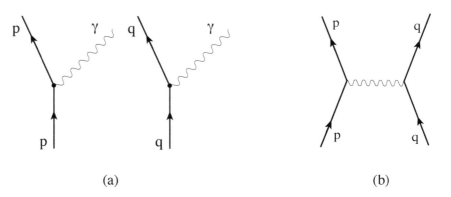

(a)   (b)

Figure 32.7: (a) The vertices of p and q interactions. (b) A typical process obtained from the vertices.

32.5. Can you make the physical process of p-$\bar{p}$ scattering into p-$\bar{p}$ by turning and twisting the diagrams of Conceptual Exercise 32.4? How about the process of q-$\bar{q}$ scattering into p-$\bar{p}$?

32.6. Can you make the physical process of q-$\bar{q}$ scattering into q-$\bar{q}$ by turning and twisting the diagrams of Conceptual Exercise 32.4? How about the process of p-$\bar{p}$ scattering into p-$\bar{q}$? How about the process of p-$\bar{q}$ scattering into q-$\bar{p}$?

32.7. Can you make p-γ scattering into p-γ from the vertices of Conceptual Exercise 32.4? How about the process of q-γ scattering into q-γ? p-γ scattering into q-γ?

32.8. Can you make the physical process of p-$\bar{p}$ annihilation into photons by turning and twisting the diagrams of Conceptual Exercise 32.7? How about q-$\bar{q}$ annihilation into photons? How about q-$\bar{p}$ annihilation into photons?

32.3.5   Numerical Exercises

32.1. Can you create a p-$\bar{p}$ pair from EM radiation by turning and twisting the diagrams of Conceptual Exercise 32.7? If so, approximately how hot should the "photon gas" be if the mass (i.e., the rest energy, mc^2) of p is 1.5 MeV? The average energy of a photon gas is related to its temperature by $2.7k_BT$, where k_B is the Boltzmann constant.

32.2. Can you create a q-$\bar{q}$ pair from EM radiation by turning and twisting the diagrams of Conceptual Exercise 32.7? If so, approximately how hot should the "photon gas" be if the mass (i.e., the rest energy, mc^2) of q is 4.5 MeV? The average energy of a photon gas is related to its temperature by $2.7k_BT$, where k_B is the Boltzmann constant.

32.3. Can you create a q-$\bar{p}$ pair from EM radiation by turning and twisting the diagrams of Conceptual Exercise 32.7? If so, approximately how hot should the "photon gas" be if the mass (i.e., the rest energy, mc^2) of p is 1.5 MeV and of q is 4.5 MeV? The average energy of a photon gas is related to its temperature by $2.7k_BT$, where k_B is the Boltzmann constant.

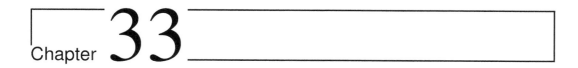

Chapter 33

Classifying Particles and Forces

The invention of the accelerators in the 1930s and their rapid development after WWII opened up a new vista for fundamental physics research. The few particles that made themselves known in the early days of nuclear physics were only the tip of an iceberg. By the mid 1950s the accelerators produced so many new particles that physicists felt like eighteenth century zoologists facing a seemingly endless variety of "animals" to study. And just like the zoologists, they embarked on their classification.

The initial key to this classification was the strength of the particles' interaction with matter. By forcefully injecting the new particles into varying thicknesses of lead, physicists could identify two major categories of particles: those that were stopped quickly (i.e., after moving a short distance) were called **hadrons**[1] because they must have interacted with matter very strongly; the remaining particles, thought to be "lighter" than hadrons, were named **leptons**—derived from the Greek word for light.

Hadrons and leptons.

Another scheme of classifying particles, which was probably more important than their stopping distance in lead, was their half-life. With the exception of the proton, electron, neutrino, and their antiparticles, all known particles were *unstable*, i.e., they did not live forever. For example, a free neutron was known to decay into a proton, an electron, and an antineutrino in about 15 minutes.[2] Based on the length of their half-lives, particles' *interactions* were divided into three categories: strong, weak, and electromagnetic.

Particles partaking in the strong interaction had a very short half-life (of the order of 10^{-20} s); those participating in the weak interaction had a long half life (of the order of 10^{-6} s and longer); electromagnetic interaction was represented by an intermediate half-life. By definition, only hadrons took part in strong interactions. They could, of course, participate in other interactions as well. [For example, the neutron beta decay is known to be a weak interaction even though neutron is a hadron.] However, particles that *did not* have strong interactions were not hadrons. Before talking about hadrons and leptons, let's see what exactly is a particle.

33.1 Spacetime Symmetry: Mathematical Poetry I

The decade of the 1860s is a landmark in the history of mathematics. In the latter part of that decade, a young Norwegian mathematician was playing with some geometric ideas, which later were developed into one of the most charming branches of mathematics. The young Norwegian was Sophus Lie (1842–1899), and his ideas culminated in what is now

---

[1] From the Greek word $\alpha\delta\rho o\sigma$ (*adros*) meaning strong.
[2] This is the prototype of the nuclear beta decay (see Section 31.3).

appropriately called the *Lie groups*. Like other elegant mathematics, Lie groups had their origin in Nature.

After Newton applied calculus to motion and wrote down the first *differential equation*, physicists and mathematicians subsequently encountered many other types of differential equations, whose solutions would describe the behavior of physical quantities. A variety of ingenious "tricks" were invented to find solutions of differential equations, but there was no *systematic* procedure for solving them. Sophus Lie discovered that through the use of group theory, which had been created to investigate the possibility of finding the roots of polynomials, he could systematically find solutions to many differential equations, and also explain why the ingenious tricks worked. Both the technique and the ideas of Lie group theory were highly abstract, to such a degree that, according to some historians, mathematicians thought that they had finally stumbled on something that physicists had no way of using. But the aesthetic appeal of Lie groups could not have escaped Nature, who found it to be the best medium in which to compose her poetry.

Lie groups and their use in solving differential equations.

Recognizing a generalization of the Lorentz transformations (called the *Poincaré transformations*) as a Lie group, Eugene Wigner (1902–1995), the Hungarian-born American mathematical physicist, set out to analyze it. In mathematical jargon, he was looking for a *representation* of the Poincaré group. A representation is a set of matrices that have identical properties as the group elements, and a set of *states* similar to the Ψ function of the quantum theory (see Section 22.3). In 1939, in a historic paper that won him the Nobel Prize, Wigner succeeded in finding the representation and the states. Of significance to us is the latter.

Wigner's work showed that Einstein's special theory of relativity, in the context of Lorentz (or rather Poincaré) transformations, classified particles using three numbers: one associated with mass, another associated with spin, and the third associated with the projection of the spin (called the *third component* of spin) along some line. Mass, m, can be any nonnegative number; spin, s, can be a nonnegative integer or a positive half-odd integer (multiple of $\hbar$, as discussed also in Section 23.3.1); the third component of spin, s_3, needs a little more explanation.

Since most particles of interest are moving (more or less on a straight line), it is natural to consider the directed line of their motion for the discussion of s_3. If the particle is massive, s_3 can take—in units of $\hbar$—any value between $-s$ and $+s$ incremented by 1 from the lowest to the highest values.

Massless particles can have only two values for their s_3: $-s$ and $+s$, with no intermediate values. The state with $+s$, i.e., with the projection of the spin along the direction of motion, is called a **positive helicity** or a **right-handed** state, and the state with $-s$, i.e., with the projection of the spin opposite to the direction of motion, is called a **negative helicity** or a **left-handed** state. These two states are related via a **mirror symmetry**. Let's see what this means.[3]

Positive and negative helicity; handedness.

Recall from our discussion in Section 8.2.2 that the angular momentum direction is associated to a rotation via the right-hand rule: point the thumb of your right hand in the direction of the angular momentum and curl your fingers; the direction of your fingers shows the direction of the associated rotation. Figure 33.1 shows a massless particle moving with velocity **v** (necessarily equal to the speed of light as pointed out in Box 28.3.3) having a projection of spin along the line of motion. The particle in Figure 33.1(a) is right handed, while that in Figure 33.1(b) is left handed.

Now suppose you look at Figure 33.1(a) in a mirror. The direction of the velocity changes (if you move your finger towards the mirror, its image comes towards you). What about the direction of the spin? Stand in front of a mirror and whirl your hand clockwise. Look at the image of your whirling hand. It is also moving clockwise! Rotations do not change

---

[3]A word of caution! Quantum spin has no classical analogue. What is to follow is a classical analysis, which, as far as mirror symmetry is concerned, also works for spin.

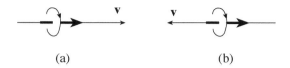

Figure 33.1: The heavy vector is the spin angular momentum vector. The light vector is the velocity vector **v** indicating the direction of motion. (a) A positive helicity state. (b) A negative helicity state.

direction in the mirror. Therefore, the image of Figure 33.1(a) is precisely Figure 33.1(b). Similarly, you can convince yourself that the mirror image of Figure 33.1(b) is Figure 33.1(a): the mirror image of a right-handed particle is a left-handed particle and vice versa. And Wigner showed that both right-handed and left-handed particles should be present in the representation of the Poincaré group. In summary,

---

Box 33.1.1. *A particle is described by its mass, m, its spin s, and the projection of its spin along motion direction. Spin can be a nonnegative integer multiple of ℏ (Bosons) or a nonnegative odd multiple of $\frac{1}{2}$ℏ (Fermions). In units of ℏ, spin projection of massive particles can be any value between −s and s incremented by one, while massless particles can have −s (left-handed state) or +s (right-handed state) for their spin projection.*

What do you know? 33.1. What drastic implication would the discovery of a spin $\frac{3}{4}$ particle have?

33.2   Hadrons

Walking in the footsteps of the eighteenth century zoologists, the mid twentieth century physicists began to divide the "kingdom" of particles into "families," "genre," and "species." Based on their spin, the hadrons were divided into **baryons** and **mesons**. The baryons are hadrons that have half-integer spins, while the spins of the mesons are integer. Examples of baryons are protons and neutrons, and of mesons, the pions.

The baryons seemed to hold a special place in the community of particles. It seemed that in any collision process, as violent as it might have been and as many different kinds of particles it might have created out of the energy of the initial colliding particles, the total number of baryons at the end was equal to that at the beginning. There was a **baryonic charge conservation** at work, very similar to the *electric charge conservation*: If you subtract the total number of negative baryons from the total number of positive baryons before the collision and do the same after the collision, the two are always equal. In other words, every time a new positive baryon is created in a collision, it should accompany a negative baryon. In general, positive baryonic charge is assigned to particles and negative to the corresponding antiparticle. Thus, protons and neutrons have positive baryonic charge; their antiparticles, negative.

Mesons, on the other hand, could be created and annihilated individually, and they were produced profusely. One group of mesons caught the attention of physicists; they were

Baryonic charge conservation.

Kaons and strangeness.

collectively called **kaons** or the **K-mesons**. Because of the mode of their production, the electrically neutral kaons were expected to participate only in the strong nuclear interactions; but they showed clear signs of weak nuclear interaction by their long half-lives. This strange behavior prompted M. Gell-Mann and K. Nishijima to introduce another "charge," similar to the electric and baryonic charge, called **strangeness**. However, unlike its predecessors, which are strictly conserved in all kinds of interactions, strangeness can change. In fact, it was soon realized that kaons could decay into nonstrange particles through the medium of the weak nuclear force.

What do you know? 33.2. What is wrong with the following reactions?

$$\Sigma^+ + \eta^0 \rightarrow K^0 + K^+ + \pi^0 \quad \text{and} \quad \Xi^- + p \rightarrow K^0 + n + \pi^0$$

Hint: Look at Figure 33.2 and invoke certain conservation law.

Kaons were only one group of hadrons that were assembled together because of their strangeness. Soon physicists discovered that using other properties, which were invented for classification purposes, they could group particles into **multiplets**. For example, the oldest particles, the proton and neutron, formed a *doublet* because of a property called **isospin**, while the three pions formed an isospin *triplet*. The grouping of particles based on their isospin was not random. Particles that fell into the same isospin multiplet shared many common properties including mass: particles belonging to the same isospin multiplet had almost the same mass, so in some sense, they were (almost) different "states" of the same physical entity.

Isospin is a property of particles used in their classification.

33.2.1   Eightfold Way: Mathematical Poetry II

Eightfold way and classification of hadrons.

In the decade of the 1960s, as if to celebrate the Lie groups' centennial, two physicists, M. Gell-Mann and Y. Ne'eman, used them to successfully classify all the known hadrons of the time into multiplets. This classification, which employed isospin and strangeness, came to be known as the **eightfold way**.

Following the mathematicians' practice of the representation of Lie groups, Gell-Mann and Ne'eman constructed two perpendicular axes. The horizontal axis was the ("third component of the") isospin I_3, and the vertical was the strangeness S. The eightfold way then more or less dictated how many and in what order the particles were to assemble themselves in the $I_3 S$-plane. It also predicted that certain quantum numbers were shared by all the particles in a multiplet. One of these quantum numbers was *spin*. For example, the three pions, all four kaons, and another meson η_0, which seemed to be a lonely *singlet*, all assembled in a spin-zero octet shown in Figure 33.2(a). Similarly, the known baryons, including our old friends the proton and neutron, formed a spin-$\frac{1}{2}$ octet[4] shown in Figure 33.2(b).

Eightfold way predicts the existence and properties of η_0.

The construction and operation of accelerators in the decade of the 1950s produced many new hadrons, for many of which the eightfold way found appropriate positions on the $I_3 S$-plane. Moreover, since none of the multiplets were completely "filled," the eightfold way *predicted* many particles in the same fashion that Mendeleev table predicted many elements a century earlier.[5] For instance, in 1961, when the eightfold way was proposed, only seven spin-zero mesons were known. The η_0 had not yet been produced. The eightfold way predicted not only the *existence* of η_0, but also its approximate *mass*. Furthermore, the shear membership of Ξ^- and Ξ^0 to the baryon octet of Figure 33.2(b) was sufficient proof that their spin, which was unknown at the time of their discovery, would have to be $\frac{1}{2}$, a

---

[4]Recall that spin is given in units of $\hbar$ (Planck constant divided by 2π).

[5]There is a fundamental difference between Mendeleev table and the eightfold way: The former is purely empirical, the latter purely mathematical.

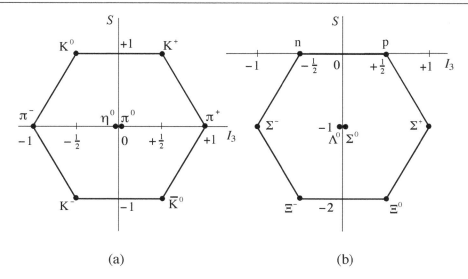

(a)   (b)

Figure 33.2: (a) The meson octet of spin 0. (b) The baryon octet of spin $\frac{1}{2}$.

prediction that was later verified by further experimental analysis. However, the greatest success story of the eightfold way was yet to come.

In the "zoo" of many particles known in 1961 were four spin $\frac{3}{2}$ baryons: Δ^-, Δ^0, Δ^+, and Δ^{++}. These particles could not fit in an octet, but either a decuplet (10-particle) or a 27-particle multiplet could accommodate them. A decuplet would form a triangle in the I_3S-plane, while a 27-particle multiplet would correspond to a large hexagon. In the same year, three new particles, Σ^{*-}, Σ^{*0}, and Σ^{*+}, were discovered, which had strangeness -1, but their spin was not known at the time. These particles could also fit in either a decuplet or a 27-member family.

BIOGRAPHY

Murray Gell-Mann was born on September 15, 1929, in New York City. Upon graduation from high school at the age of 15, he went to Yale University, where four years later he obtained his B.Sc. For graduate work, Gell-Mann attended the Massachusetts Institute of Technology, and after only three years, at the young age of 21, he was awarded a Ph.D. degree. In 1953 he and the Japanese team of Nakano and Nishijima independently proposed the notion of "strangeness" which became the cornerstone of particle classification, leading to the "eightfold way," which Gell-Mann and the Israeli physicist Ne'eman discovered independently. Gell-Mann's persistent investigation of the eightfold way led to the prediction of the omega minus, which was discovered shortly after its prediction, and the quark theory, which is at the foundation of the successful theory of strong interactions. Gell-Mann won the Nobel Prize in physics in 1969 for his classification of particles.

At a 1962 conference of particle physicists held at CERN (*Centre European pour la Recherche Nucleaure*), Geneva, now known as the European Center for Particle Physics, two new particles, Ξ^{*-} with strangeness -2 and electric charge -1, and Ξ^{*0} with strangeness -2 and electric charge 0, were reported. They fitted nicely in either the decuplet or the 27-member multiplet. However, the failure—also reported at the conference—to see any other members that would fill the 27-member multiplet made the choice clear. Here is how Ne'eman described the excitement felt in that conference:

> The creators of the eightfold way, who attended the conference, felt that this 'failure'
> clearly pointed out that the solution lay in the decuplet. They saw the [inverted]
> pyramid [of Figure 33.3] being completed before their very eyes. Only the apex was

Prediction and subsequent discovery of Ω^-.

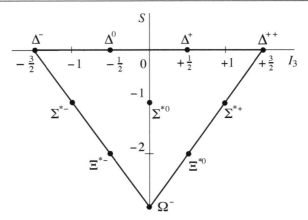

Figure 33.3: The decuplet of baryons of spin $\frac{3}{2}$.

missing, and with the aid of the model they had conceived, it was possible to describe exactly what the properties of the missing particle should be! Before the conclusion of the conference Gell-Mann went up to the blackboard and spelled out the anticipated characteristics of the missing particle, which he called 'omega minus' (because of its negative charge and because omega is the last letter of the Greek alphabet). [Pag 83, pp. 200–201]

As Figure 33.3 suggests, unlike the other members of the decuplet family which have siblings, the Ω^- is a lonely isospin singlet with strangeness -3 and spin $\frac{3}{2}$. Other properties of the Ω^-, including its mass (an important property for experimentalists, because it tells them how much energy is needed to produce the particle), could be predicted from the eightfold way formalism. After "hunting" for the particle for a couple months, the Ω^- was captured in one of about 100,000 photographs taken by a group of 33 physicists at the Brookhaven National Laboratory at the end of 1963. In February 1964, the group officially announced the discovery of the Ω^-.

33.2.2   "Fundamentalism" in Hadron Physics: The Quarks

Fundamentalism is essential for new discoveries in physics.

Whenever a phenomenological theory such as the eightfold way succeeds, physicists instinctively look for the "deeper" reason behind its success. We have already seen this instinct in the case of Planck's discovery of the electromagnetic quanta. This "fundamentalism" may annoy some philosophers of science (see Section 20.2.2), but it is the *only* effective way of discovering hidden secrets of Nature. Fortunately for physics, the creators of the eightfold way did not heed the warnings of the philosophers against such fundamentalism and sought a deeper underlying theory.

The clue to a more fundamental theory is revealed from the Lie group theory itself. It turns out that all the multiplets of the Lie group used in the eightfold way can be "built" from a single shape: a triangle. Figure 33.4 shows the construction of the octet and the decuplet from triangles. All the other multiplets can also be constructed out of triangles.

Quarks, quark model, and up, down, and strange flavors.

Murray Gell-Mann and George Zweig, working independently at Caltech in 1964, came up with the idea that the triangular mathematical building block of the eightfold way ought to be reflected in Nature as three distinct spin $\frac{1}{2}$ fundamental particles, from which all the hadrons are built. This distinction is characterized by assigning a **flavor** to each quark. Gell-Mann, borrowing from James Joyce's *Finnegans Wake*, called these particles **quarks**. On the basis of this **quark model**, all hadrons are made up of three flavors of quark and their antiparticles. The three quarks (or flavors of quark) are the **up** quark, denoted by u,

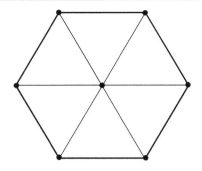 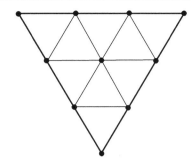

Figure 33.4: The octet and the decuplet are "made up" of triangles.

the **down** quark, denoted by d, and the **strange** quark, denoted by s. Their antiparticles are designated by putting a bar over the symbol: $\bar{u}$, $\bar{d}$, and $\bar{s}$.

Their name is not the only thing that is weird about quarks. They have unusual electric and baryonic charges as well. Every known particle has an electric charge that is an *integral* multiple of the charge of a proton e ($e = 1.6 \times 10^{-19}$ Coulomb). However, quarks have *fractional* charges. The charge of the up quark is $+\frac{2}{3}e$. The down and strange quarks have equal charges of $-\frac{1}{3}e$. Similarly, the baryonic charges of all known baryons are either $+1$ or -1, but for quarks it is $\frac{1}{3}$, and for antiquarks $-\frac{1}{3}$. The important properties of quarks are summarized in Table 33.1. The charges of antiquarks have opposite signs.

Quark	Electric charge	Baryonic charge	Strangeness charge
u	$+\frac{2}{3}$	$+\frac{1}{3}$	0
d	$-\frac{1}{3}$	$+\frac{1}{3}$	0
s	$-\frac{1}{3}$	$+\frac{1}{3}$	-1

Table 33.1: Properties of quarks.

All the hadrons can be explained by the following simple rules:

Box 33.2.1. (Rules of quark model) *All baryons are constructed out of three quarks (antibaryons out of three antiquarks). All mesons are made up of a quark and an antiquark. The spin of each quark (or antiquark) is $\frac{1}{2}$.*

For example, a proton is made up of one down and two up quarks. One up and two down quarks make a neutron. The positive pion π^+ consists of one up and one antidown quark; π^- has one down and one antiup quark; and π^0 is a hybrid of a down-antidown pair and an up-antiup pair. All K-mesons have an s quark (or its antiparticle). For instance, K^+ consists of an up and an antistrange quark; K^-, the antiparticle of K^+, consists of an antiup and a strange quark. When a down quark combines with an antistrange quark, they make a K^0; when their antiparticles combine they form a $\bar{K}^0$.

What do you know? 33.3. Look at Table 33.1 and explain why there is no meson with strangeness -1 and electric charge $+1$.

		Mesons			Baryons			
	S	+1	0	−1	0	−1	−2	−3
Q								
+2					uuu			
+1		$u\bar{s}$	$u\bar{d}$		uud	uus		
0		$d\bar{s}$	$u\bar{u},\,d\bar{d},\,s\bar{s}$	$s\bar{d}$	udd	uds	uss	
−1			$d\bar{u}$	$s\bar{u}$	ddd	dds	dss	sss

Table 33.2: All the mesons and baryons according to the quark model. Q stands for the electric charge and S for the strangeness.

Example 33.2.2. The quark model explained certain mysteries of particle production that were hitherto unexplained. To understand the mystery, let us ask the following question:

Q1: How many different kinds of meson can you get by combining the three quarks and their antiquarks? Arrange these in order of decreasing electric charge.

A1: The highest electric charge of +1 is obtained by combining u with a $\bar{d}$ or an $\bar{s}$. There are five possibilities for neutral mesons: $u\bar{u}$, $d\bar{d}$, $s\bar{s}$, $d\bar{s}$, and $s\bar{d}$. The mesons of the lowest electric charge of −1 are simply the antiparticles of the mesons of highest charge: $d\bar{u}$, and $s\bar{u}$.

Therefore, there are only nine possibilities for mesons. This explains why there are no mesons carrying two units of electric charge (while there *is* a baryon of electric charge +2, namely Δ^{++}), and why there is no meson of strangeness −1 and electric charge +1, or of strangeness +1 and electric charge −1.

Q2: How many different kinds of baryon can you get by combining three quarks? Arrange these in order of decreasing electric charge.

A2: The highest electric charge of +2 is obtained by combining three u quarks. There are two possibilities for the electric charge of +1: uud and uus; three possibilities for neutral baryons: udd, uds, and uss; and four possibilities for the electric charge of −1: ddd, dds, dss, and sss. ∎

> **What do you know? 33.4.** Is it possible for a baryon (not an antibaryon) to have strangeness +1? How about +2?

The example above illustrates the power and simplicity of the quark model. Table 33.2 shows all the hadrons predicted by the quark model. The model does not allow certain combination of charge and strangeness. For example, a baryon of strangeness −3 and charge 0 or +1 is not allowed by the quark model, while it could exist in a family of 27 members according to the eightfold way. The fact that such baryons have not been discovered, lends strong support for the quark model.

How does the quark model account for the spin of hadrons? Recall from Section 23.3.2 that there are two possible (and opposite) directions for a spin of $\frac{1}{2}$. Call these directions *positive* and *negative*. Now consider a quark and an antiquark making up a meson. If one of the spins of the quark and antiquark is positive and the other negative, they cancel; the spin of the meson is zero. If both spins are positive (or negative), they add and the spin of the meson is 1 (the −1 arising from two negative spins is the *projection* of the resulting spin; the spin itself is always positive). For baryons, if all spins align, the result is a spin $\frac{3}{2}$ baryon; if one of the spins is opposite to the other two (thus canceling one of the spins), we get a spin $\frac{1}{2}$ baryon.

Figure 33.5 shows a sample of spin possibilities for some baryons and mesons. In the first row of the figure are collected four baryons of spin $\frac{1}{2}$. The down spin cancels one of the up spins leaving the baryon with only one quark spin. The baryons of row (a) belong

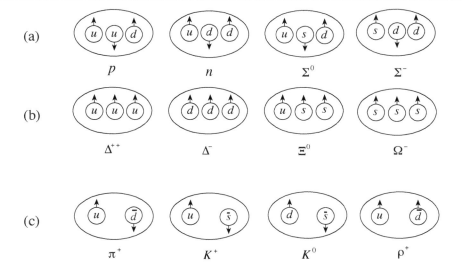

Figure 33.5: A sample of hadrons and their spins. The arrows represent the spins of the quarks. (a) Some spin $\frac{1}{2}$ baryons. (b) Some spin $\frac{3}{2}$ baryons. (c) The first three mesons have spin zero; ρ^+ has spin 1.

to the eightfold way octet of Figure 33.2(b). The second row of Figure 33.5 shows four spin $\frac{3}{2}$ baryons belonging to the eightfold way decuplet of Figure 33.3. In the last row of Figure 33.5 are four mesons, the first three of which show up in the spin zero octet of Figure 33.2(a). The last meson, ρ^+, is one of eight spin-1 mesons, which also make up an eightfold way octet (which we have not discussed).

In fact, there are many higher-spin hadrons (e.g., spin-2 mesons and spin-$\frac{5}{2}$ baryons), called *resonances*, forming octets and decuplets (which we also have not discussed). The highest spin one can get solely from the quark spins is $\frac{3}{2}$ for baryons and 1 for mesons. So where does the extra spin come from? Our discussion of spin in Section 23.3.2 showed that the total angular momentum consists of two parts: orbital and spin. Thus, if the quarks are orbiting one another inside a hadron, their orbital angular momentum could add to the total spin of the composite particle. Since orbital angular momentum takes on only *integer* values (see Box 23.3.2), it does not change the spin nature of the hadrons: mesons still have integer spins, baryons half-integer.

The quark model described in this subsection was proposed in 1964; and for a long time experimentalists tried to isolate a quark by violently bombarding hadrons. However, despite an exponential increase in the energy of the accelerators, no free quark has ever been reported. The appeal and simplicity of the quark model is too much to give up just because free quarks are not observed. In fact, in very energetic collisions, the so-called "deep inelastic scatterings," there has been clear evidence for "structure" within hadrons. If these structures are an indication of quarks, as everybody thinks, why are there no free quarks? How do quarks interact with one another? Are there really just three kinds of quark? May there be more varieties that we have not seen yet? These and other questions can be answered within the context of the forces that quarks experience from one another, and a deeper study of the other group of particles we have ignored so far, the leptons. But first some collision chemistry!

Structure within protons and neutrons is (indirect) evidence for quarks.

33.2.3 Hadron Collision "Chemistry"

With the quark model, we can understand the outcome of the collision of hadrons and other particles if we follow some simple rules.

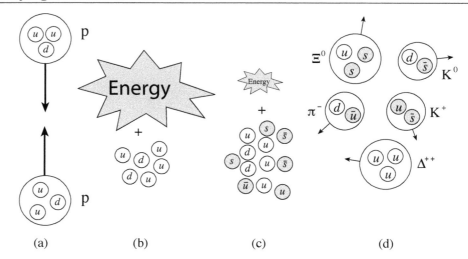

Figure 33.6: A typical proton-proton collision. (a) The two protons approach a head-on collision. (b) Schematics of the initial quarks and the energy available. (c) Schematics of the transformation of some of the energy into matter. Note that every quark created accompanies its antiquark. (d) The quarks assemble into various hadrons carrying some KE.

Box 33.2.3. (**Particle Collision Rules**) *In any particle collision, the following rules hold:*

1. *The total flavor (quark or lepton) of any kind before the collision equals the total flavor after collision.*

2. *The kinetic energy of the colliding particles can turn into matter according to $E = mc^2$. Every matter particle thus formed must accompany its antiparticle. Similarly, every particle can annihilate its antiparticle into pure energy.*

3. *No free quark can be created. All the quarks "bag" themselves into hadrons according to Box 33.2.1.*

As an example consider a proton-proton collision shown in Figure 33.6. Initially [Figure 33.6(a)] there are four u quarks and two d quarks moving at very high speed represented by the heavy long arrows attached to the two protons. Figure 33.6(b) shows schematically the quark and energy content of the collision. Some of the large energy of Figure 33.6(b) is converted into three quark-antiquark pairs (the shaded quarks) in Figure 33.6(c). Note that each quark created is accompanied by *its* antiquark. The creation shown is only one of many possibilities constrained only by energy conservation. If the available energy of Figure 33.6(b) is small, fewer quark-antiquark pairs are created; and if the energy is too small, no pairs may be created. The quarks so created cannot be free; they must assemble themselves into hadrons. Figure 33.6(d) shows one possible result of quark-antiquark assembly. The name of each hadron is shown next to its "bag." It is worth noting that there are two baryons in the final state, as required by the baryon-number conservation.

What do you know? 33.5. Suppose that in Figure 33.6(a) you had proton and antiproton colliding. What would Figure 33.6(b) look like?

The simple quark model we are considering cannot predict what the exact outcome of each collision is. In fact, such a prediction is impossible due to the statistical nature of the quantum theory. The most complete theory can only predict the probability of a particular final state out of all possible states. Our simple quark model cannot even do this. It can only say that given sufficient amount of energy, the reaction of Figure 33.6, which can be written as

$$p + p \longrightarrow \Xi^0 + \Delta^{++} + K^+ + K^0 + \pi^-, \qquad (33.1)$$

is a possible reaction. But, with exactly the same (initial plus created) quarks, the reaction

$$p + p \longrightarrow n + 2\Sigma^+ + \overline{\Xi}^0 \qquad (33.2)$$

is also possible. Furthermore, allowing for the variation in the types (flavors) of quark-antiquark pairs created, many other collision products become possible, the probability of none of which is addressed by our simple quark model.

What do you know? 33.6. Which of the following reactions are possible?

$$p + p \rightarrow K^+ + \pi^- + \Sigma^+ + \Sigma^0 + \pi^+ + K^0$$
$$p + p \rightarrow K^+ + \pi^+ + \Sigma^0 + 2K^0 + n + \bar{K}^0$$
$$p + \bar{p} \rightarrow K^+ + \pi^0 + \pi^+ + K^0 + \pi^- + K^- + \bar{K}^0$$
$$p + \bar{p} \rightarrow K^+ + 2\pi^0 + \overline{\Sigma}^+ + \Sigma^- + \Sigma^0$$

Refer to Figures 33.2 and 33.5 and Table 33.1 for quark content and quark properties.

You don't need hadrons to create hadrons. One of the most common ways of creating hadrons is colliding electrons and positrons. In a typical experiment, e^- and e^+ are accelerated to very high speeds in opposite directions and made to collide with one another. Figure 33.7 shows a typical such collision. Initially [Figure 33.7(a)] the electron and positron approach each other with high speed. The annihilation of e^- and e^+ results in the creation of pure energy. Figure 33.6(b) shows schematically the energy content of the collision. Some of the large energy of Figure 33.6(b) is converted into five quark-antiquark pairs in Figure 33.6(c). The creation shown is only one of many possibilities constrained only by energy conservation. The quarks so created cannot be free; they must assemble themselves into hadrons. Figure 33.6(d) shows one possible result of quark-antiquark assembly. Note that there is one baryon and one antibaryon in the final state, as required by the baryon-number conservation.

Hadrons are not the only particles created in either the p-p collision of Figure 33.6 or the e^+-e^- of Figure 33.7. It is quite possible that other kinds of particles, for example leptons, are also created. In fact, new leptons (discussed in the next section) were created in collisions such a the ones described above.

What do you know? 33.7. Which of the following reactions are possible?

$$e^- + e^+ \rightarrow 2K^0 + \Sigma^+ + \bar{n} + K^-$$
$$e^- + e^+ \rightarrow \Sigma^0 + K^- + \overline{\Xi}^- + \bar{K}^0$$
$$e^- + e^+ \rightarrow p + K^- + \overline{\Sigma}^0 + \bar{K}^0$$

Refer to Figures 33.2 and 33.5 and Table 33.1 for quark content and quark properties.

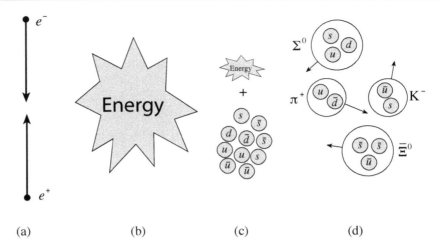

Figure 33.7: A typical electron-positron collision. (a) The two particles approach a head-on collision. (b) Schematics of the energy available after e^+-e^- annihilation. (c) Schematics of the transformation of some of the energy into matter. Note that every quark created accompanies its antiquark. (d) The quarks assemble into various hadrons carrying some KE.

33.3   Leptons

The world of leptons is nothing like the overcrowded world of hadrons. Since leptons do not participate in the strong interaction, they cannot form particles as do quarks. So there are only a handful of leptons, and despite subjecting them to violent collisions, no structure has been found inside leptons. As far as we know, all leptons are fundamental particles.

33.3.1   The Electron

The first lepton, the electron, was discovered in 1897 as one of the constituents of atoms. It is negatively charged, has spin $\frac{1}{2}$, and like proton, is absolutely stable, in the sense that it never disintegrates. There is a good reason for this stability. We know that electric charge is conserved. We also know that electron is the *lightest* electrically charged particle. Furthermore, in any disintegration, the final decay products must be lighter than the decaying particle.[6] Since there are no charged particles lighter than the electron, it becomes impossible for it to disintegrate.

Electron is the best known elementary particle partly because it was the first to be discovered, but more importantly, because we have a remarkably successful theory (the Dirac equation and its offspring quantum electrodynamics) describing the electron with unprecedented accuracy. Because of such precise knowledge of the electron, the electronic industry has been able to harness this particle to a point where it can manipulate microscopic currents, and construct miniature integrated circuits and microchips—the heart of all computers.

33.3.2   The Muon

As mentioned earlier, in the confusion ensuing Yukawa's proposal of a particle responsible for nuclear forces, muon was erroneously identified as the Yukawa particle initially. Later,

---

[6]You can always move along with the (necessarily massive) decaying particle—which, therefore, appears motionless to you—and watch its disintegration process (see Box 27.5.2). Then the mass of the decaying particle is the source of the energy before decay ($E = mc^2$), which must be distributed among the masses and the KEs of the decay products.

it was discovered that the pions were the true sought-after particles, and that muon was a decay product of pions. Muons are one of the most abundant particles in the cosmic radiation at the surface of the Earth. Each second thousands of muons pass through our bodies, both due to their speed and because they are not strongly interacting.

Except for muon mass, which is about 200 times that of the electron, there is no other appreciable difference between the two. Muon itself has spin $\frac{1}{2}$ and is negatively charged and its antiparticle has the same spin but positive charge. In fact, muon is so much like an electron that it is sometimes called a "fat electron." Its enormous mass is a good source of energy for its decay. Unlike electron, which cannot find a lighter charged particle into which to decay, the muon can easily decay into an electron plus other particles.

BIOGRAPHY

Leon M. Lederman was born in New York in 1922 of immigrant parents. After graduating from high school in 1943, Leon spent three years in the US Army. Then, in September of 1946 he entered the Graduate School of Physics at Columbia, where he participated in the construction of a 385 MeV Synchrocyclotron. After receiving his Ph.D. in 1951, he was invited to stay on, which he did, for the next 28 years. In 1958, Lederman took his first sabbatical at CERN where he organized a group to do an experiment measuring the precise magnetic moment of the electron. In 1979, he became Director of the Fermi Lab where he supervised the construction of one of the highest energy accelerators in the world. Recently Lederman has been increasingly involved in promoting science education for gifted children and public understanding of science. He helped to found the Illinois Mathematics and Science Academy, a three year residence public school for gifted children in the State of Illinois. Lederman shared the 1988 Nobel Prize in physics for his discovery of the muon neutrino.

33.3.3   The Neutrinos

An analysis similar to the beta decay of neutron (see Section 31.3.1) reveals that there are two other particles accompanying the electron in the decay of muon; and the elusiveness of those particles point to their identity as neutrinos. Neutrinos are so elusive that to stop half of the ones produced in a typical nuclear decay, one needs a slab of lead that is not a few centimeters, or a few meters, or a few kilometers, but a few *light years* thick!

The two neutrinos produced in a muon decay interact differently with matter. When the energy of one of them is transformed into matter upon impact with other particles, a positron is produced, while the other neutrino always produces a muon. Because of this distinction, particle physicists call the first one an **electron neutrino**, denoted as ν_e, and the second one a **muon neutrino**, denoted as ν_μ. Furthermore, a detailed analysis of all reactions involving muons, electrons, and neutrinos indicates that, just as in the case of quarks, there is a new kind of conserved leptonic charge called **flavor**.

Flavor, also a property of leptons.

Electron and its neutrino carry an "electron flavor," while muon and its neutrino carry a "muon flavor." Their antiparticles (and antineutrinos) carry the corresponding opposite flavor. In any reaction, the total electron and muon flavor is conserved. Thus, in the muon decay, the electron e^- accompanies an electron antineutrino $\bar{\nu}_e$ and a muon neutrino ν_μ. The flavor of e^- cancels the opposite flavor of $\bar{\nu}_e$ (so that the total electron flavor at the end becomes zero and equal to the initial electron flavor), and the initial muon flavor is carried away by ν_μ. We write this reaction as

$$\mu^- \to e^- + \bar{\nu}_e + \nu_\mu. \tag{33.3}$$

What do you know? 33.8. Which of the following reactions are possible?

$$e^- + e^+ \to \mu^- + e^+, \qquad e^+ \mu^- \to \nu_\mu + \bar{\nu}_e, \qquad \mu^- + e^+ \to \mu^+ + e^-,$$

$$e^+ \mu^- \to \nu_e + \bar{\nu}_e \qquad \mu^- + \bar{\nu}_e \to e^+ + \nu_\mu, \qquad e^- + \nu_\mu \to \mu^- + \nu_e$$

Neutrinos have another property which is unique to them. Although their spin is $\frac{1}{2}$, and they are expected to have positive and negative helicities (see Section 33.1), neutrinos have *only negative* helicity; and antineutrinos have *only positive* helicity. Since neutrinos participate only in the weak interaction, we say that weak interaction breaks the mirror-image symmetry. In the high energy physics community, this property has been summarized by saying: "God is a weak left-hander!"

That's it! Only four leptons, two charged and two neutral! At least that was the story until the 1970s when a "cultural revolution" in the physics of fundamental interactions undermined the established wisdom, both theoretically and experimentally.

33.4   End-of-Chapter Material

33.4.1   Answers to "What do you know?"

33.1. It would mean that the special theory of relativity is wrong because STR *with a mathematical rigor* implies that spin can only be integer of half integer.

33.2. They both violate baryon number conservation. The first reaction has a baryon on the left and none on the right. The second reaction has two baryons on the left and one on the right.

33.3. To have strangeness -1, the meson must have an s quark. The other constituent of a meson is an antiquark. If the antiquark partner is $\bar{u}$, then the charge of the meson will be -1; if it is $\bar{d}$ or $\bar{s}$, then the meson will be electrically neutral.

33.4. No! A baryon consists of three quarks. If the baryon is to have any strangeness, then it must have at least one s quark. But s quark has strangeness -1. So, no positive strangeness is possible for baryons.

33.5. It would be purely energy, because all the quarks in p annihilate their antiquarks in $\bar{p}$ and turn into energy.

33.6. Determine the quark content of both sides and make sure that each flavor of the quark is preserved in the reaction, allowing for the cancellation of a quark by its antiquark on the right-hand side. For example, the quark content of the left-hand side of the first reaction is $uuduud$, and the particles on the right-hand side, respectively, have the quark contents, $u\bar{s}$, $d\bar{u}$, uus, uds, $u\bar{d}$, and $d\bar{s}$. If you cancel all the quarks and their corresponding antiquarks on the right-hand side, you'll end up with $uuuudd$, which is the quark content of the left-hand side. So, the first reaction is possible. Similarly, the second reaction is impossible; third is possible; and fourth is impossible.

33.7. The left-hand side doesn't have any quark. So, all the quarks on the right-hand side must accompany their antiquarks. Refer to Figures 33.2 and 33.5 and Table 33.1 for quark content and quark properties. Following this rule, you should get the following answers: The first reaction is possible; the second and third reactions are impossible.

33.8. The first reaction has no net flavor on the left, but muon and antielectron flavor on the right. The second reaction is fine. Third reaction has anti-electron flavor on the left and electron flavor on the right, and the opposite for muon. Fourth reaction has muon flavor on the left but none on the right. Fifth reaction is fine. The last reaction is fine.

33.4.2   Chapter Glossary

Baryon A hadron with half-integer spin.

Baryonic Charge A property assigned to baryons whereby baryons are given a charge of $+1$ and antibaryons -1.

Eightfold Way A scheme of classifying hadrons based on the abstract mathematical notion of Lie groups.

Flavor A property assigned to quarks and leptons. The names assigned to quarks and leptons are also flavor assignments. Thus, we speak of up flavor, down flavor, electron flavor, etc.

Hadron A particle that interacts strongly with matter. The strength of the interaction is measured by the thickness of lead that it can traverse before stopping. Hadrons traverse only a few centimeters.

Helicity A property of massless particles with nonzero spin. The property is described by the component of spin along the direction of motion of the particle.

Left-Handed Particle A massless particle whose spin lies opposite to the direction of motion.

Lepton A particle that interacts less strongly with matter than hadrons. Leptons traverse in lead distances of the order of meters and much more.

Meson A hadron with integer spin.

Mirror Symmetry A property of a massless particle with nonzero spin. If the particle can be both right handed and left handed, then it is said to have mirror symmetry.

Quark Model A model according to which all baryons are made up of three quarks and all mesons of one quark and one antiquark.

Quarks Fundamental particles, which together with their antiparticles make up all hadrons.

Right-Handed Particle A massless particle whose spin lies in the direction of motion.

Strangeness A property associated with some hadrons, based on the unexpected relation between their modes of production and decay.

33.4.3   Review Questions

33.1. What are the two main categories of particles? What is this categorization based on?

33.2. Describe helicity. What is positive helicity? What is negative helicity? How are these related to handedness, and mirror symmetry?

33.3. What is a hadron? How many categories are hadrons divided into? What is this categorization based on? Is the number of baryons going into a reaction equal to the number coming out? Is the number of mesons going into a reaction equal to the number coming out?

33.4. What is eightfold way? What is the underlying mathematical idea behind it? Name the particle whose prediction was one of the most significant successes of the eightfold way.

33.5. After "fundamentalism" was applied to the eightfold way, what particles were predicted? What are they called? How many of them were there in 1964?

33.6. What is flavor? To which particles is it applied? Name the three flavors discovered in 1964.

33.7. What is the largest electric charge a hadron could possess? The largest a meson could possess? The largest a baryon could possess?

33.8. State the rules of particle collision. What happens to the KE of the incoming particles? Has a free quark been seen? What happens to the quarks that are produced in high energy collisions?

33.9. How do you compare the number of leptons and hadrons? How many leptons were there in 1964 when hadrons were successfully classified? How many flavors?

33.10. How many neutrinos were there in 1964? Are neutrinos right-handed, left-handed, or both?

33.4.4 Conceptual Exercises

33.1. You send a beam of particles through a piece of lead a few centimeters thick, and note that nothing comes out at the other end. Did the beam consist of hadrons or leptons?

33.2. You send a beam of particles through a piece of lead a few meters thick, and note that a fraction of the beam comes out at the other end. Does the emerging beam consist of hadrons or leptons or both?

33.3. A beam of hadrons consisting of a single species of particles is sent through an inhomogeneous magnetic field as in Section 23.3.2 and Figure 23.7. It is observed that four blobs are formed on the capturing screen. Are the particles baryons or mesons?

33.4. In Figure 33.2, the meson hexagon is symmetric about the horizontal axis but the baryon hexagon is not. Why? Hint: Think quarks!

33.5. With three quarks you can make only a handful of hadrons. Where do hundreds of hadrons discovered in the 1950s and 1960s come from? Hint: Quarks in a hadron can be in higher energy states, just like the hydrogen atom, but these energies are millions of times larger than those of H-atom.

33.6. Table 33.3 shows the quark content of some common hadrons. Using this table, decide which of the following collisions are possible. Assume that the colliding particles on the left have no shortage of energy to produce as many particles as desired. A bar over a particle symbol signifies the antiparticle of that symbol.

$$p + p \longrightarrow \Delta^{++} + \Sigma^0 + K^+ + \pi^0 + \pi^-$$
$$p + p \longrightarrow \Sigma^+ + n + \pi^0 + \pi^+$$
$$p + \bar{p} \longrightarrow \Xi^0 + \bar{n} + K^0 + K^+$$
$$p + \bar{p} \longrightarrow \Xi^0 + \overline{\Sigma}^0 + K^+ + \pi^-$$

Particle	Quark content	Particle	Quark content	Particle	Quark content	Particle	Quark content
Δ^{++} Σ^+ Ξ^-	uuu uus dss	p Σ^0 Ω^-	uud uds sss	n Σ^-	udd dds	Δ^- Ξ^0	ddd uss
K^+ η^0	$u\bar{s}$ $s\bar{s}$	π^+ π^-	$u\bar{d}$ $d\bar{u}$	K^0 K^-	$d\bar{s}$ $s\bar{u}$	π^0 $\overline{K}^0$	$u\bar{u}$ or $d\bar{d}$ $s\bar{d}$

Table 33.3: Table of common hadrons.

Chapter 34

The Standard Model

The successful classification of strongly interacting particles (or hadrons) in the scheme of eightfold way and the subsequent discovery of the quark model answered the question of which hadrons have similar properties and what they are made of. The question of the *dynamics* of the interaction of hadrons remained unanswered (or alternatively, had multiple answers).

Although the quark model explained the variety of the observed hadrons, the fact that quarks were never observed in isolation made it hard for many physicists to take the model seriously. If hadrons are made up of quarks, then by providing a hard blow to a given hadron, we should be able to "knock" one or more quarks out of it just as we can "knock" an electron—a constituent of the atoms—out of an atom. But regardless of the strength of the "blow," no isolated quark was ever observed. In all experiments, irrespective of the amount of the energy provided to the colliding hadrons, the final products were *more hadrons*. It appeared as if hadrons were simply transforming into each other rather than being composed of more fundamental particles.

An attractive idea that caught the attention of the physics community in the early 1960s was the **bootstrap hypothesis**, or "nuclear democracy." Since no structurally more fundamental particles show up in the violent collisions of hadrons, the bootstrap hypothesis argued, hadrons themselves—being the by-products of the interaction of hadrons—must be fundamental: hadrons are both fundamental and composite! And just as the legendary Baron von Münchhausen could lift himself up by his bootstraps, so does a hadron lift itself up into existence by its bootstrap. Although the bootstrap hypothesis, for lack of a better theory, remained popular into the early 1970s, it eventually lost its appeal due to its enormously complicated assumptions, its use of labyrinthine mathematics, and its inability to predict any new physics. The correct theory of hadrons found its origin in a little-noticed 1954 paper by Chen Ning Yang and Robert Mills.

34.1 Gauge Theory: Mathematical Poetry III

In the classification of particles in the last chapter, one particle, the photon, is conspicuously absent. Photon is neither a hadron nor a lepton; it belongs to a third class of particles called **gauge bosons**. These particles, just like photon (which mediates the electromagnetic force among charged particles), mediate various other forces. To understand the origin of the gauge bosons, we must learn something about *symmetry*.

34.1.1   Global and Local Symmetries

A square is a more symmetrical shape than a rectangle. Why? Is it because the sides of a square are all equal? A rhombus has four equal sides, but it is not as symmetric as a square. Neither is it because the angles of a square are all equal; a rectangle has four equal angles as well. It *is* because both the sides and the angles are all equal. But why does that make a square more symmetric than both a rectangle and a rhombus? There is an *operational* definition of symmetry, which can tell us quickly whether a shape is symmetric or not, and how symmetric it is.

Symmetry of geometrical shapes.

For geometrical shapes, the operation (also called *transformation*) is their motion, in particular their rotation. If we rotate a shape by a certain angle and it does not change,[1] we say that the shape is symmetric under that rotational angle. A rectangle, a rhombus, and a square are all symmetric under a 180° rotation about their centers, but only the square (among the three shapes) is also symmetric under a 90° and a 270° rotation. *The larger the number of symmetry angles, the more symmetric the shape.* Thus, according to this definition, a regular hexagon is more symmetric that a square, which is more symmetric than an equilateral triangle, which is more symmetric than a rectangle (or a rhombus). And the most symmetric shape becomes a circle, because it is symmetric under *any* (and, therefore, infinitely many) rotational angles.

Abstracting the idea above, mathematicians and physicists define the symmetry of a system as the invariance of that system under some abstract transformation. More precisely,

> **Box 34.1.1.** *A physical system is said to be symmetric under a transformation if it remains unchanged when that transformation is performed on it, i.e., if the system after operation coincides with the original system.*

Symmetry of the universe.

For example, a homogeneous and isotropic universe (see Section 29.4) is symmetric under translation and rotation. All (distant) points of a homogeneous universe are similar; therefore, if you *translate* one point of the universe to another, the universe does not change. Similarly, (on a large scale) all angles of an isotropic universe are similar; therefore, if you pick any point P of the universe and *rotate* other points of the universe about P, the rotated universe coincides with the original one.

> **What do you know?   34.1.** Which of the following is the most symmetric under rotation in a plane: a square, an ellipse, a parabola, or a rectangle? Which one is the least symmetric?

Rotation and translation are only two kinds of (concrete) transformations. There are other more abstract transformations that can be applied to *symmetric* physical quantities without changing that quantity. For example, the Lorentz transformation of Section F.4, when applied to space and time coordinates, leaves the spacetime distance unchanged. In general, certain physical quantities expressed in terms of physical variables are invariant under certain transformations of the variables.

Symmetry—the epitome of perfection in mathematics—has found its most natural setting in *Lie groups*, of which you heard in the context of particle classification (see Section 33.2.1). While particle classification exemplified a *static* application of Lie groups to fundamental physics, the Lie groups in the present context is applied to the *dynamics* of the fundamental constituents of matter.

---

[1] In the sense that you cannot distinguish the new shape from the original shape.

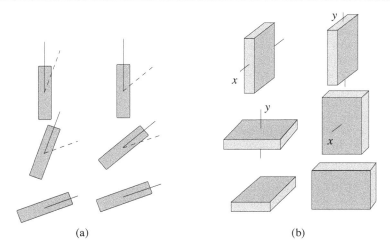

(a)   (b)

Figure 34.1: (a) Rotation in the plane. In the left column (top to bottom), a rectangle is rotated first by 20° and then by 50°. In the right column, a rectangle is rotated first by 50° and then by 20°. The end results are the same. (b) In the left column, a brick is rotated first by 90° about the x-axis and then by 90° about the y-axis. In the right column, the brick is rotated first by 90° about the y-axis and then by 90° about the x-axis. The end results are *not* the same.

What do you know?   34.2. Suppose that the symmetry operation is changing the sign of a mathematical quantity. Which expressions are symmetric under this operation? $x^2, x^3, x^2 + x^4, x^2 + x^3, |x|.$

Example 34.1.2. (This example is slightly technical, but not mathematical.) You have heard the words "Lie groups" a number of times by now. Let's delve a little deeper into this subject, as the discussion will be helpful later in this chapter. A paradigm of Lie groups is rotation, whose basic understanding will shed some light in the nature of Lie groups in general, and their application in physics, in particular.

Figure 34.1(a) shows a rectangle lying in a plane (here, the plane of the paper). Given a point in the plane such as the center of the rectangle, one can rotate the rectangle about that point. When you rotate the top figure of the left column by some amount, say 20° (the angle between the solid vertical line and the dashed line in the top figure), you get the middle rectangle; and if you rotate the middle rectangle by some other amount, say 50° (the angle between the solid line and the dashed line in the middle figure), you get the bottom figure. Now rotate the top figure of the right column (which is identical to the top figure of the left column) by 50° (the angle between the solid vertical line and the dashed line in the top figure), you get the middle rectangle; and if you rotate the middle rectangle by 20° (the angle between the solid line and the dashed line in the middle figure), you get the bottom figure. The bottom figures of the right and the left columns are identical, indicating that *the order of the two rotations is immaterial: the two rotations commute.*

Figure 34.1(b) shows a brick standing on one of its edges in space. Given a point in space (here the center of the brick), one can rotate the brick about that point. Contrary to the plane, where there is *only one axis*—the axis perpendicular to the plane—about which you can rotate, you have *three axes* passing through the point about which you can rotate.[2] Since picturing arbitrary rotations in three dimensions is hard, let's restrict rotations to 90°, which for our purposes is sufficient. Rotate the top figure of the left column by 90° about the x-axis to get the middle brick; rotate the latter by 90° about the y-axis and get the bottom figure. Now reverse the order: rotate the top figure of the right column (which is identical to the top figure of the left column) by 90°

Rotations in a plane commute.

Rotations in space do not commute.

---

[2]There are really infinitely many axes about which a rotation can take place. However, in a sense that we need not go into, only three of them are "independent."

about the y-axis to get the middle brick; then rotate the latter by 90° about the x-axis and get the bottom figure on the right. The bottom figures of the right and the left columns are *not* identical, indicating that the order of the two rotations does matter: *the two rotations do not commute.*

This commutativity (or lack thereof) resembles the property of matrices mentioned in Section 22.2.2. In fact, the above rotations can be represented by matrices. Once this is done, one finds that the matrices representing the two-dimensional rotation commute with one another, while those representing the three-dimensional rotation do not. The discussion of rotations and matrices is indeed more general: most of the abstract Lie groups encountered in physics can be regarded as "rotations" in some abstract multi-dimensional space, and therefore, represented by matrices. The rotation in the plane, having only 1 axis, has an abstract analog in Lie group theory denoted by $U(1)$. The letter U stands for **unitary**,[3] and 1 indicates that the matrix is one by one, namely just a (complex) number. The rotation in space, having 3 axes, has an abstract analog in Lie group theory denoted by $SU(2)$. Again U stands for unitary and S for **special**.[4] The "2" in $SU(2)$ is related to the three axes of rotation via $2^2 - 1 = 3$. In general, abstract multi-dimensional rotations are denoted by $SU(n)$, and possess $n^2 - 1$ axes of rotations.

Abelian and nonabelian Lie groups.

A Lie group all of whose matrices commute is called **abelian**, and if its matrices do not commute, the group is **nonabelian**. Thus, rotations in a plane form an abelian Lie group, while the Lie group of rotations in space is nonabelian. Similarly $U(1)$ is abelian, but $SU(n)$ is nonabelian if $n \geq 2$. As we shall see shortly, the commutativity property of Lie groups plays a significant role in the nature of the fundamental forces. ■

Abstract quantities require higher mathematical skills; and since the reader may not have such skills, visual aids can be useful in representing abstract quantities. For example, let a (infinite) sheet of paper represent a field as described in Box 32.2.1. Also assume that the points of the paper represent points of space (more precisely, spacetime, but we need not worry about such subtleties).[5] Since space and time do not have any color, it would be more realistic to replace the (usually white) paper sheet with a transparency sheet. A visual representation of the *invariant* physical quantity, which depends on the fields, although helpful, is not necessary because such a visualization is not crucial for the present discussion. All you need to know is that such quantities exist, and that they do not change under certain transformations of fields. Figure 34.2(a) shows a transparent sheet representing a field.

Now let's see if we can come up with a visual representation of the *transformation* of a field. What is the most natural thing one can do to a transparency sheet? Write or draw on it. And to do so, one has to pick a marker of a particular color. Pick a gray marker, and don't worry about the fact that gray markers are not easily found! For our discussion, the usual writing or drawing is too complicated (nevertheless, we'll come back to it later), because the effect of the marker is different at different points of the sheet. We want the marker to have a *uniform* effect at all points of the sheet. So instead of writing or drawing, color the sheet gray uniformly. This will give you the sheet of Figure 34.2(b). The uniform coloring of the sheet is called a **global transformation**, because the transformation is the same at all points of the sheet. Translating back into the language of mathematics and physics, a global transformation of a field is that which changes the field in the same way at all points of space (and time).

Global transformation.

What is the significance of the global transformation? There is a clear mathematical procedure for writing down a physical theory that is symmetric under a global transformation. Such a theory gives the same results whether you write it in terms of the field represented by Figure 34.2(a) or Figure 34.2(b); the theory is "color blind," or more precisely, "gray blind."

---

[3]A unitary matrix has the property that if you complex conjugate each element of one of its rows and multiply it by the corresponding element of another row and add all these products, you get zero. If you add the square of the elements of a single row, you get 1. The same two properties hold for the columns.

[4]For readers familiar with determinants, "special" refers to the fact that the determinant of the matrix is one.

[5]In reality, a field varies from point to point and from time to time, but the sheet of paper looks the same at all of its points; another subtlety that we should ignore.

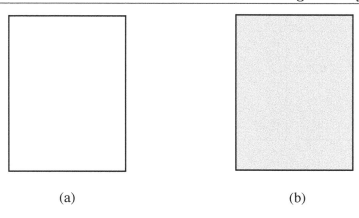

(a)   (b)

Figure 34.2: (a) The transparent sheet representing a field. (b) The "colored" sheet representing a transformed field.

Globally symmetric theories are, however, boring and do not give you interesting physics. Nevertheless, they are important because they are the starting point of the interesting theories involving **local transformations**. As you may have guessed, local transformations transform the field differently at different points. In our transparency-sheet model, a local transformation does not paint gray all points of the sheet. So when you write or draw on a transparency, you are *locally* transforming the sheet. Figure 34.3 shows some local transformations of the transparency sheet.

Local transformations.

34.1.2 Gauge Fields

In their 1954 paper, Yang and Mills considered a field theory that was symmetric under a particular group of (global) transformations.[6] The theory was, of course, not *locally* symmetric. In our analogy of transparency sheets, the theory was the same whether you used the sheet in Figure 34.2(a) or the one in Figure 34.2(b). But if you used any of the sheets in Figure 34.3, the symmetry of the theory would be lost.

BIOGRAPHY

Born on September 22, 1922, in Hofei, Anwhei, China, **Chen Ning Yang** was brought up in the peaceful and academically inclined atmosphere of the campus of Tsinghua University, just outside of Peiping, China, where his father was a Professor of Mathematics. After receiving his M.Sc. degree in 1944, Yang went to the USA, entering the University of Chicago in January 1946, where he came under the strong influence of Professor E. Fermi, and where he obtained his Ph.D. degree in 1948. Yang has contributed greatly to fundamental physics, but is mostly known for his prediction (with T. D. Lee) of the parity violation in weak interactions, and his discovery (with Robert Mills) of gauge theories. He shared the 1957 Nobel Prize in physics with T. D. Lee.

Addition of compensating fields makes a theory locally symmetric.

The next question was whether it was possible to somehow improve the theory to make it locally symmetric. And Yang and Mills found the revolutionary answer. They realized that if they added certain "compensating" fields, or **gauge fields**, which were determined by the assumed initial symmetry, then the theory would become locally symmetric. Again in the language of transparencies, this scheme corresponds to the introduction of the sheets of Figure 34.4 to "make up" for the sheets of Figure 34.3. If you "gauge" the deficiencies of any one of the sheets of Figure 34.3, you can build a corresponding sheet that covers those

---

[6]This group is the Lie group $SU(2)$, encountered in Example 34.1.2.

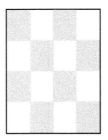

Figure 34.3: Some local transformation of the "field" represented by a transparency sheet.

deficiencies in such a way as to render the entire theory—consisting of the original fields (sheets) plus the new gauge fields (sheets)—symmetric.

What are these gauge fields good for, anyway? To answer this question, theoretical physicists simplified the Yang-Mills symmetry,[7] and discovered that the localization of a theory, which is globally symmetric under a transformation that is one step simpler than Yang-Mills', produced only one compensating field; and this single gauge field could be identified as photon! Since the photon field was known to mediate the electromagnetic interaction, it was hoped that the gauge fields of the Yang-Mills theory could be made to mediate the weak and strong interactions. However, there were serious theoretical problems that had to be overcome before gauge particles could be identified as the exchange particles of the other forces.

The simplest local symmetry has a gauge field identifiable as photon.

Example 34.1.3. The question of how many gauge fields are associated with a given Lie group is an important theoretical question. We just saw that $U(1)$ had only one gauge field. It was also pointed out in Example 34.1.2 that $U(1)$ was the analogue of the rotation in a plane, which has only one axis. Furthermore, the next Lie group, $SU(2)$, was associated with rotations in space, which had three independent axes, where the "3" was related to the "2" of $SU(2)$ via $3 = 2^2 - 1$. In any locally symmetric theory, the number of independent "axes of rotation" in the abstract multi-dimensional space in which the Lie group operates is also the number of gauge fields of the theory. Thus, a theory based on the Lie group $SU(2)$ has three gauge fields, and one based on $SU(n)$ has $n^2 - 1$. ∎

How many gauge fields are there in a given theory?

Gauge fields of a locally symmetric theory ought to be massless.

A locally symmetric theory does not allow its gauge fields to be massive: in the mathematical formulation of the theory, any "mass term" violates the local symmetry. On the other hand, weak and strong nuclear forces are *short ranged*; therefore, their exchange particles had better be either massive (see Section 31.3.2 and the following example), or possess some other property that confined their interaction to short ranges. The latter part of the 1960s and the early part of the 1970s saw the solution to the dilemma in the contexts of *spontaneous symmetry breaking* and *confinement*.

Example 34.1.4. We can understand the connection between the mass of the gauge field and the range of the force by appealing to the uncertainty principle and the Feynman diagrams. In the context of these diagrams, a particle is exchanged when it goes from one vertex to another. Let Δt denote the flight time of the particle from one vertex to the other. This flight time is related to the energy of the particle by the time-energy uncertainty relation: $\Delta E \Delta t \geq \hbar$, where $\geq$ is often replaced by $\approx$, indicating that the left side of the inequality is not *too much* greater than the right side, and the uncertainty is written as an approximate equality: $\Delta E \Delta t \approx \hbar$.

Although ΔE is the *uncertainty* in the energy of the gauge particle, its order of magnitude should not be too much different from the energy of the particle. If the particle is not moving too fast, its energy is mostly its rest energy, i.e., its mass (times c^2). We now note that if the range

---

[7]By changing the group from $SU(2)$ to $U(1)$. See the previous footnote and Example 34.1.2 for more detail.

Figure 34.4: The compensating "fields" of the local transformation of the fields of Figure 34.3.

of the force is small, the flight time Δt is also small, making ΔE, or the mass of the particle, large. Conversely, if the mass of the exchanged particle is large, Δt will be small, making the force short-ranged.

There is another aspect of the force that needs to be emphasized: its probability of occurrence. If the gauge particle is massive, the force is short-ranged, and the two vertices are close together. This means that the particles that exert this kind of force on one another must get very close to each other. And since the occurrence of a close encounter is generally rare, the interaction will be infrequent. Thus there is an inverse relation between the mass of the gauge particle and the frequency of the phenomena resulting from the exchange of that gauge particle. ∎

The more massive the gauge particles, the less frequent the interaction process.

BIOGRAPHY

Robert Mills was born in Englewood, New Jersey, and graduated from George School. He studied at Columbia College, 1944–1948, while on leave from the Coast Guard. Mills demonstrated his mathematical ability by winning the William Lowell Putnam Mathematical Competition in 1948, and by receiving first-class honors in the Tripos. He earned a master's degree from Cambridge, and a Ph.D. in Physics from Columbia in 1955. After a year at the Institute for Advanced Study in Princeton, Mills became a professor of Physics at Ohio State University in 1956 and remained there until his retirement in 1995. Mills is best known for his work with C. N. Yang on the nonabelian gauge theory.

Spontaneous Symmetry Breaking (SSB)

An important concept in any field theory is the **vacuum**. Lacking any (real) particles, it is the state of the lowest possible energy, and as such, also the most stable of all other "states" of the field theory. For our purposes, you can think of the vacuum state as the equilibrium state of the physical system. As a concrete example, consider a perfectly spherical balloon or ball (whose surface is perfectly smooth), which we keep inflating. As we add more and more air in the balloon, the pressure inside increases, but the sphericity of the balloon does not change: every point on and inside the balloon is similar to other points, i.e., the equilibrium state (the vacuum) of the balloon and its content are *rotationally invariant*.

The inflation cannot, of course, go on forever; the balloon eventually pops. Just before it pops, any part of the balloon is as likely to crack as any other part. However, the process of popping picks out a *particular* part of the balloon through which the air gushes out. Once this happens (and it happens "spontaneously"), the symmetry of the balloon and the process is broken, and the equilibrium state is no longer rotationally invariant. There is an analogous mathematical process, by which the choice of a vacuum (out of an infinite equivalent possibilities) spontaneously breaks the symmetry of a field theory.

Jeffrey Goldstone, a British theoretical physicist, showed in the early 1960s that when the *global* symmetry of a theory is spontaneously broken (due to the unavoidable choice of a vacuum), some massless particles are created. These massless particles are called the **Goldstone bosons**. This is the inverse of the process needed for short ranged interactions,

Goldstone bosons.

Higgs mechanism.

and therefore, by itself will not help us. However, soon after Goldstone's discovery, Peter Higgs, another British physicist, applied the idea of SSB to *local* symmetries, and found a remarkable property, now called the **Higgs mechanism**. Based on this mechanism, when a theory that is locally symmetric (and, therefore, necessarily contains *massless* gauge fields) is spontaneously broken, some of the gauge fields acquire mass after symmetry breaking. In fact, some Goldstone bosons of the global symmetry disappear and a corresponding number of the originally massless gauge fields become massive. Some of the gauge fields "eat up" the Goldstone bosons and gain weight! Thus

---

Box 34.1.5. *Some of the gauge fields of a locally symmetric field theory acquire mass after the spontaneous symmetry breaking.*

Confinement

More elaborate Lie groups lead to gluons.

Spontaneous breaking of local symmetry "fattens" the gauge fields and makes them good candidates for short ranged interactions. There is, nevertheless, another mechanism by which interactions can be confined to short ranges. This mechanism occurs for all but the simplest Lie group (whose gauge field is photon).[8] More precisely, when the symmetry is more elaborate than the simplest kind, the massless gauge fields interact among themselves[9] in such a way as to make the corresponding force *increase* with distance. Because of the similarity of this behavior and the (rubber) glue, these gauge fields are called **gluons**. The particles exchanging gluons behave as free particles (in the sense that they don't feel any force) when they are very close, but experience strong forces when they are pulled apart.

What is Nature's Symmetry?

Yang-Mills' idea of local symmetry and gauge fields is appealing; and Higgs' mechanism of giving mass to the gauge fields adds to the appeal. In his *Ode on a Grecian Urn*, John Keats said "Beauty is truth, truth beauty." The intrinsic mathematical charm of Lie groups lends strong support to its potential validity. The truth of fundamental interactions may, therefore, reside in the Lie groups, but not every Lie group contains the truth. Thus, physicists ventured into the enchanting land of Lie groups in search of *the* group that held the truth. The search was mainly aimed at unraveling the weak nuclear force. Among the frenzy that ensued Goldstone and Higgs' embellishment of the work of Yang and Mills, the work of three prominent physicists of the time stood the test of time: Steven Weinberg, Abdus Salam, and Sheldon Glashow. Their work not only gave a full explanation of the weak nuclear force, but, as a prerequisite of doing so, it *unified* that force with the electromagnetic force into what is now called the **electroweak** force.

34.2 Electroweak Interaction

Central to the application of Lie group theory to the interaction of particles is how the particles are grouped together. We saw this grouping in the classification scheme of the eightfold way (see Section 33.2.1). We also saw how the *correct* grouping was crucial in the placement of the two Ξ^* particles and the prediction of Ω^-. The grouping narrows down the choice of the Lie groups, and the right grouping may pick the Lie group that Nature herself has decided upon. Glashow's paper appeared in 1961, Salam's in 1967, and Weinberg's in 1968. Between these three papers, a certain choice of Lie group was made which was only one of many other choices proposed by other physicists in the 60s. However,

---

[8]In the language of Example 34.1.2, the mechanism occurs for all nonabelian Lie groups.

[9]Unlike photon, which, being neutral, cannot interact with another photon.

by the latter half of the 70s, experimental numbers were getting closer and closer to the **Weinberg-Salam-Glashow (WSG) model**.

To understand how general Lie group theory is applied to particle physics, let's examine a special example, the simplest Lie group.[10] In this simplest model, every electrically charged particle constitutes one group, a singlet.[11] To find out what kind of interaction this simple model entails, we need to examine the Feynman diagrams (see Section 32.2.1) of the theory; and since the vertices of the Feynman diagrams are the building blocks of the entire theory, we need to look at the possible vertices allowed by this simple theory. But first we need to know the *general* rule for constructing vertices. This rule, in conjunction with the other Feynman rules given in Box 32.2.2, can yield many interesting physical processes.

> **Box 34.2.1. (Feynman Rule for a General Vertex)** *A vertex consists of an initial particle drawn as a solid line directed upward, a final particle also drawn as a solid line directed upward, and a gauge particle, with all three attached at a single point. The initial and the final particles (not necessarily different) must belong to the same group of the model under consideration.*

In the simple model we are considering, for each singlet there is only one vertex, with the initial and the final particles of the vertex being the same and the gauge particle being necessarily neutral, because electric charge conservation demands that the total charge entering a vertex equal the charge leaving it. If the singlet consists of an electron, we recover the vertex of Figure 32.1, i.e., the theory of quantum electrodynamics.

BIOGRAPHY

Steven Weinberg was born in 1933 in New York City. Encouraged by his father, he developed an early interest in science, and by the time he was 15 or 16 his interests had focused on theoretical physics. Weinberg received his BS degree from Cornell in 1954, and after a year of graduate study and research at Niels Bohr Institute, he returned home to attend Princeton University, where he received his Ph.D. in 1957. Weinberg's research from 1959 to 1966—while at Berkeley—was mainly in theoretical high energy physics. His active interest in astrophysics dates from 1961–62; he wrote some papers on the cosmic population of neutrinos and then began to write a book, *Gravitation and Cosmology*. Published in 1972, the book has become one of the standard references on the subject. While he was a visitor at M.I.T. in 1967, his research turned in the direction of the unification of weak and electromagnetic interactions. In 1973 Weinberg was offered and he accepted the chair of Higgins Professor of Physics at Harvard. His work during the 1970s was mainly concerned with the implications of the unified theory of weak and electromagnetic interactions and the unification of all interactions. In 1982 he moved to the physics and astronomy departments of the University of Texas at Austin, as Josey Regental Professor of Science. Weinberg received the Nobel Prize in physics in 1979 for his work in the unification of the electromagnetic and the weak interactions.

34.2.1   The Lepton Sector

The next in complication, beyond the simplest Lie group discussed above, is a Lie group associated with *doublets*. Charged leptons with their companion neutrinos (see Section 33.3) are natural candidates for such doublets. These doublets are represented as

$$\begin{pmatrix} \nu_e \\ e^- \end{pmatrix}, \quad \begin{pmatrix} \nu_\mu \\ \mu^- \end{pmatrix} \quad \Rightarrow \quad \begin{pmatrix} \nu_l \\ l^- \end{pmatrix}, \tag{34.1}$$

---

[10]If you have read Example 34.1.2, you'll recognize this simple group as $U(1)$.

[11]The double use of the word "group" in the present context is unfortunate but not conveniently avoidable. I hope the context clarifies the distinction between "group" as an aggregate or collection of particles and "group" as a mathematical entity (as in Lie groups).

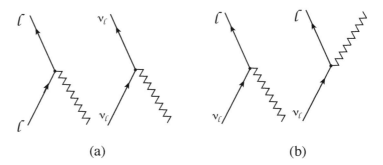

Figure 34.5: The four vertices of the electroweak interaction.

where the last doublet is a convenient generic representation of the other two with l being e or μ. In the 1960s when the Weinberg-Salam-Glashow (WSG) model was proposed, there were only two doublets of leptons: the electron doublet and the muon doublet. The number of doublets has no effect on the validity of the model. In the general discussions of the model, we use the generic doublet to emphasize the generality of the conclusions. As is common in the literature, the member with lower electric charge is placed in the bottom row of the doublet.

What kind of Feynman vertices do we get from this theory? With l representing either of the leptons, Figure 34.5 shows the four possible electroweak vertices. In Figure 34.5(a), the gauge fields, represented by a zigzagged line, are electrically neutral—because the initial and final leptons have the same charge—and it does not matter whether they are drawn in the lower part (initial state) of the diagram or in the upper part. As discussed in Section 32.2.1, the exchanged particles are drawn horizontally anyway. Figure 34.5(b), on the other hand, shows gauge particles that carry electric charge, because the initial lepton has a different charge than the final. The gauge particle on the left of Figure 34.5(b) carries a negative charge, while the gauge particle on the right is positively charged (to neutralize the negative charge of the final lepton).

BIOGRAPHY

Abdus Salam (1926–1996) was born in Jhang, a small town in what is now Pakistan. When he cycled home from Lahore, at the age of 14, after gaining the highest marks ever recorded for the Matriculation Examination at the University of the Punjab, the whole town turned out to welcome him. In 1946 Salam was awarded a scholarship to St. John's College, Cambridge, where in 1950 he received the Smith's Prize from Cambridge University for the most outstanding pre-doctoral contribution to physics. He also obtained his PhD in theoretical physics at Cambridge. Salam returned to Pakistan in 1951 with the intention of founding a school of research, but it soon became clear that this was impossible, and in 1954 he left his own country for a lectureship at Cambridge. Many years later, at the International Center for Theoretical Physics, Trieste, which he created, he instituted the famous "Associateships" which allowed deserving young physicists from the developing countries to spend their vacations there in an invigorating research atmosphere. Salam used his academic reputation to influence international scientific affairs. He shared the Nobel Prize in physics in 1979 for his work in the unification of electromagnetism and the weak interactions.

Therefore, the electroweak interaction has four gauge particles, two neutral, one positive, and one negative. Spontaneous symmetry breaking gives mass to three of these: the two charged ones, which are now denoted by W^+ and W^-, and one of the neutral ones, which is given the symbol Z^0. The other neutral gauge particle remains massless and is identifiable as the photon γ, responsible for the electromagnetic interaction.

What do you know? 34.3. The accompanying figure shows two possible vertices in the electroweak theory. Are the gauge particles electrically charged? If so, what sign do they have?

Example 34.2.2. The discussion leading to the four gauge particles of Figure 34.5 is not entirely correct. If you have followed Examples 34.1.2 and 34.1.3, you have noticed that the Lie group that is only one step more elaborate than the $U(1)$ of QED is $SU(2)$. But $SU(2)$ has only $2^2 - 1$ gauge fields. So where does the fourth gauge field come from? It turns out that the doublets of Equation (34.1) are not entirely correct. Here is why.

At the time of the proposal of the electroweak theory, neutrino's were assumed massless.[12] Therefore, based on the discussion of Section 33.1, they should be right-handed as well as left-handed. However, as early as 1955, it was known that weak interactions violate the **mirror symmetry**. This violation manifests itself in the elimination of the right-handed neutrinos. Somehow Nature has chosen the left-handed neutrinos, but not their mirror image. So how do you group the particles in the model?

Weinberg suggested splitting the massive lepton (i.e., the electron or muon) into its left-handed and right-handed components,[13] grouping the left-handed part with the lepton's neutrino into a doublet, and leaving the right-handed component as a singlet. The Lie group appropriate for this arrangement of particles is a combination of $SU(2)$ and $U(1)$ denoted by $SU(2) \otimes U(1)$. The $SU(2)$ part deals with the left-handed doublet and gives three gauge bosons; the $U(1)$ part deals with the right-handed singlet and gives one gauge boson. Altogether, the theory yields four gauge bosons.

∎

Food for Thought

FFT 34.2.3. On January 21, 1983, in a packed auditorium at CERN (now called *European Center for Particle Physics*), the group working on the detection of the W bosons announced their success. They had found a signal exactly at the predicted mass value. A few months later, the Z boson with the expected mass was also discovered at CERN. Today thousands of sightings of these particles have been reported, and their masses measured more and more accurately. In the latest *Particle Physics Booklet*, the mass of W is given as 1.429×10^{-25} kg and that of Z as 1.604×10^{-25} kg.

Although you are surely familiar with the World Wide Web, you may not know that it was invented at CERN with the sole purpose of bringing particle physicists from around the world together for faster exchange of scientific ideas. Corporate America turned WWW into a mutibillion dollar source of commercial profit, without anybody knowing that the origin of this profit is a financially struggling laboratory of scientific research of the purest kind. However, the billionaires of the "dot com" industry don't know or care about the W and Z bosons and all the other discoveries made at CERN and similar laboratories. Perhaps, if they knew that the pursuit of even the purest and seemingly most abstract research could help their industry, they would not be so indifferent to abstract science.

How idyllic our society would be if ordinary citizens showed some interest in science beyond its practical applications; followed science news with a fraction of the interest with which they follow entertainment and sports news; voluntarily donated to scientific organizations a small portion of what they donate to their institutions of faith; their topics of social conversation included the excitement over the discovery of the bending of light in the Sun's gravity or the discovery of a new particle predicted by a theory, with which they had at least some familiarity; and showed some appreciation for the power of the human mind that could create the mathematics that miraculously predicted such exotic phenomena.

Contrary to the popular culture and the advertising arm of corporate consumerism, average men and women *could* take delight in abstract thought. Politicians, administrators, army generals,

World Wide Web was invented by particle physicists at CERN.

---

[12]The newly discovered "neutrino oscillation," by which mechanism an electron neutrino can turn into a muon neutrino, implies that neutrinos have (a small) mass.

[13]Recall that a massive particle has all the possible spin projections along its direction of motion. But for spin-$\frac{1}{2}$, these are just $-\frac{1}{2}$ and $+\frac{1}{2}$, i.e., left-handed and right-handed.

and parents have not always been interested solely in entertainment, sports, money, power, and politics. Twenty-five centuries ago ordinary citizens of Greece, as well as politicians and generals, participated in vigorous intellectual arguments at dinner tables and cafes (see the quotation in Section 1.5 and page 30 of [Ham 93]). Post-Renaissance Europe also saw a brief period of such intellectual vitalization.

Alas, today's dominant pragmatic global ideology promotes larger SUVs, higher-resolution TV screens, fancier cell phones, 24-hour television programming, complex and costly movie animation and special effects, and a thousand other futilely pleasurable inventions. And in such an environment, the interest in the works of the predictors and discoverers of the bending of light, of antimatter, of Ω^- and $W^{\pm}$ and Z^0, in short, in the highest achievement of the human mind, is considered out of fashion and "nerdy."

μ-ν_e electroweak interaction.

The vertices of Figure 34.5 can be put together to create *processes*. Since we have already discussed the electric part of the electroweak interaction in Chapter 32, we ignore processes involving the exchange of photons. The simplest purely weak process is the exchange of a W boson between a muon and an electron neutrino, in which the muon turns into its neutrino and the electron neutrino turns into the electron [see Figure 34.6(a)]. The exchanged particle could be interpreted as a W^- moving to the right or a W^+ moving to the left; the two interpretations are completely identical. This diagram represents a μ^--ν_e scattering into an electron and a muon neutrino. While this process is important in its own right, it becomes more interesting when we use the rules of Box 32.2.4, and turn the electron "leg" of the Feynman diagram as shown in Figure 34.6(b), in which the sign of the W boson has been suppressed, allowing either one of the two interpretations given above.

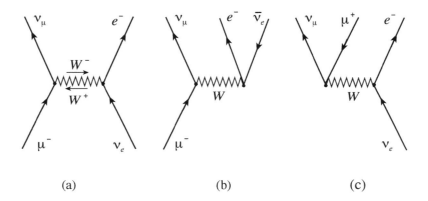

Figure 34.6: (a) A simple exchange of W^- (or W^+) in a μ^--ν_e scattering. The exchange can be considered as a W^- moving to the right or a W^+ moving to the left. (b) Muon decay process obtained from (a) by twisting a "leg." (c) Although a good diagram, this process cannot go due to energy conservation.

Figure 34.6(b) shows a muon in the initial state and three particles in the final state. It is the process of muon decay encountered in Equation (33.3). You may be tempted to turn the muon leg of the Feynman diagram of Figure 34.6(a) to obtain Figure 34.6(c), which for all reasons is a good Feynman diagram except for one: energy conservation. While Figure 34.6(b) is energetically possible due to the large muonic mass (muon's mass is almost 200 times the electron's mass), Figure 34.6(c) cannot go because the initial neutrino is (almost) massless.

What do you know? 34.4. What can be exchanged between e^- and μ^- if they, and only they, both appear in the initial state of a Feynman diagram?

Some simple vertices, which were forbidden in QED can take place in the electroweak theory. For example, the second diagram in Figure 32.5(b) is impossible, because the initial photon is massless while the final particles are massive. However, the analogue of that diagram, such as the four shown in Figure 34.7 are possible physical processes. In fact, these diagrams describe how the massive gauge bosons of the electroweak interaction decay, and all of these decay processes have been observed and measured experimentally, in perfect agreement with the theoretical predictions.

BIOGRAPHY

Sheldon Glashow was born in 1932 to a family who immigrated to the US to find freedom and opportunity denied to Jews in Czarist Russia. From an early age, partly due to the encouragement of his elder brother, Glashow knew that he wanted to be a scientist. At the Bronx High School of Science, he became friends with a group of bright students that included Steven Weinberg. The group spurred one another to learn physics while commuting on the New York subway. Sheldon and Steven attended Cornell University, where they once again became part of a talented class, including many future prominent scientists. After getting his BS from Cornell, Glashow went to graduate school at Harvard University in 1954, and wrote his PhD thesis under the supervision of Julian Schwinger, a codiscoverer of QED and a Nobel laureate. After several years at Niels Bohr Institute, CERN, Stanford, and Berkeley, Glashow returned to Harvard University in 1966 where he has remained since. He shared the 1979 Nobel Prize in physics for his work in the unification of the electromagnetism and the weak nuclear force.

> **What do you know? 34.5.** What can be exchanged between e^- and ν_μ if they, and only they, both appear in the initial state of a Feynman diagram?

🧺 Food for Thought 🌱

FFT 34.2.4. Weak interactions had been studied both theoretically and experimentally prior to the WSG model. This model (plus other models proposed at the same time) suggested new experiments which were performed in the 1970s. One such experiment, whose sheer possibility tells of the enormous progress in the physics of the twentieth century, is the reaction $\nu_\mu + e^- \rightarrow \nu_\mu + e^-$. Why?

Recall from Section 31.3.1 how the neutrinos of the 1930s β-decay baffled physicists to the point of abandoning the energy conservation. Through a persistent study of these elusive particles, we came to know them, and to realize that they can penetrate thicknesses of lead not meters, kilometers, or thousands of kilometers long but light years long! As elusive as they were, we found many of their properties whenever we could catch a fleeting glimpse of them. But there is no limit to mankind's scientific ability. By the 1970s we had the neutrinos under such a tight control that we could send them in a collision course with electrons!

From the point of view of Feynman diagrams, the reaction mentioned above is interesting because the only way that it can go is through the diagram $\rightsquigarrow$, in which the exchanged gauge boson can only be a Z^0: it can't be a photon because photons can couple only to charged particles, and ν_μ is neutral; it can't be a W^+ or W^-, because the initial and final particles at each vertex have the same charge. This reaction was one of the first strong indirect proofs of the existence of the *neutral weak gauge boson*.

The more recent data from the ν_μ-e^- scattering, as well as the data from the earlier days of the weak interactions could be used to predict the masses of $W^\pm$ and Z^0.[14] By the latter half of the 70s the masses of these particles could be predicted. Based on the available data, the model set these masses at $M_W = 1.4 \times 10^{-25}$ kg and $M_Z = 1.6 \times 10^{-25}$ kg.

---

[14]These data were not terribly accurate, as the techniques used to obtain them were at their earliest stages. The data improved considerably over time.

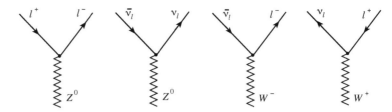

Figure 34.7: All these diagrams are possible physical processes. They are precisely how the massive weak bosons decay. The l represents either e or μ.

34.2.2   The Quark Sector

Weak interactions are not confined to the leptons. In fact, the earliest weak interactions were observed in nuclei—thus the name "weak nuclear force"—in the process of the β-decay, a paradigm of which is the disintegration of neutron into a proton, an electron, and an antineutrino (see Section 31.3.1). The quark structure of the neutron and proton suggests that in this disintegration, one of the down quarks in neutron changes to an up quark at a W vertex much like the μ-ν_μ vertex of Figure 34.6(b).

If d is to change into u at a vertex, the two must belong to the same doublet, just like μ and ν_μ of Equation (34.1). In the early 1960s there were only two lepton doublets: the electron doublet and the muon doublet. The idea of arranging the quarks into doublets seemed very attractive. However, there were only three quarks, and putting u and d in a doublet would leave the s quark without a partner. Could it be that there was a fourth quark yet to be discovered? That s indeed had a partner, but the hadron in which it was hiding had not yet been detected? This idea was very "charm"ing mathematically. Glashow and his collaborators could not resist this idea, not only from an aesthetic point of view, but also from theoretical calculations which explained some of the difficulties associated with a three-quark model of hadronic weak interactions. **Charm** was the name, and c the symbol, they gave to this hypothetical fourth quark. Since s had a negative charge, c must possess an electric charge of $+\frac{2}{3}$, just like u.

November of 1974 was an exciting month for the high energy physics community. In that month, the discovery of a new particle was reported by two groups: one on the east coast at Brookhaven National Laboratory (BNL) led by Samuel Ting who called the new particle J; the other on the west coast at Stanford Linear Accelerator Center (SLAC) led by Burton Richter who called it ψ. What was surprising about this new particle (now called J/ψ) was the "very narrow peak in the cross section," as opposed to the boring broad peaks signaling very short-lived particles that occur frequently. The narrow peak meant that J/ψ lived much longer than usual, and that the physicists could identify its mass very accurately.

Actually J/ψ had been peeping through the detectors at the two coasts months before it fully exposed itself to the particle physicists. In January of 1974, the SLAC group noticed a slight "bump" in the cross section of electron-positron scattering at around 3 GeV (giga electron volt, where giga means 10^9), while plotting the cross section as a function of energy in steps of 200 MeV. Imagine a giant walking on Earth with steps that are 20 miles apart. This giant may put one step on the foot of a mountain peak on one side and the next step on the other side. If you asked him whether he saw a mountain or not, he might say that there was just a slight bump, but nothing substantial. The peak of the mountain was missed between the two giant steps.

It turned out that the 200-MeV steps were like the giant's step: they missed the peak. By reducing the step size from 200 to 10 MeV, the SLAC group unmistakably pinpointed the very sharp peak at 3.1 GeV in June 1974. However, due to some inconsistencies, they had to refine their measurement. At the same time, the BNL group also discovered a peak,

Charm or c quark shows up in the mathematical model of the weak interaction of hadrons.

Charmed quark is discovered experimentally.

but did not announce it. By early November, the inconsistencies of the SLAC group were resolved, and the discovery announced on November 11, 1974. On exactly the same day the BNL group also announced their findings. Four days later an Italian team at Frascati confirmed the observation of the J/ψ particle. This particle was identified as a bound state of a charm and an anticharm, and was given a third name: **charmonium**.

What do you know? 34.6. Draw a Feynman diagram representing the production of charmonium.

Once charmonium was discovered and the existence of the charmed quark established, physicists ventured to construct hadrons consisting of a charm and some other quarks, a hadron with a "naked charm." By 1976 the SLAC group was able to produce mesons consisting of a charm and one of the older antiquarks: $\bar{u}$, $\bar{d}$, or $\bar{s}$. The 1976 Nobel Prize went to Burton Richter and Samuel Ting for their discovery of the charmed quark.

No sooner had the charmed quark peeped through one detector at SLAC than a new lepton started to crawl out of another. As early as 1974, a group at SLAC led by Martin Perl was seeing certain "$e\mu$" events from the collision of electrons with positrons at very high energies. These events were unique in that aside from an electron and a positive muon (or a positron and a negative muon), no other charged particle or photons were produced as end products. In a typical e^+-e^- collision a lot of charged particles and photons are also produced; but not in these $e\mu$ events. Perl suggested that these events come from the production of *heavy leptons* (leptons that are much heavier than e and μ) and the subsequent decay of these leptons into electrons and muons. For example, in the reaction $e^+ + e^- \rightarrow \ell^+ + \ell^-$, in which ℓ represents a heavy lepton, ℓ^+ could decay into a μ^+ and neutrinos and ℓ^- into an electron and neutrinos. It took Martin Perl and his collaborators four more years to convince the physics community that what they had discovered in 1974 was the τ lepton or tauon,[15] and by the end of 1978 all the properties of tauon was nailed down. One half of the 1995 Nobel Prize went to Martin Perl for his discovery of tauon.[16]

Tauon is discovered at SLAC.

The fact that the third lepton was discovered 40 years after the second, and the second 40 years after the first may be effective ammunition for a numerologist, but to a physicist it is only a coincidence. The discovery of **tauon** τ in 1977—40 years after the discovery of muon in 1937 and 80 years after the discovery of electron in 1897—was a big surprise for all physicists.[17] Despite its huge mass (a τ^- is 3500 times heavier than an e^-), it is still a point particle.

Tauon has its own neutrino ν_τ, and the two carry a **tauon flavor**. Because of its huge mass, tauon can easily decay into a muon or an electron, each reaction carrying a neutrino and an antineutrino to preserve flavor:

$$\tau^- \rightarrow \mu^- + \bar{\nu}_\mu + \nu_\tau$$
$$\tau^- \rightarrow e^- + \bar{\nu}_e + \nu_\tau. \tag{34.2}$$

In each case, there is a ν_τ to carry the initial tauon flavor, and an antineutrino to cancel the muon or electron flavor.

What do you know? 34.7. Draw the Feynman diagrams representing the tauon decays of Equation (34.2). Hint: Look at the Feynman diagram of the muon decay in Figure 34.6.

---

[15]They chose τ from the Greek $\tau\rho\acute{\iota}\tau o\nu$ for "third"—the third charged lepton.

[16]The other half went to Frederick Reines for another significant work: his detection of neutrinos.

[17]To put all numerological anticipations to rest, we should remark that there are strong theoretical arguments against a "fourth generation" lepton in 2017 or anytime in the future!

A fifth quark, called **bottom** is discovered at Fermilab.

With the word of the discovery of a heavy lepton getting around in the experimental particle physics circles, the prospect of yet another quark loomed over many accelerator labs. If charm raised the "quark sector" to the level of the then-existing two lepton pairs, the discovery of tauon (and its neutrino) demanded a new quark doublet. As early as 1974, there were mixed signals coming from detectors at Fermilab that hinted at a new hadron. However, these "mixed" signals were not untangled until 1977, when the discovery of Υ— pronounced **upsilon**—at Fermilab was announced. Υ, with the enormous mass of 9.4 GeV, was shown to be a bound state of a $-\frac{1}{3}$-quark and its antiquark. Because of its negative electric charge, this quark was thought to be the lower member of a doublet just as the strange and down quarks are the lower members of their doublets. For this reason, the quark was named **bottom** and given the symbol b.

Obviously, if you have a bottom quark, you need a **top quark** as well. The discovery of the bottom quark at Fermilab started the "race to the top" in all laboratories capable of accelerating particles (and their antiparticles) to a sufficiently high speed. After the unsuccessful searches at "lower" energies, many laboratories dropped out of the race. Only Fermilab and CERN remained the contenders. With the bottom quark mass at about 5 GeV (almost half the mass of the Υ consisting of the equally massive b and $\bar{b}$), physicists anticipated a mass of 10, 15, or 20 GeV for the top quark. However, cranking up the Υ-production energy by a factor of 4 or 5 yielded no results. Factors of 10 and 15 to production energies of 50 or 75 GeV also showed no sign of the top quark. After 18 years of experimenting, Fermilab reported the detection of a top-antitop bound state at about 350 GeV in 1995, giving an approximate mass of 175 GeV to the top quark, a 35-fold increase over the mass of the bottom quark!

After an 18-year chase, the top quark is captured at Fermilab.

How do the six quarks pair up together? While in the case of leptons it was clear which neutrino belonged to which charged lepton [see Equation (34.1)], the assignment of quark partners is not as clear-cut. How do we know that the (negative) s-quark belongs to the (positive) c-quark and not to u or t? In fact, we don't! And when we pair up the quarks, we allow this possibility and write the pairs as

$$\begin{pmatrix} u \\ d, s, b \end{pmatrix}, \quad \begin{pmatrix} c \\ s, d, b \end{pmatrix} \quad \Rightarrow \quad \begin{pmatrix} t \\ b, s, d \end{pmatrix}, \tag{34.3}$$

where it is understood that the pairings of the three negative quarks are not equal. For example, u pairs mostly with d, occasionally with s, and rarely with b. In the above equation, the dominant negative member of the doublet has been written first.

Neutrino oscillation (or neutrino mixing) points to the massiveness of neutrinos.

It turns out that this "polygammy" of the quarks is related to the fact that quarks are massive. The "monogamous" partnership of neutrinos with their corresponding negative leptons was based on the assumption that neutrinos were massless; and until recently, this assumption seemed to be consistent with all experimental observations. However, since the late 80s there have been a growing number of observations that indicate a neutrino "mixing." The earliest and strongest evidence came from the Sun, within which neutrino-producing nuclear processes take place. These processes indicate a certain ν_e to ν_μ abundance ratio, which is *not* observed on Earth. The explanation is that on the way to Earth, neutrino's transform into one another; and this can happen only if there is a mixing of ν_e, ν_μ, and ν_τ similar to the d, s, b mixing in Equation (34.3).

Is there a fourth generation of leptons and quarks? Despite an exponential increase in the energy of the accelerators, no new leptons have been discovered since 1977; and although only four years after the announcement of the discovery of charm—the hidden partner of the strange quark—the bottom member of a new generation of quarks showed up at Fermilab, no new generation of quarks has been observed ever since. In fact, there are strong theoretical arguments in favor of only three generations of quarks and leptons. Table 34.1 summarizes our present knowledge of the fundamental constituents of matter.

Leptons				Quarks			
Name	**Symbol**	**Mass**	**Charge**	**Name**	**Symbol**	**Mass**	**Charge**
Electron	e	0.5109989	-1	Up	u	1.5-4.5	$+\frac{2}{3}$
Electron neutrino	ν_e	< 0.003	0	Down	d	4-8.5	$-\frac{1}{3}$
Muon	μ	105.65835	-1	Strange	s	80-155	$-\frac{1}{3}$
Muon neutrino	ν_μ	< 0.19	0	Charm	c	1000-1400	$+\frac{2}{3}$
Tauon	τ	1777	-1	Bottom	b	4000-4500	$-\frac{1}{3}$
Tauon neutrino	ν_τ	< 18.2	0	Top	t	175000	$+\frac{2}{3}$

Table 34.1: The fundamental constituents of matter. All masses are given in MeV, and all charges as multiples of the charge of a proton.

Food for Thought

FFT 34.2.5. The dream of the alchemists of the Middle Ages was to convert base metals into gold. The grandchildren of alchemists had the less ambitious goal of combining or changing ordinary material to get more ordinary material, but in the process understand the laws of such combination. The melting of ice into water revealed a property shared by all solids; and the rust gathered on some metals pointed to a reaction of an element in the air—later identified as oxygen—with the metal. The discovery of the elements provided chemists with the powerful technique of fusing them to form molecules, old and simple as well as new and complex.

But the dream of the alchemists was still just that, a dream. Molecules, although completely different from the atoms that built them, were not *elements*. Was it possible to change one element into another? This required the identification of elements as an aggregate of atoms and the involvement of physicists. Once the atomic structure was discovered and the nature of the difference between atoms unraveled, the possibility of transforming one element into another opened up. Nuclear physics became the modern alchemy; but alas, the force of economics inhibited the production of gold out of base metals for practical and commercial use.

If chemistry is likened to the construction of buildings out of bricks, then a chemist is a builder who has no control over the kind of bricks he uses. The atoms of the elements used to form the molecules of compounds are the bricks, and a chemist has to work with these atoms given to him. The nuclear physicist is the brick maker. She takes the *naturally* available clay and molds it into different shapes and colors of bricks. The clay is the naturally occurring protons and neutrons required to form different kinds of atoms. No other form of clay exists in the universe—at least not in the present universe.

As long as protons and neutrons are the building blocks, the nuclear physicists can only transform one atom into another, perhaps even a *new kind of atom*, but nevertheless an atom with a nucleus and a number of electrons somehow configured around it. To make a different kind of clay, a new breed of builders had to be born. These were the particle physicists of the 1950s, who created new kinds of building blocks, new kinds of clay. The naturally available "clay" consists of the up and down quarks hidden inside protons and neutrons. By creating strange, charm, top, and bottom quarks, particle physicists opened the gate to the world of hadrons, objects that had not existed in the universe possibly since a tiny fraction of a second after its creation.

All this is good and exotically interesting, you may say, but what is the use of it? Clearly, we are not going to see a toothpaste made of upsilons and charmoniums in the near future! Nor will there be an opportunity for the application of hadrons made of bottom or strange quarks in our cars' engine. So why do physicists even bother with these weird particles? I have two answers to these questions.

The first answer is in the form of a question: Why do artists bother with impasto? Why do composers bother with sonatas? And poets with sonnets? Clearly, we are not going to see a toothpaste made of the Moonlight Sonata, or the portrait of Mona Lisa, or the sonnets of Elizabeth Barrett Browning? What use do these have in our lives? Isn't the answer that they are part of what makes us human? Aren't they a distinguishing trait of *homosapiens*? If so, aren't science, physics and math equally—if not better—distinguishing traits of homosapiens? Perhaps we don't question art, music, and poetry because they are vehicles of enjoyment for us. But in order to enjoy them, we must understand them; and with some training, which starts when we reach our toddling age, we do understand them. It is not, however, as easy to understand the physics behind the art of particle creation and interaction,[18] and the question of practical utility has become a comforting excuse for many people to avoid physics and other hard sciences altogether; and with that, the most important trait of our species.

The second answer is that as homosapiens we have been trying to understand our surroundings ever since we started to walk on our two feet. It is (no, it ought to be) quite natural for humans to be curious, ask questions about the universe, and try to understand it without enslaving the exercise of understanding with fetters of practical utility. While at the beginning, the evolutionary trait of curiosity was naturally instilled in every individual, today that same curiosity is present mostly in children and scientists. In a very real sense, scientists are the agents of our future evolution. If we tie this curiosity with the rope of utility, pragmatism will become the dominant ideology of our race, as it became the dominant ideology of our race influenced by the Roman empire two millennia ago; and just as then, we may fall into the abyss of another dark age.

> In a very real sense, scientists are the agents of our future evolution.

34.2.3   The Higgs Boson

We have talked about the spin-$\frac{1}{2}$ matter fields—the leptons and quarks—that participate in the electroweak interactions. We have also talked about the four gauge bosons (γ, Z^0, W^+, and W^-) that mediate the electroweak force between the matter fields. It appears that we are done. After all, what else is there except material objects and forces among them? Well, there *is* something else!

Remember the *Higgs mechanism*, whereby some of the gauge fields "eat up" the Goldstone bosons and become massive? (See Box 34.1.5 and the discussion preceding it.) In order to give mass to the gauge bosons, the Higgs mechanism introduces a *scalar* field—usually denoted by ϕ—into the theory.[19] And when you have a field, you have a particle. The particle content of the field ϕ is the **Higgs boson**; sometimes called the "God particle," because it gives mass (and somewhat erroneously, existence) to all the other particles. This is the last piece of the electroweak theory left to be discovered. It is predicted to be extremely massive, perhaps of the order of 100 to 150 GeV.

> Higgs boson is the "God particle."

With all the predictions of the electroweak having been substantiated, there is little doubt in the minds of the elementary particle physicists that the Higgs boson will eventually be captured. It is only a question of when and where. The laboratory that can pump sufficient amount of energy into its accelerator will be the first to see the "God particle." At this point, all eyes are on Geneva, where the physicists at the European Center for Particle Physics are attaching the last pieces of the largest accelerator ever, the **Large Hadron Collider** (LHC),[20] in the hope that within a few years, they can announce the discovery of the Higgs boson.

---

[18]Not because physics is intrinsically hard, but because we do not start our training in physics at an early age—if we ever do. In an unlikely world in which children learn physics and "play" in laboratories, while being completely deprived of art and music, physics is easy and enjoyable, and art and music, "hard subjects" in high school!

[19]Scalar simply means that the field (or its associated particle) has zero spin.

[20]After its first failed run!

34.3 Quantum Chromodynamics (QCD)

The last section described one of the nuclear forces, the weak one; and in the process, unified the electromagnetic force with that force. What about the strong nuclear force? What holds the protons and the neutrons inside a nucleus? More fundamentally, what holds the quarks inside protons and neutrons? To answer these questions, we reexamine the quark model.

34.3.1 The Quarks

Section 33.2.2 discussed the quark structure of hadrons. One of these hadrons, the spin-$\frac{3}{2}$ baryon we called Δ^{++}, consists of three u quarks, all of whose spins are up. Now suppose that we were to "bring" three u quarks together to form a Δ^{++}, in analogy with forming atoms by bringing electrons and nucleons together. In the latter case, Pauli exclusion principle limited the number of electrons one could assemble in the lowest energy state. For example, only two of the three electrons of a lithium atom could end up in the lowest energy state; the third one has to go to the next energy level. These energy levels differ slightly (a few eV), which is orders of magnitude smaller than the rest energy, $E = mc^2$, of the atom, which is billions of eV. In other words, the mass of a lithium atom is the same whether its third electron is in the second energy level, or the third, or any other higher level.

The world of quarks is a world of much higher energies. If a quark inside a hadron moves up to a higher "energy level," the identity of the hadron will change, because the difference in the energy of two levels can reach millions or even billions of electron volts, affecting the rest energy, i.e., the mass, of the hadron; and hadrons differing in their masses are different hadrons. For example, there is a group of baryons which have very similar properties to Δ^{++} except for their increasingly higher masses. It is believed that the quarks of each member of the group are in higher and higher energy levels, and that Δ^{++} itself is the ground state of the group with all its quarks in the lowest energy level. In the context of the atoms, it is like having the three electrons of a lithium atom in the lowest energy state *with all their spins pointing up*.

To allow the quarks to be in the lowest energy level of a baryon without violating Pauli exclusion principle, Gell-Mann introduced the notion of a **color charge**. Each quark in a baryon (including the Δ^{++}) has a different color; and a **red** quark differs from a **green** quark, and the two differ from a **blue** quark. You can place a red quark next to a green quark, next to a blue quark all in the lowest energy state of a baryon *with all their spins pointing up* without violating Pauli exclusion principle! Furthermore, it is the "force" of this color charge that holds the hadron together, just as the (electric) force of the positive nucleus and the negative electrons holds the atom together.

> Quarks have color charge!

We denote the color of a quark by a subscript under its symbol. For example, a red up quark is denoted by u_r, a green charmed quark by c_g, and a blue strange quark by s_b. The net color charge of a hadron is zero. Stated differently, a hadron is *colorless* (or *white*). The mathematical theory of colored quarks predicts two colorless combinations: a quark and an antiquark of the same color, and three quarks of different colors. The first combination gives the mesons, the second, the baryons. Thus, a positively charged pion could be formed from any of the following three combinations: $u_r \bar{d}_r$, $u_g \bar{d}_g$, or $u_b \bar{d}_b$; and the quark structure of a proton could be $u_r u_g d_b$, $u_r u_b d_g$, or $u_b u_g d_r$.

> There are two colorless combinations of quarks and antiquarks.

34.3.2 The Gluons

The same mathematics that classified hadrons and unified the weak and electromagnetic interactions also explains the strong nuclear force. The Lie group theory—in its local (gauge) version—puts the three color charges of quarks in a triplet and allows the resulting *eight*

gauge bosons to intermediate the force between them.[21] These gauge bosons are called
gluons, because they "glue" the quarks together inside hadrons; and the theory of quarks
interacting via gluons is called **quantum chromodynamics** or QCD.

BIOGRAPHY

David Gross was born in Washington, D.C., on February 19, 1941, to Bertram Meyer, son of immigrant
Jewish parents from Czechoslovakia-Hungary, and Nora (Faine), born in the Ukraine, immigrated to
the United States with her family after World War I. In David's family, the children were encouraged to
read at a very early age. His father and mother treated their children as intellectual equals, thus greatly
bolstering their self-confidence and interest in ideas of all kinds. In 1953, his father's job took the family
to Jerusalem. One of the great advantages of growing up in Jerusalem at that time was the absence
of television and the many other distractions present then and now in the United States. Thrown back
onto his own resources, he became an avid reader. From the age of 13, he was attracted to physics and
mathematics. He studied mathematics on his own and soon, exceeding the knowledge of his teachers,
was excused from mathematics classes. Determined to become a theoretical physicist, upon graduating
from High School, he entered the Hebrew University and majored in physics and mathematics. After
receiving a B.Sc., Gross applied to graduate schools in the United States, and was only accepted at the
University of California at Berkeley. In 1964, he started to do research under the supervision of Geoffrey
Chew, the charismatic leader of the S-Matrix "bootstrap" approach to the strong interactions. Gross
found this revolutionary new theory very exciting at first, but gradually became disillusioned. He rapidly
finished a thesis and spent most of his last year at Berkeley in thoughts of new directions. Concluding
his graduate studies in 1966, Gross was nominated to the Harvard Society of Fellows, where he started
on the path that led to the discovery of asymptotic freedom and QCD. In 1969, having had many offers
for faculty positions, he decided to go to Princeton, as an assistant professor, where he had a remarkable
group of graduate students, especially Frank Wilczek (codiscoverer of QCD) and Edward Witten (one
of the founders of string theory). He shared the 2004 Nobel Prize in physics with his student Frank
Wilczek and David Politzer.

Our discussion of Section 34.1.2 told of two ways of having short-ranged forces: spontaneous symmetry breaking (SSB), by which gauge bosons acquire mass, and confinement.
The latter is at work in the strong forces between quarks. In any nonabelian gauge theory—
i.e., any theory which has more than one particle in its multiplets—in which SSB is absent,
the gauge bosons carry charges (such as the color charge under present discussion), which
not only holds the particles together, but also the gauge bosons themselves. In the case of
the quarks, the color charges carried by the gluons makes them attract each other and the
quarks into a "sticky glue."

An outcome of the interaction among gluons is that the strength of the force between
quarks *increases* with distance. In contrast to the electrical force between two electrically
charged particles, which decreases as you pull them apart, the color force between two
quarks *increases* when you pull them apart: quarks strongly resist separation. This is
also true of the gluons themselves. So, while the quarks and gluons materially reside
inside hadrons, one cannot forcefully knock them out of hadrons, because any attempt
at separating them will encounter a strong resistance which only increases with separation.
That is why isolated quarks and gluons cannot be found: isolation means infinite separation,
and infinite separation means infinite resistance. The energy aimed at separating the quarks
will instead create other quarks and antiquarks which pair up with the original quarks to
form new hadrons (see Section 33.2.3).

The notion of color as a property of quarks originated in the quark model of hadrons
in the mid 1960s. But it was not until 1973 that it was used as the main ingredient
in a *dynamic* theory of hadrons, quantum chromodynamics. QCD has been extremely
successful in explaining and predicting the outcomes of hadronic collisions and structures.
It was because of this success that the 2004 Nobel Prize in physics was awarded to the three

Why isolated quarks and gluons are not found.

---

[21]If you are familiar with Example 34.1.2, then you may care to know that the Lie group associated with
the three colors is $SU(3)$. Therefore, the number of gauge bosons is $3^2 - 1 = 8$.

originators of QCD, David Gross, Frank Wilczek, and David Politzer.

34.4   The Standard Model

When the electroweak theory and quantum chromodynamics are put together, the result is called the **standard model** of the fundamental particles and interactions. It has been remarkably successful in explaining and predicting all the fundamental interactions[22] of all particles. As we have numerously seen, these interactions are best described by Feynman diagrams. Let's take a look at some.

Standard model is electroweak theory and QCD combined.

BIOGRAPHY

Frank Wilczek's grandparents emigrated to the US from Poland on his father's side and from Italy on his mother's side in the aftermath of World War I. Both his parents were born on Long Island, and he was born in 1951. His parents were children during the time of the Great Depression, and their families struggled to get by. This experience shaped many of their attitudes, and especially their aspirations for Frank. Frank went to public schools in Queens, and was fortunate to have excellent teachers. Because the schools were big, they could support specialized and advanced classes. At Martin van Buren High School there was a group of thirty or so students who went to many such classes together, and both supported and competed with one another. More than half of them, including Frank, went on to successful scientific or medical careers. Frank went to the University of Chicago with large but amorphous ambitions. He read voraciously in many subjects, but he wound up majoring in mathematics, largely because doing that gave him the most freedom. During his last term at Chicago, he took a course about the use of symmetry and group theory in physics and he felt an instinctive resonance with the material. For graduate work, he went to Princeton University's mathematics department, but kept a close eye on what was going on in physics. He became aware that deep ideas involving mathematical symmetry were turning up at the frontiers of physics; specifically, the gauge theory of electroweak interactions. He started to talk with a young professor named David Gross, and his proper career as a physicist began. His collaboration with Gross led to the discovery of QCD for which they both won the Nobel Prize in Physics in 2004.

The most common procedure for creating heavy hadrons is the collision of particles and antiparticles. Fermilab uses protons and antiprotons, while electrons and positrons are used at SLAC. In either case, a fundamental particle (electron or quark) collides with its antiparticle at very high energies, converting some of the energy of motion into matter. Figure 34.8 shows two important examples of such processes. In Figure 34.8(a) an electron collides with a positron, both of which annihilate into a virtual photon, and the energy carried by the photon turns into a charm and anticharm pair, which could bind to form a charmonium (or could bind with other quarks and antiquarks produced similarly to form a variety of hadrons). In Figure 34.8(b) an energetic up quark in a proton annihilates an equally energetic antiup quark in an antiproton into a virtual photon, whose energy turns into a bottom and antibottom pair, which could bind to form a bottomonium or—if other quark-antiquark pairs are produced—a variety of other hadrons.

The gauge boson exchanged between the participating particles determines the kind of interaction taking place. The exchange of photon in the first two diagrams of Figure 34.8 makes those two processes electromagnetic. The first process cannot be strong because electrons do not have color charge. The only other alternative is the exchange of Z^0, which although possible, contributes negligibly to the process, as detailed calculations show. On the other hand, the exchange particle of Figure 34.8(b) could be either a Z^0 or a colored gluon.

While the first two diagrams of Figure 34.8 are collision processes, the last two represent weak decays. Figure 34.8(c) is the beta decay of a d quark. This, of course cannot happen

---

[22]Gravity is an exception. It plays no role in the energies and distances available in accelerators. However, it becomes important in energies that were once available moments after the big bang.

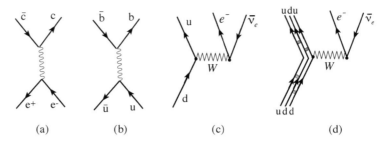

Figure 34.8: Examples of processes in the standard model. (a) A free electron collides with a free positron to produce a charm anticharm pair. (b) An up quark (in a proton) collides with an antiup quark (in an antiproton) to produce a bottom and an antibottom. (c) A weak d quark decay. (d) Neutron beta decay.

in isolation; however, if the d quark is inside a neutron (bound by the exchange of colored gluons to two u quarks), then the decays can represent a neutron beta decay as shown in Figure 34.8(d). These and other, more complicated diagrams corresponding to actual physical processes, can be translated into mathematical expressions, which theoretical physicists have calculated, and the results of calculations have been compared with high energy experiments. In these comparisons, the standard model has emerged as a remarkably successful theory in describing three (electromagnetism, weak nuclear, and strong nuclear) of the four fundamental forces of nature.

34.5   Grand Unification

Our experience with gauge theories has shown that the more elaborate the symmetry of the theory (the Lie group), the larger the number of gauge bosons, and therefore, the more forces included in the theory. For example, the electroweak theory grouped two particles together and came up with four gauge bosons, including the photon and three weak gauge fields; thus unifying electromagnetism with the weak nuclear force.

The unification of forces via the enlargement of the Lie group caught the attention of the physicists in the 70s and 80s. It was thought that one could achieve a "grand" unification by grouping more fundamental particles together and using the machinery of Lie groups and gauge theories. These **grand unified theories** or GUTs have not been as successful as the electroweak theory; however, the idea is so elegant that a brief account of grand unification is worth our while.

The fundamental particles to be grouped together are, of course, the quarks and the leptons. That there may be a kinship among these two categories of particles comes from the following observations:

- there are six leptons and six quarks coming is three doublets [see Table 34.1 and Equations 34.1 and 34.3];

- as far as we can tell, both leptons and quarks are point particles with no structure within; the radius of the electron, for example is known to be less than 10^{-18} m;

- consider the lightest pairs of leptons and quarks; $\left(\begin{smallmatrix} \nu_e \\ e^- \end{smallmatrix}\right)$ for the leptons[23] and $\left(\begin{smallmatrix} u \\ d \end{smallmatrix}\right)_r$, $\left(\begin{smallmatrix} u \\ d \end{smallmatrix}\right)_g$, and $\left(\begin{smallmatrix} u \\ d \end{smallmatrix}\right)_b$ for the quarks, where the subscripts indicate the color of the pairs; if you add all the electric charges of the quarks you get $+1$, which cancels the negative charge of the lepton pair, giving a total electric charge of zero, an indication that perhaps quarks and leptons belong to a (electrically neutral) family.

[23] Here we are ignoring the quark and neutrino mixing discussed after Equation (34.3).

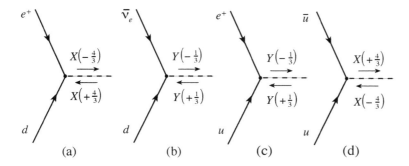

Figure 34.9: Some vertices of the GUT. (a) and (b) come from the first family of Equation (34.4) while (c) and (d) come from the second (larger) family. The quarks in the diagrams can have any one of the three colors.

The most direct generalization of the electroweak unification puts three colored quarks and an antilepton pair in one family of 5 fundamental particles and 24 gauge fields.[24] The mathematics of this particular generalization makes a larger family as well. Equation (34.4) shows two such groupings.

$$
\begin{pmatrix} d_r \\ d_g \\ d_b \\ e^+ \\ \bar{\nu}_e \end{pmatrix}, \qquad
\begin{pmatrix}
0 & \bar{u}_r & \bar{u}_g & u_b & d_b \\
\bar{u}_r & 0 & \bar{u}_b & u_g & d_g \\
\bar{u}_g & \bar{u}_b & 0 & u_r & d_r \\
u_b & u_g & u_r & 0 & e^+ \\
d_b & d_g & d_r & e & 0
\end{pmatrix}.
\tag{34.4}
$$

Note that in both cases the family is electrically neutral as a whole.

What are the 24 gauge fields? Some of them are already familiar. For example, the gauge fields that connect—in the sense of Box 34.2.1—the first three members of the first family of Equation (34.4) are the eight gluons; and the ones that connect the last two members of the family are the four electroweak bosons. The remaining 12 gauge fields, generically denoted by X (whose electric charge is $\pm\frac{4}{3}$) and Y (whose electric charge is $\pm\frac{1}{3}$) are responsible for new interactions. Some of the vertices of these new interactions are shown in Figure 34.9. These vertices promise an intriguing possibility.

Take the last vertex of Figure 34.9; rotate it 90° counterclockwise (see Box 32.2.2 to find out how processes are made out of vertices); take the first vertex of Figure 34.9; rotate it 90° clockwise; now tie the two gauge bosons together to obtain Figure 34.10(a). This figure describes a physical process in which two up quarks turn into a positron and an antidown quark.

Now imagine that the initial two u quarks are inside a hadron, say a proton, shown as an oval in the lower part of Figure 34.10(b). At the end of the process, there will be three fundamental particles: a down quark remaining intact from the initial proton; an antidown quark coming from the interaction of the initial two up quarks; a positron also coming from the two up quarks. The d and the $\bar{d}$ interact via gluons and confine themselves inside a hadron, the π^0, shown as an oval in the upper part of Figure 34.10(b). The positron, having no color charge, cannot interact strongly, and escapes to outside (to be most likely annihilated by one of the many electrons awaiting its arrival). The upshot of all of this is the intriguing decay of a proton: $p \to \pi^0 + e^+$.

---

[24] In the language of Lie groups, the symmetry group is $SU(5)$, leading to $5^2 - 1 = 24$ gauge bosons.

> **What do you know? 34.8.** Can you come up with another Feynman diagram for proton decay?

The proton decay may seem to be a drawback for GUTs, because protons are the essential ingredient of every element—including all the material *we* are made of; and the decay of protons is tantamount to the decay of everything! This clearly is not happening. So why do we even bother to talk about GUTs? The answer lies in the *rate* at which the proton decays. Recall from Section 31.2.1 that any decay process is characterized by a half-life. If the half-life is short, we can observe a lot of decays; if the half-life is long, we can observe some decays; if the half-life is very long, we may observe only a few decays; and if the half-life is incredibly long, we may have a hard time observing *any* decays.

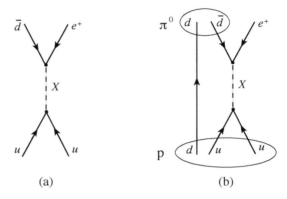

Figure 34.10: (a) Two u-quarks interact via an X particle to turn into a positron and a d-quark. (b) The proton decay $p \rightarrow \pi^0 + e^+$.

In the case of the proton, the rate of decay depends on the mass of the X boson: the more frequent the decay the less massive the X boson, and vice versa (see Example 34.1.4). Experiments have shown that the half-life of the proton is at least 10^{31} years.[25] This sets the mass of X at about 10^{14} GeV. The electroweak bosons, which took an incredible amount of resource for their discovery in 1983, are less than 100 GeV. To provide more and more energy, the circumference of the accelerators need to be increased more and more. For the electroweak bosons this circumference was a few kilometers. To produce X, the circumference has to increase to many times the Earth circumference!

While it seems impossible to observe the X particle directly, proton decay, as rare as it may appear, *is* actually observable. Equation (31.2), describing any decay process, is a statistical equation. It gives the (approximate) number of particles left after some time. To see *half* of particles decay, we, of course, need to wait one half-life. But if we are content with observing only a few decays, we don't have to wait that long. We can turn the question around: How many initial protons do we need in order to see 100 decay events in one year? With $t = 1$, $N(t) = N_0 - 100$, and $t_{\text{half}} = 10^{31}$, Equation (31.2) gives $N_0 = 1.44 \times 10^{33}$. With 3.3×10^{29} protons per ton of water, we need 4360 tons of water to see approximately 100 proton decays in one year. **Example D.34.1 on page 66** of *Appendix.pdf* gives the numerical details.

In the summer of 1982 a group of experimental particle physicists filled a pool 22.5 m long, 17 m wide, and 18 m deep in a salt mine near Cleveland, Ohio with 8000 tons of water. The mine, safely located 600 m under Lake Erie, was fairly immune from the hundreds of thousands of cosmic particles hitting the surface of the Earth every second. Furthermore, an

---

[25] For comparison, note that the age of the universe is only 1.37×10^{10} years!

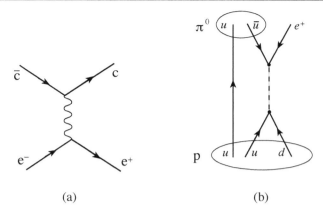

Figure 34.11: (a) Production of charmonium from electron positron annihilation. (b) Another diagram for the proton decay $p \to \pi^0 + e^+$.

extra 2 m thickness of water on all six sides of the pool was designated as a shield rather than a source of decaying protons. This left 3300 tons of water as the source, potentially yielding about 75 proton decays per year. By early spring of 1983, after 204 days of observation and hoping to see some of the 40 expected events, the group reported the absence of the decay of protons, dealing a violent blow to GUTs.

34.6   End-of-Chapter Material

34.6.1   Answers to "What do you know?"

34.1. A square is the most symmetric; a parabola is the least symmetric.

34.2. x^3 and $x^2 + x^3$ are not symmetric under change of sign. The rest of the expressions are.

34.3. The gauge particle in the left diagram must be positively charged to neutralize the negative lepton in the initial state and make the initial total charge equal to the final charge, which is zero. The gauge particle in the right diagram must be negatively charged to make the final total charge equal to the initial charge, which is negative.

34.4. Consider the μ^- vertex. You can have a ν_μ or μ^- at the final end of this vertex. If ν_μ is at the final end, then the gauge boson would be a W^- *moving away* from the vertex. But then the e^- vertex cannot accept this W^-, because, with electron having another minus charge, the final end of the electron vertex would have to be doubly charged, and there are no such particles. Therefore, the final end of the muon vertex has to be a μ^-, in which case the gauge particle has to be neutral, i.e., a photon or a Z^0. Then this neutral gauge particle can be absorbed by the e^- vertex which would have an e^- at the final end of the vertex.

34.5. Neutrino cannot have a photon at its vertex, because it is neutral. So the only exchanges could be W or Z^0, and both are possible. The W exchange would give rise to $e^- + \nu_\mu \to \nu_e + \mu^-$ reaction, and the Z^0 exchange would yield $e^- + \nu_\mu \to e^- + \nu_\mu$.

34.6. The simplest diagram has an electron and a positron in the initial state, such as the diagram in Figure 34.11(a). This is one of the ways charmonium is actually produced. Instead of electron and a positron one can use a quark and its antiparticle. For example, an up quark in a proton and an antiup quark inside an antiproton could replace the electron and positron of the figure.

34.7. On the left of the μ^--decay diagram, replace μ^- with τ^- and ν_μ with ν_τ. This will give the second decay mode in Equation (34.2). On the right of the diagram you just obtained, replace e^- with μ^- and $\bar{\nu}_e$ with $\bar{\nu}_\mu$ to get the first decay mode in Equation (34.2).

34.8. The quark content of π^0 is not only $d\bar{d}$, but also $u\bar{u}$. So, a diagram like the one in Figure 34.11(b) is another candidate for proton decay.

34.6.2 Chapter Glossary

Bottom The fifth flavor of quarks. It was discovered in 1978.

Charm The fourth flavor of quarks. It was discovered in 1976.

Charmonium A meson composed of a charm quark and its antiquark.

Color A property (charge) of quark taking part in strong nuclear interaction.

Compensating Fields Same as gauge fields.

Gauge Boson Particles introduced in gauge theories, which become ultimately responsible for intermediating various interactions.

Gauge Fields The fields introduced in gauge theories, whose associated particles are called gauge bosons, which in turn are ultimately responsible for intermediating various interactions. Gauge fields appear in a theory when the global version of its symmetry is changed to a local version.

Gauge Theory A mathematical theory based on the idea of the local symmetry of some physical theories. All fundamental forces of nature are now described by gauge theories.

Global Symmetry A symmetry whose operation is uniform across all space and time.

Gluons The massless gauge particles of quantum chromodynamics responsible for the strong nuclear force.

Goldstone Bosons Certain massless particles, which appear in globally symmetric theories after the symmetry is spontaneously broken.

Grand Unification An attempt at inventing a gauge theory which unifies the strong and the electroweak interactions.

Higgs Boson The surviving spin-zero particle of the Higgs mechanism.

Higgs Mechanism A process by which some of the massless Goldstone bosons disappear and a corresponding number of gauge bosons acquire mass.

Lie Groups A branch of advanced mathematics which combines two areas of mathematics, namely group theory and analysis. Lie groups are the means by which the idea of symmetry is implemented in physics.

Local Symmetry A symmetry whose operation differs at different points of space or different instants of time.

Quantum Chromodynamics A gauge theory based on the color charge of the quarks, whose gauge bosons, the gluons, cause a force that increases with distance, thus "confining" the quarks inside hadrons.

Spontaneous Symmetry Breaking A process through which the lowest energy state of a theory, called a *vacuum*, is chosen. As a result of this choice, the symmetry associated with the theory is spontaneously lost.

Symmetry An operation performed on mathematical objects such as figures or expressions. These objects are generally changed under the operation of symmetry. If an object does not change, it is said to be symmetric under that particular operation.

Top The sixth (and last) flavor of quarks. It was discovered in 1995.

34.6.3   Review Questions

34.1. What was the bootstrap hypothesis? Did it predict any phenomenon? How long did it survive? What was wrong with it?

34.2. What is symmetry? How is global symmetry different from local symmetry? Give an example of a symmetric figure. Under what kind of symmetry operation is it symmetric?

34.3. Suppose a sheet of transparency is your mathematical field. What would be a global symmetry operation? What would be a local symmetry operation? If a theory is globally symmetric, would a uniformly colored sheet of transparency work in that theory?

34.4. What are needed to make a globally symmetric theory locally symmetric? What are they generically called? What is the significance of these objects?

34.5. How does spontaneous symmetry breaking occur? What kind of particles are created when a global symmetry spontaneously breaks? What happens to these particles when a local symmetry spontaneously breaks?

34.6. What is confinement and how does it come about?

34.7. State the Feynman rule for a general vertex when particles are grouped together. What particles are attached at a vertex?

34.8. How are the leptons grouped in the WSG model? How many gauge bosons does the model have? How many of them are massive and how many massless? Which fundamental force(s) does the model describe? Have the gauge bosons been discovered?

34.9. What happens when you try to group the existing three quarks in the WSG model? What do you have to do to make the grouping work? What charge does the new particle have? Was it ever discovered? When?

34.10. How many pairs of fundamental particles do we have altogether? When was the last of these particles discovered? What is its name?

34.11. Have all the particles introduced in the WSG model been discovered? If not, how many are awaiting discovery? Where are they expected to be seen?

34.12. What property of quarks participates in the strong interaction? Which gauge theory explains the strong interaction of hadrons? What is the name of the gauge bosons of this theory? How does the strong force between two quarks change with the distance between them?

34.13. What is the standard model of the fundamental forces? In the context of the standard model, can you draw the Feynman diagram responsible for the nuclear beta decay (neutron decay)?

34.14. What is a grand unified theory? In the simplest GUT, how many fundamental particles are grouped together? How many gauge bosons does it have? What is the most important prediction of this GUT? Has this prediction been confirmed by experiments?

34.6.4   Conceptual Exercises

34.1. Suppose a theory is represented by a square and another by a hexagon. If these theories were made locally symmetric under the operation of rotation, which shape would give rise to more gauge bosons?

34.2. Imagine rotating a planar shape by two different angles in succession. If you interchange the order of rotations do you end up with the same final result? Now imagine rotating a *solid* shape by two different angles about two different axes. If you interchange the order of rotations do you end up with the same final result? Hint: See Figure 34.1.

34.3. With transparency sheets representing fields and coloring representing the symmetry operation, assume one theory accepts clear sheets as well as blue and green sheets and yields the same results. A second theory accepts clear, blue, green, and yellow sheets and yields the same results. When localized, which theory is more capable of incorporating several interactions?

34.4. A theory that is globally symmetric gives rise to 3 massless Goldstone bosons upon spontaneous symmetry breaking. To make the theory locally symmetric, 5 gauge bosons are needed. When this locally symmetric theory is spontaneously broken, how many massive and massless gauge fields will result?

34.5. Which of the following reactions are possible in the electroweak theory? For the ones that are possible, draw their Feynman diagrams. Remember that γ (the photon) can attach to any charged particle.

$$\gamma + \nu_e \to W^+ + e^- \qquad \gamma + \nu_e \to Z^0 + \nu_e$$

$$\nu_e + \nu_\mu \to \nu_e + \nu_\mu \qquad \nu_e + \mu^- \to \nu_e + \mu^-$$

$$W^- \to \bar{\nu}_e + e^- \qquad Z^0 \to \bar{\nu}_e + \nu_\mu$$

$$e^- + \mu^+ \to \nu_e + \bar{\nu}_\mu \qquad \mu^- + e^+ \to \nu_e + \nu_\mu$$

34.6. Charmonium is made up of a charm and an anticharm. Can a charmonium decay into two photons? Hint: See Figure 32.4(c).

34.7. Two groups of fundamental particles are assembled in a pair of columns as follows:

$$\begin{pmatrix} p_1 \\ p_2 \\ p_3 \end{pmatrix} \quad \text{and} \quad \begin{pmatrix} q_1 \\ q_2 \\ q_3 \end{pmatrix}.$$

The physical properties of these particles are summarized in Table 34.2. The gauge particles resulting from both groupings are the same, and all gauge particles with the same electric charge are identical. What are the electric charges of the gauge particles?

Particle	p_1	p_2	p_3	q_1	q_2	q_3
Electric charge	+1	0	−1	+1	0	−1
Mass in MeV	50	250	400	200	450	900

Table 34.2: Table of some hypothetical particles.

34.8. Referring to Conceptual Exercise 34.7 and Table 34.2, suppose you have a vertex with p_1 in the initial state, p_2 in the final state, and (of course) a gauge particle. What pairs of particles can be attached to the other end of the gauge particle? Draw all such Feynman diagrams.

34.9. By rotating the diagrams (or parts thereof), change the diagrams of the previous question in such a way that only one particle appears in the initial state (a decay diagram). Which of these decay diagrams can correspond to real decays?

34.10. Referring to Conceptual Exercise 34.7 and Table 34.2, suppose you have a vertex with p_1 in the initial state and p_3 in the final state.
(a) What pairs of particles can be attached to the other end of the gauge particle?
(b) By rotating the diagrams, change them into decay diagrams.
(c) Which of these decay diagrams can correspond to real decays?

34.11. Referring to Conceptual Exercise 34.7 and Table 34.2, now consider a vertex with p_1 in both the initial and final states.
(a) What pairs of particles can be attached to the other end of the gauge particle?
(b) By rotating the diagrams, change them into decay diagrams.
(c) Which of these decay diagrams can correspond to real decays?

34.12. Suppose you try to pull out one of the quarks inside a proton by colliding it with a very energetic probe (a high energy particle such as another proton). Will a quark come out of the proton? Why or why not? What happens to the energy of the incoming particle?

34.13. What theory explains the neutron decay? Which constituent of the neutron gets transformed? Which of the four fundamental forces of nature is involved in the decay?

34.14. Four of the many vertices of the GUT described by the groupings of Equation (34.4) are shown in Figure 34.9. Draw five more vertices and designate the charge of the gauge particle of each vertex. Pay attention to the direction of the arrows on the quark and lepton lines.

Chapter **35**

New Age "Physics"

The decade of the 60s was a mixture of the good, the bad, and the nonsense. The unpopularity of the American war in Viet Nam spread rapidly across campuses, into the streets, and beyond borders. The misconceived association of science with the war initiated an antiscience wave that propagated with unprecedented speed into all strata of human population, including the academia. The antidote of science very naturally emerged as the folklore and tradition of the very peoples at which the atrocities of the war were aimed.

The oriental way of life became a panacea of the New Age. The new generation of "revolutionaries" protested against everything rooted in western civilization, and strove to replace them with their eastern counterparts: scientific medicine gave way to Qi-gong, acupuncture, and a host of other alternative medicine practices; western religion and philosophy were replaced by Hare Krishna and Eastern mysticism; the mind-body exercises became increasingly popular.

The New Age Movement has a history that goes back to the nineteenth century spirituality, but it picked up momentum in the 1960s in the midst of a "cultural revolution" in which all established institutions and ideas, including religion, came under scrutiny by the new generation of students. The "establishment" became a hated word, and it eventually encompassed anything that was "old." The more authoritative western religions were replaced by the decentralized Eastern theosophy. In the absence of any authority, ideas from atheism, monotheism, polytheism, Buddhism, Chinese folk religion, Christianity, Hinduism, Islam, and Judaism, as well as astrological,[1] ecological, and archaeoastronomical ideas could coexist in the New Age spirituality.

Among the adherents of the New Age Movement were some physicists, who, in the crowded marketplace of such diverse ideas, found a kiosk to place physics. But, being a science, physics was at odds with the rest of the ideas sold in the marketplace. Was there a way of reconciling physics with the spiritual world of the New Age Movement? Many came to think so. And a pioneer was Dr. Fritjof Capra, who, "sitting by the ocean one late summer afternoon" in 1969, saw "cascades of energy coming down from outer space, in which particles were created and destroyed in rhythmic pulses; [he] 'saw' the atoms of the elements and those of [his] body participating in [a] cosmic dance of energy; [he] felt its rhythm and [he] 'heard' its sound, and at that moment [he] *knew* that this was the Dance of Shiva, the Lord of Dancers worshiped by the Hindus." With these words in the preface of the first edition of his influential book *The Tao of Physics* (abbreviated as *TP* here) Capra sets out to "establish" his claimed parallel between modern physics and Eastern mysticism.

What is it about modern physics that attracts the New Agers? Several things: it is the cornerstone of modern technology (therefore, it is the basis of the physical surrounding of

---

[1]The New Age is sometimes called the *Age of Aquarius*.

the New Agers), it has reached the smallest dimensions of matter and the farthest corners of the universe (therefore, it fits perfectly with the "universal" message of the New Age Movement), it is counterintuitive (therefore, contrasting the intuitive *established* classical physics), it is highly abstract (therefore, prone to interpretation), and, most importantly, it cannot be expressed in any human language. If you cannot "express" quantum physics and relativity, the New Agers may argue, then "any" expression (and interpretation), including mine, is viable. Modern physics, especially quantum theory, is more fictional than any speculative fiction writer could ever imagine. And the New Agers, as we shall see later in this chapter, use this fictional character to connect modern physics to their spirituality. But first, let us see how great physicists themselves put a mysterious halo around the quantum theory and paved the way for the New Agers to abuse it.

35.1   Sins of the Fathers

The founders of our civilization, whether they are scientists, mathematicians, poets, writers, musicians, or artists, share many of the same strengths and weaknesses that we possess. Outside their areas of expertise they are quite ordinary characters, who, in particular, can be poor judges of politics, religion, social structure, economic outlook, etc. Newton believed in alchemy; Fourier, one of the greatest mathematical physicists of the nineteenth century, was a close friend of Napoleon's and joined him on some of his invasions; Einstein encouraged President Roosevelt to initiate the development of atomic weapons; Brian Josephson, the discoverer of an important quantum mechanical effect bearing his name and a Nobel Laureate, is a strong believer in the paranormal; many great physicists participated in the Manhattan Project and later regretted it. But these mistakes are not made right because of the science of their makers, just as the science is not made wrong because of the mistakes of its discoverers: *It is the message that counts not the messenger.*

> It is the message that counts not the messenger.

Quantum theory has been the subject of the most cruel abuses by the outsiders as well as, sad to say, the insiders, including some of the originators of the theory. The two most important contributing factors are probably the quantum "weirdness" and quantum universality; the latter making the theory's presence known to a wide audience, some of whom use the "interpretation" of the theory to legitimize their stance.

When we put the universality of the theory in the context of its time we may be able to understand the tendency for its abuse. The 1920s were the years in which the Bolshevik revolution had just succeeded in establishing the Soviet Union as an alternative socioeconomic system which was foreseen by the believers to eventually replace capitalism. They were also the years in which Freud was enjoying the credit of another kind of revolution: a revolution in the "science" of the mind. These two revolutions offered alternatives for the beliefs of the existing society, and many, including some physicists, found it quite natural to offer theirs. These motives plus the desire of scientists to be able to communicate the excitement of their discoveries to the public, and the mythically enchanting flavor that quantum theory could add to this communication, made the situation ripe for disaster.

It is unfortunate that some of the most harmful abuses of the quantum theory were initiated by the very people who created it. Bohr, Heisenberg, Born, Pauli, and others, prompted by the power and universality of the quantum theory, and ensnared by the charm of its "interpretation," proposed some of the most fatuous ideas, and made some of the most provocative statements in the realm of politics, religion, philosophy, and psychology.

These physicists' utterance of such amateurish statements is an unfortunate circumstance spurred by an enthusiasm for their revolutionary discoveries. Woefully, these statements have been the starting points of the most strident attacks on physics, and the most bizarre applications in the New Age Movement. To expose the nature of these attacks and applications, it is therefore important to bring into the open the flaws of these hasty statements and to separate them from the scientific greatness of their originators. The following is a sample of this regrettable chapter in the social impact of the quantum theory.

35.1.1   Physics and Politics?

Max Born, one of the venerable founders of the quantum theory, the same Max Born that paved the way for the rapid development of the theory with his introduction of probability into the Schrödinger equation, had an interest in philosophy and politics as well. This interest, however, was marred by a serious confusion that stretched the relevance of the quantum theory so far as to apply it to social phenomena. At a time when "Communism versus Capitalism" was the topic of debate in practically all intellectual circles of Europe, Born writes:

> The thesis "light consists of particles" and the antithesis "light consists of waves" fought with one another until they were united in the synthesis of quantum mechanics. ... Only why not apply it to the thesis Liberalism (or Capitalism), the antithesis Communism, and expect a synthesis, instead of a complete and permanent victory for the antithesis? There seems to be some inconsistency. But the idea of complementarity goes deeper. In fact, this thesis and antithesis represent two psychological motives and economic forces, both justified in themselves, but in their extremes, mutually exclusive. ... There must exist a relation between the latitudes of freedom Δf and of regulation Δr, of the type $\Delta f \cdot \Delta r \approx p$. ... But what is the "political constant" p? I must leave this to a quantum theory of human affairs. [Bel 98]

We can probably accept the "fighting" between the particle and wave nature of light as a metaphor, but to apply the quantum wave-particle duality to the human society makes absolutely no sense. The wave-particle duality was achieved after a long theoretical and experimental struggle stretched over many generations of physicists starting with Newton and Huygens in the seventeenth century. The physicists did not invent the wave-particle duality out of their whims; they had to accept it (somewhat reluctantly) after nature forcefully imposed it on them. Is there any empirical evidence that Capitalism and Communism are "dual?" If so, of what? Human society? But human society has gone through many different phases: hunting-gathering, tribal communes, slave-owning states, feudalism, etc. Why not pick any two of these stages and proclaim them dual (or thesis and antithesis)?

Born then invokes the complementarity principle, and once again, without any justification, proclaims Capitalism and Communism as "two psychological motives and economic forces" that only in the extremes are mutually exclusive, otherwise they are justified in themselves. Then, associating freedom to Capitalism and regulation to Communism, he writes an uncertainty principle for them. He does not say how to quantify freedom (one has to if one wants to give meaning to Δf) and regulation. He does not say how to assign a matrix to each of these concepts—matrices are *required* for the formulation of uncertainty principle. Finally, as though the rest of the concepts were clear and justified, he leaves the "political constant" p to a quantum theory of human affairs.

The truth is that Born's idea can be (ab)used in practically all situations of conflict. When there is *no* empirical basis for an argument, *anything* can be a basis for that argument. We could just as well (mis)apply the uncertainty principle to the present (2002) conflict between the US-UK coalition and Iraq: let ΔU be the "latitude" of the US-UK interest in the war and ΔI that of Iraq; then $\Delta U \cdot \Delta I \geq p$. Similarly, let ΔW be the uncertainty in Wall Street and ΔI the indecision of the investors, then $\Delta W \cdot \Delta I \geq E$, where E is the "economic constant" to be discovered in a future quantum theory of economics.

When there is no empirical basis for an argument, any wild assumption can be a basis for that argument.

35.1.2   Physics and Mysticism?

If Born tried to merely connect (quantum) physics with the mundane politics, Pauli, the great Austrian physicist, who has been compared to no less a giant than Einstein himself, wanted to go beyond the secular affairs of mankind. In a letter to Markus Fierz on 12 August 1948, Pauli writes:

> Science and religion must have something to do with each other. (I do not mean "religion within physics," nor do I mean "physics inside religion," since either one would certainly be one-sided, but rather I mean the placing of both of them within a whole.) I would like to make an attempt to give a name to that which the new idea of reality brings to my mind: the idea of reality of the symbol. ...It contains something of the old concept of God as well as the old concept of matter (an example from physics: the atom. The primary qualities of filling space have been lost. If it were not a symbol, how could it be "both wave and particle?"). The symbol is symmetrical with respect to "this side" and "beyond" ...the symbol is like a god that exerts an influence on man. [Bel 98, p. 29]

In essence, Pauli is saying that because we cannot picture the electron as both a wave and a particle, we have to accept the fact that it is only a symbol, and this symbol is a manifestation of some kind of a god.

Pauli's mysticism is also demonstrated in a series of letters he wrote to Jung, the Swiss psychologist, in which not only did he report of his numerous personal dreams anonymously, but also participated in a physics-psychoanalysis dialog with Jung. Pauli adopts the Jungian concepts of "synchronicity" and "collective unconscious" and proposes that the simultaneous appearance of the physical concept of "field" and the psychological concept of "unconscious" in the nineteenth century is a manifestation of this "collective unconscious" [Lak 70, p. 84].

The "collective unconscious" and the "symbol" are different manifestations of a Neo-Platonic spirit or soul that Pauli advocated because of his "interpretation" of the statistical nature of the quantum theory. His argument goes something like this: Quantum physics says that (microscopic) objects obey a fundamentally probabilistic law. Therefore, the microscopic *and macroscopic* world must be statistical; and in a statistical world one cannot have causality. If causality is to be abandoned, we also have to abandon "rationality." If rationality and reason cannot explain the natural phenomena, there must be a universal entity that governs the behavior of objects of reality. This entity is what Plato called the Idea, or Form, or Soul (or *anima mundi* in Latin).

The flaw of Pauli's argument is, of course, to extrapolate a statement that applies to individual atoms to a collection of trillions and trillions of atoms. We have already seen what a *qualitative* difference exists between a few random events and a large collection of such events, the difference that we stated as the law of large numbers (see Section 16.3). Although atoms and their constituents obey the probabilistic laws of the quantum theory, and their individual behavior is unpredictable and "irrational," *any macroscopic* collection of these atoms behaves quite predictably and rationally. That is the difference between physics and philosophy: physics distinguishes between quantity and quality, philosophy does not.

Distinction between physics and philosophy.

35.1.3   Observer-Created Reality?

The most damaging "interpretation" of the quantum theory is the abolition of objectivity, the tenet of science. The alleged downfall of objectivity arises for two reasons: the probabilistic nature of the theory and the interaction of the measuring device with the object of measurement. The first reason was advocated most strongly by Pauli, while Bohr and Heisenberg promoted the second reason. Objectivity, the probability argument goes, means the existence of the outside world independent of the observer. This, in turn, means that if different observers measure the same phenomenon they get the same result. For example, if there is a total eclipse of the Sun in a region of the globe, everybody in that region will see it the same way. The statistical nature of the quantum theory violates this notion of objectivity. Two different observers measuring, say, the position of an electron may obtain two different answers. It therefore appears that knowledge of the outside world depends on the observer.

The second reasoning goes something like this: since any measurement is an *interaction* between the observer and the observed, one can no longer say that the outside world exists independent of the observer. This line of argument is best described by Heisenberg:

> We can no longer speak of the behaviour of the particle independently of the process of observation. As a final consequence, the natural laws formulated mathematically in quantum theory no longer deal with the elementary particles themselves but with our knowledge of them. Nor is it any longer possible to ask whether or not these particles exist in space and time objectively...
>
> When we speak of the picture of nature in the exact science of our age, we do not mean a picture of nature so much as a *picture of our relationships with nature.* ...Science no longer confronts nature as an objective observer, but sees itself as an actor in this interplay between man and nature. The scientific method of analysing, explaining and classifying has become conscious of its limitations, which arise out of the fact that by its intervention science alters and refashions the object of investigation. In other words, method and object can no longer be separated. [Hei 58]

It is not a big logical leap from these remarks to conclude that reality (the outside world), being dependent on the observer, is actually *created by the observer* [Pag 83, pp. 47–48]. An observer-created reality, of course, undermines the very foundation of science and promotes a variety of pseudoscientific quackery. "Physics, the most exact science, tells me," a touch therapist may argue, "that I, as an observer, can create the reality. And since my interaction with reality is controlled by my brain, I can influence the outside world through my mind. With some training, therefore, I should be able to influence the body of a sick person through my mind, and extract the disease out of his/her body. Of course, I need to bring my hands close to the sick person to transfer the 'healing energy' into his/her body."

How a touch therapist may take advantage of the observer-created reality!

Both arguments against objectivity are wrong. The probability argument is wrong, because it *extrapolates* the behavior of an *individual* subatomic particle to the macroscopic world. The law of large numbers tells us that "individual chaos leads to collective determinism." Therefore, different observers making measurements on identical *macroscopic* samples obtain identical results, *as predicted by the quantum theory.* Quantum theory has as much predictive power as the classical theory, if not more. The reader recalls from our discussion of the law of large numbers in Section 16.3, that the prediction of the outcome of a mere trillion *random* events (tossing of black and white coins) is so exact that to see a violation of the prediction one has to wait more than the age of the universe! Even at the individual level, the theory is *predicting* the probability of outcomes; the problem is that to test this prediction, you need a large number of outcomes. The statement, "The probability of getting a head is 50%" is not violated by throwing two coins and observing that both are head. And reality does not become dependent on the observer simply because a second observer's coins showed up as two tails.

The second argument, in which the interaction between the observer and the observed is used to undermine objectivity, is also wrong. The interaction is *not* between the observer (a person) and the observed, but between the observed and an inanimate and consciousless *apparatus*. This observation is completely observer-independent: two (or two thousand) observers using the appropriate instruments for measuring a particular phenomenon will get identical (statistical) results. For instance, suppose the theory predicts that the probability of a radioactive decay of a particular material is 0.02 and two observers look at a sample of a million atoms of that material. If one observer sees 19,800 nuclei decaying and the other 20,200, are we to say that the theory is wrong? Or that the reality of nuclear decay is observer-dependent? Of course not. Both these numbers lie within 1% of the "exact" prediction of 20,000; and the verification of the prediction of any theory to within 1% is a remarkable success. The two observers can be completely detached from observation by going home at night and let the counters, recorders, and computers register the results. The will of the observer has absolutely no effect on the outcome, as disappointing as this may be for touch therapists and "quantum healers!"

35.2   Union of Philosophy, Science, and Religion

Two trademarks of the New Age Movement are its alleged "universality," and its abhorrence of reductionism, a characteristic of science and Western philosophy. This universality also defines Eastern mysticism, and forges a strong tie between the New Age Movement and Eastern mysticism. But is universality a merit or a drawback? Is it a mark of modernism and sophistication, or a sign of stagnation in some primitive belief system? Let history tell us what it knows!

The ever increasing reliance of the population of ancient civilizations on agriculture created a caste in society whose task it was to find means to guarantee sufficient irrigation and good crops. The members of this caste had to make observations of the sky to predict seasonal changes, and to follow the movement of "gods," upon whose mercy so much of the fate of the earthlings depended (see Section 1.1). And to communicate with them, the caste constructed images of these gods in their temples. A member of this caste was a scientist, a priest, and a philosopher all wrapped up in the same person. The priest's main goal was to befriend gods to win their compassion. Science (astronomy) was a peripheral tool forced on the priests because of their desire to understand the motion of gods in heaven. It was the infancy and the primitiveness of science that led to its union with religion and philosophy.

This process is known to have happened in Egypt and Babylon where all knowledge was monopolized by the priesthood, and temples were centers of higher learning much the same as universities are today. In fact, we know that the early Greek thinkers frequently visited temples in Babylon and Egypt to acquire knowledge, which was a mixture of science, religion, and philosophy. To New Agers, this is an ideal situation. *The Tao of Physics* glorifies the Greek philosophy of the sixth century BC, "a culture where science, philosophy, and religion were not separated." "The sages of the Milesian school in Ionia," the book goes on to say "were not concerned with [the] distinction [between science, religion and philosophy.]" For these thinkers, the important task was to see "the essential nature, the real [alive and growing] constitution of, things which they called 'physis' " [Cap 84, p. 6].

Eastern philosophy is entirely based on the unity of all things including mind and matter, religion and science, body and spirit. And in this respect, early Greek philosophers were not much different from the Eastern mystics. The later Greek philosophers called these early thinkers "hylozoists," meaning "those who think matter is alive." As much as New Agers lament the later development of Greek philosophy, such a development was as similar in character and as inevitable as the much earlier development of language in homosapiens. The criticism of the early philosophers by the later Greek thinkers was a natural evolution of superior thought from the outmoded and inferior beliefs.

Partly because of the mysticism of the early sages, superstition reigned uncontrollably among the Greek populace. The hylozoism of the early philosophers translated into a variety of "spirits" for inanimate objects in the minds of ordinary people. In response to this widespread superstition the later philosophers developed more rational methodologies. In the fight against superstition two prominent figures deserve our attention: Plato (427–347 BC) and Epicurus (341–270 BC).

35.2.1   Plato

Plato believed that the cause of superstition and "wrong" knowledge was the false impressions our senses convey to us. To correct this, he recommended to completely do away with sense impressions and rely primarily on the mind. He doubted whether sight or hearing has

The union of science, religion, and philosophy in early civilizations.

Fascination with the personification of the inanimate.

Plato combats superstition with the power of the mind.

any truth, and quoted the poets as saying that we neither see nor hear anything accurate or clear [Whi 76, pp. 63–65]. He argued that since unequal things could suggest the same idea,[2], the idea must have resided within us. Since our senses cannot distinguish between equals and unequals, they cannot apprehend the "truth." Therefore, Plato concluded, the judgments made by thought are "accurate" and "true" whereas those made by sensation are not. In fact, the emphasis on the "Ideas" (or "Forms") was so great that Plato arrived at the doctrine that sensible objects were in some sense copies, or poor imitations, of "Ideas."

The supremacy of the mind in Plato's doctrine had some profound effect on the later generation of thinkers. For example, it greatly promoted mathematics, and paved the way for the genius of the golden age of Greece. However, it also had its drawbacks. The primacy of the "Idea" led Plato to argue that since the ideas cannot come from our senses, they must have existed in us from our birth, even possibly from a previous life [Ros 72, pp. 23–27]. This led Plato to the notion of the soul and suggested that the philosopher ought to pay more attention to the desires of the soul than those of the body.

Plato was so enchanted by the power of the mind that he refused to allow anything that originated from outside the mind to be used in his body of knowledge, in particular mathematics. It is well known that his attack on the then current state of geometry was twofold. In the first place, he took geometers to task for relying on unsupported assumptions or axioms, for which they were unable to provide any grounds. In the second place, he maintained that it was the fault of their procedures that they relied on sensible figures such as those drawn in the sand, whereas what they should have been talking about were not those sensible figures but rather certain geometrical entities which were apprehended *not by the senses, but by the mind* [Whi 76, p. 96].

Plato chides geometers for using drawings on sand.

Geometry was a fairly advanced science at the time of Plato. It had gone through a long experimental stage in the hands of Egyptians and Babylonians, who had discovered many rules of thumb concerning plane and solid figures. By the fifth century BC a trend of axiomatization of geometry had been established which tried to incorporate the rules of thumb in a more systematic analysis of geometrical figures. Plato's contemporary geometers were caught up in this transition period from experimentation to pure axiomatization. And, Plato, being a staunch Idealist, wanted to completely do away with any further experimentation and replace the process by pure speculation.

35.2.2   Epicurus

The second philosopher combating superstition was Epicurus. However, his fight against superstition came from a completely opposite direction than Plato's. Epicurus argued that the weird imagery that was at the root of all superstitions originated from our mind; that superstition was human thought run amok. Therefore, we should not trust our mind, but rely solely on the experience that our senses impart on us. Our senses are the only trustworthy means that connect us to the outside world and are in immediate contact with it.

Epicurus's emphasis on the interaction between the senses and the outside world led him naturally to the concept of the atom. To him, atoms were the exclusive "messengers" that emanated from the objects in the outside world and impressed the senses. Although Epicurus's atom was far different from modern atom, its *idea* was truly revolutionary. Based on this idea Epicurus was able to explain many natural phenomena, which were otherwise hard to explain. For example, at a time when the cosmos was assumed to be finite and made of crystals, Epicurus advocated an infinite universe full of atoms moving in the pervasive vacuum.

Epicurus combats superstition by relying on the senses.

The *materialistic* philosophy of Epicurus,[3] although far reaching, had its drawback. Epi-

---

[2]For example the word (or the Idea of a) "tree" can refer to many "unequal" things that have branches, leaves, stems, etc.

[3]See Section 42.1.1 for a more thorough description of the materialistic philosophy.

curus was weary of anything that employed the mind even to the slightest degree, including mathematics, especially geometry, which by his time had turned highly axiomatic. Being a staunch experimentalist, he saw in the axiomatization of geometry another attempt at perpetuating nonmaterial and speculative apprehensions. It was in reaction to such trends that Epicurus attacked both Plato and mathematics. A contemporary of Euclid, he considered the latter's work on the axiomatization of geometry a fulfillment of Plato's dream of the supremacy of Ideas over the senses.

In the struggle between the senses and the mind, the latter won. In fact, Plato's influence on philosophy and mathematics was so great that Archimedes, the forerunning implementor of modern scientific methodology, who combined the senses (in the form of experimentation with fluids and their containers) with the mind (in the form of conjuring mathematical ideas from the experiments and subsequently proving the ideas), kept his methodology a secret for fear of being branded as one who "dirties his hands." The primacy of the mind over the senses was in tune with Christian ideology, and the two dominated philosophy for almost two millennia, bringing about a mental stalemate that was terminated only by returning to the Archimedean methodology, namely by combining the senses (observation) with the activity of the mind (theory). This was done by Galileo and Newton, and has persisted ever since.

Archimedes discovers the scientific methodology by combining the senses with the power of the mind.

Scientific progress requires a decisive break from superstition in any shape and form. In the history of our race this breakage happened in Greece—and not in any other part of the world—in a dual mode, Platonic and Epicurean. When Archimedes realized that these two seemingly opposite approaches actually complimented each other, it was too late, because the Romans, who cared only about politics, law, and rhetorics, took charge of leading humanity. The revival of scientific methodology in its modern form fell upon the post Renaissance scientists of Europe, and not in any other part of the world, including the Orient.

Fascination with the personification of the inanimate.

This scientific development is not welcomed by New Agers, who are more in tune with the early Greek philosophers such as Anaximander because he "saw the universe as a kind of organism which was supported by 'pneuma,' the cosmic breath, in the same way as the human body is supported by air" [Cap 84, p. 6]. Clearly, there is a connection between this kind of philosophy and the Eastern mysticism in which the cosmos "is seen as one inseparable reality—forever in motion, alive, organic; spiritual and material at the same time" [Cap 84, p. 11]. The difference is that the Greeks discovered rational philosophy and the scientific methodology (albeit ephemerally), but the Eastern mystics did not.

35.3   Physics-Eastern Thought "Parallelism"

Capra is one of the first to attempt to establish a connection between modern physics and Eastern philosophy or other New Age ideas, but he is not the only one. There is a plethora of authors having written on the parallel between modern physics and some of the New Age notions. Such parallelism, of course, does not exist, and it is the task of this chapter to demonstrate this. How could the authors show the parallelism, then? By relying on some faulty statements made by famous physicists; driving a wedge between classical and modern physics and between classical physics and Eastern mysticism; attributing inaccurate statements to physics; making universal analogies between New Age and modern physics that could hold for any two activities; and presenting failed ideas as if they were at the forefront of research, while refusing to acknowledge the successes of the current physics because it contradicts the New Age philosophy.

35.3.1   Quotations from Famous Physicists

The opening section of this chapter contained numerous examples of unorthodox statements made by some of the greatest physicists of the twentieth century. These and other similar statements have become valuable tools in the hands of the New Agers to advance their mystical ideas. Some New Age authors do not stop at suchlike statements. They take even an innocent statement by a well-known physicist and, through a labyrinth of interpretations, give it the appearance of supporting their views. Here are some illustrative samples.

Dr. Fred Alan Wolf, in his book, *The Spiritual Universe: How Quantum Physics Proves the Existence of the Soul*, narrates a story about a wanderer who visits the Buddha and asks if there *is* a soul; the Buddha remains silent. The wanderer asks if there is *no* soul; again the Buddha remains silent. After the wanderer leaves, Ananda, the Buddha's disciple inquires about the master's silence. The Buddha replies that if he had answered yes, he would be siding with brahmanas who hold to the eternalist theory, and if he had answered no, he would be siding with those who hold to the annihilationist theory.

Wolf then quotes Robert Oppenheimer[4] (when a student asks him about the existence and movement of an electron in an atom) as saying:

> If we ask, for instance, whether the position of the electron remains the same, we must say "no."
> If we ask whether the electron's position changes with time, we must say "no."
> If we ask whether it is in motion, we must say "no."
> If we ask whether it is standing still, we must say "no." [Wol 96, p. 176]

Wolf then concludes that Oppenheimer's remarks and the Buddha's answers "point to the same thing. For in both Buddhist logic and QT [quantum theory], it is necessary not to hold any fixed opinion but to see things as they are without mental projections."

On page 224 of the same book, there is a quote from Einstein:

> As far as the laws of mathematics refer to reality, they are not certain; and as far as they are certain, they do not refer to reality.

Wolf takes this profound statement, and through an ambiguous and meandering argument involving order and chaos and the ancient Chinese system of thought and ch'i, connects it to how the soul talks to us!

The Dancing Wu Li Masters, by Gary Zukav, is full of quotations and New Age conclusions drawn from them. Page 29 of this book starts with a statement (of the sort sampled in Section 35.1) by John Wheeler, a well-known American physicist:

> May the universe in some strange sense be "brought into being" by the participation of those who participate? ... The vital act is the act of participation. "Participator" is the incontrovertible new concept given by quantum mechanics. It strikes down the term "observer" of classical theory, the man who stands safely behind the thick glass wall and watches what goes on without taking part. It can't be done, quantum mechanics says.

Then Zukav predictably concludes that the language of Eastern mystics and Western physicists are becoming very similar.

Pursuing this line of thought, the participator cannot separate himself from the world, the concept of the outside world loses it meaning, and since the existence of the outside world is the tenet of objectivity, the notion of objectivity also loses its meaning.[5] We then read:

> ... We cannot eliminate ourselves from the picture. We are a part of nature, and when we study nature, there is no way around the fact that nature is studying itself. Physics has become a branch of psychology, or perhaps the other way round.

---

[4]A prominent American physicist noted for his leading role in the Manhattan Project.
[5]See Section 35.1.3 for a discussion of reality and objectivity.

Zukav then gives a quote, in which Carl Jung, the Swiss psychologist and psychoanalyst, connects the psyche of an individual and the outside world. Finally, the following quote by Wolfgang Pauli (who corresponded extensively with Jung),

> From an inner center the psyche seems to move outward, in the sense of an extraversion, into the physical world ...

seals the conclusion that Zukav is trying to reach: "If these men are correct, then physics is the study of the structure of consciousness."

The preceding quotes are just representative samples. If you take any New Age book and flip through it, chances are that you'll see many quotes from Bohr, Heisenberg, Born, Pauli, and many other modern physicists. These quotes, the fault of some of which we analyzed in Section 35.1, are the indispensable basis of the arguments used to connect modern physics with Eastern mysticism.

35.3.2   Separating Classical and Modern Physics

While Plato's rational philosophy, with its denial of the senses, promoted the idea of a Soul, albeit of a different kind than the soul or spirit of the early Greek philosophers, Epicurus, with his emphasis on the senses and his hypothesis of atoms, drew a clear line between matter and spirit. This distinction became the cornerstone of Western thought, including both religion and science. Western religion separated matter and spirit and concentrated on spirit; post-Renaissance science also separated matter and spirit but focused on matter. Both science and Western religion broke away from their early Greek roots. And since Eastern mysticism is very similar to the early Greek philosophy, science (in particular, classical physics) and Western religion have nothing in common with the Eastern philosophy.

If Eastern mysticism is to parallel modern physics, and if the former has nothing in common with classical physics, then it becomes essential for the New Agers to separate classical physics from modern physics so they can tie the latter with Eastern mysticism. To do so, they *only* look at the *difference* between the two. Consider the following passage from Zukav's book:

> Li ... means "organic pattern." The grain in a panel of wood is Li. The organic pattern on the surface of a leaf is also Li, and so is the texture of a rose petal. In short, Wu Li, the Chinese word for physics, means "patterns of organic energy" .... This is remarkable since it reflects a world view which the founders of western science (Galileo and Newton) simply did not comprehend, but toward which virtually every physical theory of import in the twentieth century is pointing! [Zuk 80, p. 5]

The words "organic" and "energy," the telltale signs of the New Age Movement, are both embraced in Wu Li, the Chinese word for physics. Do we need any stronger evidence for the connection between (modern) physics and Eastern thought? And it has to be *modern* physics, because there are no "patterns of organic energy" in classical physics.

Here is an argument for separation found in *TP*:

> The concept of matter in subatomic physics, for example, is totally different from the traditional idea of a material substance in classical physics. The same is true for concepts like space, time, or cause and effect. These concepts, however, are fundamental to our outlook on the world around us and with their radical transformation our whole world-view has begun to change. [Cap 84, pp. 3–4]

After this quote, Capra claims that such changes brought about by modern physics lead toward a worldview similar to those held in Eastern mysticism.

Alternative medicine is one of the hallmarks of the New Age Movement, and if there is a "scientific" explanation for this kind of medicine, it has to be groped for in the quantum theory and relativity. An alternative medical doctor, who associates quantum physics with the healing process of his patients, first has to dispose of classical physics as a science of his medicine. Deepak Chopra, MD, uses some diagrams in his bestselling book, *Quantum Healing: Exploring the Frontiers of Mind/Body Medicine*, to differentiate between classical and quantum physics, and to affiliate the latter with Ayurveda, the ancient Indian medicine.

In these diagrams Chopra draws a horizontal line and speculates that everything that we can explain in terms of cause and effect occurs above the line. If event A causes event B to happen, he puts A and B above the line and connects them with a straight arrow, presumably showing that B follows from A. This, he claims, is the classical world of Newtonian physics, the world of billiard balls, in which the collision of the moving ball A with the stationary ball B will surely cause B to move, the world of the senses with which we are familiar.

But Chopra wants to "explain" the connection between mind and body, between the activity of the brain and the chemical and physical bodily processes that follow. To accomplish this, he takes advantage of the major findings in neurobiology, and reduces this process down to its molecular level, saying that neuropeptides are chemicals that are produced and transmitted between neurons in many of the brain's activities. But how are the neuropeptides produced? How does one explain the creation of neuropeptides out of thoughts?

> We need to consult the quantum to really understand how the mind pivots on the turning point of a molecule. A neuropeptide springs into existence at the touch of a thought, but where does it spring from? A thought of fear and the neuro-chemical that it turns into are somehow connected in a hidden process, a transformation of nonmatter into matter. [Chop 90, p. 95]

Once again, quantum physics comes to the rescue of New Age mysticism. The region underneath Chopra's horizontal line is a mysterious region where inexplicable phenomena take place. So if B follows from A and one cannot explain how A causes B, one simply connects A to B via a U-shaped curve—a detour from the Newtonian straight line—the lower half of which is in the region below the horizontal line. This region is outside space and time, and unreachable by any of our senses or any physical or chemical device, so Chopra says. And it is in this region where thought turns into neuropeptide, where nonmatter turns into matter. What does all this have to do with quantum physics? We'll come back to this question later.

Remark 1. It may be helpful to recount a bit of background to *Quantum Healing*. The hardcover edition of the book came out in 1989 and its paperback edition in 1990. The book was dedicated, "With a full heart and deepest thanks to Maharishi Mahesh Yogi." In the Introduction, the author narrates his meetings with Maharishi, in which the latter talks about some special ancient techniques. Then he tells Dr. Chopra, "Now I want you to learn them, and at the same time I want you to explain, clearly and scientifically, how they work"[6] [Chop 90, p. 3]. And Chopra sets out to find the answer.

In the next few hours of their meeting, Maharashi teaches Chopra some "primordial sound" techniques prescribed for specific illnesses, including some incurable ones. These techniques are of such immense value that Chopra repeatedly mentions their source in his book: the book is warmly and thankfully dedicated to Maharishi, there are many places in the book where he is mentioned with reverence, and the index has his name as an entry with numerous reference pages. It appears that all this takes place up to the 14th printing[7] of the book! In the 15th printing, all citations of Maharishi's name are erased; the meetings with him, which were the starting point of the quantum healing, are not mentioned at all; the crucial "primordial sound" techniques which were "the strongest healing therapies in *Ayurveda*" and were prescribed for incurable diseases like cancer, are gone. ∎

---

[6] A technique that is in any way related to science comes *after* the prerequisite scientific knowledge has been gained. One doesn't start with a technique and *then* ask for (or order) a scientific explanation!

[7] Printing, not edition! As far as the publication date and publisher's name are concerned, there is only one edition of the book. All our references to *Quantum Healing* are made to the printings that came before the 14th.

That classical and modern physics are different is hardly debatable. However, what the New Agers are completely avoiding is how much *modern physics is dependent* on classical physics. Take the master equation in (nonrelativistic) quantum theory, the Schrödinger equation. As every student of physics knows, this equation uses the concepts of momentum, kinetic energy, and potential energy, all of which were directly linked to classical physics. Furthermore, the mathematics used in writing the Schrödinger equation (differential equations) was *invented* by classical physics itself. The commonality does not stop here. Practically any concept (e.g., angular momentum and spin) introduced in the quantum theory has a classical counterpart. The same is true of the relativity theory. (See Section 42.3.) When Einstein tried to write the equation of the general relativity that bears his name, in one crucial step, *he had to rely on the Newtonian gravity.*

> Modern physics has a lot in common with classical physics, contrary to what New Agers want to make us believe.

> Without Newtonian gravity, Einstein would not have succeeded in creating his general theory of relativity.

Modern physics evolved from classical physics, and when applied to circumstances in which classical physics is valid, both give identical results. Classical physics, which explains the motion of planets and moons beautifully, is not designed for atoms and subatomic particles, just as quantum mechanics, which explains atoms and molecules, is not designed for planets and moons. There is, of course, a difference in this dichotomy of the large and the small. Quantum physics can, *in principle*, be used to describe planets and moons, but its usage is so unwieldy, it is almost impossible.[8] As an analogy, consider trying to measure the distance between New York and Los Angeles. One could use a jet plane flying at a certain speed and time its flight, thereby measuring the distance (by taking the product of the time and speed). Another alternative is to use a ruler and measure the distance in increments of one foot! Although we get practically the same result for the distance, it is obvious which method is more feasible. Applying quantum physics to the motion of planets and moons is as unwieldy (many times more so) as measuring the NY–LA distance with a ruler. Similarly, applying classical physics to the motion of subatomic particles is like trying to measure the length of a sheet of paper by sending an airplane from one side of it to the other and timing the plane!

35.3.3   Customized Physics

Physics is the antipode of mysticism, and if the New Age authors took physics as is, they would not be able to reconcile it with their mystical beliefs. So, they have to change it to fit their need. They do not present their altered customized physical ideas to physicists, who can immediately identify their flaws. They write their ideas for other New Agers and the novice. These authors are, therefore, not fettered by accuracy and correctness. They can take any one of their experiences and connect it to modern physics, once the latter has been interpreted (im)properly.

Consider the notion of wave-particle duality. Some New Agers think that waves are nonmaterial and wave-particle duality becomes nonmaterial to material transformation. We have already seen how Chopra considers the area below the horizontal line of his diagrams a region where nonmaterial thought can transform into material neuropeptide. With this kind of transformation and ascribing materiality to particles and non-materiality to waves, the connection between mind/body and quantum physics becomes irresistible. In the context of this peculiar interpretation, light, which has both wave and particle properties, takes a U-shaped detour whenever it transforms from wave (nonmaterial) to photon (material).

> Here lies the connection between mind/body and quantum physics!

As discussed in Section 23.2, light consists of photons (particles), whose *probability distribution* on a screen, after passing through a double-slit, looks like a wave.[9] It is not a single photon that can be a wave or a particle; a single photon is *always a particle*. However, under certain circumstances, a *macroscopic collection* of photons can assemble into a wave.

---

[8]Classical physics, however, cannot—even in principle—be used for atoms and subatomic particles.

[9]Although Section 23.2 discussed the interference of electrons, all subatomic particles, including photons, exhibit interference in a double-slit experiment.

Therefore, if anything, an electromagnetic wave is *more* material than a single photon! An electromagnetic ***wave cannot turn into a photon*** either above or below the "line!"

Many people are intrigued by the masslessness of the photon. They equate the massless to the nonmaterial. However, in the theory of relativity, mass has no significance in determining a "concrete" particle. In fact, the equivalence of mass and energy through $E = mc^2$, allows particles to be massless, but possess energy. Photons, either individually as particles, or in multitude as waves, are as real and as material as any other particle, large or small. (See Food for Thought 28.3.4 for further detail.)

Remark 2. It is instructive to generalize the nonmaterial-to-material transformation occurring under Chopra's horizontal lines to a (more feasible) transformation of microscopic energy to macroscopic motion, which is as valid as the one he uses, because they are both conjectures. So, let's propose the following: there is no straight-line connection between (the energy of) a molecule and the (motion of) tectonic plates. Therefore, we draw a U-shaped arrow, passing through the region below the horizontal line, to connect the molecule to the tectonic plates. The whole area below the line, as Chopra suggests, is not a region to be visited in space and time; it just stands for "wherever it is you go" when you turn the energy of a molecule into the motion of tectonic plates.

Now consider the process of lifting a finger. Here is how Chopra gives an "explanation" of this process:

> If you want to lift your little finger (point A), a physiologist can trace the neurotransmitter (B) that activates an impulse that runs down the axon of the nerve (C), causing a muscle cell to respond (D), resulting in the lifting of your little finger (E). However, nothing a physiologist can describe will get him from A to B—it requires a detour. [Chop 90, p. 100]

A good example of illustrating the flaw of this sort of reasoning is to utilize the molecule-tectonic transformation to show how a sneeze in the American Midwest can cause an earthquake in Mexico City. We use the argument verbatim, but change the physiological events to geological events. Here is how it goes: the energy of one of the molecules exhaled in the sneeze (A) of a flu-stricken person in the American Midwest penetrated the Earth crust and turned into the motion of a tectonic plate (B), carrying a seismic wave which ran to the fault under Mexico City (C), causing a displacement of the Earth crust there (D), resulting in the collapse of a building (E). A seismologist can trace the motion of the tectonic plate (B) that produces a seismic wave that runs to the fault under the city (C), causing an earthquake there (D), leading to the collapse of the building (E). However, nothing a seismologist can describe will get him from the sneeze A to B—it requires a detour. ■

The "explanation" of how a sneeze in the American Midwest can cause an earthquake in Mexico City!

We have so far seen how the New Agers have connected modern physics with Eastern mysticism and Ayurveda. But New Agers are spiritual people, so they must have found a way of connecting modern physics to spirits. Indeed Wolf has found the connection in his book mentioned earlier. The connection takes us to the beginning of time:

> ... about 15 billion years ago, give or take five billion years, the physical universe was created from nothing. It seems there was a big bang and some time later ... the universe will come to an end in a big crunch.....

> How about the spiritual universe? Did it have a beginning? Will it come to an end? ... we might wish to inquire how the soul fits into this big bang and big crunch universe .... What about Aristotle's concept of the soul's being a physical substance? Is the soul a *physical* process ...? Or can it be simultaneously real and not physical? Does the universe gush forth soul as it brings forth matter and energy? Does the soul require energy? Or is the soul some form of energy itself?

> .... If it turns out that the soul is physical, a new vision of the universe may appear— nature not only produces matter and energy from nothing, but also creates soul. But where is the soul? And how could nothing just produce something, anything at all? ... Perhaps it's because nothing is really something after all.

> ... According to quantum theory, a vacuum ... is not empty, but consists of a vast amount of positive and negative fluctuating energy. Thus, out of a vacuum can be

derived a number of unusual phenomena, including matter, antimatter, energy, and now, as $\underbrace{I\ suggest,\ even\ spirit\ and\ soul.}$ [Wol 96, pp. 124–126]

                                    [Our emphasis.]

The author then goes on to introduce the concept of zero-point energy, quantum vacuum, and the fact that vacuum, in quantum field theory, contains (virtual) particles and antiparticles. All these become a seemingly convincing tool to prove the existence of soul. Let's see how convincing the argument *really* is.

First, the universe did not come about from nothing. Once you say *from nothing*, even if you mean vacuum, you are implying that space and time existed before the big bang. It did not! Big bang not only created matter and energy, but also space and time. So, the notion of "before big bang," even if only implied, is wrong! Present physics is incapable of describing the very moment of big bang. So, any statement made about the moment of creation or before it, is pure speculation.

Second, the positive and negative energy that the author talks about are all *virtual* in the context of Feynman diagrams [see Figure 32.2]. For example, the photon exchanged between two electrons is virtual. Virtual photons are merely mathematical entities whose pictorial representations are found to be convenient in theoretical calculations. Virtual particles and processes could have observable effects only if they interact with an external field. And this observable effect is *not* seeing particles and antiparticles or positive and negative energies, but a small perturbation in an existing physical quantity. We have already discussed one such effect (see Lamb shift on page 488) resulting from vacuum polarization and its interaction with the Coulomb force between the electron and proton of a hydrogen atom.

Third, all the theoretical discussion in the quotation above seems to be a distraction for the insertion of the last six words. We could replace the words "spirit" and "soul" with any other pair of words, and nobody can tell us that our choice is less appropriate than "spirit" and "soul," because, ultimately, they are all "suggestions."

Some authors take the most radical approach. They change the very *methodology* of science. This approach is particularly emphasized by Fritjof Capra, who divides the scientific methodology of physics into stages, and correctly identifies the first two stages as making observations and developing mathematical theories. But he adds a third stage on which he dwells a great deal, because it is at the heart of his argument. Capra believes that the crucial third stage of the methodology of physics is the physicists' desire to convey their results to nonphysicists, which requires them to express their ideas and theories in plain language. Capra even require such a "translation" for physicists as well. That is, that physicists, even when communicating their ideas among themselves, should not be content with the formulation of their theories in the mathematical language, which nature itself has forced upon them, but they ought to translate them into plain human languages [Cap 84, pp. 17–18].

Changing methodology by introducing a third stage beyond observation and theory!

Physicists do feel an obligation to express their discoveries to the public. However, this obligation stems not from a "third stage" in the scientific process, but from a desire to share the excitement of understanding nature, and to arouse interest in a difficult and mathematical field among the youth, a minority of which will (have to) eventually constitute the next generation of physicists. The physicists themselves have a perfectly suitable language in which to communicate: mathematics. That is why they have to go through a rigorous training lasting almost a decade after high school to learn this language. No doubt, they will use a human language to communicate as well; but that is secondary to mathematics, which contains the essential knowledge. One needs only to attend a physics conference to see how crucial mathematics is in the communication among physicists.

To say that physicists are *required* to express their findings in human languages is like saying that Chinese poets are *required* to translate their poetry into English. Not only is the translation unnecessary, but the poetry often loses its meaning in the process. It would, of course, be a great service to the English-speaking people if good Chinese poems were translated into English. But the *requirement* of such a translation makes absolutely no sense. It has been established ever since Pythagoras that Nature speaks in the language of mathematics, and as physicists kept trying to understand Nature, they were *forced* to learn this language ever so increasingly. The *requirement* of the translation of mathematical laws of physics into human languages makes no sense either.

Modern physics, in particular, can *only* be spoken in the mathematical language. While classical physics, which deals with familiar objects, is more tuned to our intuitive understanding of nature, relativity and quantum physics are so counter-intuitive and "strange" that their comprehension *demands* the medium of mathematics. How else can one speak of four-dimensional spacetime, of the creation of space and time at the big bang, and of the infinite-dimensional vectors?

> To say that physicists are *required* to express their findings in human languages is like saying that Chinese poets are *required* to translate their poetry into English.

35.3.4 Universal Analogies

Physics is a very precise, objective, and verifiable description of reality. Eastern mysticism is a very vague, subjective, and personal experience of spirituality. They have nothing in common. But the New Agers claim that the two are "parallel." How do they substantiate their claim? One way is to find a commonality between the two. But their version of commonality could, in principle, be used for *any* two activities. For example, Capra talks about the long training period required of an experimental physicist before he/she can repeat an experiment in subatomic physics. Then he goes on to say:

> Similarly, a deep mystical experience requires, generally, many years of training under an experienced master, and, as in the scientific training, the dedicated time does not alone guarantee success. If the student is successful, however, he or she will be able to "repeat the experiment." The repeatability of the experience is, in fact, essential to every mystical training and is the very aim of the mystics' spiritual instruction. [Cap 84, p. 23]

Just because two activities require long periods of training and have the property of repeatability, it does not follow that there is a parallel between them. Any skill, even the most primitive one like the punching of a boxer, requires "many years of training under an experienced master, and ... the dedicated time does not alone guarantee success." And the fact that the boxer has to "be able to 'repeat the experiment'" is self-evident. This similarity implies absolutely no connection between the experience of a boxer and that of a physicist or a Zen student.

Capra introduced his third stage in the methodology of physics (see page 552) for a purpose. Physics, especially modern physics, is highly mathematical, and as long as mathematics is the medium of the communication of physics, nothing unusual happens. But if you try to translate modern physics in ordinary languages, peculiar things emerge which could be made parallel to the "riddles" of Eastern mysticism.

The third chapter of *TP* opens with two quotations. The first, by D. T. Suzuki, a Zen master, states that since our "inner experience" transcends any kind of language, talking about that experience leads to puzzling contradictions. The second quotation is by Werner Heisenberg, who talks about the serious problems of language, and the fact that one cannot talk about quantum mechanical entities such as atoms in ordinary human languages. Then Capra argues that, since both the structure of the atom and the Zen experience cannot be spoken in ordinary language, they must be related [Cap 84, p. 33].

The inability of two people or two groups of people to do the same thing does not mean that those people or groups have any common characteristics. If neither Karl nor Emmy can play tennis, do we conclude that there is some kind of sameness between Karl and Emmy? That perhaps they have the same sport interest? It may be hard to translate Bhojpuri poetry to English. It may be just as hard to translate Nahuatl poetry to English.

Does that mean that there is a similarity, a common characteristic, a parallelism, between Bhojpuri and Nahuatl?

Some New Agers do not feel obligated to show any evidence for the mystic experience which is allegedly parallel to modern physics. For example, a New Ager may claim that a certain mystical experience jumps from a lower level to a higher level without offering any proof of the jump. He can then parallel the "jump" to a quantum jump and thus "establish" the connection between mysticism and quantum physics. That is precisely how Chopra draws a connection between Ayurveda and quantum physics. The Ayurvedic method of curing any disorder requires a level of total, deep relaxation. The fundamental assumption in Ayurveda is that the human body knows how to maintain balance. A disease throws off this balance. Therefore, to cure the disease, the lost balance should be restored [Chop 90, p. 14].

How does one restore the balance? According to Chopra, some patients have an innate ability[10] to restore the lost balance—even if the disease that threw it off is as deadly as cancer—and in the process, they experience a dramatic shift in awareness.

> At that moment [just before the cure] such patients apparently jump to a new level of consciousness that prohibits the existence of cancer. ... This leap in consciousness seems to be the key. ... The word that comes to mind when a scientist thinks of such sudden changes is *quantum*. [Chop 90, p. 17]

Any "jump" involves the quantum theory?

Thus, according to this quote, a sudden jump from an existing level of anything to a new level is quantum mechanical: a jump up (or down) the stairs, a burst into laughter (a jump in the volume of our voice), or a cough (a jump in the the amount of air coming out of our mouths)! While the jumps just mentioned do actually exist and are verifiable, the "jump to a new level of consciousness" cannot even be verified,[11] let alone connected to the quantum theory.

35.3.5   Oblivion to True Progress

The investigation of combustion started modern chemistry in the eighteenth century. A flame rises *away* from a burning substance. So it was quite natural to think of the burning process as the removal of a substance from the burnt material. *Phlogiston* was the name given to this substance. The phlogiston theory of burning was so predominant in the early eighteenth century that when it was found that objects *gain* weight after combustion, the advocates of the theory gave phlogiston a negative weight! The whole idea of phlogiston was eventually abandoned in light of strong experimental evidence against it.

Heat was another concept that started on the wrong track. It was assumed to consist of a "fluid" called *calorie*, which flows in and out of substances making them hot or cold. At some point, two kinds of calorie were in existence: positive (causing a rise in temperature) and negative (causing a drop in temperature). The caloric theory of heat eventually fell on its face when it was irrevocably shown that heat was a form of energy.

Today only the historians and philosophers of science speak of the phlogiston theory of combustion and the caloric theory of heat. Some of them take such rejected theories seriously enough to challenge the existing theories of science. They argue that since phlogiston and caloric theories were at some point in the past "acceptable" theories, they are as important as, say the relativity theory "accepted" today. They completely overlook the fact that the rejected theories were either short-lived or proposed without any experimental support, and as soon as the theories were confronted with experiments, they proved wrong and had to be discarded.

---

[10]Nowhere in his argument does Chopra say how to generate this innate ability in those patients who don't *naturally* have it. The cure, therefore, becomes the responsibility of the patient!

[11]The author himself uses the word "apparently."

At the beginning of high energy physics in the early 1960s, numerous theories were proposed, among which were some phlogiston-like ideas that, for lack of observational disapproval, survived for a few years. One such theory was the *bootstrap* hypothesis mentioned at the beginning of Chapter 34. Bootstrap hypothesizes that hadrons are both fundamental and composite. And if this sounds as if it's coming straight from an Eastern mystic philosophy book, it should be of no surprise. Bootstrap theory is a favorite among the New Agers. For example, almost a quarter of *TP*, is devoted to this and the related *S-matrix* theory.

Even though they are as dead as the phlogiston theory of combustion and the caloric theory of heat, the New Agers adhere to the bootstrap and S-matrix theories. Why? Because they both existed in the physics community in the recent past, and because they do not contradict New Agers' beliefs. This adherence is so strong that in the eyes of some New Agers, any other theory is old-fashioned and futile. For example, Zukav divides theoretical physics into two schools, one following the old way of thinking, the other the new way. He associates physicists who are in search of the fundamental building blocks of the universe such as quarks and leptons with the old school. As for the new school, he notes the impossibility of presenting all the different approaches pursued by the physicists belonging to this school, mentions the ideas of a couple of physicists in this school, and concludes:

> These theories, at the moment, are speculative. ...
>
> The most successful departure from the unending search-for-the-ultimate-particle syndrome is the S-Matrix theory. In S-Matrix theory, the dance rather than the dancers is of primary importance. [Zuk 80, pp. 244–245]

Bootstrap's relation to Eastern mysticism.

Zukav dismisses theoretical physicists who study proven quarks and leptons.

Zukav admires theoretical physicists who propose unproven speculative ideas.

These words were written in 1980, four years after the charm quark, a "new building block," was discovered, three years after tauon, another "new building block," was produced in the laboratory, two years after the bottom quark, yet another "new building block," was created, and one year after the originators of the electroweak theory, a theory for the "building blocks," were awarded the Nobel Prize for physics. While all of these *real* discoveries are branded "old school" and dismissed, some "speculative" theories are given legitimacy, and, of course, the S-matrix theory gets the fullest attention.

Despite all this progress, and the recognition of the potential of the standard model by the physics community in 1980, page 314 of the *Dancing Wu Li Masters* contains the following quote from Geoffrey Chew, the founder of the bootstrap hypothesis:

> Our current struggle [with certain aspects of advanced physics] may thus be only a foretaste of a completely new human intellectual endeavor, one that will not only lie outside physics but will not even be describable as "scientific."

On page 280 of *TP* Capra writes:

> In the Eastern view then, as in the view of modern physics, everything in the universe is connected to everything else and no part of it is fundamental. The properties of any part are determined, not by some fundamental law, but by the properties of all the other parts.

Then on page 290, we find the following quote also from Chew:

> Carried to its logical extreme, the bootstrap conjecture implies that the existence of consciousness, along with all other aspects of nature, is necessary for self-consistency of the whole.

From this quote, Capra predictably concludes: "This view, again, is in perfect harmony with the views of the Eastern mystical traditions which have always regarded consciousness as an integral part of the universe."

The first edition of *TP* came out in 1975. By that time, elementary particle physicists had discovered a promising venue to pursue their interest: gauge theories. Bootstrap and S-matrix had left the land of physics and entered into the land of history, philosophy, and mysticism. Only some New Agers elevated the bootstrap hypothesis and S-matrix to the level of a viable theory. The fourth edition of *TP* came out in 2000. In the intervening 25 years, a lot of progress was made in the world of *real* physics. The charm quark, which was proposed as part of the electroweak unification, was discovered in 1976 (see Section 34.2.2). The heaviest lepton, tauon, was found in 1977, and a year later the fifth quark, bottom, appeared in a particle detector at Fermilab. By 1983, enough energy could be transferred to colliding particles at CERN that the heavy gauge particles of the electroweak theory ($W^{\pm}$ and Z_0) could be produced there. And the last and heaviest piece of the puzzle of the standard model of fundamental particles and interaction, the top quark, was discovered at Fermilab in 1995. Yet there is absolutely no mention of any of these tremendous successes in the fourth edition of *TP*!

Similarly, there is no mention of the successes of quantum chromodynamics, the correct theory of the strong interaction of hadrons. Although we see the acronym QCD in an afterword to the second edition of the book, there is a considerable amount of misinformation about its success. In this afterword, Capra talks about the fact that quarks have never been observed; that physicists have invented various mechanisms to explain their confinement; that the existence of quarks "would lead to severe theoretical difficulties;" and that QCD can be applied only to a very limited range of phenomena. Then he laments that "in spite of all these difficulties, most physicists still hang on to the idea of basic building blocks of matter which is so deeply ingrained in our Western scientific tradition" [Cap 00, p. 316].

This portrayal of QCD conveys the impression that physicists have invented "various mechanisms" to ensure that quarks are confined, as if they knew that quarks were not free and *then* came up with a theory that confined them. This is completely inaccurate. QCD is the application of local Lie group theory (gauge theory) to the color charge. The confinement of quarks comes as a surprising *by-product* of this application, as explained at the end of Section 34.1.2. The discoverers of QCD did not say to themselves, "We need to confine quarks, so let's invent QCD!" Just as Einstein didn't say to himself, "I need to make light bend in the gravitational field, so let me invent the general theory of relativity!"

Capra disparages QCD as being applied only to a very limited range of phenomena. It is a pinnacle of either lack of knowledge or dishonesty to characterize QCD in this way while glorifying bootstrap and S-matrix as the theories of hadrons. The fact is that QCD has been successful in every situation in which hadrons interact strongly. Like any other good theory, it has explained many hadronic phenomena, predicted many outcomes, and suggested new experiments for its confirmation. QCD has been so successful that the discoverers of the theory were given the Nobel Prize in physics in 2004. On the other hand, bootstrap has produced nothing but the satisfaction of some New Age mystics.

Capra's lament as quoted above says it all. Holistic philosophy is opposed to any notion of "basic building blocks of matter." This notion is a *reductionist* notion, and reductionism is in conflict with Eastern mysticism. Physicists, nevertheless, pursue the path that started with Archimedes and Galileo and continues even today. At the heart of this pursuit is specificity and reductionism, the nemesis of Eastern mysticism. Starting with the year in which Capra conceived the unification of Taoism and modern physics and ending with the publication year of *From Atoms to Galaxies*, the following lists the Nobel Prizes in physics awarded to those who "still hang on to the idea of basic building blocks of matter."

1969 Murray Gell-Mann "for his contributions and discoveries concerning the classification of elementary particles and their interactions."

1976 Burton Richter and **Samuel Chao Chung Ting** "for their pioneering work in the discovery of a heavy elementary particle of a new kind."

The author of TP does not accept the observationally verified quarks and gluons!

1979 Sheldon Lee Glashow, **Abdus Salam**, and **Steven Weinberg** "for their contributions to the theory of the unified weak and electromagnetic interaction between elementary particles, including, inter alia, the prediction of the weak neutral current."

1980 James Watson Cronin and **Val Logsdon Fitch** "for the discovery of violations of fundamental symmetry principles in the decay of neutral K-mesons."

1984 Carlo Rubbia and **Simon van der Meer** "for their decisive contributions to the large project, which led to the discovery of the field particles W and Z, communicators of weak interaction."

1988 Leon M. Lederman, **Melvin Schwartz**, and **Jack Steinberger** "for the neutrino beam method and the demonstration of the doublet structure of the leptons through the discovery of the muon neutrino."

1990 Jerome I. Friedman, **Henry W. Kendall**, and **Richard E. Taylor** "for their pioneering investigations concerning deep inelastic scattering of electrons on protons and bound neutrons, which have been of essential importance for the development of the quark model in particle physics."

1992 Georges Charpak "for his invention and development of particle detectors, in particular the multiwire proportional chamber."

1995 "for pioneering experimental contributions to lepton physics."

   Martin L. Perl "for the discovery of the tau lepton."

   Frederick Reines "for the detection of the neutrino."

1999 Gerardus 't Hooft and **Martinus J.G. Veltman** "for elucidating the quantum structure of electroweak interactions in physics."

2004 David J. Gross, **H. David Politzer**, and **Frank Wilczek** "for the discovery of asymptotic freedom in the theory of the strong interaction."

2008 Yoichiro Nambu "for the discovery of the mechanism of spontaneous broken symmetry in subatomic physics."

   Makoto Kobayashi and **Toshihide Maskawa** "for the discovery of the origin of the broken symmetry which predicts the existence of at least three families of quarks in nature."

35.4 End-of-Chapter Material

35.4.1 Chapter Glossary

Ayurveda is a system of traditional medicine native to India, and practiced in other parts of the world as a form of alternative medicine. Evolving throughout its history, Ayurveda remains an influential system of medicine in South Asia.

Bhojpuri is a regional language spoken in parts of north-central and eastern India.

Collective Unconscious is a term of analytical psychology, coined by Carl Jung. It is a part of the unconscious mind, shared by a society, a people, or all humanity, that is the product of ancestral experience and contains such concepts as science, religion, and morality.

Hemophilia is a group of hereditary genetic disorders that impair the body's ability to control blood clotting or coagulation, which is used to stop bleeding when a blood vessel is broken.

Materialistic Philosophy A branch of ancient philosophy that puts primary emphasis on matter. The concept of atoms is the tenet of this philosophy.

Nahuatl is one of the Native American languages spoken in Mexico.

Neuron is an excitable cell in the nervous system that processes and transmits information by electrochemical signaling.

Neurotransmitters are chemicals which relay, amplify and modulate signals between a neuron and another cell.

Tao is a concept found in ancient Chinese philosophy. The character itself translates as "way," "path," or "route," or sometimes more loosely as "doctrine" or "principle." However, it is used philosophically to signify the fundamental or true nature of the world.

Zen is a school of Buddhism referred to in Chinese as *Chan*, which is itself derived from the Sanskrit *Dhyana*, meaning "meditation." Zen emphasizes a form of meditation known as *zazen*—in the attainment of awakening, often simply called the path of enlightenment.

35.4.2 Review Questions

35.1. Name some of the "social mistakes" that some great scientists made.

35.2. Which two social developments occurred at about the same time that the quantum theory and relativity were being discovered?

35.3. Who applied the uncertainty principle to Communism vs Capitalism? What was his contribution to the quantum theory? Was he right in applying physics to human society? Why?

35.4. What argument did Pauli use to connect science and religion? What is the role of electron and quantum physics in this connection? Can you reconstruct Pauli's argument for the existence of a Soul or *anima mundi*? What is the flaw in his argument?

35.5. How do some people use the probability argument to deny the objectivity of the outside world? Why is their argument wrong? What does the law of large number in probability theory have to do with this?

35.6. How can the nature of measurement as an interaction between the measuring device and what is to be measured lead erroneously to the notion of "observer-created reality?" How can a touch therapist or a faith healer use this to defend his/her practice?

35.7. What are the trademarks of the New Age Movement? How is the New Age Movement tied to Eastern mysticism?

35.8. Why did religion, science, and philosophy start together and were unified at first? Is there any historical evidence for this? Why did the early Greek philosophers and scientists visit the temples of Egypt and Babylon?

35.9. How did Plato combat superstition? What did he think of the senses? Of the mind? What did he think of mathematics? What did he think of geometry? Did he agree with his contemporary geometers? What did they do that he did not like? What were the benefits of Plato's way of thinking? What was its harm?

35.10. How did Epicurus combat superstition? What did he think of the senses? Of the mind? What did he think of mathematics? Why was he Euclid's axiomatization of geometry? What were the benefits of Epicurus' way of thinking? What was its harm?

35.11. Who combined Plato's and Epicurus' philosophy? Did he publicize his way of thinking? Whose philosophy won in subsequent centuries, Plato's or Epicurus'?

35.12. How do the authors of the New Age Movement show the the parallelism between their beliefs and modern physics? Name some of the approaches they use.

35.13. How do the Buddha's and Oppenheimer's statements point to the same thing? Is that really the case?

35.14. How does the notion of a "participator," rather than observer, lead to the alleged downfall of objectivity?

35.15. In which culture is spirit separate from matter? In which culture are they not separate? Does Western religion separate the two? Does science separate the two? What is the difference between the religious and scientific separation?

35.16. How do you translate physics in Chinese? What is the literal meaning of the translation? Is that a good reason to call physics "organic?"

35.17. When Chopra meets Maharishi Mahesh Yogi, the latter tells him "I want you to explain ... scientifically, how they [Ayurveda] work." Is this how scientific investigations work? Do scientists receive orders or requests to discover the laws of nature?

35.18. Using U-shaped paths, we gave an example of how a sneeze can cause an earthquake in Mexico City. Give other examples that use the same argument to "explain" some really weird phenomena.

35.19. In terms of mathematics, how is modern physics different from classical physics? How is it the same? Give some examples of how quantum physics and relativity are crucially tied to classical physics.

35.20. When does modern physics give a different prediction from classical physics? When do they give the same prediction? Which one is more convenient to use when they give the same prediction? The book gives measuring the distance between NY and LA as an example of the comparison of classical and quantum physics. Can you think of other examples?

35.21. How is the wave-particle duality of quantum physics used in *Quantum Healing*? Why is it *needed* to think that waves are nonmaterial and particles are? How is this weird interpretation connected with the notion that thought can create a molecule? What is the correct way of connecting the wave nature and particle nature of light? What is the reason that people think that photons are nonmaterial? Explain why photons are as material as a baseball or an electron.

35.22. How does *Quantum Healing* "explain" the intelligence of the brain? How far does the chain of intelligence go in the structure of the brain cells? What is the weird conclusion? Instead of intelligence take another property such as "living." If you start with a cell which is living and follow the same line of reasoning, what conclusion do you reach?

35.23. How are the creation (and therefore the existence) of spirit and soul "explained"? Take any other similar words and insert them in place of "spirit and soul" in the last line of the quotation on page 551. Haven't you just as much proved the existence of your choice of words as "spirit and soul?"

35.24. How many stages are there in the scientific methodology? Is the physicists' desire to talk about their results to nonphysicists another stage of scientific methodology? How is this requirement different from requiring poets of one language to translate their poems into another language? Is that feasible?

35.25. Because modern physics and Eastern mysticism both require years of training and they both involve repeatability, they have a lot in common. What is wrong with this argument? Give a few examples that require training and involve repeatability and have nothing to do with modern physics.

35.26. Because modern physics and Eastern mysticism cannot be expressed in ordinary language, they have a lot in common. What is wrong with this argument? Give a few examples in which two things have the same *negative* property and conclude some arbitrary parallelism between them.

35.27. Give some (mundane) examples in which the "jump-to-a-new-level" argument can be used to apply quantum physics to very ordinary situations.

35.28. What are the phlogiston and caloric theories? Are they still used? Who cares about them?

35.29. What are the bootstrap and S-matrix theories? Are they still used? Who cares about them?

35.30. Why don't the New Age authors mention the current successful theories of physics such as the standard model of quarks and leptons? Name some of these successes.

36

Epilogue: The Eye of Physics

The discovery of the atomic nucleus opened the gates to a completely new territory, an utterly unexpected entity that took all scientists by surprise. While the idea of an atom had been in the making since the ancient times, and modern chemists had been dealing with them for over a hundred years, Rutherford's experiments of 1911 ushered the physicists into unchartered terrains. The nucleus showed the theoretical physicists the right way of tackling the problem of the structure of atoms, discovering along the way quantum physics, one of the most encompassing theories ever.

36.1 Modern Experimental Techniques

Rutherford started an ingenious method of experimental investigation suited for the study of atoms and subatomic particles. Prior to 1911, scientific observations were essentially of two kinds. First there were those observations that required direct participation of the human senses: looking at an object either with a naked eye or through a telescope or a microscope; putting two charged objects next to each other and "seeing" how they attracted or repelled one another; pushing on a cylinder full of gas and "seeing" how the thermometer's reading changed. In all these cases, the instruments, as well as the systems under study were in the domain of the human senses. The experimenter could see and feel the telescope, the microscope, the thermometer, the gas, the cylinder, etc. And although the instruments had gotten sophisticated, the reliability of their operation could be traced back to the human senses: telescopes could see unseen planets and faraway galaxies, but one could rely on their accuracy by comparing their performance with the performance of the rigorously tested previous generation of telescopes in the domain of the applicability of the previous generation. And the latter were trusted because they too were successfully compared with *their* previous generation of telescopes. These comparisons went all the way back to Galileo's telescope. And one could rely on that telescope and its discovery of Jupiter's moons by comparing what it saw with what human eye saw when there was grounds for comparison: Galileo's telescope saw the same Jupiter that we saw, only bigger; Galileo's telescope saw the same Moon that we saw, only bigger; Galileo's telescope saw the same Milky Way that we saw, only in more detail. So, if Galileo's telescope saw some moons around Jupiter, they must be there.

The second kind of observation was done mostly by atomic chemists. Although the outcome of their experiments impacted human senses (change of color or smell or other physical properties of the products of a chemical reaction), the chemical processes of the experiment involved hypothetical atoms and were therefore invisible. Chemists had to *assume* certain hypotheses, test the hypotheses by mixing chemicals and see if the outcome

Involvement of human senses in pre-1911 experiments and observations.

Involvement of human
senses in pre-1911
chemical experiments
and observations.

was as expected. If so, the hypotheses were confirmed, and eventually became part of the chemical lore. If not, new hypotheses had to be proposed and tested, until observation confirmed them. Because one could see and smell chemicals, one was comfortable with the experiments, although the invisible atoms made many people uneasy.

In Rutherford's experiment no human sense was involved. Both the probes and the targets were invisible! To rely on the experiment took a great leap of faith; faith in the laws of physics and our theoretical understanding of nature. Whereas Galileo's observation could be "confirmed" by our ability to see at least Jupiter, if not its moons, and atomic hypothesis could be "confirmed" by the visible change in chemical substances, Rutherford's experiment could not be "confirmed" by our senses! Here lies the source of some philosophers' mistrust of modern experimental techniques of fundamental processes. The laws of physics are used both in setting up the experiment and in interpreting the outcome. Can we trust such experiments and the knowledge gained by them?

Can we trust
experiments that involve
no human senses?

ooooooooooooooooooooooooooooooooooooooooo

The allegory of the
green seers of Arud.

In the ancient village of Arud, due to some genetic disorder, the overwhelming majority of the population was "green blind." Anything that was green appeared gray to them. The few who could see the color tried in vain to convince the rest of the village that the color of trees, bushes, and grass was different from the color of the sky just before a rainstorm. The "green seers" would put a green leaf next to a gray picture of the leaf and would ask the Arudians to look carefully at the two to find the difference in color. Every time the green seers would fail. The villagers could not tell the difference. This went on for many years. Finally the green seers gave up, and the villagers accepted the fact that a few among them had the extraordinary power of seeing a color that did not exist, and after a while they just did not care about the green seers.

Not all the villagers forgot about the issue. Arudian *philosophers* were deeply interested in the "epistemology of green," as they called it. They argued that the existence of something is proved only if all (or overwhelming majority of) people agreed on it. We all agree that horses exist, because we have all seen a horse and perhaps even ridden it. We all agree that trees exist, because we have all seen them in our gardens or out in the country. We have no disagreement about the blue color of the sky, or the orange color of the sunset. But when a few people claim that the color of a tree is green rather than gray, and no one else can see that color, then the existence of green should be questioned.

The arguments of the philosophers were so convincing that many villagers believed that "green" did not exist. The green seers suggested a "test" for the existence of green. They compiled a list of common green objects, gave it to a skeptic messenger, and told him to go to the farthest village where no Arudian had gone before. Once there, the messenger was to ask the green seers of that village to make a list of some common green objects without showing the list he was carrying. If the two lists had a lot of overlap, then "green" must exist. Although the test was successful, the philosophers were so eloquent and convincing that the villagers remained skeptic, and never accepted the color "green."

ooooooooooooooooooooooooooooooooooooooooo

> But nobody ...seriously questions our knowledge of medium-sized physical objects like trees and tables. In mainstream epistemology, sceptical arguments about trees and tables reflect back on our assumptions: since we obviously do know about tables, an argument that such knowledge is impossible challenges us to find the flaw in our reasoning.
>
> By contrast, scepticism in the epistemology of science is by no means just a philosophical exercise. For it is not at all obvious that we know about the entities postulated by modern scientific theories, such as gravitational waves and neutrinos. After all, we never have any direct sensory evidence for these entities. [Pap 97, p. 3]

This is a quote from a book on the philosophy of science, and the last sentence says it all: Since we have no sensory evidence for neutrinos, electrons, protons, neutrons, atoms, molecules, genes, etc., we have all the rights to question their existence. By the same token, since the Arudians did not have any sensory evidence for the color green, they had all the rights to doubt its existence. But let's go deeper into the reason that the villagers could not see the color green: They were not equipped with the cone-shaped photoreceptors that were sensitive to the green wavelength. On the other hand, the green seers had the *equipment* or the *instruments* with which they could detect the green color. Looking from outside, we can realize how silly the village philosophers were for denying the existence of the color green simply because they could not see it.

Physicists and other scientists are the few "green seers" of our world, not because their *physiology* permits them to see what others can't, but because their science and their equipments are sensitive to what none of our senses can detect. An invisible alpha particle emerging out of an invisible radioactive nucleus obeys the laws of physics as it approaches and consequently scatters from an invisible gold nucleus all the way to the counter at which it is detected. It's quite natural to extend these laws, which have passed all the tests in the visible world, to the invisible domain of the nuclei and the subatomic particles, *unless* a clear and detectable contradiction results.

Why do we have to trust the click of a counter or the reading of an apparatus? Remember the green seers' test given to the messenger? Ask a physicist about the properties of a proton: its charge, its mass, its spin, etc. He'll give you a list of numbers, some to many significant figures. Go half way around the world; find a qualified physicist; and ask her for the list of the same quantities. Compare the two lists, and find that they match exactly! On your way back, stop at a third "village" and ask a third physicist. The third list matches exactly with the other two. Just because we cannot see subatomic particles and processes, it does not mean that they don't exist. There *are* people who can see them. And the test of their ability is the unprecedented accuracy with which they agree with one another. These people are the physicists and chemists and biologists, who through laborious training and unselfish devotion to science have *developed* the ability to invent instruments and theories that truly *can* see the domain that is invisible to ordinary human beings.

> Scientists are the "green seers" of our race!

36.2 Modern Mathematical Techniques

Modern experimental techniques showed a whole new world at the heart of an atom. The inhabitants of this world spoke in the same language that the heavenly bodies were discovered—by Greeks and later by Galileo and Newton—to speak, only in a much more sophisticated and poetic tone. The relativity of 1905 and 1916, and the quantum theory of 1925–26 had already employed some of the most elaborate mathematics of their time. Little wonder that the names of the noted mathematicians of the time appear alongside the physicists'. Henri Poincaré, the brilliant French mathematician, codiscovered the special theory of relativity (STR); Marcel Grossmann, the Hungarian mathematician and a classmate of Einstein, discovered the connection between relativity and differential geometry, a connection that Einstein used to develop his *general* theory of relativity (GTR); David Hilbert, the great German mathematician, laid the mathematical foundation of the quantum theory half a century before the theory was discovered.

When Paul Dirac considered combining the quantum theory and STR, he had to first (re)invent some mathematical tools. He discovered the version of the language in which the dialogue with nature became much more transparent; the version in which nature spoke of antiparticles and the mysterious spin. When asked about his philosophy of physics in Moscow in 1955, Dirac responded "Physical Laws should have mathematical beauty."

> "Physical Laws should have mathematical beauty."
> Paul Dirac

Mathematical beauty is a reflection of natural *symmetry*; and symmetry is studied in geometry and another branch of mathematics known as *group theory*. When geometry and group theory are combined with the common language of *all* branches of physics, namely differential equations, a dialect of mathematics emerges called the theory of *Lie groups*.

Eugene Wigner, Dirac's brother-in-law, was a pioneer in recognizing the necessity of Lie groups in the manifold of relativistic spacetime. A quarter of a century later, when the classification of the "zoo" of particles created in the growing accelerators became a necessity, Gell-Mann and Ne'eman found that these particles assemble themselves remarkably in the *representations* of a specific Lie group. And Gell-Mann suggested that the most *fundamental* representation of that Lie group must correspond to the most fundamental entities of nature. That's how he predicted the existence of quarks.

If particles assemble themselves in the representations of Lie groups, and if we know that fundamental forces are the results of the exchange of particles, could the fundamental forces also be connected to Lie groups? Once it was realized that the electromagnetic force was associated with the *simplest* (local, gauge) Lie group, it was natural to try the next Lie group—that was one step more complex—as a candidate for other forces. The first attempt by Yang and Mills was unsuccessful. However, the attempt by Weinberg, Salam, and Glashow, which carried along the *unification* of the electromagnetic and weak nuclear forces, achieved miraculous success.

36.3   An Endangered Species?

The modern experimental and mathematical techniques have opened up a physical vista comparable to that opened up four centuries ago by Galileo and Newton. And giants like Einstein, Bohr, Schrödinger, Dirac, and others have made tremendous progress in unraveling many of the secrets hidden inside atoms. But there are many unanswered questions and many undiscovered pieces of the puzzle.

Today, a vanishingly small fraction of our race is pursuing the footsteps of Copernicus, Galileo, Newton, Maxwell, Einstein, and Schrödinger. They work in fields such as *high energy physics*, *fundamental particles*, *cosmology*, and *string theory*. In short, they are the physicists who investigate the topics discussed in this (and the next) part of the book. Engaged in finding solutions to some fundamental questions of the kinds that those intellectual giants asked, they seek answers to such questions as "How are the fundamental forces of nature operating?" "How can we unify them?" "What is the nature of the fundamental particles?" and "How did the universe begin and what happened subsequently?"

The pursuit of these questions have no foreseeable practical applications, just as the pursuit of the puzzle of the hydrogen atom in the 1920s had no obvious potential for industrial utility. However, because these questions are posed, and because they are the unanswered questions *derived* from the natural pursuit of the earlier inquiries, it is the duty of our race to expend some of its resources to support those who are asking them, even though there is no immediate return for the expenditure. This is how it has been ever since the people of Egypt and Babylon allotted part of their communal wealth to the investigation of the sky by the priesthood.

Alas, the largest depositories of human resources, the profit-hungry Wall Streets and the practical-minded Sputnik-driven governments show little or no interest in supporting such a noble pursuit. Because of the dwindling funds for fundamental research, many of the brightest minds have left this pursuit for other occupations in finance and industry, for which there are better financial rewards. This "brain drain," though unnoticeable now, will be fatally damaging in the long run, because if continued, it will hamper an evolutionary trend that started millions of years ago, a trend that has set us apart from the rest of the animals.

36.4   Continuity and Specificity of Physics

Quantum theory of Schrödinger and Heisenberg, borrowing concepts and procedures from the classical physics of Galileo and Newton, and evolving them to the level applicable to the atomic domain, brought physics to a higher stage at which it could explain the hydrogenic nonrelativistic phenomena. Similarly, Einstein's special relativity, also borrowing concepts and procedures from the physics of Galileo and Newton, advanced physics to such a level that it could explain the motion of objects that moved close to the speed of light. Classical physics was a tree that lived for over 300 years with a single stem, growing taller and stronger, until at the turn of the twentieth century it sprouted two new branches.

The discovery of the atomic nucleus, which used theoretical concepts entirely borrowed from classical physics, and many experimental gadgetry also borrowed from classical experimental tools, became the playing field for the two new branches of the tree of physics. While the hydrogen atom was mostly explained in terms of the quantum theory alone, the hydrogen *nucleus*, the proton, could not be understood but by a merger of the special theory of relativity and quantum theory, a step taken by Paul Dirac in 1928, without being aware of its use in the structure of the proton 45 years later.

Inseparability of modern and classical physics.

All the progress in the understanding of the fundamental forces and particles has been made by tackling *specific* problems. Quarks were the by-products of the mathematics of Lie groups used in the specific task of classifying hadrons, whose prolific production was the result of the specific desire to learn more about the atomic nucleus; and the latter was discovered when Rutherford sought an experimental verification of Thomson's specific model of the atom. Every step that brought us closer to an understanding of the microworld, was taken in response to a specific question, a specific problem, or a specific experiment.

Part VIII

Twentieth Century Physics: Macrocosmology

Chapter 37

Physics of the Cosmos

The material of the previous part of the book, once put in proper order, becomes the storyline of the universe. How can the tiniest subatomic particles tell us about the vast panorama of the cosmos? How is it that the ultimate "whole" can be understood in terms of its paltry "parts?" If posed as philosophical questions, we may never know the answers. But physics, in its quest for increasingly fundamental relations among natural phenomena, has been able to give a rational and observationally confirmed answer to these questions.

Briefly stated, general theory of relativity (GTR) gives a picture of an expanding universe, which has been confirmed by observation; the same theory predicts a time-reversed contraction, which at a finite time in the past puts the universe at a "singularity," an event with no spatial or temporal extension, a point in the four-dimensional spacetime representing the universe itself at infinite energy, infinite temperature, and infinite density. This event has come to be known as the big bang.

Singularities are mathematical "diseases" that nature cannot tolerate. However, the GTR *does* yield such a singularity. This simply means that GTR alone is not suited for the extreme conditions of the big bang, although it predicts it. Since small objects are studied in the quantum theory, only a consistent combination of GTR and quantum physics, a theory of **quantum gravity**, can remove the singularity of spacetime. Unfortunately, such a theory is, as of now, lacking.

A combination of GTR and quantum physics, a theory of **quantum gravity** may hold the answer to the big bang singularity.

Although today's physics cannot explain the very onset of the big bang, as the universe expanded and cooled down, the heat, energy, and temperature "reduced" to the amount that the physics studied in the last part of the book becomes relevant. And that is where we start our story of creation in the next chapter; but first we have to look at how the two constituents of the universe, namely matter and radiation, are held together.

37.1 The Friedmann Equation

Grasping the dynamics of the universe requires an understanding of how its constituents interact among themselves. The dominant force being gravity, we examine its role in holding the universe together. For a complete understanding of the gravitational dynamics of the universe, we need the full machinery of GTR. However, a (slightly modified) Newtonian theory of gravity can also describe the behavior of a simple model of the universe surprisingly well. The modification involves giving mass to the energy of EM radiation and making it a source of gravity. It is this modified Newtonian cosmology that will be the guiding light in our search for the history of the universe.

To this end, consider two (distant) points P_1 and P_2 of our universe separated by a distance R. Construct a sphere with its center at P_1 and its radius equal to $\overline{P_1P_2}$ as shown

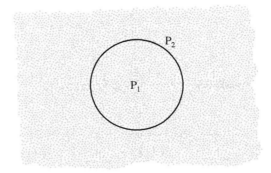

Figure 37.1: Two points P_1 and P_2 and the sphere filled with matter and radiation between them.

in Figure 37.1, and concentrate on the motion of an object of mass m at P_2. The sphere is filled with matter and radiation exerting a force of gravity on m. The rest of the universe has no effect on m, because m happens to be in a spherical cavity as far as the rest of the universe is concerned, and no force exists in such a cavity (see Math Note E.9.2).

As the mass m moves away from the center of the sphere due to the expansion of the universe, it possesses both kinetic energy and potential energy, much like a projectile fired from the surface of a planet. Adding the two energies and writing the mass of the sphere in terms of the density of the material filling it, **Math Note E.37.1** on **page 140** of *Appendix.pdf* derives the **Friedmann equation**, which is the fundamental equation of cosmology:

$$\left(\frac{v}{R}\right)^2 = \frac{8\pi G}{3}\rho - \frac{kc^2}{R^2}, \tag{37.1}$$

where v is the speed of m, G is the universal gravitational constant, and ρ is the density of matter and radiation filling the sphere of radius R.

There are three kinds of universes: closed, open, and flat.

As explained in detail in Math Note E.37.1, k determines the kind of universe we live in. If $k > 0$, the expansion of the universe will eventually halt and the *contraction* begins. We call this kind of universe **closed**. If $k < 0$, the universe will expand forever. We call this kind of universe **open**. If $k = 0$, the universe will expand forever, but the expansion constantly slows down, completely stopping only after an infinite amount of time. We call this kind of universe **flat**. It is this latter kind of universe on which we shall concentrate, because it is simpler, and observation points strongly in its favor.

The quantities in Equation (37.1) all vary with time (and only with time, as the universe is assumed homogeneous and isotropic, barring any physical quantity from varying with location or angle). For example, we know that ρ is a *decreasing* function of time, because the amount of material in a given large volume of the universe remains the same but the volume itself increases due to the universal expansion. Different constituents of ρ may (in fact, as we shall see, do) vary differently, and ρ may consist mostly of matter or mostly of radiation, depending on the particular epoch of the universe under consideration.

*Estimate for the **critical density** of the universe.*

We can obtain an estimate for ρ for a flat universe. Setting $k = 0$ and noting that $v = HR$ by the Hubble law, Equation (37.1) yields $\rho_c = 3H^2/(8\pi G)$, where the subscript under ρ stands for "critical," because ρ for a flat universe has come to be known as the **critical density**. Using the two extreme values of H given in Section 29.4.2, **Example D.37.1** on **page 67** of *Appendix.pdf* calculates the current critical density and finds that it lies between 8.9×10^{-27} kg/m^3 and 1.16×10^{-26} kg/m^3 equivalent to 5.3 and 6.9 nucleons per cubic meter.

> **What do you know? 37.1.** For a flat universe, $k = 0$ and the left-hand side of Equation (37.1), the square of the Hubble parameter, is proportional to the density. How do you explain the fact that H grows with ρ, while we know that larger density means stronger attraction of mass towards the center and, thus, its slowing down?

The *observed* nucleon number density n_b (the "b" stands for *baryon*, as protons and neutrons are the prototypes of this group of particles) is approximately 0.25 per cubic meter. If the universe is approximately flat, then its constituents cannot be just nucleons and radiation, because they account for *at most* $0.25/5.3 = 0.047$ or 4.7% (or less, if 6.9 nucleons per cubic meter is used for the critical density) of the critical mass. But we already know that ordinary matter is not the only constituent of the universe. Dark matter, whose existence is demonstrated unequivocally by the larger-than-expected speed of stars in galaxies (see Section 9.1.3 and Box 9.1.8), is a viable candidate for—and its abundance can account for much of—the missing mass required to "flatten" the universe.

We need dark matter to "flatten" the universe.

BIOGRAPHY

Aleksandr Friedmann was born to a father who was a ballet dancer and a mother who played professional piano. He began as a very ordinary student, but soon rose to the top of his class. Upon graduation from gymnasium in 1906, he entered the University of St Petersburg, finishing his undergraduate studies in 1910. He began his MS degree in 1911, but due to the turmoil in the prerevolutionary Russia, did not submit his Master's dissertation until 1922. By this time, he had taken up a new interest. Einstein's general theory of relativity was not known in Russia and by late 1920, Friedmann had already been working intensively on the theory. In an article sent to Zeitschrift für Physik in 1922, he showed that the radius of curvature of the universe can be either an increasing or a periodic function of time. Einstein quickly responded to Friedmann's article, saying "the solution given in it does not satisfy the field equations." Upon reading this comment, Friedmann sent Einstein the detailed calculations of his article and asked Einstein to examine it. After analyzing Friedmann's calculations, Einstein wrote to Zeitschrift für Physik: "In my previous note I criticised [Friedmann's work On the curvature of Space]. However, my criticism, as I became convinced by Friedmann's letter communicated to me by Mr Krutkov, was based on an error in my calculations. I consider that Mr Friedmann's results are correct and shed new light." Modern cosmology was born.

An accurate treatment of the Friedmann equation requires a knowledge of the exact proportion of the matter and radiation contribution to the density. This makes the solution to the equation prohibitively complicated. Fortunately, during various epochs of the universe, one of the two components was distinctly dominant, and therefore, the density in the Friedmann equation can be assumed to be composed *either* entirely of matter *or* entirely of radiation. The distinction between matter and radiation becomes fuzzy when the universe is so hot that even matter particles may be moving close to light speed, in which case they behave a lot like a photon. Thus, sometimes it is more convenient to classify constituents as **relativistic** and **nonrelativistic**.

What are relativistic and nonrelativistic constituents of universe?

The dominance noted above refers not to the number, but the *equivalent mass*, because it is this mass that gives rise to gravitational forces. In the case of matter, density, which is denoted by ρ_m, is simply the mass density. In the case of radiation, it is the *energy density*—for which there is an exact formula in terms of temperature as we shall see later—divided by c^2 (remember that $E/c^2 = m$). When ρ consists mostly of matter, we refer to the universe as **matter-dominated**, and if it consists mostly of radiation, we call it **radiation-dominated**.

37.2   Matter Dominance

Let's assume that the universe is composed mostly of matter, with negligible contribution from radiation. Let ρ_{m0} and R_0 denote the present matter density and "size" of the universe, and $\rho_m(t)$ and $R(t)$ its density and "size" at time t after the big bang.

First, what exactly *is* the "size" of the universe? It does *not* refer to the actual extent of the universe. By "size" is meant the scale of the universe, which is measured by the separation of two "typical" distant galaxies. The best way to understand R is to concentrate on two *specific* galaxies, say the Milky Way and another one far enough for homogeneity and isotropy to hold, and let R be their separation. The variation of R tells of the variation of the entire universe. For example, if the Milky Way is 500 Mly away from this reference galaxy now (so that $R_0 = 500$ Mly), and at some time t in the past their separation was 250 Mly, then we know that at time t the entire universe was half the present universe.

> **What do you know? 37.2.** Does it make sense to say that the universe was "half" its present size if the distance between two distant galaxies was half its present size?

Consider the sphere centered at the Milky Way (or any other galaxy) with radius R_0. The matter inside this sphere does not change as the universe expands. Using this fact, **Math Note E.37.2 on page 141** of *Appendix.pdf* finds R, ρ_m, and H as a function of time.[1] The results are displayed in the first row of Table E.3.

Example D.37.2 on page 67 of *Appendix.pdf* looks at the universe when its size was 1% of its current size. It shows, for instance, that it was between 14 and 18 million years old, that its matter density was about a million times the current density, and that its Hubble constant was approximately 11,000 km/s per Mly, or about 600 times the current Hubble constant.

As mentioned earlier, and as discussed in detail in the next section, the universe is—and has been for most of its existence—dominated by matter. Therefore, the formulas in the first row of Table E.3 are valid for our present universe. In particular, the last formula, when evaluated today, connects the value of the Hubble parameter to the age of the universe. Since H lies between 21 and 23 km/s per Mly, or between 2.23×10^{-18} s^{-1} and 2.44×10^{-18} s^{-1}, that formula gives an age of the universe between $t_0 = \frac{2}{3H} = 3 \times 10^{17}$ s (9.49 billion years) and 2.73×10^{17} s (8.67 billion years), both of which are noticeably different from the measured age of 13.7 billion years.

Part of the discrepancy is due to the assumption of a perfectly flat universe. What if the universe were not flat? A closed universe is ruled out, because it would be younger than the two estimates calculated above. Here is why: A closed universe will eventually stop

Accelerated expansion! expanding in the future. This implies that it is (and has been) slowing down, and therefore, to have the present rate of expansion, it must have been expanding more rapidly than a flat universe at the earlier times. Therefore, it must have taken it less time to reach this stage. An open universe is a more likely candidate. However, the recent observation of an *accelerating* expansion casts doubt on the conventional standard cosmology. It appears that the **cosmological constant** Λ, which Einstein introduced in his equation to "stop" the expansion of the universe, plays an important role in modern cosmology and could explain the strange acceleration.

At any rate, even if the simple flat universe that we are studying is not a sufficiently accurate description of the *present* universe (although there are strong pieces of evidence that it is), it turns out to be a good theory for the very early universe, which is physically

Horizon and the flatness of the early universe. more interesting. This has to do with the notion of the **horizon**.

An observer at any point P of the universe can see events whose light signals have had time to travel to P. The farthest of these events form a horizon. Any event farther than the horizon from P is unobservable by P, because its light signal has not had enough time to reach P. As the universe expands, the horizon expands with it, but *at a faster rate*. In fact, while the scale of the universe increases in proportion to $t^{2/3}$ [see Equation

---

[1]Instead of expressing ρ_m as a function of t, it is more common to write t in terms of ρ_m.

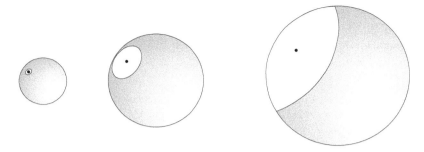

Figure 37.2: A point (the dot) and its horizon (the white area) as the universe (sphere) expands. Notice how the horizon covers a larger and larger fraction of the universe as the universe expands.

(E.125) or the first entry in the first row of Table E.3] when matter dominates and to $\sqrt{t}$ [see Equation (E.140) or the first entry in the second row of Table E.3] when radiation dominates, the horizon grows proportionately with time, as Example D.37.3 illustrates. It follows that the horizon covers a larger and larger portion of the universe as time passes. The horizon is a small part of a young universe; and the younger the universe, the smaller the fraction of the universe covered by the horizon. Figure 37.2 illustrates the expansion of the horizon relative to a two-dimensional spherical "universe." The smaller the relative size of a portion of a curved surface, the flatter it looks (a football stadium looks completely flat, while a continent does not). Therefore, the horizon, which borders the only part of the universe about which we can talk sensibly, covers an almost flat portion of the universe at the beginning, and the approximation of a flat universe becomes especially valid for a very young universe.

37.3   Radiation Dominance

When electromagnetic radiation is in thermal equilibrium with matter at a particular temperature, it is described by the black-body radiation function (20.5). In its early stages of evolution, when the universe is a hot plasma (a collection of positive ions and negative electrons), radiation and matter are indeed in thermal equilibrium because of the interaction of photons with the charged particles,[2] and therefore, the EM radiation obeys Equation (20.5). However, as the universe cools and neutral atoms are formed, radiation no longer interacts with matter, and the thermal equilibrium is broken. We say that radiation **decouples** from matter. Two questions arise: (1) What happens to the wavelength and temperature of the radiation? (2) How is the radiation described after decoupling?

Decoupling of radiation from matter.

Math Note E.37.3 on **page 143** of *Appendix.pdf* shows that the wavelength of the radiation increases in proportion to the scale of the universe, and that the radiation is *still a black-body* radiation—even though it is no longer in equilibrium with matter—with a temperature inversely proportional to that scale. This is a significant conclusion, because it tells us

Box 37.3.1. *To look for the radiation left over from the big bang, watch for a black-body radiation, including its characteristic curve.*

With T_0, λ_0 and R_0 denoting the present temperature, wavelength, and scale of the universe, respectively, Equation (E.134) of Math Note E.37.3 yields the third row of Table E.3.

---

[2]Recall from Chapter 32 that the EM force—represented by the exchange of photons—exists between charged particles only.

Our next task is to find the equivalent mass density of radiation. This quantity, denoted by ρ_γ, is derived in **Math Note E.37.4** on **page 144** of *Appendix.pdf* and displayed in the following equation and in the fourth row of Table E.3.

$$\rho_\gamma(T) = 8.36 \times 10^{-33}T^4 \text{ kg/m}^3. \tag{37.2}$$

Combining the third and fourth rows of Table E.3 gives the density of radiation in terms of the scale of the universe, displayed in the last row of Table E.3. This last expression, which shows a decrease in ρ_γ proportional to the inverse *fourth* power of scale size, seems strange! After all, when we discussed the matter density, we argued that the increase in volume—which is proportional to R^3—and the constancy of the number of nucleons give rise to an inverse R^3 dependence for ρ. Why can't we argue similarly and arrive at the same inverse R^3 dependence for ρ_γ? The answer lies in the difference between the energy of a nucleon and that of a photon. While the *number* density varies as inverse R^3 for both cases, the energy of a nucleon is mc^2 and is *independent of the scale* of the universe. For photons, the energy is hc/λ, which *decreases* with the increasing scale. This decrease introduces another factor of R in the denominator.

With ρ_γ—in terms of R—at our disposal, we can solve the Friedmann equation and obtain the size of the universe, the density of the universe, and the Hubble parameter as functions of time for a radiation-dominated universe. Math Note E.37.4 has all the details and arrives at the formulas displayed in the second row of Table E.3.

The matter mass density is directly related to the matter number density, because the two are related by a multiplicative constant (mass of the matter particle). This is not the case with the *radiation* energy and number densities, as we saw above. Is there a formula that connects the *number density* of photons to the temperature? **Math Note E.37.5** on **page 146** of *Appendix.pdf* gives the answer in the affirmative and also finds the average energy of a photon as a function of the temperature. These results are:

$$n_\gamma(T) = 2 \times 10^7 T^3 \text{ photons/m}^3, \qquad \langle E_\gamma \rangle = 2.7k_BT = 3.726 \times 10^{-23}T \text{ J}. \tag{37.3}$$

What do you know? 37.3. (a) What do you get when you multiply the two formulas of Equation (37.3)? (b) What do you get when you divide the result by c^2?

Suppose we are located at a point in the universe and we look out to see points as far away from us as possible. Since the universe is 13.7 billion years old, any signal that we receive from any point in the universe can be at most that old. This means that the signals can come from within a radius of 13.7 billion light years. Therefore, we say that the horizon is 13.7 billion light years now. A billion years ago the horizon was 12.7 billion light years; 10 billion years ago it was 3.7 billion light years, and so on.[3]

Starting at the big bang, we can also argue that the horizon radius one second after the big bang was 300,000 km, two seconds after the big bang it was 600,000 km, etc. This argument assumes that the horizon radius r_h is speed of light times the time elapsed: $r_h = ct$. In the absence of gravity and universal expansion, this is true. But, as **Math Note E.37.6** on **page 147** of *Appendix.pdf* shows, expansion increases the spread of horizon further so that $r_h = 3ct$ for a matter-dominated universe, and $r_h = 2ct$, when radiation is dominant.

The scale of the universe grows in proportion to $t^{2/3}$ when matter dominates and to $\sqrt{t}$ when radiation dominates. In either case, the horizon "catches up" with the universe: it grows at a faster rate than the universe itself, and therefore, covers more and more fraction of the universe. **Example D.37.3** on **page 68** of *Appendix.pdf* compares the fraction of the universe covered by horizon one hour, one week, and one year after the big bang.

---

[3] As the next paragraph makes it clear, these statements are not entirely correct.

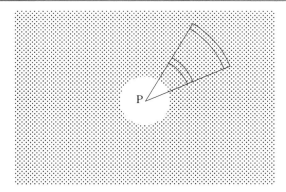

Figure 37.3: The larger shell is twice as far as the smaller shell. Its volume is *four* times as large as the smaller volume. Therefore, the number of "stars" in the larger shell is four times larger than the number in the smaller shell.

37.4 Expansion and Olbers' Paradox

When the expansion of the universe was established, one of the mysteries of astronomy could be explained. The mystery is this: "Why is the night sky dark?" Although the question may sound childish, there is a deeper motivation for asking it, and no less an astronomer than Kepler was one of the first to recognize the significance of the question. If the universe is infinite, homogeneous, and isotropic, then the number of stars in a spherical shell subtended by a cone should grow in proportion to the distance squared. Figure 37.3 shows the cross section of two spherical shells subtended by a cone centered at P, the observation point. The bigger shell is twice as far as the smaller shell; therefore its area is *four* times as large. Assuming the same thickness for both shells, the volume of the big shell is also four times the smaller volume. Since the number density of the stars is assumed uniform, there are four times as many stars in the larger shell than in the small one.

> Why is the night sky dark?

How do the amounts of light received by P from the two shells compare? Since there are four times as many stars in the big shell, its light intensity is four times the intensity of the smaller one. But intensity decreases as the *square* of the distance. This means that a typical star in the big shell shines four times weaker than a typical star in the small shell. Therefore, the overall intensity received at P from the big shell *is equal to* the overall intensity from the small shell. This independence of the intensity from distance and the assumption that the universe is infinite (and therefore, that there are infinitely many stars in the universe) implies that we should be receiving infinitely intense light from the sky day and night!

What do you know? 37.4. Knowing that intensity or brightness is the amount of power received (or emitted) per square meter, explain why the intensity of a source decreases as inverse square of the distance.

The above paradox prompted Kepler to assume that the universe was finite. But a finite universe, as Newton showed, was gravitationally unstable, and would collapse on itself. To overcome these difficulties, Loys de Chesaux, a Swiss astronomer, attributed the darkness of the night sky to the absorption of light by a hypothetical fluid that supposedly filled the entire universe. Almost 80 years later, in the 1820s, the German astronomer Heinrich Olbers picked up on the same fluid theory, and the entire paradox became known as the **Olbers' paradox**. But the fluid theory did not solve the problem, because the absorption of the starlight would heat up the fluid to the point that it would also shine as brightly as stars.

Today, with the evidence for the expansion of the universe we can explain the darkness of the night sky: When the universe expands, the wavelength of EM radiation expands with it. The stretch of the wavelength changes the visible light to longer wavelengths such as microwaves and radio waves, which are not visible, rendering the sky dark.

Expansion of the universe explains Olbers' paradox.

37.5   Echo of the Big Bang

The runaway distant galaxies tell of an expanding universe, an expansion that is a whimpering reminder of a once colossal explosion. Galaxies are made of nucleons and other forms of matter, and their running away tells us the story of the big bang as told by only one component of the universe. Are there other observable remnants of the big bang coming from the other component of the universe, radiation?

In 1964, two radio astronomers, Arno Penzias and Robert Wilson, at Bell Laboratories were using a horn antenna, built earlier for communication via the *Echo* satellite, to measure the intensity of the radio waves coming from the Milky Way. The radio signals from all astronomical objects come in as "noise," much like the statics picked up by radios during a thunderstorm. Distinguishing the signal noise from other spurious noises is not trivial, although it is much easier if the source is small, such as a star. In this case, one can switch the antenna beam back and forth between the source and the empty sky. If there is a detectable difference, it must be due to the source. The enormous size of our galaxy makes such a directional distinction difficult. In order to observe any signal from the galaxy, the antenna had better be as "noise free" as possible.

Role of pigeons in the discovery of the big bang!

By a technique using liquid helium, Penzias and Wilson could reduce the expected spurious noise considerably. They started their observation in the spring of 1964 using a relatively short wavelength of 7.35 cm, where the radio noise from the Milky Way should have been negligible. To their surprise, they detected a strong signal. They changed the antenna direction; the noise was still there. It appeared that the noise was coming from practically every direction. To make sure that the fault was not of the antenna, they dismantled the 20-foot horn, and discovered that some pigeons had nested in the antenna and deposited a "white dielectric material" there! After cleaning the mess and pointing the antenna to the sky in early 1965, they observed very little difference in the level of the noise. The noise did not want to quit.

Theoretical argument points to the existence of a background radiation in the universe.

Puzzled by the persistent noise, Penzias contacted some colleagues, who eventually directed him to Princeton University. It turned out that a group of physicists there had been working on the formation of nuclei at the early universe. James Peebles, the theorist of the group, had argued that the observed structure of the visible universe, indicating a composition of about 99% hydrogen and helium, was a strong evidence for an intense radiation at the early universe. Peebles calculation revealed that during the first few minutes of the evolution of the universe, the nuclear processes would take place at such a rapid pace that a large fraction of the nucleons would "cook" into the nuclei of heavy elements. The present absence of such elements must be the result of a mechanism that prevented their formation. The only candidate for such a prevention is a very dense and hot background radiation. Peebles estimated the present temperature of this radiation to be around 10 K.

In a subsequent meeting, the Princeton group and Penzias and Wilson decided to publish a pair of companion letters in the *Astrophysical Journal*, in which Penzias and Wilson would announce their observation, and the Princeton group would explain the cosmological implications. The radiation has come to be known as the **cosmic microwave background** (CMB) radiation. It is the same black body radiation that was discussed in Section 37.3.

Penzias and Wilson could detect only a small portion of the curve characteristic of a black body radiator. As it reaches the Earth surface, CMB loses most of its content to the

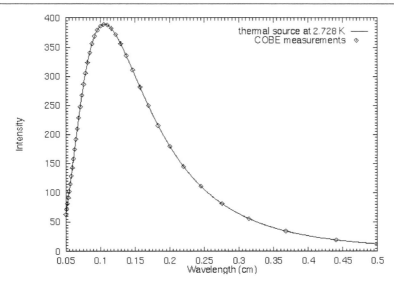

Figure 37.4: The black body radiation curve as detected by COBE. The squares are the data points, and the solid curve is the BBR curve corresponding to a temperature of 2.725 K.

atmosphere. Nevertheless, Penzias and Wilson could determine the temperature of CMB from that small portion. To see the entire spectrum, and to determine the temperature more accurately, the antenna has to be lifted above the Earth's atmosphere. And that is where the COsmic Background Explorer (COBE) comes in.

After the discovery of Penzias and Wilson, and the realization of the limitation of their observation, physicists started to plan ways of getting their instruments above the atmosphere. The obvious choice was a satellite, and by 1974 three groups of physicists from east and west coasts submitted proposals to NASA. As the project was complicated and NASA experienced the *Challenger* disaster in 1986, the COBE satellite was not launched until November 18, 1989.

Less than two months after its launch, COBE had faithfully sent enough information that the investigators could construct the shape of the radiation curve. Was the curve a black body radiation curve as predicted by the big bang theory? In the winter meeting of the American Astronomical Society, held outside Washington DC, on January 13, 1990, one of the principal investigators put up on the screen an image of the plotted data points as well as a theoretical black body radiation curve of the type introduced in Equation (20.5). An eerie silence fell over the audience, which immediately turned into a frenzied applause and a standing ovation. The data points fell exactly on the theoretical curve! The CMB *was* a black-body radiator, and its temperature was 2.725 ± 0.001 K. Figure 37.4 shows this historically significant plot.

CMB *is* a black body radiator, and its temperature is 2.725 K.

If you have been following the math notes, you have no doubt noticed that many of the formulas derived so far are dependent on the temperature alone. With COBE's precise measurement of the present temperature of the CMB radiation, we can determine the present numerical value of many quantities. Let's start with number densities. Substitute 2.725 for T in Equation (37.3) to obtain 4×10^8 photons/m^3 for photon number density. On the other hand, the present nucleon *number* density is about 0.25 per cubic meter. Thus, there are $4 \times 10^8/0.25$ or about 1.6 billion photons for every nucleon. Radiation clearly out*numbers* matter. However, superiority is not measured by numbers; it is the density that counts.

With 0.25 nucleons per cubic meter and 1.67×10^{-27} kg as the mass of a nucleon, the

The present universe is matter-dominated.

present density ρ_{b0} of *baryonic* matter is found to be 4.18×10^{-28} kg/m³. The present radiation density is obtained from the fourth row of Table E.3, which yields $\rho_{\gamma0} = 4.61 \times 10^{-31}$ kg/m³. We see that the present ρ_b is almost 1000 times larger than the present ρ_γ, verifying our earlier assertion that the present universe is matter-dominated. Moreover, matter is not just baryonic. In fact, as already mentioned, it is mostly dark. If we include the dark matter, the density jumps to about 2.5×10^{-27} kg/m³, and the ratio $\rho_{m0}/\rho_{\gamma0}$ jumps to about 5400. The universe is definitely matter-dominated.

But has it always been that way? To answer this and other related questions, we need to know how to express ρ_m as a function of temperature. **Math Note E.37.7 on page 147** of *Appendix.pdf* derives an approximate formula, which we reproduce here:

$$\rho_m(T) = 1.24 \times 10^{-28} T^3 \text{ kg/m}^3. \tag{37.4}$$

Universe became matter-dominated at $T_{\text{eq}} = 14,800$ K.

The earlier epochs of the universe witnessed a faster rise in radiation density than in matter density: ρ_γ increases as the fourth power of T while ρ_m does so as the third power. In fact, you can easily find the temperature at which the two become equal: just equate (37.4) and the fourth row of Table E.3 and solve for T. This procedure yields $T_{\text{eq}} = 14,800$ K, where the subscript is a reminder of the equality of densities. When the universe was hotter than about 14,800 K, it was radiation-dominated. As it cooled down below this temperature, it became matter-dominated.

Universe became matter-dominated about 25,000 years after the big bang.

An interesting question is "How long was the universe dominated by radiation?" Since 14,800 K is the transition from radiation to matter dominance, the question can be rephrased as "How long after the big bang did the universe cool down to 14,800 K?" With $T = 14,800$, the second equation of the second row of Table E.3 gives $t = 1.05 \times 10^{12}$ s, or about 33,000 years. However, this is not quite right, because the calculation ignores the neutrinos which also contribute to the density. It turns out [see Section 38.1, especially Equation (38.1)] that neutrinos increase the density by a factor of 1.681 and reduce the time by $\sqrt{1.681}$. So, instead of 33,000 years, we get $33000/\sqrt{1.681}$, or 25,500 years. Thus, for over 25 millennia (a tiny fraction of its age), the universe was dominated by radiation. However, those "few" years were crucial in determining the fate of the universe as we shall see in the next chapter.

Now we can explain where the numbers used in Example D.37.3 came from. **Example D.37.4 on page 68** of *Appendix.pdf* provides the details.

When did radiation say "goodbye" to matter?

Another interesting question is "When did the CMB radiation separate from the matter?" As long as the universe had positive ions and negative electrons, i.e., as long as it was a plasma, photons could not escape. Only after the atoms formed, did the radiation decouple from matter. And for atoms to form, the universe must be so cool that the photons lack enough energy to ionize them. As an example, take the hydrogen atom, which is mostly found in its *ground state*, i.e., what we called the $n = 1$ electron orbit in Section 21.3. The energy associated with this state is -13.6 eV, and a photon with an energy of 13.6 eV or more can kick the electron out of the atom, that is, ionize it. What is the temperature at which the photons have this kind of energy?

Equation (37.3) gives the average energy of a photon as a function of temperature. For this energy to be 13.6 eV$= 2.18 \times 10^{-18}$ J, the temperature of the photon "gas" has to be 58,500 K. This is, however, a huge overestimation, because when the average energy is 13.6 eV, there are a lot of photons (too many for our purpose) that have much higher energies. Considering the fact that photons outnumber the hydrogen atoms by a factor of 1.6 billion, it is possible to have enough photons of 13.6 eV or higher energy to ionize the atoms at a much lower temperature. Math Note E.37.5, using a little calculus, shows that at about 5900 K, $1/(1.6 \times 10^9) = 6 \times 10^{-10}$ of the photon population has energies of 13.6 eV or higher; i.e., for each H-atom there is a photon that has an energy of at least 13.6 eV. Therefore, a black body radiator at 5900 K has sufficient number of photons at high enough energy to ionize the available atoms, although the average energy of its photons—by the second equation of (37.3)—is only 1.4 eV.

The consideration above is not the only argument reducing the overestimated ionization temperature. Other factors, such as the probability of a photon interacting with the electron of the hydrogen atom (the so-called scattering cross section) reduce the temperature to about 3000 K, called the **decoupling temperature** and denoted by T_{dec}. With this temperature, we can now find the age of the universe when radiation separated from matter. Since T_{dec} is smaller than T_{eq} found above, the decoupling occurred *after* matter dominated the universe. Using the third row of Table E.3, we first find the ratio of the scales: $R(t)/R_0 = 2.725/3000$, which is 0.00091; then use the first row to get $t/t_0 = 0.00091^{3/2} = 2.74 \times 10^{-5}$, so that $t = 2.74 \times 10^{-5} t_0$ or $t = 375000$ years, assuming that $t_0 = 13.7$ billion years. This time is denoted by t_{dec} and is called the **decoupling time**.

375,000 years after the big bang, radiation decouples from matter.

> **What do you know? 37.5.** What is the earliest time that you can obtain information from the universe using electromagnetic signals?

37.6   End-of-Chapter Material

37.6.1   Answers to "What do you know?"

37.1. Larger ρ is achieved at earlier times, and at earlier times, the force of the big bang explosion overcomes the gravitational attraction.

37.2. Yes! Look at the balloon analogy: When the radius of a (spherical) balloon doubles, the distance between points on the surface of the balloon also doubles.

37.3. (a) When you multiply the number of photons per cubic meter by the (average) energy of each photon, you get the amount of energy the collection has per cubic meter, i.e., the energy density. (b) When you divide energy by c^2, you get the equivalent mass. In this case, you get the equivalent mass *density* ρ_γ.

37.4. Consider two spheres centered at the source, one having a radius twice as big as the other. They both receive the same amount of power (the total power of the source) on their entire surface area. Since area goes as radius square, the bigger sphere has four times as much area as the smaller one. Therefore, the amount of power it receives per square meter (the intensity) is one fourth the power received by each square meter of the smaller surface. Thus, increasing the distance by a factor of two reduces the intensity by a factor of four.

37.5. The earliest time would be when the background radiation was no longer trapped in the plasma of positive and negative charges, i.e., when atoms started to form 375,000 years after the big bang. No electromagnetic information about the universe is available less than 375,000 years after the big bang.

37.6.2   Chapter Glossary

Closed Universe A universe which expands for a while and then starts to contract.

CMB Cosmic Microwave Background Radiation.

Cosmic Microwave Background Radiation The electromagnetic radiation present in all regions of the universe. It is the remnant of the very hot and energetic EM radiation present from the very moment of the big bang. It has a BBR curve corresponding to a temperature of 2.725 K.

Cosmological Constant A constant multiplying a certain term which is added to Einstein's equation to prevent the universe from expanding.

Critical Density The total density (including matter, radiation, dark matter, etc.) required to make the universe flat.

Decoupling A process by which one or more particles stop interacting with the rest of the constituents of the universe.

Decoupling Temperature The temperature (about 3000 K) at which radiation stopped interacting with matter because of the formation of hydrogen and helium atoms.

Decoupling Time The time (about 375,000 years after big bang) at which radiation stopped interacting with matter because of the formation of hydrogen and helium atoms.

Flat Universe A universe which expands for ever, but the expansion keeps slowing down without any minimum limit.

Horizon The boundary beyond which the universe is not observable.

Matter-Dominated Universe A universe whose constituent is almost entirely matter.

Nonrelativistic A term describing a constituent of the universe which moves slow compared to light speed, so that its KE is much smaller than its rest energy mc^2.

Olbers' Paradox The puzzle of the darkness of the night sky. If the universe is homogeneous, isotropic, and infinite, then it can be shown that night should be as bright as day.

Open Universe A universe which expands for ever, and if the expansion slows down, it will never slow down below a certain minimum limit.

Quantum Gravity The Holy Grail of theoretical physics. The intensely sought after theory which combines Einstein's general theory of relativity and quantum mechanics.

Radiation-Dominated Universe A universe whose constituent is almost entirely radiation.

Relativistic A term describing a constituent of the universe which moves close to light speed, so that its KE is much much bigger than its rest energy mc^2.

37.6.3   Review Questions

37.1. Is quantum gravity a combination of quantum theory and Newtonian theory of gravity? Why is quantum gravity necessary?

37.2. Who was Aleksandr Friedmann, and what was his contribution to physics?

37.3. How many kinds of universe did Friedmann obtain when he applied GTR to the entire universe? Describe each one in detail. What is the role of the critical density in this classification?

37.4. How do you describe a relativistic particle? A nonrelativistic particle?

37.5. What is a matter-dominated universe? A radiation-dominated universe?

37.6. What is the cosmological constant? What is the horizon?

37.7. What does it mean in physics to say that something decouples from the rest of the universe?

37.8. If the universe is homogeneous and isotropic and infinite, do you expect dark nights or bright nights? Why? What is Olbers' paradox?

37.9. Why is it necessary to have very energetic EM radiation at the beginning of the universe? What would happen if this radiation were not present? Would that contradict the present observation regarding the components of the universe?

37.10. What is cosmic microwave background radiation? Where is its source in the sky? Is it coming from a particular direction? What kind of graph do you get when you plot its intensity versus the wavelength of the radiation? Is there a temperature associated with that graph? If so, what is that temperature?

37.11. Is the universe matter-dominated or radiation-dominated now? Has it always been this way? How long after the birth of the universe did matter dominate? How hot was the universe then?

37.12. What is the decoupling temperature? How old was the universe when the decoupling took place? What did the universe consist of before that? After that? Why is it not possible to see how the universe was prior to the decoupling?

37.6.4   Conceptual Exercises

37.1. Is the density ρ in Friedmann equation (37.1) the density of ordinary matter? If not, what is it?

37.2. Suppose we could monitor the speed of a distant galaxy continuously. In which case would we see the biggest change in the speed of the galaxy, a closed universe, an open universe, or a flat universe? In which case would we see the smallest change in the speed?

37.3. For the universe to be flat, its density should be about 10^{-26} kg/m^3. The observed density of the universe (mostly in the form of hydrogen and helium) is around 4×10^{-28} kg/m^3. Does this mean that the universe is not flat?

37.4. An electron has a KE of 2 keV. Is it a relativistic or a nonrelativistic particle? A proton has a KE of 5 MeV. Is it a relativistic or a nonrelativistic particle? An electron has a KE of 5 MeV. Is it a relativistic or a nonrelativistic particle?

37.5. The universe today may not be flat. Could you say that it was flat 13 billion years ago? When was it flatter, 12.5 billion years ago or 13.5 billion years ago?

37.6. Suppose you could plot the BBR curve of the CMB radiation 7 billion years ago. Would the peak of the curve shift to the left or to the right of its current peak?

37.7. Suppose you go back in time to when the temperature of the CMB radiation was twice its present value.
(a) Would the matter density increase or decrease? By what factor?
(b) Would the radiation *equivalent mass* density increase or decrease? By what factor?
(c) Would the radiation *number* density increase or decrease? By what factor?
(d) Would the average energy of each photon increase or decrease? By what factor?

37.8. What is the relation between the two formulas of Equation (37.3) and the formula in Equation (37.2)?

37.9. Which direction in the sky do you have to point your radio telescope to see the CMB radiation? What wavelength do you have to tune your telescope to get the maximum intensity?

37.6.5   Numerical Exercises

37.1. Assume that the Hubble constant is 22 km/s per Mly.
(a) What is the critical density of the universe?
(b) Assuming that the universe is made solely of nucleons, how many nucleons (on the average) do you expect to find in a cubic meter of the universe? A nucleon has a mass of 1.67×10^{-27} kg.

37.2. What is the intensity (in Watts/m^2) of a 100-Watt light bulb a meter away from it? What is its intensity 10 meters away from it? The area of a sphere of radius r is $4\pi r^2$.

37.3. Use Wien's displacement law to find the wavelength at which cosmic microwave background radiation peaks. To what category of the EM spectrum does this wavelength belong?

37.4. A second after the big bang, the temperature of the universe is about 2 billion Kelvin, and it is dominated by radiation.
(a) How many photons are there per cubic meter?
(b) What is the radiation density of the universe?
(c) How much does each liter of the universe weigh?
(d) What is the matter density of the universe?

Chapter 38

Early Universe

The last chapter studied a universe consisting of matter and radiation, with matter dominating the universe (but radiation out*numbering* matter particles) at the present time, and for almost the entire history of the universe, except for the first few dozen millennia when radiation ruled the universe. What about the earlier times? Did photons outnumber the nucleons by the same factor of 1.6 billion? Did the short era of radiation dominance start with the big bang, or was there a period in the early universe when matter's importance equated or surpassed radiation? These are important questions, because as we have seen, the content of the universe determines its fate. The following example sets the stage for the story of our very young universe.

Example 38.0.1. The average energy of a photon increases proportionately to the temperature. As there is no upper limit to how hot the universe was at the very beginning, it is possible that the early photons were so energetic as to be able to create fundamental particles and their corresponding antiparticles. For instance, quantum electrodynamics allows the possibility of the creation of an electron-positron pair through the collision of two photons. The Feynman diagram in Figure 32.4(d)—reproduced in Figure 38.1 for convenience—describes precisely this process.

Not *any* two photons can create an e^+-e^- pair. The conservation of energy requires that the initial photon pair have a total energy at least equal to the rest energy of the e^+-e^- pair; or that each photon have energy at least equal to the rest energy of an electron. Putting 9.1×10^{-31} kg for the mass of an electron and 3×18^8 m/s for the speed of light in $E = mc^2$, we obtain 8.2×10^{-14} J as the required energy of each photon. Equating this to the average energy of a photon, given in Equation (37.3), yields a temperature of 2.2×10^9 K.

We shall see later that, at temperatures well above 2.2 billion K, the electrons and positrons move close to the speed of light, and therefore, behave much like photons. Other particles (and their antiparticles) may also be present at higher temperatures; and if the temperature of the universe is high enough, all these particles move close to light speed. So there is not much difference between radiation and matter, and the radiation formulas apply to all components of the universe. In particular, to find the approximate age of the universe at this temperature, you can use the second row of Table E.3. However, the contributions of different constituents to the density of the universe differ slightly from the photon contribution. The next section is devoted to this topic. ■

38.1 Content of the Early Universe

The example above showed that a black body radiator, the average energy of whose photons is equal to the rest energy of an electron can create electron positron pairs. The minimum temperature of such a radiator is 2.2 billion K, and is called the **threshold temperature** for electrons. The e^+-e^- pairs so created have a tendency to annihilate each other back into two photons. Thus the two processes of creation and annihilation are in equilibrium:

Threshold temperature of electrons.

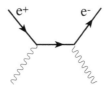

Figure 38.1: The electron-positron creation from two photons.

$e^+ + e^- \leftrightarrow \gamma + \gamma$. At temperatures above 2.2 billion K the equilibrium exists; below that temperature, only annihilation takes place because now the photons are too "cold" to produce e^+-e^- pairs.

Other particle-antiparticle pairs have their own threshold temperatures. For example, protons, with a mass 1836 times larger than the electron mass, have a threshold temperature of 4 trillion K. Lighter particles such as muons, pions, kaons, etc., have threshold temperatures lying between 2.2 billion and 4 trillion K. Heavier particles such as neutrons, tauons, $W^\pm$, Z^0, the Λ's, the Σ's, etc. have threshold temperatures larger than 4 trillion K. Thus, as we get closer and closer to the moment of the big bang, crossing the threshold temperatures of various particles, the content of the universe becomes more and more diversified. This scenario gives a very complicated picture of the early universe, while, in reality, the early universe was relatively simple, for the following reason.

> **What do you know? 38.1.** $W^\pm$ and Z^0 are, respectively, 156,750 and 178,470 times heavier than an electron. What are the threshold temperatures of $W^\pm$ and Z^0?

At the earliest stages of the universe, all particles are relativistic.

Although there is a large number of different kinds of particles (and their antiparticles) in the early universe, because of the tremendous heat present, all particles are *relativistic*; that is, their rest energy (mc^2) is negligible compared to their kinetic energy. Therefore, their masses could be ignored: they become *massless*, just like photons (see Section 28.3.3). And the universe can be treated as a hot gas of different kinds of "photons." The contribution of these different massless particles to the density of the universe is distinguished by the product of three numerical factors which depend, respectively, on the number of orientations (projections) of the spin of the particle, on whether the particle has an antiparticle or not, and on whether the particle is a boson or a fermion.

The factors above stem from the calculation of the contribution of different *species* of particles to the total energy density. For example, since there are two spin states (up and down) for a spin-$\frac{1}{2}$ particle, one has to introduce a factor of two, one for spin up and the other for spin down. Similarly, if the particle has an antiparticle, then the two contribute equally to the energy density, and one has to introduce another factor of two. In general, for a massive particle of spin s, the first factor mentioned in the last paragraph is $2s + 1$; for a massless particle it is 2 (see Box 33.1.1). The second factor is 1 if the particle does not have a distinct antiparticle, and 2 if it does. These plus the last factor, which is 1 for bosons and 7/8 for fermions—and is harder to derive—are explained in some detail in **Math Note E.38.1** on **page 148** of *Appendix.pdf*.

Thus the recipe for writing down the expression for the density of the universe is to add all the densities of the different (relativistic) particles in thermal equilibrium, which, except for the numerical factors above, are identical to the photon density given in Equation (37.2). For example, to find the density ρ_e of the electron species, multiply ρ_γ by $\frac{7}{8}$ because electron is a fermion, and also by 2 because electron has an antiparticle. The factor of 2 for the spin of the electron is already included in ρ_γ, because photon also has two spin orientations.

What do you know? 38.2. What is the density of the electrons in the universe when its temperature T is well above 2.2 billion K?

Later, we will be interested in the relation between the temperature of the universe and the time elapsed since the big bang. This relation is given in the second row of Table E.3 for a photon gas. To find the same relation for a gas consisting of a variety of "photon" species, we need to replace ρ_γ with the *total* density $\rho = \rho_\gamma + \rho_e + \rho_\nu + \ldots$, where ρ_γ is the photon density, ρ_e the electron density, ρ_ν the neutrino density, etc. Since all of these densities are multiples of ρ_γ, the total density ρ is also a multiple of ρ_γ. Call this multiple α, so that $\rho = \alpha \rho_\gamma$. Then the middle equation of the second row of Table E.3 in Appendix E will have a $\sqrt{\alpha}$ in the denominator:

$$t = \frac{2.3 \times 10^{20}}{\sqrt{\alpha}\, T^2}. \tag{38.1}$$

What do you know? 38.3. The present universe consists mostly of photons and neutrinos. For such a constituency $\alpha = 1.681$. What would you get (in years) if you use this and the current temperature of the universe in Equation (38.1), which gives the time elapsed since the big bang? Why is the answer so wrong?

38.2   The Genesis as Told by Physics

The history of the early universe cannot be told chronologically. The reason is that events proceed with extraordinary speed at the beginning. The number of things that happen in the first second may be far more than what happens in the next second. Our analysis so far has shown that the important parameter is the temperature T, as it determines the content of the universe, and therefore, the processes that take place in it.

It is convenient to divide the course of the development of the universe into epochs, each epoch specified by a range of values of the temperature. Starting with a very simple and symmetric universe, you will see how the laws of physics developed in the previous parts can predict the content of each epoch, and ultimately of the present universe. It is remarkable that the predictions of these laws so closely match the data collected in numerous observations of the universe.

38.2.1   The First Epoch $10^{14} < \mathrm{T} < \infty$

The present laws of physics cannot explain how the big bang occurred. Such an explanation requires a quantum theory of gravity, which as of this writing, is lacking. Ultimately, such a theory combines the general theory of relativity—which works so well in the large scale structures—with the quantum theory, which is the theory of the microscopic phenomena. So we don't know what happened at the *start* of the first epoch, but it is reasonable to assume that sometime during this epoch, the universe consisted of all the known fundamental particles and gauge bosons. In fact, let's assume complete symmetry among all particles and their antiparticles, and let physics explain the ensuing asymmetry, i.e., the presence of matter and the absence of antimatter.

What do you know? 38.4. Why is it reasonable to assume complete symmetry between matter and antimatter at the beginning of the creation?

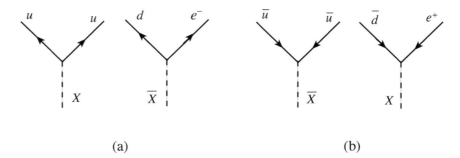

Figure 38.2: (a) Creation of pure matter. (b) Creation of pure antimatter.

Consider a very hot universe consisting of six quarks and six antiquarks, six leptons and six antileptons, and all the gauge bosons. All particles are assumed to populate the universe in equal numbers. At the beginning of the first epoch, when the temperature is sufficiently high, every process takes place in both directions. For example, if a particle annihilates its antiparticle to produce two photons, two (other) photons collide and produce the original particle antiparticle pair.

As the universe expands and cools down, certain processes, whose threshold temperatures lie above the prevailing temperature, take place only in one direction. For example, $W^{\pm}$ annihilate each other completely once the universe cools down below their threshold temperature of about 4×10^{14}. This kind of annihilation raises the question: If all particles are accompanied by exactly the same number of their antiparticles, why did they not annihilate each other completely? After all, our existence indicates that matter survived!

CP violation is the same as matter-antimatter asymmetry.

The matter-antimatter asymmetry is a phenomenon for which we don't have a complete explanation. However, there is a strong experimental evidence for the asymmetry, and the grand unified theories give a reasonable theoretical argument in favor of the phenomenon. The experimental evidence, known as the **CP violation**, was discovered in 1964 by James Cronin and Val Fitch, who were awarded the Nobel Prize in 1980 for this discovery. The "C" stands for **charge conjugation** and "P" for **parity**. These are abstract mathematical symmetry operations performed on fields: charge conjugation changes the sign of the charge of the particle associated with the field and parity reflects the position of the particle about the origin.

Decay of GUT gauge particles and the origin of matter-antimatter asymmetry.

To see how the grand unified theories (GUTs) can explain the absence of antimatter in the present universe, consider the decay of the X particle and its antiparticle shown in Figure 38.2. Recall from Section 34.5 that the X gauge boson of the grand unified theories attaches to two quarks or a quark and a lepton (see Figure 34.9). This allows the possibility of the decay of X into two up quarks and its antiparticle[1] $\overline{X}$ into a down quark and an electron [Figure 38.2(a)]. This may give the impression that we have created pure matter (u, u, d, e^{-}) out of an equal amount of matter and antimatter $(X\overline{X})$. But that is not true, because for every pair of processes depicted in Figure 38.2(a), there is a pair depicted in Figure 38.2(b), which creates pure antimatter out of an equal amount of matter and antimatter. Thus the decay of an $(X\overline{X})$-pair by itself cannot create pure matter.

However, the existence of the CP violation *can* produce pure matter out of an equal amount of matter and antimatter. Suppose that—due to the CP violation—the processes of Figure 38.2(a) take place slightly more frequently than those in Figure 38.2(b). For example, assume that for every billion occurrences of Figure 38.2(b), there are a billion and one occurrences of Figure 38.2(a). Then matter particles outnumber the antiparticles by

---

[1]We are labeling the X with a positive charge as particle and the one with the negative charge as antiparticle.

one part in a billion. Thus GUTs *plus* the CP violation can explain the matter-antimatter abundance asymmetry. Although the universe is created with equal number of particles and antiparticles, by the end of the first epoch particles slightly outnumber antiparticles.

38.2.2   The Second Epoch 10^{12} K $< T < 10^{14}$ K

The second epoch starts a few nanoseconds after the big bang (see **Example D.38.1** on **page 69** of *Appendix.pdf*). The universe consists of all the leptons and antileptons, five quarks and their antiquarks, eight gluons and the photons. All the very heavy particles, such as the weak bosons and the top quark, have disappeared because 10^{14} K is below their threshold temperature. At the start of the epoch, all particles come in equal abundance, but the heavier particles and their antiparticles start to annihilate, so that by the end of the epoch only the lighter particles survive.

🧺 Food for Thought 🌱

FFT 38.2.1. As we have seen in the last chapter, and will see in the present one, physics can explain fairly accurately the course of the development of the universe starting only a moment *after* the big bang, and many observations testify to that explanation. At the present time, however, physics cannot explain the act of creation itself. That is why we had to start a few nanoseconds after the big bang. This is still a remarkable achievement for science; but some pseudoscientists argue (and many people buy the argument) that since physics cannot explain the act of creation, their particular story of creation is *the* story.

The realm of the unknown is continuously diminished by the advancement of science, but it will never be completely annihilated. There was a time when the appearance of a comet in the sky was a sign of the "wrath of God." Now that Newtonian physics can explain the motion (and predict the appearance) of all known objects in the solar system, no one panics when Halley's comet visits us every 76 years.[2] There was a time when the synthesis of organic material was considered interfering with the act of God, but now the production of synthetic substances is a routine laboratory procedure.

Now, there are people, including scholars, who think that the use of embryonic stem cells in research is committing murder; who insist on including the "intelligent design" hypothesis in school curricula as an alternative to evolution; who question the very notion of objectivity of science because they disapprove of its historical and cultural development. All this makes the science awareness of the public more necessary, and the task of the science educators more urgent.

Due to the small scale of the universe, the quarks are so close to each other at the beginning of the second epoch that they do not feel any strong force.[3] The quarks are the "free" roamers of the universe. But as the universe expands, the interquark distance increases and the quarks begin to feel the strong force of gluon exchanges and bag themselves and the gluons inside hadrons. **Example D.38.7** on **page 72** of *Appendix.pdf* gives a very crude estimate of the **quark confinement temperature**.

Quark confinement temperature.

> **What do you know? 38.5.** Given that $\alpha_q = 1.75$, (a) what was the quark density ρ_q at the end of the second epoch? (b) What was the quark *energy* density in eV/m^3 then? (c) If the average energy of each quark was 10 MeV, what was the quark number density then? (d) What was the volume (of the cube) occupied by each quark? (e) On the average, how far apart were the quarks from each other? (f) How does this compare with the size of a nucleon, 10^{-15} m?

---

[2]However, there are people who think that their souls can ride to heaven with a hitherto-unknown comet if they commit suicide when it journeys through the sky.

[3]Recall from Section 34.3.2 that the strong force between two quarks *increases* with increasing distance.

38.2.3   The Third Epoch 10^{10} K $<$ T $< 10^{12}$ K

The universe is about a tenth of a millisecond old now and consists almost entirely of electrons, muons, neutrinos and their antiparticles, and, of course, photons. Hadrons, which are heavier and which were formed out of quarks in the second epoch, have started annihilating their antiparticles, and will complete this annihilation soon after the start of the third epoch. A small number (small compared to the number of the other constituents) of protons and neutrons—the minute "extra" matter particles produced in the first epoch—are immersed in the sea of other particles.

In the middle of this epoch, when the temperature is 10^{11} K, the muons disappear (μ^+ and μ^- annihilate each other), and the only survivors will be electrons, positrons, neutrinos and antineutrinos, the photons, and a small "contamination" of protons and neutrons. The small "extra" number of muons (created in the first epoch due to matter-antimatter asymmetry) decay very rapidly due to their very short half-life.

The density of the universe after μ^+-μ^- annihilation is approximately 4.5×10^{12} kg/m^3, as shown in **Example D.38.2** on **page 69** of *Appendix.pdf*. This is so large (a grain of sand made of this material would weigh several tons!) that even neutrinos, which can easily pass through layers of lead that are light years thick, are trapped.

Reactions such as $p + e^- \leftrightarrow n + \nu$ and $n + e^+ \leftrightarrow p + \bar{\nu}$ keep the number of protons and neutrons equal. Here is why. Suppose there are twice as many protons as neutrons. Then—as there is no shortage of the electrons, positrons, and neutrinos—it is twice as likely for the first reaction to go from left to right (than right to left) and the second reaction to go from right to left (than left to right), causing more protons to turn into neutrons than vice versa. Ultimately, the numbers become equal. Therefore, the third epoch starts with equal population of protons and neutrons.

Equality of proton and neutron populations at the beginning of the third epoch.

However, as the end of the third epoch approaches, the *p-n* conversion becomes asymmetric: Because *n* is heavier than *p*, more neutrons convert to protons than vice versa. **Example D.38.3** on **page 70** of *Appendix.pdf* gives another reason for the smaller abundance of neutrons. There is also the decay of neutron into proton [what is responsible for beta decay in nuclei as discussed in Sections 31.2.1 and 34.4 and Figure 34.8(d)]. However, this decay is too slow to have any effect on the *p-n* imbalance, at least in the present epoch. By the end of the third epoch, the protons outnumber the neutrons by a factor of three, i.e., out of all the nucleons, 25% are neutrons and 75% protons.

38.2.4   The Fourth Epoch 10^{9} K $<$ T $< 10^{10}$ K

About a second has passed since the big bang. The universe is still composed of electrons, positrons, three kinds of neutrino, photons, and a small "contamination" of protons and neutrons. The density has fallen down to the point that the neutrinos are no longer trapped. They *decouple* from the rest of the universe.

What do you know? 38.6. For the constituents of the universe at the beginning of the fourth epoch, $\alpha = 5.375$. Using Equation (38.1), can you estimate the age of the universe at the beginning of the fourth epoch?

A little after the neutrino decoupling, at a temperature of about 2.2×10^9 K, the electrons and positrons annihilate each other, and the energy released as a result of the annihilation heats up the photons, making the photon gas slightly hotter than the decoupled neutrino gas. **Math Note E.38.2** on **page 148** of *Appendix.pdf* shows that the ratio of the photon temperature T_γ to the neutrino temperature T_ν is $\sqrt[3]{\frac{11}{4}}$. The end of this epoch coincides with the end of the antimatter era. From now on, the matter of the universe is the "contamination" of protons and neutrons formed in the second and third epochs out of quarks, and the "contamination" of electrons (which were produced in the first epoch along with the extra u and d quarks that later formed the protons and neutrons) that survived the e^+-e^- annihilation in the present epoch.

What do you know? 38.7. How long after the big bang did antimatter completely disappear from the universe? Hint: Right after electron-positron annihilation, $\alpha = 1.681$.

Since the electric charge is strictly conserved, and no physical process can create a positive charge without an equal negative charge, the number of electrons equals the number of protons, and this equality will hold for eternity. Any process that changes the number of protons will also change the number of electrons by the same amount. One such process is the neutron decay, which becomes more important as the age of the universe becomes comparable to the half life of a neutron, 614 seconds.

Number of protons is equal to the number of electrons ... always!

38.2.5   The Fifth Epoch 10^8 K $<$ T $< 10^9$ K

The dominant constituents of the universe are now photons and the neutrinos. The universe is about three minutes old (see **Example D.38.4** on **page 70** of *Appendix.pdf*), and the time span is long enough that neutron decays play a significant role in the abundance of p and n. In fact, **Example D.38.5** on **page 71** of *Appendix.pdf* shows that during this time, about 20% of the neutrons turn into protons via beta decay. The nuclear particles now consist of approximately 87% protons and 13% neutrons.

Shortly after the start of this epoch, 3 minutes and 16 seconds after the big bang, something remarkable happens. Every couple of neutrons pair up with a couple of protons and form a helium nucleus. This means that 26% of the nucleons end up inside the helium nuclei, with free protons constituting the remaining 74%. Since nucleons have almost equal masses, the mass of all nuclear matter in the universe is now 26% helium nucleus and 74% proton.

The formation of helium nuclei, although quick, is not an instantaneous process. There are intermediate steps that need to be taken before helium can be created. First a proton and a neutron must fuse together to form a deuteron: $p + n \rightarrow D + \gamma$. Then two deuterons fuse into a helium nucleus $D + D \rightarrow ^4$He$+\gamma$. Helium is a very tightly bound nucleus; once formed it is very hard to break it apart. Even though the protons and neutrons had a lot of opportunity to fuse at earlier times, they did not form helium. Why? The answer is the **deuteron bottleneck.**

Helium nuclei are formed 3 minutes and 16 seconds after the big bang.

In order to make helium, first we must make deuterons. And while ^{4}He is very tightly bound, deuteron is not. It takes only 2.224 MeV to break up the deuteron into a proton and a neutron.[4] At earlier times, although deuterons *were* formed, they were immediately broken apart by the energetic photons, i.e., deuteron production went both ways: $p + n \leftrightarrow D + \gamma$. What exactly is the temperature at which deuterons are made and not broken apart? What is the threshold temperature for D-formation? **Math Note E.38.3** on **page 149** of *Appendix.pdf* shows that it is 950 million Kelvin, which is just under the initial temperature of this epoch.

What do you know? 38.8. If you assume that the average energy of the photons in the early universe is 2.224 MeV, the binding energy of a deuterium, what kind of temperature would you get? Is that anywhere near the temperature you expect?

---

[4]This is about a third the energy needed to pull a single nucleon out of ^{4}He.

Because ^{4}He is so tightly bound, it is energetically a more favorable final product than other multinucleon elements, which are also formed, though in much smaller quantities. For example, some deuterons may survive; or a proton can fuse with a deuteron to form ^{3}He; or a D can fuse with a ^{4}He to form a lithium. Although these other elements do form at the temperature of 950 million Kelvin, most of them disintegrate immediately, because of their small binding energies, leaving only a minute trace behind.

The formation of ^{4}He via the deuteron bottleneck—as predicted by the laws governing nuclear interactions—points to two important cosmological predictions. First, that a universal radiation must exist to prevent the formation of deuterons at earlier times, and second, that the ratio of the hydrogen mass abundance to helium mass abundance in the universe should be 74% to 26%. If radiation did not exist, neutrons and protons would form He as soon as the temperature fell to the point that the average KE of the particle content of the universe (nucleons and electrons) was not sufficient to—through violent collisions—break apart the deuteron that was being formed. This would change the H-He abundance ratio. **Example D.38.6** on **page 71** of *Appendix.pdf* estimates this ratio to 60%-40%, in complete violation of observation!

38.2.6   The Sixth Epoch $15,000 \text{ K} < \text{T} < 10^8 \text{ K}$

Approximately 4 hours and 56 minutes have passed since the big bang. Ignoring the neutrinos, which have their separate existence, the ^{4}He nuclei, protons, and electrons are immersed in a sea of photons. For every nucleon, there are 1.6 billion photons, and this ratio persists for the rest of the life of the universe. The temperature of radiation (and matter, which is in thermal equilibrium with it) is $\sqrt[3]{11/4} = 1.401$ that of neutrino, and this ratio will also persist forever. An analysis similar to Example D.38.4 shows that $\rho_\gamma = 0.836 \text{ kg/m}^3$, $\rho_\nu = 0.57 \text{ kg/m}^3$, and $\rho_m = 0.000028 \text{ kg/m}^3$. Thus, the universe is still dominated by radiation and neutrinos, however matter has gained some grounds in the competition for dominance: while radiation was 300,000 times more dense than matter at the beginning of the fifth epoch, it is only 30,000 more dense now.

Not much will happen during the sixth epoch. The universe keeps expanding, the wavelength of the photons and neutrinos keep stretching, and because of the interaction between the photons and the charged particles of the universe (the protons and the electrons), electromagnetic radiation is trapped until the end of this epoch.

38.2.7   The Seventh Epoch $3000 \text{ K} < \text{T} < 15,000 \text{ K}$

Twenty-five millennia have passed since the big bang. The radiation and neutrinos have cooled down to the point that matter is beginning to be the dominant contributor to the density.[5] And for another 300-400 millennia nothing of interest occurs. The photons' and neutrinos' wavelengths keep increasing, and the photons are trapped in the plasma of the positive ions (the helium nuclei and protons), and the negative electrons.

However, at the end of the seventh epoch, about 400 millennia after the big bang, the photons reach a temperature that is no longer sufficient to ionize the helium and hydrogen atoms that kept being formed all along. The bound electrons and ions can no longer interact with the photons, and the latter are decoupled from matter, roaming the universe unnoticed until 1964, when it is picked up by the horn-shaped antennas of two radio astronomers on Earth. The helium and hydrogen gas mixture is now ready to form stars, galaxies, and other structures.

---

[5]The baryonic matter (protons and neutrons) is not as dense as radiation at the temperature of 15,000 K. Only if you include the dark matter—which must creep up in the universe at some stage—does matter become dominant.

38.3   End-of-Chapter Material

38.3.1   Answers to "What do you know?"

38.1. Example 38.0.1 shows that the threshold temperature of a particle is proportional to its mass. Since the mass of $W^{\pm}$ is 156,750 times the mass of the electron, the threshold temperature of $W^{\pm}$ is 156,750 times the threshold temperature of the electron, which is 2.2 billion Kelvin. This gives $156,750 \times 2.2$ billion K or 3.45×10^{14} K as the threshold temperature of $W^{\pm}$. Similarly, the threshold temperature of Z^0 is 3.93×10^{14} K.

38.2. For electron, the constant multiplying ρ_γ is $\alpha_e = (7/8) \times 2 = 1.75$. Therefore, $\rho_e = 1.46 \times 10^{-32} T^4$ kg/m^3.

38.3. With $T = 2.725$ K and $\alpha_e = 1.75$, Equation (38.1) yields $t = 2.4 \times 10^{19}$ s. Since there are 3.15×10^7 seconds in a year, t would be 758 billion years, instead of 13.7 billion years! The answer is wrong because Equation (38.1) assumes a radiation dominated universe, a dominance which stopped after a mere 15000 years after the big bang.

38.4. In many respects matter and antimatter are very similar. For example, the most important property that distinguishes one particle from the rest, i.e., its mass is the same for a particle and its antiparticle. Therefore, there is no reason to assume that at creation, preference of any sort was given to either matter or antimatter.

38.5. (a) $\rho_q = 1.75\rho_\gamma$, and with ρ_γ given by Equation (37.2) and $T = 10^{12}$, we get $\rho_q = 1.46 \times 10^{16}$ kg/m^3. (b) To get the energy density, multiply result of (a) by c^2. This yields 1.32×10^{33} J/m^3. Converting this to eV gives 8.23×10^{51} eV/m^3. (c) The number density n_q is just the energy density divided by the average density of each quark, which is 10^7 eV. So, $n_q = 8.23 \times 10^{44}$ quarks/m^3. (d) If each cubic meter is shared by 8.23×10^{44} quarks, then the volume occupied by each quark is $1/8.23 \times 10^{44}$ or 1.22×10^{-45} m^3. (e) For this volume, the side of the cube should be $\sqrt[3]{1.22 \times 10^{-45}}$ or 1.07×10^{-15} m. (f) Extremely well!

38.6. With $T = 10^{10}$ K and $\alpha = 5.375$, Equation (38.1) gives $t = 0.99$ s.

38.7. With $T = 2.2 \times^9$ K and $\alpha = 1.681$, Equation (38.1) gives $t = 36.65$ s.

38.8. The average energy of photons is related to the temperature by the second formula in Equation (37.3). Convert the binding energy of the deuteron to Joules to get 3.56×10^{-13} J. Then the second formula in Equation (37.3) yields a temperature of 9.55 billion K, which is over a hundred times larger than the expected temperature. The reason is that when the *average* energy is 2.224 MeV, there are a lot of photons that have energies much higher than this. We don't need this many photons to ignite the reaction. Furthermore, the photons outnumber the nucleons by a ratio of 1.6 billion to 1, which introduces another overestimating factor. Math Note E.38.3 provides the technical reasoning for getting 950 million K as the desired temperature.

38.3.2   Chapter Glossary

CP Violation A phenomenon in which matter and antimatter behave differently while expected to behave in the same way.

Deuteron Bottleneck The process of the formation of deuteron, a necessary prerequisite for helium nucleus production.

Quark Confinement Temperature The temperature below which the inter-quark distances are so large that the strong force causes the quarks to form hadrons.

Threshold Temperature The temperature associated with a particular particle above which the EM radiation is so energetic that it can create that particle and its corresponding antiparticle.

38.3.3   Review Questions

38.1. What is the threshold temperature of electrons? Is there anywhere in the universe now which is this hot? What is the threshold temperature of protons?

38.2. What is the content of the universe at the beginning of he first epoch? What is the most significant event of the first epoch? What kind of asymmetry causes this event to happen?

38.3. When did the second epoch start? How hot was the universe at the beginning of this epoch? What is the content of the universe at the beginning of this epoch? What is the status of quarks at the beginning of the second epoch? Are they free? Do they remain free until the end of the epoch?

38.4. When did the third epoch start? How hot was the universe at the beginning of this epoch? What is the content of the universe? What happens to the hadrons in this epoch? Are there any hadrons left by the end of this epoch? Which hadrons are they? How many of them? Do muons survive this epoch? How dense is the universe? How much would a grain of sand of this period weigh? What happens to the protons and neutrons? What is the proton-neutron ratio at the end of the third epoch?

38.5. When did the fourth epoch start? How hot was the universe at the beginning of this epoch? What is the content of the universe? What happens to the neutrinos? What happens to the electrons and positrons? Do neutrinos have the same temperature as photons? If not, which one is hotter? By how much? What caused this difference in temperature? Are there any antimatter left at the end of the fourth epoch? How does the number of protons compare with the number of electrons?

38.6. When did the fifth epoch start? How hot was the universe at the beginning of this epoch? What is the content of the universe? How does the number of protons compare with the number of neutrons? What happens shortly after the beginning of this epoch? What percentage of the universe becomes He nucleus and what percentage protons? What is the deuteron bottleneck? What are the two important cosmological predictions of the deuteron bottleneck and the formation of helium? Have these predictions been verified observationally?

38.7. When did the sixth epoch start? How hot was the universe at the beginning of this epoch? What is the content of the universe? What is the photon-nucleon ratio? How much is the matter-radiation mixture hotter than neutrinos? Is the universe matter-dominated or radiation-dominated?

38.8. When did the seventh epoch start? How hot was the universe at the beginning of this epoch? What is the content of the universe? What important thing happens at the end of this epoch? How old is the universe then?

38.3.4   Conceptual Exercises

38.1. Can GUT alone explain matter-antimatter asymmetry? If not, what else is needed?

38.2. In which epoch do tauons, with a mass of 1777 MeV, disappear?

38.3. In which epoch do the charged pions, with a mass of 140 MeV, disappear?

38.4. The density of the universe in the middle of the third epoch is 4.5×10^{12} kg/m^3. How many kilograms does a cc (cubic centimeter) of this universe weigh? Put one cc of this material on one side of the scale. How many trucks (30 tons each) do you have to put on the other side of the scale to balance the scale?

38.5. At the beginning of the third epoch, the number of protons is equal to the number of neutrons. What fraction of the neutrons turn into protons by the end of the third epoch?

38.6. If neutrinos could be used to get information from the cosmos, what would the earliest time be that we could "see" the universe?

38.7. By what percentage is the temperature of photons higher than the temperature of the neutrinos in the universe?

38.8. The hydrogen mass in the universe is three times larger than the helium mass. How many times is the hydrogen *number* larger than the helium number in the universe?

38.9. What is the significance of 15,000 K as the temperature dividing the sixth and seventh epochs? Hint: See the discussion after Equation (37.4).

38.3.5 Numerical Exercises

38.1. From the content of the universe at the beginning of the second epoch, find α. Then use it and Equation (38.1) to calculate the age of the universe at the beginning of the second epoch.

38.2. The temperature of the universe is now 2.725 °K, and its scale size is 500 Mly.
(a) How hot was the universe when its length scale R was 100,000 ly, the size of the Milky Way?
(b) What was $\lambda_{\max}$ then?

38.3. Consider the universe one minute after the big bang. Assume a radiation dominated universe.
(a) What was the temperature then?
(b) What was the density then?
(c) Assuming that the current temperature of the universe is 2.725 °K and its scale size is 500 Mly, what was the scale size of the universe one minute after the big bang?
(d) How do you describe the rate of expansion of the universe during the first minute of its creation?

38.4. The mass of the proton is 1.67×10^{-27} kg.
(a) What is the minimum energy E_{avg} required of photons for protons and antiprotons to exist in equilibrium with photons in the universe?
(b) What is the corresponding temperature?
(c) What is the corresponding mass-equivalent density of radiation?
(d) Assuming a radiation dominated universe, how long after the big bang did protons and antiprotons start to annihilate themselves?

Chapter 39

The Last Epoch

A few hundred thousand years after the big bang, when the universe is about a thousandth of its present size, and the radiation—which kept the electrons away from the helium nuclei and protons—has receded into the emptiness of space, atoms are formed. These atoms will be the ingredients of all structures that take shape later. Matter, which has long fought for a universal dominance, is now completely in control. Radiation and neutrinos have very little say in the fate of the universe from now on. And the electromagnetic interaction, which trapped the radiation in a soup of negative electrons and positive ions, is itself trapped inside the atoms; and the stronger nuclear force is even more tightly imprisoned inside the nuclei of the helium atoms. All forces have given way to gravity, which now commands the universe.

All interactions are trapped except gravity, which rules the cosmos.

39.1 Cosmic Structure Formation

In the global smoothness of the hydrogen-helium gas, here and there, a spot became slightly denser than its surrounding. The slightly larger gravitational force of this spot attracted the nearby atoms, becoming even denser and heavier. The atoms in the proximity felt even a stronger gravitational attraction, and rushing to the site, made the spot even larger. This snowball effect continued at a faster and faster rate until a huge blob of H and He atoms was formed. If the universal gas consisted of around 74% hydrogen and 26% helium, then this blob, as well as the structures that followed from it, such as stars, galaxies, and globular clusters, should also have the same abundance ratio. And they indeed do!

The H-He abundance persists in the interior of stars and galaxies.

39.1.1 Jeans Mass

The accretion of atoms towards the middle of the blob creates an outward pressure that resists the rush to the center. If the clumping is to take place, there must exist sufficient amount of helium and hydrogen, whose gravitational attraction can overcome the pressure resistance. Therefore, there must be a minimum mass at which the clumping begins. This mass is denoted by m_J and is called the **Jeans mass**—after Sir James Jeans who calculated it in 1902 for the conditions under which stars are formed—and is given by $M_J = (p/G)^{3/2}/\rho^2$, where G is the universal gravitational constant, p is the pressure, and ρ the density of the gas.

For gravitational clumping to occur, the mass of the blob must be *larger* than the Jeans mass; and if the Jeans mass is too large, the clumping will be very unlikely, because a gigantically massive blob is not a common occurrence in the universe. **Example D.39.1**

Radiation decoupling was important for the formation of structure in the universe.

on **page 72** of *Appendix.pdf* shows how important the decoupling of radiation was for the formation of structures.

> **What do you know? 39.1.** Except for a numerical factor, pressure is just the energy density. (Can you show this?) Compare the pressure before radiation decoupling, when the radiation pressure dominated, to the pressure after decoupling, when the matter pressure dominated (in the form of an ideal gas). For matter pressure, multiply its number density by the average KE (17.1). To find the number density, divide Equation (37.4) by the mass of a nucleon, 1.67×10^{-27} kg.

39.1.2   Wrinkles in Spacetime

We learned in Example D.39.1 that matter clumps only after radiation decoupling, about 350,000 years after the big bang. But in order for the clumping to begin, some parts of the universe must first become denser than their surroundings, providing the "seeds" of the cosmic structures we see today such as galaxies. Putting aside the question of how this could have happened (after all the universe was supposedly homogeneous and isotropic, and any variation in density would violate that), we can ask if there is any *direct* evidence for this variation.

If the seeds were present 350,000 years after the big bang, then they may have been present even earlier, before the radiation decoupling. Any variation in the density accompanies a variation in temperature: the denser spots are hotter than their surroundings in proportion to their density. The radiation, being in contact with matter, would also be hotter at some spots and colder at others. And when radiation decoupled from matter, it should have carried this variation with it. However, COBE's measurement of the background radiation (CMB) showed an almost perfect agreement with a black body radiation curve of temperature 2.725 K, with little room for variation. That was in January 1990, when the full capacity of COBE was yet to be used.

Looking at a full Moon you see a perfect circle with a perfect contour. Look through a pair of binoculars, and you will detect some small raggedness on the contour of the Moon. Use a powerful telescope, and you will discover mountains and valleys spreading all over the Moon. In January 1990 COBE saw the CMB through its "binoculars." It took another two years to get its powerful "telescope" ready. What mankind saw on April 23, 1992 was nothing less than the universal genesis. On this day, in a meeting of the American Physical Society, the lead investigator of the COBE team presented to the audience the image shown in Figure 39.1.

The "blobs" in Figure 39.1 are variations in the temperature of the CMB as small as 0.00001 K. These variations, too small to be seen in the black body radiation curve of CMB, are remnants of the variations when CMB was in contact with matter before it decoupled from it. Figure 39.1 is a solid proof that matter indeed had areas of concentration (of possibly dark matter), which later turned into structures such as globular clusters and galaxies that we see today.

39.2   Problems with Standard Cosmology

The standard big bang cosmology has been very successful in describing the content and development of the universe. And many observations vindicate its validity: the darkness of the night sky; the hydrogen and helium abundance, with H outweighing He three to one, and with practically no other primordial element in the universe; the expansion of the universe; and the background radiation. However, all is not well with the standard model of cosmology. There are three main problems with the model having to do with the **flatness**

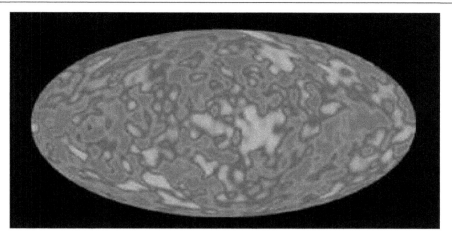

Figure 39.1: The blobs indicate variations in the temperature of the CMB as small as 0.00001 K.

of the universe, with the causality and the influence of the **horizons** of far observers, and with the distribution of **structures** in the universe. The following subsections address these problems one by one.

39.2.1   The Flatness Problem

The flatness problem is best discussed in terms of a parameter denoted by Ω_{tot}. It is the ratio of the actual density to the critical density: $\Omega_{\text{tot}} = \rho/\rho_c$. The contribution of different components to the actual density carry a subscript on their corresponding Ω. For instance, $\Omega_\gamma = \rho_\gamma/\rho_c$ is the contribution of radiation to Ω_{tot}; $\Omega_m = \rho_m/\rho_c$ is matter's contribution to Ω_{tot}; $\Omega_\nu = \rho_\nu/\rho_c$ is neutrino's contribution, etc. Therefore, Ω_{tot} can be expressed as $\Omega_{\text{tot}} = \Omega_m + \Omega_\gamma + \Omega_\nu + \dots$ A flat universe has a total density equal to the critical density, i.e., an Ω_{tot} equal to 1.

To understand the problem associated with the flatness, let's go back to the very beginning of our discussion of cosmology, to the Friedmann equation (37.1). **Math Note E.39.1** on **page 149** of *Appendix.pdf* shows that the Friedmann equation can be written as

$$\Omega_{\text{tot}}(t) - 1 = kc^2/(R^2 H^2), \tag{39.1}$$

where R and H are, of course, functions of time. For a perfectly flat universe, $k = 0$ and $\Omega_{\text{tot}}(t) = 1$, meaning that the universe must have *always* (i.e., for all t) been flat.

What do you know? 39.2. If the universe was perfectly flat at some time, could it become closed or open at some other time?

Now suppose that the universe is *not* flat, say $\Omega_{\text{tot}} = 1.5$. Then, as Math Note E.39.1 demonstrates, at the time of the radiation decoupling Ω_{tot} must have been 1.000908, at the onset of matter dominance it must have been 1.00018, and at the time of the helium formation it must have been 1.0000000000000353. Thus a big deviation from flatness now, diminishes to a very small deviation at earlier times. Turning the argument around, if shortly after the big bang, the universe was even minutely nonflat, it would have evolved into a universe completely different from the very nearly flat universe we observe. In fact, the latest (July 2008) measurement of Ω_{tot} puts it between 0.999 and 1.023!

A minute non-flatness early on grows into a huge non-flatness in the present universe.

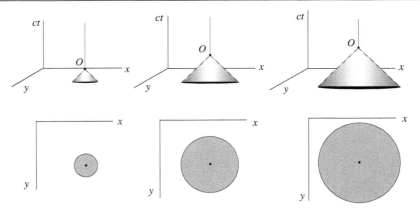

Figure 39.2: The top graphs show how the light cone gets bigger as time passes. The bottom graphs show how the horizon in the xy-plane also increases in size. The dot at the center of the bottom circles is the location of the observer O. The vertical line passing through O is his world line.

The flatness problem can be posed as the question: Why out of an infinite number of possibilities available to it at the time of creation, did the universe choose to be flat, i.e., with a constant Ω_{tot} of exactly 1? Of course, one can retort the question with another question "Why not?" But such a very special initial condition begs for an answer; and there is one, as we shall see later.

39.2.2   The Horizon Problem

Uniformity of CMB indicates a violation of causality in standard cosmology.

The horizon problem is even more serious than the flatness problem. It has to do with the very notion of causality and the amazing uniformity of the CMB radiation. COBE measurements have revealed that the temperature of the background radiation is remarkably uniform across the sky:[1] regardless of the angle of observation, CMB has a temperature of 2.725 K. On the other hand, uniform temperature of a system is achieved only when different parts of the system come in contact with one another or somehow communicate information with each other. The following argument shows that points far away from each other in the sky could never have communicated with one another!

Any observer O has two light cones, one that connects him to the future and one that connects him to the past (see Figure 27.4). Concentrate on the past light cone. All events that are inside or on the light cone are causally connected to O; i.e., O can receive, observe, interact, etc. only with those events. Now suppose that O is located in the xy-plane, not moving. All the points (events) in the xy-plane of the spacetime coordinate system have $t = 0$. Figure 39.2 shows the spacetime geometry appropriate for this discussion. In contrast to the discussion of Chapter 27, we have added an extra space dimension to make the graphs more realistic.

Now let's ask the question: "Which events occurring at some time t are causally connected to O?" The answer: All the events that lie *both* inside O's past light cone *and* in the xy-plane. These are points that lie inside and on a circle in the xy-plane with O at its center, the region that we have called the *horizon*. The size of the horizon depends on t. This is because, although O is stationary in *space* (the xy-plane), it *moves* in spacetime (on its vertical world line). So at later times, the vertex of the past light cone is farther away from the xy-plane, cutting a larger circle out of it. Therefore, O can communicate only with points in the xy-plane that lie within the horizon. At earlier times, only a small portion of space (the xy-plane) have had sufficient time for their "signals" to reach O; at

---

[1]Except for the minute variations shown in Figure 39.1

later times more and more points become "observable" to O. It should be clear that the radius of the horizon at time t is ct.

All discussions and graphs above pertain to spacetime geometry without gravity and universal expansion. When we include gravity, the only thing that changes is that the horizon radius becomes not just ct, but a multiple of that ($3ct$ when matter is dominant, and $2ct$ when radiation). Therefore, in an expanding universe, a point can be in causal contact only with points in the universe that lie inside its horizon, which grows in proportion to time. Now go back to Figure 39.2 and let $t = 0$ be the time of the big bang. Any time after the big bang, the horizon covers only a finite portion of the universe. Thus, O can communicate only with a finite portion of the universe. So, how did one point of the CMB "know" what the temperature of the other points was?

What do you know? 39.3. What was the horizon radius right after the dominance of radiation ended when the temperature of the universe was 14,800 K? (Hint: Use Equation (38.1), with $\alpha = 1.681$ to account for the presence of neutrinos. Then use $r_h = 2ct$ to find the horizon radius.) What was the distance between P_1 and P_2, two diametrically opposed points of the horizon? Could two points, which were farther apart than this, communicate with each other?

The immediate answer may be "Radiation was in contact with matter; so different parts just picked up the corresponding temperature of matter, which was after all uniformly hot." This is not a good answer, because then we have to answer the question of how one part of *matter* knew the temperature of the other parts that were causally disconnected from it, a precondition for the uniformity of the temperature. (If the horizon is finite for radiation, it is also finite for matter.) Another answer may be that the entire universe was inside the horizon, and therefore, all points of the universe were causally connected. This is not a good answer either, because the horizon is only a fraction of the universe as Example D.37.3 showed. In fact, **Example D.39.2** on **page 73** of *Appendix.pdf* illustrates that points of the present sky that are separated by very small angles could not have known of each other ever since the domination of matter. Once the radiation decouples, different regions of it will develop independently, as radiation does not interact with itself. Yet COBE's measurements clearly indicate that over the entire sky, the CMB radiation is remarkably uniform; and the only way that could happen is for different parts of the radiation to have communicated with one another. This is the horizon dilemma that standard cosmology cannot resolve.

What do you know? 39.4. In the previous **What do you know?**, you calculated the distance between P_1 and P_2, two diametrically opposed points of the horizon. As the universe expands, this distance increases. What is this distance now? The light coming from those two points have traveled almost 13.7 billion light years, i.e., the distance between us and those two points is 13.7 billion light years. What is the current angular separation between P_1 and P_2? Hint: Divide $\overline{P_1 P_2}$ by the distance from us to P_1 and P_2 to get the angle in radian. Each radian is 57.3 egrees.

39.2.3   The Structure Problem

The problem associated with structure is twofold. The first has to do with the sheer existence of structure. If the universe is homogeneous and isotropic, why are there regions in which structures exist? These regions must have been the site of "seeds" (concentration of dark matter, perhaps) that have existed since very early times. As we saw in Example

D.39.1, the amount of material needed to clump matter together just before decoupling was enormous. So much, in fact, so that different parts of it could not have been causally connected and could not have coherently collapsed to form structure.

The second problem has to do with the homogeneity and isotropy of the distribution of structures in the universe. Once you go beyond the local structures, the universe looks very similar at different angles and at different distances. Concentration of varieties of distant galaxies look very much the same. You don't find a crowd of one galaxy shape in one region of the sky and another in a different region. How did this homogeneity of galaxy distribution take place, when we know that regions of the sky only a few degrees apart could not have communicated with one another?

39.3   Inflationary Cosmology

The solution to the shortcomings of the standard cosmology came from a completely unrelated field. Grand unified theories (GUT), of which we spoke briefly in Section 34.5, although appealing, had some serious flaws. One of these flaws was the unobserved prediction of the decay of proton, mentioned at the end of Section 34.5. The other was the existence of **magnetic monopoles**, which was also not observed. Prior to the downfall of the GUT in the mid 1980s, Alan Guth, an elementary particle theorist, was trying to explain the scarcity of the magnetic monopoles in the universe. The standard cosmological model predicted that there should be as many monopoles as there are nucleons! Since all attempts at finding a monopole had failed, Guth suggested that the universe must have had an **inflation** phase during which it expanded *exponentially*. This exponential growth was so rapid that it diluted any existing monopoles so much so that today there may be only a few in our observable universe.

Connection between magnetic monopoles and cosmology.

BIOGRAPHY

Alan Guth was born in New Brunswick, New Jersey, in 1947. He attended the public schools in Highland Park, NJ, but skipped his senior year of high school to begin studies at MIT, where he remained for all his degrees, including Ph.D. Guth then held several postdoctoral positions including one at Cornell, where he was approached by a fellow postdoctoral physicist, Henry Tye, who persuaded Guth to join him in studying the production of magnetic monopoles in the early universe. This work changed the direction of Guth's career. The following year at SLAC, he continued to work with Tye on magnetic monopoles. They found that standard assumptions in particle physics and cosmology would lead to a fantastic overproduction of magnetic monopoles, contrary to observations, which have failed to detect even a single magnetic monopole. Guth and Tye began a search for alternatives to avoid monopole overproduction, and from this work Guth invented the inflationary universe.

The idea of inflation, designed to solve the magnetic monopole problem, gave an unexpected explanation for the three problems of standard cosmology in one gigantic swoop. Once the details of the *microscopic* physics of the grand unified theories are worked out, the *macroscopic* effect is the introduction of the notorious *cosmological constant* Λ in Einstein's GTR, and—from there—in the Friedmann equation, which is modified to

$$\left(\frac{v}{R}\right)^2 = \frac{8\pi G}{3}\rho - \frac{kc^2}{R^2} + \frac{\Lambda}{3}. \tag{39.2}$$

The expansion of the universe reduces both the first (because ρ decreases with expansion) and the second (because R increases with expansion) terms. Therefore, after a while the third term becomes dominant, and noting that the left-hand side is just the Hubble parameter, one gets $H = \sqrt{\Lambda/3}$. **Math Note E.39.2** on **page 151** of *Appendix.pdf* shows that in such a case, R increases exponentially.

This exponential increase solves the flatness problem immediately. In fact, Math Note E.39.2 also derives Equation (E.162), which describes an exponential *decay* of Ω_{tot}. Even if

Ω_{tot} is not equal to 1, the exponential factor makes the right-hand side equal to zero after a while, forcing Ω_{tot} to become 1 for quite a long time. **Example D.39.3** on **page 74** of *Appendix.pdf* illustrates that if the universe inflated between 10^{-34} s and 10^{-32} s after the big bang (i.e., if inflation operated for a millionth of a millionth of a millionth of a millionth of a millionth of a hundredth of a second), then, regardless of how nonflat it might have been earlier, it becomes so nearly flat that it would take 450 million million million million million times the present age of the universe for it to have an Ω_{tot} equal to 1.0001! In other words, the universe will never become unflat if inflation operates for only a (incredibly) small fraction of a second after the big bang.

> Universe will never become unflat if inflation operates between 10^{-34} s and 10^{-32} s after the big bang.

The horizon problem is also solved beautifully with inflation. Math Note E.39.2 derives the horizon radius as a function of time during the inflation, and **Example D.39.4** on **page 74** of *Appendix.pdf* uses some reasonable numbers to estimate the rate of expansion of a horizon. It is shown in that example that, during the extremely brief period of inflation (a millionth of a millionth of a millionth of a millionth of a millionth of a hundredth of a second), the horizon grows from a subnuclear distance to over 85 light years! This means that a tiny region of the universe, small enough to achieve thermal equilibrium before inflation, can expand to be much larger than the size of the presently observable universe.

Finally inflation gives a solution to the structure problem as well. At the beginning of the inflation, the radius of the horizon in our part of the universe was subnuclear, at which distance the laws of quantum mechanics were relevant. In particular, statistical fluctuations, inherent in the probabilistic nature of the quantum theory, could have caused a slight variation in the density—the seeds of the future structures—of that region of the universe. Because of the symmetry of the universe, the distribution of these seeds must have been more or less uniform. Upon expansion of that subnuclear region, the seeds and the uniformity of their distribution were carried over to the inflated universe.

> Quantum fluctuations give rise to the seeds of cosmic structures and explain the uniformity of structure distribution in the universe.

What do you know? 39.5. Would it be possible to explain the structure formation in the universe if we had not discovered the quantum theory?

39.4   Birth of a Star

The study of the formation of structure is challenging and not fully understood. Large structures such as galaxies are the subject of intense observational and theoretical investigations. One of the things (among many others) that make galaxies so hard to study is the existence of the mysterious dark matter within them. We still don't know what exactly dark matter is. There have been numerous candidates for it (including massive neutrinos), but a satisfactory explanation is yet to come. For this and other reasons, we avoid any discussion of galactic formation and evolution. Instead, we focus on the formation and evolution of much smaller structures, stars, which are far better understood.

The birth of a star starts with a large, cold, and diffuse cloud of a hydrogen-helium gas mixture. A tiny region around a point P of this cloud is slightly denser than its surrounding. The mutual gravitational attraction between this region and its nearby regions pulls them together; and since P is more massive, it moves less than the surrounding, attracting atoms, becoming heavier, and therefore resisting motion even further. P becomes the center of accretion. As the atoms get attracted towards P, they accelerate, their (average) kinetic energy increases, i.e., the region becomes hotter (see Box 17.2.5).

> Minimum mass required for a star to be born. Gestation period of a star is a million years!

Further increase in the mass strengthens the gravitational field, attracting more and more atoms toward the center, heating up the core. Once sufficient quantity of hydrogen and helium atoms is accumulated at the core (about a million years after the start of the gravitational contraction), its temperature reaches about ten million Kelvin, and, as Example 31.2.6 showed, nuclear fusion, mostly in the form of **proton-proton cycle**, sets in. A star is born. What is the smallest mass at which the nuclear fusion starts? **Math Note E.39.3** on **page 152** of *Appendix.pdf* estimates this minimum mass to be about half a solar mass $M_\odot$, concentrated in a sphere with a radius of $0.13R_\odot$.

A star lives, so long as the fusion process, which transforms hydrogen into helium, continues. The life of the star is a balance between the gravitational force trying to collapse the star and the outward pressure of the gas resisting this collapse. This pressure is caused by the increase in the gas temperature, which in turn is a result of the interaction of radiation (a by-product of fusion) with the atomic nuclei in the star. What is the largest mass that the radiation pressure can support? **Math Note E.39.4** on **page 153** of *Appendix.pdf* estimates this maximum mass to be about $7.5M_\odot$. The life-span of a star is determined by its mass. A star as massive as the Sun is expected to live for over 10 billion years. More massive stars (10 to 100 solar masses) cannot provide sufficient radiation resistance to the strong gravitational force, and collapse much sooner, perhaps in a mere 10 million years.

It takes light about a million years to go from the core to the surface of a star!

The energy of the fusion is carried mostly (about 95%) by photons and the neutrinos (about 5%). The very energetic photons (in the form of gamma rays) travel from the core to the surface, undergoing multiple scattering from the dense material along the way, and losing considerable amount of energy to them, thereby heating them up and increasing their pressure. By the time a photon reaches the surface, it has lost so much energy that it is now visible. The photons emerging from the surface of a star are not the same photons that started at the core. In the process of scattering, a photon is absorbed by a charged particle, and another photon with a different wave length is emitted, which in turn gets absorbed by yet another charge particle, and a third photon is emitted, etc. Between the time that the gamma ray starts at the core and the visible photon emerges at the surface, almost a million years passes!

What do you know? 39.6. Estimate the speed of "light" as it goes from the core of the Sun to its surface. The radius of the Sun is 700,000 km. Does this mean that a photon travels this slow in the Sun?

The solar neutrino oscillation (or mixing) is proof that neutrinos are massive.

The neutrinos, being extremely weakly interacting, penetrates easily through the layers of matter and escape into outer space unscathed. The **solar neutrinos** have been detected on Earth with a puzzling twist. The neutrinos produced in the Sun are all electron neutrinos, and based on the kind of reaction that takes place in the interior of the Sun, a certain amount of electron neutrino flux is expected on Earth. However, experiments since the early 1990s have detected a weaker electron neutrino flux than expected. An explanation that has been gaining grounds is **neutrino oscillation**, which states that it is possible for one neutrino flavor to transform (oscillate) to another. This can happen only if neutrinos, which have been assumed massless, are actually massive. Now there is strong evidence that neutrinos indeed have mass, and that the flavor of the electron neutrinos coming from the Sun transform into other flavors such as muon and tauon on their way to Earth.

39.5   Death of a Star

Hydrogen is the first kind of fuel that a star uses. Once it is used up[2] and the photons that heated the gas surrounding the core and kept the star from collapsing are no longer produced, gravity gets an upper hand again and the star begins a new phase of collapse. Depending on the mass of the star, this phase goes through various stages in which lighter elements are used up in the process of fusion, and heavier elements are formed until the most stable nuclei like iron are reached, at which point the fusion stops. Why does it stop?

---

[2]The hydrogen is not completely used up in this stage. When only about 10% of the hydrogen is burned, this stage comes to an end, and a new phase starts.

Because iron is at the peak of the stability curve of the nuclei (see Figure 31.1 and the beginning of Section 31.2.3). No more fusion means no more radiation pressure. The star has now reached the end of its life. Gravity, encountering no resistance, pushes the entire stellar matter towards the core, and depending on its mass, the corpse of a star can take three different shapes: white dwarf, neutron star, and black hole.

39.5.1   White Dwarfs

For an ordinary star like the Sun, the gravitational collapse continues until a temperature of a hundred million Kelvin is reached. At this point, helium[3] starts to diffuse to form carbon. The photons that are released in the fusion of helium are much more energetic than those produced in the hydrogen fusion. Therefore, after the hydrogen supply ends and helium burning begins, the energetic photons create a much larger pressure, pushing the outer layers of the star outward until the star becomes a hundred or even a thousand times larger. At this stage the star is called a **red giant**; red, because the spread of the huge energy across its enormous area reduces the amount of power output of each square meter of its surface, resulting in a lower temperature (longer wavelength) than a normal star. When the Sun turns into a red giant, it will gobble up Mercury and Venus, possibly Earth and even Mars.

> Fusion of He produces more energetic photons; the higher pressure pushes star outward . . . to a **red giant**.

> **What do you know? 39.7.** Suppose that when the Sun turns into a red giant, its power output will be 250 times its current power output of 4×10^{26} Watts, and its radius 100 times its current radius of 7×10^8 m. What is the brightness J_e (i.e., power per square meter) of the red giant? Using Equation (20.1), estimate the surface temperature of the red giant. Using Equation (20.2), estimate the wavelength at which the BBR curve of the red giant peaks. Recall that the area of a sphere of radius r is $4\pi r^2$.

The final product of the collapse of a star similar to the Sun is a **white dwarf**, with a density of about a million tons per cubic meter, a radius of a few thousand kilometers (about the Earth radius), and a surface temperature of about 10,000 K. **Math Note E.39.5** on **page 154** of *Appendix.pdf* outlines a very crude approximation to the physics of a white dwarf. The white dwarf eventually releases its energy to the empty space and becomes a **black dwarf**. This is the fate of a star like the Sun, and that is what will happen to it 6 to 8 billion years from now.

> **What do you know? 39.8.** What is the gravitational acceleration at the surface of a white dwarf? How much does the light bend when grazing a white dwarf? Use Equations (9.2) and (29.1) with $M = 10^{30}$ kg and $R = r = 10^7$ m. The angle comes out in radian.

Why doesn't gravity push the white dwarf even further? What kind of resistance does gravity encounter now? The answer comes in an amazing interplay of microscopic physics and macroscopic world. The electrons have now been pushed so much that the Pauli exclusion principle (see Box 23.3.4) becomes relevant:[4] electrons cannot get any closer together, because two or more of them would have to occupy the same quantum state, violating Pauli's principle. A *quantum* principle, applicable to a strange quantum property called spin, prevents a *gigantically massive* object from collapsing under gravity. That's astonishing!

> The amazing union of microworld and macroworld!

---

[3]When we mention an element, we mean its nucleus, of course. No electron can be permanently captured to form atoms at the extreme temperature of the core of a star.

[4]Although we used Pauli's exclusion principle in Math Note E.39.3 for an ordinary star, it is in the description of a white dwarf that the principle reveals its dramatic effect.

39.5.2   Neutron Stars

If a star has a mass larger than the **Chandrasekhar mass**, it turns into a neutron star.

If a quantum *principle* prevented a white dwarf from collapsing, a quantum *process* opens a way for further collapse. However, for the process to take effect, the star ought to be more massive. If the mass of a star is larger than the **Chandrasekhar mass**, which is about 1.4 solar mass, the gravitational force becomes so strong that, at the end of the fusion cycle, the electrons and protons come in contact, and through the electroweak process of $e^- + p \to n + \nu_e$, turn into neutrons and neutrinos. The electrons and protons almost completely annihilate each other; neutrinos escape into outer space, and only neutrons remain. The electronic pressure due to Pauli's exclusion principle—also called *electron degeneracy pressure*—having terminated, the star collapses again until the neutron degeneracy prevents it from further collapse. The star has turned into a **neutron star**, a gigantic (atomic) nucleus composed of only neutrons, having a radius of about 10 km and a density of over 10^{17} kg/m^3. The end of Math Note E.39.5 derives some properties of neutron stars.

A **pulsar** is a spinning neutron star in whose magnetic field are trapped electric charges, which spin with it.

When a rotating star larger than the Sun collapses to a chunk of nuclear matter only 10 km in radius, it will spin unimaginably fast; and if the electric charges secreted during the collapse are trapped in the gigantic magnetic field of the neutron star, they produce an electromagnetic pulse that rotates with the star, sweeping the sky much like a searchlight. When one of these pulses was first discovered in 1967, it was thought to be coming from an intelligent civilization. After all, how can anything else produce such regular pulses with a period of exactly 1.34 s? Since 1967 hundreds of these **pulsars** have been observed, and their origin is clearly understood to be the collapse of a stellar mass into a neutron star.

Supernova, a cataclysmic explosion before the formation of a neutron star.

The violent collapse to a neutron star accompanies a **supernova**, a cataclysmic explosion that sometimes outshines the galaxy in which the collapsing star was once located. Chinese astronomers have recorded a supernova explosion in 1054 that was visible in the daytime for many days. The expanding shell of that explosion is what today's astronomers call the *Crab Nebula*. At the center of the Crab Nebula is a pulsar with a period of 0.033 s.

One pulsar, discovered in 1974, appeared to have a strange behavior. Its period of 0.059 s—corresponding to a frequency of 17 Hz—seemed to decrease slightly for a while and then increase by the same amount a little later; and this change occurred regularly. The discoverers of the pulsar quickly realized that the shift in the frequency was due to the Doppler effect, one caused by the motion of the pulsar towards and the other due to its motion away from the Earth. For this to happen, the pulsar must be orbiting around an unseen companion.

Change in Doppler shift points to the existence of a "partner" of a pulsar.

From the change in the Doppler shift, the period of the orbit was estimated as just 8 hours. This puts the pulsar very close to its companion (which is believed to be another neutron star). If a planet were orbiting the Sun with such a period, it would have to be at a distance of only 1.4 million kilometer from the Sun's center. Mercury, the closest planet, is 58 million kilometers away from the Sun, over 40 times farther than the pulsar is from its companion. And Mercury is already feeling the effects of the general relativistic curvature. With the companion having a mass larger than the Sun, and the pulsar orbiting it at such a close distance, the general relativistic effects are even more noticeable. In fact, the curvature is so large that the rotation of the pulsar produces gravitational *waves*. We have already seen how these waves affect the period of the pulsar in Section 29.3.2.

39.5.3   Black Holes

No physical process can stop a supermassive collapsing star.

Stars more massive than a few solar masses do not possess sufficient neutron degeneracy pressure to prevent further collapse. No physical process can stop a collapsing star that is heavier than a few suns; it collapses until it becomes a single point, a singularity of spacetime, a *black hole*. The earlier discussion of black holes concentrated on the inability

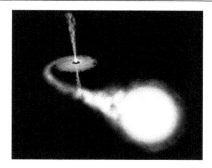

Figure 39.3: The artist's rendition of an accretion disk around a black hole.

of light to escape a spherical object whose radius was smaller than

$$r_s = \frac{2GM}{c^2} \tag{39.3}$$

called the **Schwarzschild radius**. From a general relativistic point of view, r_s is the distance—from the center of a spherical mass distribution—at which the geometry of space-time breaks down. Any spherical object whose actual radius is smaller than its Schwarzschild radius is a black hole.

Black holes abound in the universe. As they cannot emit any form of EM radiation (the primary source of cosmic information), black holes cannot be seen directly. To "see" a black hole, one has to rely on observing the theoretical predictions applied to objects that are under its influence. Black holes come in three general varieties depending on their origin and their mass scale: black holes in X-ray binaries, black holes in galaxy centers, and exploding primordial black holes.

We can't see a black hole, but physics can!

Black Holes in X-Ray Binaries

Most of the stars are members of binary pairs in orbital motion about one another. One member of some of these binaries may be a compact star—many times more massive than the Sun—that has undergone a supernova explosion and turned into a black hole. If the orbit of the normal partner is small enough, it may shed material and form an **accretion disk** orbiting around the black hole (see Figure 39.3). The material in the disk may lose energy due to various dissipative mechanisms and spiral closer to the center of the black hole. The energy released can heat up the inner regions of the disk to such an extent that X-rays are produced. [Wien's displacement law (20.2) tells us that for λ_{max} to have a typical X-ray value of 10^{-10} m, the temperature must be about 30 million Kelvin.]

X-ray binaries, detectable by X-ray telescopes in orbits around Earth, are some of the brightest X-ray sources in the sky. These X-ray sources can be identified by the Doppler shift of the optical spectrum of the normal partner. Thus, if *X-ray* telescopes detect a source at a certain location of the sky, and *optical* telescopes detect a normal star in the same location whose spectrum is Doppler-shifted, then there must exist an X-ray binary in that location, with the invisible partner being a black hole.

Black Holes in Galaxy Centers

Centers of galaxies, around which billions of stars orbit, are good places to look for black holes. There is convincing evidence for the existence of black holes at the centers of a number of galaxies, including the Milky Way. In fact, there may well be a black hole at the center of *every* galaxy. These **supermassive black holes** weigh anywhere from a million to a billion solar masses.

Supermassive black holes hold stars of galaxies together.

As in the case of the X-ray binaries, supermassive black holes are not directly observable. Aside from their entrapment of light, these black holes comprise a tiny spot in the middle of a galaxy. The Schwarzschild radius of even the most massive black holes is only a billion kilometers, the size of the solar system, while the central bulge of a galaxy has a radius of several thousand light years, or 10 million times larger than the black hole: spotting something that is 10 million times smaller than the galaxy is not easy.

Identification of supermassive black holes at galactic centers results from the observation of the motion of the nearby matter, just as in the case of dark matter discussed in Section 9.1.3. From the rotational speed of the nearby matter and $v = \sqrt{GM/r}$ [Equation (9.3)], the mass M can be estimated; and if the estimate indicates a huge mass in a small region at the center of a galaxy, that is evidence for a black hole.

Active galactic nucleus (AGN) is a supermassive rotating black hole.

Some otherwise ordinary galaxies emit intense radiation from their centers which is more luminous than all the rest of the stars of the galaxy put together. The radiation is not starlight, which is almost like a black body radiator. In fact, the centers of such galaxies emit X-rays, optical radiation, and radio waves with almost equal intensity. These are galaxies with an **active galactic nucleus** (AGN). Almost all AGNs can be detected at distances where the host galaxy is too dim to be observed. The brightest AGNs are 10,000 times brighter than a typical galaxy at the same distance. The size of a typical AGN is several hundred thousand times smaller than a galaxy. All these properties point to the possibility that AGNs are supermassive **rotating black holes**, and this rotation can be one of the sources of the intense energy of the released radiation.

BIOGRAPHY

Stephen Hawking was born in 1942 in Oxford, where his mother had moved from London because of the war. After the war the whole family moved to St Albans, where Stephen attended school. Hawking wanted to specialize in mathematics in his last couple of years at school, but his father was strongly against the idea and Hawking was persuaded to make chemistry as his main school subject. In March 1959 Hawking was awarded a scholarship to Oxford University, where he specialized in physics in his natural sciences degree. During his last year at Oxford, Hawking noticed that he was becoming rather clumsy, and when he moved to Cambridge to take up research in general relativity and cosmology, his health deteriorated so much that he had to spend two weeks in a hospital, where he was eventually diagnosed with the motor neurone, or Lou Gehrig's disease in early 1963. The doctors predicted that he would not live long enough to complete his doctorate. However, not only did he finish his doctorate in 1966, but he became a Research Fellow, and then a Professorial Fellow at Gonville and Caius College, Cambridge. In 1973 he joined the Department of Applied Mathematics and Theoretical Physics at Cambridge, becoming Professor of Gravitational Physics in 1977. Two years later Hawking was appointed Lucasian Professor of Mathematics at Cambridge, a post once held by Newton. Between 1965 and 1970 Hawking worked on singularities in the theory of general relativity in collaboration with Roger Penrose. From 1970 Hawking began to apply his previous ideas to the study of black holes. Combining quantum theory and general relativity, he discovered the remarkable property that black holes can emit what is now called *Hawking radiation*. Today, over 40 years after the doctors gave him only a few years to live, Hawking, although confined to a wheelchair and talking through an electronic device, is as alive and active as ever.

Primordial Black Holes

The fluctuations in the density of the early universe depicted in Figure 39.1 could have attracted enough material to form a collapsing star with a black hole as the end product. These black holes would have a mass of about 10^{11} kg, and therefore, a Schwarzschild radius of only 10^{-16} m, smaller than a proton! How does one look for these black holes? Quantum-sized black holes require quantum physics.

Section 32.2.2 introduced the idea of *vacuum polarization*, in which a virtual electron-positron pair is created and immediately annihilated. Although such pairs are not directly observable, and have no observable effect in an infinite space with no external fields, they have observable effects when external fields—such as the electric field of an atomic nucleus—are present. *Lamb shift*, already discussed in Section 32.2.2, is precisely the vacuum polarization effect applied to the Coulomb field of the hydrogen nucleus. Another effect, relevant to our present discussion is the following.

Near the Schwarzschild radius of a black hole, the gravitational field is so strong that it can lend energy to the creation of an electron-positron pair. One of the particles could get sucked into the black hole and the other escape away from it. The latter is then part of the **Hawking radiation**. The escaped particle could be an electron or a positron. Furthermore, since photons are their own antiparticles, a pair of photons could also be produced. Hawking radiation therefore consists of matter, antimatter, and photons. The energy taken away by Hawking radiation reduces the gravitational energy of the black hole, diminishing its size until it *evaporates*.

Hawking radiation.

How long after its formation does a primordial black hole evaporate? The formula for the lifetime (in years) of a primordial black hole is $t_{\text{evap}} = 10^{10} \left(M/10^{12} \right)^3$, where M is the mass of the black hole in kilogram. For an ordinary black hole with a mass of a few, say 10, solar masses, this formula gives $t_{\text{evap}} = 8 \times 10^{67}$ years. Even if such a black hole were formed immediately after the big bang, we would *now* have to wait $8 \times 10^{67} - 1.37 \times 10^{10}$ or 8×10^{67} years for it to evaporate! However, if a black hole has a mass of 10^{12} kg, its lifetime is 10^{10} years, and if it was formed right after the big bang, it would be evaporating now. Due to severe theoretical constraints on the number of such black holes, it is doubtful that they would ever be detected.

39.6 End-of-Chapter Material

39.6.1 Answers to "What do you know?"

39.1. The *energy* density of radiation is $\rho_\gamma c^2$. At the decoupling temperature of 3000 K and ρ_γ given by Equation (37.2), this gives a pressure P_γ of 0.06 Pa. For matter, as an ideal gas, the energy is just KE, and the energy density is number density times average KE. The average KE is just $k_B T$ [see Equation (17.1) and ignore the numerical factor], and the number density is $0.074 T^3$ nucleons/m^3. Multiplying these together and putting the numbers in gives a P_m of 8.3×10^{-11} Pa, which is a billion times smaller than P_γ.

39.2. If the universe is perfectly flat at any time during its history, then $k = 0$, and Equation (39.2) yields $\Omega_{\text{tot}}(t) - 1 = 0$, making $\Omega_{\text{tot}}(t) = 1$ for *all* time t. Therefore, the universe cannot become open or closed.

39.3. Equation (38.1), with $\alpha = 1.681$ gives the time as 25,700 years. The horizon radius is $2c(25700 \text{ years})$. The c and the "years" give light years. So, the radius is 51,400 light years. The distance between P_1 and P_2 is just the diameter of the horizon or about 100,000 light years. Any two points whose distance was larger than 100,000 light years lay outside the horizon, and therefore, could not communicate with one another.

39.4. The separation has increased in inverse proportion to temperature, i.e., by a factor of $14800/2.725$ or 5431. So, $\overline{P_1 P_2}$ has grown to 543,100,000 light years. Dividing this by 13.7 billion light years gives an angle of 0.04 radian or 2.27 degrees.

39.5. No! Only the probabilistic nature of the quantum theory allows for uniformly random distribution of the seeds of structure right after the creation of the universe.

39.6. Dividing the distance of 7×10^8 m by the time of 1 million years or 3.15×10^{13} s, we obtain 0.00002 m/s, or 0.002 cm/s, or 0.02 mm/s! This *does not* mean that photons travel this slow in the Sun. Each photon gets absorbed by matter and *after a while* a different photon is emitted. Between emission and absorption, the photon travels with the speed of light.

39.7. The total power output of the red giant is 10^{29} Watts. Its area is $4\pi(7 \times 10^{10})^2$ or 6.16×10^{22} m^2. Therefore the brightness (power divided by area) is 1.6×10^6 W/m^2. Use this in Equation (20.1) and get $T^4 = 2.86 \times 10^{13}$ and $T = 2313$ K. Now substitute this in Equation (20.2) and obtain $\lambda_{\max} = 1.25 \times 10^{-6}$ m or 1.25 μm, which is in the infrared region of the EM spectrum.

39.8. Put the numbers in the formula for gravitational acceleration and get 667,000 m/s². Put the numbers in the formula for bending of light and get 0.000148 radian. Each radian is 57.3 degrees. So, the angle is 0.0085 degrees, which is about 40 times larger than the light-bending of the Sun. Note that Equation (29.1) gives half the actual angle because it does not incorporate the curvature of spacetime.

39.6.2   Chapter Glossary

Active Galactic Nucleus A source of extremely powerful radiation in some galaxies. These sources are believed to be supermassive rotating black holes at the center of some remote galaxies.

Chandrasekhar Mass The critical mass above which a star undergoes a supernova explosion and turns into a neutron star or a black hole.

Flatness Problem A problem with the standard cosmology whereby a slight nonflatness at the beginning of the universe turns into a huge unobserved flatness now.

Hawking Radiation Radiation caused by the vacuum polarization in the presence of the intense gravitational field at the Schwarzschild radius of a black hole.

Horizon Problem A problem with the standard cosmology whereby even points separated by a very small angle could not have communicated causally, and therefore, could not have attained the uniform temperature observed in the present universe.

Inflationary Cosmology The assumption, introduced to reduce the number of monopoles in the early universe, also solved the problems associated with the standard cosmology. According to this assumption, the universe expanded at an exponentially rapid pace for a minute fraction of a second after the big bang, and then expanded at the pace set by the standard cosmology.

Jeans Mass The minimum mass required for a gas of particles to clump together gravitationally.

Magnetic Monopoles Magnets with only south or north poles. These were predicted by the GUTs to be as abundant as the nucleons in the universe.

Neutrino Oscillation The transformation of one neutrino flavor into another. This oscillation depends crucially on the massiveness of neutrinos.

Neutron Star The celestial body left over from a star a few times more massive than the Sun after it reaches the end of its life.

Primordial Black Hole A hypothetical black hole with a Schwarzschild radius of only 10^{-16} m, formed right after big bang.

Proton-Proton Cycle A set of nuclear reactions (fusion processes) which turn four protons into a helium nucleus.

Red Giant The stage just before a star like the Sun turns into a white dwarf. In this stage, the star grows to a hundred or a thousand times its original size.

Schwarzschild Radius The radius characterizing the ability of a black hole in capturing light. If light gets closer to the black hole than this radius, it cannot escape the gravity of the black hole.

Solar Neutrinos Neutrinos received on Earth and produced in the core of the Sun via fusion processes.

Structure Problem A problem with the standard cosmology whereby even points separated by a very small angle could not have communicated causally, and therefore, could not have attained the uniform structure distribution observed in the present universe. Furthermore, the assumption of isotropy and homogeneity does not allow the formation of structure like stars and galaxies, which breaks isotropy and homogeneity.

Supernova The explosion of a star a few times more massive than the Sun just before it reaches the end of its life.

White Dwarf The celestial body left over from a star such as the Sun after it reaches the end of its life.

39.6.3   Review Questions

39.1. What do you expect the hydrogen-helium abundance ration to be in the stars and galaxies? Why do you expect that?

39.2. If a star is to be formed, there should be a minimum amount of matter present. Why? What prevents the formation of the star if there is not enough matter? What is the name of this minimum amount?

39.3. Enumerate the problems with standard cosmology and explain why they are problems.

39.4. What is a magnetic monopole and what theories predict it? How was the riddle of the magnetic monopole solved and what implication the solution had for cosmology? What is the name of this theory and who proposed it? Explain how inflation solves the three problems of the standard cosmology.

39.5. Consider a region of the universe which, at the beginning of inflation, is about the size of the nucleus of an atom. How big does this region become at the end of the inflation?

39.6. How is a star formed? Why does gravity not cause a complete collapse of a star? What mechanism prevents it from collapsing? What is the source of this mechanism? Is the source present forever? If not, what happens to the source, and what is the outcome?

39.7. What happens to a star like our own Sun at the end? What is a red giant and when, during the life cycle of a star does it take place? What is a white dwarf? How dense is it; i.e., what is its mass and approximately how big is it?

39.8. What happens to a star two or three times more massive than the Sun at the end? What is a supernova? What is a neutron star? How dense is it; i.e., what is its mass and approximately how big is it?

39.9. What happens to a star whose mass is more than a few times the solar mass? How is the size of the final remnant characterized? What happens to light if it gets closer to the remnant of the star than this characteristic size?

39.10. How many kinds of black hole are there? What is an active galactic nucleus? What sort of radiation does it emit? How many times brighter is it than a typical galaxy at the same distance?

39.11. What is a primordial black hole? How big is it? How heavy is it? What is Hawking radiation and how is it created?

39.6.4   Conceptual Exercises

39.1. Why does a neutron star rotate so fast while the star that collapses into it rotates so slowly?

39.2. Why are atomic and nuclear physics so important in the physics of the stars?

39.3. What is so "super" about a supernova?

39.4. Explain how Doppler effect can be used to determine the period of a star revolving about another star.

Chapter **40**

Epilogue: The Closing Dialogue

It was his last lecture of the semester. The professor had been teaching PHY 101 for over four months, and in doing so, he not only covered some hard-core physics, but also implications that the laws and methodology of physics may have on students' worldview. He had emphasized the continuity of science, the pattern of universal discoveries that always stem from the study of specific problems, the interplay between observation and theory, and the connection between terrestrial physics and the celestial objects.

Now the professor wanted to wrap up the course in the most succinct way possible. He wanted his students to leave the class taking the most important message of the course; not necessarily in the form of the technical knowledge that they had gained, but in a more subtle way. Since his students were not science majors, and his course was possibly the first and last physics course they would have taken, he wanted to convey to them the essence of science, and he wanted to do so by a Socratic participation of the class. So, he started by bluntly posing a question.

PROFESSOR: What is the single most important lesson you learned in this course?

CHAD: That physics is the hardest subject in the universe?

PROFESSOR: (*Clearing his throat after the collective chuckle of the class subdued*) Chad is right in the sense that playing piano is hard, baseball is hard, hockey is hard, and writing is hard. In fact, what skill is not hard, if you want to really master it? The reason that some skills seem to be easy is because you have started them at an early age. If you had waited until college to play baseball while in the first grade you had started learning physics and continued it year after year, physics would have been much easier for you in college than baseball. Anyway, let's not concentrate on what you learned *about* the course, but *from* it.

DAN: I liked how physics is relevant to our everyday lives.

PROFESSOR: That is definitely one aspect of physics, although not the one that I wanted to emphasize in this particular course. What, beside the application of physics to industry and technology, have you learned from the course?

KIMBERLEY: I liked how physical laws applied to the whole universe.

PROFESSOR: That's good. Let's concentrate on the laws. Kimberley, can you name a few laws that apply to the whole universe?

KIMBERLEY: The law of gravity. The law that showed us how to calculate Sun's temperature; I forgot the name of that law.

PROFESSOR: You mean Wien's displacement law.

KIMBERLEY: Yes, that law. Also the relativity theory and the big bang.

PROFESSOR: Any other lessons you learned?

ANDY: How observation and theory go hand in hand?

PROFESSOR: That's an important lesson. Can you give us some examples?

ANDY: I thought how Galileo came up with the first law of motion was really cool.

PROFESSOR: Do you remember how he came up with the first law?

ANDY: Yes. He used inclined planes and made them smoother and smoother, and he used smaller and smaller angles.

PROFESSOR: That's great. To what kind of motion does the first law apply? Anybody.

JULIE: To the motion of any object that is left alone.

PROFESSOR: Does it apply to objects in outer space?

JULIE: As long as the objects are not disturbed.

PROFESSOR: Objects in any part of the universe?

JULIE: I think so.

PROFESSOR: Would you say that, as long as objects are not disturbed, the first law is universally true?

JULIE: Yes.

PROFESSOR: Yes indeed! The law is valid in the entire universe, yet it was discovered on an inclined plane here on Earth. Kimberley mentioned Wien's displacement law, which was also discovered on Earth, yet it could measure the surface temperature of Sun. While we are talking about it, can anybody think of some other earthly physics applied to Sun?

MAT: The nuclear fusion inside Sun that makes it shine.

PROFESSOR: Excellent! Note that Wien's displacement law, which was related to the black body radiation, which was the start of *atomic physics*, gave us the *surface* temperature of Sun; and *nuclear physics* gave us the physics of Sun's *interior*. What did the *sub*nuclear physics, the physics of quarks and leptons give us?

ANDY: Last week you were talking about the universe right after the big bang, and you started with quarks, leptons, and some other particles. I have forgotten their names.

PROFESSOR: You mean the gauge bosons? The particles responsible for fundamental forces?

ANDY: Yes.

PROFESSOR: Good! Is there a lesson to be learned here?

CHUCK: This is going to sound weird, but it looks like if we want to figure out the universe, that is so big, we need to know how the tiniest fundamental particles and forces work.

PROFESSOR: Chuck, it is not weird at all. In fact, that is the essence of physics. Philosophers call it *reductionism*, and some of them have made a hated word out of it. Just like *liberalism*, which is a hated word among conservatives, and *religious right*, which is a hated word among liberals. Some call it "fundamentalism;" and they are spared the attempt at making it a hated word. But regardless of what philosophers call it or think of it, physics has always been after fundamental processes.

JIM: But if you concentrate on the tree, wouldn't you miss the forest? I had a class in the philosophy of science, and my teacher taught us that most systems that we are in contact with are complex, and they have to be studied as a whole. For example, even a cell, although it may be made up of billions of molecules, should be studied as a whole, because only when its separate parts work together do we have a *live* cell. How can dead atoms lead to a living cell?

PROFESSOR: I'm glad you brought this up, Jim. The philosophy of science you just mentioned is called **holism**. It is popular among many philosophers, all alternative medical doctors, a few biologists, even fewer chemists and physicists. According to this philosophy, science should not break down the object of its study into smaller and smaller parts, because in so doing it loses the quality that made the object a whole. A cell, which Jim brought up is a good example. But to understand the holism of a cell, it is instructive to start with a simpler example. Take water, the most abundant material on Earth. How would a holistic philosopher advise a chemist to study water? Anybody.

JULIE: By feeling it, smelling it, maybe weighing it and heating it?

PROFESSOR: That's about it. Holistic philosophers don't believe in the idea of breaking water down to its smallest part, a molecule consisting of two hydrogen and one oxygen atoms. But faced with the undeniable fact that water consists of countless molecules, what would be the attitude of a holistic philosopher toward water now?

KIMBERLEY: Just study the molecule without worrying about what it is made of?

PROFESSOR: That's what it looks like, doesn't it?

JIM: But that is not a fair evaluation of what holistic philosophers say. They don't apply their philosophy to water molecules. They are concerned mainly with biological systems. After all it *is* true that science has not been able to explain life from a combination of atoms and molecules.

KIMBERLEY: Just because scientists can't explain life *now*, it doesn't mean that they will never be able to do it. Who knows? Maybe 100 years from now, they will be able to make living things out of nonliving molecules.

JIM: How do we know that they *will* be able to do that?

KIMBERLEY: A hundred years ago there were probably people who said that scientists would never be able to synthesize organic material, but they are doing it now. So, why shouldn't we believe that 100 years, or 200 or 2000 years from now science will not only explain, but even create life in the lab?

PROFESSOR: Let's get back to the water molecule and the cell, and let's not worry about the cell's property we call "life." I can also point to some properties of water, say its liquidity, and claim that because hydrogen and oxygen are both *gases*, their assemblage cannot explain the fact that water is *liquid*. A holistic philosopher would then argue that water molecule cannot be studied by breaking it up into atoms. "Only by studying a water molecule as a whole," he might say, "can we understand its liquidity, just as life can only be studied as a whole." As a physicist should I or shouldn't I try to understand atoms?

KIMBERLEY: Of course you should!.

PROFESSOR: And if physicists discover that atoms consist of nuclei and electrons, should they or shouldn't they try to understand nuclei and electrons?

JIM: But understanding nuclei and electrons will not help us understand life, will it?

PROFESSOR: Perhaps not directly. But history has shown that only by understanding the parts, can we understand the whole. Only because we had had a good understanding of molecules, could we have a better grasp of the unit of life, DNA. Besides, physicists have no choice. Their pursuit of truth has led them to more and more fundamental entities. Any holistic approach is against the entire tradition of not only physics, but all sciences. Remember that chemistry started the atomic theory, and benefited tremendously from it *especially* when physicists discovered the structure of the atom and a theoretical understanding of it.

CHAD: I don't know much about philosophy, but I think it makes no sense to not try to understand the parts whenever they are discovered. What is wrong with trying to see what is inside the nucleus of an atom? What is wrong with looking inside protons and neutrons to see if there are more fundamental particles there? It seems to me that the more we understand about these tiny particles, the more we understand the bigger picture, the universe itself. I thought it was so cool to learn last week that the interaction among electrons, positrons, neutrinos, protons, and neutrons explained the helium/hydrogen abundance in the universe. I can't understand what the holistic philosophers have against this. Unless they are against science itself, they should see the power of basic science.

JIM: I don't know much about philosophy myself, and all I know about holism is what I learned in the philosophy of science course I took last semester. But I think there is a deeper issue here. Science cannot explain everything; and when it can't, maybe other methods should be used to explain things. For example, going back to a cell's life, if fundamentalism or reductionism does not work, maybe holism should be given a chance.

PROFESSOR: If we allow holism—which as we just saw, is opposite to science—to replace science in explaining the life of a cell, then we should allow any other nonscientific ideas to do so as well. I hope that you have learned in this course that if an explanation is not scientific (i.e., supported by evidence and careful observation), then it is only an opinion, a story, a fable, a conjecture, a hypothesis, or a myth that may or may not be true. As for life, it is true that science cannot *now* make life by assembling nonliving molecules; but, as Kimberley said, it is possible that if we wait long enough, it will eventually do it, unless we fall into a long future Dark Age as our ancestors did for almost 20 centuries.

The hour came to an end sooner than the professor expected. The students were already reaching for their backpacks. So, he wished them a good semester break, and started packing his notes.

ooooooooooooooooooooooooooooooooooooooooooooooooooo

The last four parts of the book summarize the last 100 years of physics. Built on the classical notions of the previous three centuries, modern physics has been unimaginably successful in explaining hundreds of puzzles, and opening new vistas of investigation. The objects of the investigation of physics have been shrinking in size: from atoms, to atomic nucleus, to protons and neutrons, all the way to quarks and leptons. At the same time the domain of its application has been expanding in size: from earthly material, to stars, to galaxies, all the way to the universe itself.

> **Box 40.0.1.** *Without a comprehensive knowledge of the fundamental particles and forces, no understanding of stars, galaxies, and the universe is possible.*

If we don't follow the path of successive reduction, on which nature itself has placed us, we have to artificially stop at a certain stage and unscientifically attribute qualities to that stage that science, if given a chance, has the capability to explain. The following is a prime example of this artificial brake applied to the natural motion toward increasing reduction, and the bizarre conclusion drawn from it.

Remember Deepak Chopra from Chapter 35? He is a holistic medical doctor who was charged by the Indian guru Maharishi Mahesh Yogi to find a scientific explanation of the ancient Indian medicine Ayurveda, which uses a holistic approach of treating the mind and body together. We saw in Chapter 35 how Chopra assigned nonmateriality to waves and materiality to particles, and used the quantum mechanical wave-particle duality to transform nonmaterial thought into material bodily action, thus "explaining" the mind-body connection "scientifically."

Mind is the residence of intelligence, and any "scientific" treatment of mind/body holism is incomplete unless it includes an explanation of the source and the mechanism of the emergence of intelligence. Modern biology has been able to come very close to unraveling the molecular basis of memory. A chemical called neuropeptide is produced and transmitted from neuron to neuron in any recorded brain activity. The details of the production and transfer of neuropeptides and the exact nature of the connection between them and thought are as yet unknown, but under investigation. A further reduction, an examination of the peptide at a more fundamental level, may be necessary to unlock its function.

However, like a genuine holist, Chopra chides modern biologists for trying to reach finer and finer levels of physiology to explain intelligence. If reductionism is the wrong way to go, what is the alternative? How does a holist "explain" intelligence? Chopra claims that everything that makes up our body, large or small, has intelligence as a built-in feature. Here are his own words:

> You may find it easy to think of DNA, with its billions of genetic bits, as an intelligent molecule; certainly it must be smarter than a simple molecule like sugar. How smart can sugar be? But DNA is really just strings of sugar, amines, and other simple components. If these are not "smart" to begin with, then DNA couldn't become smart just by putting more of them together. Following this line of reasoning, why isn't the carbon or hydrogen atom in the sugar also smart? Perhaps it is. [Chop 90, pp. 65–66]

Sugar molecules are intelligent because they make up the intelligent DNA!

Atoms are intelligent because they make up the intelligent sugar molecule!

Based on this line of reasoning, we are intelligent because all the atoms and molecules that make up our body are intelligent! But if we accept the intelligence of all atoms and molecules, how do we account for the fact that a piece of wood, also made up of atoms and molecules, is not intelligent?

Part IX

Nature of Science

Chapter **41**

Misconceptions about Science

We have made quite a journey through the world of physics, and gained a substantial amount of knowledge along the way. At the end of each part you saw a summary of what the part taught us. This part of the book assembles these teachings and talks about the *nature of science* from a physicist's perspective. There has been a lot of discussion about this topic recently, and a number of "theories" have been proposed as to "what science is" and how it works. Our experience with the physical ideas and methodology exposed throughout the book should have given us enough background to draw our own conclusion about "what science is and how it works," the subject of this part of the book. But before doing that, let's dispel some misconceptions about science.

There are a lot of activities that are not science, and their very nature reveals their nonscientific character. No one confuses art, music, and poetry with science. But there are activities which are wrongly taken to be science. Three of these activities, which are most commonly confused with science are mathematics, technology, and medicine, discussed in the next few sections of this chapter.

41.1 Mathematics

Nature writes her poems in the language of mathematics. This statement crystallizes the relation between mathematics and nature (science, especially physics). Our long voyage through the vastness of physical ideas has revealed that the most fundamental laws of nature incorporate the most abstract mathematics. The deeper we cut through matter, the more abstract and elegant the mathematics needed to explain it.

The analogy of language and poetry for mathematics and physics goes beyond a mere statement of similarity. There is a strong tie between physics and mathematics as we have repeatedly observed in the book. While nature speaks in the language of mathematics, it constantly *creates* the language. And in this sense, the analogy becomes even stronger: poetry uses language, but some of the most effective linguistic phrases are *created* by poetry.

Mathematics and language, while similar, are also different. Mathematics, once abstracted from its natural origin, can grow on its own (through the activity of the human mind). Language does not have this property; if it grows, it is because of the advancement of human society. It is a remarkable phenomenon that the abstract mathematics developed by the mind of humankind finds its way back in nature.[1] Ignoring this difference, mathematics

Language, poetry, mathematics, and physics.

---
[1] We saw striking examples of this in the application of Lie groups to the connection between spacetime and matter (Section 33.1), to the classification of hadrons (Section 33.2.1), to the unification of electromagnetism and the weak nuclear force (Section 34.2), and to the theory of strong nuclear force QCD (Section 34.3).

is to science as language is to poetry.

Application of statistics in a discipline or a trade does not make it "scientific."

And just as the use of a language does not imply poetry, so does the use of mathematics not imply science. There are many academic (as well as nonacademic) disciplines and trades that use mathematics, *mostly in the form of statistics*. Are we to call those disciplines "science?" Language is used in almost all human activities. Are we to call these activities "poetry?" What is the difference between the usage of the language in poetry and in other human activities? Poets have a better command of the language and are a lot more creative in its usage, while the usage in many of the other activities is very limited. Similarly, physicists use the same mathematics that is used by pollsters, insurance companies, and marketing firms. The difference is that the latter use mathematics in a very limited way (mostly in the form of statistical data analysis), while physicists *are forced by nature itself* to use it much more extensively, creatively, and abstractly.

Physics *also* uses statistics and probability as we saw in Sections 17.1 and 22.3. The difference is that physics *predicts* the probability it uses; so it *knows* the probability of the sample with which it is dealing. The other disciplines have no trace of the probability of their sample. They have to find this probability empirically by taking "statistically significant" samples of participants. Each time a new question is posed or a new product comes out, the pollsters have to find a new "statistically significant" sample of participants.

News media and their misuse of the word "science."

It is an unfortunate fault of the scientifically uninformed media that any procedure that uses statistical analysis is called "scientific." Let us repeat loud and clear:

> **Box 41.1.1.** *The use of statistics or statistical analysis in a discipline or profession makes that discipline or profession neither a science nor scientific.*

41.2 Technology: Application of Science

The rapid development of thermodynamics and electromagnetism in the nineteenth century opened up a whole new world of technology in the twentieth century. One could say that the efforts of the physicists of the late eighteenth and mid nineteenth centuries paved the way for the technicians and inventors of the second half of the nineteenth and the first half of the twentieth centuries to create many novel and ingenious devices. This close connection between the physicists and inventors has become a source of confusion in the sociology and politics of science. And the disentanglement of this confusion is a necessary prerequisite for a true appreciation of the nature and the workings of the scientific enterprise.

41.2.1 Medicine

Our discussion of technology and its relation to science ought to start with distinguishing between science and medicine: *Medicine is a technology not a science; and as such, it is the application of science to the treatment of illnesses. In fact, medicine is to biology (and biochemistry) what industrial technology is to physics (and chemistry).* This is hard for the layman to grasp because the coverage of "science" in the major source of public information, the media, is restricted to medicine.

Medicine may have *motivated* some research, which has resulted in scientific breakthroughs, just as technology (such as steam engines) motivated some branches of physics (e.g., thermodynamics). But motivation differs substantially from the end product. If Ponzi scheme results in the accumulation of wealth, we do not identify wealth with Ponzi scheme. If a novel brings fame to an author, we do not equate fame with writing novels. And if sheer curiosity brings about scientific discovery, we should not identify curiosity with science.

Despite the fact that many biological (biophysical, biochemical, biomolecular, ...) discoveries may have taken place in medical and health institutions, medicine remains an art, a technology. And the historical fact that medicine may have been the starting point of modern biology does not change anything.

41.2.2   The Role of Culture

Science discovers the laws of nature blindly, purposelessly, and without regards to its potential human use. If it has an eye, it is to see the hidden secrets of nature. If it has a purpose, it is to connect what is known to what is unknown. If it is human, it is only due to its apparent confinement to the planet Earth. Any other intelligent species, regardless of its location in the universe will discover the same science. The fact that scientific laws are discovered by scientists, who happen to be human beings, does not make science itself human-dependent.

Science is detached from culture.

That science is void of any humanistic trait is evident from its history of development. It started in Egypt and Babylon, moved to Greece and India, went back to the Middle East, and finally landed in the West. As diverse as these civilizations were, and as violently as they clashed, the transfer of science was always inevitable, because it empowered the conquering civilizations with new means to rule more effectively. Unlike other culturally motivated characteristics of humans such as language, which has numerously been trampled to extinction by history, science has always strengthened on passage from one culture to the next.

Technology, on the other hand, applies science sightedly, purposefully, and humanly, and as such, is very much dependent on culture, politics, economy, and all the other characteristics of the human society. In fact the very word "technology" comes from the Greek word τεχνικοσ (*technikos*), meaning *art* or *artifice*. Just as a sculptor uses the raw material such as stone and clay to create a statue for *human* pleasure and *human* consumption, so does a *techn*ician use the raw material such as wires and circuits to create a television set for *human* pleasure and *human* consumption.

Technology is attached to culture.

Once the discovered laws of nature are put to application, they put on a human face with all the cultural, political, and social cosmetics fashionable at the time. The same engine that drives a truckful of food to a flood-stricken community can propel a tank to bulldoze dwellings, schools, and people. It is technology that takes electricity, a fundamental force of nature, and either warms houses in the winters of Alaska, or electrocutes a death row inmate in Texas. The laws of chemistry will not change whether we use them to make aspirin or nerve gas.

Difference between science and technology.

> **Box 41.2.1.** *The task of science is solely the discovery of the laws of nature. Technology (including medicine) uses these laws to meet human needs. Science is independent of any social, political, or economic conditions. Technology is driven by these conditions.*

41.2.3   Science Is Blind to Future Applications

Science is that human activity whose sole purpose is to understand nature by observation and experimentation, and subsequent condensation of this understanding in theories **with no consideration of the possibility or the consequences of any application**. This last statement requires some explanation. A scientist (when not commissioned by a government or industry) simply tries to solve a mystery or answer an unanswered question. Whether this answer will have any future application or not is far removed from his or her mind. History is filled with examples. In fact, all great theories of physics (including those upon which our current civilization is based) have been discovered as answers to purely intellectual questions, satisfaction of childlike curiosities, or reinterpretation of answers in a more elegant and mathematical form. Applications, if they ever came, appeared much later

Separation of science from its potential applications.

in the ventures of entrepreneurs and inventors who had little knowledge of the deep science involved. Below are some post-Renaissance examples of scientific discoveries and how they relate to technology.

Motion and Gravity

All contributions of the versatile **Galileo** to the study of motion were the result of his insatiable desire to know the universe. His experiments with inclined planes, leading to the first law of motion, were done with only one goal in mind: to unravel the secret of motion. And when he turned his telescope to the sky, his only intention was to observe the detailed features of stars, planets, and Moon.

If **Newton** had been after fortune, he would have tended his family farm. Instead he turned to pure thought and mathematics, because his interest was in understanding the solar system for the sake of understanding the solar system. This led him to the investigation of motion (thence to the invention of differential calculus) and the study of gravity (thence to the invention of integral calculus). His physics, although a monumental system of thought with many social, political, and philosophical ramifications, had very limited technological applications until very recently when man started to explore the outer space. Newton could not have dreamed of a day when a nation would have a space agency, which launched orbiting rockets. shuttles, and satellites using the laws he had discovered.

Newton's law of gravitation was truly universal. However, its theoretical utility was hampered by a lack of knowledge of the *universal gravitational constant*, which eluded all experimental detections until the brilliant experimenter, **Henry Cavendish**, measured it in 1798. Cavendish, an exceedingly shy and retiring person—so much so that he neglected to publish many of his important discoveries—was in no way interested in commercial applications of his discovery. In fact, being one of the richest men in England, he was in no need of any financial reward.

Electromagnetism

Perhaps the best examples of curiosity-driven discoveries are the electric and magnetic phenomena. From a childlike game of kites and lightnings to the discovery of the law of electrical forces, to the accidental discovery of the production of magnetic fields from electric currents, and finally to the production of electricity from the motion of wires in magnetic fields, the science of electromagnetism grew as an intellectual toy in the hands of some nineteenth century grizzled children.

The first application of electromagnetism was telegraph, whose rapid development came with the discovery that electric impulses could be used to transmit signals along a wire. Among the many electric systems attempted was the needle telegraph, based on Hans Christian Örsted's discovery in 1820 that an electric current caused an adjacent magnetized needle or pointer to deflect.

To further differentiate between science and technology, consider the following two biographies.

- The development of the electromagnet in the late 1830s provided the American inventor **Samuel Morse** with a way to transmit and receive electric signals. The son of a clergyman, he attended Yale and graduated with a degree, not in any branches of science, but in art. After some failed attempt at finding patrons for his imported historical canvases, he turned to politics, running for mayor of New York on the Nativist ticket, advocating racism, anti-Catholicism, and anti-Semitism.

  After failing in politics, in 1837, Morse solicited the partnership of Joseph Henry and Alfred Vail, and, in 1838, of a US Congressman, in the belief that he could secure federal funds for the construction of a telegraph line in the US. But when Congress turned the group down, Morse's partners gave up and Morse went on alone. In 1843, he finally won Congressional support for a 41-mile telegraph line from Baltimore to Washington, D.C.

- **Thomas Alva Edison** entered school in Port Huron, but because of hearing problems, he had difficulty following the lessons and his school attendance became sporadic. Eventually, at the age of 12, Edison took a job as a trainboy on the Grand Trunk Railway. He also began printing a weekly newspaper, which he called the Grand Trunk Herald.

  From 1862 to 1868, Edison worked as a roving telegrapher in the Midwest, the South, Canada, and New England. During this time, he began developing a telegraphic repeating instrument that made it possible to transmit messages automatically. By 1869, Edison's inventions, including the duplex telegraph and message printer, were progressing so well, he left telegraphy and began a career of full-time inventing and entrepreneurship.

  Edison moved to New York City and within a year, he was able to open a workshop in Newark, New Jersey. He produced the Edison Universal Stock Printer, the automatic telegraph, the quadruplex, as well as other printing telegraphs, while working out of Newark. Later, with financial help from his widowed father, he built a new laboratory and machine shop in Menlo Park, New Jersey, where he invented his incandescent electric light bulb.

Compare the two brief biographies above (or the biography of any other inventor) with that of any physicist (or chemist or biologist), and you will see the essential difference between science and technology. With only a few exceptions, scientists have had to master their scientific fields before they could contribute to them. This typically—but not exclusively—involves attending universities and obtaining the highest possible degree in their chosen field. Inventors, on the other hand, have very little or no training in their "field" of interest. In fact, there is probably as much training in "How to become an inventor" as there is in "How to make a fortune."

The difference between inventors and scientists tells of the difference between technology and science.

If the discovery of electric currents and their magnetic effects gave us the telegraph and the light bulb, what came later—when electricity and magnetism were truly combined—was nothing less than the founding of our modern civilization. And the bulk of the inventions were made not by the scientists who wrote down the mathematical equations, but by entrepreneurs who had little knowledge of the science but a lot of desire to get rich. The interplay between scientists' discoveries and entrepreneurs' inventions is crucial for the advancement of both. The technology that is developed as a result of new scientific discoveries is inevitably used in the instrumentation of later scientific investigations.[2] However, this essential interplay between science and technology should not cause muddling the two. On the contrary, it is more reason to make a clear demarcation between them *and advertise this demarcation*, especially when the policy makers and funding agencies mix them up, and in the long run, hurt both enterprises.

The true unification of electricity and magnetism by the purely mathematical arguments of Maxwell led to the prediction and subsequent production of electromagnetic waves. As a theoretical physicist, Maxwell could not have dreamed of the use of his discovery in telecommunication, electronics, computer industry, and a host of other enterprises on which our modern civilization is built. As he was struggling with the mathematical inconsistency of the existing electromagnetic equations under candle light in the darkness of many nights, his motivation was to find a solution, not to replace his candles with light bulbs!

Even **Heinrich Hertz**, who produced the first man-made electromagnetic wave in his laboratory, was not driven by any potential commercial application. His primary objective,

---

[2]Imagine a chemical laboratory without light bulbs, refrigerators, or computers!

like all experimental physicists, was to prove (or disprove) the prediction of a theory; in this case, Maxwell's prediction.

Nowadays the funding agencies of science and their decision makers expect practical applications (or potential for quick applications) from any scientific investigation. That is why the National Institutes of Health receives very generous financial allocations from the US government, while the handful of theoretical physicists and mathematicians investigating the fundamental forces of nature are struggling for financial survival. Yet the history of physics reveals that the most drastic changes in our civilization arising from applied technology were the progeny of some seemingly fruitless exercise of human curiosity.

Who could have predicted that the tinkering of the lodestone and amber of the sixteenth century would give rise to the transmission of information over hundreds of miles in a fraction of a second? Yet the technology of modern telecommunication is based entirely on electric charges, conductors, wires, and magnets, the toys of the earlier practitioners of seemingly futile exercises of human curiosity.

Quantum Physics

We now enter the twentieth century, and right at the gateway we find the quantum idea and a few years further the theory of relativity. As all great theories before them, these two were molded by the human desire to answer a question, to solve a puzzle. Neither of the two promised any entrepreneurial application. These were puzzles of the purest kind that demanded solutions which could only come about as a result of the purest mental activity.

When **Max Planck** introduced quanta in 1900, he did it quite reluctantly. The sheer incongruity of the idea, the reluctance of both Planck and the rest of the physics community, the awe with which the truthfulness of the idea struck all the physicists of the time should be enough evidence to convince everybody that at the time of its proposal, the *quantum idea could not and did not have any promise of application* in it!

The quantum idea found its way into the structure of atoms in the work of **Niels Bohr** who extended the discoveries of Planck and Einstein to an explanation of the behavior of the hydrogen atom. Bohr's work was so ad hoc and so strange that even the most imaginative science fiction writer could not have concocted any application for it. Bohr's only desire was to resolve a contradiction suggested by the newly discovered nucleus of the atom.

Nuclear industry grew at a rapid pace during World War II both in Germany and in the US. Although this growth was driven by the war efforts, the discovery of the nucleus itself was motivated by sheer curiosity. In an attempt to verify a theoretical model of the atom, **Ernest Rutherford** bombarded gold atoms with highly energetic alpha particles in 1911. He was dumbfounded when his experiment showed that almost the entire weight of the atom is concentrated in a very small region at the center of the atom. This discovery raised some very deep questions about the stability of matter which led to one of the highest human intellectual achievements ever.

The brain child of an international cooperation of theoretical physicists, quantum theory, was developed in stages, and at each stage a great leap of counter-intuitive faith had to be taken. By 1927, all the mysterious pieces of the puzzle came together to form a mathematical structure which was logically consistent and observationally flawless. However, the entire theory was based on the notion of probability, and, as such, was immersed in a controversy of which Einstein said "God does not play dice." Amidst such confusion the last thing *anybody* was thinking about was application! This newly opened window into the secrets of the atomic world was showing too magnificent a scenery for one's mind to get distracted by any entrepreneurial applications. The unraveling of this new mathematical theory and its complete ramification was too important for physicists to divert their thoughts to industrial consequences of the theory. Today, as we have just crossed the exit gate of the century, almost everything we touch in our daily lives runs on the idea of quantum.

41.3   Science and Values

The confusion caused by identifying science and technology has prompted some science critics to ridicule the notion of value-free science.[3] Thus, in the introduction of his book *Value-Free Science?* Robert Proctor writes:

> Still we often hear that however foul its application, science itself is pure. Science may be political in its application, but not in its origin and structure. And certainly it is true that science and technology alone are hardly a threat to world peace. Politics and moralities stand behind our sciences and give them life; science can be used for good or evil. This is one sense of the "neutrality" of science—that science (or technology) "in itself" is neither good nor bad; that science may be *used* or it may be *abused*. [Pro 91, p. 2]

Note how the two words "science" and "technology" come together. Although the author starts the paragraph with "science" alone, he immediately conjoins "technology" to it, and in the last sentence, he practically equates the two by using the parenthetical phrase "or technology." This allows the author to shift the attention from science to technology and to cite numerous examples of its destructive (and certainly value-laden) application. The good or bad and *use* or *abuse* are *already* incorporated in "technology." Nuclear physics is the *science* behind both the MRI of medicine and the bomb of the military. The latter two are *technologies*, and they are by no means "neutral": MRI is good, bomb is bad. Any attachment of "neutrality" to technology opens the possibility of looking for something good in a nuclear bomb!

> *Science and technology are **wrongly** being interchanged!*

Proctor goes on to say:

> Yet this supposed neutrality describes only the simplest technologies, the most abstract principles. The seven simple machines, perhaps, or the rules of arithmetic, may be neutral in this sense. But an abstract truth often conceals a concrete lie. "Guns don't kill people, people kill people." Yet is it surprising that a society that surrounds itself with guns will use them? ... Tools, we realize have alternative uses; the knife bought for cooking might be used for killing. Yet knives or levers are not what modern science-based technology is all about. A nuclear power plant, cruise missile, or linear accelerator can hardly be used for ends other than those for which they are designed. Science-based technologies are increasingly *end-specific:* the means constrain the ends; it is no longer easy to separate the origins of a tool from its intended use. What does it mean to "abuse" a cruise missile or a neutron bomb? [Pro 91, pp. 3–4]

Since these are typical assertions in the war against value-free science, let us analyze them in detail so that the reader can judge whether the quote implies a value-laden science. First let us correct the author in his favor and emphasize that *even the simplest machines are value-laden*, because they are built with a human need in mind. A lever and a catapult are technologies that are based on the simplest science,[4] arithmetic. Yet one is used to lift a boulder blocking the irrigation route of a village, and the other to hurl fire to kill the enemy. So, it is wrong to ascribe any neutrality to even the simplest machines. The neutrality should correctly be ascribed to the *science* (arithmetic or algebra) behind the technology and *not* to the technology itself.

> *Even the simplest machines, such as a lever or a catapult, are value-laden!*

Section 41.2 showed that any technological invention applies the knowledge gained through science to create things specifically designed for human use. This principle applies to all technology, past, present, and future. The only difference is that the inventions of thousands of years ago (simple machines) are so simple that nonscientists can understand them, and, therefore, can separate them from the science of that time (arithmetic).

---

[3]In the present context, "value" refers to moral and ethical nature of science.

[4]The differentiation between science and math, although emphasized at the beginning of this chapter, need not be highlighted here, as in ancient times, neither had any distinctive feature to set it apart from the other.

However, modern technology, which is based on such highly mathematical disciplines as electromagnetic theory and quantum mechanics, is much harder to separate from the complicated science that went into it.

It is not the simplicity of science that makes it neutral. General theory of relativity, quantum field theory, electromagnetism, and Newtonian mechanics are as neutral as arithmetic. Just as arithmetic, they are laws stated in the language of mathematics, generally in the form of differential equations, that describe the behavior of (the constituents of) the universe. This fact is extremely hard to understand for somebody who has not studied these equations and their meaning, and only sees their technological applications.

Second, the often quoted sentence, "Guns don't kill people, people kill people," is so obvious a tautology that it is pointless. It is as pointless as saying: "Cars don't drive, people do," or "Knives don't cut, people do," or "Pencils don't write, people do." It merely states the simple fact that any machine needs an operator to operate it. There is no "abstract truth" in this. It is as concrete as one can get! Abstract are the laws of science, and there are no concrete lies in them. There is no concrete lie in the universal law of gravitation, or in the four Maxwell's equations of electromagnetism, or the Schrödinger equation, or Einstein's equation of the general theory of relativity. Or the laws of thermodynamics and chemistry. Concrete are the guns and the people who use them.

In the construction of a gun—the barrel, the ignition mechanism, the path of the bullet, and the fabrication of its parts—the laws of motion, thermodynamics, and chemistry may have been used, but these same laws are also used to operate a respirator or a pacemaker. Guns, being the product of a technology motivated, financed, and encouraged by war, were made with no other purpose in mind than to kill. We did not invent the gun with which to brush our teeth! A lot of values, ethics, morality (or lack thereof) has gone into the building of a gun. An abstract truth which conceals no concrete lie is the following statement:

> **Box 41.3.1.** *The laws of physics and chemistry don't kill people, guns do.*

This statement captures the essence of the difference between technology (guns) and science (the laws of physics and chemistry), a difference that is overlooked by most critics of science.

Third, the relation between ancient science and ancient technology is *identical* to the relation between modern science and modern technology. Tools, whether invented by a *homo erectus*, an ancient Egyptian, or a Medieval artisan, are as science-based as the laser used in LASEK; and they are as much a part of technology as cruise missiles and power plants. Knife, at the time of its invention in the Bronze Age, required as much cutting-edge[5] science as laser did a few decades ago. And as it was built to replace the old Stone-Age sharp stones, knife was as *end-specific* as a nuclear power plant, a cruise missile, or a linear accelerator: it was meant to kill and cut with not to cook with!

A nuclear power plant is a "tool" that uses the *abstract* principles of physics—such as $E = mc^2$—to produce electricity for consumption. It comes out of a branch of technology, and, therefore, *is not science!* A cruise missile is nothing but a glorified gun. To it also applies the statement of Box 41.3.1. It is a destructive power which was developed with the clear intention of annihilating buildings and human beings. We can blame science for its production as much as we can blame arithmetic for the production of simple "war" machines of 2000 years ago such as catapults.

A linear accelerator is of a completely different nature. Although a substantial amount of advanced technology is used in its construction, it is not designed to produce anything for human consumption, and certainly not for his annihilation. It is a machine that probes the structure of matter to a deeper and deeper level. One can say that it is a huge microscope

GTR, quantum field theory, electromagnetism, and Newtonian mechanics are as neutral as arithmetic.

We can blame science for the production of a cruise missile as much as we can blame arithmetic for the production of catapults of 2000 years ago.

---

[5]No pun intended!

capable of "seeing" subatomic particles. Why the author puts linear accelerators alongside power plants and cruise missiles is unclear.

We have already seen how the author starts with a sentence containing "science" alone, but, to show its value-ladenness, he switches to "technology" in the next sentence. The farther we read in the book, the more it becomes clear that science and technology are used interchangeably. The following quote is an extreme case of this mix-up, in which the word "science" is used *instead of* "technology:"

> According to one common sense of science, however, politics or values are only to be found external to the practice of science—in its uses but not its origin, in its failures but not its triumphs, in the exceptional or the peripheral but not in the everyday and fundamental. This is the ideology of *pure* or *value-free* science, the belief that science "in itself" is pure and that values or politics enter only as contamination.
>
> ... And yet the myth confronts a reality: science throughout the modern world is recognized as a vital part of industrial and military production. [Pro 91, pp. 3–4]

The "science" in the second paragraph is really "technology!" String theory, gauge theory, quantum chromodynamics, and inflationary cosmology, some of the sciences of our day, are not, by any stretch of imagination, "part of industrial and military production."

Box 41.3.2. *The complexity of both modern technology and modern science has caused many critics, unfamiliar with either, to confuse the two, equate science with technology and assign "value" to both. Our discussion in this section has shown that* **while technology is loaded with value, science is completely value-free**.

Chapter 42

Characteristics of Science

One of the goals of this book, as implied by its subtitle, is to promote scientific awareness and literacy and to combat irrationality. To do so, it has to fully explain the inner workings and characteristics of rationality. As the epitome of rationality, science can be an effective sword in the battle against all forms of irrational thinking. And as the epitome of science, physics can make this sword razor sharp. What are some of the characteristics of science?

42.1   Science Studies Matter

Materiality is one of the most distinguishing characteristics of science. Starting with Galileo and his inclined planes and blocks or balls, meeting along the way Maxwell and his electromagnetic fields, and finally ending with the most exotic subatomic particles such as quarks and gluons or the gigantic black holes billions of times heavier than Sun, physics *exclusively* studies material objects. And yes, fields are as material as atoms, molecules, billiard balls, and SUVs (see Section 15.4)! Take the EM field, which consists of photons; the materiality of photons has been established not only by the kicks and nudges they impart on other particles, but also by their capacity to turn into other particles such as electrons and positrons [see Figure 32.4(d)]. Despite the claim of some advocates of the alternative medicine

> **Box 42.1.1.** *Everything that physics—both classical and modern—studies, including fields and waves, is material. Nonmaterial objects have no place in physics.*

Materiality is not confined to physics. In fact, one can say that the form of matter determines the branch of science: atoms and subatomic particles are studied in atomic, nuclear, and particle physics; molecules are studied in chemistry; the matter of life, DNA, is the subject of biology; a piece of bone in the hands of a paleontologist tells a story of our prehistoric past; a pot or a tablet in the hands of an archaeologist tells of our past civilizations; geology studies Earth; astrophysics studies stars; cosmology studies galaxies and the universe itself.

Every branch of science studies a particular form of **matter**.

No science deals with nonmaterial notions such as the id, ego, and superego of psychoanalysis, the subluxation of chiropractic, the Qi of acupuncture, or the energy field of therapeutic touch. The popularity of these pseudoscientific disciplines is a sad testimony to the preponderance of irrationality in our modern society. To see the power of rational thinking, let us go back to ancient Greece.

42.1.1 Materialistic Philosophy

Science is the *modern* discipline in which matter has a primary position. There was also a time in antiquity when some *philosophers* insisted on the primacy of "matter" and on the role of the senses as the gateway to knowledge. This **materialistic philosophy** led them to conclude that the outside world *has to* stimulate the senses by some kind of *material* agent originating from the source of the stimulation. And since we don't see these agents, they must be so tiny as to be invisible to the human eye. They called them *atoms*.

Materialistic philosophy leads to the idea of *atoms*.

The first person to systematically analyze and popularize the theory of atoms is **Democritus of Abdera** (ca. 460 BC–370 BC). However, Democritus' materialistic philosophy was incomplete because it only considered atoms. A true philosophy was to put mankind into the equation as well. It became the task of one of the greatest philosophers of antiquity, **Epicurus** (341–270 BC), to dovetail the atoms and the human experience and create the most remarkable philosophical system of his time.

Although he wrote more than any philosopher before him, very few of Epicurus' writings have come down to us in their original form. Aside from a number of fragments, only three letters to his disciples have survived. The best introduction to Epicurus' *Canonic*—a set of rules by which Epicurus formulated his philosophy—is a passage by Cicero, the Roman statesman, orator, and philosopher, who was a hostile critic of Epicurean philosophy. He abolishes definitions, Cicero complains, he has nothing to say about how reasoning is carried out and conclusions reached; he has nothing to say about how sophisms may be resolved and ambiguities cleared up. *He places the criteria of reality in the senses* [Use 87].

How Epicureanism and science are alike!

But it is precisely this placement of "the criteria of reality in the senses" that makes the Epicurean doctrine so akin to modern science. Listen to what Epicurus says in a letter to Herodotus: "For the similarity between the things... which we call real and the images received as a likeness of things and produced either in sleep or through some other acts of apprehension on the part of the mind, ... could never be, unless there were some effluences of this nature actually brought into contact with our senses" [Wea 87, p. 312]. Then a little after he demonstrates that hearing is caused by currents which "split up into particles," he gives this remarkable passage, which might as well have been written by a modern chemist: "Furthermore, we must suppose that smell too, just like hearing, could never bring about any sensation, unless there were certain particles carried off from the object of suitable size to stir this sense-organ ...."

🧺 Food for Thought 🍖

FFT 42.1.2. In the Epicurean materialistic philosophy, our senses become simple *detectors*, and very crude ones at that. These detectors are tuned to the outside world by which they are constantly impressed. And since *no detector is perfect*, the picture our senses convey to us of the outside world is by no means accurate. Similarly, the impressions we get from our social and political environment may not be accurate. If we truly grasp this simple fact, it will become much easier to overcome the prejudices and preconceptions we so tightly hold on to. It will make us wonder whether the picture of people we have constructed in our mind—because of the color of their skin or the language they speak or the culture in which they are raised—is the real picture. The history of the world is filled with occurrences of prejudice and injustice, sometimes even backed by the residing government. Here is one example.

Our senses are nothing but detectors!

For almost 300 years the political structure in South Africa fed its citizens only information suited for the survival of apartheid. Thus, a white child "learned" from the beginning that she was superior to all blacks, that it was quite natural for blacks to have separate buses, that it was unthinkable for blacks to participate in any decision-making process. Only a costly upheaval, a renaissance in the mentality of the people of South Africa, both white and black, a revolution in the political order was capable of turning the society around and teaching the population of all colors that the picture drawn in the minds of most people by the apartheid was a distorted picture, and that the real picture gives blacks, whites, and people of different colors equal weight in social status.

Relation between sense perception and prejudice.

Therefore, we ought to take this basic lesson of scientific methodology to heart and remember that just as our senses are imperfect instruments and cannot reveal the true picture of reality, so are our impressions of the society in which we live. The picture of the latter is even more distorted because it passes through numerous filters including tradition, media, advertising, and entertainment industry, many of which have a single goal: to make a profit.

By far, the most comprehensive account of Epicurean philosophy is a book of Latin poems called *De rerum natura*—or *The Nature of Things*—written by **Titus Lucretius Carus** who lived from about 95 to 55 BC There have been many translations of these poems in English, of which the most noteworthy is that written by Lucy Hutchinson (1620–1664), a seventeenth century woman of British nobility, who lived in an era when the science of physics was beginning to emerge, and the notion of atoms and their motion became popular again. Lucretius, as the only authentic Epicurean philosopher, came to be the hero of this popular culture. However, due to the intense negative publicity of the Church, Lucretius was made into a blemished hero. The rumors, spread by the foes of Epicureans, were so strong that even those who admired Lucretius routinely accepted his "madness." It is in this vein that Lucy Hutchinson, who so marvelously translated Lucretius' poems, casually adds a marginal note to her translation informing the reader: "Here is one of the Poets abrupt Hiatus for he was mad with a Philtrum his wife gave him and writt this booke but in the intervalls of his phrenzie" [Hut 96, p. 55].

Let us see what Lucretius has to say about "things" in his poems. We quote a few passages from Hutchinson's translation [Hut 96] because of its clarity, its sincerity, and its poetic character. On page 33 of this book, we find the following passage, which could be thought of as a version of the modern scientific principle of conservation of energy and matter:
"God never aniething of nothing made;
But soe are mortall men restreind with dread,
As seing severall works in heaven and earth,
And ignorant of the cause that gives them birth,
They thinke a power devine brings forth those things;"

To convince the reader of the existence of atoms, Lucretius gives the following persuasive argument on page 36:
"Yet least my sayings find no faith with you,
Because first bodies[1] are from outward view
Concealed, attend and you your selfe shall learne
There are some bodies eies cannot discerne.
The wind rufles the clouds, beats on the sands,
Overwhelms tall ships, and passing through the land
Strews it with torne up trees, the groves destroys
And rages through the hills with horrid noyse
......
And yet the winds concealed bodies be
Which passe unseene through heaven, earth and sea;"

The Epicurean material analysis of the universe led to some uncomfortable (but accurate) interpretations of nonmaterial phenomena. Epicureans felt that superstition was the main cause of human misery, and did everything to eliminate it. This brought them face to face with the established religion of the time, and on a collision course with the Greek gods. Likewise, in his material analysis of love, Lucretius was led to human physiology, and came frighteningly close to the modern chemical and hormonal explanation of that feeling. Such an interpretation was, of course, dangerously ahead of its time and was considered

While 2000 years ago Lucretius tried to explain "love" materially . . .

---

[1]Lucretius, in order to maintain poetic rhymes, refers to atoms in many different ways including *primordia, elementa, corpora prima*. The last one is translated by Hutchinson as "first bodies."

offensive, obscene, and mad. For a translation of the last 250 lines of Lucretius' Book 4—in which he describes the material cause of love—see [Hut 96, pp. 237–240].

...New Age doctors try to cure diseases by necromancy!

It is regrettable that, while a Roman poet of over 2000 years ago searched to find a *material* explanation of the seemingly nonmaterial feeling of love, modern day New Agers attack the contemporary scientific medicine for its attempts at understanding the *material* causes of diseases. By undermining the efforts of scientific medical doctors in finding cures, these New Agers are turning the wheel of progress back to pre-golden era of the Greek civilization; to a time when the "good spirits" in the teachings of Qi and Ayurveda were summoned to get rid of the "bad spirits" in a patient.

42.2   Science Trusts Only Observation

The study of matter starts with observation. Scientific statements are summaries of observations made (sometimes over many generations). Never in the history of science does one encounter a "decree" that is not suggested by observation. A claim that connects a disease with an imbalance in some unobservable "energy field," or with a pinched nerve in the spine, or an invisible imponderable Qi, or an ad hoc nonmaterial triad of id, ego, and superego is as scientific as the tribal claim that connects a disease with the evil spirits. Neither has any observational support whatsoever.

Food for Thought

FFT 42.2.1. Over 23 centuries have passed since the great Epicurus took his last breath, and less than five centuries since his method of thinking and discovery were revived, and with this revival, an explosion of human knowledge ensued. One can only wonder why humankind abandoned such a natural habit of placing *observation* (or sensation) on top of all other means of gathering information. Some people hold the Medieval Church responsible for the "darkness" of that Age. This, then, begs the question "Why did people allow this to happen?" The answer may lie in what happened *between* the Golden Age of Greece, and the Dark Ages of Europe.

With the death of Epicurean philosophy scientific method also dies.

Whatever the reason for this long stupor of humankind, our present world might have been unrecognizably more advanced had the teachings of Epicurus been the guiding light of our ancestors. It is quite possible that we would have had a Newton by the fourth century AD and an Einstein by the sixth. Then today we would be living in a society that will be possible only 14 centuries from now—if we don't fall into another Dark Age!

History can teach us a lot of good lessons only if we can recognize the present analog of forces that hindered the progress of the past. These forces come in shapes and forms that may appear completely different from their historical ancestors. During the Dark Ages, the center of hindrance was cathedrals and royal palaces. Where do we have to look for these centers today? Are they still in the more decentralized institution of religion? Or in the corporate headquarters of giant broadcasting companies that decide what to tell the society and what not to? Or in the educational system where "social intelligence" and "social promotion" of pupils replace their satisfactory performance in the lower grade? Or in the schools of education whose faculty's response to the repeated failure of the nation's students in international mathematics and physics competition is "[students] need to know how to decide which life insurance to buy, how pesticides affect their food and how interest rates affect home mortgages?" [NYT 98] Or in the community of "science educators," whose attempt at "integrating" all sciences, willy-nilly leads to the smearing of the border of science and non-science, and as a result, some authors find common grounds for physics and Eastern mysticism? (See Chapter 35.) Or in the system of government that cannot control any of the above?

A device constructed to test a theory may have used the theory in its construction.

Scientific observation is necessarily based on the existing proven theories. There cannot exist an observation that is detached from the repertoire of valid scientific theories. Some philosophers of science believe in the complete separation of observation and theories. They think that "theory-laden" observations are invalid. But that is not how science works.

Science is full of examples in which *the very device that is used to make an observation is based on the earlier theoretical works.* For example, a NMR (nuclear magnetic resonance) spectrometer that chemists use to analyze the molecular structure of substances (thereby, testing the quantum theory that describes the molecules) is built using the guidelines of the quantum theory.

Many observations are indirect. To test a theory by experimentation or observation, one may look for evidence and clues related to (but not necessarily identified with) the prediction of the theory. It is impossible to test the theory of the creation of Sun, because it occurred only once. And if we insist that "unless science can create a Sun, anything it says about the Sun is just speculation," we are closing our eyes to all the advances we have made in the science of astrophysics. If the theory predicts a multitude of features and properties of Sun that are observable *now*, and if all *present* observations agree with those predictions, then the theory is observationally tested.

You don't have to create a Sun to verify a theory of stellar creation.

42.3   Continuity of Science

No science is the creation of a single man or a single age. Newton could not have solved the mystery of gravitation without the works of Galileo and Kepler; and the latter could not have made their contributions without the observations of Tycho Brahe and the theoretical work of Copernicus. And Copernicus could not have revolutionized human thought had he not studied the work of ancient astronomers. Einstein's theory of relativity of 1905 depends crucially on Maxwell's electromagnetic theory of 1865, which in turn was based on the work of such people as Coulomb, Gauss, Örsted, and Faraday, spanning over five decades of hard work. And the earliest work on electromagnetism by Coulomb is directly connected to Newton's laws of motion of 1666.

Any scientific discipline has a body of interconnected knowledge accumulated over many generations of scientists, on which all specialists of that discipline agree; and any new knowledge that is to be added to that body uses all or part of the existing knowledge. No discovery or advancement of science exists in history that is isolated from the existing body of knowledge. A discoverer of DNA must know the structure of the chromosomes; the discoverer of chromosomes must know the structure of the nucleus of a cell; the discoverer of the nucleus of the cell must know the structure of the cell; the discoverer of the cell must have a microscope in his possession. Thus, the science of the 1950s becomes continuously connected to the science of the 1660s.

> **Box 42.3.1.** *A discipline in which the previous generation of its practitioners propose theories which are trampled by the current generation, whose theories are in turn trampled by the future generation, is not science.*

In the popular expositions of physics it is often asserted that modern physics has replaced or undermined the classical physics of Newton and Galileo. They give the impression that modern and classical physics have nothing in common, or worse, that they are in conflict with one another. Nothing is further from the truth! A new theory, as revolutionary and as radical as it may seem, simply *extends* the old theory into domains where the latter fails. In fact, in arriving at the new theory, the old theory is always a crucial guideline; and the commonality of the two theories can be shown by applying them to the domain of the validity of the old theory and seeing that both give identical results.

Relativity could not have been discovered without classical physics.

Consider, as an example, the Newtonian and the relativistic theories of motion. The latter is dependent on the former in two ways. First, relativity is a logical necessity of Maxwell's prediction of the electromagnetic waves (see Section 25.2 for details); and Maxwell's equations were derived using experiments involving forces on electric charges and currents *that*

Both special and general theories of relativity have their origins in the classical physics.

obeyed Newtonian laws of motion. So in this sense, there is a direct link between relativity and Newtonian physics.

Second, when Einstein tried to write equations governing the motion of relativistic particles under the influence of forces, he realized that all he had to do was to *copy the Newtonian second law of motion* in a way that was consistent with relativity. And in his general theory of relativity, at a crucial step in deriving Einstein's equation, he *had to use the Newtonian law of gravity.*

Relativity becomes identical to classical physics when it is applied to objects that move slowly!

As we have seen in Part VI, relativity is the physics of objects that move close to light speed. For such speeds, Newtonian theory fails while Einstein's theory of relativity prevails. This is as expected, because at the time of Newton, no observational evidence pointed to the uniqueness and universality of the speed of light. Furthermore, all observable speeds were significantly smaller than the speed of light, whose magnitude was not within the reach of the technology of the time. The theory of relativity, on the other hand, was *built* on the assumption that the speed of light is a special speed, beyond which no object is capable of moving. Nevertheless, when the objects under consideration by relativity are moving with very small velocities (small compared with the speed of light), the theory turns into the Newtonian theory, and the two theories become one and the same!

The development of the quantum theory is also vitally tied to the Newtonian physics. When Bohr applied the quantum hypothesis of Planck and Einstein to the hydrogen atom, *he used Newtonian second law of motion verbatim!* (See Section 21.3 and Math Notes E.21.1 and E.21.2 for details.) Bohr model of the hydrogen atom was indispensable for the quantum theory of Schrödinger and Heisenberg. Moreover, the master equation of quantum theory, the Schrödinger equation, owes its existence to such quantities as momentum, kinetic energy, and potential energy, *all of which are genuinely Newtonian concepts.* For example, we have seen (see Section 8.1.1 and Math Note E.8.1) how the concept of kinetic energy is *a direct consequence of Newton's second law of motion.* Therefore, we can say without any hesitation that

> **Box 42.3.2.** *If Newtonian theory had not existed, the two pillars of modern physics, relativity and quantum theory, could not have been discovered.*

This captures the essence of the continuity of science.

A number of authors have tried to take advantage of the difference between the Newtonian and quantum theory to advance their philosophical agenda. They have attributed to the old theory such pretentiously negative qualities as "mechanistic," "Cartesian," and "linear," while the new quantum theory is given such New Agedly positive attributes as "holistic" and "universal." They hide the fact that the "holistic" theory *could not have been discovered without the aid of the "mechanistic" theory,* and that when applied to the same objects for which the Newtonian theory is good, "holistic" theory becomes just as mechanistic and Cartesian.

Scientific theories of an era usually have multiple discoverers, because scientific ideas are interconnected.

The continuity and interconnectedness of scientific ideas is the reason that **they are usually discovered simultaneously and independently** by more than one scientist. The body of theories and experimental results is available to all practitioners of a scientific field, and it is very common for a few people trying to solve the unique enigma of the time to hit upon the same idea. The universal law of gravitation was suggested not only by Newton, but also by some of his contemporaries. The special theory of relativity, credited to Einstein, was also discovered by the great French mathematician Henri Poincaré. Likewise, in a recently discovered manuscript of the noted German mathematician David Hilbert, there appears an equation, which we now call Einstein's equation of the general theory of relativity. In fact, the date of the manuscript was so close to the date of Einstein's GTR paper that it took historians many days to settle the question of "Who discovered GTR first?" in favor of Einstein.

42.4   Science Is Detached from the Scientist

Because science is based on (and explains the results of) observations of nature, and since nature is detached from the scientist, scientific ideas and discoveries are also detached from the discoverers. The universal law of gravity, although discovered by Newton, has absolutely no mark of Newton's personality on it. It is an abstract mathematical statement about the behavior of the universe under the influence of one of the fundamental forces of nature that we call gravity. Students learning about gravity may know nothing (in fact, almost always they do not) about Newton's biography or the original work in which gravity was explained.

A remarkable example of this detachment of the discovery from the discoverer is the Schrödinger equation. This equation was discovered in January of 1926 by the Austrian physicist Erwin Schrödinger, who also solved it to obtain a successful description of the hydrogen atom. As a result of this solution, Schrödinger found a mathematical entity whose correct interpretation demanded some very intense intellectual exercise. Schrödinger himself wanted to interpret this mathematical entity—what we called the Ψ function in Section 22.3—as a so-called "matter wave." This interpretation, however, could not explain the observational results satisfactorily. The correct interpretation was given in June 1926 by the German physicist, Max Born, who proposed that the mathematical entity should be interpreted as "probability waves." Schrödinger, a determinist who abhorred the notion of probability and indeterminacy, is said to have remarked that had he known the consequences, he might not have written his paper. We now know that Born was right and the Schrödinger equation has been completely detached from Schrödinger!

How Schrödinger equation is completely detached from Erwin Schrödinger!

The detachment of science from the scientist completely eliminates the possibility of an "authority figure" in science. There are no "gurus" and "followers" in science. If physicists followed Einstein, it was because his theories were correct, and he happened to have a remarkable affinity with nature and its secrets. As soon as he proved to be wrong (as in his insistence on keeping probability out of quantum theory) his followers declined considerably. Because of this lack of authority figure, when giants of science die, there is no dispute, disagreement, feud, and factional in-fighting among the scientists.

The independence of science from the scientist is also called the **objectivity** of science. However, many "postmodern" philosophers of science argue that such an objectivity does not exist; that not only is science attached to the scientist that discovers it, but also to the "disciples" of the discoverer; that different groups of scientists believe in different scientific knowledge, and there is no way to tell who is right. Such an attitude towards scientific knowledge started in the 1960s, and one of the most influential proponents of this attitude is Thomas Kuhn.

In his trend-setting book *The Structure of Scientific Revolutions*, Thomas Kuhn argues that scientists accept new theories[2] only through persuasion and argumentation. Kuhn believes that the only way that a theory could be accepted unequivocally is by a formal "proof;" and since no theory can be proven (all theories are "underdetermined" as discussed in Section 5.3), the acceptance of a theory by the scientific community must be done by persuasion [Kuh 96, p. 152]. And since persuasion is a subjective act, science cannot be objective.

Thomas Kuhn's "paradigm shift" and his attempt at destroying objectivity.

This, like any other philosophy of science, completely overlooks the role and the convincing power of *observation*. Philosophers of science scarcely mention observation and experimentation in their analyses. Yet it is *only* observation that can "persuade" a scientist to accept a theory (or paradigm, as Kuhn calls it). A thousand scientists have less persuasive power on a fellow scientist than a single observation. Of course, a trillion observations cannot form a "proof" demanded by underdeterminists. Nevertheless, a single experiment, such as Rutherford's 1911 experiment showing the existence of the atomic nucleus, is sufficient proof for Niels Bohr to change his mind about the nature of the hydrogen atom and suggest

---

[2]In Kuhn's jargon, they are "converted to new paradigms."

a new theory for it. This is objectivity in action: Nature reveals itself through a myriad observational *machines*, and chooses *mathematics* as her language of communication.

Box 42.4.1. *No emotion, no ethics, no morality, no value, no politics, no persuasion ... and yes, no subjectivity plays any role in science!*

42.5   Reductionism and Science

The act of observation with the aim of gathering useful information is, of necessity, also an act of reduction. In the course of our journey through physics, we learned at every step that great and universal discoveries were made when the discoverer studied a *very specific system*:

- Aristarchus discovered the *universal* heliocentrism by asking a very specific question: "What is the distance between the Earth and the Sun?"

- Galileo discovered the *universal* first law of motion by studying a very specific system: an inclined plane with blocks sliding on it. This law now applies to the motion of any object from a molecule to a galaxy.

- Kepler discovered the *universal* laws of planetary motion by studying a very specific system: Mars and Sun. His laws apply not only to *all* planets in the solar system, but to anything that moves around a center of gravity such as the dark matter, which was discovered using these laws.

- Newton discovered the laws of motion and the *universal* law of gravity by studying a very specific system: Earth, Moon, and an apple. Newton's laws are valid not only for the motion of the Moon and the apple, but for *any* motion.[3]

- Maxwell discovered the *universal* electromagnetic waves by studying the very specific equations of electricity and magnetism. The EM waves are now the only source of information we obtain from every cosmic object.

- Planck discovered the *universal* quantum nature of the electromagnetic waves by studying the very specific process of EM wave production through heat.

- Einstein discovered the truly *universal* theory of relativity by asking a very specific question: "How do Maxwell's equations look like in a moving reference frame?"

- Schrödinger and Heisenberg discovered the quantum theory by studying a very specific and very simple system: the hydrogen atom. Quantum theory is now so *universal* that solids, liquids, gases, plasmas, or any other form of matter are accurately described by it; the entire discipline of chemistry is based on it; even the secret of life itself may eventually be unraveled by a "quantum mechanical" treatment of DNA.

- Dirac discovered the *universal* notion of antimatter by asking a very specific question: "How do we combine special relativity and quantum theory?"

- Einstein and Friedmann discovered the *universal* big bang by asking a very specific question: "What happens if we apply the general theory of relativity to an isotropic and homogeneous gas?"

---

[3] As long as the objects moving are larger than molecules and do not move close to light speed.

- Guth discovered the inflationary universe by asking a very specific question: "Why are there no magnetic monopoles?"

Is the specificity illustrated in the items above—and in many other instances too numerous to list—on purpose? Do physicists have a "holistic choice" but they just ignore it? No! Physicists *are forced* to study *specific systems* and seek answers to *specific questions*, because a single physicist—even a single generation of physicists—is incapable of studying the entire holistic universe. Every era concentrates on some specific unanswered questions left behind by the discoveries of the previous era.

> Every era concentrates on some **specific** unanswered questions left behind by the discoveries of the previous era.

Reductionism[4] is a hallmark of not just physics, but of all sciences. It gave us the amazing materialistic philosophy and its ancient atomic theory 25 centuries ago. At the turn of the nineteenth century it showed up again as modern atomic theory, when chemists tried to make sense out of chemical reactions. The success of modern atomic theory in chemistry prompted many physicists to take the notion of an atom seriously and to "reduce" a thermodynamic system to a collection of "particles" obeying the laws of mechanics. This led to the new branch of physics and chemistry, statistical mechanics, which clarified many of the hitherto mysterious notions of thermodynamics.

The physicists' scrutiny of the simplest atom, hydrogen, led to the discovery of the quantum theory, which in principle could explain the behavior of more complex atoms and molecules. Chemistry has become the science of molecules using the quantum theory. All the miraculous advances in modern chemistry are the results of the concentration of chemists on the study of molecules.

Biology has also been "reducing." The true study of life started with the discovery that all living objects are made up of cells. Once the nucleus of the cell was discovered and the complex structure inside it revealed, *it* became the subject of biologists' study. Today, biologists agree that the "unit" of life is a very large *molecule* called DNA, and to unravel the secret of life we have to understand this molecule. So chemistry, and ultimately physics, become the tools for understanding life.

> Reductionism *does not mean* that fundamental physics alone can explain life.

Some science critics take the last sentence above to mean "physics alone can explain life;" that if you find the laws of physics governing the fundamental particles, you don't need to do anything else to understand DNA. This is, of course, a completely erroneous statement. Physics to a biologist or chemist is like a ship to a sailor. The ship by itself is not sufficient to discover new lands. You have to be an expert navigator, with a lot of knowledge about maps, geography, astronomy, and of course, how to sail a ship, to be able to discover new lands. Similarly, physics is only the beginning—but a *necessary* beginning—of the understanding of molecules. A (bio)chemist has to know the structure of the atoms, the configuration of their constituents, any special features they may possess, how they interact with one another, and of course how to use quantum physics, to be able to understand the molecule formed from the atoms. Although the study of life can be reduced to the study of the atoms making up the DNA molecule, it is not reduced to such a study *in isolation*.

> Reducing study of DNA to physics and chemistry is **necessary**, but not sufficient.

42.5.1 The Whole and Its Parts

The recent holistic "school" of science has not come out of science itself, but *imposed* on it from outside. Certain philosophers and mystics, borrowing from Eastern mysticism, combined with a misinterpretation of modern physics (see Chapter 35), have come up with a "holistic" prognosis for science. They argue that "the whole is more than just the sum of its parts," and conclude that reductionism is wrong; that science should abandon its reductionist approach. A scientist (a reductionist) does not refute the statement; he refutes the conclusion. As mentioned above, understanding a molecule involves more than just a

> Holism has been imposed on science from outside.

---

[4]Reductionism is sometimes referred to as "fundamentalism" by some philosophers of science. We have already pointed out how important "fundamentalism" has been in the development of science (see, e.g., Sections 20.2.2 and 33.2.2).

knowledge of the atoms of which it consists. But *without an understanding of the atoms making up a molecule, you cannot understand the molecule itself.* Atoms form the main part of the many ingredients that are combined to form a molecule.

Statistical mechanics can further elucidate the difference between a holist and a reductionist take on the above statement. When you put a large number of particles together to form a thermodynamic system, statistical mechanics assigns a probability to every state of that system, and if you know this probability, you can *predict every property of the system*; i.e., you know the "whole." But, this probability depends crucially on the energy of each *individual particle* and how it interacts with the rest of the particles in the system (see Sections 17.2.2 and 19.3). In other words, if you don't know the properties of the individual "parts," you *cannot* know the bulk properties of the "whole." Therefore,

> **Box 42.5.1.** *Although the whole is more than the sum of its parts, if you don't know the parts, you cannot know the whole.*

Electrical conductivity of the "whole" can *only* be explained by the quantum physics of its "parts."

A good example of the relevance of the statement in Box 42.5.1 is electrical conductivity. In relation to this property, materials are divided into three major categories: **conductors** offer very little resistance to the motion of charges; **insulators** do not allow charges to move in them; and **semiconductors** conduct electricity only when heated up. For a long time the conducting property of materials remained a mystery until the discovery of the quantum theory and the understanding of the *atomic* structure of materials. *Only through a good understanding of the atoms have physicists been able to explain electrical conductivity.*

Here is another "wholistic" question that can be answered only through an understanding of the parts: Why does water dissolve salt but not oil? A water *molecule* consists of one oxygen and two hydrogen atoms. The chemists have shown that because of the atomic structure of H and O, the most stable configuration of the water molecule is when the three atoms form a triangle. This leads to a *polarization* of water molecule (i.e., the center of the positive charges in the nuclei and the negative charges of the electrons are separated). Because of this polarization, certain molecules like salt split into positive and negative ions when mixed with water, and the ions are attracted to the two ends of the water molecules due to the electrical attraction. Other molecules like oil are too tightly bound for water molecule to split them.

42.6   Commonality of Instrumentation

If science deals with matter and is based on observation, there should be a similarity among the instruments used for observing the material structure of the objects of various branches of science, as long as the circumstances of observation are the same. For example, the same kind of machine that was used in the early 1900s to discover the arrangement of atoms in a crystal was also used in the mid 1900s to discover the arrangement of atoms in a DNA molecule.

The era of (crude) observation, which started the scientific age and was done by human senses, is over. Science has become exceptionally refined, demanding extraordinary means for the perusal of matter. If you go to *any* scientific laboratory, as large as a particle accelerator—measuring kilometers in diameter—or as small as a high school chemistry lab, you will see microscopes, telescopes, recording machines, etc., of various sizes and shapes built for a single purpose: to "see" matter at different levels of structure.

There is no instrument that can observe id, Qi, or Ayurveda.

A non-scientific discipline has no (does not need) instruments; and if it does, they have no resemblance to the instruments used in various branches of science. There is no instrument that can observe the id of psychoanalysis, the Qi of ancient Chinese medicine, or the Ayurveda of ancient Indian medicine. Of course, there may be a "decorative" instrument in the office of an alternative medical practitioner—such as the X-ray machine of a chiropractor—to give the *appearance* of a modern facility, but the outcome of the use of the machine has little influence on the course of action taken by the practitioner.

42.7   Reproducibility?

In the nineteenth century, when chemistry became the dominant science of the time, and most of physics could be performed on a table top, the notion of reproducibility—that all scientists equipped with the same apparatus, and arranging the same experimental set-up, should be able to get the same result out of the same table-top experiment—became a primary characteristic of science.

Although all scientists still embrace the opportunity to reproduce scientific results, they realize that the narrow-minded interpretation of reproducibility is not (cannot be) possible. By the narrow-minded interpretation we mean the kind of reproducibility advocated by the creation "scientists." To be convinced of the theories of evolution and big bang, creationists settle for nothing less than the recreation of man or the universe in a laboratory!

Reproducibility, in its more sensible connotation, refers to the reenactment of observation. If one observer looks at Sun and draws scientific conclusions about it, the same conclusions should also be the result of the observation by another observer on the other side of the globe. And if the theory tells us that at such and such stage of the development of Sun, such and such phenomena must have taken place, a rational person cannot expect the claimer to create a Sun to confirm his claims! There are many other ways of testing the claim; for example, by creating—on a small scale in the laboratory—the physical conditions prevalent at that particular stage of Sun's development; or by looking for stars similar to Sun, but younger in age. This is precisely what is being done in laboratories and observatories around the world; and this is how the entire fabric of science is so strongly held together.

Chapter 43

Science versus Pseudoscience

There are many disciplines that use techniques that *appear* to be scientific. For example, we have already mentioned the use of statistics in a business adventure or a survey. Although the media report these as "scientific," neither the surveyors nor the actuaries of the business consider themselves scientists. Such endeavors, while resulting in useful knowledge, have no claim to science.

However, there are other activities that have nothing in common with science, yet their practitioners insist or imply that those activities are scientific. What is worse, somehow the society has accepted their claims, and the practitioners obtain licenses for the practice of their activities! Examples of these activities are all branches of alternative medicine,[1] psychoanalysis,[2] and intelligent design (formerly called creation science). There are also activities that, although not classified (or claimed) as science, have implications that trespass into the scientific territories. Examples of this category of activities are the claim that we have been visited by aliens riding unidentified flying objects, all psychic phenomena, and faith healing. We study the nature of all these activities under the general heading of **pseudoscience**, and in the next section list some of its characteristics.

43.1 Some Characteristics of Pseudoscience

Just as science has characteristics that distinguish it from other human enterprises, pseudoscience also has certain characteristics that, in comparison with those of science, will help us identify it. We shall first list these characteristics and then look at some pseudoscientific disciplines in some detail.

43.1.1 Observation and Reinterpretation

A common characteristic of all forms of pseudoscience is that unlike science, their claims are not based on observation and material evidence, but on authority, hearsay, and unauthenticated reports. Take the example of UFOs. Although there have been countless reports of UFO sightings, and many "first-hand" observers of them, not a single one of them has been accompanied by hard evidence: an extraterrestrial vehicle, a watch, a camera, a piece of clothing, or a toy. All stories of the encounters of various kinds involve one or two

---

[1] While most practitioners of alternative medicine are silent about the scientific character of their practice, others find some scientific basis for it: [Kea 97] identifies some factions of chiropractic as scientific, and [Osc 00] finds the scientific basis of the energy medicine, including touch therapy.

[2] [Ged 05] claims that psychoanalysis is a biological science.

persons who are somehow abducted by aliens, subjected to some extraordinary experience conveniently distant from all potential witnesses, and later released.

Pseudoscience relies on **authority**.

An astrologer or a palm reader relies only on a set of rules and a dogma that all fortune tellers adhere to. Her forecast is based entirely on these rules, and no amount of observation will change them. These rules may have been originated decades or even hundreds of years ago. In fact, the older these rules are the more "authentic" they appear to be. Psychoanalysts refer to the "original" books written by the gurus of psychoanalysis, Sigmund Freud and Carl Jung. They are the masters, the experts, the "authority." Creation scientists refer to the Bible for clues. This book is the authority, the ultimate source of knowledge. Everything we want to know and will ever want to know is in there.

But why do pseudoscientists have to keep returning to the authority? The answer is simple: In light of new observation and evidence, the "rules of the game" need to be *reinterpreted*. And what better place to look for new interpretation than the original "authoritative" source. *While for scientists observation becomes the ultimate authority and theories are modified or replaced to agree with observation, pseudoscientists maintain the old statements of their authorities, but give new meanings to those statements as the occasion arises.*

Reinterpretation is the key for the survival of pseudoscience.

Thus, if the book says that the universe was created in six "days," and modern evidence points to the formation of stars and galaxies that take millions of years, then the word "days" becomes symbolic and could be equated to millions and billions of years. And if the psychoanalytic theory of *Oedipus complex* fails the test of observation and experimentation conducted on many children, psychoanalysts ascribe new meaning to Freud's writings on this subject.

43.1.2   Falsifiability

Perhaps the most important difference between science and pseudoscience is the testability of their claims. *The claims of science are testable (or falsifiable); those of pseudoscience are not.* A very simple-minded (but illustrative) example of a nonfalsifiable statement, which—when cast in a more elaborate form—is not unlike statements made by pseudoscientists is: "Ghosts exist, but cannot be observed or studied by a scientist or a disbeliever." Obviously, such a statement cannot possibly be tested.

Here is a more subtle example of the unfalsifiability of pseudoscientific statements. Uri Geller, a psychic claiming to have supernatural power in the 1970s, was once invited to appear on the Tonight Show to demonstrate his paranormal ability. He had claimed that he could read people's minds, bend spoons and keys by merely rubbing them, determine the content of a container without looking inside, to name a few. On the advice of James Randi, a magician and debunker of claims of paranormal, Johnny Carson, the host of the show, took every precaution to ensure that the psychic had no access to the props prior to his appearance. Thus, the crew of the show had put some drawings in sealed envelopes, and had filled a few containers among many with water and covered them tight before Uri Geller appeared on the show. When presented with this set of objects, Uri Geller was incapable of predicting the subjects of the drawings or identifying the containers with water. Feeling embarrassed, his only response was "I am always capable of mind reading, spoon bending, and predicting the contents of containers without looking at them. However, *tonight* I feel pressured, and only momentarily I have lost my powers." There is simply no way that this statement can be tested or falsified.

43.1.3   Abusing Limitation of Science

Another important difference between science and pseudoscience, related to falsifiability, is that *science carries the burden of proof of its claims while pseudoscience places this burden on science.* And because the proof entails demonstrating that something does not exist, it becomes *logically impossible.* Let us dissect this a little further.

To prove that an apple *exists*, all I have to do is fetch one from my refrigerator. Science works this way: it proves the positive. But to prove that Elvis does NOT visit Graceland every midnight is an impossible task. If I say that nobody has seen him, many people may come forward and claim that they have seen Elvis at midnight. If I suggest camping at Graceland at night to catch Elvis when he shows up, and the attempt fails, I will be told that I did not look hard enough, or that he came but he hid behind trees and bushes. If I suggest illuminating the entire Graceland with powerful searchlight at midnight, and the experience fails again, I'll be told that I ruined everything, that he does not like bright light. And this can go on and on. In fact, there are an *infinite number* of "explanations" for why Elvis does not show up when I am looking for him in Graceland at midnight.

Elvis in Graceland!

> **Box 43.1.1.** *While for the proof of the <u>existence</u> of something you need only one instance of that thing, to prove the <u>nonexistence</u> of something, you would have to eliminate an <u>infinite</u> number of possibilities—an impossible task!*

The two camps of theorists and experimentalists in science are constantly watching each other. If theorists claim something, the experimentalists immediately go to their labs to test the theorists' claim. This professional competition is a beneficial symbiotic relation that is advantageous to both camps and to the rest of humanity, and drives science to newer and newer frontiers.

Science watches over itself!

In complete contrast, pseudoscience is *characterized* by its exploitation of the limitations of science. In fact, the very existence of pseudoscience depends on this limitation. The smaller this limitation, the less chance for the survival of pseudoscience. Prior to the advent of modern science—in the Dark Ages—pseudoscience was rampant. Many natural phenomena were associated with the supernatural because there was no scientific explanation for them. An eclipse would be interpreted as the wrath of god. A comet would be the messenger of pestilence and destruction. Today, when we can predict the occurrence of an eclipse or the arrival of a comet to within a fraction of a second, pseudoscience and superstition have a much smaller space in which to maneuver. But space they have! Knowledge is like a huge ocean which we have just begun to explore. Scientists have discovered only a small portion of a small region close to its shore. Pseudoscience has the rest of the ocean to exploit.

Pseudoscience exploits the limitations of science.

43.1.4   Exploiting Mystery

Pseudoscience thrives on mysteries while science tries to solve them. As science expands the borders of knowledge, pseudoscience gropes for new territory. At one time astrology had a huge base of support among the public. In fact, as late as the seventeenth century, well known astronomers such as Copernicus and Kepler studied the stars to control and influence earthly phenomena. Over the course of many decades, it became clear that there was no scientific basis for a connection between heavenly motions and earthly happenings. So, astrology was confined to narrower and narrower spheres until it shrunk to the confines of the psychics, fortune tellers, and some politicians!

However, our understanding of the universe opened up new territories for pseudoscience. If our Sun is only a star among billions of other stars, then what prevents us from assuming that there are other planets revolving around other suns and supporting intelligent creatures who may have a civilization far more advanced than ours. And if these creatures are advanced enough in technology, is it not reasonable that they have ventured out of their planet in search of other civilizations? Is it not sensible to assume that this search has brought them to our planet?

Scientists do not deny the possibility of alien visits; they deny the actuality of such visits due to lack of any material evidence.

All these claims may be true. In fact, many scientists believe in the possibility of intelligent life somewhere in the universe, maybe even in our own galaxy. As scientists, they cannot outright deny the *possibility* of the visitation of Earth by those intelligent beings. What they *are* denying is the *actuality* of such a visitation due to the nonexistence of any *material evidence.*

It is in the nature of science to open up new questions every time it stumbles over a new discovery or a new world outlook. Sometimes these questions could remain unanswered for decades. In the mean time, it does not take a whole lot of imagination for the scientifically uninformed—but otherwise motivated—people to fabricate stories, forge theories, and manufacture far-fetched ideas that can live in the world of unanswered questions until those questions find answers and the cycle starts all over again.

Box 43.1.2. *As long as science has not found all the answers (will it ever?), as long as even a smidgen of the unknown survives (will it ever not?), there will be stories of UFOs, of extraterrestrial creatures, of ghosts and poltergeists, and of faith healing and psychic surgery.*

43.2   Intelligent Design

One of the biggest public debates in America, next to abortion, is evolution vs creation. Creation "scientists"[3] believe that there is scientific support for the biblical story of creation and that the theory of evolution is wrong. They overlook the fact that there are as many "stories" of creation as there are religions and ancient—sometimes extinct—civilizations. If creation is to be a scientific theory, it has to be universal. The theory of relativity is as valid in the US as it is in Argentina, Brazil, Yemen, or Zimbabwe. But the story of creation in which a Buddhist, a Christian, a Jew, a Moslem, a Zoroastrian, or a Hindu believes is different from the rest. Why should a particular story be singled out as *the* "scientific" creation?

There are *many different* stories of creation.

Any science starts with observation. Is it not, therefore, logical to ask whether the biblical stories have any observational truth to them? If these stories are truly based on observation, how long ago were these observations made? Two thousand years ago? Four thousand years ago? If so, should we rely on the conclusions derived from such ancient observations? Should we, therefore believe in the (3000-year-old) conclusion that Sun and all the (five!) planets move around Earth? That there are only four elements: water, Earth, air, and fire? That the heaven is made of crystal spheres to which celestial bodies are attached, and that the comets must shatter these crystals to enter our view?

Creationism (or intelligent design theory) claims to have the answer to the riddle of creation. It aims at replacing the scientific answer, i.e., the theory of evolution. It is therefore instructive to examine these two theories in some detail.

43.2.1   What Is Evolution?

The story of creation is without a doubt the oldest story of humankind, and the question of how humans came into existence is probably the first one asked by the conscious man. We now have a surprisingly successful *scientific* story of creation: the **big bang** theory of Part VIII, which originated in Einstein's general theory of relativity. The answer to the question of the creation of man came in 1859 when **Charles Darwin** published one of the

---

[3]Intelligent design is a new "cover" for creationism in that it does not directly refer to the Bible, but to the intervention of some form of intelligence in the formation of life on Earth.

masterpieces of scientific literature, *The Origin of Species*, in which he put forth the theory of evolution, one of the most revolutionary scientific theories ever.

Darwin did not hypothesize evolution in the form of a dogma or a random wild hypnotic dream. Darwin's theory, like all scientific theories, was based on a large number of *observations* he had made over 20 years earlier. Examination of the flora and fauna of the Galapagos islands, which Darwin encountered in 1835 on the voyage of *H.M.S. Beagle*, initiated his speculation of the evolution theory.

Evolution is based on observation.

Although the seed of the theory of evolution had been germinating in Darwin's mind as early as 1835, and he pondered over them for 24 years, it was only after he realized that he was not alone in thinking in terms of evolution—that other naturalists, notably Alfred Russel Wallace, had also come up with the same theory—that Darwin hastened to publish his ideas in 1859.

Like a legitimate science, evolution was discovered simultaneously and independently by at least two scientists (Section 42.3).

The essence of evolution is that different species have a common ancestor. The mechanism operative in this theory is, according to Darwin himself, "modification of the descendants of any one species" and "[n]atural selection [which] leads to divergence of character and to much extinction of the less improved and intermediate forms of life" [Rus 88, p. 97].

With the advent of genetics, the dynamics of evolution now possesses a molecular explanation. In fact, Darwin himself thought of a *material* transmission of traits from parents to offsprings. It is now well established that genes are the agents of this transmission. Although all the characteristics of the parents are more or less preserved in the transmission, once in a while a **mutation** takes place inside the genes causing the offspring to be different from the parent. If the mutation assists the survival of the offspring, it will persist in future generations and new species will evolve. This is the "natural selection" component of evolution.

Darwin thought that traits are *materially* transmitted from parents to offsprings. Genetics provides the dynamics of this transmission.

The fact that genes mutate has been proven both in the lab and in the birth defects of the offsprings of humans as well as other animals. The fact that *selection* is operative is evident from the variety of "pets" and cattle that humankind have bred. Even a superficial look at the different breeds of dogs around us reveals the astonishing power of (artificial) "selection." All these breeds have been "created" out of a single variety, which thousands of years ago was what we now call wolf.

43.2.2   What Is Creationism?

Creation "science" states that all the variety of animals were created exactly in the form in which they exist now by a single all powerful creator. They insist that evolution is not a science because "no one observed the origin of the universe, the origin of life, the conversion of a fish into an amphibian, or an ape into a man" [Gis 73, p. 3].

As we believe in Tutankhamen because of the (indirect) evidence provided by pyramids and tablets, so do we have to believe in evolution because of the (indirect) evidence of variety and similarity.

What are the implications of this statement? Has anyone "observed" King Tutankhamen, or Hammurabi? No. Yet creationists have no qualms about the fact that they lived 3000 or 4000 years ago. Why does anybody believe in these historical facts? It is not that someone who can speak English has survived 30 or 40 centuries and has given us an eye-witness account of the lives of the Egyptian and the Babylonian kings! *It is because of the evidence!* This evidence is in the form of pyramids, tablets of wall paintings, sculptures, and hieroglyphic writings or cuneiforms.

To understand these tablets, one inevitably *ought to interpret them*. One interpretation is that some alien civilization has placed them on Earth. (There *are* some who believe this "creationist" version of the evidence!) The other interpretation—based on countless observations by many generations of archaeologists—would argue that primitive humans, who at first had no writing techniques[4] settled in the valleys of the Nile and Tigris-Euphrates

---

[4]Because radioactive carbon dating proves that the oldest tablets with writing on them go as far as 5000 or 6000 years ago while the existence of primitive human beings capable of drawing goes back as far as 15,000 years ago.

rivers, and gradually developed the skill of writing, and these tablets are their ways of telling us their stories.

Every observation depends on the existing theories. So, when we observe the Egyptian hieroglyphic tablets or the Babylonian cuneiforms, we have to accept the linguistic "theory" gained by a careful and critical examination of the tablets by experts, in the process of which hieroglyphics gets translated into modern languages. Then, we can safely say that the tablets relay the story of a king or a queen or a prince that lived 4000 years ago. And no aliens will be required to place the tablets and the pyramids on Earth.

There are countless pieces of evidence, both live and fossilized, pointing to the validity of evolution, just as there are countless pieces of evidence pointing to the validity of the theory that Hammurabi ruled Babylonian empire about 4000 years ago. Scientists explain the evolutionary pieces of evidence in terms of mutation and natural selection, just as historians explain the archaeological pieces of evidence in terms of social and political progress. And just as there are people who believe that pyramids were *created* by aliens from outer space, so do creationists believe that species were created by an intelligent designer.

How do creationists "explain" the origin of species, once they discard evolution? If evolution cannot create a reptile out of a fish, they argue, then creation is the only alternative. However, the same logic that disposes of evolution is used in *support of* creation: "We do not know how God created, what processes He used, *for God used processes which are not now operating anywhere in the natural universe*.... We cannot discover by scientific investigations anything about the creative processes used by God" [Gis 73, p. 25].[5] The italic, by the way, is original!

43.2.3 The Design Argument

One of the oldest arguments against evolution is the impossibility of the occurrence of complicated design such as life forms without a "designer."[6] In his *Natural Theology*, the eighteenth century theologian, William Paley writes:

> In crossing a heath, suppose I pitched my foot against a *stone*, and were asked how the stone came to be there; I might possibly answer, that ... it had lain there for ever: nor would it perhaps be very easy to show the absurdity of this answer. But suppose I had found a *watch* upon the ground, and it should be inquired how the watch happened to be in that place; I should hardly think of the answer which I had before given, that ... the watch might have always been there.

What are the odds that a monkey at a typewriter types a single sentence from Hamlet?

Paley then concludes "that the watch must have had a maker." If the watch has a maker, shouldn't such qualitatively and quantitatively more intricate a structure as the human eye, or ear, or heart, or brain?

The impossibility of a complicated structure to come together on its own is best described by the famous example of a monkey and a typewriter. How likely is it that a monkey bashing away on a typewriter can produce Shakespeare's Hamlet? To grasp the smallness of the likelihood of this occurrence, let us drastically reduce our expectation, and look into the probability that the monkey can produce a small—very small—phrase from Hamlet. In a conversation with Polonius over the shape of a cloud, Hamlet says: "Methinks it is like a weasel." Including the spaces between words and excluding the period at the end, the phrase consists of 28 letters. Let's ignore the complication arising from the case of each letter and assume that only capital letters are available. What is the probability that a monkey at a typewriter can produce this sentence? **Math Note E.43.1 on page 154** of *Appendix.pdf* shows why it would take a monkey 12 trillion times the age of the universe to do this *even if it had the speed of a computer*!

---

[5]This is a masterpiece of unfalsifiable claims! The statement says: All creatures were created by a creator but there is no way that science can discover this "truth."

[6]The discussion of the design argument is taken from [Daw 87, pp. 43–63].

It is these staggering numbers that the advocates of the design argument use to refute evolution. However, there is a flaw in this kind of argument. We have to distinguish between *single-step* selection and *cumulative* selection. In the single-step selection, the random choices made in each trial are completely independent of the previous trials. This is what we calculated in Math Note E.43.1. The selection involved in evolution is not single-step. The "letters" of life replicate themselves and the end product of one generation of selection is the starting point of the next generation.

To mimic the cumulative selection, let us assume that our computer monkey begins by choosing a random sequence of 28 letters as before. However, it now "breeds from" this random phrase. It duplicates it repeatedly, but with a certain chance of *random error*—"mutation"—in the process of copying. The computer examines the mutant nonsense phrases, the "progeny" of the original one, and chooses that which, *however slightly*, resembles the target phrase. This goes on generation after generation. In actual runs, the target is usually reached in less than 100 generations in at most a few minutes. A far cry from trillion ages of the universe!

Other computer simulations mimicking the action of genes have shown a more intriguing results. In one computer model, Richard Dawkins[7] chose nine genes, each represented by a number in the computer. The simulation draws one line segment (or branch) at a time, and the genes tell the computer how to draw this branch. For instance, one gene might influence the angle of branching, another, the length of some particular branch, and a third, the depth of the recursion, the number of successive branchings.

In every generation, the program takes the genes that are supplied to it by the previous generation, and hands them on to the next generation but with minor random errors—mutations. A mutation simply consists in +1 or −1 being added to the value of a randomly chosen gene. Although the mutations are random, the cumulative change over the generations is not random. The progeny in any one generation are different from their parent in random directions. But which of those progeny is selected to go forward into the next generation is not random; it is selected by the eye of the person at the computer. This is where Darwinian selection comes in.

Mutations are random, but the selection of the progeny is not.

The human tells the computer which one of the current litter of progeny to breed from. The genes of the chosen one are passed across to the simulation, and a new generation begins. This process, like real-life evolution, goes on indefinitely. Each generation of these so-called "biomorphs" is only a single mutational step away from its predecessor and its successor. But after 100 generations of *evolution*, much can happen. How much? Richard Dawkins writes:

"Evolution" of insects on the computer screen!

> When I wrote the program, I never thought that it would evolve anything more than a variety of tree-like shapes. I had hoped for weeping willows, cedars of Lebanon, Lombardy poplars, seaweeds, perhaps deer antlers. Nothing in my biologist's intuition, nothing in my 20 years' experience of programming computers, and nothing in my wildest dreams, prepared me for what actually emerged on the screen. I can't remember exactly when in the sequence it first began to dawn on me that an evolved resemblance to something like an insect was possible. With a wild surmise, I began to breed, generation after generation, from whichever child looked most like an insect. My incredulity grew in parallel with the evolving resemblance. .... I still cannot conceal from you my feeling of exultation as I first watched these exquisite creatures emerging before my eyes. I distinctly heard the triumphal opening chords of *Also sprach Zarathustra* in my mind. I couldn't eat, and that night "my" insects swarmed behind my eyelids as I tried to sleep.

In the Appendix of *The Blind Watchmaker*, Dawkins shows a collection of the bimorphs that he created on his computer desktop using a software developed by himself and included on the disk accompanying the book. The collection sample shapes have a striking

---

[7]See [Daw 87, pp. 50–74].

resemblance to a jumping spider, eagle, ant, planktonic larva, moth, and a host of other shapes.

43.3 Psychoanalysis

Sigmund Freud, the founder of psychoanalysis, is probably the most influential figure of the twentieth century. Psychoanalytic approaches have been applied to such widely diverse fields as history, political science, literature, music, and the arts. Nonetheless, the basis of psychoanalysis is pseudoscientific. This observation came about very slowly and, unfortunately, very ineffectively.

The dogma of psychoanalysis.

The entire discipline of psychoanalysis is based on the *dogma* of the division of consciousness into three levels. First, there is the conscious level; "below"[8] this lies the preconscious. "Below" the preconscious lies the unconscious. According to Freud, unconscious contains memories, desires, and feelings that have been repressed by the individual because they would be too traumatic or painful to face directly due mostly to sexual inhibition.

Freud's theory also deals in great detail with the development of personality and specifically with the development of sexual behavior and sexual identity. He divided personality into three "structures:"

The dogma of the division of personality.

- The **id** is thought of as the "seed" of personality. The newborn has only an id and the other structures of the personality develop from it. The id is also the most animalistic part of the personality, seeking only to obtain pleasure and avoid pain.

- The **ego** develops as the child grows. It is reality-oriented and modifies or controls the desires of the id by taking into account the possible consequences of an action.

- The **superego** is the conscience of the individual. It judges whether an action is right or wrong, according to whatever set of moral standards the child has been taught.

These three structures of personality interact in complicated ways. For example, the ego may postpone the gratification of the id, and the superego may "battle" with both the id and the ego because behavior often falls short of the moral code it represents. But more often in the normal person, the three work as a team, producing integrated behavior.

What evidence dictates this kind of division and the interaction among them is not clear. Why not add another structure between id and ego, or between ego and superego? In fact, in order to allow for more possibilities of interactions, perhaps we should let personality have a multilevel structure. Divide the personality, in order of decreasing "animalisticity," into id, idd, iidd, subego, ego, superego! Or some equally arbitrary multilevel structure!

Two major characteristics of science, materiality and observation, are lacking in psychoanalysis.

Two major characteristics of science are emphasis on *observation* and the *material* nature of its objects of study. These two characteristics of science are conspicuously absent in psychoanalysis. Freud never made any observation of any material object from which to deduce psychoanalysis. His only object of "study" was personality and mind.

When a discipline starts with a dogma that is remote from and insensitive to observation and experimental data, interpretation and reinterpretation becomes an essential part of the discipline as we saw in Section 43.1.1; and psychoanalysis is no exception. By constantly reinterpreting its content, psychoanalysis can accommodate every truth (or falsehood). Every new case, regardless of its nature, is a new confirmation of the psychoanalytic theory. While a valid scientific theory, such as the general theory of relativity, makes the *risky* prediction of the bending of light as it grazes the Sun, psychoanalysis is risk free, because it makes either no prediction, or every possible outcome can be *interpreted* as a prediction after the fact!

---

[8]It is not clear whether the word "below" has any real spatial connotation!

43.3.1 Neo-Freudians

Let us put psychoanalysis to the test of another characteristic of science: its detachment from the scientists. The great ideas of science are so impersonal that scientists rarely refer to the original writings of the discoverers. Because of this detachment, the idea will live a long time after the discoverer dies. The theory of relativity is just as valid today as it was in 1905 when Einstein discovered it. Neither the "charisma" of Einstein nor the "cult of personality" of Newton has had any effect on the validity and acceptance of relativity or the laws of motion.

What happened to psychoanalysis after (even at the time of) Freud? In *The History of Psychoanalysis* [Fin 90, p. 77], by Reuben Fine, himself a psychoanalyst, we read "From the very beginning of psychoanalysis, schisms and dissensions within its ranks have been notorious." Should any reader with even a slight familiarity with the workings of science not wonder "Why the schisms and dissensions?" Can the opposing parties not bring their disagreements to the court of observation and let this ultimate judge of all sciences rule out all claims except one to which all must adhere? We don't hear anything about the "schisms and dissensions" in the ranks of relativity theory! The history of relativity has no chapter on the split among its rank due to disagreements on the power of c in $E = mc^2$! Or on what Einstein meant by the letter c in that formula!

Disagreement, schisms, and dissensions occur only in disciplines in which observation, the most fundamental ingredient of science, is absent. This frequently occurs in politics and religion where *dogma* replaces theories and *interpretation* of dogma replaces observation. The similarity between religion and psychoanalysis is very well described in the following statement:

> The historian must offer some rational explanation of these innumerable splits and divisions, which are paralleled perhaps only in the history of religion. The analogy is by no means inappropriate, for psychoanalysis has also been dominated by a long series of charismatic figures, like the religious sects. When the works of these charismatic figures are carefully pursued, they often make little or no sense .... [Fin 90, pp. 78]

Charisma playing a dominant role in science? Are relativists to be mesmerized by the charm of a "charismatic figure" into believing that $E = mc^3$? or $E = mc^4$? Or perhaps $E = mc^{2.5}$? And why are these charismatic figures allowed to make statements that "make little or no sense"? Why aren't their works "carefully pursued" to begin with? Is there no yardstick in psychoanalysis with which to measure the validity of psychoanalytic statements? Psychoanalysis is so arbitrary that "neither Jung nor Adler grasped the bases of the first psychoanalytic system" [Fin 90, p. 78].

Here is a historian of psychoanalysis, himself a psychoanalyst, who proclaims that even Jung and Adler, the two most prominent psychoanalysts after Freud, did not grasp "the bases of the first psychoanalytic system." In other words, only Freud could grasp psychoanalysis, i.e., *psychoanalysis cannot be detached from Freud.*

Psychoanalysis lacks another characteristic of science: its detachment from the scientist.

The history of psychoanalysis after Freud is a history of sectarianism. We read about the hatred that Freud developed toward Adler after the latter showed a mild independence in the discipline; about the love-hate, father-son, and "unanalyzed homosexual transference" relation between Freud and Jung [Fin 90, pp. 79–87].

But the best story is that of Wilhelm Reich, who "up to 1933 was one of the leading figures" in psychoanalysis, and whose book *Character Analysis* "made a significant contribution to the understanding of character." In his attempts to combine psychoanalysis and Marxism, Reich developed a theory which he called "orgonomy," and for which he found some followers. In later years he saw "orgone energy" as the antidote of nuclear energy and envisioned himself as the savior of mankind whom the FBI was protecting because of his knowledge of secrets. He was convicted of fraud regarding the "orgone box," leading to his incarceration in a federal prison in which he died in 1957 [Fin 90, pp. 108–109].

We are not attacking Reich's personality here. Even a true scientist may be attracted to the realm of pseudoscience as Newton was to alchemy. However, Newton's attraction to alchemy was completely detached from his scientific discoveries, and no other physicist followed him into alchemy. On the other hand, the very fact that Reich attracted some followers means that his "weird" ideas were the natural evolutionary products of his earlier (more widely accepted) theories.

The discipline of psychoanalysis allows so much religion-like interpretation that "whenever anything out of the way is presented in analysis, the originator avoids criticism by saying that he represents a different 'school'" [Fin 90, p. 87]. Adler, Jung, and Reich were no exceptions; they were the rule. Nowhere in the history of science does one read about scientists creating "schools." You don't read about "schools of light power," each school "believing" in a different power than 2 in $E = mc^2$! We see this kind of flourishing of schools over and over again, not only in psychoanalysis, but in any "scientific" field founded by a single person and based on a "dogma." We see the same phenomenon in chiropractic, founded at the end of nineteenth century by Daniel David Palmer, an American grocer, phrenologist, and magnetic therapist.

As long as the "science" is attached to the personality of its founder, as long as one has to go back to the founder to verify the correctness of "scientific statements," as long as students of the discipline—even a hundred years after its "founding"—are constantly referred to the work of the founder, and as long as there is no solid observational yardstick to check the claims of the "leaders," the word "science" is only a decorative emblem attached to the discipline to attract new "disciples." At no time in the education of a physicist is he/she referred to the original work of Newton, Planck, or Einstein.[9] Why? Because gravity, quantum theory, and relativity are completely detached from Newton, Planck, and Einstein. And all students who study, say, relativity deal with exactly the same set of concepts. It does not matter whether these students speak the same language, live in the same place, and are close or far from where Einstein lived. There is no need to refer to Einstein's original work on relativity (and if you do not know German you cannot anyway), because the theory of relativity can be taught in *any* language.

The dogmatic nature of psychoanalysis, as in all other dogma-driven disciplines, leaves its door open to every possible unchecked whim. Since the "master" has founded the entire discipline on fancy, why should the loyal disciples refrain from proposing genuine fanciful "theories." When Freud himself arbitrarily divides human psycho into id, ego, and superego, it is only natural for Jacques Lacan, one of his most prominent French followers, to claim that the diversity in the geometrical and topological structures of sphere, Möbius strip, and the Klein bottle,[10] "is very important as it explains many things about the structure of mental disease" [Sok 98, p. 19].

Now, a Möbius strip, and a Klein bottle are geometrical constructs—invented by mathematicians after whom the objects are named—with some strange and counter-intuitive properties, for the demonstration of which the invention took place to begin with. These constructs are utterly products of imagination which, so far, have not found applications in even the most mathematical of all sciences, physics. When mathematics find its way into physics, it is in a most natural and compulsory manner. If Newton used calculus in his description of motion, it was because it was natural (even necessary) to use it. If Maxwell used vector analysis in his study of electromagnetic theory, it was because nature practically forced it on him. Einstein did not use differential geometry for his general theory of relativity simply because differential geometry was "fashionable." In fact, if anything, it was the general theory of relativity that made differential geometry fashionable by heralding its relevance to the workings of nature.

---

[9]Unless, of course, he/she is interested in the history of science.

[10]The use of the vocabulary of modern mathematics and physics has become fashionable in some postmodern disciplines. It is not uncommon to encounter words like "nonlinear," "topology," "quantum," and "field" in the texts of postmodern writers.

To cast aside any doubts that Lacan's claims may be only metaphors, we quote his response to a question suggesting that the mathematics and the topology used by Lacan may be "an analogy for an explanation of the life of the mind." Lacan replies: "...It is not an analogy. It is really in some part of the realities, this sort of torus.[11] This torus really exists and it is exactly the structure of the neurotic. It is not an analogon; it is not even an abstraction, because an abstraction is some sort of diminution of reality, and I think it is reality itself" [Lac 70, pp. 195–196].

So, why is Lacan using *the language* of topology, geometry, and mathematics in his psychoanalytic study of the mind? Is there some rational necessity for its usage? Is psychoanalysis forcing these mathematical concepts on him? Does he gain simplicity and clarity by their usage? Obviously not! Lacan's use of mathematics as his medium of caprice may be his futile ambition of bestowing exactness to the pseudoscience of psychoanalysis. After all, physics uses mathematics, and it is considered an exact science. So, why not *impose* mathematics on psychoanalysis and give *it* the appearance of an exact science as well!

43.3.2 How Can Psychoanalysis Become a Science?

Psychoanalysis analyzes (i.e., studies) the "psyche" or the mind. Is there any "matter" associated with the mind? Although some self-proclaimed "scientists" try to separate the mind from any form of matter, a proper scientific investigation of the mind first identifies its material basis. Obviously, the matter of the mind is the brain. Therefore, if psychoanalysis is to become a science, it has to study the brain.

Even a cursory look at the brain reveals that it is a very complicated "whole" consisting of several equally complicated structures. When each of these structures is dissected further and further, it is seen that the building block of the brain is a **neuron**, a nerve cell. Thus, psychoanalysis is "reduced" to the understanding of neuron.

Before the "holistics" snap at this conclusion, we have to emphasize immediately that understanding neurons is only the first step. As clearly indicated in Box 42.5.1 and the discussion before and after it, to understand the brain and its functions (mind), you have to understand the neuron. But that is not the end! After (and *only after*) you understand the neuron, you have to see how they interact, communicate, send chemicals, etc. among one another. Psychoanalysts do not follow this procedure; that is why they are not scientists. But there *are* scientists who are actively pursuing the understanding of neurons; these are the neurologists, the neurobiochemists, the molecular biologists, the biophysicists, and others.

Box 43.3.1. *A truly scientific investigation of the "mind" will take the **material** organ of the mind—the brain—and concentrate on its building blocks, the nerve cells. After (and **only after**) the nerve cells are understood, can one put them together to get a complete picture of the "whole" brain.*

A neuron is a very complex arrangement of molecules, and its understanding will not come easy. It will probably be decades maybe even centuries before we completely fathom the workings of this (or any other) cell. *Only then* can we ask how they are put together to form the brain and how this formation constitutes the mind. But the difficulty of the task ahead of us should not be a reason for hastily inserting pseudoscientific techniques into the science of the brain.

Should we wait centuries for a full comprehension of the brain before we treat the mentally ill? Of course not! The absence of the scientific knowledge of the human body

---

[11]A geometrical object in the shape of a doughnut. It is obtained by gluing the two ends of a cylinder. If the two ends are "twisted" before gluing, a Klein bottle is obtained.

did not prevent medical doctors from treating patients throughout history. Any technique *proven to be effective* in the treatment of mental diseases should not be prevented simply because of our lack of knowledge of how it affects the neurons. However, when there is a choice between a drug that is based on the scientific research of the chemistry of the brain and an herb or a procedure that is based on the practice of an alternative medical doctor, the former should be selected.

43.4   Alternative Medicine

In 2007, 85 million Americans spent some $34 billion for therapies ranging from acupuncture to zone effect.

One of the first items on the priority list of the American public is physical and emotional well-being. The appetite and the desire for good health and good life has made the public alarmingly vulnerable to medical quackery. In 2007, 85 million Americans spent some $34 billion for therapies ranging from acupuncture and Ayurveda to yerbera and yoga [Nah 07]. In 1993 the US Senate created the Office of Alternative Medicine in the National Institutes of Health, and in 2000 it promoted the "Office" to the National Center of Complementary and Alternative Medicine and increased its budget to $68.7 million [NIH website].

According to Wayne Jonas, one of the past directors of OAM, there are "people in both extremes, advocates who believe in their alternative therapy wholeheartedly and not in science, and skeptics who disbelieve it wholeheartedly and believe in science. The truth is that reality is somewhere in between." In the August 1997 issue of *Nature: Medicine* he gets a chance to respond to critics of the OAM.[12] Referring to the homeopathic claim that biological information can be stored and transmitted by water and wires, he asks, "Even though this concept is implausible, the potential implications it holds for understanding basic biological and cellular communication are enormous. Can we not afford to invest a small amount in pursuit of this question?" If we were to apply this principle to the claim that gold could be produced by puffing at dirt, it would read something like this: "Even though the concept of producing gold by puffing at dirt is implausible, the potential implications it holds for economy and commerce are enormous. Can we not afford to invest a small amount in pursuit of this question?" If we were to apply this principle to the concept of flying it would read something like this: "Even though flying by flapping ones arms is implausible, the potential implications it holds for human transportation and urban traffic are enormous. Can we not afford to have a few people jump from Empire State Building while flapping their arms?"

The popularity of alternative medicine is a reflection of the scientific unawareness of the American public, its inability to distinguish between real and false medicine, and its hearty reception of pseudomedicine "doctors." As part of the task of this book, we have chosen to expose only three forms of the alternative medicine: chiropractic, therapeutic touch, and magnetic therapy. A fourth form, quantum healing was discussed in Chapter 35. Lack of space does not allow consideration of other forms of alternative medicine, but the pseudoscientific bases of all branches of alternative medicine are so similar that our handful of examples suffice to show the true nature of the entire discipline.

Before going into the details of the specific branches, we need to understand the nature of the human body's reaction to drugs and drug substitutes. In particular, we have to know something about the so-called *placebo effect*.

43.4.1   The Placebo Effect

A lot more science is used in medicine today than a hundred years ago.[13] Nevertheless, remnants of the old unscientific practices still linger on. For example, doctors and their

---

[12]The National Center of Complementary and Alternative Medicine has turned away from its earlier practice of advocating alternative medicine. It is now investigating the efficacy of alternative practices, and, in the process has undermined certain claims of alternative practitioners.

[13]This subsection is taken from [Bro 98].

patients continue to ascribe healing powers to pills and procedures that have no intrinsic therapeutic value for the condition being treated. Consider, for example, the widespread—and medically pointless—use of antibiotics to fight colds and flu caused by viruses. It is not that these treatments do not offer benefits: most of them do. But in some cases, the benefit may come from the **placebo effect**, in which the very act of undergoing treatment—seeing a medical expert or taking a pill—helps the patient to recover.

A typical medical research proceeds with a so-called **double-blind study**: in a group of patients participating in the study, some are treated with a placebo (a pharmacologically inert capsule or injection), and neither the doctors nor the patients (thus the term double-blind) know who receives the placebo and who the drug. A landmark study in the early 1950s by a Harvard University team, and several studies of depression, suggested that

Double-blind study.

> **Box 43.4.1.** *For a wide range of afflictions, including depression, pain, high blood pressure, asthma, and cough, roughly 30–40% of patients experience relief after taking a placebo.*

In some cases, the response can be even more pronounced: researchers of the University of Kansas Medical Center in the late 1950s investigated the effectiveness of the then routine arterial ligation surgery to treat angina pectoris (chest pain caused by insufficient blood supply to the heart). The doctors performed the surgical procedure in one set of 13 patients; with a second group of five patients, they made only a chest incision but did no further surgery. Among the patients who received the actual surgery, 76% improved. Notably, 100% of the placebo group got better! (Arterial ligation surgery is no longer performed.)

Placebo effect shows itself even in surgical procedures!

The healing environment is a powerful antidote for illness. The decision to seek medical assistance restores some sense of control. The symbols and rituals of healing—the doctor's office, the stethoscope, the physical examination—offer reassurance. And merely the act of taking a pill can have a therapeutic effect. For example, the drug propranolol is often prescribed after a heart attack to regulate the heartbeat and prevent further damage. In a study of more than 2000 patients, the death rate was cut in half among patients who took propranolol regularly compared with those who took the medication less regularly. But in the same study, *patients who took placebos regularly also had half the death rate of those who took them less regularly.*

Notably, placebos seem to be most reliably effective for afflictions in which stress directly affects the symptoms: in certain forms of depression and anxiety, for example, distress is the illness. And conditions such as pain, asthma, and moderate high blood pressure can become worse when the patient is upset. Indeed, placebos may work in part by lessening the apprehension associated with disease. Studies of both animals and humans have shown that the functioning of the immune system falters under stressful conditions. For instance, stress increases the secretion of hormones such as cortisol, which in turn lowers resistance to disease. It is not inconceivable that by reducing anxiety, placebos could influence countless diseases, including some that we do not usually think of as subject to psychological influence.

A patient's expectation of improvement is also crucial. Researchers know that across a wide range of illnesses, patients who think they will feel better are more likely to do so. Expectation operates more specifically as well. A 1968 study done by a team of researchers at the State University of New York Downstate Medical Center in Brooklyn showed that patients with asthma who were given an inhaler containing only nebulized saltwater but were told they would be inhaling an irritant or allergen displayed more problems with airway obstruction. When the same group was told that the inhaler had a medicine to help asthma, their airways opened up!

Dramatic illustration of the power of placebo effect.

The placebo effect is now understood to be due to the release of endorphins, the same substances responsible for pain reduction caused by the physiological stress of excitement. That the placebo effect is a real chemical reaction has been demonstrated by eliminating it with the injection of the drug naloxone which neutralizes endorphins. It has also been shown that the release of endorphins in the placebo situation is a classically conditioned response that can be taught to rats.

It may well be that in all "alternative" treatments, it is the placebo effect at work rather than the procedure. Very few practitioners of alternative medicine have been willing to subject their practice to a double-blind study. In the few cases that they have,

> **Box 43.4.2.** *Double-blind studies have shown that "alternative" treatments are no more effective than placebo.*

Acupuncture: it doesn't matter where you stick the needles!

For example, the May 4, 2005 issue of the Journal of the American Medical Association (JAMA) reports a randomized, controlled trial, in which the effectiveness of acupuncture is compared with sham acupuncture in treating migraine, which is touted as one of the great successes of acupuncture. There were 302 patients in the study. In 12 sessions over 8 weeks, sham acupuncture, in which the needles are inserted in the "wrong" points, was just as effective as inserting them in the "correct" points.

43.4.2 Chiropractic

A very popular branch of alternative medicine, one that has exhibitions in the malls, advertises in rented booths in county fairs, and has multiplied like mushrooms in the yellow pages of all urban communities in the US, is chiropractic. The entire profession of chiropractic is based on the *assumption* that every pain and disease (or "dis-ease," as chiropractors call it) is caused by the nerves pinched between the bones of the spinal column.

A grocer in Iowa becomes a phrenologist and later, the founder of chiropractic.

The founder of chiropractic is Daniel David ("D. D.") Palmer (1845–1913).[14] Palmer started as a grocer in Davenport, Iowa, but soon was attracted to phrenology,[15] and eventually practiced as a full-time "magnetic healer." To attract patients, D. D.'s brother ran articles in newspapers claiming that D. D. cured patients simply by the motion of his hands, and that he could cure tumors and cancers without medicine. His technique was to locate the dysfunctional organs and to impart a "life force from his hands into that dormant organ, thereby assisting it to throw off the unnatural condition."

Chiropractic dates its origin to September 18, 1895, when Palmer claims to have manipulated a spinal bone of Harvey Lillard, a janitor where he had his office, and curing him of a 17-year-old deafness instantly. "Shortly after this relief from deafness, I had a case of heart trouble which was not improving. I examined the spine and found a displaced vertebra pressing against the nerves which innervate the heart. I adjusted the vertebra and gave immediate relief" [Mag 95, pp. 10–11]. From these two instances of "cure," Palmer concluded that other diseases were caused by the same "pressure on nerves." In Palmer's own words, "the science (knowledge) and art (adjusting) of Chiropractic were formed at that time."

Palmer referred to the "displaced vertebrae" as "luxations." Then, shortly after the turn of the century, one of his disciples began calling the alleged problem areas "subluxations." The term *subluxation* soon became central to chiropractic theory and is still used by chiropractors today.

To base the entire profession of chiropractic on "subluxation" and to diagnose all the diseases and pains as pinched nerves in the spinal disks is nothing but pseudoscience, namely, the replacement of observation and experimentation with the rule of authority and unproven ideas. There is nothing wrong with stating a theoretical idea. Science is full of theories,

---

[14]For a detailed account of the controversial discipline of chiropractic, see [Mag 95]. Unless otherwise referenced, what follows in this section is taken from this book.

[15]Phrenology was a pseudoscience, very popular in the nineteenth century, based on the assumption that one can diagnose diseases by analyzing the bumps on the head.

principles, and laws. Scientific theories, however, are results of thousands of painstaking observations and hundreds of unsuccessful or partially successful "little" theories and conjectures. The predictions and explanations of these theories (principles, laws, conjectures) have been tested over and over again. When conventional medicine claims that penicillin kills bacteria, it can prove it in an overwhelming majority of cases. But when chiropractors claim that they can cure pain by manipulating the spine, the only "proof" they offer is word of mouth, claims of other practitioners, and in rare cases, patients who have been "cured"—very likely due to other mechanisms such as the placebo effect (Section 43.4.1).

In 1896, with the help of a local minister, Palmer coined the name *chiropractic* from Greek words meaning "done by hand." That same year, he incorporated his first school which was renamed Palmer Infirmary and Chiropractic Institute in 1902. Several years later, D. D. was jailed for practicing medicine without a license. His son B.J. took charge of the Institute, and refused to give his father access to school grounds after the latter's release from prison. An arbitration committee resolved the dispute between father and son by allowing the son to purchase the school upon which he renamed it Palmer School and Infirmary of Chiropractic (or Palmer School of Chiropractic) in 1907.

And this is how chiropractic started in America. From its inception, the discipline was factionalized. There were those so-called "straights," such as B. J. Palmer, who believed in the centrality of the spine. There were also "mixers," such as Willard Carver, who advocated other modalities in addition to the spinal manipulation. Today, the same factionality still exists among chiropractors.[16] Some chiropractors practice under the assumption that the dogma of the founders—D. D. and B.J. Palmer—is the truth and everything else follows from that [Kea 97, p. 41]. Others rely on spiritual inspiration (Innate Intelligence), empirically testable but untested (and uncontested) hypothesis such as subluxation, and uncritical rationalism. An example of this mentality is the claim of some chiropractors that "chiropractic works because the nerve system is the master switchboard of the body."

The existence of a wide spectrum of beliefs in the chiropractic community should be an evidence of its pseudoscientific character. The factions among chiropractors resemble the denominations of a religion. When there is such a wide variety of beliefs and methodologies in any discipline, one has to conclude that the discipline has nothing to do with science. We have to distinguish between the temporary polarization of a scientific community upon the introduction of new radical ideas[17] and the permanent division of the chiropractic community which has lasted over 100 years.

> The factions among chiropractors resemble the denominations of a religion.

What about subluxation? Is it still the central theme of chiropractic? Is there any experimental proof of its role? Although in 1975 the American Chiropractic Association (ACA) "disaffirmed the [monocausal] doctrine that holds to a singular approach to the treatment of disease," the ACA's current "Chiropractic: State of the Art" booklet states that "classical subluxation" theory and the "nerve compression hypothesis" still occupy a "central place in the chiropractic rationale." The policy handbook of the International Chiropractors Association (ICA) states that "subluxation is a reasonable and credible diagnosis" [Mag 95, pp. 31–32]. In 1980, a prominent chiropractic educator asked 1000 chiropractors on the ACA's mailing list whether they agreed with various statements related to such beliefs. Of 268 respondents, 4% agreed that subluxation is the cause of *all diseases*, and a whopping 70% agreed that "the chiropractic subluxation may be related to the cause of most diseases." To the question of whether they thought that the subluxation hypothesis was scientifically supported 95% said that it was only "partially" supported or not supported at all! In other words, chiropractic "doctors" are *consciously* practicing a theory that has,

---

[16]A standard joke among chiropractors is: For every "DC" ("doctor" of chiropractic) there is an equal and opposite DC.

[17]When Einstein introduced the notion of the relativity of time in 1905, many physicists opposed him, partly because they did not understand him and partly because it was undermining the cherished ideas with which they had grown up. The correctness of the relativity of time was quickly verified and the whole physics community immediately rallied behind Einstein.

at best, "partial" scientific support.

When, in 1971, a medical doctor challenged his local chiropractic society to produce ten sets of "before and after" X-rays that demonstrate the effect of chiropractic treatment, the society refused and referred him to the Palmer School of Chiropractic. When he contacted the school, the vice president of the school replied:

> Chiropractors do not make the claim to be able to read a specific subluxation from an X-ray film. [They] can read spinal distortions, which indicates the possible presence of a subluxation and can confirm the actual presence of a subluxation by other physical findings. [Mag 95, p. 36]

43.4.3   Therapeutic Touch and Magnetic Therapy

Recently Therapeutic Touch (TT) has gained a powerful foothold in the American nursing profession.[18] According to Dolores Krieger, R.N., its leading proponent, "Therapeutic Touch is a healing practice based on the conscious use of the hands to direct or modulate, for therapeutic purposes, selected unphysical human energies that activate the physical body" [Kri 93, pp. 3–4]. By centering your consciousness, "you will have stepped into another, often unrealized dimension of yourself" [Kri 93, p. 19]. Thusly prepared, through the palms of the hands held slightly above the body, the Therapeutic Touch practitioner can scan the body for every deficit or illness. The evidence that one has detected such "living currents of energy flow" comes through sensations such as tingling, heat, pressure, or elasticity. Following assessment of the patient's energy field, the practitioner, palms facing the patient and about two or three inches from his body, "rebalances the patient's energies." The practitioner sends energy from her hand chakras (energy centers) through the patient's energy field, thereby inducing a health-promoting energy flow.

In defining this "nonmaterial energy field that I cannot see or taste or smell," Krieger (p. 16) draws upon metaphors such as that of steam power and electricity. Yet she admits that heat, one of the sensations practitioners sense through their hands, cannot be detected by scientific instruments such as the thermocouple. This is an excellent example of the prime characteristic of pseudoscience: sensation of heat exists, but no scientific instrument can detect it. (Remember the paradigm statement: "Ghosts exist, but no scientist can observe them"?) Krieger goes on to say, "Obviously Therapeutic Touch deals with a very different aspect or conception of temperature differential than the one we currently understand in biophysics" (p. 31).

In the hands of touch therapists, the scientific ideas of heat, temperature, and energy take on new undetectable and unmeasurable meaning!

The idea of treating patients with nonmaterial "fields" is not anything new. For thousands of years, wonder and magic were associated with the mysterious forces exerted by lodestones. Paracelsus (1493–1543), a physician and alchemist reasoned that since magnets have the power to attract iron, perhaps they can also attract diseases and leach them from the body. But Paracelsus was also aware of the important role of the patient's mind in the process of healing. He wrote, "The spirit is the master, the imagination is the instrument, the body is the plastic material. The moral atmosphere surrounding the patient can have a strong influence on the course of the disease. It is not the curse or the blessing that works, but the idea. The imagination produces the effect."

From magnets to animal magnetism!

The development in eighteenth century England of carbon-steel permanent magnets more powerful than lodestones brought renewed interest in the possible healing powers of magnets, and among those interested was Maximilian Hell, a professor of astronomy at the University of Vienna. Hell claimed several cures using steel magnets, but he was rapidly eclipsed by a friend who borrowed his magnets to treat a young woman suffering from a severe mental illness. The friend was Franz Anton Mesmer (1734–1815), and Mesmer's success with the "magnets from Hell" led directly to the widespread promotion of his theory of

---

[18]This subsection is adopted from [Bal 98] and [Liv 98].

"animal magnetism." Although he first used actual magnets, he later found he could "magnetize" virtually anything—paper, wood, leather, water—and produce the same effect on patients. He concluded that the animal magnetism resided in himself, the various materials simply aiding the flow of the "universal fluid" between him and the patients.

In 1784 Mesmer sought the French government approval of his practice of a treatment based on animal magnetism. He considered his Animal Magnetism a physical force of supreme interest to the sciences. He had only to point a finger toward his patient to induce a therapeutic "crisis," such as this one: "Bodies would begin to shake, arms and legs move violently and involuntarily, teeth chatter loudly. Patients would grimace, groan, babble, scream, faint, and fall unconscious." With repeated provocation the attacks would gradually become less severe and eventually disappear, and recovery would follow.

Possibly in response to the turmoil generated by Animal Magnetism, the French King Louis XVI appointed a commission to investigate the practice. The four commissioners appointed five additional commissioners including the American ambassador, Benjamin Franklin, and Lavoisier, the founder of modern chemistry. Avoiding a blind alley taken by several contemporary researchers and funding groups, they rejected Mesmer's demand that they focus on cures.[19] The commissioners considered that their first duty was to find out whether animal magnetism existed; the question of its utility could be taken up only after the question of its existence had been answered affirmatively.

In one experiment, a woman who claimed an experience of warmth in every part of her body that received magnetism, was blindfolded. The parts of her body which were exposed to magnetism without her knowledge did not respond, but when she was made to *believe* that she was being magnetized while blindfolded, she felt the usual sensations, although nothing whatever was being done. From this study it was concluded that imagination controlled these sensations. In another experiment, the commission sought to test the power of imagination in the production of convulsions. A particularly susceptible subject was blindfolded and asked to touch four trees in succession, one of which had been charged with Animal Magnetism. At the fourth tree, 24 feet from the magnetized one, he fell in a convulsion, losing consciousness. Similarly, several nonmagnetized cups were presented in succession to another subject. The second cup agitated her somewhat and the fourth provoked a crisis. She was calmed afterwards by drinking from an additional cup that had, in fact, been magnetized. From their series of experimental findings the commission concluded that "imagination without magnetism produces convulsions and ... magnetism without imagination produces nothing" and "that this fluid without existence is consequently without utility."

The inspiring words of an eighteenth century commission on animal magnetism.

The report of the scientific commission on animal magnetism damaged Mesmer's reputation and he eventually faded from public view. However, "magnetizing" persisted in various forms. Many early magnetizers evolved into students of hypnosis and developed various forms of hypnotherapy.[20] Daniel David Palmer, of whom we spoke earlier, developed chiropractic (see Section 43.4.2). Others focused on hand gestures without actual touch, an approach recently reborn as therapeutic touch. Mary Baker Eddy was "cured" by a magnetizer, but she later became convinced that cures could best be achieved through prayer, and founded Christian Science.

In the twentieth century, materials scientists and engineers developed stronger and stronger permanent magnets—alnico magnets in the 1930s, ferrite (ceramic) magnets in the 1950s, and rare-Earth magnets in the 1970s and 1980s. The latest rare-Earth magnets, neodymium-iron-boron, are over a hundred times more powerful than the steel magnets available in the nineteenth century. Both ferrite magnets and the latest "neo" magnets have had a tremendous impact on modern technology, but they have also restimulated interest

---

[19]Although Mesmer might not have known of the placebo effect, his practice clearly showed him the power of this—to him unknown—effect. Thus his insistence on cures.

[20]The trance induced in many of Mesmer's patients is thought to be what is now called a hypnotic trance, and most dictionaries today list mesmerism as a synonym for hypnotism.

in the use of permanent magnets for magnetic therapy. Unlike earlier magnetic materials such as steel and alnico, the new magnets have great resistance to demagnetization. While earlier magnets had to be long to avoid being demagnetized by the internal fields produced by the poles at the ends, new magnets can be mounted in a variety of thin products that can be applied to the body with the magnetic field emanating from the surface.

Magnetic products are typically of two kinds: those whose north and south poles alternate, and those with only one pole facing out. Many magnetic therapy products have alternating arrays of north and south poles facing the patient. Some have detailed explanations of why a circular array of poles is optimal, while others offer poles in checkerboard or triangular arrays. Nikken, the Japan-based firm that has used a multilevel marketing scheme to expand from an annual business in the US of $3 million in 1989 to $150 million in 1998, primarily offers products with alternating poles.

The difference between such multipolar and unipolar magnetic devices is the "reach" of the magnetic field. The field from even unipolar magnets decreases very rapidly with increasing distance from the magnet, but the field from multipolar magnets decreases much more rapidly. If multipolar magnets really have any effects on the human body, they will be limited to depths of penetration of only a few millimeters. (Many refrigerator magnets are multipolar, which limits the thickness of paper they can hold to the refrigerator, but also limits the damage they can do to nearby credit and ATM cards.)

Most promoters of magnetic therapy recognize the need for offering some plausible explanation. The broadest explanation has been presented by Kyochi Nakagawa, MD, of Japan, who claims that many of our modern ills result from "Magnetic Field Deficiency Syndrome." The Earth's magnetic field is known to have decreased about 6% since 1830, and indirect evidence suggests that it may have decreased as much as 30% over the last millennium. He argues that magnetic therapy simply provides some of the magnetic field that the Earth has lost.

Which one is more fatal, Magnetic Field Deficiency Syndrome or Mosquito Deficiency Syndrome?

In response to this claim we can also say that over the past 10,000 years (the last Ice Age), the Earth temperature has risen considerably, and this may be the cause of "our modern ills." Maybe we are suffering from "Ice Deficiency Syndrome," and we need to put ice cubes around our neck and ankles! Or perhaps due to the eradication of malaria in most parts of the world, many of our modern ills result from "Mosquito Deficiency Syndrome." To cure this, we may have to take a sizable dose of mosquito larva every day! When we accept—without any scientific evidence—the effect of one thing on another, any arbitrary phenomenon can be the cause of any other phenomenon. (See Food for Thought 13.3.3 for further discussion of magnetic therapy.)

43.5   A Project for the Reader

We have dissected psychoanalysis and some branches of alternative medicine with the scalpel of scientific characteristics to further clarify the difference between science and what *claims* to be science. You can take any other discipline claiming to be scientific (e.g., if it ends in *"-ology"* or *"-ics"* or has the word "science" in its name), put it to the test of some of the characteristics of science, and decide whether or not that discipline is indeed science. As a first step, go through its literature and history as much as you can; find out about some of the major scholars of the discipline; compare their writings and look for opposing views. Then go back to Chapter 42 and put the discipline to the test of the characteristics in that chapter. For example, ask yourself the following questions:[21]

1. Does the discipline study matter?

2. Does observation play a fundamental role in the discipline?

---

[21]This list is not necessarily exhaustive.

3. Does the discipline have historical continuity?

4. Does the discipline study the parts to understand the whole?

If the answer to *any* of the above questions is "no," then the discipline is not science. The first two items are sometimes hard to evaluate. A discipline whose practitioners do nothing but put a hamburger on the table and stare at it may be argued to affirm the first two questions. The first two questions are more useful when their answers are "no." For example, in the case of psychoanalysis, the answers are "no," and one can conclude without asking any further question that psychoanalysis is not science.

The third and fourth items are easier to evaluate, and usually give away the true nature of the discipline. If upon examination of the history of the discipline you find that generations of "scientists" simply trampled the "theories" of the older generations, or if the "theorists" keep going back and forth between contradictory ideas, you are dealing with a discipline that is not scientific. Similarly, if the discipline studies a "whole" but does not care about the parts, their structure, what *they* are made of, how they interact among themselves, etc., you have to raise doubts about the scientific nature of the discipline.

Glossary

Absolute Zero The zero of the universal temperature scale.

AC Alternating current.

Acceleration The change in velocity divided by time. Velocity is speed to which is assigned a direction. So a change in velocity can be made by changing its magnitude as well as its direction (see Chapter 6).

Accelerator Machines designed to accelerate particles to very high energies for the purpose of colliding them with target nuclei and other particles to probe what is inside the targets.

Action at a Distance The notion that a body exerts a force on a test object instantaneously, no matter how far the object is from the source of the force.

Active Galactic Nucleus A source of extremely powerful radiation in some galaxies. These sources are believed to be supermassive rotating black holes at the center of some remote galaxies.

Alpha Ray A by-product of nuclear radioactivity. It carries two units of positive charge, and was later identified as the nucleus of the helium atom consisting of two protons and two neutrons.

Alternating Current (AC) A source of electricity producing a current that changes periodically with time.

Amber A fossilized tree resin found in the Baltic region. When rubbed with a cloth it could make chaff, bits of thread, and other light particles jump and stick to it.

Ampere (amp) The unit of the electric current named after the French physicist André–Marie Ampère. It is equal to one Coulomb of charge passing through a wire in one second.

Amplitude The property of a wave corresponding to its strength. For water waves, for instance, it is the height of the wave.

Amplitude Modulation (AM) A technique of transmitting radio waves by modulating the amplitude of the carrier wave.

Angular Momentum The analog of ordinary momentum in rotational motion. One can call it the "quantity of rotational motion." If the rotation takes place about a fixed axis, then angular momentum of a system is the sum of the product of the momentum of each constituent of the system and its distance from the axis.

Anomalous Magnetic Moment A tiny correction to the magnetic moment of an electron as predicted by quantum electrodynamics.

Archimedes' Principle A principle—derivable from Newton's laws of motion—stating how a liquid exerts an upward force on an object immersed in it.

Aristotelian Dynamics A set of philosophical statements laid out by Aristotle without any experimental verification of those statements. Do not confuse the word "dynamics" used here with the same word used in today's physics where observation and experimentation is crucial.

Astrology The study of the sky with the purpose of understanding and controlling what happens on Earth. Astrology is not a science, because there is absolutely no connection between what humans do on Earth and what happens in the sky.

Astronomy The *science* of the study of the objects in the sky, without any intention of relating them to what is happening on Earth.

Atmospheric Pressure The pressure at some point on the surface of the Earth due to the weight of the column of air (extending over the thickness of the atmosphere) located at that point.

Atomic Clock A type of clock that uses an atomic resonance frequency standard as its timekeeping element. They are the most accurate clocks in existence.

Atomic Number The number of protons in a nucleus. This determines what element the nucleus belongs to.

Atomic Weight (really, atomic mass) The weight of a given collection of atoms or molecules of a substance. For hydrogen this is (approximately) one; all other substances are an (almost) integer multiple of the hydrogen weight.

Average Speed Distance traveled in some time interval divided by that time interval. Do not confuse average speed with average velocity.

Average Velocity A vector quantity defined as displacement of an object in some time interval divided by that time interval. Do not confuse average velocity with average speed.

Avogadro's Number The number 6.02×10^{23}. The number of atoms in 12 grams of carbon.

Ayurveda is a system of traditional medicine native to India, and practiced in other parts of the world as a form of alternative medicine. Evolving throughout its history, Ayurveda remains an influential system of medicine in South Asia.

Baryon A hadron with half-integer spin.

Baryonic Charge A property assigned to baryons whereby baryons are given a charge of $+1$ and antibaryons -1.

Bernoulli Principle A principle in fluid dynamics relating the pressure of a fluid to its speed.

Beta Rays A negatively charged ray discovered at the end of the nineteenth century, later identified as a beam of electrons.

Bhojpuri is a regional language spoken in parts of north-central and eastern India.

Big Bang The initial moment at which the entire universe, including matter, space, and time was created under extreme conditions of temperature, pressure, and energy. See Part VIII for a full discussion.

Binding Energy The absolute value of the total energy (which is negative) of a bound system.

Black Body An idealized black object, one that absorbs 100% of the EM waves that fall on it. Being a perfect absorber, it is also a perfect radiator.

Black Hole A celestial body which captures light, when the light passes the body at a close enough distance. A black hole is characterized by the fact that its escape velocity is larger that the speed of light.

Blue Shift A decrease in the wavelength due to the Doppler effect.

Blurred Quantity A physical observable not appearing in the labeling of a quantum state. A measurement of this quantity does not give an exact result, but several results with different probabilities.

Bohr Model A model of the atom which mixes the quantum ideas of Planck and Einstein with classical mechanics, leading to quantized orbits and quantized energy.

Boson A particle whose spin is a multiple of $\hbar$.

Bottom The fifth flavor of quarks. It was discovered in 1978.

Bound System A system consisting of two or more bodies held together by the force of gravity. A characteristic of a bound system is that its total energy is negative.

Brightness The flux of electromagnetic radiation energy.

Calorie The (old) unit of heat. It has been replaced with Joules, since heat is just another form of energy.

Cathode Rays Same as beta rays.

Causally Connected Refers to two events. If an observer or a light signal can be present at both events, then the events are causally connected.

Center of Gravity Same as center of mass.

Center of Mass A point of a rigid body whose motion under the influence of the external forces describes the motion of the rigid body if the entire body were concentrated at that point. Sometimes called **center of gravity** because it is the point at which you can balance the rigid body on a pivot.

Centripetal Acceleration The acceleration of an object moving *with constant speed* on a circle. Its direction is toward the center (thus the name *centri*petal).

Centripetal Force The force on any object that moves on a circle. It is the force that keeps the object on the circle.

Chain Reaction A nuclear process in which the newly formed neutrons are absorbed by heavy nuclei to produce lighter nuclei *and some neutrons*, which are absorbed by more heavy nuclei to produce even more neutrons to be absorbed by even more heavy nuclei, etc.

Chandrasekhar Mass The critical mass above which a star undergoes a supernova explosion and turns into a neutron star or a black hole.

Charm The fourth flavor of quarks. It was discovered in 1976.

Charmonium A meson composed of a charm quark and its antiquark.

Closed Universe A universe which expands for a while and then starts to contract.

CMB Cosmic microwave background radiation.

Coherent Sources Sources of waves which oscillate in unison and maintain the relative motion of their oscillation.

Collective Unconscious is a term of analytical psychology, coined by Carl Jung. It is a part of the unconscious mind, shared by a society, a people, or all humanity, that is the product of ancestral experience and contains such concepts as science, religion, and morality.

Color A property (charge) of quark taking part in strong nuclear interaction.

Compatible Observables Two or more observables which can be measured simultaneously without any restriction on the accuracy of their measurement.

Compensating Fields Same as gauge fields.

Conductor A material in which electric charges are free to move.

Conservation of Angular Momentum A principle similar to the first law of motion. It states that if a system is isolated, then its total angular momentum does not change.

Conservation of Energy A deep fundamental principle of physics, which states that energy cannot be created or destroyed. When restricted to kinetic and potential energy, it becomes **conservation of mechanical energy**, which states that, in the absence of other forms of energy, the sum of the potential and kinetic energies is constant.

Conservative System A system for which conservation of mechanical energy holds.

Constant-Volume Gas Thermometer A thermometer based on the physical principle that at constant volume, the pressure of a gas is proportional to its temperature. When the pressure of the working gas of the thermometer is reduced, the reading of the temperature becomes less and less dependent on the kind of gas used. This reading is the universal temperature of the substance being measured.

Constructive Interference Regions of space where the waves of two coherent sources add to oscillate with twice the amplitude of the wave of each source.

Coordinate Time Time kept by a clock that is moving relative to the clock that keeps the proper time.

Coordinate Velocity The rate at which the coordinates of a moving object change with time.

Cosmic Microwave Background Radiation The electromagnetic radiation present in all regions of the universe. It is the remnant of the very hot and energetic EM radiation present from the very moment of the big bang. It has a BBR curve corresponding to a temperature of 2.725 K.

Cosmological Constant A constant multiplying a certain term which is added to Einstein's equation to prevent the universe from expanding.

Coulomb The unit of electric charge.

Coulomb's Law A mathematical statement, discovered by Charles Augustine de Coulomb, expressing the electrostatic force between two charges.

CP Violation A phenomenon in which matter and antimatter behave differently while expected to behave in the same way.

Critical Density The total density (including matter, radiation, dark matter, etc.) required to make the universe flat.

Cycle The motion undergone by a simple harmonic oscillator during one period.

Cyclotron One of the first particle accelerators in which an alternating electric field accelerated charged particles, while a magnetic field kept them in circular orbits.

Dark Matter An abundant *invisible* constituent of the universe whose existence is vindicated through its gravitational effect on the *visible* part of the universe.

Daughter Nucleus One of the lighter nuclei produced in a nuclear (usually fission) reaction.

DC Direct current.

de Broglie Relation A simple mathematical formula relating a particle's momentum to its wavelength.

De Magnete The first scientific book on electricity and magnetism. It was written by William Gilbert.

Decoupling A process by which one or more particles stop interacting with the rest of the constituents of the universe.

Decoupling Temperature The temperature (about 3000 K) at which radiation stopped interacting with matter because of the formation of hydrogen and helium atoms.

Decoupling Time The time (about 375,000 years after big bang) at which radiation stopped interacting with matter because of the formation of hydrogen and helium atoms.

Deferent A large circle centered at the center of Earth on which the center of another circle, called epicycle, moves.

Density Mass divided by the volume it occupies. Density measures the concentration of mater in a region.

Destructive Interference Regions of space where the waves of two coherent sources cancel (destroy) each other.

Deuteron Bottleneck The process of the formation of deuteron, a necessary prerequisite for helium nucleus production.

Diffraction Interference of waves passing through a single aperture. When a wave diffracts, it bends as it passes through the aperture.

Direct Current (DC) A source of electricity producing a current that does not change with time.

Displacement A directed line segment (arrow) drawn from the initial position of an object in motion to its final position. The initial and final positions are determined by the beginning and end of a time interval.

Displacement Current The term which Maxwell added to the fourth equation of electromagnetism, and which is the key to the prediction of the EM waves.

Distance The length of the path taken by an object in motion in some time interval.

Divergence A technical term referring to the total flux through the bounding surface of a small volume divided by the volume.

Doppler Effect The change in the frequency (or wavelength) of a wave when its source moves towards or away from the detector or the detector moves towards or away from the source.

Ea The Babylonian supreme god of Mesopotamia.

Eightfold Way A scheme of classifying hadrons based on the abstract mathematical notion of Lie groups.

Electric Dipole Two opposite charges of equal magnitude separated by a small distance.

Electric Field The physical entity surrounding any (source) charge. A test charge introduced in this field experiences the electric force exerted by the source charge.

Electric Induction The process of producing electricity from magnetism by changing the magnetic flux through a loop of wire.

Electric Potential The difference between electric potential energy of a charge at two points divided by the charge.

Electric Potential Energy The energy stored in a charge when it is moved from one point to another in an electric field.

Electromagnetic Spectrum The range of electromagnetic waves having different frequencies (or wavelength), all traveling at the speed of light.

Electromotive Force (emf) The ability of a battery, or any other generator of electricity, to move the carriers from the attractive pole to the repulsive pole. It is simply the voltage of the battery or the generator.

Electron Microscope A microscope using the wave property of electrons and the fact that electric and magnetic fields could act as lenses for the electrons.

Electron Volt (eV) A very small unit of energy used in atomic interactions.

Elektron The Greek word for amber.

Engine A device which converts heat into mechanical energy.

Entropy A thermodynamic quantity which measures the messiness of a system and how much useful energy is available when that system undergoes a process. The law of increase of entropy determines the arrow of time.

Epicycle The circle (or sphere) on which a planet moves. The center of this circle moves on a larger circle. In the simplest model, this larger circle, called *deferent*, has the Earth at its center. The combination of the two motions results in a retrograde motion as well as a change in the brightness of the planet (due to its approach to Earth).

Equivalence Principle The principle that identifies the acceleration of an RF with the gravity felt in that RF.

Escape Velocity The minimum velocity—associated with a celestial body—which, when given to a projectile, sends the projectile farther and farther away from the body in such a way that the projectile never returns.

Ether A hypothetical medium whose undulation was believed to manifest itself as light. Relativity theory did away with ether altogether.

Euclidean Distance The distance with which the reader is familiar. It is the notion that describes how far one house is from another or one city is from another.

Event Something that happens at a point in space and at a single instant in time. A point in the four-dimensional spacetime. A point with 4 coordinates, one of which is time.

Exclusion Principle A principle stating that two electrons (more generally, fermions) cannot occupy the same quantum state.

Fermion A particle whose spin is an odd multiple of $\hbar/2$.

Fictitious Force A force in a reference frame created solely due to the acceleration of the RF.

Field Lines Streams of directed curves whose direction at a point gives the direction of the field at that point, and whose density represents the strength of the field at that point.

Final State The configuration of real particles after all interactions among them has taken place. It is used in the context of Feynman diagrams.

First Law of Motion One of the three laws of motion, a limited version of which was discovered by Galileo. He discovered that on an ideal infinitely smooth surface, an object keeps moving without having to be pushed.

Fission A nuclear interaction in which a heavy (parent) nucleus absorbs a slow neutron and disintegrates into lighter (daughter) nuclei plus some extra neutrons.

Flat Universe A universe which expands for ever, but the expansion keeps slowing down without any minimum limit.

Flatness Problem A problem with the standard cosmology whereby a slight nonflatness at the beginning of the universe turns into a huge unobserved flatness now.

Flavor A property assigned to quarks and leptons. The names assigned to quarks and leptons are also flavor assignments. Thus, we speak of up flavor, down flavor, electron flavor, etc.

Flux (of a physical quantity) The amount of a physical quantity crossing a square meter per second. Flux is a local quantity: Pick a point; pick a small loop perpendicular to the motion of the quantity; see how much of that quantity passes through the loop in one second; divide this by the area of the loop. That's how you find the flux of the quantity in question.

Focus One of the two points of an ellipse having the property that if you connect any point on the perimeter of the ellipse to these points, the sum of the resulting line segments is a constant, independent of the point chosen on the perimeter (see Figure 3.1).

Force Anything that changes the momentum of a system. Everything else being equal, the force that changes the momentum faster is stronger.

Free Fall Any motion that is caused by gravity *and gravity alone.* A projectile or an object dropped from a height is a good approximation to free fall. The reason that it is not *exactly* a free fall is the presence of atmosphere which affects the motion of most (but not all) objects only slightly. A small stone dropped from a height is almost in free fall, but a parachute is not.

Frequency The number of cycles a simple harmonic oscillator undergoes in one second. Frequency is the inverse of the period.

Frequency Modulation (FM) A technique of transmitting radio waves by modulating the frequency of the carrier wave.

Friction A force that acts against the motion of an object. It is present whenever the surface of the object is in contact with another surface.

Fusion A nuclear reaction in which lighter nuclei fuse together to form heavier nuclei and in the process release energy.

Galilean Relativity A restricted version of relativity which applies only to mechanics and motion. On a ship moving smoothly, all physical experiments look identical to those done on land.

Galvanometer A sensitive device—built on Örsted's discovery—to measure small electric currents.

Gamma Factor A quantity defined for a moving object. If the object has speed v, its gamma factor is $1/\sqrt{1-(v/c)^2}$.

Gamma Ray Another by-product of nuclear radioactivity. It is a very energetic photon, carrying an energy of the order of a few MeV or more.

Gauge Boson Particles introduced in gauge theories, which become ultimately responsible for intermediating various interactions.

Gauge Fields The fields introduced in gauge theories, whose associated particles are called gauge bosons, which in turn are ultimately responsible for intermediating various interactions. Gauge fields appear in a theory when the global version of its symmetry is changed to a local version.

Gauge Theory A mathematical theory based on the idea of the local symmetry of some physical theories. All fundamental forces of nature are now described by gauge theories.

Gauss A unit of magnetic field. Magnetic field of Earth is about 1 Gauss.

Geocentric Model A model of the solar system according to which the Earth is at the center and the Moon, the Sun, all the planets, and the stars move around it.

Global Symmetry A symmetry whose operation is uniform across all space and time.

Gluons The massless gauge particles of quantum chromodynamics responsible for the strong nuclear force.

Goldstone Bosons Certain massless particles, which appear in globally symmetric theories after the symmetry is spontaneously broken.

Grand Unification An attempt at inventing a gauge theory which unifies the strong and the electroweak interactions.

Gravitational Constant The universal physical constant G setting the scale of the strength of the gravitational force.

Gravitational Field The physical entity surrounding any massive body. A test object introduced in this field experiences the gravitational force exerted by the massive body.

Gravitational Lens is formed when the light from a very distant, bright source is bent around a massive object between the source and the observer.

Gravitational Potential Energy The energy related to the work done by gravity. More precisely, the work done by gravity in taking a test object from an initial position to a final position is the difference between the potential energies at those two points.

Hadron A particle that interacts strongly with matter. The strength of the interaction is measured by the thickness of lead that it can traverse before stopping. Hadrons traverse only a few centimeters.

Half-Life Describes radioactivity. It is the time in which half of the initial (large) number of radioactive nuclei disintegrate.

Hawking Radiation Radiation caused by the vacuum polarization in the presence of the intense gravitational field at the Schwarzschild radius of a black hole.

Heat A form of energy associated with the overall mechanical energy of the particles making up a substance.

Heat Pump A device, which by using up some mechanical energy (work), extracts some heat from a cold reservoir (outside in winter and inside in the summer) and delivers some more heat to a hot reservoir (inside in winter and outside in summer).

Helicity A property of massless particles with nonzero spin. The property is described by the component of spin along the direction of motion of the particle.

Heliocentric Model A model of the solar system in which the Sun is assumed at the center while the planets, including Earth, revolve around it on circular (later modified to elliptical) orbits. Although Copernicus is associated with heliocentrism, Aristarchus, a third century BC mathematician was the first to propose the model based on his measurement of the sizes and distances of Moon and Sun.

Hemophilia is a group of hereditary genetic disorders that impair the body's ability to control blood clotting or coagulation, which is used to stop bleeding when a blood vessel is broken.

Hertz The unit of frequency.

Higgs Boson The surviving spin-zero particle of the Higgs mechanism.

Higgs Mechanism A process by which some of the massless Goldstone bosons disappear and a corresponding number of gauge bosons acquire mass.

Homogeneity The property of being the same at all points.

Horizon The boundary beyond which the universe is not observable.

Horizon Problem A problem with the standard cosmology whereby even points separated by a very small angle could not have communicated causally, and therefore, could not have attained the uniform temperature observed in the present universe.

Horse Power A unit of power defined as the amount of work an ideal horse performs in one minute. Officially, a horse power is equivalent to 745.7 Watts.

Huygens' Principle states that the motion of a wave can be determined by assuming that each wave front is composed of infinitely many point sources each producing spherical waves.

Hydrodynamics The study of fluids (liquids and gases) in motion.

Hydrostatics The study of fluids (usually liquids) at rest.

Ideal Gas A gas in which the constituent particles do not interact, so that their energy is only kinetic. Most gases behave like an ideal gas under normal conditions.

Incompatible Observables Two observables which cannot be measured simultaneously with unlimited accuracy.

Inertia The property of an object that maintains its state of motion. Objects with large inertia tend to resist any change in their motion more than objects with smaller inertia.

Inertial Frame A reference frame in which the first law of motion holds. Usually it is an RF with no acceleration.

Inflationary Cosmology The assumption, introduced to reduce the number of monopoles in the early universe, also solved the problems associated with the standard cosmology. According to this assumption, the universe expanded at an exponentially rapid pace for a minute fraction of a second after the big bang, and then expanded at the pace set by the standard cosmology.

Initial State The configuration of real particles before any interactions among them takes place. It is used in the context of Feynman diagrams.

Instantaneous Speed Distance traveled in some time interval divided by that time interval when the time interval is taken to be as short as possible.

Instantaneous Velocity A vector quantity defined as displacement of an object in some time interval divided by that time interval when the time interval is taken to be as short as possible.

Insulator A material which cannot conduct electricity.

Interference A property of waves in which two specially prepared sources (coherent sources) construct a pattern at some points of which the waves oscillate with double amplitude (constructive interference) and at other points the wave disappears (destructive interference).

Irreversible Process A thermodynamic process which takes a system from an initial state to a final state, after which it is impossible to take the system from the final state to the initial state.

Isotope A variation of the nucleus of an element determined by its neutron number. While the nucleus of an element is *determined* by the number of protons it contains, the number of neutrons is not. Different isotopes of the nucleus of an element have different number of neutrons.

Isotropy The property of being the same in all directions.

Jeans Mass The minimum mass required for a gas of particles to clump together gravitationally.

Josephson Junction Two superconductors separated by a thin insulating layer. Under certain conditions, it is possible to tunnel a pair of electrons from one superconductor to the other through the insulating barrier.

Joule Scientific unit of work and energy, named after the English physicist who showed that heat is a form of energy.

Kelvin The scientific temperature scale, equal to Celsius scale plus 273.16.

Kelvin Temperature The scale of the universal thermometer, named after Lord Kelvin, who suggested the thermometer for scientific measurements.

Kilo Watt Hour A unit of energy used commercially, especially by electric power companies. It is the energy consumed or produced in one hour by a consumer or producer whose power is 1000 Watts.

Kilogram (kg) The scientific unit of mass.

Kinetic Energy The energy associated with the motion of an object. It is *defined* in such a way that, when the work of the force on the left-hand side of the second law of motion is calculated, the right-hand side gives a change in the kinetic energy.

Lamb Shift A tiny correction to the Coulomb's law due to the self interaction of a charged particle as it occurs in quantum electrodynamics.

Laser (Light Amplification by the Stimulated Emission of Radiation) A purely quantum mechanical invention, allowing the production of highly intense and monochromatic EM waves.

Law of Large Numbers A very general law in probability theory stating that when the sample size in an experiment gets larger and larger, the chance of the outcome of the experiment being anything but the average (expected, mean, most probable value) becomes smaller and smaller.

Left-Handed Particle A massless particle whose spin lies opposite to the direction of motion.

Length Contraction A relativistic effect whereby a length shrinks in the direction of its motion.

Lepton A particle that interacts less strongly with matter than hadrons. Leptons traverse in lead distances of the order of meters and much more.

Lie Groups A branch of advanced mathematics which combines two areas of mathematics, namely group theory and analysis. Lie groups are the means by which the idea of symmetry is implemented in physics.

Light Cone (of an event) is the collection of all events causally connected to that event.

Liquid Crystal A quantum mechanical substance that flows as a liquid but maintains some of the ordered structure characteristic of a crystal.

Local Symmetry A symmetry whose operation differs at different points of space or different instants of time.

Lodestone Pieces of magnetite which are permanently magnetized.

Longitudinal Wave A wave for which the medium oscillates along the direction of wave motion.

Lorentz Transformation A mathematical rule that gives the coordinates of an event as measured by one observer in terms of those measured by a second observer.

Magnetic Field The "imaginary" lines along which iron filings would line up when placed in the vicinity of a magnet. Invented by Faraday, these lines—and the corresponding magnetic field—became the cornerstone of Maxwell's mathematical theory of electromagnetism and led to the prediction of electromagnetic waves.

Magnetic Flux A mathematical quantity which is qualitatively described as the "number" of magnetic field lines crossing an area.

Magnetic Moment A physical quantity associated with some particles proportional to their angular momenta. A magnetic moment placed in a magnetic field aligns itself with that field.

Magnetic Monopoles Magnets with only south or north poles. These were predicted by the GUTs to be as abundant as the nucleons in the universe.

Magnetite A crystalline iron ore which has magnetic properties.

Mass Measures the inertia of an object. The ratio of the speed of the standard of mass (kept in Paris) to the speed of the object whose mass is to be determined, in an experiment in which the standard mass and the object are at first stationary and then fly apart due to an internal mechanism.

Mass Excess The difference between the mass of a nucleus and thee number of nucleons times the unified atomic mass unit.

Materialistic Philosophy A branch of ancient philosophy that puts primary emphasis on matter. The concept of atoms is the tenet of this philosophy.

Matrix Mechanics The study of microscopic systems employing Heisenberg's idea that all physical quantities are represented by matrices.

Matter-Dominated Universe A universe whose constituent is almost entirely matter.

Mechanical Energy Energy associated with the mechanical motion and forces. It is the sum of the potential and kinetic energies.

Mechanical Equivalent of Heat (MEH) A conversion factor relating the old unit of heat (calorie) to the unit of energy, Joule.

Meson The name given initially to the particle responsible for the short-ranged nuclear force. A hadron with integer spin.

MeV Million electron volt, is a convenient unit of energy for nuclear interactions. It is also the unit in terms of which the masses (times c^2) of some particles are given.

Microprocessor Another device, used frequently in computer technology, whose invention relied heavily on the quantum theory.

Mirror Symmetry A property of a massless particle with nonzero spin. If the particle can be both right handed and left handed, then it is said to have mirror symmetry.

Mole A collection of 6.02×10^{23} (Avogadro's number of) atoms or molecules of a substance.

Momentum Also known as the *quantity of motion*, is a vector quantity in terms of which the first law of motion is stated. An isolated system retains its momentum forever.

Momentum Conservation The essence of the first law of motion. The momentum of an isolated system is conserved (i.e., it does not change).

MRI Magnetic resonance imaging. See *nuclear magnetic resonance*.

Nahuatl is one of the Native American languages spoken in Mexico.

Natural Radioactivity Radioactivity that occurs naturally (as opposed to artificial radioactivity).

Neuron is an excitable cell in the nervous system that processes and transmits information by electrochemical signaling.

Neurotransmitters are chemicals which relay, amplify, and modulate signals between a neuron and another cell.

Neutrino A "little" neutron. A very weakly interacting particle produced in some nuclear processes.

Neutrino Oscillation The transformation of one neutrino flavor into another. This oscillation depends crucially on the massiveness of neutrinos.

Neutron The neutral particle found in all nuclei (except hydrogen). All nuclei are made up of protons and neutrons.

Neutron Star The celestial body left over from a star a few times more massive than the Sun after it reaches the end of its life.

Newton (N) The scientific unit in which force is measured.

NMR See *nuclear magnetic resonance*.

Nonrelativistic A term describing a constituent of the universe which moves slow compared to light speed, so that its KE is much smaller than its rest energy mc^2.

Normal Force The force felt by any object that is placed on a surface. It is usually perpendicular to the surface of contact.

Nuclear Decay A process in which a nucleus disintegrates into other nuclei. Same as nuclear radioactivity.

Nuclear Magnetic Resonance The nuclear process used in medicine to image pars of the body and diagnose diseases. Replaced by "magnetic resonance imaging" due to the public fear of the word "nuclear."

Nucleon The constituent of all nuclei. Nucleon refers to either a proton or a neutron.

Observable Used in matrix mechanics to describe any physical quantity.

Observer A point with respect to which the motion of an object is considered.

Ohm The unit of electric resistance.

Olbers' Paradox The puzzle of the darkness of the night sky. If the universe is homogeneous, isotropic, and infinite, then it can be shown that night should be as bright as day.

Open Universe A universe which expands for ever, and if the expansion slows down, it will never slow down below a certain minimum limit.

Orbital Angular Momentum The angular momentum resulting from the motion of a particle around a center of force.

Oscillation A motion that repeats itself.

Pair Production A process in which a particle and its antiparticle are created out of pure energy (usually the energy of a pair of photons).

Parallax The change in the angle of the line of sight of an object in motion relative to an observer.

Parent Nucleus The heavy nucleus that undergoes a nuclear (usually fission) reaction and disintegrates into some lighter (daughter) nuclei.

Pascal Unit of pressure. Same as N/m^2.

Period (oscillation) The time required for a simple harmonic oscillator to return to its "original" position, which could be any position during the course of its motion.

Period (planet) The time it takes a planet to make a complete revolution around the Sun.

Photoelectric Effect An effect whereby electric current is emitted from the surface of certain metals when light shines on that surface.

Photon The particle of light. Also the quantum of the electromagnetic field (see quantum field theory).

Planetary Model A model of the atom whereby the nucleus is considered a Sun and the electrons move around it as planets.

Plasma Confinement A process referred to fusion reactors in which the deuterons are confined for a long enough period so that the fusion process can take place.

Polarization A property of transverse waves whereby certain materials block the wave when held in a certain orientation in front of the wave, and allow the wave to pass when rotated 90 degrees from the blocking orientation.

Position Vector A directed line segment (arrow) drawn from the observer to the object in motion.

Positron The antiparticle of the electron predicted by the relativistic quantum mechanics and Dirac equation.

Potential Energy Energy which has the "potential" of turning into kinetic energy. Potential energy is defined for the so-called conservative forces, of which gravity is one.

Power The rate of production or consumption of energy. The number of Joules produced or consumed in a second.

Precession of Perihelion A slight rotation of the major (or minor) axis of the elliptical orbit of a planet each time it completes its revolution around the Sun.

Pressure Normal force exerted on a surface divided by the area of that surface.

Primordial Black Hole A hypothetical black hole with a Schwarzschild radius of only 10^{-16} m, formed right after big bang.

Probability Amplitude The wave function Ψ; the solution of the Schrödinger equation.

Probability Distribution The plot of probability as a function of a random variable.

Proper Time The proper time of an observer is the time kept by his/her clock.

Proton The nucleus of hydrogen atom. It carries one unit of positive charge, and together with neutron they constitute all nuclei. An element is determined by the number of protons in its nucleus.

Proton-Proton Cycle A fusion reaction taking place in the core of stars to convert hydrogen to helium and in the process create energy necessary for the sustenance of the star.

Pulsar A collapsed star, whose small size endows it with an extraordinarily fast rotation around its axis.

Pulse A single disturbance that travels in a medium.

Quantization The process of reconciling a classically observed force with the quantum theory.

Quantum The smallest unit of electromagnetic radiation carrying an amount of energy proportional to its frequency.

Quantum Angular Momentum The quantum analogue of classical angular momentum. Unlike its classical counterpart, no two components of the quantum angular momentum can be measured simultaneously.

Quantum Chromodynamics A gauge theory based on the color charge of the quarks, whose gauge bosons, the gluons, cause a force that increases with distance, thus "confining" the quarks inside hadrons.

Quantum Dot A quantum box in which one can trap an electron.

Quantum Electrodynamics A quantum field theory describing the interaction of charged fundamental particles with photons.

Quantum Entanglement A quantum phenomenon involving two subatomic particles in which the measurement of a property of one subatomic particle influences such measurement of the other even though the two particles may be completely separated.

Quantum Field Theory A mathematical theory in which every interaction and particle is considered as a field, which is quantized in the sense that it represents a specific particle. The matter particles interact via the exchange of particles representing the force.

Quantum Gravity The Holy Grail of theoretical physics. The intensely sought after theory which combines Einstein's general theory of relativity and quantum mechanics.

Quantum Tunneling A quantum phenomenon by which the probability of finding a particle in a classically forbidden region is nonzero.

Quark Confinement Temperature The temperature below which the inter-quark distances are so large that the strong force causes the quarks to form hadrons.

Quark Model A model according to which all baryons are made up of three quarks and all mesons of one quark and one antiquark.

Quarks Fundamental particles, which together with their antiparticles make up all hadrons.

Quasar (*Quasi-stellar radio source*) A powerfully energetic and distant galaxy with an active galactic nucleus. Quasars were first identified as being high redshift sources of electromagnetic energy, including radio waves and visible light, that were point-like, similar to stars, rather than extended sources similar to galaxies.

Ra The Egyptian Sun God.

Radiation-Dominated Universe A universe whose constituent is almost entirely radiation.

Radioactivity A process in which a heavier nucleus disintegrates into lighter nuclei, and in the process, emits one of three radioactive decay products: alpha, beta, or gamma radiation.

Radioisotope An isotope of an element which is radioactive.

Raisin-Pudding Model A model of the atom whereby the negatively charged electrons (the "raisins") are embedded in a positively charged background (the "pudding").

Random Event Any event whose outcome cannot be predicted by the laws of physics.

Red Giant The stage just before a star like Sun turns into a white dwarf. In this stage, the star grows to a hundred or a thousand times its original size.

Red Shift An increase in the wavelength due to the Doppler effect.

Reference Frame The collection of all objects (including people) which do not move relative to one another; i.e., the position vector of each object relative to any other object does not change.

Refrigerator A device, which by using up some mechanical energy (work), extracts some heat from a cold reservoir (inside of the refrigerator) and delivers some more heat to a hot reservoir (outside of the refrigerator).

Relativistic A term describing a constituent of the universe which moves close to light speed, so that its KE is much much bigger than its rest energy mc^2.

Renormalization A mathematical technique which tames the nonsensical infinities that show up in the calculations associated with higher order Feynman diagrams (diagrams with more than two vertices).

Reservoir A large body with which a (small) system can exchange energy. A reservoir is characterized by the fact that its temperature does not change due to the exchange of energy with a (small) system.

Resistance The property of a medium, in which electric current flows, which inhibits the flow of the charges.

Retrograde Motion The slow-down, reversal of direction, another slow-down, and another reversal of the direction of motion of planets. This is most conspicuous for Mars.

Reversible Process A thermodynamic process which takes a system from an initial state to a final state, after which it is possible to take the system from the final state to the initial state.

Right-Hand Rule (RHR) A rule that associates a straight-line direction with a rotational motion: Curl the fingers of your right hand along the direction of rotation, your thumb points in the direction of the straight line. For example, when rotation is about a fixed axis, the rule gives a direction along that axis.

Right-Handed Particle A massless particle whose spin lies in the direction of motion.

Rigid Body An object the distance between whose constituents does not change when (mild) forces are applied to it. Stated differently, a rigid body does not change shape under the application of (not too strong) forces.

Root Mean Square (rms) The square root of the average of the square of a quantity.

Scanning Tunneling Microscope (STM) A microscope, whose construction depends on the quantum phenomenon of tunneling. STM's magnification is so large that with them, one can see atoms.

Scattering A process in which two particles interact (usually with very high kinetic energy) to produce two or more particles. It often involves the conversion of the initial KE to mass via $E = mc^2$.

Schrödinger Equation A fundamental equation describing the behavior of subatomic particles under the influence of (usually electrical) forces. This equation is based crucially on the assumption that particles have wave properties.

Schwarzschild Radius The radius characterizing the ability of a black hole in capturing light. If light gets closer to the black hole than this radius, it cannot escape the gravity of the black hole.

Semi-Major Axis Half of the major axis of an ellipse. The major axis is the length of the longer axis of the ellipse (see Figure 3.1).

Sharp Quantity One of the physical observables appearing in the labeling of a quantum state. A measurement of this quantity yields an exact result.

Short-Ranged Force A force that operates only at short distances (i.e., distances smaller than the nuclear size).

Simple Harmonic Motion (SHM) An oscillatory motion described mathematically in terms of trigonometric functions. A mass attached to one end of a spring while the other end is held fixed, describes a simple harmonic motion when the mass is displaced slightly and then released.

Simple Harmonic Oscillator (SHO) An object undergoing simple harmonic motion.

Simple Harmonic Wave (SHW) A wave produced in a medium whose source undergoes a simple harmonic motion.

Solar Neutrinos Neutrinos received on Earth and produced in the core of the Sun via fusion processes.

Solenoid A long coil made to produce magnetic field by passing a current through it. An electromagnet.

Spacetime Distance A property associated with two events that is independent of the coordinates of the two events.

Spacetime Momentum Mass times spacetime velocity.

Spacetime Velocity The rate at which the spacetime coordinates of a moving object changes with its proper time.

Specific Heat The amount of heat needed to raise the temperature of one kilogram of a substance by one degree Celsius.

Spectral Lines The lines appearing in the spectroscopy of gases of elements. Each element has its unique spectral lines, making them like a finger print of that element.

Spin (classical) The angular momentum or the rotational motion of a rigid body about an axis (usually an axis of symmetry) passing through the body.

Spin (quantum) The angular momentum intrinsic to a particle. It is named so, because it was (erroneously) thought that particles have an intrinsic rotational motion similar to the Earth spinning about its axis.

Spontaneous Symmetry Breaking A process through which the lowest energy state of a theory, called a *vacuum*, is chosen. As a result of this choice, the symmetry associated with the theory is spontaneously lost.

Standard Model The theory that explains three of the four fundamental forces of nature: electromagnetism, weak nuclear force, and strong nuclear force.

Statistical Mechanics The branch of physics which considers the bulk matter as an aggregate of microscopic particles obeying the laws of physics and applies these laws as well as statistical techniques to them to predict their bulk properties.

Strangeness A property associated with some hadrons, based on the unexpected relation between their modes of production and decay.

Structure Problem A problem with the standard cosmology whereby even points separated by a very small angle could not have communicated causally, and therefore, could not have attained the uniform structure distribution observed in the present universe. Furthermore, the assumption of isotropy and homogeneity does not allow the formation of structure like stars and galaxies, which breaks isotropy and homogeneity.

Super Force The Holy Grail of fundamental physics. It is the single force which encompasses all the fundamental forces of nature, including gravity.

Supernova The explosion of a star a few times more massive than the Sun just before it reaches the end of its life.

Superposition The property of waves whereby two waves reaching a single point add to give the oscillation of the medium at that point.

Symmetry An operation performed on mathematical objects such as figures or expressions. These objects are generally changed under the operation of symmetry. If an object does not change, it is said to be symmetric under that particular operation.

Synchronicity was coined by Carl Jung to express a causal connection of two or more psycho-physic phenomena. This concept was inspired to him by a patient's case. One night, the patient dreamt a golden scarab. The next day, during the psychotherapy session, a real insect hit against Jung's cabinet window. Jung caught it and discovered surprisingly that it was a golden scarab.

Synchrotron The second-generation cyclotrons which incorporated the effects of relativity, which became important when the accelerated particles achieved a speed close to light speed.

Tao is a concept found in ancient Chinese philosophy. The character itself translates as "way," "path," or "route," or sometimes more loosely as "doctrine" or "principle." However, it is used philosophically to signify the fundamental or true nature of the world.

Tension (As in a rope) is the force that a rope sustains when it is stretched. The harder the stretching, the larger the tension.

Tesla A unit of magnetic field. Magnetic field of Earth is only 0.0001 Tesla.

Thermodynamics The branch of physics studying the bulk properties of matter.

Thermometry The art and science of making and using thermometers.

Thermonuclear Fusion Nuclear fusion taking place in the core of stars due to the extremely high temperature.

Threshold Temperature The temperature associated with a particular particle above which the EM radiation is so energetic that it can create that particle and its corresponding antiparticle.

Time Dilation The slowing down of clocks (including biological clocks such as aging) in motion.

Top The sixth (and last) flavor of quarks. It was discovered in 1995.

Transformer An electric device which increases (step-up transformer) or decreases (step-down transformer) an AC voltage.

Transistor A device whose invention depended crucially on quantum mechanics. It replaced vacuum tubes which were used in the construction of many electronic gadgets.

Transverse Wave A wave for which the medium oscillates perpendicular to the direction of wave motion.

Tunnel Diode A semiconductor device consisting of two oppositely charged regions separated by a gap. Electrons can tunnel through this gap, and their numbers can be controlled by varying the potential between the two regions.

Uncertainty Principle A physical principle, discovered by Werner Heisenberg, stating that two incompatible observables cannot be measured simultaneously with unlimited accuracy.

Unified Atomic Mass Unit Denoted by u, is a convenient unit in which to measure nuclear masses. It is $1.6605388 \times 10^{-27}$ kg, very nearly equal to a nucleon mass.

Uniform Motion A motion in which velocity of the moving object does not change. This motion necessarily takes place on a straight line.

Uniformly Accelerated Motion A motion in which acceleration of the moving object does not change.

Universal Law of Gravitation The mathematical formula giving the gravitational force between two objects in terms of their masses and the distance between them.

Universal Temperature The temperature scale derived from the constant-volume gas thermometer.

Universal Thermometer The constant-volume gas thermometer when the pressure of the working gas is extrapolated to zero.

Vertex Used in conjunction with Feynman diagrams. It is a point in the diagram where three or more particles meet. Vertex is the most fundamental part of a Feynman diagram and represents a particular kind of interaction.

Volt Unit of the electric potential.

Watt Unit of power, named in honor of the British inventor James Watt who improved the steam engine.

Wave A continuous succession of pulses traveling in a medium.

Wave Equation A mathematical equation obeyed by all waves. Maxwell showed that electric and magnetic fields obey this equation and therefore they can propagate as waves.

Wave Mechanics The study of microscopic systems employing the Schrödinger equation.

Wave-Particle Duality The idea that all particles have wave property and all waves have particle property.

Wavelength The distance between two successive similar points of a wave. Denoted by λ, wavelength is measured in meters.

Weight The force of gravity exerted on an object is its weight.

Weightlessness The condition prevailing in a freely-falling enclosure under which objects in the enclosure float.

White Dwarf The celestial body left over from a star such as the Sun after it reaches the end of its life.

Wien's Displacement Law A law stating that the wavelength at which the maximum of a black body radiation curve occurs is inversely proportional to the temperature (in Kelvin).

Work A physical quantity defined for a force. It is the product of that force and the displacement of the object on which the force acts.

Worldline A curve in a spacetime coordinate system.

Zen is a school of Buddhism referred to in Chinese as *Chan*, which is itself derived from the Sanskrit *Dhyana*, meaning "meditation." Zen emphasizes a form of meditation known as *zazen*—in the attainment of awakening, often simply called the path of enlightenment.

Answers to Selected Exercises

Here we have collected the answers to selected conceptual and numerical exercises in the textbook.

Chapter 1

Conceptual Exercises

1.1 The Earth should spin once as it makes a complete revolution around the central fire. Greece would be on the dark side of the Earth.

1.3 Argentina is in the Southern Hemisphere. So, regions in the south have larger angles.

1.5 (a) 12 o'clock. (b) 12 o'clock. (c) 6 o'clock. (d) Between 11 and 12 o'clock. (e) 9 o'clock. (f) Between 6 and 7 o'clock. (g) Between 5 and 6 o'clock. (h) Between 1 and 12 o'clock.

Numerical Exercises

1.1 (a) 2000 miles. (b) 318 miles.

Chapter 3

Conceptual Exercises

3.1 (a) They are equal. (b) Equals major axis. (c) Equals major axis.

3.3 (a) 9 o'clock. (b) 3 o'clock. (c) 7 o'clock. (d) 5 o'clock. (d) Between 1 and 2.

Numerical Exercises

3.1 (a) 2.7 billion km. (b) 5.4 billion km.

3.3 (a) 2.39×10^{-19}. (b) 12.6 million km.

Chapter 4

Conceptual Exercises

4.1 The second.

4.3 (c) Less than.

Numerical Exercises

4.1 (a) 0. (b) 3 m/s. (c) 6 m/s. (e) Yes. 0.75 m/s^2. (f) 0. (g) Yes. -2 m/s^2. (h) 6 m. (i) 24 m. (j) 18 m. (k) 36 m. (l) 9 m. (m) 69 m.
4.3 (a) 6.84 s. (b) 230 m.
4.4 (a) 2 m. (c) 2.5 s. (d) 25 m. (e) 27 m.
4.5 1.43 s.

Chapter 6

Conceptual Exercises

6.1 Yes.
6.3 See Figure 6.3 and compare the initial position of the near tree relative to the far tree with its final relative position.

Numerical Exercises

6.1 (a) 1860 miles. (b) 8 hours. (c) 232.5 mph, 104 m/s. (d) 40 mph, 17.9 m/s. (e) 30 mph, 13.4 m/s. (f) 0. (g) 0.
6.2 (a) 40,000 m, east. (b) 62832 m. (c) 11 m/s, east. (d) 17.45 m/s. (f) 0.015 m/s^2. (g) At C it is southeast; at D it is south.

Chapter 7

Conceptual Exercises

7.1 No.
7.3 (a) No. (b) Same speed.
7.5 You and Earth move with the same momentum as your initial momentum.
7.7 They are both zero.
7.9 The change in your momentum takes place slower with a (thick) haystack, making the force smaller.
7.13 (a) 0. (b) Force of the remaining portion of the liquid.
7.14 (a) Upward force. Smaller than weight of metal. (b) Same. (c) Down. Down.
7.15 (a) Upward force. Larger than weight of wood. (b) Same. (c) Up. Up.
7.17 (a) No. (b) Weight of monkey (down) and tension in rope (up). (c) Equal.
7.18 (a) Yes. (b) Weight of monkey and tension in rope. (c) Larger than weight.
7.19 (a) No. (b) Weight of monkey and tension in rope. (c) Equal.
7.20 (a) Yes. (b) Weight of monkey and tension in rope. (c) Smaller than weight.
7.21 (a) 0. (b) The floor force is bigger. (c) Floor force is the reaction to your push on the floor. (d) Third law.
7.23 (a) Yes. They are equal. (b) No. They are equal.
7.25 (a) 1 m/s^2. (b) With a force of 3 N. Same. (c) With a force of 2 N. (d) With a force of 1 N.
7.27 (a) Larger. (b) Force of person on scale. (c) Larger.
7.28 (a) Down. (b) No. It is the ceiling.

Numerical Exercises

7.1 (a) 8 kg·m/s. (b) 5 kg·m/s. (c) 13 kg·m/s. (e) 2.56 m/s.
7.2 (b) Away. (c) 9 kg·m/s. (e) 15 kg·m/s. (f) 6 kg·m/s, toward spaceship. (g) 0.067 m/s.
(h) Yes. 447.5 s.
7.3 (a) 7550 N, up. (b) 30.2 m/s^2, up, slowing down. (c) 2550 N, up. (d) 10.2 m/s^2, up,
slowing down. (e) 4.9 m/s.
7.5 (a) 5 m/s^2. (b) 6000 N. (c) The road.

Chapter 8

Conceptual Exercises

8.3 If force is perpendicular, no work is done, No work means no change in KE.
8.6 Bicycle.
8.9 Turns into heat.
8.11 (a) Yes. (b) No. (c) A force is required to overcome the drag force. This force does
work, and fuel is needed to provide the work. (d) No.
8.13 Same.
8.15 Total momentum is zero; total KE is not.
8.17 Too much time in chimney means hopping faster and using more energy there. Too
little time in chimney means moving faster in chimney and using more energy *there*.
8.19 (a) No. (b) Yes. (c) The farther from the axis, the faster the part moves.
8.21 (a) No. (b) No.
8.23 With He in the balloon, the weight of the balloon is less than the buoyant force. With
air in the balloon, the weight of the balloon is more than the buoyant force.
8.25 Small area means large pressure.
8.27 Larger speed accompanies smaller pressure according to the Bernoulli principle.

Numerical Exercises

8.1 (a) 240,000 J. (b) $-240,000$ J.
8.2 (a) 196,000 J. (b) $ME = 196000$ J, $PE = 0$, $KE = 196000$ J, $v = 39.6$ m/s. KE and
v do not change. (c) $ME = 196000$ J, $PE = 159250$ J, $KE = 36750$ J, $v = 17.15$ m/s.
(d) $ME = 196000$ J, $PE = 73500$ J, $KE = 122500$ J, $v = 22.14$ m/s.
8.3 $v = 3.81$ m/s; $KE = 15.24$ J; $KE_{\text{bul}} = 320$ J. The difference is the heat that is
produced when bullet penetrates the block.
8.5 (a) 2×10^7 J. (b) 4×10^7 J. (c) 10 kg.

Chapter 9

Conceptual Exercises

9.1 Decreases by a factor of 4. Decreases by a factor of 9.
9.3 Slow it down.
9.5 Straight down towards the center of Earth.
9.6 Yes. Kepler's law.
9.7 Yes. Absolute value decreases, so PE increases.
9.9 Earth's escape velocity.
9.11 They "fall up!"

Numerical Exercises

9.1 (a) 1.88×10^{11} m. (b) 4.32×10^6 s. (c) 43634 m/s. (d) 0.0635 m/s^2. (e) 8.56×10^{29} kg.
9.3 8.9 mm.

Chapter 11

Conceptual Exercises

11.1 Speed doesn't change; wavelength and period halve.
11.3 The intensity decreases because it spreads overs a larger circle.
11.5 (a) Constructive. (b) Destructive. (c) Constructive.
11.7 Increase.
11.8 One edge of Sun moves away from us while the opposite edge moves towards us. If we measure the wavelength of identical light from these two edges and see a difference, it indicates that Sun spins.

Numerical Exercises

11.1 (a) 1.7 cm to 17 m. (b) 4.3×10^{14} Hz to 7.5×10^{14} Hz.
11.3 385 m.
11.4 (a) No. (b) Yes.
11.5 (a) Yes. (b) No.
11.6 (a) 0.85 m. (b) 0.79 m. (c) 428 Hz. (d) 0.91 m. (e) 375 Hz.
11.7 (a) Approaching. (b) 5×10^{-9}. (c) 6.7 mph. (d) 88 mph.

Chapter 12

Conceptual Exercises

12.1 (a) Northeast. (b) (South of) West. (c) Southeast. (d) (North of) West. (e) Southwest.
12.5 (a) C (or D?) (b) B (c) Left. (d) Up. (e) Left is positive, right negative. Equal in magnitude.
12.4 (a) Second from left.
12.6 (e) Left charge is stronger.

Numerical Exercises

12.1 5.7×10^{13} kg.
12.2 (a) 98,000 N. (b) 0.000033 C.

Chapter 13

Conceptual Exercises

13.3 The source of all magnetism is electric current. There is no other source.
13.7 (a) No. (b) Yes. (c) No.

13.9 (a) Smaller loop. 11/4. (b) Smaller loop. (c) Larger loop. (d) Yes. (g) Smaller loop. (h) Larger loop.

Chapter 14

Conceptual Exercises

14.1 Positive charge.
14.3 (a) Yes. (b) Yes. (c) Electric.

Numerical Exercises

14.1 (a) 3×10^7 m. It is 3/4 the circumference. No. (b) 60,000 m. Still wavelength is too long! (c) 60 m. Yes, this is a typical radio wave.

Chapter 16

Conceptual Exercises

16.1 Slightly less than 0.5; some accurate measurement of the areas of the graph leads to the probability being approximately 0.48.
16.3 (a) 250,000 (b) No. (c) Very unlikely.
16.4 (a) 1000 times. (b) 100.
16.5 (a) Very likely. (b) Almost impossible. (c) Almost impossible. (d) Almost for certain. (e) Almost impossible.

Numerical Exercises

16.2 (a) 0.2461. (b) 0.1172. (c) Over twice as likely. (d) 11,720; 24,610.

Chapter 17

Conceptual Exercises

17.1 12; 60.
17.3 KE is the same. He has larger average speed.
17.5 Pressure increases. Volume increases. Density decreases.
17.6 Remain the same.
17.7 (a) 16. (b) 0; 4; 6.
17.11 No. Think of open and closed systems.

Numerical Exercises

17.1 (a) KE is the same. (b) 2.24. (c) They don't change. (d) 1419 m/s; 1718 m/s; 635 m/s and 768 m/s.
17.3 The smaller system will have 4 positive coins and the larger system 6.

Chapter 18

Conceptual Exercises

18.3 Swimming pool.
18.5 Fall.
18.7 If the weather gets hot, water can absorb some of the heat (lowering the temperature of the air) without changing its temperature a lot. If the weather gets cold, water can give off some heat (raising the temperature of the air) without changing its temperature a lot.

Numerical Exercises

18.2 712 m. Body does work internally.
18.4 (a) 60.8%. (b) 1216 Watts.

Chapter 20

Conceptual Exercises

20.1 (a) Microwaves. (b) No. (c) Large. (d) X-rays. (e) Gamma rays.
20.3 The first one.
20.4 True.
20.6 Blue and violet.
20.8 None.

Numerical Exercises

20.1 (a) 0.24 μm. (b) 12,000 K. (c) 1.18×10^9 W/m^2. (d) 3.32×10^{28} W. (e) 3.7×10^{11} kg or 370 million tons.
20.3 10,000 K; 5.67×10^8 W/m^2.
20.5 (a) 1.77 eV. (b) Yes. (c) 0.27 eV.

Chapter 21

Conceptual Exercises

21.3 No.
21.5 From Box 21.3.1, you can get $E_n = -13.6a_0/r_n$.
21.7 The dark lines correspond to the wavelengths of photons that were absorbed and sent the electrons of the H atoms to higher energy levels.

Numerical Exercises

21.1 (a) 3 eV. (b) 7.3×10^{14} Hz. (c) 0.41 μm. (d) Yes; violet.

Chapter 22

Conceptual Exercises

22.1 Electron B; electron.
22.5 Yes; yes.
22.7 Between zero and 25.
22.9 0.33; 0.33.
22.11 No. Electron would have a chance of escaping from within the H atom.

Numerical Exercises

22.1 (a) 2.65×10^{-16} m. (b) No. Diameter of the aperture should be comparable to the wavelength, in which case the bacterium will not fit in the hole.
22.3 6.6×10^{-14} m.
22.5 0.16.

Chapter 23

Conceptual Exercises

23.1 KE would be negative!
23.3 No. Some bullets go through neither. They are stopped.
23.5 $I = (a_1 + a_2)^2 = I_1 + I_2 + 2I_1I_2$, a_1 and a_2 are amplitudes; I_1, I_2, and I intensities. If both crests or troughs meet, amplitudes have the same sign; their product is positive, and I will be larger than the sum of the intensities. Do the rest yourself!
23.9 (a) $n = 2$. (b) -3.4 eV.

Numerical Exercises

23.1 10^{13}.
23.3 $210 = 14 \times 15$. So, angular momentum is $15\hbar$; it has 31 projections.

Chapter 25

Conceptual Exercises

25.1 Northwest.
25.3 It is not moving.
25.5 300,000 km/s.
25.6 Into your hand.
25.11 If both observers move on the perpendicular bisector of the events, they both see the two events at the same time.
25.13 You have to be in the middle of the two events. It doesn't work if the train passes you.

Chapter 26

Conceptual Exercises

26.1 Clock on spaceship.
26.3 No. No.
26.5 Yes! 2.35×10^{12}.
26.7 0.31 ly.
26.9 About half the speed of light.
26.11 (a) 1 year. (b) 10 years. (c) 10 years.
26.13 (a) 1 ly. (b) 1 year old. (c) 13 years old.

Numerical Exercises

26.1 (a) 2. (b) 44.7 years. (c) 90 years old; 40 years old. (d) No. They will be dead. (e) 44.7 ly. (f) 2 ly.
26.3 1.34×10^7 m/s.

Chapter 27

Conceptual Exercises

27.1 (a) No. (b) No. (c) Yes.
27.3 Yes.
27.5 Positive.
27.7 (b).

Numerical Exercises

27.1 (c) $x'_A = -4.47$ m, $x'_B = 4.47$ m. (e) $ct_A = 99.9$ m, $ct_B = -99.9$ m. (h) 4.47 m. (k) $cT_A = 199.9$ m, $ct_B = 0.1$ m.
27.2 (c) $\beta = 0.9375$, $\gamma = 2.87$. (d) 16 ly. (e) 15 ly. (f) 0. (g) 5.57 ly. (h) 14.35 y in the future. (i) -19 ly. (k) No, would have to cover 19 ly in 14.35 years.
27.3 (b) 201 ly. (c) 200 ly. (d) 0. (e) $\beta = 0.995$, $\gamma = 10$. (f) 20.1 ly. (g) No.
27.5 3.1×10^7 m/s.
27.7 (a) 20.001 years. (b) 9.998 years. (c) 20.001 years. (d) 50 years. (e) 0.2 ly. (f) 9.998 ly. (g) 0.2 ly. (h) 10.4 ly.; 10.4 years. (i) Both. (j) 2.99985×10^8 m/s; 0; 2.99985×10^8 m/s.

Chapter 28

Conceptual Exercises

28.1 (a) 50 mph. (b) 30 mph. (c) 40 mph.
28.3 Yes, by calculating Δs. They all read the same time.
28.5 $9.95c$ and $10c$.
28.7 29.67 GeV.
28.9 938.27 MeV.

Numerical Exercises

28.1 (a) $7.937c$ and $8c$. (b) $254.74176c$ and $254.74373c$.
28.3 (a) 1.1×10^{20} J. (b) 1.15×10^{21} J. (c) 1.15×10^{22} J. (d) 1.15×10^{23} J. (e) 1.15×10^{25} J; 1.15×10^{16} gallons. (f) 1.8 gallons.
28.4 (a) 1.8×10^{17} J. (b) 1.8×10^{10} s; 571 years.
28.5 (a) 2 kg. Yes. (b) 10 kg·m/s. (c) 5 m/s. (d) 25 J; no.
28.6 (a) 1.49×10^{9} kg·m/s. Yes. (b) 5.85×10^{17} J. (c) 4.2 kg; no. (d) $\gamma = 1.55$; $\beta = 0.76$.

Chapter 29

Conceptual Exercises

29.1 Either.
29.3 (c).
29.7 Yellow.
29.9 The one on Moon runs the fastest. The one on Earth runs the slowest.
29.11 Mercury is closest to Sun; therefore gravity is strongest there, and GTR effects more pronounced. The effect for other planets is too small to measure.
29.17 First the gas is isotropic and homogeneous. Later it is not.
29.19 5 Mly is not large enough to call the galaxy "distant," and Hubble law does not hold for this galaxy. A galaxy cannot be 20 billion ly away, because then it would have to move faster than light.

Numerical Exercises

29.1 $v = 31.3$ m/s; $T = 20$ s.
29.3 (a) 0.53 rad. (b) 30 degrees. GTR gives twice this deflection.
29.5 (a) 1.73×10^{7} m/s. (b) 4.1×10^{17} s. (c) 13 billion years.

Chapter 31

Conceptual Exercises

31.3 (a) 938.78 MeV. (b) 938.779986 MeV.
31.5 Stability means how hard it is to break the nucleus apart. The minimum energy required to break the nucleus apart is the energy needed to separate *one* nucleon from the nucleus. The larger the binding energy per nucleon, the larger this minimum energy.
31.7 No.
31.11 Carbon 14.
31.13 Decays take place because of neutron excess. The larger the ratio N/Z, the less stable the nucleus and thus more likely the decay. Alpha emission increases this ratio, especially when N and Z are small.
31.15 In a nucleus the process can take place because of the existence of the binding energy: the proton can borrow some BE and create a neutron. A free proton cannot do that.
31.17 No.
31.19 With too many protons and neutrons, the nucleus becomes less stable. So breaking into smaller nuclei is energetically favorable.
31.21 No. Splitting a light nucleus (to the left of the peak of the stability curve of Figure 31.1) would create less stable nuclei.

31.23 Palladium; About 200 MeV.

31.25 Vanadium; About 250 MeV.

31.27 Iron is at the peak of the stability curve. You cannot fuse elements at the peak of the curve, because the end product would be less stable than the initial participating nuclei.

Numerical Exercises

31.1 (a) 469.3 MeV. (b) 8.5 MeV.

31.3 (a) 1.485%. (b) 199.65 days.

31.5 4.2 MeV.

Chapter 32

Conceptual Exercises

32.3 Clockwise rotation gives electron-positron scattering. Counterclockwise rotation also gives electron-positron scattering. The other graphs give the same thing.

32.5 Yes; yes.

32.4 Yes; yes; no.

Numerical Exercises

32.1 Yes. 6.4 billion K.

32.3 No.

Chapter 33

Conceptual Exercises

33.1 Hadrons.

33.3 Baryons.

33.5 A new configuration of quarks with a new (large) energy leads to a different mass and therefore, a different hadron.

Chapter 34

Conceptual Exercises

34.1 The hexagon.

34.3 The second theory.

34.5 All reactions in the left column are possible. Only the second reaction in the right column is possible.

34.7 0, ± 1, and ± 2.

34.9 A diagram that can correspond to a real decay is $q_3 \rightarrow q_2 + p_2 + \bar{p}_1$.

34.13 Electroweak theory; the d quark turns into a u quark; weak force.

Chapter 37

Conceptual Exercises

37.1 No.
37.3 No.
37.5 Very closely so. 13.5 billion years ago.
37.7 (a) Increase. 8. (b) Increase. 16. (c) Increase. 8. (1) Increase. 2.
37.9 Any direction. 1 mm.

Numerical Exercises

37.1 (a) 9.7×10^{-27} kg/m^3. (b) 5.8.
37.2 7.96; 0.000796.
37.4 (a) 1.6×10^{35}. (b) 133760 kg/m^3. (c) 133.76 kg. (d) 0.992 kg/m^3.

Chapter 38

Conceptual Exercises

38.1 No. You also need CP violation.
38.3 Third.
38.7 40%.

Numerical Exercises

38.1 $\alpha = \underbrace{\frac{7}{8} \times 2 \times (3 + \frac{3}{2} + 5)}_{\text{leptons and quarks}} + 8 + 1 = 26.75.$

38.3 (a) $T = 1.96 \times 10^{10}$ K. (b) 121376 kg/m^3. (c) 660,000 km (smaller than Sun!).

Chapter 39

Conceptual Exercises

39.1 Angular momentum is conserved. Angular momentum in the product of angular speed and moment of inertia. Moment of inertia is very large for an ordinary star, but very small for a neutron star.

Bibliography

[And 94] Anderson, G. *Sage, Saint and Sophist*, Routledge, 1994.

[Apo 69] Apostle, H. *Aristotle's Physics*, Indiana University Press, 1969.

[APS 98] Extracted from November 30, 1998 issue of *What's New* on the American Physical Society web site: http://aps.org/WN/index.html.

[Arch 74] Archambault, R. D., Ed. *John Dewey on Education*, The University of Chicago Press, 1974.

[Bal 98] Ball, T. and Alexander, D. "Catching Up with Eighteenth Century Science in the Evaluation of Therapeutic Touch," Skeptical Inquirer (July/August), pp. 31–34, 1998.

[Bec 64] Beck, F. *Greek Education 450–350 B.C.*, Barnes and Noble, 1964.

[Bel 98] Beller, M. Physics Today, September 1998, p. 29.

[Bey 96] Beyerstein, B. and Sampson, W. "Traditional Medicine and Pseudoscience in China I." Skeptical Inquirer (July/August), pp. 18–26, 1996.

[Blo 83] Bloor, D. *Knowledge and Social Imagery*, Routledge and Keegan Paul, 1976.

[Boo 83] Boorstin, D. *The Discoverers*, Random House, 1983.

[Bor 82] Bordeau, S. *Volts to Hertz*, Burgess Publishing Company, 1982.

[Bre 91] Brecht, B. *Galileo*, Grove Weidenfeld, 1991.

[Bro 91] Brooke, J. *Science and Religion*, Cambridge University Press, 1991.

[Bro 62] Brophy, J. and Paulucci, H. *The Achievements of Galileo*, Twayne Publishers, 1962.

[Bro 98] Brown, W. "The Placebo Effect." Scientific American (January), pp. 90–95, 1998.

[Bur 98] Burke, P. *The European Renaissance*, Blackwell, 1998.

[Can 90] Canfora, L. *The Vanished Library*, University of California Press, 1990.

[Cap 83] Capra, F. *The Turning Point*, Bantam Books, 1983.

[Cap 84] Capra, F. *The Tao of Physics*, 2nd ed., Bantam Books, 1984.

[Cap 00] Capra, F. *The Tao of Physics*, 4th ed., Shambhala, 2000.

[Cha 61] Charlesworth, M. *The Roman Empire*, Oxford University Press, 1961.

[Chop 90] Chopra, D. *Quantum Healing*, Bantam Books, 1990.

[Cre 86] Crews, F. *Skeptical Engagements*, Oxford University Press, 1986.

[Daw 87] Dawkins, R. *The Blind Watchmaker*, W. W. Norton and Company, 1987.

[Dew 87] Dewey, J. *Democracy and Education*, The Free Press, 1944.

[DOE 01] See http://www.eia.doe.gov/oiaf/ieo/world.html, the US Department of Energy web site.

[Dra 70] Drake, S. *Galileo Studies*, University of Michigan Press, 1970.

[Dre 53] Dreyer, J. *A History of Astronomy From Thales to Kepler*, Dover, 1953.

[Ein 52] Einstein, A. et al. *The Principle of Relativity*, Dover, 1952.

[Eis 82] Eisenberg, D. (with T. L. Wright) *Encounters with Qi: Exploring Chinese Medicine*, W. W. Norton, 1982.

[Eve 66] Everson, T. and Cole, W. *Spontaneous Regression of Cancer*, W. B. Saunders, 1966.

[Eys 73] Eysenck, H. and Wilson, G., Eds. *Experimental Study of Freudian Theories*, Methuen and Co., 1973.

[Far 67] Farrington, B. *The Faith of Epicurus*, Basic Books, 1967.

[Fes 69] Festugière, A. *Epicurus and His Gods*, Russell & Russell, 1969.

[Fey 63] Feynman, R. *Feynman Lectures on Physics*, vol. I, Addison-Wesley, 1963.

[Fin 90] Fine, R. *The History of Psychoanalysis*, Continuum, 1990.

[Fre 50] Freud, S. *The Interpretation of Dreams*, Modern Library, 1950.

[NYT 98] Gardner, H. *New York Times*, March 2, 1998 op-ed section.

[Ged 05] Gedo, J. *Psychoanalysis as Biological Science: A Comprehensive Theory*, The Johns Hopkins University Press, 2005.

[Gis 73] Gish, D. *Evolution: The Fossils Say No!*, ICR Publishing Company, 1973.

[Ham 93] Hamilton, E. *The Greek Way*, W. W. Norton & Company, 1993.

[Ham 93] Hamilton, E. *The Roman Way*, W. W. Norton & Company, 1993.

[Har 91] Harding, S. *Whose Science? Whose Knowledge?*, Cornell University Press, 1991.

[Har 83] Harding, S. and Hintikka, M., Eds. *Discovering Reality*, D. Reidel Publishing Company, 1983.

[Hei 58] Heisenberg, W. *The Physicist's Conception of Nature*. Translated by Arnold J. Pomerans, Harcourt, Brace, 1958.

[Hin 88] Hines, T. *Pseudoscience and the Paranormal*, Prometheus Books, 1988.

[Hol 01] Holton, G. and Brush, S. *Physics, the Human Adventure*, Rutgers University Press, 2001.

[Hui 92] Huizenga, J. *Cold Fusion: The Scientific Fiasco of the Century*, University of Rochester Press, 1992.

[Hut 96] Hutchinson, L. *Lucretius: De rerum natura*, University of Michigan Press, 1996.

[Hut 98] Huth, J. "Latour's Relativity." In *A House Built on Sand*, Koertge, N., Ed., Oxford University Press, 1998.

[Jac 75] Jacobs, D. *The UFO Controversy in America*, Indiana University Press, 1975.

[Kea 97] Keating Jr., J. "Chiropractic: Science and Antiscience and Pseudoscience Side by Side." Skeptical Inquirer (July/August), pp. 37–43, 1997.

[Kli 97] Kline, P. "Obsessional Traits, Obsessional Symptoms and Anal Eroticism." *British Journal of Medical Psychology*, pp. 299–305, 1968.

[Kri 93] Krieger, D. *Accepting Your Power to Heal: The Personal Practice of Therapeutic Touch*, Bear, 1993.

[Kuh 96] Kuhn, T. *The Structure of Scientific Revolutions*, The University of Chicago Press, 1996.

[Lac 70] Lacan, J. "Of Structure as an Inmixing of an Otherness Prerequisite to Any Subject Whatever." In *The Languages of Criticism and the Sciences of Man*, pp. 186–200. Macksey, R. and Donato, E., Eds., Johns Hopkins University Press, 1970.

[Lat 88] Latour, B. "A Relativistic Account of Einstein's Relativity," *Social Studies of Science* 18 (1988): 3–44.

[Lau 70] Laurie, S. *Historical Survey of Pre-Christian Education*, Scholarly Press, 1970.

[Lak 70] Laurikainen, K. *Beyond the Atom*, Springer–Verlag, 1988.

[Liv 98] Livingston, J. "Magnetic Therapy: Plausible Attraction?," Skeptical Inquirer (July/August), pp. 25–30, 1998.

[Mag 95] Magner, G. *Chiropractic: The Victim's Perspective*, Prometheus, 1995.

[McF 01] McFadden, J. *Quantum Evolution: A New Science of Life*, W. W. Norton, 2001.

[Mey 71] Meyer, H. *A History of Electricity and Magnetism*, MIT Press, 1971.

[Nah 07] Nahin, R. et al. *National Health Statistics Reports*, #18, July 30, 2009.

[Nee 89] Ne'Eman, Y. and Kirsh, Y. *The Particle Hunters*, Cambridge University Press, 1989.

[NIH website] http://www.nih.gov/about/almanac/organization/NCCAM.htm

[Nol 74] Nolen, W. *Healing: A Doctor in Search of a Miracle*, Random House, 1974.

[Oku 89] Okun, L. Physics Today, June 1989, p. 31.

[Osc 00] Oschman, J. *Energy Medicine: The Scientific Basis*, Elsevier, 2000.

[Pag 83] Pagels, H. *The Cosmic Code*, Bantam Books, 1983.

[Pap 97] Papineau, D., Ed. *The Philosophy of Science*, Oxford University Press, 1997.

[Par 02] Park, R. *Voodoo Science: The Road from Foolishness to Fraud*, Oxford University Press, 2000.

[Pel 94] Pellegrino, C. *Return to Sodom and Gomorrah*, Random House, 1994.

[Pro 91] Proctor, R. *Value-Free Science?*, Harvard University Press, 1991.

[Rad 75] Radice, B. *Letters of the Younger Pliny*, Penguin, 1975.

[Ran 80] Randi, J. *Flim-Flam!*, Lippincott and Crowell, Publishers, 1980.

[Ran 87] Randi, J. *The Faith Healers*, Prometheus Books, 1987.

[Ran 93] Randi, J. *Secrets of the Psychics*, Videotape recording, PBS Nova Series, 1993.

[Ras 96] Raso, J. "Alternative Health Education and Pseudocredentialling," Skeptical Inquirer (July/August), pp. 39–45, 1996.

[Ris 72] Rist, J. *Epicurus: An Introduction*, Cambridge University Press, 1972.

[Ron 64] Ronan, C. *The Astronomers*, Evans Brothers, 1964.

[Ros 72] Ross, D. *Plato's Theory of Ideas*, Cambridge University Press, 1951.

[Rus 88] Ruse, M., Ed. *But Is It Science?*, Prometheus, 1988.

[Sag 96] Sagan, C. *The Demon-Haunted World*, Ballantine Books, 1996.

[Sam 96] Sampson, W. and Beyerstein, B. "Traditional Medicine and Pseudoscience in China II." Skeptical Inquirer (September/October), pp. 27–34, 1996.

[Sel 96] Selby, C. and Scheiber, B. "Science or Pseudoscience? Pentagon Grants Funds Alternative Health Study," Skeptical Inquirer (July/August), pp. 37–43, 1996.

[Sok 98] Sokal, A. and Brichmont, J. *Fashinable Nonsense*, Picador, 1998.

[Sta 98] Starr, C. *A History of the Ancient World*, Oxford University Press, 1991.

[Tip 94] Tipler, F. *The Physics of Immortality*, Doubleday, 1994.

[Tue 97] Tuerkheimer, A. and Vyse, S. "The Book of Predictions: Fifteen Years Later," Skeptical Inquirer (March/April), pp. 40–42, 1997.

[Use 87] Usener, H. *Epicurea*, Leipzig, 1887.

[Wal 81] Wallechinsky, D., Wallace, A., and Wallace, I. *The Book of Predictions*, Morrow, 1981.

[Wea 87] Weaver, J. *The World of Physics*, vol. 1, Simon and Schuster, 1987.

[Wel 84] Wells, C. *The Roman Empire*, Stanford University Press, 1984.

[Whi 76] White, N. *Plato on Knowledge and Reality*, Hacket Publishing Company, 1976.

[Wol 96] Wolf, F. A. *The Spiritual Universe*, Simon & Schuster, 1996.

[Zuk 80] Zukav, G. *The Dancing Wu Li Masters*, Bantam Books, 1980.

Index

Milton Keynes UK
Ingram Content Group UK Ltd.
UKHW050441111024
44932UK000050B/443